THE SOCIO-ECONOMIC IMPACTS OF ARTISANAL AND SMALL-SCALE MINING IN DEVELOPING COUNTRIES

For Sophia, Chris, and our beloved Spike, who we all miss ever so dearly

The Socio-Economic Impacts of Artisanal and Small-Scale Mining in Developing Countries

Edited by

GAVIN M. HILSON
Environmental Policy and Management Group (EPMG)
Imperial College of Science, Technology and Medicine, UK

A.A. BALKEMA PUBLISHERS / LISSE / ABINGDON / EXTON (PA) / TOKYO

Library of Congress Cataloging-in-Publication Data
Applied for

Typesetting: Charon Tec Pvt. Ltd, Chennai, India
Printed by: Krips, The Netherlands

Published by: A.A. Balkema Publishers, a member of Swets & Zeitlinger Publishers
www.balkema.nl and www.szp.swets.nl

ISBN 90 5809 615 7

Contents

Acknowledgements

In preparing this book, I received inputs from a number of individuals. My special thanks goes to Dr Clive Potter, Professor John Monhemius, Professor Don Huisingh, Dr Paolo Galizzi, Professor Barbara Murck, Dr Janjaap Blom and all of the A.A. Balkema staff, the Institution of Mining and Metallurgy (IMM) staff, Mr. B.R. Yakubu, and Mr. Joseph Eyison. My special thanks also to all of the contributors for their patience and commitment to the project, their tolerance of my frequent email reminders, and their ability to meet my tight deadlines within their own very hectic schedules. Finally, I would like to give thanks to all of my colleagues at the Environmental Policy and Management Group (EPMG), Imperial College for their support and encouragement, especially Konstantinos Evangelinos and Daniel Weisser, who provided regular motivation and timely inspiration. This book is dedicated to Sophia Bains, Chris Hilson, and our beloved Spike, who we all miss ever so dearly.

Foreword

The world is big and diverse. Few of us ever see more than a small part of it. None of us see all of it. Yet, the temptation to generalize from our own limited experience is almost universal.

When institutions that project themselves globally, such as international financial organizations, United Nations' bodies, or aid agencies try to devise effective strategies for coping with problems, they are usually based on generalizations from the experience of a limited number of all too human policy makers. They may be very experienced, but have not been everywhere, or seen everything. There is therefore a very real risk that the resulting policies will not fit all sizes, and may, in some circumstances, produce pernicious results.

"Small scale mining" or "artisanal mining" mean very different things to different people, depending on their experience and what they have seen. In some places, these are virtually all-male activities. In others, they are overwhelmingly female. Sometimes, they are the voluntary choice of people who see them as the best available strategy for building their futures. Sometimes, participants are indentured, or even enslaved.

There are small-scale operations in which no children work; others that are dominantly staffed by children. Some forms of small-scale mining are reasonably safe and healthy activities; others are extraordinarily dangerous and unhealthy. Some governments support and encourage small-scale mining, while others prohibit it and try to stamp it out. Some artisanal miners reap most of the benefit of their labour by selling their products into relatively free markets, while others are, in one way or another, forced to sell at a fraction of the market price into captive markets, to criminal gangs, or even to armed insurgents who use the revenues to fuel civil conflicts. Some forms of this activity are a seasonal supplement to livelihoods earned in other ways; sometimes, it is the sole support of participants.

Inherent in much that has been written about artisanal and small-scale mining is an urgent desire to "fix" what is perceived as being wrong with this activity. But what is not working with ASM, and what needs to be done, depends very heavily on whose perspective we choose to adopt. Almost all of those concerned are subject to contradictory pressures, which lead to vacillating policies, or policy paralysis.

Governments often see this sector as a source of problems: earnings on which taxes are unpaid, occupation of lands to which there is no title, revenues flowing outside the banking system, and even as an economic power base for criminals or revolutionaries. From this perspective comes a desire simply to stamp out the activity. But this desire is often tempered by the realization that many governments, particularly the poorest ones, have precious few economic alternatives to offer. Thousands of artisanal miners may be less of a threat to the public order than thousands of the unemployed. The logic of these contradictory urges has led many to see the ideal policy as one that supports artisanal miners by helping them become more productive, better educated, healthier, and less environmentally damaging, ultimately with more opportunities both inside and outside the mining sector. But these aspirations without the resources to back them have proven a hollow alternative.

While more than a few mineral deposits have first come to the attention of large mining companies because their geologists see artisanal miners in the area, most of these companies have also seen small miners as a problem to be solved. Sometimes, many thousands of them live and work on ground subject to formal mining concessions, which cannot be developed until a way is found to deal with these occupants. Sometimes, keeping them off the concession is harder than getting them off in the first place. Increasingly, companies are seeing their interest best served by helping to develop a positive future for artisanal miners if, for no other reason, than it being the most effective acceptable means of avoiding conflict with them. As in the case of government, this requires resources that are not always there.

Environmental organizations, appalled by the environmental consequences of this activity, have sometimes supported its abolition. But in most cases, such prohibitions have simply driven the activity from the formal economy, where it tends to fall into the hands of the lawless, and becomes even more difficult to manage. And environmental organizations have increasingly made common cause with artisanal miners against the large companies that may be seen as the common foe.

Increasingly, there is an awareness on the part of all of these actors that change in this sector is very hard indeed to achieve without the support and cooperation of the small-scale miners themselves. This requires engaging directly with small-scale miners and their families as humans, and finding out more about who they are, how they operate, what they want, and what they see as the problems needing to be fixed.

In short, promoting positive change in this sector is very much a matter of respect for the humanity and dignity of the people who participate in it. With their dependents, this is a group of people more numerous than the French, the British, or the Germans. Moving forward requires a willingness to engage with them, listen to them, and learn from them.

Respect is a two-way street, and lest we fall victim to ethno-romantic dreams about small-scale miners, we must remember they have sometimes failed to respect others. They have participated in massacres of indigenous people in the Amazon Basin, have fallen in behind questionable political leaders in a number of countries, and have overwhelmed local villages with in-migration following discovery of new gold deposits. Part of the question about small-scale mining is whether participants are capable of reaching new levels of political and social maturity.

This is, in part, because global policy processes, the decisions of global institutions, and the forces of global markets increasingly affect the lives and futures of small-scale miners. Yet, they generally have difficulty finding ways to participate. Encouraging the emergence of leadership and finding ways to give it a space at the table should be a high priority for international organizations whose work affects small-scale miners. The International Labour Organisation has long shown leadership in this area; others need to do more. But while development of that leadership can be encouraged externally, it can only be done internally. It must emerge from within the community of small-scale miners itself.

The Socioeconomic Impacts of Artisanal and Small-Scale Mining in Developing Countries is an ambitious attempt by some of the world's most knowledgeable researchers and experts in the field of small-scale mining to portray the complex reality of this activity.

As the former Director of the Mining Minerals and Sustainable Development Project, I can say that we much wish that we had had this very valuable resource on hand as we were preparing our project report, *Breaking New Ground.* A new MMSD working paper, bringing together our project's work on artisanal and small-scale mining, will be issued by the International Institute for Environment and Development in late 2003, and we hope to provide additional support and background for the conclusions reached and documented so clearly here.

There are vastly different concepts of where small-scale village-based economic activities – not just mining but artisanal fishing, wood cutting, and peasant agriculture – fit into the economy of the future. In the view of some, they are remnants of antiquated ways of doing things, swept away by the irresistible forces of corporate capitalism. Others point out that artisanal modes of production offer something global capitalism seems not to provide: large numbers of livelihoods for the poor. Or they argue that these village-level activities are an alternative to even further mass migrations to the deeply troubled megacities of the developing world. And sometimes artisanal modes of production have proven their durability: they have been able to continue for centuries, without destruction of the resource base, perhaps some demonstration of their compatibility with ideas of sustainable development. Large-scale corporate capitalism has yet to prove this.

To these arguments in the case of mining, we can add that most artisanal and small-scale miners exploit deposits that are not of much interest for the large-scale techniques employed by the world's mining giants. These may be located over or adjacent to the kind of large-scale, low-grade deposits that large mining companies like to exploit, but are often deposits that would go unexploited by corporate techniques. But despite all of these arguments, resources continue to flow away from small-scale production and into a decreasing number of increasingly larger corporations. Whether this trend eventually leads to a sustainable economy, or is driving us all in a dangerous direction, is perhaps a question it is extremely difficult to answer and one perhaps not appropriate for a book such as this one.

In general, prices of most mineral commodities have been declining for a very long period of time, with occasional upward spikes. This means that if small-scale mining remains static, it will yield less and less reward. The pressure is on small-scale miners to become more productive or to suffer ongoing impoverishment. The ways they are succeeding – or failing – in coping with these challenges will determine the future of this activity and its value as a route out of poverty.

Luke J. Danielson

About the Contributors

John S. Andrew is an Assistant Professor in the School of Urban and Regional Planning at Queen's University in Kingston, Ontario, Canada. His core area of research is conflict management in planning. He practices as an independent facilitator and mediator, and consults on public consultation and conflict management. He teaches in a graduate program, primarily in the areas of environmental and social planning. He holds a Ph.D. in Geography and a Master of Science and Planning from the University of Toronto, and a B.Sc. in physical geography from The University of Western Ontario. Email: andrewj@post.queensu.ca

Philip Andrews-Speed is Director of the Centre for Energy, Petroleum and Mineral Law and Policy at the University of Dundee, Scotland. He spent fourteen years as a geologist in the international mining and petroleum industries before coming to the Centre in 1994, gaining an LLM in Energy Law and Policy, and joining the academic staff. Dr Andrews-Speed leads the Centre's China Programme, which covers research, consultancy and professional training in the petroleum, electricity, mining, and water sectors. The focus of his research is policy, regulation and reform especially in the gas, electricity and coal sectors. Email: c.p.andrewsspeed@dundee.ac.uk

Benjamin Nii Ayi Aryee is currently the Chief Executive of Ghana's Minerals Commission, the country's governmental agency responsible for promoting the mining sector as well as managing and regulating the utilization of Ghana's mineral resources. He graduated from the University of Ghana with a Bachelor's degree in Economics and Geography in 1984, and, in 1992, earned a M.Eng degree in Mineral Economics from McGill University in Montreal, Canada. He is currently pursuing an Executive MBA at the Ghana Institute of Management and Public Administration. Email: b_aryee@mincomgh.org

Jasminda P. Asirot is a Research Assistant in the Small-Scale Mining Project at the Philippine Institute for Development Studies. Email: asirotj@esb.dlsu.edu.ph

Christian Beinhoff is an extractive metallurgist and has extensive experience in planning, erection, commissioning and rehabilitation of mineral industries operations in the iron and steel, non-ferrous and precious metals extraction sectors. Dr Beinhoff has 15 years of expertise in consulting work for multi-national companies related to sustainable mineral resource development and environmental protection. He is presently working as Chief Technical Advisor at UNIDO for the GEF-financed Global Mercury Project. Email: cbeinhoff@ unido.org

Antonio Bermeo is a Civil Engineer, with a Masters degree in environmental engineering from the Escuela Politecnica in Quito, Ecuador. In his career, he has worked within academia and Project Management in both national and international projects. Since 1995, he has worked as a consultant and as a project coordinator in the PRODEMINCA mining sector support project in Ecuador.

Edmund Bugnosen is perhaps the only European-based consultant specializing in small-scale mining and mining community development projects. He also coordinates the modest operations of Small Mining International, a corporation-not-for-gain, which maintains, and continuously collects, information on small-scale mining, and, at the same time, provides technical enquiry services on the same subject. Email: AYBAN@aol.com

Sergio Castro graduated from the University of Chile in 1972, with postgraduate specialization in mineral processing. In 1974, he joined the Department of metallurgical Engineering of the University of Concepción. He is currently Titular Professor, and his main research areas are: Surface Chemistry, Flotation, Mineral Processing, and Effluent Treatment in the Mining Industry. Professor Castro is the author of numerous scientific and technological publications, and Editor of 10 books published by the University of Concepción, and one published by Elsevier. He has dedicated 30 years of his professional career to the copper, gold and molybdenum mining industry. He has been a visiting professor at several foreign Universities, including the University of British Columbia and Universidad Industrial de Santander in Colombia. Email: scastro@udec.cl

Geoff Crispin has been working with indigenous people in Australia and around the world since the 1970s. He has worked for various agencies of the UN, CFTC, the Australian Government, private companies and private donors, focusing on small-scale income generation projects. He has been stationed in Africa, The Caribbean, the Middle East, South East Asia, and, most recently, Papua New Guinea as an AusAID Small-Scale Mining Advisor. He holds a M.Sc. degree from the University of New South Wales and a M.Ed. degree from the University of Southern Queensland. Email: geoff.crispin@bigpond.com

Florence Luzia Bukali da Graça is a managing partner, policy researcher and consultant at Social Development Consultants in Maputo, Mozambique. She specializes in environmental management, development, social and cultural issues, and HIV/AIDS. Email: florence@teledata.mz

Hamid Etemad is an Associate Professor of Marketing and International Business at the Faculty of Management, McGill University. He holds a Master of Engineering from the University of Tehran in Mechanical Engineering, a M.Sc. in Mechanical Design, an MBA in International Business and Operations Management, and a Ph.D. in International Business from the University of California, Berkeley. Email: hamid.etemad@mcgill.ca

Ajoy K. Ghose is a Professor Emeritus and former Director of the Indian School of Mines in Dhanbad. He is currently a Consultant, and an Adjunct Professor at the Department of Mining and Geology at Deemed University. He is editor-in-chief of the *Journal of Mines, Metals & Fuels*. Email: ajay_g@vsnl.net

Mrinal K. Ghose earned a M.Sc. in chemistry from Rajshahi University in 1969, after which, he became a lecturer at Chakaria College, Chittagong, Bangladesh. He carried out doctoral research in chemistry between 1972 and 1975, and post-doctoral research between 1976 and 1979 in environmental engineering, both at the Jadavpur University in India. Dr Ghose joined the Environmental Engineering Division of Central Mine Planning and Design Institute Ltd. in 1987, and, in 1989, joined the Centre of Mining Environment at the Indian School of Mines. Email: ghosemrinal@eudoramail.com

Rosalind P. Harris is an Associate Professor of Sociology, Department of Sociology, University of Kentucky, Lexington. Her research focuses on postcolonial development;

rural sociology; race, gender and class; and qualitative methodology. Email: rharris@pop.uky.edu

Marieke Heemskerk is an anthropologist and a research associate at the University of Wisconsin. Dr Heemskerk has conducted research with Maroon gold miners and their families in Eastern Suriname since 1996. Email: mheemskerk@wisc.edu

Gavin M. Hilson (Editor) is a member of the Environmental Policy and Management Group (EPMG), Imperial College of Science, Technology and Medicine. He is author of over 30 refereed publications on the environmental/socioeconomic impacts of mining. He is currently undertaking research on the impacts of artisanal gold mining in Ghana. Email: g.hilson@imperial.ac.uk

Jennifer Hinton holds a Bachelor of Applied Science in Geological Engineering (Environmental and Geotechnical), and a Masters of Applied Science in Mining and Mineral Process Engineering from the University of British Columbia. She is currently a doctoral candidate and Bridge Strategic Fellow at the Department of Mining Engineering, University of British Columbia. Ms. Hinton has most recently undertaken research on the impacts of artisanal and small-scale gold mining in the Brazilian Amazon, focusing on technical and social measures to mitigate risks from mercury pollution in artisanal mining communities, and has also conducted a major review of 'clean' alternative artisanal gold mining technologies. In addition to her work in artisanal mining, she has worked, researched and published in areas ranging from the biological indicators of metals pollution in the environment, innovative mine closure alternatives, sustainability issues in mining, and the remediation of contaminated sites, to the incorporation of environmental and social factors into mine planning. Email: jhinton_crm@telus.net

Rediar Hoffner is a Mining Engineer, trained at the Royal Institute of Technology, Stockholm, Sweden. Mr. Hoffner has worked as a consultant within the mining sector for more than two decades. He is a member of the Expert Advisory Group of the "Collaborative Group on Artisanal and Small Scale Mining (CASM)". Currently, he is the Project Coordinator for the EDF-funded Mining Diversification Programme in Zambia.

Danilo C. Israel is a Senior Research Fellow at the Philippine Institute for Development Studies. He holds a Ph.D. in Resource Economics. Email: disrael@pidsnet.pids.gov.ph

Norman S. Jennings is a Senior Industrial Specialist in the Sectoral Activities Department, International Labour Office in Geneva. Email: jennings@ilo.org

Stephens M. Kambani is a Senior Lecturer and Head of the Department of Mining Engineering, School of Mines, University of Zambia, Lusaka. He holds B.Min.Sc. (Mining Engineering, UNZA, Zambia), M.Eng. (Mineral Economics, McGill University, Canada), and Ph.D. (Mineral Economics, Montan University, Austria) degrees. Email: skambani@mines.unza.zm

Esther Kazilimani is a Freelance Health and Gender Consultant in Mozambique. Email: esther@zebra.uem.mz

Crispin Kinabo holds the position of Senior Lecturer in the Department of Geology, University of Dar-es-Salaam, Tanzania. Email: crispin@uccmail.co.tz

Beatrice Labonne is presently a Senior Adviser for partnership and network development for seven UN economic and social agencies with global and regional focus. She is former Director of the Division for Natural Resources Management and Economic and Social

Development in the UN Department of Economic and Social Affairs in New York. Ms. Labonne has more than 30 years of experience in the field of mineral exploration, economic and social development, and is the author of several publications on mineral/natural resources management and sustainable development, corporate responsibility, conflict prevention, gender dimension, and artisanal mining and sustainable livelihoods. She holds a Masters degree in geology. Email: blabonne@unog.ch

Luiz Drude de Lacerda is presently Head of the Graduate Program in Marine Sciences at the Universidade Fedral do Ceará at Fortaleza, Brazil. Since 1985, Professor Lacerda has researched the biogeochemistry of mercury in tropical ecosystems, particularly in the Amazon Region. His work has dealt with the atmospheric deposition of mercury, the transfer of mercury through the tropical forest ecosystem, the contamination of biota and humans, and the control and abatement of mercury contamination. Professor Lacerda has been actively involved in the organizing committees of the "Mercury as a Global Contaminant" conference series, and has been a member of many international panels on Hg contamination. Email: ldrude@fortalnet.com.br

Kuntala Lahiri-Dutt is a Research Fellow at the Australian National University. She specializes in the socio-economic impacts of mining projects on indigenous/rural people in developing countries. Her research has a strong focus on gender. Email: kuntala@coombs.anu.edu.au

Bo Lundberg holds undergraduate and doctoral degrees from the University of Lund in Sweden. He has worked as a consultant in international mining projects for more than 20 years, and his expertise includes institutional development, sectoral environmental assessments, and the elaboration of environmental policies for the mining sector.

Guo Ma is a Lecturer in the Marketing Department of the College of Economics and Business Administration, Chongqing University in China. Her current research interests include marketing in the retailing and energy industry. Email: maguocq@yahoo.com

Maideyi L. Mabvira-Meck is a Lecturer in Environmental Geology at the University of Zimbabwe in Harare. Email: maideyimeck@yahoo.com

Oliver Maponga is a graduate of the University of Zimbabwe, McGill University, Montreal, Canada and Curtin University's Western Australian School of Mines (WASM). Dr Maponga is currently Chairman of the University of Zimbabwe's Institute of Mining Research, and heads a team undertaking research and offering consultancy services to the minerals industry. Dr Maponga has held short-term appointments in Canada and the Western Australian School of Mines. His areas of expertise include small-scale mining, mineral project development, mining environmental management, mining policy, mineral commodity analysis and mining investment analysis. Email: mapongao@science.uz.ac.zw

Munyindei Masialeti is a Lecturer in Mining Engineering at the University of Zambia School of Mines in Lusaka, Zambia. Email: mmunyi@yahoo.com

Gary McMahon is a Principal Economist, Global Development Network in Washington DC. Email gmcmahon@gdnet.org

Maideyi Meck holds a Bachelor of Science Honours Degree in Geology from the University of Zimbabwe, and recently completed an MPhil in Geology from the same University. Apart from teaching, his professional experience includes a six-year appointment as Assistant

Mineralogist at the Institute of Mining Research in Zimbabwe. His other area of expertise is environmental issues in mining, including the effects of Acid Mine Drainage.

Winnie V. Mitullah is a Senior Research Fellow and Lecturer at the Institute for Development Studies, University of Nairobi, Kenya. She has researched, consulted and published in areas of provision and management of urban services, institutions and governance, community development and the role of stakeholders in development.

Salvador Mondlane Jr. earned his first degree in Geology at Eduardo Mondlane University in 1990, his MSc. in Mineral Exploration and Mining Geology from the University of Leicester, UK in 1993, and is currently a Ph.D. candidate at the University of Zimbabwe. He is currently a lecturer at Eduardo Mondlane University, and his research interests include the socioeconomic and environmental impacts of small-scale mining. Email: mondlane@science.uz.ac.zw

Bill Murray is a development economist who has undertaken many projects in Mongolia over the last decade for the World Bank, the Asian Development Bank and bilateral agencies. He has also worked in more than two dozen other countries, and is Managing Director of Murray Harrison Ltd, a consulting company established in 1981, with offices in London, Hong Kong and Canberra. Email: bill.murray@murrayharrison.com

Jason S. Ogola is a Professor of Mining and Environmental Geology and Head, Department of Mining and Environmental Geology as well as Deputy Dean in-charge of the School of Environmental Sciences, University of Venda for Science and Technology, South Africa. He has wide experience spanning over 25 years in research and teaching in the fields of mining/economic and environmental geology. His research includes evaluation of mineral resources and their rational utilization, integrated artisanal gold mine management, and impacts of mining on the environment. Email: ogolaj@univen.ac.za

Monica A. Omulo is a Lecturer in Environmental Sciences, Department of Environmental Studies, Maseno University, Kenya. She holds a B.Sc from the University of Nairobi, and a M.Sc. Degree in Environmental and Ecological Sciences and an M.Phil. in Environmental Management from Lancaster University, UK. She has published in the areas of environmental resource management, groundwater quality, integrated artisanal gold mine management, and the impacts of mining on the environment.

Meredith M. Redlin is an Assistant Professor of Rural Sociology, Department of Rural Sociology, South Dakota State University, South Dakota, USA. Her research focuses on social theory; sociology of agriculture; environmental sociology; rural diversity; and stratification in the United States and internationally. Email: meredith_redlin@sdsu.edu

Michael Renner is a Senior Researcher at the Worldwatch Institute, a non-profit research organization in Washington DC concerned with sustainable development. He serves as Project Director for the Institute's annual "Vital Signs" book, a collection of critical global trends and indicators. His research and writing has focused on the linkages between the environment, natural resources, and conflict. Email: mrenner@i-2000.com

Malebabo Sakoane is a Mining Policy Analyst at the Minerals and Energy Policy Centre in Johannesburg, South Africa. He holds a B.Sc. degree in Applied Environmental Science, a post-graduate diploma in Environmental Engineering, and a Masters degree in Environmental Analysis. Email: Malebabo@mepc.org.za

Kamaleddin Sheikholeslami Salmasi is a Course Lecturer in the Faculty of Management and a Research Fellow in the Department of Mining, Metals and Materials Engineering, McGill University, Canada. He holds a B.Sc. in Industrial Management and an MBA from the University of Tehran, a D.P.A. (Graduate Diploma in Public Administration) from Carleton University, Canada, and a Ph.D. in Mineral Economics and Mining Engineering from McGill University. Email: salmasi@management.mcgill.ca

Andrea Seeling is a consultant at the Asistente Codirección, Proyecto "Apoyo a la Pequeña Explotación Minera" (APEMIN), Avenida del Minero in Bolivia. Email: aseeling@yahoo.com

Bingjia Shao is an associate professor and the vice director in the Marketing Department of the College of Economics and Business Administration, Chongqing University in China. His research interests are e-commerce and agricultural economic management.

Xunpeng Shi is an economist and policy researcher at the Coal Research Institute (CCRI) in China. His current research focus is coal and energy economics and policy, mineral law and policy in China. He is currently studying for an LLM in Mineral Law and Policy at the Centre for Energy, Petroleum and Mineral Law and Policy, University of Dundee, Scotland. Email: shixunpeng@yahoo.com

Dennis Shoko holds a Ph.D. in Sedimentology from the University of Zimbabwe. He is currently a lecturer at the University of Zimbabwe's Geology Department. His research interests include the socioeconomic and environmental aspects of small-scale mining.

Suzanne E. Tallichet is an Associate Professor of Sociology, Department of Sociology, Social Work, and Criminology, Morehead State University, Kentucky, USA. Her research focuses on gender and work; social psychology; rural sociology; and qualitative methodology. Email: s.tallic@moreheadstate.edu

Håkan Tarras-Wahlberg holds an undergraduate degree in Geology from the University of Cape Town and Masters and Doctoral degrees in Geography from Cambridge University. He works as a consultant within the wider field of natural resource management with many of his assignments related to the mining sector. An important part of his research focuses upon metal contamination in water and water quality policy issues, and the environmental impact of mining activities. Email: hakan.tarras-wahlberg@hifab.se

Rachael van der Kooye works as a free-lance journalist with Suriname's national newspaper *de West* and for the environmental NGO, Conservation International. Since 1991, Ms. van der Kooye has followed the Suriname mining boom closely as an investigative journalist, investigating and reporting on diverse social and environmental issues related to small-scale gold mining, including environmental pollution, Brazilian gold miners, and conflicts between and among multinationals, local gold miners, and the national government.

Marcello Veiga has worked for the past twenty five years as a metallurgical engineer and environmental geochemist for mining and consulting companies in Brazil, Canada, the US, Venezuela, Chile and Peru. Dr Veiga has worked extensively on environmental, social and economic issues related to artisanal gold mining in numerous Latin American countries. As a professor at the Department of Mining Engineering, University of British Columbia, Canada from 1997 to 2002, his research topics included sustainable development in mining, mine closure and reclamation, remedial procedures for metal pollution, in particular mercury pollution, bioaccumulation, and adverse effects of metals in the environmental, Acid Rock Drainage and mineral processing. He is currently working for the United Nations Industrial Development Organization (UNIDO), in Vienna, looking into issues related to artisanal mining in Asia, Africa and South America. Email: m.veiga@unido.org

Zhong Ziran joined the former Ministry of Geology and Mineral Resources in 1991, and took positions of Deputy Chief for the Division of Mining Law Affairs from April 1994 to June 1996, Division Chief from July 1996 to March 1997, and Deputy Director General for the Bureau of Mineral Development Management from April to August 1997. He was Deputy Director General for Oil and Gas Administration at the National Commission of Mineral Resources from September 1997 to June 1998. He holds a Bachelor of Engineering degree from the Department of Geology, Hefei University of Technology, and M.Sc. and Ph.D. degrees in mineral economics from the Graduate School of the Chinese Academy of Geosciences. Email: zrzhong@mail.mlr.gov.cn

General Introduction

GAVIN M. HILSON

What is "artisanal" and "small-scale" mining? For decades, experts have debated this question but have been unable to reach a compromise. The efforts that have been made at various international conferences and associated workshops to develop plausible universal definitions have complicated matters even further.

In fact, the disproportionate amount of time spent trying to define "artisanal" and "small-scale" mining over the years has, in turn, precipitated their gradual acceptance as mainstream classificatory terms for the most rudimentary branches of the mining sector. This is largely because of increasing realization that no one criterion is sufficient to define either term. Using examples of ore grade and market price, Hollaway (1997, p. 35) illustrates this point in the following passage:

> A gemstone mine treating one ton a day, a gold mine treating ten tons a day and a coal mine producing a hundred tons a day can all be artisanal/small-scale mines. Such mines may employ, or attract, a handful or tens of thousands of workers, so the size of the workforce is also an unsatisfactory criterion. The value of production may be enormous, as in Brazil's Serra Peleda which generated over $200 million in one year at its peak, or tiny, as with an isolated African woman panning a few specks of gold on a remote stream.

This book, however, is unconcerned with crafting definitions of artisanal and small-scale mining. Nor does it attempt to further arguments in favour of developing baseline interpretations of either term. Rather, it seeks to provide an overview of the impacts mining activities at the grassroots have had on rural communities and rural development. In doing so, the authors that have contributed to this edited volume treat both "artisanal" and "small-scale mining" as the low-tech, labour intensive branches of the mining industry prevalent in developing world economies.

SCOPE

Artisanal and small-scale mining is more than simply an industry with the potential to contribute positively to foreign exchange earnings and employment; it is a way of life. Its participants, most of whom were lured into work by the prospect of gaining wealth quickly, include rural community dwellers, nomadic peoples, seasonal subsistence farmers, and retrenched large-scale mine workers. However, despite providing higher wages than comparative rural sectors of industry, artisanal and small-scale mining is generally associated with a deteriorated quality-of life. Its operators are often malnourished and overworked because they are subjected to harsh conditions and tedious labour; prospective mining regions generally lack adequate sanitary facilities, and thus propagate a plethora of diseases; and the majority of the industry's operations are haphazard, threatening the livelihoods of both workers and surrounding villagers. Outside of a small group of initiatives undertaken by the World Bank, the International Labour Organization, and various branches of the United Nations, little research has been carried out to gain an improved understanding of the key anthropological characteristics of the industry, and its socio-economic impacts.

Yet, governments around the world continue with reckless abandon to regularize resident artisanal and small-scale mining operations, despite clearly possessing insufficient knowledge of its mechanics and socio-cultural characteristics. As one African mineral policy-maker put it during a personal interview, "almost every government in the Third World lacks the very information needed to implement effective artisanal and small-scale mining policies and regulations … but most still go ahead and implement them anyway". It is safe to say that most governments have inadequate data concerning the locations of scores of relevant artisanal mining operations; what each is extracting and in what quantities; the nature of the technology being deployed at sites; and the groups of people involved.

The purpose of this book, therefore, is to examine the socio-economic impacts of artisanal and small-scale mining, and, in the process, address the important issues that should be taken into account during important industry policy-making activities.

CONTRIBUTORS

The contributors to this volume are based in 22 countries – Australia, Austria, Bolivia, Brazil, Canada, Chile, China, Ghana, India, Indonesia, Kenya, Mozambique, the Philippines, South Africa, Suriname, Sweden, Switzerland, Tanzania, the UK, the USA, Zambia, Zimbabwe – and work in academia, government, the international agency and NGO sector, consultancies and industry. They have a wealth of experience and interest in different facets of artisanal and small-scale mining. Some have overlapping experience, which adds to the strength of this book.

ORGANIZATION OF THIS BOOK

The book is divided into five sections. The first section explores the policy and legal environment for artisanal and small-scale mining. To effectively regulate, and support, such an industry, a plethora of issues must be addressed. The fact that artisanal and small-scale mining has long been treated as an informal sector of the economy requires that governments initially establish sound regulatory regimes for operations. This involves, *inter alia,* drafting pertinent legislation, creating procedurally simple licensing systems for small-scale miners, and analyzing the composition of the industry.

Once the constituents and significance of a de facto and de jure legislative framework have been realized, governments should begin studying more in depth the nature of resident artisanal and small-scale mining operations, and use the recovered information to refine and strengthen existing regulations and policies, and establish robust industry support services. The second section of the book – "Artisanal and Small-Scale Mining, Labour and the Community" – examines the chief characteristics of artisanal and small-scale mines, providing policy-makers with pertinent information concerning the state of the industry today.

The remainder of the book is divided into three sections – "African Case Studies of Artisanal and Small-Scale Mining", "Asian Case Studies of Artisanal and Small-Scale Mining", and "Latin American Case Studies of Artisanal and Small-Scale Mining". The collective aim of these sections is to provide the reader with up-to-date information on the state of artisanal and small-scale mining within the three continents where it occurs. The case studies presented provide an interesting and varied selection of perspectives concerning the socio-economic importance of the industry and the problems it has caused.

The book aims to facilitate a much-needed exchange of ideas between international agency researchers, academics, and government policy-makers, and seeks to initiate further research in an industry in dire need of improvement and support.

REFERENCE

Hollaway, J. (1997). Policies for artisanal and small scale mining in the developing world – a review of the last thirty years. In A.K. Ghose (Ed.) *Mining on a Small and Medium Scale* pp. 35–42. Intermediate Technology Publications, UK.

Part I

Policy and Regulatory Issues in the Small-Scale Mining Industry

1

Introduction Part I

GAVIN M. HILSON

A report produced by the South African Branch of the recently-concluded Mining, Minerals and Sustainable Development (MMSD) project, states that, artisanal and small-scale mining is "typically practised in the poorest and most remote rural areas by a largely itinerant, poorly educated populace, men, and women with few employment alternatives" (MMSD, 2002). In short, artisanal and small-scale mining activities are now widespread simply because there are few alternative employment opportunities in many parts of the developing world. An increasing number of rural inhabitants are turning to artisanal and small-scale mining in order to feed their families, further reinforcing assertions that the industry is "poverty-driven".

Some 10–15 years ago, many governments, amid claims of having finally recognized the potential economic value of the mineral outputs derived from artisanal and small-scale mines, began implementing a series of policies that both legalized resident activities and regulated their production. Regional definitions of "artisanal" and/or "small-scale mining" were soon crafted and eventually embodied into national mineral legislation. Licensing and registration schemes were then created and enforced upon all existing and new mine operators; those choosing to function outside this newly-forged "legalized" industry segment were viewed as "clandestine" and "illegal".

The impulsive decisions to license this rudimentary sector of industry have produced mixed results. On the one hand, by better regulating production, governments have inevitably improved control of mineral outputs that contribute positively to both national export earnings and foreign exchange. For example, gold and gemstones reportedly valued at US$1 billion a year are produced and collected by governments in sub-Saharan Africa. Moreover, in China, gold production from small-scale mining is currently worth about US$200 million a year; in Bolivia and Brazil, US$180 million; US$140 million in Indonesia; and about US$250 million in Peru (ILO, 1999). Artisanal and small-scale mining also provides employment to millions world wide; it has been conservatively estimated that between 11.5 and 13 million people are employed in the industry directly, and that as many as 100 million depend on its existence for their livelihoods.

On the other hand, by rapidly implementing a series of policies and laws for artisanal and small-scale mines, governments have, in effect, introduced formalities to what has long been treated as an informal sector of industry; many miners have been unable, or are unwilling, to adapt to new regulatory regimes. As a result, authoritative bodies now face a plethora of *unanticipated* problems, such as rampant "illegal" (transient) mining, illicit mineral marketing activities, and land use disputes between small- and large-scale mining parties. These problems have been exacerbated by consistent failure on the part of governments to provide resident artisanal and small-scale miners with adequate support and extension services. The industry's operators frequently lobby for increased financial assistance, improved equipment, and training, and regularly voice complaints over how they receive secondary treatment to large-scale mine operators.

The primary objective of this introductory section is to examine many of the important regulatory issues and concerns in the artisanal and small-scale mining industry. It serves as an important backgrounder on the industry, profiling key policy and legislative-related issues.

The first chapter, by Edmond Bugnosen, provides pointers to governments keen on implementing sector-specific legislation for small-scale mining. Apart from reviewing existing small-scale mining legislation in selected developing countries, the author uses the results and lessons learned to develop an appropriate small-scale mining legislative framework for Mozambique (as an illustrative case study).

In Chapter 3, John Andrew and Gavin Hilson describe the pervasiveness of the land use conflicts that often occur between small- and large-scale mining parties. The authors explain how, as a direct consequence of governments prioritizing the infusion of foreign large-scale mining activity, thousands of indigenous artisanal miners have been displaced. The authors outline potential mediation procedures for resolving the inevitable ensuing land use disputes.

Chapter 4, by Stephens Kambani, describes the characteristics of the "illegal mining segment", a label that has been a part of the mainstream mining lingo since the first efforts were made to regularize artisanal and small-scale mining activities. The author begins by examining the major issues concerning illegal mining and marketing in the small-scale mining sector in developing countries. The causes of these undesirable activities are then analyzed, and possible interventions to minimize or eradicate them are presented.

In Chapter 5, Hamid Etemad and Kamaleddin S. Salmasi critically examine Foreign Direct Investment (FDI) theories in relation to mining policies in developing countries. The authors first review FDI theories from a mining perspective, arguing that it provides inadequate guidelines for decision-making and policy formulation in the sector. They further argue that the general theory of FDI is, in part, responsible for the misguided inadequacies of mining policies, particularly those pertaining to the small-scale mining industry. The chapter confirms that the theoretical underpinning of operating decisions in mining policy formulation is in dire need of modification.

The concluding chapter of the section, by Michael Renner, profiles many of the wars that have occurred over mineral deposits – particularly, diamonds. Case studies are presented from Africa and Asia that illustrate how, in many states with weak authority, rebel parties have seized control of mineral resources; this has precipitated war between anarchic groups, some of which have been sponsored by outside governments.

The overarching aim of this introductory section is twofold. First, it describes many of the challenges associated with regularizing artisanal and small-scale mining. Secondly, it illustrates the problems that can come about if cavalier approaches are taken to regulate the industry.

REFERENCES

International Labour Organisation (ILO) (1999). *Social and labour issues in small-scale mines*. Report for discussion at the Tripartite Meeting on Social and Labour Issues in Small-scale Mines. International Labour Organization, Sectoral Activities Programme, International Labour Office, Geneva.

MMSD (2002). MMSD *Southern Africa Draft Report*. Prepared for a meeting of the MMSD Southern Africa Steering Committee held in Johannesburg, 11–12 February, 2002.

2

Small-Scale Mining Legislation: A General Review and an Attempt to Apply Lessons Learned

EDMUND M. BUGNOSEN

Every country, including those in the developed world, has its own small-scale mining sector. It is a basic and inherent component of national mineral industries, as well as the global mining business as a whole. In developed countries, this sector has evolved into a formal and normal enterprise bearing no significant difference – apart from size – to large-scale mineral operations and projects. On the other hand, the small-scale mining sectors of most developing countries are still in an infancy stage in terms of formalization. While evidently employing, and providing a means of livelihood to, a large number of people, the small-scale mining sectors of the developing world are also beset with the numerous negative characteristics and traits that are generally prevalent in any informal undertaking.

Regularization of the small-scale mining sectors of the developing world is a key to addressing the negative impacts of the sector, and, at the same time, optimizing its potentially positive contributions; legislation is an initial and vital step in the formalization process.

This chapter provides a general review of existing small-scale mining legislation in selected developing countries, and uses the results and lessons learned to develop an appropriate small-scale mining legislative framework for Mozambique. The research reported in this chapter is based on the results of the following two projects: (1) a small-scale mining legislation study funded by British Department for International Development (DfID); and (2) a World Bank assignment in Mozambique. The central aim of the chapter is to provide reference material for governments keen on developing and/or amending small-scale mining laws and regulations.

GENERAL REVIEW OF SMALL-SCALE MINING LEGISLATION

The need for small-scale mining legislation

Several international and regional seminars and conferences have helped to underscore the importance of implementing legislative measures for small-scale mining, and have made a number of recommendations for implementing relevant laws and regulations (Table 2.1). Since at least 1978, at the time of the UNITAR International Conference on the Future of Small-Scale Mining, there has been a call for bilateral and multilateral agencies to support government requests for assistance to their small-scale mining sectors through the strengthening of services related to legislation. More recently, at a UNIDO meeting in 1997, it was specifically recommended that the acquisition of a fully transferable mining title be made available to artisanal miners; it was noted that the process should be simple, quick and transparent.

Many countries have reacted to the perceived need to enact small-scale mining legislation, and have introduced laws and regulations (see Table 2.2). Some countries (*e.g.* Zimbabwe, Ghana, and the Philippines) have introduced special legislation to cover the sector, while others have elected to incorporate small-scale mining provisions within general mining acts or laws. Few countries have made an effort to distinguish between "artisanal" and "small-scale" mining,

Table 2.1 Conference recommendations on small-scale mining legislation.

Bilateral and multilateral agencies to support government requests for assistance to their small-scale mining sectors through strengthening of government services relating to legislation. (UNITAR International Conference on the Future of Small-Scale Mining, Mexico, 1978.)
Existing mining legislation applicable to the mining industry should be reviewed to support small-scale mining. (Workshop on Mineral Policy for Small-Scale Mining, India, 1984.)
Attention of participants was drawn to the limits and dangers related to small-scale mining and proposed legislative measures to counter the real dangers to the social and physical environment. (UNECA Workshop on the Enhancement of the Contributions of the African Non-fuel Mineral Sectors Towards the Region's Economic Advancement, Zimbabwe, 1990.)
Government and their agencies should endeavour to provide simple, clear, understandable and stable sets of laws and regulations on small/medium scale mining. (UN Interregional Seminar on Guidelines for the Development of Small/Medium Scale Mining, Zimbabwe, 1993.)
Governments must move towards legalising artisanal mining and streamlining registration and licensing procedures. (World Bank Roundtable on Artisanal Mining, USA, 1995.)
Government should address small-scale mining in their mining codes. (Global Conference on Small/Medium Scale Mining, India, 1996.)
Acquisition of a fully transferable mining title to discoverers (artisanal gold miners) of a deposit should be a simple, quick and transparent process. (Expert Group Meeting on UNIDO High Impact Project: Introducing New Technologies for Abatement of Global Mercury Pollution from Artisanal Gold Mining, Austria, 1997.)

Table 2.2 Small-scale mining legislation in selected developing countries.

Country	Law & Comments
Brazil	General mining code has provisions for small-scale mining; some individual States also have their own small-scale mining legislation
Burkina Faso	Mining Legislation (Kiti No ANVIII-0328TER/FP/PLAN-COOP), has provisions for artisanal exploitation
Chile	Mining Code, 1983, includes small-scale mining provisions
Ethiopia	Mining Proclamation No 52/1993 includes provisions for small-scale mining and artisanal mining
Indonesia	Basic Mining Law No. 11, 1967, has provisions for peoples' mining, which refers to small-scale mining activities of local people and co-operatives
Ghana	Minerals and Mining Law, 1986, has provisions for small-scale mining; Small-scale Gold Mining Law, 1989, represents specific small-scale mining legislation
Guinea	Mining Law, 93/025/CTRN, has provisions for artisanal mining
Guyana	Mining Act, 1992, as amended provides claim system for small-scale mining operations
Lao PDR	Draft mining legislation has provisions for small-scale mining and artisanal mining
Mozambique	Mining Law, 1986, include provisions for acquiring Small-scale Mining Certificates
Malawi	Mines and Minerals Act, 1981, does not specifically mention small-scale mining, but contains provisions applicable to small-scale mining
Namibia	Mining Act, 1994, includes provisions for pegging and the registration of claims as a mechanism for acquiring mining rights by small-scale miners
Papua New Guinea	Mining Act, 1992, as amended, has provisions for small-scale mining
Philippines	General Mining law (RA 7942), 1995, has provisions for small-scale mining; has also specific small-scale mining legislation (RA 7076, 1991 and PD 1899, 1984) and special safety rules and regulations (RD 97-30, 1997) purposely for small-scale mining
Tanzania	Mining Act, 1979, provides provisions for small-scale mining opportunities; legislative framework under review
Uganda	Mining Act, 1964, mentions small-scale mining rights
Zambia	Mines and Minerals Act, 1995, includes provisions for small-scale mining operations
Zimbabwe	Mines and Minerals Act (Chapter 165, provides simple mechanisms of pegging and registration of claims) Mining (alluvial Gold) (Public Streams) Regulation, 1991, is specific to small-scale alluvial gold extraction

wever. Interestingly, some countries do not make specific mention or refer-ce to small-scale mining, but provide licensing and registration mechanisms t actually cater to the requirements of small-scale mine operators.

Reasons for enacting small-scale mining laws and legislation

There are a number of reasons for enacting small-scale mining legislation. They are as follows:

- To curb the illegal mining and illegal trading (smuggling) of metals and mineral products;
- To stop supply of gold to the black market;
- To address environmental destruction or specific environmental concerns such as erosion and siltation of water courses, as in the case of the Zimbabwe Mining (Alluvial Gold) (Public Streams) Regulations;
- To develop and exploit existing small mineral deposits;
- To generate employment opportunities, thereby improving living conditions in rural areas;
- To generate additional foreign exchange earnings;
- To protect and rationalize viable small-scale mining activities;
- To provide mechanisms for collecting government revenues from operations;
- To control the mining rights of cultural minorities within their ancestral lands, as in the case of the *Peoples Mining Act* of the Philippines;
- To provide equitable sharing of a nation's wealth and natural resources;
- To address the exploitation, by the small-scale mining sector, of a specific mineral product, as in the case of gold in Ghana;
- To encourage prospecting and mining of minerals by methods not involving substantial expenditures, or the use of specialized techniques (*e.g.* Ghana); and
- To encourage private citizen prospectors to increase mineral production, and to liberalize the marketing of minerals (*e.g.* Zambia).

Unfortunately, many of these reasons are seen as reactive measures by governments, and do not necessarily match the priorities and immediate concerns of the majority of small-scale miners. Consequently, some of the resultant provisions do not address specific conditions, limitations, and the long-term aspirations of small-scale mining communities.

Defining small-scale mining

Defining small-scale mining can be difficult, and it is perhaps for this very reason why some legislation has avoided providing specific definitions altogether. However, the absence of a universal definition of small-scale mining has not hindered the development of relevant policies and legislation to cover the sector.

Different countries have simply adopted their own definitions (Table 2.3). Some fail to distinguish between "artisanal" and "small-scale" mining, whereas others have adopted more than one definition. It is, however, apparent that the

Table 2.3 Selected definitions of artisanal and small-scale mining.

Surinam: The exploitation of mineral deposits which, due to their mode of occurrence and size, can be mined economically by simple means and techniques.

Mexico: Mines with annual production values not exceeding US$3.00 million, provided that their daily production capacity is less than 200 tons per day (for metal mines) and 300 tons per day (for non-metal mines).

Chile: Mining operations by a person who works a mine property or process plant by himself with or without the family support, a maximum number of five salaried workers, or by legal society with no more than six partners. It also includes operations by a mining co-operative with partners who are actual artisan miners. Small-scale mining is also defined as that producing up to 2,000 metric tons yearly of fine copper, or equivalent.

Philippines: Small-scale mining refers to mining activities, which rely heavily on manual labour using simple implements and methods, and do not use explosives or heavy mining equipment. Also defined as a single unit of operation involving an annual production not exceeding 50,000 metric tons of run-of-mine ore with the following requisites: (a) working is artisanal, either open cast or shallow underground mining without the use of sophisticated mining equipment; and (b) minimal investment on infrastructure and processing plants (not exceeding 10 million pesos).

UN, 1972: Any single unit mining operation having an annual production of unprocessed materials of 50,000 metric tons or less as measured at the entrance of the mine is a small-scale operation.

Brazil: Individual or collective extractive work, using rudimentary tools, manual devices or simple portable machinery – for immediate exploitation of a mineral deposit which, by its nature, dimension, location and economic use, can be worked, independent of previous exploration work, according to criteria set by the National Department of Mineral Production. Also provides a separate definition for "garimpagem" (artisanal mining) as individual work performed by panners with rudimentary forms of mining using manual or portable equipment, and applied only to alluvial, colluvial and eluvial deposits.

Guinea: Small-scale mining refers to exploitation of precious materials, in this case, gold, diamonds and other gemstones found in primary or alluvial deposits, outcrops or sub-outcrops.

Burkina Faso: Provides a definition for artisanal exploitation; refers to activities conducted on ore bodies or deposits by natural or legal persons using traditional techniques or low mechanization levels.

Ethiopia: Small-scale mining are operations to be designated as such by the Minister of which the annual run-of mine ore does not exceed a certain limit, which differs from one mineral product to another and on the nature of mineral occurrence. Also provides a definition for artisan mining, which refers to non-mechanized mining, operations of gold, platinum, precious minerals, metals, salt, clay and other similar minerals, as essentially manual nature carried out by Ethiopian individuals of groups of such persons.

Ghana: Small-scale mining refers to operations of individual Ghanaians or organized groups of Ghanaians (4–8 individuals), or a co-operative of ten or more individuals, entirely financed by Ghanaian resources at a certain limit, and carried out on a full time basis using simple equipment and tools. It also refers to prospecting and mining in an area designated, and which uses specialised technologies and methods not involving substantial expenditure.

majority of country-specific definitions have some similarities. The common features among most definitions include the following:

- The proponents or stakeholders are usually limited to citizens of the country concerned, consequently prohibiting the participation of foreign nationals and companies;
- The use of sophisticated equipment is restricted, with only simple means and techniques normally allowed; and
- Set limits on the level of output and infusion of capital, though these limits differ from one country to another and even by mineral product.

General policy directions of small-scale mining legislation

Nearly all of the small-scale mining legislation implemented to date recognizes and considers the difference between the extraction of mineral resources for commercial motives and personal use. Consequently, appropriate provisions – in the form of informal licensing or gratuitous permits – are incorporated to allow free access to mineral products by landowners and government agencies, when extracted for their own use. This particularly applies to operations engaged in the extraction of non-metallic and building materials. The Philippines has adopted a similar policy for mineral fertilizers (*guano*).

There are conflicting concepts or policies regarding the transformation and upgrading of small-scale mining to larger-scale operations. On the positive side, the legislation of Zambia requires holders of a small-scale mining licence to upgrade to a large-scale licence when mining operations reach a substantial scale. Guinea has adopted a similar policy. However, other countries have adopted completely different policies. Ethiopia, for example, provides provisions to stop artisanal mining operations when it is seen that a subject mineral deposit requires a more advanced method of development. The proposed legislation of PDR Lao has incorporated a similar provision. Areas designated for small-scale mining in both Tanzania and Indonesia may also be declassified or abolished and opened for large-scale mining operations, provided the area or deposit is known to have potential for large economic exploitation.

It appears that most small-scale mining legislation has the general tendency to limit the technological development of small-scale mining operations. This is also manifested in such (small-scale mining) licensing provisions as (a) workings to be limited to a certain depth; (b) prohibition of the use of explosives; (c) prohibited use of mechanised equipment; and (d) non-application of advanced processing technologies.

GENERAL PROVISIONS OF SMALL-SCALE MINING LEGISLATION

Environmental protection

Environmental protection concerns, and pollution control measures, are addressed in different legislation in various ways. Some countries (*e.g.* Zambia,

Ghana, Papua New Guinea, Philippines, Bolivia and Ethiopia) have made the preparation and approval of environmental protection plans and programmes a mandatory procedure in the obtaining of small-scale mining rights. These environmental plans, however, are not necessarily as detailed as those required for large mining operations.

Other countries have targeted specific environmental issues such as mercury pollution (Ethiopia, Tanzania, and Philippines) and river siltation (Zimbabwe), by providing specific provisions in legislation itself.

Certain countries, such as Guinea and Zambia, also require licence holders to post surety bonds to ensure compliance with environmental protection and pollution control plans. In Ghana and the Philippines, provisions for an environmental tax (*Land Reclamation Fund of Ghana* and the *Special Fund of the Philippines*) exist. In both cases, funds are retained from the sales of output derived from small-scale mining production, and then used to rehabilitate small-scale mining areas.

Health and safety

Safety concerns in small-scale mining appear to be generally covered by overall health and safety legislation. In effect, small-scale mining operations are normally treated in the same manner as large mining operations. There are, however, cases like India, where small-scale mines are exempt from the application of certain safety provisions altogether. The Philippines appears to be the only country to have separate safety regulations for small-scale mining. Its regulations (Small-Scale Mine Safety Rules and Regulations) were only recently adopted in 1997.

Marketing

There are no common approaches taken by governments towards the marketing of mineral products produced at small-scale mines. Typical practices include: (a) separate licensing of mineral traders, as in Uganda, Ghana and Zambia; and (b) government regulated buying schemes, particularly for gold and precious stones (Burkina Faso, Republic of Guinea, Zimbabwe and the Philippines). Otherwise, permission to market or export mineral products is generally part of a small-scale mining licence or permit.

Penalties and incentives

Under the legislation of various countries, penal provisions such as fines, imprisonment and cancellation of licences or permits are provided for violations.

On the other hand, incentives for small-scale mining are also provided in some of the legislation. Examples of incentives include: (a) tax exemptions (Philippines), (b) exemption from royalty fees (Zambia), and (c) provisions for prospecting grants from governments (Zimbabwe).

Decentralised regulation

Participation of local government in the regulation of small-scale mining activities is provided for in some of the legislation. It takes the form of decentralized licensing (Indonesia, Ghana, Philippines), and in the actual management of the resource, as in the case of Zimbabwe and Sierra Leone.

LICENSING SMALL-SCALE MINES

There are several types or procedures involved with licensing small-scale mining operations, including the following:

1) *Informal or undocumented license*: This kind of small-scale mining "license" does not provide a written document *per se*, nor does it involve any formal procedures of application and processing. It is generally in the form of pronouncements or specific provisions and incorporated into the relevant mining legislation. Apparently, this provision or scheme is provided to accommodate specific groups, particularly indigenous members of the community and landowners. One example of this type of practice is the right of *libre approvechamiento* or free digging. This is extended to Venezuelan small-scale miners, to allow the exploitation of alluvial minerals on unappropriated lands or in rivers. Land owners in Papua New Guinea are also allowed to extract gold from their lands without formal licensing, provided that only panning or other primitive procedures are used. The same policy has also been adopted in the Dominican Republic. The principle of informal licensing is also embraced by other countries (Zambia, Uganda, Philippines) to legalise the extraction of mineral products by landowners and government agencies for their own use, particularly in cases involving the extraction of building materials.
2) *Strata licensing*: This system provides mining rights over a certain vertical extent/depth of a mineral deposit or a given area. Countries that embrace this concept include Ethiopia, Indonesia, Papua New Guinea, Guinea and the Philippines. The depth set by Ethiopia is 15 metres, and in the Philippines, it is 50 metres. In the case of Papua New Guinea, the depth is determined on a case-by-case basis.
3) *Group permitting*: This system, which is applied under the Zimbabwean *Alluvial Gold Mining Act* and the Philippine *People's Small-Scale Mining Law*, provides a permit over a specified area for a group of individuals. It simplifies the procedures, makes monitoring much easier, and, at the same time, encourages the grouping of small-scale miners into cooperatives and similar associations.
4) *Licensing by type or name of minerals*: This is the most common designation of small-scale mining licenses. Apart from separate licences for quarry resources and building materials, specific minerals such as gold, diamonds, and precious stones, are licensed separately by some countries. Gold, diamonds and gemstones are the most common minerals targeted for small-scale mining licensing. Other mineral products include building materials,

industrial minerals and mineral fertilizers. In Mongolia, the following three categories of minerals are recognised: strategic (including precious metals, gemstones, oil and coal), specific (ores and some non-metallic minerals) and common. Licenses for strategic and specific minerals are issued by the Minister for Geology and Mining, while licenses for common minerals are issued by local governors. Indonesia has adopted a similar three-part categorisation scheme, based on "strategic", "vital" and "others".

5) *Staggered and single licence*: Single licensing is a system that provides one overall licence to cover all related activities, including exploration, development, mining extraction, processing, and the marketing of mineral products. It is a simple process, and is considered the most appropriate licensing method for small-scale mining operations. A staggered licensing system, on the other hand, provides separate licences for each stage of mining – *i.e.* a separate licence or permit for prospecting, exploration and extraction. This is normally used in the licensing of large-scale mining operations, but has also been adopted for licensing small-scale mining, albeit with certain variations, in certain developing countries.
6) *National government licence or local government licence*: This is a case exclusive to Brazil, where small-scale mining (*garimpagem*) is regulated by the national government via the National Department of Mineral Production. However, individual States can also have regulations and licensing procedures for the same sector, but within their jurisdiction. Otherwise, the normal practice involves one licensing system for the whole country, which is usually enacted at the national level. It is only the implementation that is localised or devolved to local government, and this is not even common practice.

Most countries provide short-term (*i.e.* one to two years' duration) small-scale mining licenses. The notable exceptions are Ethiopia and Zambia, where 10-year licenses are provided but these countries also have separate provisions for a shorter-term artisanal mining licence.

In terms of the areas covered by small-scale mining licences, based on available information, PDR Lao provides the largest coverage (five square kilometres) for a small-scale mining licence, followed by Zambia, where a coverage of 400 hectares is provided. Otherwise, most countries provide less than 20 hectares of area per concession. Some countries also provide different areas of coverage based on the type of mineral (Ethiopia, Kenya, Uganda), and on the legal personality of the applicant, as in Ghana and Indonesia.

Most small-scale mining licences are treated as transferable assets. The notable exceptions are the small-scale gold licences of Ghana, the claim offered by the Tanzanian Government, and the artisanal licence of PDR Lao. However, Lao also provides a separate and transferable small-scale mining licence.

DIFFICULTIES WITH LEGISLATING SMALL-SCALE MINING

The main problem in legislating small-scale mining operations appears to be in situations where mining areas are already covered by other mining rights, such as prospecting permits and exploration permits related to large mining projects. The same problem also occurs when prospecting or exploration permits for large mining operations are granted over areas where there are existing small-scale mining operations. The Philippines tried to address this problem through special legislation (PD 1150) to regulate gold panning and sluicing inside existing mining claims. In a way, the legislation allows double licensing over the same area, but proved ineffective due to non-issuance of consent by concession holders.

There are, however, other mechanisms available to legalise small-scale mining within existing mining concessions. These include tributing arrangements, operating agreements and contract mining. Most are initiated by the parties themselves, or are performed with the assistance of authorities, as in the case of Papua New Guinea, where the government, through the Department of Mines, has actually prepared standard tributing contracts.

This "privatised licensing scheme" of small-scale mining has been reportedly adopted in Mongolian legislation as well; in Mongolia, licence holders are encouraged to contract out (to small-scale miners) portions of the deposit that may not be viable to exploit using conventional (large-scale) industrial techniques. On the other hand, it is mandatory in many countries for large-scale concession holders to gradually reduce the size of their concessions as exploration progresses. By this process, small deposits that are of no interest to the licence holders are effectively relinquished. These areas should then revert back to the government, and opened for location by small-scale miners.

IMPACT OF SMALL-SCALE MINING REGULATIONS

A recent ECOSOC report (ECOSOC, 1996) discusses in detail the socio-economic impacts of small-scale mining. However, each is treated in general terms; the report fails to provide a clear relationship between these impacts and small-scale mining legislation. There is very little documented or anecdotal information available on the impact of small-scale mining legislation. Of the material that does exist, most generally originates from narrative accounts, and, collectively, covers only a small group of countries.

In Ghana, the promulgation of the *Minerals and Mining Law* in 1986, and the establishment of the Minerals Commission to regulate and manage the utilization of the country's mineral resources and co-ordinate policies in relation to them, has been instrumental in the rapid growth of the gold-mining sector (Acquah, 1996). The later adoption and implementation of small-scale mining legislation has reportedly resulted in the provision of systematic support to the small-scale mining sector. Between 1991 and 1994 exports of gold and diamond increased by 61% and 142% respectively, largely because of reforms made to small-scale mining policy (Ofei-Aboagye, 1995). It has also been observed that the legislation resulted in better and realistic pricing of minerals, the collection of a land reclamation fund through buyers, and the creation of over ten thousand jobs after the first year of implementation (correspondence from Dr KA Kwakye).

In Zimbabwe, the Mining (Alluvial Gold) (Public Streams) Regulation has been seen as a genuine attempt to control panning problems. It has decentralized power to local authorities, helped to stem the problem of migration, and attempts to reduce and control environmental degradation. However, the regulation has failed to stem the rise in illegal panning (Maponga & Musabayana, 1996), and there is lack of evidence showing practical rehabilitation of permitted sites. The regulation has been described as lacking a long-term vision for the activities of alluvial gold panners (Mohiddin, 1997).

In Brazil, although the need for effective licensing and regulatory procedures is recognized, "the great majority of mining operations are unauthorized and unregulated and the environmental and social consequences are serious" (Bezerra et al., 1996). The impact of the Brazilian small-scale mining policy has been described by Brazil Mining Associates Consultants (BRMINE), who claimed that the legislation is inadequate and inappropriate. The legislation and constitutional reforms in relation to small-scale mining (*garimpo*) were intended to promote mining co-operatives by creating *garimpo* reserves. Although many co-operatives were licensed and are functioning, most small-scale miners prefer an individual panner's license (*lavra garimpeira*). They can then associate themselves as a civil association rather than a co-operative that must comply with co-operative requirements (BRMINE, 1997).

In the Philippines, the licensing of small-scale sand and gravel extraction activities has provided local governments with a new source of revenue, particularly in the form of royalty fees. Some local governments have even instituted checkpoints to ensure that all extraction is performed legally and that royalty fees are paid. This has created several small business opportunities such as trucking, manufacturing of building materials (hollow blocks), and the trading of sand and gravel concessions.

A REVIEW OF CONCLUSIONS AND LESSONS LEARNED

The following conclusions can be drawn from this work:

1) There is wide recognition of the importance of small-scale mining industries and the need for an appropriate regulatory framework. Many countries have enacted laws and regulations to cover the sector. Even the few countries that do not appear to treat small-scale mining as a special area of legislative concern have mechanisms in their general mining legislation that allow for the licensing of small-scale mining operations.
2) It is important to cover all the major issues related to small-scale mining, such as the illegal trading of mineral products, health, safety, environmental degradation, and taxation, in the legislative framework.
3) The promotion of the mining sector in developing countries as a whole remains biased towards large-scale mining with the aim of attracting foreign investment. Thus, there is a need to balance this trend by implementing policies for small-scale mining that promote locally-initiated and owned mining enterprises, and by emphasizing its integration into the formal mining industry.

4) A large proportion of small-scale miners worldwide continues to operate outside of the regulatory framework. In many cases, this is because of a lack of area available to small-scale miners after vast mining concessions are granted to large mining companies. However, it may also suggest that (i) the legislation is not perceived by the miners as being advantageous to them, (ii) the provisions of the legislation are inappropriate, and (iii) there is a lack of institutional and administrative capacity, and capability to implement the legislation. Therefore, it is important to consider the particular conditions, needs, and limitations of miners, as well as the institutional and financial resources required to implement legislative measures.
5) Government motivation to legislate small-scale mining does not necessarily match the priorities of small-scale miners. It is therefore important to also consider the aspirations and principal priorities of miners (refer to Part 2 of this book).
6) The various restrictive provisions of many existing small-scale mining licences, such as: (i) a ban on the use of explosives and machinery, (ii) their short-term duration, (iii) the limited depth of underground workings allowed, and (iv) lack of security of tenure, which hinders the development and inhibit the efficiency of small-scale mining. It also leads to illegal practices in an otherwise legalised activity. These provisions, therefore, should not be adopted or amended accordingly.
7) It is important to provide tradable and long-term small-scale mining licences, which help create an entrepreneurial atmosphere in the small-scale mining sector. This also allows for proper mine planning and closure, and supports the implementation of environmental protection programmes.

Based upon these conclusions, the next section of the chapter sketches a plausible small-scale mining regulatory framework for Mozambique.

ANALYSIS AND PROPOSAL FOR MOZAMBIQUE

Background

Small-scale mining is believed to provide a means of livelihood, jobs and business opportunities to about 500,000 persons in Mozambique. The activity is mainly poverty driven; hence, it is people-initiated and a direct poverty alleviation measure for the country as a whole, with little cost, and limited intervention, on the part of the government. The small-scale miners of Mozambique, however, do not only generate cash income for themselves, their families and their respective communities. They also serve as grassroots prospectors; for a country where there is little formal exploration activity by large mining companies, this is vital to the existence and development of the country's mining sector. (See "African Case Studies of Artisanal and Small-Scale Mining" for more in-depth analysis of artisanal and small-scale mining in Mozambique.)

On the other hand, small-scale mining in Mozambique – as in many other countries – has had a wide range of adverse impacts on people, communities and the environment. It also contributes to a number of unwanted economic activities such as the illegal trading and smuggling of mineral products, non-payment of taxes, and corruption.

The government of Mozambique is fully aware of both the positive and negative contributions of the small-scale mining sector to its people and the economy, and, therefore, has acted accordingly. It has already instituted a licensing system (mining certificates), which is issued at the provincial level, thereby making it more accessible to miners. A Department of Small-Scale Mining within the National Mining Directorate was also established to provide support and technical assistance. In addition, a Mining Development Fund (Fundo de Desenvolvimento Miniero) has also been established to provide financial support to the sector. However, these measures have proven inadequate, in effect, underscoring the need to reform and improve the general legislative framework for this sector of industry. The proposals in the succeeding paragraphs, which are influenced by the foregoing review, are being presented for use in the on-going reform.

Small-scale mining legislative proposal for Mozambique

It is clear from the foregoing review of small-scale mining legislation, and from the background conditions of the small-scale mining sector of Mozambique, that any reform of the legislative framework on small-scale mining will need to address specific aspects (both positive and negative) of operation and effects. It is also important to note that the desired effects and results of any laws and regulations can only be attained if they are properly and efficiently implemented; this is only accomplishable if appropriate and effective institutions are established.

Each country, however, will also have specific priorities that need to be given particular consideration. Unfortunately, priorities can change with development of the sector, potentially making specific proposals inappropriate, untimely or obsolete; this appears to have been the case of Mozambique. Therefore, the proposals being presented mainly cover the principal components or factors believed to be necessary to improve and transform the country's small-scale mining legislative framework.

Access to mineral properties

Small-scale miners either compete with everybody else to obtain mineral properties, or, as in the case of Mozambique, receive (affirmative) assistance from governments through identification and segregation of areas solely for small-scale mining. There are few cases wherein the practice of segregating areas for small-scale mining appears to have been effective and appropriate. Based on the general review, notable cases include pebble extraction in the Philippines, emerald mining in Zambia, and alluvial gold extraction in Zimbabwe. The practice (of segregating), however, is dependent on first establishing the mineral potential of prospective small-scale mining regions. Segregated areas with undetermined mineral potential will do little to sway miners to operate in "designated" regions. The drawback of this practice, which is also seen in Mozambique, is that it requires governments to mobilize significant resources in order to accurately determine mineral potentials; in many cases, however, governments lack the resources to undertake rigorous prospecting activities.

It is therefore proposed that Mozambique continue to pursue its current practice of segregating areas for small-scale mining, but only as a reactionary measure to legitimize or legalize operations in existing small-scale mining areas and other regions expected to be discovered by bands of artisanal miners. The government should not take the initiative of prospecting and exploring areas to be segregated for small-scale mining, but rather provide other licensing schemes (discussed in the succeeding paragraphs) to allow small-scale miners to compete with other parties interested in acquiring mineral properties.

Licensing

The current legislation in Mozambique provides a system of licensing small-scale mining operations through mining certificates over segregated areas, as mentioned above. However, the practice appears to only address the concerns of artisanal miners who are mainly involved in mining as a means of livelihood, and does not cater to the needs and requirements of small-scale mining entrepreneurs. The system, therefore, deters grassroots prospecting by local inhabitants who have proven to be effective discoverers of mineral deposits. Similarly, the system does not encourage the development of a local mining-business culture in the country, which is very important towards developing a sustained mining industry in Mozambique.

Consequently, it is proposed that Mozambique defines, adopts and issues three types of mining licenses/permits for small-scale mining operations. These licenses or permits should be based on the main objectives of the applicants undertaking the mining activity, and should include the following:

1) *Mining certificate* – an improved or amended version of the existing mining certificates, to be issued to artisanal miners who are mainly involved in mining as a means of livelihood.
2) *Gratuitous small-scale mining license* – to be issued to applicants who wish to extract mineral products (especially building materials, decorative stones and mineral fertilizers) for their own use, and not for commercial purpose.
3) *Commercial small-scale mining license* – to be issued to applicants who wish to extract mineral products for business and commercial purposes.

It is further proposed that the general features and provisions of the abovementioned licenses as shown in Tables 2.4, 2.5 and 2.6 be considered and/or adopted.

Administrative and institutional management

At present, there are two separate government institutions dealing with small-scale mining in Mozambique. They are: (a) the Small-Scale Mining Department within the National Mining Directorate, and (b) the Mining Development

Table 2.4 Elements of a *mining certificate* that should be addressed in Mozambican small-scale mining legislation.

Items	Provisions/features
Purpose	To legalize operations in small-scale mining segregated areas
Rights covered	Prospecting, exploration, extraction, processing and marketing
Who can apply	Mozambican nationals (individuals, partnerships, cooperatives and associations)
Minerals covered	All minerals and metals, including sand and gravel and other building materials
Size/area	To be determined during segregation process
Duration	To be indicated during the segregation process
Transferability	Transferable and tradable to qualified holders (Mozambican nationals)
Operational restrictions	None
Tax provisions, fees	Limited to registration fees and occupational tax, if any
Issuing office	MINERE, Provincial office

Table 2.5 Elements of a *gratuitous mining certificate* that should be addressed in Mozambican small-scale mining legislation.

Items	Provisions/features
Purpose	To legalize extraction of mineral products for non-commercial use
Rights covered	Extraction and processing
Who can apply	Mozambican individuals and government agencies
Minerals covered	Building materials, decorative stones, and mineral fertilizers
Size/area	To be specified upon application
Duration	To be specified, or upon extraction of volume required; which ever comes first
Transferability	Non-transferable, non-tradable
Operational restrictions	To be specified
Tax provisions, fees	Limited to registration fees
Issuing office	MINERE Provincial office

Fund (Fundo de Desenvolvimento Miniero). These offices should be drastically reformed and their functions should be streamlined to provide distinct and separate responsibilities. It is proposed that the Small-Scale Mining Department be developed purely as a regulatory body within the office of the National Mining Directorate, while the Fundo de Desenvolvimento Miniero be strengthened to serve as an independent institution with its own legal personality, and

Table 2.6 Elements of a *commercial small-scale mining license* that should be addressed in Mozambican small-scale mining legislation.

Items	Provisions/features
Purpose	To legalize commercial small-scale mining operations
Rights covered	Prospecting, exploration, extraction, processing and marketing
Who can apply	Mozambican nationals (individuals, partnerships, cooperatives, associations and companies), partnerships and companies with at least 50% ownership by Mozambican nationals
Minerals covered	All minerals and metals, including sand and gravel, and other building materials
Size/area	Dependent on the possible extent and occurrence of the mineral deposit
Duration	10 years, renewable for the same period
Transferability	Transferable and tradable to qualified holders
Operational restrictions	None
Tax provisions, fees	Usual taxation of mining operations
Issuing office	MINERE, National Directorate for Mines offices

tasked mainly to provide technical assistance and extension services to small-scale miners.

The main function of the regulatory body should include processing and the issuance of licenses, maintenance of statistical records, and ensuring compliance with other regulations (health & and safety, environmental protection, tax/rentals, etc.). The technical assistance institution, on the other hand, should be mandated to provide free and/or affordable services to small-scale miners, particularly helping miners to comply with required regulations. The reform may involve the movement and re-assignment of existing personnel in both institutions to ensure that the right persons are appropriately placed. The staffing pattern of the regulatory unit should follow the usual government set-up, but the technical assistance institution should be composed of a multi-disciplinary team (mining engineers, geologists, processing engineers, economists, social scientists, etc.) to ensure that it can easily and fully address all of the basic demands of miners.

Marketing

Marketing provisions should ensure that miners receive a fair price for their mineral and metal products, and that illegal trading (black-market) and smuggling activities are eliminated or minimized. The marketing system should also be linked to the collection of any related government revenues and taxes. A separate licence for mineral traders should also be developed and issued to interested applicants to foster competition among buyers. Similarly, other forms of transparent marketing of mineral products, such as auctions and open-bidding, should be promoted and supported. A government or private monopoly in the marketing and purchasing of mineral products (from small-scale miners) should be avoided (See Chapter 4 for a more comprehensive discussion on the marketing of minerals).

Environmental protection
In the absence of appropriate environmental legislation, it is proposed that environmental protection and pollution control policy in small-scale mining should follow the concept of "the polluter pays". Applicants for a small-scale mining licence, therefore, should be required to prepare simplified pollution control and environmental protection plans backed with performance bonds to ensure that resources will be available to remediate any adverse effects of the operations.

Health and safety
Monitoring of small-scale mining operations for health and safety matters is generally difficult because of a lack of resources, and problems related to accessibility. However, the safety of miners should not be compromised. It is therefore proposed that health and safety violations be linked to the small-scale mining license, such that continuous violations of safety regulations and/or the occurrence of unavoidable accidents should be grounds for suspension and even cancellation of a license or permits.

REFERENCES

Bezerra, O., Verissimo, A. & Uhl, C. (1996). The regional impacts of small-scale gold mining in Amazonia. *MERN Bulletin*, 10.

Brazilian Mining Associates Consultants (BRMINE) (1997). *Mining in Brazil, Legal Framework and Taxation and Panners Co-operatives and their Predictable Developments, Legal Aspects of Mineral Activity in Brazil*, Brazil.

ECOSOC (1996). *Recent Developments in Small-scale Mining*, UN Secretary-General Report (E/C.7/1996/9), prepared for ECOSOC Committee on Natural Resources.

Maponga, O. & Musabayana, D. (1996). Socio-economic and environmental impacts of alluvial gold panning, Zimbabwe. *MERN Bulletin*, 9.

Mohidden, H. (1997). *Small-scale Mining and Gold Panning in Zimbabwe*. Thesis, Institute of Ecology and Resource Management, University of Edinburgh.

Ofei-Aboagye, E.C. (1995). *Women's Participation in Artisanal Mining in Ghana*, conference paper, World Bank Roundtable on Artisanal Mining, USA.

3

Land Use Disputes between Small- and Large-Scale Miners: Improving Conflict Management

JOHN S. ANDREW AND GAVIN M. HILSON

Small-scale mining in developing nations is routinely associated with land use conflicts with other stakeholders, primarily large mining companies. The scale of these disputes (which occasionally involve armed conflict) is usually sufficient to have significant adverse impacts on the natural environment and the local population. These conflicts have proven very difficult to manage, and have imposed great costs on a broad range of stakeholders.

This chapter explores intense land use competition as a source of conflict between small- and large-scale mining parties, using several case studies to illustrate its points. It then proposes mediation as a promising approach to resolving these serious disputes. Finally, it generates recommendations for how this approach to conflict resolution could be tailored to this context in order to make it an effective and efficient process, and increase its potential to produce lasting, consensus-based settlements.

LAND USE COMPETITION IN DEVELOPING COUNTRIES

Throughout history, different parties have competed for land plots. In tropical countries such as Brazil and PNG, where resources are rich and diverse, groups such as loggers, farmers and ranchers are constantly competing for land, contesting for it to be utilized differently. In areas containing minerals, however, parties are in competition for the same resources but deploy different methods for their extraction. In many cases, the ensuing conflicts between the parties are exacerbated by harsh climatic conditions and population pressures.

The poorly enforced legislative processes and accompanying monitoring activities prevalent in the developing world are the main reasons why a number of land use conflicts have occurred in recent years. Specifically, many indigenous claims to land have been quashed by recent government initiatives undertaken to promote foreign investment and regulate industrialization in rural regions. Unruh (2002, p. 275) puts the issue into perspective:

> The legal problem concerns the ongoing disconnect between formal state law, and customary or traditional law, which governs how a great deal of the world's poor intersect with property. The former allows assets to be fungible and used, as such, by individuals; but the latter – the maintenance and security of community and lineage connections to land in an often risky physical, social, and political environment.

An estimated 20–80% of land delivery in developing countries is informal, and, therefore, does not conform to the legal cadastral system and land use controls (Fourie, 1998). Thus, the actions that have been taken to better regulate rural lands have posed a major problem for locals, who have been forced, often for the first time, to adhere to a more controlling, legislated environment.

LAND USE CONFLICTS BETWEEN SMALL AND LARGE-SCALE MINING PARTIES

In recent years, in an attempt to alleviate financial crises and curb national inflation, several governments have restructured their economies under the tutelage of the World Bank and the IMF. In fact, structural adjustment and stabilization has become the norm across Sub-Saharan Africa and Latin America, where virtually every country has undergone, or is in the process of undergoing, some form of adjustment in line with plans set out by the Bretton Woods Institutions. As Jackson (1999, p. 281) explains, the majority of these programs aim to reduce short-term *macroeconomic equilibria* – "getting the prices right" – but also involve medium-term adjustment of the main productive elements of the economy.

Economic restructuring depends heavily upon inputs from the private sector. In the majority of cases, governments have decreased ownership in domestic industries, and have drafted policies and implemented legislation for the purpose of promoting increased foreign investment. This, however, has had pervasive effects on the livelihood strategies of many people, as their established means of income generation have been seriously disrupted. The decision to promote increased foreign investment has therefore put many developing countries in a vulnerable, more destabilized, socio-economic state.

The mining and minerals sector has been mainly targeted by governments striving to promote foreign investment in their economies. The main strategies pursued have emphasized the divestiture and subsequent privatization of state-owned operations, and the provision of mineral exploration and mining licenses at a minimal expense. In order to *guarantee* adequate availability of land resources to foreign mineral exploration and mining companies, most governments have legalized informal artisanal and small-scale mining industries with the intention of organizing and localizing activities.

However, despite repeated contentions of the need to improve the organization of resident artisanal and small-scale operations, and to ensure that output does not escape through illegal channels, governments appear to have been quick to ignore artisanal and small-scale miners in favour of large companies. It is therefore no surprise that the "relationship between large mining companies and smaller-scale operations has often been characterized by tension and mistrust" (MMSD 2002, p. 324). Many countries are dependent on substantial foreign investment in the mining sector, and, in many cases, this has contributed to conflict. A 2001 survey of mining companies revealed that when such conflict is violent, it often serves as a significant disincentive for mining companies to maintain and make further investments in their operations in the area of unrest (MMSD, 2002).

Many natural resource conflicts are typically severe and debilitating, resulting in *inter alia* violence, resource degradation, the undermining of livelihoods, and the uprooting of communities (Castro & Neilsen, 2001); each party simply wishes to pursue its own interests to the fullest. Such circumstances characterize the mining sector in a number of developing countries, where conflicts have been exacerbated by the cavalier approaches governments have taken toward exploration geology and the demarcation of land plots. Concessions of land are generally awarded to large-scale mining companies, which are then prospected to ascertain the locations of prospective mineral deposits. Often, numerous illegal artisanal operators are found working the land, and, when asked to relocate, disputes occur.

The problem is not only confined to large-scale miners, however. Licensed small-scale miners often "acquire" concessions from the government without any knowledge of mineral content. In many cases, following extensive pitting and trenching, miners realize that awarded areas contain minimal mineral deposits, and are therefore forced to abandon concessions outright. The inevitable encroachment (on neighbouring large-scale mine plots) that follows induces further conflict.

CASE STUDY ANALYSIS

According to the United Nations publication, *Recent Developments in Small-Scale Mining* (United Nations 1996: s. 43), "recently, many large companies seeking to establish operations in developing countries have concerned themselves with the small-scale mining issue, establishing specialized divisions that deal with community relations". Although in certain instances, working partnerships have successfully been forged between large- and small-scale miners, there remains a disproportionately greater percentage of cases of strained relations between the parties. As much of the improvement that has been achieved in this area is now well documented in the literature, it has created the misconception that in most parts of the world, large- and small-scale miners are coexisting in harmony.

The main antagonists continue to be international mining companies, which, despite operating under extremely favourable economic conditions, repeatedly take advantage of the fact that developing world governments are cash and resource-strapped. More specifically, rather than turning over portions of an awarded concession – that have proven unsuitable for large-scale activity – to artisanal and small-scale mining parties, the management of international mining corporations typically elects to withhold land, and governments, most of which are operating at the mercy of these companies, are forced to comply with their demands.

In fact, in the majority of cases where compromises have been reached between large- and small-scale mining parties, (large-scale) companies have either released land reluctantly, have turned over largely spent and "mined out" portions of a concession, and/or have freed up portions of a concession in exchange for compensation. For example, in the Philippines, the Benguet Corporation has allowed a group of small-scale gold miners to operate legally on a certain portion of its concession largely because of a forged agreement that gives the company exclusive rights to the tailings of small-scale miners (Bugnosen, 2001). Similarly, in the Great Dike area of Zimbabwe, "companies have allowed artisanal miners to *rework* tailings and operate in *abandoned* sections of functional mines, as well as *marginal* areas of the concession, with agreements to sell at least part of the production to the company" (United Nations 1996: s. 45). Ghana Goldfields, which is operating in Tarkwa, western Ghana, has not only been involved in the relocation of local villagers – the associated schemes for which have generated considerable controversy in various circles – but have also established purchasing services for illegal miners operating within their concession, equipping them with identification cards and mandating that they serve as a "police" force to prevent additional artisanal miner encroachment.

CONFLICT PREVENTION *VS.* CONFLICT RESOLUTION

Many researchers and scholars have proposed proactive measures that should be taken to reduce the risk of conflicts arising between small- and large-scale miners. Most of this work to date has placed the onus for taking action on large-scale mining companies. Among the more comprehensive and practical recommendations directed at these firms are the following identified by the International Labour Organization (1999):

1. Provide affordable assaying services.
2. Share geological and other technical information with small-scale miners.
3. Provide practical training and technical advice.
4. Help to establish or sponsor small-scale central processing plants.
5. Provide purchasing services, tools and equipment to local communities.
6. Assist with the procurement and storage of explosives.
7. Provide custom milling services and workshop facilities.
8. Buy and treat tailings.
9. Release land that is sub-optimal for large-scale mining.
10. Provide emergency assistance and mine rescue.

While many companies have implemented a number of these, and related, measures, most have done so strictly for the benefit of the company, rather than small mining parties. Moreover, mining companies have nearly always made these decisions unilaterally, with little, or no, meaningful consultation with key stakeholders, such as small-scale miners, non-governmental organizations (NGOs), foreign aid agencies, environmental groups, human rights groups, and local and national governments.

In principle, strategies to proactively attempt to prevent disputes from arising are more valuable and worthy of attention and resources than methods of conflict *resolution*. However, the intensity of the land use competition and the extreme volatility of the conflicts that ensue make the prevention of these disputes a Herculean task. Until these approaches have had time to evolve and mature, we must be realistic in our expectations of them. At present, it would be irresponsible to devote all of our energy and attention to the prevention side of the equation. Rather, it is essential to address these serious disputes along two parallel tracks: conflict prevention and conflict resolution. Given the considerable attention that the literature has directed at the former (especially in the area of community consultation), this chapter concentrates on the latter.

The remainder of this chapter introduces mediation as a conflict resolution process, establishes the need for mediation in disputes between small- and large-scale mining parties, and discusses features that should be incorporated into a mediation process in order to increase its effectiveness at resolving these conflicts.

INTRODUCTION TO MEDIATION

Mediation is a voluntary, negotiation-based process in which parties involved in a current or potential dispute meet together with the assistance of a neutral mediator for collaborative problem solving and consensus building, with the goal of achieving a mutually acceptable resolution. At the heart of mediation is the process of negotiation or bargaining between stakeholders, in an attempt to resolve issues on which they disagree. Taylor (1992) defines negotiation as "resolving conflict through a process of communication, exchange, and commitment to a course of action. It is intended to reach an agreement that benefits all parties while recognizing that each side will protect and promote its own self interest."

The participation of disputants in mediation is voluntary, including their ability to withdraw from the process at any time. The process is confidential and without prejudice to the legal rights of any party. By entering into mediation, disputants do not surrender their right to later pursue a different conflict resolution channel. Nor do the proceedings have any legal influence on a concurrent or subsequent process. The mediator has no decisionmaking or adjudicatory authority to impose a settlement on the parties. Disputes are resolved only when the parties themselves reach what they consider to be an acceptable resolution. The settlement of issues is based on a consensus of all of the parties, rather than a majority vote.

The mediator is a completely independent, neutral and impartial party, who is normally jointly selected by the disputants. The mediator works with the parties to design a fair process, helps them to obtain the resources they require, organizes and manages the meetings, assists the parties to set and adhere to realistic deadlines, maintains minutes of each session, and coordinates the exchange of information between the parties. In addition to managing the process, the mediator contributes to discussions about the substance of the conflict, shuttles ideas and offers back and forth between the parties, helps each party to formulate proposals that are more likely to be acceptable to the other parties, conducts private caucus sessions with fewer than the full complement of parties (if all parties agree), participates in the generation of creative options, and assists in the writing of the final agreement.

THE NEED FOR MEDIATION IN SMALL-SCALE MINING DISPUTES

A review of the literature on small-scale mining revealed no examples of attempts to resolve land use disputes through mediation, although it is possible that there are cases which have not been documented (Solomon, 2001). Ayling and Kelly (1997) observe that little attention has been directed to the development of mechanisms for managing natural resource conflicts in general, and argue that these are urgently needed to equitably distribute resources and lessen the risk of violence. Epps and Brett (2000) recommend the creation of mechanisms to resolve local mining disputes, although they do not specifically address small-scale mining.

Hilson (2002a) calls for large mining companies to improve their communication with communities (including small-scale miners). Opportunities are needed for a local community to learn information from a company, express its concerns and ideas, have its questions answered, and provide input about various phases of mine development. Hilson also recommends that mining companies provide appropriate compensation packages to local communities adversely impacted by their activities. Since it is essential that compensation meets the specific needs of each community (which has seldom been achieved to date), this requires the company to have clearly identified those needs. Mediation is capable of achieving each of these objectives.

Representatives of local communities in which small-scale mining is prevalent often express frustration with their attempts to negotiate compensation with large, multi-national mining companies (Mining Watch Canada, 2000). In one case in Ghana, consultants working for a Canadian mining company negotiated with local residents being displaced by its mining operation. However, the negotiations occurred despite a tremendous power imbalance; a history of repression of local rights and coercion by the company; and a lack of any alternative process available to the residents (Mining Watch Canada, 2000). It is likely that a skilled mediator could have managed the process to ensure that this power differential, although very real, would not handicap any party at the bargaining table. The issue of the distribution of power among participants in mediation will be further explored later in this chapter.

Tyler (1999) stresses the importance of making use of a completely neutral, outside mediator in the resolution of natural resource disputes. Epps and Brett (2000) and MMSD (2002) each argue that a neutral party is needed in disputes between small- and large-scale miners, especially when an impasse has been reached. The *Action Plan for Change* developed by MMSD North America establishes the need for dispute resolution mechanisms that can be applied at the project/operation level of mining (MMSD, 2002).

Mediation should also play an important role in the efforts of many countries to develop more comprehensive systems of regulating all mining activities. The advantage of greater regulation of small-scale mining is well documented in the literature, including its potential to reduce conflicts between various types of mining operations (MMSD, 2002). To date, many international initiatives for the regulation of small-scale mining have been designed, yet very few have been successfully implemented by governments (Andrews-Speed et al., 2003). Legislation and licensing, in concert with strict enforcement, would represent significant progress toward preventing future land use conflicts between large mining companies and small-scale miners. However, such efforts have, in the recent past, proven to be catalysts for considerable conflict among stakeholders, especially when companies already hold permits, when there are already conflicting pieces of legislation, or when there are illegal mining operations (Bugnosen et al., 1999).

Employing the services of an independent mediator would be particularly beneficial for bringing about positive changes to the regulatory environment. The effectiveness of any regulatory program will depend on whether the process used to develop it is accountable and transparent, avoids the conflicts of interest associated with local government decisions, and involves the meaningful participation and accommodates the key interests of all stakeholders (Andrews-Speed et al., 2003). As Danielson (2003, p. 98) points out: "Successful approaches will require cooperative and sympathetic methods of solving problems, rather than harsh solutions". Mediation will lessen the risk that small-scale miners will perceive regulation as unfairly prohibitive or punitive, and be further marginalized in an underground economy, exacerbating illegality, land use conflicts and environmental degradation.

DESIGNING EFFECTIVE MEDIATION FOR MINING DISPUTES

Andrew (2003) evaluated land use conflicts arising from small-scale mining with respect to how well they satisfy a set of 19 characteristics of disputes. These are prerequisites for mediation, in the sense that they increase the likelihood that mediation will produce a satisfactory outcome. This analysis found that these disputes would probably satisfy 10 of the 19 preconditions, while the remaining nine may or may not be satisfied. For none of the preconditions could it be predicted with any certainty that these disputes would fail to meet its requirements. The article concluded that mediation holds enough potential to resolve land use conflicts associated with small-scale mining that it warrants use in this setting, at least on an experimental basis.

Andrew (2001) employed various statistical techniques to test the influence that 17 factors (based on a review of the literature) had on the success of Alternative Dispute Resolution (ADR) processes in 54 waste management disputes. The ADR processes included negotiation, facilitation and mediation. Success was measured using the following four criteria: whether a final settlement of the conflict was achieved; whether the conflict was resolved more quickly and at less cost than a conventional conflict resolution process; participant satisfaction with the process; and the duration of the process. Unlike Andrew (2003), this article included not only characteristics of the conflict, but also characteristics of the ADR process. Of the 13 characteristics of either the ADR process alone (10) or of the process and the conflict combined (3), seven were found to influence the outcome of the process. They were: the number of people directly involved in the process, the participation of all stakeholders in the dispute, the participation of the government (if it was required to approve a final settlement), the type of representatives, the overall effectiveness of party participation (a blend of nine sub-factors), the joint design and control of the ADR process, and the neutrality of the facilitator or mediator.

Andrew (2001) concluded that fewer characteristics of ADR processes are actually important to successful outcomes that are widely claimed in the literature. For those characteristics found to influence ADR success, the degree of that influence was usually less than expected. Nevertheless, these findings, many of which contradicted previous research by others (most of which was not empirically based) were used to generate recommendations for designing more effective ADR processes. These recommendations pertain to good ADR practice, and are easily adapted to the mediation of land use disputes between large- and small-scale miners. This study suggests that for mediation to be effective in addressing this type of dispute, its process must be designed with the following points in mind.

While it is essential to include representatives of all parties that hold a stake in the outcome of the dispute, it is also important to balance this with the need to limit the number of individuals present at the bargaining table. This may involve an early identification by the mediator of the parties which represent genuinely different interests, as well as opportunities for coalitions to be formed to allow multiple parties to be represented by a single negotiator. In many cases, the number of representatives per party may need to be restricted. The participation of any government department or agency that will be responsible for approving any settlement reached must also be directly involved in the mediation process. Contrary to a conventional viewpoint in the mediation literature, stakeholders should be represented by professional advocates trained and experienced in negotiation (such as lawyers or technical consultants), rather than by the principals themselves.

Overall effectiveness of party participation was comprised of nine components in the Andrew (2001) study. Adapted to mining disputes, the findings suggest that all parties should have:

- Adequate financial resources to hire any professional assistance required;
- Sufficient opportunities to express their opinions and to actually influence decision;
- Adequate access to usable information about the conflict and the mediation process;
- A good understanding of the process itself;
- Representatives with the full authority to represent their constituencies;
- Representative with strong negotiation skill;
- Flexibility and willingness to negotiate in good faith; and
- Cohesiveness within the constituency itself.

The mediation process should include provisions that permit the early and direct participation of the disputants in jointly designing the process itself, as well as the cooperative management of the process throughout its duration. Finally, the complete neutrality and impartiality of the mediator is essential to an effective mediation process.

In addition to the factors identified in the research by Andrew (2001) as being important features of an effective ADR process, the authors believe that there are a number of characteristics of mediation that will increase its effectiveness in attempts to resolve land use disputes between small- and large-scale miners in developing countries. These have been adapted from an extensive review of the conflict management literature and from the authors' professional experience. The following discusses these 10 key features of mediation and elaborates on some of the characteristics of the process that have already been identified. We do not include in this discussion most of the fundamental rules of mediation that are frequently discussed in the literature and widely accepted. Many of these (*e.g.* consensus agreements rather than majority rule, the mediator having no adjudicatory authority, etc.) have been mentioned in the earlier section of this chapter that introduced mediation. Those that are discussed in the following (*e.g.* voluntary participation, confidentiality, etc.) are included because they are not always observed in practice and/or may not be matters on which there is currently a consensus of opinions among scholars and practitioners. The following also intentionally omits any discussion of characteristics of the mediator (other than neutrality and impartiality), an area of analysis which is beyond the scope of this chapter.

1. Role of government

The unique position of various levels of government in developing nations as regulators of the mining sector carries with it the responsibility to effectively manage small-scale mining disputes. These governments also have a moral obligation to encourage and make the necessary arrangements for key stakeholders to participate in an equitable and efficient mediation process. The principle of *subsidiarity* holds that any governance function is best carried out by the lowest level of government capable of doing so. However, in the case of resolving mining disputes, this is superseded by the greater importance of the accountability, neutrality and impartiality of the process. Consistent with this are Epps and Brett's (2000) recommended mechanisms for handling mining conflicts, which involve local government initiating and managing mediation processes, and national government stepping in when disputes cannot be locally managed. Hilson (2002b) cites the example of Ghana in recommending that national governments expand their role in resolving land use conflicts. While not specifically addressing conflict resolution, Andrews-Speed et al. (2003) propose for China the creation of an agency of the central and provincial governments to provide "one stop" regulation and administration services to small-scale mining. Regardless of which level of government oversees the process, it is important that the mediation sessions be held in the local community or communities where the dispute exists.

2. Complete neutrality and impartiality of the mediator

The typical conflict characteristics of contentious interparty relationships and an imbalance of power between stakeholders make it essential that mediation services be provided by a completely independent neutral entity. A mediator must be entirely neutral and impartial, and be perceived as such by each stakeholder. Neutrality requires that the mediator has no past, present, or likely future relationship with any of the parties, and does not stand to gain anything from any possible outcome. Impartiality depends on the mediator having no bias or preference for any party or position. While this does not imply that the mediator cannot and will not hold personal opinions on the matters in dispute, he or she must separate those from the management of the process and the substance of the discussions. This does not necessarily require that the disputants jointly pay for the mediation process. There are various types of innovative financial mechanisms that can make it possible for one type of party to pay a greater proportion of these costs, without compromising the integrity of the process. In some cases, it is appropriate for mediation processes to be financed by a government or international agency, with the disputants bearing none of the cost burden.

Despite their important functions as proponents and managers of conflict resolution processes, it is nearly always inappropriate for governments to assume the role of mediator (MMSD, 2002). Governments are usually stakeholders in these disputes. In some cases, they enforce claims granted to mining companies at the expense of small-scale miners. In many developing nations, governments fail to provide equitable justice and legal systems, and may be associated with corruption and human rights violations. In short, parties in mining disputes often have little trust or confidence in their government. Government officials rarely have formal training or experience in conflict resolution. Finally, governments may also lack sufficient resources or legal authority to provide effective mediation services (MMSD, 2002).

In many cases, it will be necessary to contract with an independent organization offering professional conflict management services, in order to ensure the independence and impartiality of the mediator. In some cases, the need for independence will require the employment of foreign-based expertise. One such organization is Oxfam, which, in 2000, established the position of Community Aid Abroad Mining Ombudsman for Australia-based mining companies (MMSD, 2002). One of the roles of the Ombudsman is that of a mediator of mining disputes. This is really just an extension of one of the functions that Oxfam Community Aid Abroad was already carrying out in a few cases, including the following two recent disputes in Indonesia: the Indo Muro Gold Mine in Central Kalimantan, and the Kelian Gold Mine in East Kalimantan (Oxfam Mining Ombudsman Annual Report, 2000). In other mining conflicts, Oxfam was less directly involved, observing negotiation meetings, encouraging parties to negotiate, and providing support to communities and non-governmental organizations (NGOs).

In some conflicts, it may prove difficult to find a neutral organization that is acceptable to all of the parties, which may provide further need for a mediator from outside of the country (MMSD, 2002). In the longer term, the mining industry should establish an international agency to co-ordinate and supervise all conflict resolution processes world-wide. This agency would operate at arm's length of any country, company or mining group. It would be funded by all governments and mining companies, and governed by a board with representation from all stakeholders in the global mining community. Such an agency could provide mediation services based at the regional small-scale mining support centres, as suggested in Chapter 8 of this volume.

There is sufficient evidence from other conflict settings that the use of an entirely independent mediation service can help to provide many of the other conditions necessary for effective mediation that are discussed in the following sections, including (*inter alia*) the willingness of key stakeholders to participate in the process and negotiate in good faith, a balance of relative power between the parties, and the ability of all participants to negotiate effectively.

3. Inclusion of all stakeholders

It is essential that any mediation process applied to a mining conflict be open to the participation of any parties with a stake in (*i.e.* affected by) the outcome (Johnson & Duinker, 1993; Australian EPA, 1995; Epps & Brett, 2000). This is important because it allows the interests, values and concerns of all relevant parties to be incorporated into the decisionmaking (MMSD, 2002). Suliman (1999, p. 290) asserts that "in localized conflicts local leaders should be the major actors in conflict resolution." Commonly involved stakeholders include mining companies, small-scale miners, local communities, governments (local, regional and national), NGOs, foreign aid agencies, international governmental organizations (*e.g.* specialized agencies of the United Nations and World Bank) and industry associations.

In some disputes it may be difficult for the mediator to identify all of the relevant parties, and to then ensure their participation. While the known parties may assist in this regard, in some cases, they will intentionally fail to identify other stakeholders that should be present. Due diligence in identifying stakeholders before commencing the mediation process is an essential responsibility of the mediator and the parties that have already committed to their participation. Although it is advisable to include *bona fide* stakeholders that may emerge after the mediation process has begun, this may be quite disruptive. Epps and Brett (2000) address the significant long term costs that may be associated with failing to include stakeholders. Perhaps the most significant of these is the considerable risk that a (voluntarily or otherwise) excluded party will obstruct the implementation of any settlement reached.

4. Voluntary participation of stakeholders

It is critically important that the participation of all parties be completely voluntary. Parties must be free to choose whether it is in their best interests to enter into the process, and must be at liberty to exit the process at any time without suffering any sanctions or repercussions. In order to believe that mediation is the option most likely to meet its interests (including achieving a settlement), a stakeholder must believe that there is, in fact, a conflict that needs resolving. A dispute must have "matured" to the point where the stakeholders are sufficiently motivated to negotiate; yet, have not become so volatile (or even violent) that calm, rational negotiations in a "safe" environment with a fair balance of power are impossible (Castro & Nielsen, 2001).

In many types of disputes for which mediation is an alternative to a more traditional and formal legal adjudication process (often litigation), the disputants are required to attempt mediation before they can gain access to the legal process. However, mandatory mediation is inappropriate for most disputes between small- and large-scale miners in developing countries, for several reasons. First, in many of these disputes, there is no institutionalized legal system to deal with claims. In fact, in some situations the only alternative is a violent confrontation. Second, if mediation is mandatory, there is less incentive for all of the parties and the mediator to establish the kinds of favourable conditions for reluctant disputants to voluntarily participate. In other words, if disputants are aware that the mediation process is a good one (in terms of its efficiency, equity, accountability, integrity, etc.), most will freely decide to participate. If, at any time, they believe that the mediation process is unlikely to protect their rights and meet their main interests, they will withdraw from it. An effective, fair mediation process need not be mandatory to ensure participation. Third, mandatory mediation programs are often unsuccessful, simply because there is never any effective way of forcing a reluctant party to negotiate earnestly, and in good faith.

Similarly, it is inappropriate to require that disputants attempt to resolve a dispute on their own prior to requesting mediation. In many conflicts, factors such as power imbalances, a history of violence, or a weak regulatory/legal environment preclude the existence of any other reasonable, productive means of parties negotiating on their differences.

5. Disputant involvement in the design of the mediation process

It is important that all of parties jointly design all aspects of the mediation process, with the assistance of the mediator. This includes planning the format of the negotiation sessions (which may become fairly detailed), the location of the sessions, scheduling and setting deadlines for various phases of the process, and establishing the ground rules for conducting the process and for participant behaviour (Johnson & Duinker, 1993). All parties must agree to all aspects of these logistical matters before proceeding with the mediation process. There must also be a consensus on the need for a mediator; and who that should be and the nature of their role and responsibilities.

There is considerable evidence in the literature that parties are more likely to negotiate productively, and to respect a final settlement, when they are able to adopt a sense of ownership of both the process and its outcome (Australian EPA, 1995). Consensus-based process design also has a powerful side benefit – by tackling this relatively easy challenge at the beginning, the parties gain confidence in their ability to work together and make mutual decisions, which establishes a positive environment for the remainder of the process.

6. Confidentiality of the process

At the beginning of a mediation process, the participants must agree (by signed consent) on the degree of confidentiality of the process, and what sanctions will be brought to bear on violators of this policy. Decisions about confidentiality include whether the sessions will be open to the public and/or the media or closed (*in camera*), whether the participants will be allowed to discuss the substance of what is discussed with persons not involved, whether the deliberations will remain confidential after the process has ended, whether a final settlement will be confidential, and whether official minutes will be kept (and who will have access to them). In most cases, it is beneficial for the parties to keep all discussions confidential, at least until the process has ended (hopefully with a signed agreement). Parties should normally conduct closed mediation sessions (if permitted by law), and appoint a single spokesperson (usually the mediator) as the only point of contact for the media. It is often advisable to avoid any official record of what is discussed, and for the mediator (and maybe even the negotiating parties) to destroy their notes after each session. We also recommend that all parties agree that all discussions will be without prejudice, meaning that no party can be held to anything they said in mediation (even in a subsequent legal process should mediation fail) until a final agreement has been signed. Finally, the participants should agree that if they are successful in achieving a final signed agreement, all of the signatory parties will be legally bound to it.

7. Balance of power

The effectiveness of a mediation process relies in part on there being a reasonably balanced distribution of power among the stakeholders (Epps & Brett, 2000; Castro & Nielsen, 2001; MMSD, 2002). The perception of the participants in mediation about a balance or imbalance of power is more important than the actual distribution of that power (if the latter could even be determined). Parties that feel they are at a significant power imbalance will often be unwilling to enter into a voluntary conflict resolution process, sceptical about its ability to produce an outcome that is acceptable to them (MMSD, 2002). However, in some cases, a party may feel so impotent that it believes it has no other reasonable opportunities to have its interests heard, and will therefore elect to participate in mediation in spite of its perceived lack of power.

One of the important functions of the mediator is to work to "level the playing field" as much as possible, by reducing any perceived power imbalances. A skilled and trusted mediator running a sound conflict resolution process can lessen many parties' concerns about power imbalances (MMSD, 2002) and establish conditions for more productive negotiations. The subject of power is complex and well beyond the scope of this chapter. However, in considering techniques that mediators may use to try to improve the balance of power between parties, it is helpful to consider some of the commonly accepted sources of that power that may be relevant in any mining dispute. The freedom of a party to either not participate in mediation or to "walk away" if it chooses is often suggested as the single greatest provider of power. This is the important concept of the *Best Alternative To a Negotiated Agreement* (BATNA), first put forth by Fisher and Ury (1981). A mediator can often reduce disparities in parties' perceived power by separately assisting each in determining their BATNA. Parties often do not realize going into a mediation how much having their own interests met depends on the cooperation of the other parties. They often overestimate their ability to achieve a good outcome for themselves through some other channel. The mediator may be able to help them adopt a more realistic assessment of their BATNA, and therefore increase their incentive to negotiate in a collaborative manner.

One source of power that may be particularly relevant to land use disputes between small- and large-scale miners is access to good technical information and the ability to make use of it in order to make decisions. Based on mediation experience in other types of disputes, it is clear that the mediator can play an important role in ensuring that parties share this type of information with each other, and that all parties have access to similar levels of hired technical expertise. The important role of information in mediation will be addressed in a subsequent section below.

These are just two examples of how the power balance issue may be effectively managed, often through the actions of the mediator. Each of the following other potential sources of perceived power suggests other measures that the mediator and the parties may take to ensure that an imbalance of power does not compromise the consensus-building process.

Other potential sources of party power:
- Having the ability to allow other parties to meet their needs (increases their cooperation).
- The ability to impose sanctions or costs on other parties.

- The ability to make credible threats to other parties (especially violence).
- The level of negotiation skill of the representative(s) (e.g. persuasiveness, charisma, etc).
- The ability to exert control over the mediation process.
- The ability to control information (and/or professional expertise).
- Having political influence (*e.g.* by monetary contributions, providing employment, etc.)
- Possessing authoritative power (*e.g.* legislation, policy, international support, etc.).
- Possessing moral power (*e.g.* accepted norms, the status quo, appeal to principles, etc.).

8. Party representation and negotiation skill

A good mediation process will include at the table negotiators who truly represent the interests of the identified stakeholders and who have the skill and experience to advance the interests of their constituencies. Parties must be allowed to select their own representatives, and should employ democratic methods for doing so (Castro & Nielsen, 2001; MMSD, 2002). Each representative must have full authority to speak for its party and commit it to agreements (ideally without need for ratification). However, this does not supersede the need for representatives to remain accountable to their constituencies at all times (MMSD, 2002), and to communicate regularly and openly about the substance of the negotiations.

In disputes involving large- and small-scale mining, the level of experience and skill with negotiation-based processes usually varies greatly between parties. These disparities are especially large when the stakeholders include indigenous peoples. Parties that are politically marginalized or which lack resources (as commonly found in developing nations) are often unable to represent their interests effectively (MMSD, 2002). In such situations, the mediator should assist disadvantaged parties to better understand the conflict resolution process and their legal rights, to communicate their interests to the other parties, and to understand other parties' interests. In principled negotiation (on which good mediation is based), it is in all of the participants' best interests for stakeholders to be skilfully represented. Since there are limits to a mediator's ability to correct deficiencies in the representation of some parties, it may be necessary for more advantaged parties to pay for the others to hire professional representation.

Marieke Heemskerk and Rachael van der Kooye observe in this volume (Chapter 36) that small-scale miners sometimes form negotiating coalitions to properly represent their collective interests. The example they provide is of a dispute in Suriname between Maroon miners and the Canadian mining company Golden Star Resources.

In the case of the Kelian Gold Mine in Indonesia, the Oxfam Mining Ombudsman was acting as a facilitator of negotiations between the mining company (Rio Tinto Indonesia/PT Kelian Equatorial Mining) and the local community. Late in the process, a government official became involved as a party, thereby violating an earlier agreement between the other two parties.

The company also began negotiating with a separate party (Team Murni), which claimed to represent the community. However, Team Murni lacked the endorsement of the Council for People's Prosperity, Mining and Environment (LKMTL), an organization formed to represent local interests in a meeting of more than 2,000 people from affected communities; this was extremely disruptive to the negotiations. Eventually, the company brought in an independent mediator from the Australian Federal Court, and negotiations resumed. Oxfam recommended (*inter alia*) that the company recognize LKMTL as the only *bona fide* representative of local community interests (Oxfam Mining Ombudsman Annual Report, 2000). These two examples illustrate the important role that representation issues play in the mediation of mining disputes.

9. Information

As was the case for party representation and negotiation skill, disputes between small- and large-scale miners often involve problematic differences in the abilities of stakeholders to manage information pertinent to the situation and proposed solutions. This often proves to be a significant impediment to productive mediation. Enabling all of the parties to understand and make effective use of the information that is collectively available increases the probability that the disputants will be successful in resolving their differences. The mediator and the more sophisticated parties share the responsibility of ensuring that parties with less experience and fewer resources (typically small-scale miners and indigenous communities) are not disadvantaged in the process by their lesser capacity to utilize information.

Parties in possession of relevant information must be required to share it with others, in a format that is useful to them. The mediator is responsible for ensuring that proper disclosure and exchange of information occurs (Australian EPA, 1995). In many cases, it is necessary for technical consultants to be made available to disadvantaged parties, even if this cost is met by other parties (Castro & Nielsen, 2001). The same applies to legal expertise, since legal information is often of great value. It may also be prudent to begin the mediation process with a program to jointly educate all of the parties about certain issues, especially those concerning the environmental impacts of mining activities. This may be one of the functions of the government or the mediator (Epps & Brett, 2000). Fostering collaborative behaviour with respect to collecting and analyzing information is in the best interests of all parties. As Epps and Brett (2000, pp. 5–21) point out: "… the more people co-operate with each other in dealing with uncertainty, sharing information and committing themselves to reciprocal plans of action, the less uncertainty everyone has to face".

10. Understanding party values, rights and interests

Recognizing that differences in fundamental values and principles contribute to many conflicts between large- and small-scale miners, Suliman (1999) warns that "outsiders", such as foreign-based mining companies, must understand that in some cultures (*e.g.* traditional African societies), the right to use land is more important than the right of formal ownership of that land. Epps and Brett (2000) discuss the importance of protecting community rights when dealing with mining disputes. This is particularly important for indigenous peoples, who are often important stakeholders in disputes between large- and small-scale mining (MMSD, 2002). The 1992 United Nations Draft Universal Declaration of the Rights of Indigenous Peoples (Paragraph 20) states that: "Indigenous peoples have the right to require that States and domestic and transnational corporations consult with them and obtain their free and informed consent prior to the commencement of any large-scale projects". The Mining and Indigenous Peoples Consultation held in 1996 advanced the United Nations' opinion by "demand[ing] that Indigenous Peoples be consulted with, and full and comprehensive information be provided in a timely manner, when mining activities are being considered for sites located on Indigenous Peoples' lands ..." (in Epps & Brett, 2000).

One of the tasks of the mediator at the outset of the mediation process is to ensure that all stakeholders recognize the legitimacy of the values, rights and interests of the other parties (Epps & Brett, 2000; MMSD, 2002). This may be particularly challenging in small-scale mining disputes, in which the stakeholders often hold conflicting values and opinions regarding fundamental issues such as land ownership and access, traditions, ancestral rights, human rights, environmental protection, sustainability, and means of resolving conflict.

CONCLUSION

There is a clear need for mediation as a process to attempt to resolve land use disputes between small- and large scale miners, and often also involving other stakeholders. Caused chiefly by competition for the use of land, these conflicts have proven to be particularly troublesome, and typically have a great impact on the local population and environment. Mediation has sufficient potential in this challenging context that it should be strongly encouraged by governments, mining companies and international agencies with a stake in the mining industry and the well-being of local populations in areas where small-scale mining occurs. In addition to reiterating many of the recommendations of Andrew (2001), this chapter has generated recommendations for designing mediation to maximize its effectiveness and efficiency, and its likelihood of producing good, enduring settlements.

While governments have a responsibility to encourage and provide logistical support to mediation, they are seldom in an appropriate position to assume the role of mediator. It is often necessary to employ the mediation services of an independent (often foreign-based) organization. The mining industry should work toward the establishment of an international, independent agency to co-ordinate and supervise all conflict resolution processes globally.

The mediator and the key disputants share the responsibility of ensuring that all stakeholders in the conflict have the opportunity to participate in the mediation. However, it is essential that all parties participate voluntarily, with no sanctions brought to bear if they choose either not to enter or to withdraw during the process. All parties must agree that there exists a conflict requiring resolution, and its stage of development must be amenable to mediation. The disputants having attempted to resolve the dispute on their own should not be a prerequisite for mediation. The parties should all be involved in the planning of all aspects of the mediation process, including the need for, and selection of, a mediator. There also needs to be consensus on a number of decisions concerning the confidentiality of the process.

A reasonable balance between stakeholders in the distribution of power, levels of negotiation skill and experience, and ability to manage information is necessary for productive mediation. Although these can vary considerably between parties (especially when indigenous peoples are involved), an effective mediator has many tools at his or her disposal to assist disadvantaged or marginalized parties, and help to "level the playing field." This chapter has made a number of recommendations in this area. Negotiators at the table must be truly representative of the interests of their constituencies, have open lines of communications with them, and be fully authorized to commit their parties. Because of its importance to productive mediation, the ability of each party to negotiate effectively is in the best interests of everyone involved in the mediation process. More sophisticated parties should help to ensure that parties with less experience and fewer resources (typically small-scale miners and indigenous communities) are not disadvantaged in mediation. This may involve contributing to the cost of providing weaker parties with professional expertise. Finally, all participants must recognize the legitimacy of the values, rights and interests of the other stakeholders.

REFERENCES

Andrew, J.S. (2001). Making or breaking alternative dispute resolution? Factors influencing its success in waste management conflicts. *Environmental Impact Assessment Review*, 21, 23–57.

Andrew, J.S. (2003). Potential application of mediation to land use conflicts in small-scale mining. *Journal of Cleaner Production*, 117–130.

Andrews-Speed, P., Minying, Y., Lei, S. & Cao, S. (2003). The regulation of China's township and village coal mines: A study of complexity and ineffectiveness. *Journal of Cleaner Production*, 11, 185–196.

Australia Environmental Protection Agency (1995). *Community Consultation and Involvement*. Australian Federal Environment Department.

Ayling, R.D. & Kelly, K. (1997). Dealing with conflict: Natural resources and dispute resolution. *Commonwealth Forestry Review*, 76(3), 182–185.

Bugnosen, E., Twigg, J. & Scott, A. (1999). Small-scale mining legislation and regulatory frameworks. *Minerals and Energy*, 14(2), 35–38.

Bugnosen, E. (2001). *Country Case Study on Artisanal and Small-Scale Mining: Philippines.* Working Paper No. 83, Mining, Minerals and Sustainable

Development (MMSD) Project, International Institute for Environment and Development (IIED), London.

Castro, A.P. & Nielsen, E. (2001). Indigenous people and co-management: Implications for conflict management. *Environmental Science and Policy*, 4, 229–239.

Danielson, L. (2003). Editorial: Artisanal and small-scale mining from an NGO Perspective. *Journal of Cleaner Production*, 11, 97–98.

Epps, J. & Brett, A. (2000). Engaging Stakeholders. In J.M. Otto & J. Cordes (Eds.), *Sustainable Development and the Future of Mineral Investment*, Paris: United Nations Environment Programme. 5–1 to 5–38.

Fisher, R., Ury, W. & Patton, B. (1991). *Getting To Yes: Negotiating Agreement Without Giving In, Second Edition*. New York: Penguin Books.

Fourie, C. (1998). The role of local land administrators: An African perspective. *Land Use Policy*, 15(1), 55–66.

Hilson, G.M. (2002a). An overview of land use conflicts in mining communities. *Land Use Policy*, 19(1), 65–73.

Hilson, G.M. (2002b). Land use competition between small- and large-scale miners: A case study of Ghana. *Land Use Policy*, 19(2), 149–156.

International Labour Organization (ILO) (1999). *Social and Labour Issues in Small-Scale Mines*. Report for discussion at the Tripartite Meeting on Social and Labour Issues in Small-Scale Mines, Geneva, 17–21 May 1999. Geneva: International Labour Organization Office.

Jackson, P. (1999). New roles of government in supporting manufacturing: The capabilities of support agencies in Ghana and Zimbabwe. *Public Administration and Development*, 19(3), 281–298.

Johnson, P.J. & Duinker, P.N. (1993). *Beyond Dispute: Collaborative Approaches to Resolving Natural Resource and Environmental Conflicts*. Thunder Bay, ON: School of Forestry, Lakehead University.

Mining Watch Canada (2000). *Mining Watch Canada Newsletter*. Number 3, Spring.

Mining, Minerals and Sustainable Development (MMSD) Project (2002). *Breaking New Ground: The Report of the MMSD Project*. International Institute for Environment and Development. London: Earthscan Publications Ltd.

Oxfam (2000). *Community Aid Abroad Ombudsman 2000 Annual Report*. Oxfam.

Solomon, M. (2001). E-mail correspondence. November 19.

Suliman, M. (1999). Conflict resolution among the borana and the fur: Similar features, different outcomes. In M. Suliman (Ed.), *Ecology, Politics and Violent Conflict* (pp. 286–290). London: Zed Books.

Taylor, W.H. (1992). Planners as negotiators: Practice and preparation. *Plan Canada*. March: 6–11.

Tyler, S.R. (1999). Policy implications of natural resource conflict management. In D. Buckles (Ed.), *Cultivating Peace: Conflict and Collaboration in Natural Resource Management* (pp. 263–280). Washington, DC: World Bank Institute.

United Nations (1996). Recent developments in small-scale mining; A report of the Secretary-General of the United Nations. *Natural Resources Forum*, 20(3), 215–225.

Unruh, J.D. (2002). Poverty and property rights in the developing world: not as simple as we would like. *Land Use Policy*, 19(4), 275–276.

4

Key Issues in Illegal Mining and Marketing in the Small-Scale Mining Industry

STEPHENS KAMBANI

This chapter discusses major issues concerning illegal mining and marketing in the small-scale mining (SSM) sector in developing countries (DCs). Causes of these undesirable activities are analyzed, and possible interventions to minimize or eradicate them are presented.

Illegal mining can best be described as any form of mineral extraction without legal title to the prospect being worked where required by the authorities. The major target of illegal miners are high unit value minerals offering a prospect of high returns, such as diamonds, colored gemstones, gold, and, to a much lesser extent, relatively low value minerals. An added advantage of high unit value minerals is that their transportation does not require an elaborate infrastructure. Thus, many illegal mining operations are found in locations lacking basic infrastructure (Cramer, 1990).

The illegal marketing of minerals involves the selling of product outside legally accepted channels. The predominant sources of minerals sold illegally originate from illegal mining activities. However, as is discussed later in the chapter, because of a variety of reasons, some legal mines are also involved in illegal trading. Available information indicates that most gold and gemstones produced in DCs by small-scale miners are sold illegally.

With regard to the distribution of illegal activities by mineral commodity, it is apparent that high unit value minerals such as gold and gemstones (including gem diamond) have the highest concentration of illegal mining and marketing activities, both by value of output and numbers of people involved.

Therefore, the common feature of minerals mined and traded illegally by small-scale miners is their high unit value. Because of this, they require less transportation infrastructure. In addition, these minerals generally require minimal processing technology, an attribute that does not offer a barrier to illegal miners. For instance, gold is recovered using simple mercury amalgamation processes, while gemstones are sold as unprocessed rough material upon recovery from rock matrixes, using simple techniques of chiseling, hand sorting, washing or a combination of these methods.

CAUSES OF ILLEGAL MINING

Before analyzing the causes of illegal mining and marketing, it is important to first provide essential background information on small-scale mining (SSM) to reveal its inherent characteristics, some of which contribute to these activities. SSM is a highly heterogeneous sector, and, from a technical and structural perspective, is conducted at three different levels (Kambani, 2000).

The first, and lowest, level is artisanal mining (or micro-scale mining), which features the simplest of operations, and is characterized by the use of simple tools and the absence of a formal enterprise. Due to the simplicity of operations and absence of meaningful investment, the activity can easily be started and discontinued. For the same reason, artisanal mine sites can develop rapidly, as migrating miners are attracted in large numbers to a reported new discovery of precious minerals. For instance, in the early-1990s, Brazil's gold rush involved an estimated one million *garimpeiros* (artisanal miners) working more than 2,000 prospects; the activity supported close to 4.5 million people overall (Davidson, 1990).

Mobility and use of simple tools are therefore the main characteristics of artisanal mining. Because of these attributes, this segment of industry features numerous illegal mining activities.

The second level comprises traditional SSM operations. More specifically, it consists of both licensed and unlicensed non-mechanized and/or semi-mechanized activities. It has a basic management structure and features hired labor. Equipment is usually antiquated. This group also contributes significantly to the illegal trading of gold and gemstones.

The third group consists of licensed SSM operations using highly mechanized methods and which features formal management structures. Although there is generally no illegal mining in this category, similar to the second level, there is a noticeable level of illegal trading within this group. This may be due to various reasons, including tax evasion, marketing constraints, and pre-financing of mining operations by illegal buyers.

The causes of illegal mining and marketing are discussed in detail in the subsequent sections.

Causes of illegal mining

The following factors are considered to be the major causes of illegal mining:

- Lack of access to mineral rights
- Weak institutional framework support
- Poor security at mines

The lack of entitlement to mineral rights is a result of many factors. One of the major causes is excessive state control of mining operations. Prior to the economic liberalization that had proliferated most DCs during the past decade, many countries in the developing world embraced the socialist type of economy, which, to a great extent, influenced the types of policies applied to various sectors of their economies.

For many sectors, especially mining, which was dominant in the economies of many DCs, state control through direct ownership of production and marketing facilities was common. DCs pursued this strategy in asserting sovereignty over their natural resources, and aimed to maximize foreign exchange earnings and revenues through the exploitation of mineral resources.

However, complete state control of the sector meant the exclusion of small miners from mining, which left artisanal parties with no option but to mine illegally in areas where they could not be easily controlled. Once this happened, it became extremely difficult to police the numerous small-mining operations countrywide that would periodically spring up.

The lack of access to mineral rights may also be caused by difficult, and often bureaucratic, procedures to obtain licenses for mineral exploration and mining. In some cases, the prohibitive license fees have also played a role, especially for marginal artisanal miners.

Weak institutional framework support from mining authorities is another major cause of illegal mining. Mining authorities are supposed to provide mineral rights in a transparent and efficient manner. In addition to this service, they are required to monitor the sector to ensure that proper mining methods, safety, and, in some cases, adherence to environmental regulations, is maintained and unlicensed operations are eradicated.

Unfortunately, the operational budgets of such institutions are usually inadequate, resulting in low staffing and poor field inspection funds. Moreover, these institutions are often highly centralized in major towns, making it difficult for small miners based in rural areas to access them for the purpose of securing appropriate licenses.

The resulting poor security further propagates illegal mining, and if an existing mine is left unguarded, illegal miners may invade it (Fig. 4.1).

Impacts of illegal mining

Illegal mining has social, economic and environmental ramifications. Because the activity is conducted illegally, operators do not normally adhere to existing health, safety and environmental regulations.

There are four major negative consequences of illegal mining. These are identified as:

- Uncontrolled environmental degradation;
- Poor health and safety;
- Loss of tax revenues from the activities by the government; and
- Lawlessness, especially crime, in high density gold mining areas.

Generally, SSM has a poor track record with regard to degradation of the environment. The large numbers of small miners involved, and their widespread distribution, makes the monitoring and control of the sector difficult. Illegal miners further worsen the situation since they are even more difficult to monitor.

Figure 4.1 Arrested illegal miners on the ground in Ndola Rural Emerald Mining area, Zambia.

For instance, uncontrolled alluvial gold mining along some of Zimbabwe's rivers involving an estimated 50,000 artisanal miners has, in the past, caused substantial environmental degradation (Davidson, 1992).

Figure 4.2 shows land degradation in a prime land area of Lusaka caused by illegal artisanal miners producing stone aggregates used in home construction.

Since illegal miners are aware that they are unmonitored by the mining authorities, they do not maintain acceptable safety and health standards. The mining methods used are usually unsafe, leading to serious fatalities. For instance, many injuries and deaths in Zimbabwe have been attributed to mining gold at depths of five to eight metres in unsupported pits and tunnels (Davidson, 1992).

Illegal miners enter underground workings using crude shafts. The entrance may be concealed during the day, thereby posing an additional danger to local inhabitants. Ventilation and lighting in such workings are also generally poor.

Of serious health concern to miners is mercury poisoning. Mercury emissions into the environment are a specific problem of small-scale gold mining in DCs due to careless processing of gold amalgam (Priester, 1993). Reduced life expectancies have been reported among gold miners as a result of mercury poisoning.

One other negative effect of illegal mining is the loss of government revenues. Specifically, considerable revenues are lost through non-payment of mining license fees by illegal operators.

Lawlessness in high-density mining areas is sometimes a major problem. Police forces may be absent, or present but helpless. For instance, in the Philippines, armed groups are known to be involved in lucrative mining and protection businesses, while others resort to highway robbery and extortion.

Figure 4.2 Degraded land by stone aggregates artisanal miners in city of Lusaka.

CAUSES OF ILLEGAL MARKETING

There are many causes of illegal marketing. These are summarized below:

- State control of marketing
- Marketing constraints
- Pre-financing of mining operations by illegal buyers
- Illegal mining
- Local currency overvaluation
- Tax evasion

State control of marketing

In most DCs, marketing legislation has established state marketing agencies for mineral products. These special laws require that all gemstone and gold sales, which are the main products obtained from SSM activity, be channeled through government marketing agencies.[1]

Under the state control of marketing arrangements, a number of marketing strategies have been pursued, including the following:

- Establishing state marketing agencies to market mineral products from state-owned mining companies and production from some private mining

[1] Section 319 of the Mines and Minerals Act Chapter 165 of Zimbabwe is a typical example of such legislation. It states, "No person other than the Precious Stones Board shall sell or otherwise dispose of precious stones" (Mines and Minerals Act, 1974, p. 646).

Table 4.1 Examples of state marketing agencies.

Marketing Agency	Country
The Minerals Marketing Corporation of Zimbabwe for the sale of minerals other than gold and gemstones	Zimbabwe
The Metal Marketing Corporation of Zambia for the marketing of all Zambian metals	Zambia
Reserved Minerals Corporation for the marketing of minerals other than emeralds.	
Zambia Emerald Industries Limited for cutting, polishing and marketing of emeralds	
Tanzania Gemstone Industries	Tanzania
Pakistan Gemstone Corporation	Pakistan
Gems e Pedras Lapidary de Mozambique	Mozambique
Myanmar Gemstones Enterprise[2]	Myanmar

enterprises. Table 4.1 gives examples of government mineral marketing agencies that have been established.

- Government mining banks have also been established to buy products from small-scale miners especially precious minerals like gold, and gemstones. Examples include gold sales to reserve banks in Zimbabwe, Tanzania, Philippines and Ghana.

Under this controlled system of marketing, in which the state monopolizes the buying and selling of these high-unit value minerals to international markets, the system, by design, does not capture production from illegal mining activities, thereby encouraging illegal marketing.

The basic operational characteristics of state controlled marketing agencies are as follows (Kambani, 1995):

- The state decides what price to pay the producer. Pricing is fixed arbitrarily; it is not based on market competitive bidding. Since the state has to resell the mineral products on international markets, producer prices are considerably low in order to cover their own costs. From a producer point of view, it is, in essence, an under-pricing of their commodities. The effect of under-pricing is discussed in greater detail later.
- Producers are normally paid in local currency. In countries where there are foreign exchange shortages, this impacts negatively on the sector because equipment and spares have to be imported in foreign currency. Furthermore, where the local currency is overvalued, a payment in local currency results

[2] Gemstone Mining in Myanmar (formerly Burma) was nationalized in 1963. Myanmar Ministry of Mines still controls all legal mining. In 1976, the Myanmar Gems Corporation was established and later renamed "Myanmar Gems Enterprise" in 1989 to mine and market gemstones. In an attempt to drastically reduce illegal mining, the government announced joint venture mining contracts with nationals in 1990.

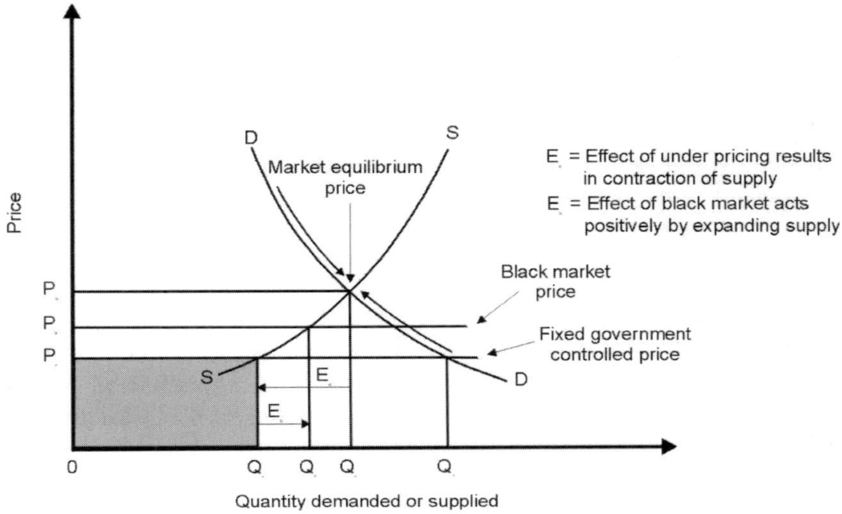

Figure 4.3 Effect of under-pricing.

in under-pricing of the commodities. In conjunction with the arbitrary pricing discussed above, producer prices may further deteriorate. Because of these factors, miners may be tempted to sell illegally to raise the necessary foreign currency with the added advantage of securing realistic market prices. The effect of local currency overvaluation is also discussed in detail below.

- The state decides when to pay the miner. Often, the miner must wait long periods before payments can be made because marketing agencies can only pay them after their commodities have been sold. However, most small miners operate on a shoe string budget, and therefore, cannot withstand the financial strain of delayed payments.
- Other arrangements have involved payment in local currency but pegged at realistic market prices. This method has been used in the marketing of gold in Ghana, Tanzania and, more recently, Zimbabwe, with spectacular results.

From the above discussion, it is clear that unfair pricing policies, delays in payments, and local currency overvaluation, contribute to illegal mineral marketing. The effect of unfair pricing is now discussed.

Figure 4.3 demonstrates the effect of under-pricing policies. The diagram shows the aggregate supply and aggregate demand schedules for any given commodity with the equilibrium market conditions established at price P_e. It is important to note that the supply curve represents the cost function of the commodity under consideration. This function is established by arranging the individual quantities of the mineral produced by all active operations in an ascending order of production costs. It is obvious that at the price level P_e, miners will supply exactly the quantity Q_e, since additional quantities cannot be mined profitably. At this price, consumers are also willing to buy exactly the quantity Q_e and the market is said to be at equilibrium. The price P_e is the competitive market price determined by conditions of supply and demand.

The effect of under-pricing is shown by the introduction of a lower fixed government price depicted as P_o below the market equilibrium price P_e. Similarly, the price obtained on the illegal market (black market) is shown as P_b on the diagram. There are three important observations to be made at this point. They are as follows:

(1) At government controlled price P_o

As previously discussed, government marketing agencies are forced to set their producer prices P_o below the competitive world market P_e at which they can resell the minerals because they have to cover their own operating costs. At the low fixed government price P_o, only a small quantity Q_o can be mined and marketed at a profit, while all other producers would incur a loss due to production costs exceeding this price level.

In theory, if the market can be fully controlled by the authorities, the mining sector will sooner or later shrink to Q_o. In this case, the country will lose export earnings equivalent to:

$$[(P_e * Q_e) - (P_o * Q_o)]$$

(2) At intermediate black market price P_b

However, in practice, the market cannot be fully controlled. Illegal trading will inevitably develop at a black market price level somewhere between P_o and P_e. For producers with marginal costs greater than the fixed government price P_o, the decision to sell on the black market is a matter of economic survival. With lower cost producers, it would be normal business behavior to maximize profits by taking advantage of a higher market price.

Empirical evidence indicates that the quantities supplied on the black market ($Q_2 - Q_o$) are much greater than the quantities traded officially (Q_o). It can also be observed in Figure 4.3 that the quantities supplied on the black market depend on the shape of the supply curve. The quantities increase with a flatter supply curve presented by S_1 with:

$$(Q_2 - Q_3) > (Q_2 - Q_o)$$

A flatter supply curve is an indication of the price elasticity of the supply side, or that supply increases substantially with small price increments. It is also apparent that at the black market price P_e, the volume of profitable production, and thus the income contribution to the economy as a whole from the sector, will be greater than at a low fixed government price level. That is:

$$(P_e * Q_2) > (P_o * Q_o)$$

(3) At market equilibrium price P_e

The greatest volume of production and SSM sector income will, however, be achieved if the market is allowed to operate freely without unnecessary government interference, whereby an equilibrium price close to the world price would develop. The total income earned in the sector will be highest at $P_e * Q_e$. Obviously, this is the most desirable situation for the economy and its producers.

However, evidence from some countries that liberalized their economies and markets more than a decade ago – such as Zambia – still indicates the continued presence of significant levels of illegal marketing. This provides a sufficient conclusion that there are other important underlying factors contributing to the scourge. Among these are overvaluation of local currencies, tax evasion, marketing and financial constraints. These are now discussed.

(4) Over-valuation of local currency

In a number of DCs, mineral products are largely for export, and therefore, the trade is sensitive to changes in foreign exchange rates. Where local currencies are overvalued, it may lead to a similar situation to that discussed above, whereby an under-pricing of minerals occurs if marketing is controlled by government marketing agencies.

Figure 4.4 demonstrates the effect of overvaluation (Kambani, 1995). The figure shows the demand and supply schedules for convertible currency (*e.g.* US dollar) with an equilibrium price or exchange rate established at P_e. This is the market-clearing rate as determined by market forces of supply and demand.

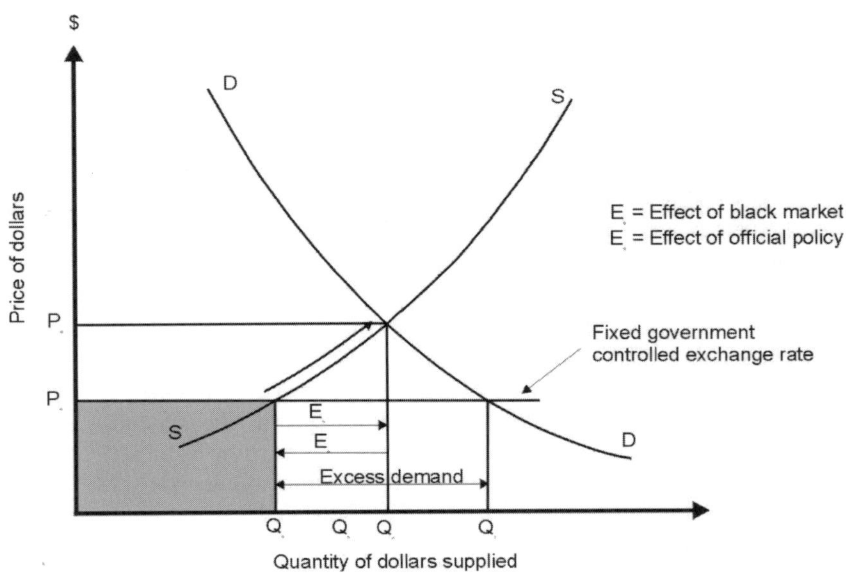

Figure 4.4 Effect of local currency overvaluation.

However, in some countries like Zimbabwe, where there are foreign exchange shortages, governments may intervene and manage foreign exchange rates artificially by lowering the exchange rate to P_o. Reasons for doing so are discussed later in this section. This has the net effect of under-pricing the available foreign exchange. With $P_o < P_e$, this situation represents the case of local currency overvaluation (*i.e* paying less for the available foreign exchange).

At this low price, P_o, supply of foreign exchange will slide from Q_e to Q_o, while demand will expand to Q_1, resulting in excess demand for hard currency in the amount of $Q_o - Q_1$. The fact that the market clearing rate P_e is higher than the government fixed rate P_o creates an opportunity for a parallel market (or black market) for hard currency to set in. Thus, DCs, which have experienced foreign exchange shortages and an overvalued local currency, have also attracted a strong black market presence for hard currencies. Those with access to officially available foreign exchange can buy it at this low fixed rate of P_o for resale at a market rate of P_e. Exporters of mineral commodities will be attracted by the higher black market rate when converting their hard currency earnings into local currencies.

It can also be observed in Figure 4.4 that the low fixed rate results in a lower supply of hard currency needed by the economy for investment and development than would prevail under competitive conditions. This is indicated as a decline in the supply of hard currency from Q_e to Q_o. Apparently, the black market tries to restore the imbalance in the market by expanding the supply to Q_e from Q_o.

Since the black market rate for foreign exchange is much higher than the official rate, it correspondingly results in a higher general price level of goods and services procured using this rate. It is largely because of this that governments do not tolerate the foreign exchange black markets.

Some DCs have adopted a stabilization program where the exchange rate is fixed. The objectives of these exchange-rate-based stabilization programs are said to have been justified on the grounds that only under a regime of fixed nominal exchange rates can the latter fulfil its traditional role as a nominal anchor for the domestic price level. Examples of countries that have pursued this approach include the high inflation Latin American countries of Brazil and Mexico in the late-1970s, Zambia during the mid-1980s, and, more recently, Zimbabwe (Montiel & Ostry, 1993).

Currency devaluation, on the other hand, has the effect of reducing local costs. In many DCs, the international competitiveness of the mining industry has been maintained through a policy of active periodic devaluation. For instance, in the case of copper, the average total costs fell from US$0.82 in 1981 to US$0.66 in 1986. While industrial countries such as Canada and Australia were assisted by productivity improvements, DCs such as Zambia, Peru and the Philippines experienced exchange rate adjustments (Kumar, 1989; Auty, 1991). Generally, devaluations benefit exporters while hurting importers. However, Kumar (1989) argues that while exchange rate adjustments may benefit the industry in the short run, it is not a viable strategy in the long run without real improvements in productivity.

From the foregoing discussion, it is apparent that state controlled marketing and an overvaluation of local currencies has generally contributed to illegal trading because they lead to low producer prices. However, where payments by state marketing agencies have been prompt, and where a real exchange rate has been used, smuggling activities have been observed to diminish. For instance, Sri Lanka saw the declared value of gemstones rise from US$550,000 in 1977 to US$26 million in 1982, following the institution of various liberalization measures, including the adoption of a local price level determined by the market; spectacular results were also achieved in gold production with similar experiences. Kenya solved its gold smuggling problem by paying a market price plus a 20% premium as an export incentive, while Congo DR had an increase from 2.9 tons of gold in 1982 to six tons in 1983 when the currency was devalued to the level of convertibility (Walrond, 1989). In Tanzania, with the introduction of measures to pay competitive prices to those on the black market, there was a dramatic increase in the amount of gold declared through the Central Bank (Walrond, 1989). However, the inability of the government to periodically review the prices to take into account price erosion due to inflation, in turn, resulted in a significant drop in the amount of gold declared, with most of it again being sold on the black market (Kambani, 1994). More recently, Zimbabwe adopted similar measures with the Central Bank, paying a premium price on gold at above market price.

Marketing constraints

Despite market liberalization in some countries, illegal trading still continues. With a few exceptions like Thailand, Sri Lanka, and Brazil, most DCs do not have well-developed local markets for gemstones and gold. Therefore, small miners usually lack access to markets. This has created an opportunity for illegal buyers to provide product and commonly visit the miners to buy the minerals directly; miners therefore do not have to travel abroad to seek markets, which can be expensive.

Financing constraints

Because of the high risk associated with SSM, the securing of adequate financing has proven to be a challenge for most of its operators. However, illegal buyers undertake some form of pre-financing of mining operations. More specifically, illegal buyers provide the finances to the legal mine owner against future production. This form of pre-financing is very common in Zambian emerald mining and has provided more than 50% of emerald production to foreign illegal buyers (Bull, 2002).

Illegal mining

Illegal mining is a natural partner of illegal trading. Where markets are government controlled, there exists a problem of how to transfer production from numerous unregistered operations into legal channels. Government buying agencies are not in a position to deal directly with illegal miners. Consequently, in the absence of a free market, illegal production will automatically find its way into illegal markets, thus worsening the already-existing illegal trading position. Marketing liberalization policies would help to bring illegal production into legal channels.

Tax evasion

Some miners are driven by a desire not to pay the necessary taxes to the government, and therefore do not declare their production. Small-scale miners of precious minerals are well known for under-declaring output, and, in certain cases, not declaring output at all.

IMPACT OF ILLEGAL MARKETING

Rampant illegal marketing has serious economic ramifications in any given country. Available data indicate that about 50% of Brazil's estimated 120–140 tonnes of gold annual output is undeclared (Davidson, 1990). In the Philippines, the Central Bank was buying an average of 8.5 tonnes per year during the period 1985–1990 as declared production, which was estimated at between 33–40% (Muyco, 1993; Cramer, 1990). In Zambia, it is estimated that more than 50% of emerald production valued at more than US$200 million is undeclared annually (Kambani, 1994). More than 50% of Angola's diamond production valued close to US$1 billion produced in originally rebel-held areas is mined and sold illegally.

The most obvious direct economic loss from illegal trading is taxes. Significant government tax revenues are lost, making the SSM sector one of the smallest contributors to the fiscus. Furthermore, illegal trading denies a country valuable data on its value of production, and exports that may be required to design appropriate policies for the sector. For instance, strategies to establish local processing facilities may not be possible without key data.

Virtually all production from SSM is sold in raw form, especially gemstones and gold. The inability to add value to minerals, in turn, results in loss of potential employment and taxes that may be generated by forward integration into lapidary processing and jewelry making, and backward integration from the establishment of industries to supply the sector with various inputs.

Although a country may earn some invisible income from illicit trading, this type of trade is often characterized by a large involvement of aliens, which implies that substantial incomes accruing to these foreigners are repatriated abroad, resulting in weakened final demand linkages locally.

Generally, it can be concluded from the discussion in this chapter that illegal mineral exports diminish the SSM sector's contribution to the national economy.

REFERENCES

Auty, R.M. (1991). Mismanaged mineral dependence: Zambia 1970–90. *Resources Policy*, 17(3), 170–183.

Bull, T. (2002). Jewels in a copper crown. *Profit – Zambia's Business Magazine*, 1(3), 24.

Cramer, S. (1990). The legal battle for small-scale mining. *Appropriate Technology*, 17(2), 5–6.

Davidson, J. (1992). Zimbabwe – legal notes. *Small Mining International Bulletin*, No. 4, February 1992. SMI SMI-8.

Davidson, J. (1990). Gold mining and Garimperios. *Small Mining International Bulletin*, No. 2, June 1990, SMI-6.

Kambani, S. (2000). Policy and strategy options for small-scale mining development in Zambia. *Minerals and Energy*, 15(3), 25–28.

Kambani, S. (1995). The illegal trading of high unit value minerals in developing countries. *Natural Resources Forum*, 19(2), Elsevier Science Limited.

Kambani, S. (1994). *Strategies for the Enhancement of the Contribution of Gemstone Mining to the Economies of Developing Countries*. Ph.D thesis, Montan University, Leoben, Austria.

Kumar, R. (1989). Government Fiscal strategy for mining to 2000. *Natural Resources Forum*, 10(4), 275–284.

Montiel, P.J. & Ostry, J.D. (1993). Targeting the real exchange rate in developing countries. Finance and development. *International Monetary Fund and World Bank*, 30(1), 38–40.

Muyco, J.D. (1993). *Small-scale mining in the Philippines: status, developments and policy directions*. Interregional Seminar – Guidelines for Development of Small/Medium-scale mining, Harare, 15–19 February 1993.

Priester, M. (1993). *Environmental Protection Measures for the Reduction of Hg-Emissions from Gold mining Activities: A Case Study from Colombia*. Guidelines for development of small/medium-scale mining, Harare, 15–19 February 1993.

Walrond, G.W. (1989). Small gold mines – The production and declaration problem: another dilemma for a small developing country. *Natural Resources Forum,* 10(4), 343–349.

5

The Policy Environment and Foreign Direct Investments in Mining Ventures in Developing Countries: Implications for Small-Scale Mining

HAMID ETEMAD AND KAMALEDDIN S. SALMASI

This chapter critically examines Foreign Direct Investment (FDI) theories in relation to mining policies as a context for small-scale mining in developing countries. It first reviews relevant FDI theories from a mining perspective, arguing that the general theory of FDI has not been, and is no longer, rich enough to provide adequate guidelines for decision-making and policy formulation in the mining sector. It further argues that the general theory of FDI is, in part, responsible for the misguided inadequacies of mining policies, particularly those pertaining to small-scale mining (SSM).

The chapter introduces a family of moderating factors that can potentially improve the efficacy and value of mining ventures from the perspective of both investment decisions and host country mining policies. Based on the proposed modifications, a set of research questions were devised, which were, in turn, subjected to testing and systematic empirical verification with worldwide-mining executives through a survey questionnaire; the survey results are presented and discussed. This research confirms that the theoretical underpinning of operating decisions in mining and mining policy formulation, especially in developing countries, are in need of change and due for modification.

LIST OF ACRONYMS

DCs	Developed Countries
EPU	Export Promoting Unit
FDI	Foreign Direct Investment
FSAs	Firm Specific Advantages
LDCs	Lesser Developed Countries

LMEs	Large Mining Enterprises
LSAs	Location-Specific Advantages
MNEs	Multinational Enterprises
OSA	Ownership-Specific Advantages, and LSA and Internalization
SSM	Small-Scale Mining
SSMs	Small-Scale Mines
SMEs	Small-Scale Mining Enterprises

BACKGROUND

Globalization has affected all facets of mineral commodities and mining profoundly. Consider the following:

1) On the demand side, the markets for mineral commodities have been globalized for some time. Most minerals are trading globally within a very narrow price band on the international market, and even in cases where they are not traded internationally, the narrow price band prevails locally.
2) Large mining enterprises have reduced their dependence on local suppliers by adopting multi-point, multi-employer, and multi-input supply chains, characterized by high flexibility and redundancy. In the process, the relative bargaining of the supplier (*i.e.*, the host country) is reduced dramatically (Porter, 1980, pp. 4–8).
3) On the supply side, market drivers and consumer behavior characteristics are also moving away from distinct local trends; the new trend points toward convergence, rather than the divergence of the past.
4) Global competition has forced Large Mining Enterprises (LMEs) to compete on both the supply and demand sides; they must continuously compete in order to become, and remain, globally competitive.
5) The industry support system, which was formerly monopolized by a small group of large enterprises, is now open to providers of all sizes since barriers to entry and exit are removed by globalization.

The overriding consequence of the foregoing processes is that barriers to movements of most, if not all, factors – input and output alike – are dramatically reduced, with little place for the collection of economic rent for location, size, affinity, or political stripe. Small-Scale Mining Enterprises (SMEs) must increasingly abide by the same rules and compete with LMEs in order to survive. Global competition will soon force local governments to revise their SME-oriented policies, and stop the implicit flow of subsidies from other economic sectors to their local SMEs under different industrial policy banners. FDI provided a framework for discussions in the past but is no longer sufficient. More specifically, there is a need for a critical examination of FDI theories in relation to mining policies, as a context for small-scale mining.

A number of scholars, such as Radetzki (1985a), Johnson and Pintz (1985), Brower (1987), Walde (1988), Bomsel (1990), Claessens (1993), Aldous (1993), Frecker and Sharwood (1993), O'Neill (1993), Ritchie (1993), Brown et al.

(1994), Strongman (1994) and Mikesell (1997), have discussed the role of FDI and multinational mining operations in both Lesser Developed Countries (LDCs) and Developed Countries (DCs). Few, however, have analyzed investment in the mining and mineral sector from the perspective of FDI theory and frameworks. In fact, this extant literature runs counter to prevailing globalization trends. Investments are becoming integral parts of the broader global markets, sharing the same cost of capital and subject to the common risk-adjusted return expectations. Most discussions in the received literature are from the viewpoint of LDCs (or, to a lesser extent, DCs), and are set within a practical-operational framework in terms of incremental improvements, and cost/benefits in specific mining projects without due regard for their implicit linkages to other mining investments. In spite of their dominant impact, both the theoretical (*i.e.* those of FDI) and practical (*i.e.* those of Multinational Enterprises) viewpoints are seriously missing.

REVIEW OF RELEVANT FDI THEORIES AND ADAPTATION OF THE MODERN THEORY FOR INVESTMENTS IN MINING

Monopolistic Advantage Theory

Hymer (1976) demonstrated that FDI occurred largely in oligopolistic industries. Hymer's work implied that the small number of dominant firms held advantages unavailable to others. He argued that their advantages might include economies of scale, superior technology, or superior knowledge in marketing, management, as well as ready-access to sources of finance largely unavailable to others. Firms capable of readily taking advantage of such imperfections develop further and maintain their competitive advantages, which, in turn, can enhance their operations globally. It had been reported as early as 1980 that superior knowledge alone was sufficient for an investing firm to produce differentially superior product(s) leading to some measure of control over the price, and, inevitably, an advantage over indigenous firms (Caves, 1982). Access to sources of more inexpensive capital could also provide a cost advantage over the indigenous firm, if, and when, they are able to acquire the necessary financial resources to sustain their local operations.

The conditions required by LMEs to gain an advantage over local firms are present in the mining sector of LDCs. The limited number of firms usually involved typically have access to superior inputs, including technology, information, sources of finance, as well as some degree of control over international markets. It is such Firm Specific Advantages (FSAs) that give these firms a *differential advantage* over local operators. The replication of these same advantages in several locations provides the basis for the economies of scale needed to further enhance market power. By definition, local mining firms have *location-specific advantages* (LSAs), including a local presence, access to local knowledge of subsoil resources, and close relations with local governments. LSAs can provide a basis for an independent differential advantage, which, in turn, can provide local mining firms with a basis to compete with LMEs.

However, in spite of their relatively superior local knowledge, indigenous firms are typically inferior to Multinational Enterprises (MNEs) in terms of FSAs. In the case of mining, although such FSAs cannot alone lead to a successful operation over the longer term, the implication of Monopolistic Advantage Theory is that the survival of SMEs depends on management finding ways to become competitive at a smaller scale.

Oligopolistic Reaction Theory

Knickerbocker (1973) postulated that in oligopolistic industries, companies *react* to each other's moves, and imitate each other in order to reduce the risk of being out-competed. His theory was based on the following two findings: (i) the tendency for firms in a number of industries to cluster together their direct investments in foreign countries (herding-type behaviour); and (ii) that most companies at the forefront of international expansion are typically in industries dominated by oligopolies. The objective of Knickerbocker's exposition was to make a case for a relationship between the clustering together of foreign investments, and the desire of these firms to react to the moves of their rivals. In the case of mining, both of Knickerbocker's conditions are satisfied when a foreign firm independently, or in joint venture with local firm(s), establishes a successful local operation. The success of the operation has a signalling effect – "the inherent risks are manageable" – which, after the first successful foreign-based operation, forces others into the "herding behaviour and oligopolistic reaction" in order to re-establish the old equilibrium. As a direct result, competing foreign firms hastily emulate the leader and secure a market position before the scale operations of the first movers can become entry barriers.

The key question, therefore, is the extent to which SMEs are impacted by such LME behaviour. United Nations' studies indicate that a relatively small number of MNEs based in industrialized countries dominate the sector through the control of both the supply chain and the marketing of minerals. This confirms the existence of the requisite conditions under which MNEs can operate oligopolistically, take advantage of market imperfections, and further influence market conditions in order to enhance their own firm-specific advantages (*e.g.*, market power, economies of scale, etc.) and create barriers for other firms, including SMEs. Naturally, the challenge for SMEs is to generate a family of advantages of their own to counter the above dynamics.

Internalization Theory

As explained by Buckley and Casson (1976), and elaborated by Rugman (1979) and others, MNEs are assumed to hold control over their own internal markets and resources, including technologies, managerial expertise, and marketing know-how. When MNEs expand into international markets, they can potentially earn additional profits in their own "internal" markets (*i.e.* the networking of sister-subsidiaries and affiliates), where their relatively superior resources and assets, such as knowledge, intellectual prosperity and specific competencies, are further leveraged.

Logically, MNEs should decline to transfer their FSAs to non-affiliated firms through contractual agreements, such as licensing, for a number of reasons:

- It may result in the MNE giving away its technological and managerial know-how (FSA) to a potential competitor;
- Licensing does not often provide the MNE with adequate control over manufacturing and managerial decisions in the foreign country;
- The MNE's know-how may not be easily transferable to a third party, thereby requiring higher costs and further adaptations to the local market conditions without commensurate returns; and
- A transfer to a third party/agent over which the MNE has less than full control may compromise the MNE's FSA, leading to greater long-term cumulative losses.

In response to foregoing, and related, costs, the Theory of Internalization states that MNEs internalize their capabilities within their internet markets, populated by their own network of sister-subsidiaries. Internalization allows for the replication, further development and exploitation of FSAs within the MNE's internal markets in a virtuous cycle, without the risk of dissipation or loss. The clear implication of this theory is that independent SMEs cannot gain easy access to the MNE/LME's pool of advantages.

The Eclectic Theory

Dunning (1973, 1977, 1980, 1988) extended Internalization Theory by incorporating LSAs, arguing that they can augment firm-specific or Ownership-Specific Advantages (OSA), and that the interactions of these factors influence the direction of international investments and enhance the power of FDI. Dunning's location-specific advantages referred to those derived from the use of resource-endowments or assets tied to a particular foreign location, such as mines and minerals. Dunning argued successfully that it is the combined **O**wnership-specific advantages, and **L**SA and **I**nternalization (OLI) that influence MNE's FDI decisions, and further enhance (or detract from) both the associated long-term viability and profitability of these decisions.

Although Dunning's Eclectic Theory is the most encompassing explanation, it faces several challenges in its application to mining. Dunning acknowledges the heavy dependence of FDI on all three pillars of the Eclectic Theory – namely OSA/FSA, LSA and Internalization. He does not, however, clearly specify their interdependencies, their bilateral inter-linkages, and, more importantly, the conditions under which these three pillars must be integrated to lead to a competitive FDI in general, in the mining sector in particular. A host of mining-related factors impact – *i.e.* add to, or even deduct from – the objective attractiveness of the location-specific advantages in FDI decisions. The mirror image of such impacts portrays the condition under which the local authorities, and SMEs by extension, will be operating.

THE POTENTIAL IMPACT OF THE LOCAL GOVERNMENT

Governments or local authorities possess the capability of further empowering all firms, including MNEs, or can inhibit their activities with their actions (or inaction) either selectively (*e.g.*, targeted to a specific sector and category of firms) or generally (*e.g.*, applied to all investments). The ultimate impact may, however, be very different. The following theoretical scenario briefly highlights the impact of government policy on the LSAs in the mineral sector of LDCs, and points to inter-linkages beyond the local market. Consider the following simple FDI decision by a typical MNE, using Dunning's Eclectic Theory of the "three pillars" (Dunning, 1986, 1988) of FDI:

1) The OSA/FSA: The choice and the extent of using a particular factor or, a combination of factors, from the firm's portfolio (*e.g.*, skills, expertise, proprietary knowledge, rights, etc.) operationalized to a particular ownership or firm-specific advantage (OSA or FSA), is entirely an internal decision. However, the quality of these decisions will undoubtedly impact the efficacy of MNE's OSA. The firm need not secure extraneous rights or prior authorizations from external agents, including a potential government, to make such decisions. All of the necessary rights and authorization for enabling the firm to use all facets of its own FSAs are, by definition, within the firm's internal domain of decision-making. One obvious criterion is that the net returns to combinations of selected firm-specific factors involved in a given investment cannot fall much below the returns from similar investments in other locations or markets.
2) The LSA: Three aspects distinguish LSA, especially in mining, from FSA. First, in the case of mining, mines, subsoil resources, and all advantages associated with them are only available at the specific mine site or resource location. While an FSA is mobile, a LSA is not. Secondly, such LSAs are under the control of the authorities with jurisdiction over the specific location (the federal and the local governments, the landowner, license or right holder, etc.). While a typical MNE can independently decide on FSA, it cannot make any decision regarding LSA. Thirdly, any entity wanting to benefit from a particular LSA must cooperate with the local authority in order to secure the necessary authorization to exploit the potential LSA. The quality of that authorization is, therefore, bound to impact the LSA's attractiveness, value, potency and efficacy.
3) The FSA-LSA interdependencies: The authorization to exploit is not usually issued automatically to all general applicants. A list of preconditions is normally imposed and must be met before a specific authorization is considered. Most specify a set of operating conditions, restrictions, and requirements; expect certain performance standards to be met; and demand performance guarantees. Furthermore, the process is time consuming, not without cost, nor is it automatic. Therefore, the authorization becomes both firm- and location-specific.

A practical example of such authorization in mining is "the right to mine." The local authority in charge of the specific LSA may exercise several choices, including the following: (i) simply exercising the necessary "*rights to mine*" to a portion of a given geographical location for a specified time period and a pre-specified fee; (ii) soliciting the calling for bids from other interested companies; and (iii) offering a partial or time-based authorization for a certain time period with, or without, the right to extend or renegotiate tenure. Based on a firm's specific expertise, or as a part of its pre-qualifying conditions, the local authority may even favour certain firms, thereby imposing implicit costs on others. Naturally, *without such authorization*, the specific LSA *is of no use* to the firm, regardless of the objective market value of that resource or the associated benefits that the firm can derive from it through applying the full force of its own FSAs. Furthermore, the imposition by the authority may make the LSA less (or even more) favourable to a firm than the objective value because of restrictions on access (or even further enabling it by offering, for example, tax holidays and financial incentives). Such impositions may favour a local firm with relatively lower FSAs; it may also do the reverse – *i.e.* favour a Junior Mining Company (with inferior FSA as compared to major mining firms with superior FSA). Such complicating factors, particularly with regard to mining, were not envisioned by the general FDI theories, and therefore, further developments are required.

As there is an absence of sound theoretical frameworks, the following brief theoretical assessment is designed to shed further light on the situation.

FURTHER THEORETICAL DEVELOPMENT

General FDI theory suffers from certain deficiencies in terms of its application to mining. These fall into the following two categories: (i) the evaluation of LSA; and (ii) the LSA-OSA interactions beyond those envisioned by this theory.

The LSA and the LSA-OSA interactions

As previously stated, the intrinsic and objective value of a LSA, as a proxy to its usefulness to a firm, is not of practical value, and, therefore, is not applicable to FDI decisions on mining projects. As a theoretical notion, however, any location-specific resource, such as a mine, has an objective value independent of both the strength of the prospective firm's FSA and the intentions, if not the actions, of the local authorities. We refer to this objective value as Location-Specific Resource (LSR).

In terms of LSA-OSA/FSA interactions, although the local authority's decisions are outside of the MNE's sphere of control, its knowledge of an MNE's FSA in terms of its potential applicability, efficiency, as well as its brand equity and past behaviour in similar situations, may influence the local authority in its degree of authorization, thereby impacting the useful value of the LSR to the firm.[1] Although the local authority does not actually decide

[1] Therefore, LSA can be schematically represented as: $LSA = LSR * LA$, where LA can assume a range of positive numbers. For values greater than 1, LA enhances the value of LSR; and for values less than 1, it reduces the value of LSR. It is neutral when $LA = 1$.

jointly with the MNE, its authority in granting the local authorization to the MNE resembles a joint one, and may or may not involve implicit negotiations for arriving at the final values. In the absence of explicit negotiations, the provision of information, whether offered by the firm to the local authority voluntarily or collected by the authority for its own assessment, will necessarily link LSA to FSA implicitly (which is not envisioned by general FDI theory and constitutes a violation of both the Eclectic and Internalization Theories), and may furthermore expose the applicant's FSA to higher dissipation risks than expected by the MNE (a risk that MNEs rather avoid). Based on its own objectives, the local authority may demand, as precondition to its granting the authorization, FSAs that have not been deployed elsewhere, and which would not necessarily optimize MNE's internal use of assets and resources in its internal markets. It should not, therefore, be difficult to see that the complexities of most mining FDI decisions can complicate matters. However, although FDI theory remains underdeveloped, it can still provide rich insight.

Implications for the firm

In accordance with the above discussion, a firm is required to estimate the sum-total of all combined advantages associated with investment in a given location. This value should serve as the initial basis from which the firm's decision to apply for local authorization, submit a bid to invest and operate, or alternatively, to withdraw from further participation, is based. In fact, the following three separate determinations must be made before making a decision: (i) the choice and composition of LSA; (ii) the most objective estimate of LSR; and (iii) the most realistic value for LA.

A noteworthy point is that it is not the objective value of the LSA (*e.g.*, the presence of a rich ore deposit that is extremely economical to exploit) that should influence a firm to decide in favour of, or against, operating in a given location. Rather, it is the nature of the authorization, by the local or national authorities, that modifies that value. The local authorization can act as a modulator, which can either be highly positive[2] (provides a set of incentives that enhance the absolute value of the resource), or highly negative (sets debilitating conditions, such as higher taxes, risks, and disincentives, thereby reducing the expected absolute value of the resource) in terms of determining the practical value of the resource. It is this LA modulator that transforms the absolute value of the LSR into a highly valuable location-specific advantage on one extreme (*i.e.* when LA is much greater than one and thus LSA becomes much greater than LSR) and a location-specific disadvantage on the other extreme (*i.e.* when LA is much smaller than one and thus LSA becomes much smaller than LSR).

[2] The term "highly positive" should be interpreted as LA being greater than 1 (*i.e.*, $1 < \mathrm{LA} < \infty$), thereby further enhancing the objective value of the resource (LSR). Similarly, the term "highly negative" should be taken as LA being smaller than 1 ($0 < \mathrm{LA} < 1$), thereby decreasing the objective value of the resource.

As stated previously, the eventual value of this modulated LSR for a given firm may heavily depend on the local authority's assessment of the firm's FSAs, market position, relations with the local authorities, and its control over the market structure, including its own internal markets. Such assessment would, naturally, impact the ultimate returns of a mining project to both the firm and the local authority. However, the ultimate returns from a given investment in a given location (or subsidiary, once established) cannot be much inferior to similar investments elsewhere, regardless of location of the industry, especially in integrated and diversified MNEs. This provision pits returns from a mining investment against all others in the firm, if not the industry and the global market as a whole, for similar investments.

The implications of the foregoing discussion are re-articulated as follows. First, before arriving at a mining decision, not only must the value of the resource be estimated objectively, but the *nature* of a local authority's intentions, or actions, must also be assessed. The current policy orientation in a location is a good indication of its immediate directions. In light of this insight, the gradual withdrawal of FDI from developing countries in the 1970s, in spite of much announcement about improved investment environments in LDCs and DCs alike, can be easily attributed to the worsening operating conditions resulting from the lingering bureaucratic policies of the 1960s. Although this was not the intention of the policy makers at the time, the results were predictable and inevitable.

Secondly, a critical distinction in the case of the mining industry must be made – namely, that the local authority's decisions moderate local assets. This is neither recognized in FDI theories nor in the work of mining scholars. The immediate implication is that even passive local authorities – those who do not seek equity participation and control – impact the total operation to a higher degree than their actual authority. Naturally, their impact is much higher for SSM than their larger counterparts, regardless of nationality and actual size.

Thirdly, the local authority is capable of tilting decisions in favour of one firm as opposed to the other, as if they were active participants in the market place. This explains why certain Junior Mining Companies have been more successful than their larger counterparts in securing certain privileged positions.

Fourthly, the most profound implication is that they transform mining FDI decisions into a set of interactive, and related, decisions between the local authority and the potential investor. Thus, the local authority's relatively positive (or relatively negative) enactments can impact the eventual profitability accordingly, and thus influence the investor's decisions. This aspect is neither recognized nor anticipated by the general FSA-LSA Internalization-based models; it has also been overlooked by the mining investment scholars.

Fifthly, another clear implication is that the pattern of decisions is much closer to joint decisions made interdependently and interactively, rather than independently and sequentially, and by at least two independent decision-making agents (*e.g.*, the Government and the MNE), rather than one (*e.g.*, the MNE or the government alone).

A review of mining policy transformation in the decades of the 60s, 70s, and 80s, indicates that the actual decisions of both the local governments and

MNEs were taken independently and sequentially, one reacting to the decisions of the other, and with less than optimum results for both; some vestiges of that legacy still remain. However, one can also observe that the newly-emerging models (*i.e.* interactive and interdependent decisions) are setting the trend and becoming a much-practiced case for successful FDI decisions in mining. The evolving pattern of the Contracts of Work in Indonesia, for example, suggests strongly that the mining industry has recognized the efficacy of the cooperative model, and, as a result, the *modus operandi* of the mining industry is gradually evolving in that direction (See Chapter 32).

One critical question is how, and why, should LDC governments increase the efficacy of their authorization to allow for higher utilization of their mine resources? The "why" part of the question can be answered with relative ease. In light of the transparency of operations, and the abundance of information in the context of an exceedingly globalized environment, less competitive LSAs will be ignored. Because of secondary and tertiary value-adding activities, the opportunity costs of such glanced-over resources will quickly rise to much higher values than the actual value of the resource. Thus, it is theoretically possible that the reform of the mining code, fiscal regime and foreign investment policies may provide other value-adding benefits to a MNE's operation, thereby making an objectively inferior LSR attractive.

The how part, once the justification is in place, can be partly answered by the reform of these codes, as most LDCs had done in the late-1980s and 1990s.[3] Moreover, the moderating character of the local authority can further enhance or inhibit the value of LSA. Conversely, the related critical question for MNEs is as follows: are "reforms' overall impact" enhancing, or still inhibiting, relative to other opportunities?

RESEARCH METHODOLOGY

Motivations and research questions

As already explained, the general theory of FDI does not explicitly describe decision processes; this implies that the theory is independent of the process itself. The actual practice of FDI follows an implicit, if not, explicit, process, with clearly observable path/process dependency. The theory appears to be at variance with the prevailing practice, at least in the context of mining. The emerging practice is unclear, and, at times, troubling. This research, therefore, aims to shed some light on both the theoretical and practical aspects of FDI. It constitutes an empirical examination of the prevailing reality of MNE-LDC's FDI relations, focusing on both the pattern of host country policies and regulations, as well as the firm's pattern of behavior in the mining sector.

[3] According to Eggert (1997), some 90 countries have revised their mining codes during the 1980s and 1990s, or are now considering major changes to attract local and foreign investment.

The basic theme of this research is the FDI's path/process dependency. It is against this background that an effort is made to address the following broad research questions:

1) What are the influential components of host country-centered advantages (LSA), and how do they impact the investing firm's screening and selection of a location?
2) How do these LSAs interact with both the FSAs and the inherently larger risk factors associated with mining operations as determinants of FDI decisions?
3) How has the modern practice of FDI in the mining and extraction sector evolved over time, and is that evolution in full concordance with the general theory, or is it diverging from it?

The design of the research instrument

A part of a wider empirical study that included in-depth interviews, detailed longitudinal case studies, as well as a survey through a questionnaire, the present investigation focuses primarily on the feedback obtained from surveys. The design of the questionnaire instrument in this study followed several steps based on prior studies and research. The factors appearing most frequently in the literature and in previous surveys formed the initial basis. Secondly, five longitudinal country case studies have been conducted, of which, three have developed considerable depth and breadth on FDI in mining (see Etemad & Salmasi, 2001). These cases served to confirm and verify the cross-country importance of the selected factors for further examination and inclusion in this study. Thirdly, in the first phase of design, semi-structured interviews were conducted with academics and mining company executives, who were asked to examine the nature of each specific research question (to be included in the survey questionnaire). The questionnaire was then prepared, and pre-tested using a scheme of five-six point semantic differential scales, in order to ensure maximum clarity to respondents. To facilitate a high response rate, the questionnaire design was optimized for visual impact and ease of answering. Questionnaires were designed for, and sent to, the executives of major mining companies with worldwide operations. Prior to mailing the questionnaire, the executives had been contacted.

The questionnaire was sent to most major mining companies based in North America, Australia, Europe, South Africa, Indonesia and Japan (a total of 70). After two follow-up correspondences with the recipients through both regular mail and e-mail over the span of four months, 47 questionnaires were received, of which, 42 were deemed usable; the response rate was therefore 60%. These usable questionnaires constitute the research database from which the results reported in this chapter are based. Overall, respondents provided invaluable information, which serves to portray both the mineral investment policies and MNE practices in LDCs.

Experiences and involvement of respondents

In terms of the scope of activities and involvement, 76% of companies are active in exploration, mine development and mineral processing, and 24% of

Table 5.1 Positions of the respondents in the company (% of total).

President		Executive Manager		Consultant		Mid-level Manager	
5	12%	23	55%	9	21%	5	12%
Average Work Experience of the Respondents: 21.7 years							

Table 5.2 Reasons/incentives of mining companies for investing in a developing country.

	Degree of Importance •		
The Element/Issue	Rank	Average	S.D.*
Profitability of the Project	1	4.80	0.5
Availability of Good Mineral Resources and Higher Ore Grades	2	4.46	1
More Favourable Fiscal Regime	3	4.37	0.7
Entry into the Market for Long-term Concerns	4	3.66	1.1
Market Expansion (Host country's market size)	5	2.89	1.2

• Degree of Importance of Factors on a Scale of 5 = most important to 1 = least important.
* Standard Deviation.

are engaged in the provision of technical and managerial services. Table 5.1 provides details concerning respondents' positions and experiences within the sampled companies.

SURVEY RESULTS

Reasons cited by mining companies for investing in LDCs

The reasons behind multinational mining companies' decisions to invest in the mineral sector of developing countries have been well discussed in the literature (*e.g.*, Selassie, 1995; Strongman, 1994; Kumar, 1990). Incentives such as an abundance of resources, profitability of the project, maximized profitability over time through sustained long-term presence in a country (*i.e.* granting long-term claims tenure), and the possibility of investing in minerals with strategic importance, have all been cited as primary factors in many studies in the past. This background served as a basis upon which this study relied.

The first question of the survey sought to determine reasons for investing in a developing country. The most important reasons in the literature were summarized and listed in a 12-item list, and respondents were asked to *"indicate the degree of importance and also rank in order of importance your reasons for investing in a developing country"*. The questionnaire design provided a semantic differential 5-point scale, with its two anchor points of 5 = "very important" and 1 = "not important at all," for each listed reason. The results for the five top ranked items are reported in Table 5.2.

Table 5.3 Satisfaction of foreign companies with mineral investment environment of LDCs.

The Element/issue	Degree of satisfaction • Rank	Average	S.D.
Repatriation of Capital	1	3.65	1.63
Equity Ownership	2	3.53	1.65
Absence of Price Control on Final Product (s)	3	3.44	1.78
Access to Foreign Currency	4	3.42	1.34
Repatriation of Profits	5	3.42	1.41
Mining Rights	6	3.33	1.35
Land Ownership	7	3.29	1.43
Hiring of Foreign Specialists	8	3.28	1.5
Pricing of Inputs	9	3.24	1.66
Government Incentives (tax holidays & exemptions, cheap loans, etc.)	10	2.95	1.58

• Degree of Satisfaction of elements on a Scale of 5 = most satisfactory to 1 = least satisfactory.

As can be inferred from Table 5.2, the *"profitability of the project"* (which is an overall measure), *"the mineral potential of the country"* (a measure of LSA given the firm's FSA) and *"more favourable fiscal regime"* of the host country (a measure reflecting the friendliness of the local authority or the operating environment) were reported as the three single-most important reasons for mining companies to enter into a mining agreement with a LDC. The next two most-reported reasons dealt with entry and expansion into the host country market. Although most large mining projects take a global, rather than local, market prospective into consideration over the long term, the size of a local market seems to offer an easier short-term proxy for profitability and expansion.

The importance of the operating environment

Respondents were asked, "*In your opinion how satisfactory is (was) the investment environment in the developing country, where your most recent operation is (was), with respect to each of the following?*" (4 = Very Satisfactory, 3 = Satisfactory, 2 = Unsatisfactory, 1 = Completely Unsatisfactory, 0 = Not Applicable). Results are reported in Table 5.3.

The factors viewed as "very satisfactory" in developing countries, and therefore, ranked highly, include incentives related to the fiscal policies of the host government (*e.g.*, the right to repatriate capital, access to foreign currency, tax holidays and exemptions), the regulatory system (*e.g.*, equity ownership, mining rights, land ownership), economic policy (*e.g.*, free market pricing of inputs and final products) and administrative policies (*e.g.*, allowing for the hiring of foreign specialists). A careful re-examination confirms that the firm executives recognize that the attraction to local resources is enhanced by enabling policies. In our theoretical conceptualization, all of the above items constitute integral

Table 5.4 Degree of importance attached to mineral investment decisions in LDCs.

	Degree of importance		
Element/issue	Rank	Average	S.D.
Right to Mine (Security of Tenure)	1	4.85	0.4
Stability in Laws, Regulation and Government Policies	2	4.80	0.4
Repatriation of Profits	3	4.78	0.5
Mining Code	4	4.54	0.7
Fiscal and Financial Incentives (income taxes, government grants, etc.)	5	4.37	0.7
Access to Foreign Currency	6	4.22	1.1
Repatriation of Capital	7	4.2	1.1
Political Stability in the Host Country	8	4.10	1
Land Ownership	9	4.10	1
Bureaucratic Procedures Involved in Doing Business	10	3.97	1
Access to International Arbitration	11	3.95	1.2
Hiring of Foreign Specialists	12	3.8	1
Labor Law, Work Ethics and Work Condition	13	3.73	0.8
Equity Restrictions (Less Than × Percent)	14	3.68	1.1
Adequate Infrastructure	15	3.44	0.9

parts of the moderating factor of the LA. Results shown in Table 5.3 also indicate that host country governments have understood, and responded to, MNEs' demands by improving their investment.

Major factors that attract FDI to the mineral sector of LDCs

In the context of mining, many location-specific factors are now used as screening criteria for both investment and modes of entry. Notable examples include security of tenure, availability of land for exploration, and policies impacting their efficacy, such as stable and attractive fiscal and political regimes, progressive mining codes, economic and currency stability, simplified administrative procedures, contractual stability, and international arbitration (Eggert, 1997; Selassie, 1995; Strongman, 1994; World Bank, 1992; Kumar, 1990).

The survey question dealing with this issue was phrased as follows: "*How important are the following factors for you in formulating contracts concerning mineral projects in a developing country?*" A list of 20 items helped to capture the details and operationalize this question. Results are reported in Table 5.4.

Issues such as local demand for product(s), financial contribution of the host country, accessibility to local expertise, availability of inexpensive inputs, and infrastructure of the host country, all of which are influential in attracting general FDI, were identified as less important to mining companies. Table 5.4 clearly shows that the concern of the mining sector is different from that of general foreign direct investment. Important factors to general FDI, such as inexpensive inputs, *do not* assume a high degree of importance in mining companies relative to right to mine and stability. Furthermore, the degree of importance of factors

Table 5.5 Comparison of the results of the present survey with previous surveys (Important factors in selecting LDCs for mineral investment).

Elements/issues	EWC 1989	CHL 1990	WB 1990	ONL 1992	ESCAP 1993	S&E 2001
Right to Mine	1		1	1	2	1
Stability in Laws, Regulation and Government Policies	7			10	8	2
Repatriation of Profits (and capital)	2		5	2	3	3
Mining Code	6		4	8	4	4
Geological Potential of Host Country		1		3	1	5
Fiscal Incentives (tax rates and terms)	5		2	6	9	6
Access to Foreign Currency			3	9	7	7
Political Stability		2		5		8
Access to International Arbitration	8	4				9
Equity Control	4			7	6	10
Availability of Adequate Infrastructure		3				11
Management Control	3			4	5	12

reported in Table 5.4, as compared to previous surveys, has changed over time. This suggests that mining and extractive FDI has begun to attribute different importance to factors impacting the long-term health of their operation and on-going profitability when compared to the past.

Trend over time (a comparison of the present and previous surveys)

As reviewed earlier, a number of major studies reporting the results of comprehensive surveys concerning FDI in the mineral sector of LDCs served as benchmarks for the present study. These are identified below, and again highlighted in Table 5.5.

- A survey was undertaken by the Economic and Social Commission for Asia and the Pacific (ESCAP-1993) among 39 mining companies headquartered in North America, Europe, Asia, Australia and Africa, the title of which was "the factors being important in considering mineral investment" (See "ESCAP" in Table 5.4).
- O'Neill (1992) surveyed Australian Mining Companies to determine their "reasons in selecting a specific African country for mineral investment" (See "ONL" in Table 5.4).
- The World Bank's Mining Unit (July 1990) surveyed International Mining Companies for "determinations of mineral exploration and investment in developing countries" (See "WB" in Table 5.4).
- Charles (1990) studied Multinational Mining Companies: "Determination of important factors in selecting countries for mineral exploration," (See "CHL" in Table 5.4).

- The East-West Centre (1989) surveyed 32 International Mining Companies to determine "critical and negotiable factors for major mineral exploration in a country" (See "EWC" in Table 5.4).
- The present study (See "S&E" in Table 5.4).

As can be seen in Table 5.5, there is a general agreement between results of the present survey and previous surveys regarding critical issues. Moreover, the right to mine remains as the most important factor in all surveys. Mineral potential of the host country, stability, repatriation of profits, mining code, access to foreign currency and fiscal incentives are among the most frequently mentioned factors in all surveys. Our theoretical conceptualisation appears to be very consistent with the results of all surveys, including the present study, and provides a realistic reflection of the prevailing reality encompassing FDI in mining.

Although the above-mentioned factors do not *guarantee* an inflow of foreign capital, the strong competition for FDI among many LDCs has simply forced many to adopt conducive and competitive mineral policies. They are also custom-tailoring them further to offer more favourable conditions to potential investors; however, investors are not flocking in. A complex decision framework, as conceptualized theoretically and also discussed earlier, appears to be influencing foreign investment in the mineral sector of the LDCs. The general theory of FDI has not provided for the particular refinements presented above and already in action in the mining-oriented FDI.

CONCLUDING REMARKS

This chapter has argued that the globalization of the mining and operating environment has had a lasting impact on SMEs and LMEs alike; host governments are not spared. In fact, they are forced to enact both national and international policies that align mining with other sectors of the economy. Furthermore, the previous guidelines provided by the general theory of FDI have become inadequate, and may, in part, be responsible for the misguided inadequacies of the mining policies in the recent past, which has particularly impacted SSM. This chapter introduced the concept of *moderating factors* that impact the efficacy and value of mining assets from the perspective of both investment decisions, and host country mining policies and operations. The empirical testing in this research confirms the presence of such moderating factors in both operating decisions and mining policy formulation in the mining industry, especially in developing countries. The results of this research, combined with those of five previous comprehensive surveys on this general topic, point to a newly-emerging global perspective. This perspective is exceedingly influenced by the globalization of the environment, and may require the attention of both decision and policy makers.

Two direct implications of this chapter's theoretical development impact mining decisions and operations. First, in a mining investment decision, not only must the value of the given resource (*i.e.*, the LSR) be assessed objectively, but the nature of a local authority's treatment of that resource must also be evaluated.

A critical distinction in the case of the mining industry is that, local authorities behave as if they own the resources, are a partner in the venture, and are entitled to make decisions on behalf of the investment. Their decisions inevitably moderate the value of local resources and assets as if they were indeed active partners in the mining investment decision. This mining-specific characteristic is not recognized in general theories of FDI. The immediate implication of such behaviour is that even passive local authorities impact the nature of a total operation as if they were indeed partners in the decision process. This adds a new dimension and transforms most theories of FDI into a set of interactive decisions between the local mining authority and the MNE prospecting for a potential investment. Therefore, the local authority's relatively positive or negative actions can easily impact the eventual outcome, and can range from further empowering and enticing (favouring the LME/MNE) to hindering to MNE decisions (in favour of smaller mining firms, local or international alike).

Secondly, the nature and the potency of local authorization may also depend on the prospecting nature of the MNE's actual or perceived profile with respect to a given location-specific resource, and the lobbying actions of other competing firms, including SMEs. A more co-operative and conciliatory MNE, for example, may receive a higher-valued moderator than a less co-operative MNE for the same resource. Relatively more cooperative SMEs can, for example, form a coalition with local SMEs in order to influence a local authority's moderating action in their favour. Such cooperative and influential strategies can explain how an MNE with relatively inferior FSA (*e.g.*, technology) may receive a higher multiplier than its counterpart with relatively superior FSAs. This is a plausible explanation for cases in which Junior Mining Companies have managed to win the ultimate bidding war against larger, well-endowed but less flexible counterparts. The general theory of FDI fails to explain why larger firms with superior FSA (and even broader and richer network of internal markets), given the same LSA, loose out to firms with inferior FSAs. Both of these aspects are not recognized, and, therefore, not discussed by the general theory of FDI. The above explanation implies that decisions can be made interactively and cooperatively by two parties, rather than one decision agent (*e.g.*, the MNE), which was the pattern throughout the 1980s and 1990s, when transformations in mining policy had not yet come about.

In a truly interactive case, the government(s), the investing firm(s) and its potential support team may agree, at the early stages of development, on a set of regulatory guidelines jointly, based on the particulars of the mine, along with national industrial policies; the jointly-agreed principles would then govern the consequent operation. Many countries have already adopted and implemented such adaptive models, the Indonesian approach being a prime example. Such approaches allow for both parties to change their positions interactively in response to changes in the environment.

In the sequential gaming approach, the regulatory restrictions set by the government limit the LSA. MNEs are then forced to evaluate location-specific benefits as characterized by the local authority in light of applicable regulations (*e.g.*, mining codes, mining regulations, tax regimes, and foreign exchange and repatriation parameters) before making a decision. During the 1950s, 1960s and 1970s, the policies of developing countries made the investment climate unappealing, if not, hostile, to potential investors. As a result, many mining MNEs decided against mining investment in LDCs, whilst comparative ventures in DCs flourished. Faced with these results, LDCs began reforming their operating environments in the 1970s and 1980s through progressive revisions of their mining codes, fiscal regimes and policies on foreign investment. Additional reforms in the 1980s, in a sense, increased the empowering magnitude of the local authority multiplier and reduced the inhibiting/taxing impact of their regulations, although the actual value of the resource (LSA) had deteriorated. Faced with the improved and more conducive operating environments in LDCs, MNEs began to reconsider mining and reinvestment in LDCs.

RECOMMENDATIONS FOR POLICY FORMULATION

Small-scale operators may, perhaps, be considered the only purists of the mining industry in many developing countries, operating within inhospitable terrains, with much little financial backing, and greater exposure relatively high risks. They are often exposed to the uncertainties of inspections, fines and closure. Many SSM operators are discontent with the status quo and hence, have created conditions of their own that isolate them from competition; this enables them to operate successfully and contribute to economic development (Etemad & Salmasi, 2001; ILO, 1999). However, in light of globalization of the environment, such practices are not sustainable. The following recommendations are therefore made.

Employment-oriented policies

Although the employment impact of SSMs is considerable and has been discussed in the past (see ILO, 1999), LDC governments must re-examine the conditions under which job creation by the small and medium-sized mining companies takes place and ensure that they are sustainable over the long-term. The theoretical developments, confirmed by empirical results, point to the potency of cooperative and interactive systems. Within such systems, SMEs can play much more significant roles than in the past. This suggests that governments should review their key policies and make plans with a view to creating meaningful roles for small and medium-sized mining companies. The overriding conclusion of this research is that global competitiveness has removed the barriers that shielded SMEs from the brutal impact of LMEs and global competition. Furthermore, the theory and practice point clearly to a cooperative policy option, whereby SMEs can become a part of large mining investment. Therefore, local governments must encourage SMEs by providing an increased supply of loanable funds to reduce risks and financial pressures, thereby easing regulatory and administrative burdens; promoting the development of strong regional economies; upgrading physical infrastructure; providing technical and managerial advisory services; and improving operational and financial management capabilities.[4]

The following are other potentially valuable policy measures capable of assisting SMEs to increase their efficiency and competitiveness in the global market:

1) The introduction of modern mining and processing techniques by local governments can increase productivity and mineral recovery; furthermore, model mines and training centres could help SMEs and SSMs undergo an easier transition into modern mining.
2) Governments should encourage the participation of active and successful mining co-operatives in large mineral projects.
3) In countries where exploration permits cannot be freely traded in the free market, there are no effective means of preventing this occurrence. The elimination of restrictions on the transfer of exploration (and also mining) permits is therefore recommended. This recommendation is consistent with our theoretical developments, as it allows for a division of labour based upon the inherent special capabilities of local SMEs.
4) Departments of Mining and Mineral Processing should focus on issues related to small-scale mining and offer services such as information, technical assistance and advisory services, development (or transfer) of equipment and technology, low-interest loans, and advice on the purchasing of supplies and equipment, among others.

Increased competitiveness and participation within global markets

Many LDCs, despite possessing good mineral potential, have yet to play a significant contribution to national exports. Serious effort must therefore be made to develop commodities for which the sector has comparative advantages, and can join the league of important producers and exporters of industrial and mineral products. The following highlights some relevant policy measures:

1) The adoption of appropriate policies for foreign exchange, credit facility, tax incentives and ease of customs formalities for exporting mineral and mining goods, and elimination of wasteful bureaucratic procedures, are requirements for both policy effectiveness and promoting mineral exports.
2) An Export Promoting Unit (EPU) should be activated within the departments responsible for SSMs, which would provide market and commodity information to exporters and potential exporters, and assist SSM exporters with advertisement of their products in international markets. Similarly, the investment in marketing research, product development, and short-term

[4] The ability of SSMs to finance development can be improved if they form groups and establish micro-finance mechanisms. A market for properties can also be established to allow owners of small mines to sell their properties to, or enter into joint ventures with, people who have the resources and skills to develop them (Barry, 1996)

loans to SSM exporters through the Export Promotion Bank should be a part of the EPU's agenda.

3) Increasing the productivity of the mining sector, in particular SSMs, requires harnessing modern expertise and equipment. The provision of credit to assimilate modern technology and expert human resources should therefore be a high priority in the mining policies of LDCs.
4) As transportation of mineral and mining products requires costly expenditures, improving transportation through the acquisition of greater knowledge of shipping and logistics, the utilization of the shipping industry's full potential, and the improvement of the efficiency of LDCs' port facilities can dramatically help to reduce costs.

Other policy measures, such as supporting export unions and co-operatives, the establishment of export insurance facilities, the forming of joint export ventures with foreign companies, supporting the Export Promotion Bank, and the introduction of export incentives for completely processed goods, are also recommended.

Technology, research and development

The importance of research and development (R&D) and technological innovation for the future well-being and long-term competitiveness of the mineral industry is strikingly obvious. For the mining sectors of LDCs to be technologically sound, the implementation of the following recommendations is necessary:

1) Local processing of minerals and smelting increase local value-added and may add to foreign exchange earnings, and contribute to industrial development. This also has an important impact on the demand for local goods, services and employment, and on self-sufficiency of the processed products. This, in turn, requires the creation and promotion of further processing, manufacturing, and maintenance facilities for mining equipment and machinery. To facilitate such developments, companies should be encouraged to create strategic alliances for manufacturing with domestic manufacturers (with the participation, or under the supervision, of credible foreign manufacturers, where required) as much as possible.
2) Governments should encourage small mining companies to work with existing research and development facilities. Collaborative technology networks between the universities and mining companies could provide smaller mining enterprises with easy access to the expertise, advice, and information technology needed to increase their competitiveness through technological augmentation.
3) Governments should establish a policy to reward mining companies that: (a) minimize their energy consumption; (b) maximize the use of recycled materials in the production of metal parts; (c) lower production costs, and, in particular, the use of foreign exchange; (d) develop new products; and (e) develop technologies to improve manufacturing processes for mineral- and metal-based products.

4) Companies involved in basic and precompetitive research should be encouraged to liaise and co-operate with government-sponsored organizations to expedite exploitation, and avoid the duplication of research results.

REFERENCES

Aldous, R.T.H. (1993). *Changing patterns of world mining investment and the implications for Australia and her mining industry*. Conference Series, Published by the Australian Institute of Mining & Metallurgy, August 1993, pp. 267–274.

Barry, M. (Ed.) (1996). *Regulating Informal Mining: A Summary of the Proceedings of the International Roundtable on Artisanal Mining*. Industry and Energy Department, World Bank, Occasional Paper No. 6, April 1996, pp. 1–14.

Bomsel, O. (1990). *Mining and metallurgy investment in the Third World: The end of large projects*. Organization of Economic Co-operation and Development (OECD), 1990, p. 13.

Brower, J. (1987). Conflicting goals of mining companies and host governments. In H.E. Johansson, O.P. Matthews & G. Rudzitis (Eds.), *Mineral Resource Development: Geopolitics, Economics and Policy* (pp. 22–50). West View Press.

Brown, R., Faber, M., & Sisulu, M. (1994). Policies towards inward foreign investment. *IDS Bulletin*, 25(1).

Buckley, P. & Casson, M. (1976). *The future of multinational enterprise*. London: Macmillan.

Caves, R.E. (1982). *Multinational enterprise and economic analysis*. Cambridge: Cambridge University Press.

Charles, J. (1990). Ranking Countries for Minerals Exploration. *Natural Resources Forum*, 14(3).

Claessens, S. (1993). Alternative Forms of External Finance: A Survey. *World Bank Research Observer*, 8(1), 91–117.

Dunning, J.H. (1988). The Eclectic Paradigm of International Production: A Restatement and Some Possible Extensions. *Journal of International Business Studies*, 19(1), 1–31.

Dunning, J.H. (1980). Toward an Eclectic Theory of International Production: Empirical Tests. *Journal of International Business Studies*, 11(1), 9–31.

Dunning, J.H. (1977). *Trade, Location of Economic Activity and MNC: A Search for an Eclectic Approach*. In the International Allocation of Economic Activity: Proceedings of a Nobel Symposium held at Stockholm (pp. 395–418). London: Macmillan.

Dunning, J.H. (1973). *The determinations of international production* (pp. 289–302). Oxford Economic Press.

Eggert, R.G. (1997). National mineral policies and the location of exploration. *Proceedings of the Sixth Annual Professional Meeting of the Mineral*

Economics and Management Society (MEMS), Colorado School of Mines, USA, March 1997, p. 1.

Etemad, H. & Salmasi, K.S. (2001). The rugged entrepreneurs of Iran's small-scale mining. *Journal of Small Business Economics*, 16(2), 125–139.

Frecker, D. & Sharwood, M. (1993). *Joint ventures between the state and mining companies*. In Foreign Investment and Joint Ventures in the Mining Sector, Selected Papers, United Nations, pp. 3–21.

Hymer, S. (1976). *International operations of national firms: A study of direct foreign investment*. Cambridge; Massachusetts: MIT Press.

International Labour Organization (ILO) (1999). *Social and labour issues in small-scale mines*. Report for discussion at the Tripartite Meeting on Social and Labour Issues in Small-Scale Mines, Geneva, 17–21 May 1999.

Johnson, C. & Pintz, W. (1985). *Minerals and the developing economies*. In Economics of the Mineral Industries, American Institute of Mining, Metallurgical, and Petroleum Engineers, pp. 22–24.

Knickerbocker, F.T. (1973). *Oligopolistic Reaction and Multinational Enterprise*. Boston: Harvard University.

Kumar, R. (1990). Policy reform to expand mining investment in sub-Saharan Africa. *Resources Policy*, 220–244.

Mikesell, R.F. (1997). Explaining the resource curse, with special reference to mineral exporting countries. *Resources Policy*, 23(4), 191–199.

O'Neill, D.R. (1993). *Junior companies: Investment strategies for developing countries*. In Foreign Investment and Joint Ventures in the Mining sector, Selected Paper, United Nations, 1993, p. 43.

Porter, M. (1980). *Competitive strategy: Techniques for analyzing industries and competitors*. Free Press.

Radetzki, M. (1985a). *State mineral enterprises: An investigation into their impact on international mineral markets*. Resources for the Future, p. 28.

Ritchie, D. (1993). *Mineral sector investment criteria: An industry perspective*. In Foreign Investment and Joint Ventures in the Mining Sector, United Nations, 1993, pp. 47–51.

Rugman, A.M. (1979). *International diversification and the multinational enterprise*. Farborough: Lexington.

Selassie, H. (1995). *International joint venture formation in the agribusiness sector*. Averbury, pp. 2, 10.

Strongman, J. (1994). *Strategies to attract new investment for African mining*. The World Bank, p. 9.

Walde, T. (1988). Third World mineral investment polices in the late 1980s: From restriction back to business. *Mineral Processing and Extractive Metallurgy Review*, 3, 123–126, 174–177.

World Bank. (1992). *Strategy for African Mining*. World Bank Technical Paper, No. 181. Mining Unit, Industry and Energy Division, World Bank.

World Bank, Occasional Paper No. 6, *A Summary of the Proceedings of the International Roundtable on Artisanal Mining*. Organized by the World Bank, Washington, D.C. May 17–19, 1995 Industry and Energy Department, Barry, M. (Ed.), April 1996.

6

Violent Mining Conflicts and Diamond Wars

MICHAEL RENNER

Abundant natural resources have helped fuel conflicts in a broad range of countries in the developing world. At issue are a range of minerals and gemstones – discussed in this chapter – but also other resources, including oil and natural gas, timber, and agricultural products such as coffee. Altogether, about a quarter of the roughly 50 wars and armed conflicts active in 2001 had a strong resource dimension – in the sense that legal or illegal resource exploitation helped trigger or exacerbate violent conflict, or financed its continuation. The human death toll of these resource-related conflicts was horrendous. Rough estimates suggest that more than five million people were killed during the 1990s. Close to six million fled to neighboring countries, and anywhere from 11 to 15 million people were displaced inside the borders of their home countries.[1]

However, some people – warlords, corrupt governments, arms traffickers, and unscrupulous corporate leaders – benefited enormously from the pillage. The Angolan rebel group UNITA, for instance, is estimated to have earned more than US$4 billion from diamond sales during the 1990s. At the height of the diamond wars in Africa, conflict diamonds are believed by some sources to have accounted for as much as 20 percent of the world's rough diamond production. Governments, rebels, and warlords have used the money derived from conflict commodities to arm themselves and to line their pockets, even as their

[1] One-quarter share of all conflicts having a resource dimension is the author's assessment based on existing literature. Number of deaths derived from data in Milton Leitenberg, "Deaths in Wars and Conflicts Between 1945 and 2000," Center for International and Security Studies, University of Maryland, May 2001 (Paper prepared for Conference on Data Collection in Armed Conflict, Uppsala, Sweden, 8–9 June 2001). Refugee numbers derived from UN High Commissioner for Refugees, <www.unhcr.ch>. Number of internally-displaced persons derived from U.S. Committee for Refugees, <www.refugees.org>, both viewed 25 August 2002.

societies at large suffered terribly from warfare and the lack of essential goods and services.[2]

Since the late-1990s, awareness of the close links among illegal resource extraction, arms trafficking, violent conflict, human rights violations, humanitarian disaster, and environmental destruction, has grown rapidly. Expert panels established by the United Nations have investigated cases in Angola, Sierra Leone, and the Democratic Republic of Congo. Civil society groups have launched a campaign against "conflict diamonds" from these countries, and have shed light on other conflict resources as well. Company and industry practices are coming under greater scrutiny. Media reports have helped carry these concerns from activist and specialist circles, to a broader audience.

In contrast to the cold war era, today's conflicts are less about ideologies and seizing the reins of the state, than the struggle to control or plunder resources – capturing sites rich in minerals, timber, and other valuable commodities, or controlling points through which they pass to markets. There is a self-sustaining vicious cycle at work in which the spoils of resource exploitation fund war; such war provides the means and conditions that allow continued illegitimate access to these resources.

In some places, the pillaging of raw materials prolongs wars initially triggered by other factors, such as unresolved grievances or ideological struggles. When the cold war rivalry came to an end in the late-1980s, much of the support previously extended by the two superpowers to governments and rebel groups among their Third World allies disappeared. External patrons (either governments or nationals living outside the country) have not vanished completely, but warring factions are increasingly relying on a variety of criminal means, including extortion, pillage, hostage-taking, monopolistic control of trade, drug trafficking, exploitation of coerced labor, and commandeering of humanitarian aid within their borders. Perhaps the most important revenue source is the often-illicit extraction and trading of natural resources.[3]

---

[2] Estimates from United Nations, Security Council, "Report of the Panel of Experts Appointed Pursuant to Security Council Resolution 1306 (2000), Paragraph 19, in Relation to Sierra Leone" (New York: 20 December 2000), from Christine Gordon, "Rebels' Best Friend," *BBC Focus On Africa*, October–December 1999, cited in Ian Smillie, Lansana Gberie, and Ralph Hazleton, *The Heart of the Matter: Sierra Leone, Diamonds and Human Security* (Ottawa, ON, Canada: Partnership Africa Canada, January 2000), from Fatal Transactions Campaign, op. cit. note 3, and from U.S. Government Accounting Office, *Critical Issues Remain in Deterring Conflict Diamond Trade*, GAO-02-678, (Washington, D.C.: June 2002), p. 7.

[3] Switch from superpower patronage to resource exploitation from Mark Duffield, "Globalization, Transborder Trade, and War Economies," in Mats Berdal and David M. Malone, *Greed and Grievance: Economic Agendas in Civil Wars* (Boulder, CO: Lynne Rienner Publishers, 2000), p. 73, and from Richard Dowden, "War, Money and Survival: Rounding Up," <www.onwar.org/warandmoney/index.html>. Alternative revenue streams from Mary Kaldor, *New and Old Wars: Organized Violence in a Global Era* (Stanford, CA: Stanford University Press, 1999), pp. 102–103, and from David Keen, "Incentives and Disincentives for Violence," in Berdal and Malone, op. cit. in this note, pp. 29–31.

However, some resource-related wars are of a different origin. In a number of cases, nature's bounty attracts groups that may claim they are driven by political oppression or the denial of minority rights, but are, in effect, predators seeking to enrich themselves through illegal resource extraction. They initiate violence not necessarily to overthrow a government, but to gain and maintain control over lucrative resources – typically one of the few sources of wealth and power in poorer societies. They are greatly aided by the fact that many countries are weakened by poor or repressive governance, deteriorating public services, a lack of economic opportunity, and deep social divides.

A third dimension of the relationship between resources and conflict concerns the repercussions from resource extraction itself. In many developing countries, the economic benefits of mining and logging operations accrue to a small business or government elite, and to foreign investors. However, in case after case, an array of burdens – ranging from the expropriation of land, disruption of traditional ways of life, environmental devastation, and social maladies – are shouldered by the local population. Typically, these communities are neither informed nor consulted about resource extraction projects. This has led to violent conflict in a number of places. Rather than full-fledged war, these conflicts usually involve smaller-scale skirmishes, roadblocks, and acts of sabotage, even though state security forces, and, occasionally, rebel groups, engage in major human rights violations. A number of these conflicts have evolved into secessionist struggles.

These and other cases discussed here are all "civil" conflicts in the sense that the violence takes place within a given country. Nevertheless, there are important global and regional connections – through the world market for illegal resources and the supply of arms, and through spillovers into neighboring countries.

THE NATURE OF RESOURCE CONFLICTS

Lootability and obstructability

Governments tend to have the capacity to extract whatever resource is found on their territory (if need be, relying on the technical expertise and capital of multinational companies). By contrast, rebel or warlord forces often have limited technical and financial capabilities. Large-scale mining operations are generally beyond their abilities and are hence less lootable, but alluvial diamond deposits, for instance, are accessible even to those without sophisticated technologies and high tech equipment. But obstructability is an additional important factor. Rebel or predatory groups may be able to block the shipment of a natural resource to processing plants or markets (hence, denying the government a source of income), or may extract ransom before allowing the passage of commodities to their intended markets.[4]

---

[4]Philippe Le Billon, "The Political Ecology of War: Natural Resources and Armed Conflicts," *Political Geography*, No. 20 (2001), pp. 561–584.

Non-state groups – including secessionist movements and rebels fighting to overthrow a government, as well as regional warlords and predatory groups that are more criminal than political in their motives – rely on a range of means to "loot" mineral resources. They seek to pillage existing stocks of raw materials, and have, at times, awarded illegal concessions to companies willing to contract with them. They have also pressed large numbers of civilians into mining operations (through force or indentured labour), or put some of their own combatants to work.

Although governments' reliance on natural resource deposits to fund military operations would seem to be a simple exercise of sovereign right, there are often questionable or even illegal aspects to such actions. In the first place, the democratic legitimacy of some regimes is in question. More specifically, concessions are occasionally awarded in ways that "bend" or circumvent applicable laws. Revenue streams are kept off the books, as governments end up not only purchasing arms and military equipment but also enriching corrupt elites. In some cases, government troops use extreme violence to depopulate resource-rich areas and keep them securely in their hands.

A perverse logic

Most of the violence in resource-related conflicts is directed against civilians. Since establishing undisputed control over resources is a key objective, armed groups seek to intimidate the local population into submission, or use terror to drive people away. "Hence the importance of extreme and conspicuous atrocity," observes Mary Kaldor of the University of Sussex, including directly expelling people, rendering an area uninhabitable by the indiscriminate spread of landmines, shelling houses and hospitals, chopping off people's limbs, imposing long sieges and blockades to induce famine, and inflicting systematic sexual violence. Unlike ideologically-based movements, those pursuing resource wealth do not compete for the hearts and minds of the local population. Young boys are often turned into child soldiers, and girls into sex slaves for older fighters. Many combatants are forced to commit atrocities, often against their own relatives, in order to traumatize them and to spread a sense of complicity that will prevent them from being accepted back into their communities.[5]

Actions that are often described as chaos, collapse, and senseless violence in media reports actually flow from a certain – albeit, perverted – logic. David Keen, a lecturer at the London School of Economics, argues that violence serves an economic function, maintaining a conflict economy that benefits certain groups – government officials, warlords, combatants, arms smugglers, and unscrupulous traders and businesspeople. Those who benefit from this violent "mode of accumulation" derive profit, power, and status, even as it spells impoverishment, broken lives, and death for society at large. Groups living off a lucrative resource have a vested interest in maintaining the status quo and, if

[5] Kaldor, op. cit. note 3, pp. 90, 98–100; Dowden, op. cit. note 3.

need be, in prolonging conflict. They are likely to find this to be a more attractive choice than settling conflict because it allows them to maintain their privileged position and bestows a quasi-legitimacy on their actions.[6]

The resource curse

Why are some countries susceptible to resource-based conflicts? While the availability and lootability of natural resources are key factors, they alone do not explain the issue. Where conflict does break out, it is typically the result of a combination of political, social, economic, and military factors.

Ample resource endowments can have negative economic consequences, as countries grow overly dependent on these resources, allocate inadequate capital and labor to other sectors – agriculture, manufacturing, and services – and under-invest in critical social areas such as education and health. The result is a failure to diversify the economy, and to stimulate innovation and the development of human skills.

Resource extraction industries tend to have "enclave" characteristics, that is, they create only small pockets of wealth and have few linkages to the rest of the national economy, particularly if the resources are exported before any processing takes place. The benefits to the economy and population at large are therefore quite limited. Frequently, enclaves are even physically separated, as mineral deposits are often found in remote areas.[7]

A statistical analysis by Prof. Michael Ross of the University of California at Los Angeles shows that the more a country depend on exporting minerals, the worse it scores on UNDP's Human Development Index. Specifically, they underachieve in terms of under-5 mortality rates, life expectancy at birth, and child education; they also experience significantly higher levels of inequality between the rich and poor than other countries with comparable levels of income.[8]

This outcome is not mere coincidence. Societies whose main income is derived from resource royalties instead of value-added product seem prone to develop a culture with widespread corruption. Resource royalties enable political leaders to maintain a stranglehold on power by funding a system of patronage that rewards followers and punishes opponents. Moreover, because such regimes rely less on revenues derived from a broad-based system of taxation, they also have less need for popular legitimacy and feel less pressure to be accountable.[9]

---

[6] David Keen, "Incentives and Disincentives for Violence," in Berdal and Malone, op. cit. note 3, pp. 22, 24, 27; David Keen, "The Economic Functions of Violence in Civil Wars," *Adelphi Paper 320* (Oxford: Oxford University Press for the International Institute for Strategic Studies, 1998); Kaldor, op. cit. note 3, pp. 110–111.

[7] Michael Ross, *Extractive Sectors and the Poor*, Oxfam America, October 2001, pp. 5, 7; Indra de Soysa, "The Resource Curse: Are Civil Wars Driven by Rapacity or Paucity?," in Berdal and Malone, op. cit. note 3, pp. 120, 121, 125, 126.

[8] Ross, op. cit. note 7, pp. 8, 11, 12.

[9] De Soysa, op. cit. note 7, pp. 120, 121, 125, 126; Le Billon, op. cit. note 4.

Such societies suffer from extremely poor governance: corruption and patronage are rife, public goods and services are withheld from most people, and state institutions are deliberately weakened to thwart potential challengers to the ruler, while a parallel network outside these formal institutions is created for the benefit of leaders and their cronies. State revenues are diverted to generate huge illicit fortunes for rulers that are used for patronage payments to key regime supporters. The kleptocracy under Zaïre's dictator Mobutu serves as a prime example of this phenomenon.[10]

The allocation of mining and other resource concessions to regime supporters is both a scheme for corruption and private enrichment, and a crucial mechanism to turn resources into cashable wealth to help prop up the existing regime through the purchase of arms and the maintenance of armed forces. Many governments of resource-rich countries spend a very high portion of state income on internal security to suppress democratic movements or other challenges to their power.[11]

The proliferation of arms and combatants

In such situations, rulers often foster and manipulate conflicts among different communities, factions, and ethnic groups as a means to maintain control. However, this intensifies friction within society: discontented and aggrieved groups turn increasingly to protest and violence; rivals rise to challenge the discredited leadership; and ruthless political-criminal entrepreneurs who sense an opportunity for pillaging resources use violence to achieve their objective. In a country with a poorly developed and diversified economy, and an absence of democratic rules, seizing control of a prized resource is the most likely ticket to wealth and power.[12]

In many developing countries, particularly in sub-Saharan Africa, government forces are in decay and private security organizations are on the rise, including forces loyal to regional warlords, citizens' self-defense groups, corporate-sponsored forces, foreign mercenaries, and criminal gangs. This is happening for a number of reasons. Without cold war-motivated sponsorship and under increasing pressure from western donors to tighten belts, many governments can no longer maintain large armies. Soldiers go unpaid or underpaid and often turn to other sources of funding, including plunder and extortion. Such fragmentation is even more likely where rulers have deliberately created rival security forces that keep each other in check, which poses a serious challenge to central control. Some military commanders become de facto local

---

[10] William Reno, "Shadow States and the Political Economy of Civil Wars," in Berdal and Malone, op. cit. note 3, pp. 45, 46, 56, 57. Zaïre from Michela Wrong, *In the Footsteps of Mr. Kurtz. Living on the Brink of Disaster in Mobutu's Congo* (New York: HarperCollins, 2001).

[11] Ross, op. cit. note 7, pp. 13–14.

[12] Reno, op. cit. note 10, pp. 47–53; Smillie et al., op. cit. note 2, p. 15; Le Billon, op. cit. note 4.

warlords, establishing quasi-commercial mining, logging, or drug-producing fiefdoms.[13]

During the 1990s, a number of private military firms rose to prominence – companies like Executive Outcomes, Sandline International, Defense Systems Ltd., and Ghurka Security Guards. They offer a range of services, including training, consulting, and guarding of facilities, as well as mercenary activities, such as procuring or brokering weapons, and running combat operations. Several beleaguered governments, including those of Angola, Sierra Leone, and Papua New Guinea, turned to these parties for assistance with fighting rebel groups, paying them with revenues derived from natural resources or, in some cases, granting them (or affiliated companies) concessions to diamonds and other resources.[14]

Multinational mining corporations often rely on private security forces to guard their operations and facilities. Some companies, such as Freeport-McMoRan in Indonesia, have subsidized or helped to train and arm government security forces, or have made equipment and facilities available. These units have been involved in severe human rights violations.[15]

The massive proliferation of small arms and light weapons plays a key role in all of this. Resource-based conflicts are primarily carried out with such weapons because they are cheap, widely available, easy to conceal and smuggle, and easy to use and maintain. There is considerable uncertainty about the numbers but an estimated eight million pistols, revolvers, rifles, submachine guns, and machine guns were manufactured in 2000 (of these, just under one million were military-style weapons, the balance, commercial firearms). Estimates are evolving, but it is thought that at least 638 million small arms and light weapons exist worldwide. At least 15 billion rounds of ammunition were produced in 2000 alone.[16]

Because many activities along the resource-conflict spectrum are illicit and involve actors of questionable legitimacy, grey- and black-market transfers carry special significance (See Chapter 4). The trafficking of arms is closely

---

[13] Kaldor, op. cit. note 3, pp. 92, 93; Le Billon, op. cit. note 4.

[14] Project Underground, "Militarization & Minerals Tour," <www.moles.org/Project_Underground/mil/intro.html>, viewed 6 July 2001; Kim Richard Nossal, "Bulls to Bears: The Privatization of War in the 1990s," <www.onwar.org/warandmoney/index.html>; International Alert, *The Privatization of Security: Framing a Conflict Prevention and Peacebuilding Policy Agenda* (London, May 2001); Chaloka Beyani and Damian Lilly, *Regulating Private Military Companies* (London: International Alert, September 2001).

[15] Freeport-McMoRan from Abigail Abrash, "The Amungme, Kamoro & Freeport," *Cultural Survival Quarterly*, Spring 2001, p. 40.

[16] Ease of use of small arms and other attributes from Michael Renner, *Small Arms, Big Impact: The Next Challenge of Disarmament*, Worldwatch Paper 137 (Washington, DC: Worldwatch Institute, October 1997), pp. 10–12; statistics and estimates from Small Arms Survey, *Small Arms Survey 2002* (New York: Oxford University Press, 2002), pp. 5, 6, 13, 14.

linked to illegal trade in raw materials such as minerals and diamonds. Arms and commodities often travel on the same routes, in opposite directions. Revenues from commodity sales finance the purchase of arms, ammunition, military equipment, uniforms, and other items; weapons are often directly bartered for natural resources, drugs, animal products, and other commodities.[17]

The environmental toll

Mining is highly destructive to the environment, both because of the methods of extraction used and the fact that operations mainly take place in ecologically-fragile areas. Mining operations involve the removal of vegetation and the destruction of habitat for many plants and animals. Toxic chemicals, either intentionally dumped or accidentally released, often contaminate rivers and lakes, in the process imperiling the livelihoods of local (frequently indigenous) communities.[18]

These impacts are felt even under the best of circumstances. They are worse in conflict zones where resource extraction is typically performed in a rapacious fashion that makes light of the survival interest of local communities and discounts biodiversity. As much resource extraction occurs illegally, and because miners are intent on extracting resources before they potentially lose control over an area, they have no incentive to conduct their operations in a responsible, let alone truly sustainable, manner. Their primary interest is raising funds for weapons purchases or self-enrichment and they try to extract as much, and as fast, as possible. In these circumstances, the enduring value of preserving biological treasure troves and ecosystems crucial to human well-being all too easily loses out to the urge to exploit the narrow spectrum of natural wealth – minerals, metals, and gemstones – that can be turned into cash.[19]

The eastern part of the Democratic Republic of the Congo, for instance, has been ravaged by a series of civil wars since the early-1990s. Although it is impossible to assess the full consequences as long as war and insecurity prevail, there is little doubt that a number of factors – including battles over control of mineral deposits, widespread illegal mining (and logging and poaching), the overall state of anarchy in much of the country, and refugee movements – have had devastating environmental impacts.[20]

---

[17] Renner, op. cit. note 16, pp. 33, 34; Small Arms Survey, *Small Arms Survey 2001* (New York: Oxford University Press, 2001), pp. 107–108.

[18] For an overview, see John E. Young, *Discarding the Throwaway Society*, Worldwatch Paper 101 (Washington, DC, January 1991), pp. 7–12; and Gary Gardner and Payal Sampat, *Mind Over Matter: Recasting the Role of Materials in Our Lives*, Worldwatch Paper 144 (Washington, DC, December 1998), pp. 17–20.

[19] This juxtaposition of the immediate cash-in value of resources versus the long-term preservation of biodiversity is made by Terese Hart and Robert Mwinyihali, *Armed Conflict and Biodiversity in Sub-Saharan Africa: The Case of the Democratic Republic of Congo (DRC)* (Washington, D.C.: Biodiversity Support Program, 2001), p. 12.

[20] Ibid., pp. 14, 19, 21, 25.

Congo's national parks – Kahuzi-Biega, Salonga, Virunga, Maiko, Garamba – and the Okapi Reserve have been severely affected. The lure of resource wealth drew not only a variety of armed factions but also some 10,000 miners, with calamitous consequences. Kahuzi-Biega National Park and the Okapi Wildlife Reserve are both UNESCO World Heritage sites, a status that extols their unique value to all of humanity. Severe environmental degradation as a result of "gold rush"-like events has landed them on the organization's list of sites in danger. Poaching of elephant tusks left only two out of the 350 elephant families in Kahuzi-Biega in 2000. Likewise, the number of eastern lowland gorillas has been reduced to the point where the species is now threatened with extinction. Moreover, coltan miners strip off the bark of eko trees to fashion troughs in which they flush out coltan from ore-bearing mud, and thousands of trees have been destroyed, undermining the livelihoods of the local indigenous people, the Mbuti, who use the eko trees for gathering honey.[21]

HOW CONFLICTS ARE FINANCED BY NATURAL RESOURCE PILLAGE

The pillaging of commodities has made possible the continuation of several violent conflicts in developing countries. Diamonds have been of particular concern in the conflicts discussed in some detail here: Sierra Leone, the Democratic Republic of the Congo, and Angola. It is a bitter irony that these glittering stones, which advertisers work hard to associate with the idea of love and personal commitment, are also connected with gruesome violence.

Sierra Leone

Diamonds played a central role in the conflict that devastated Sierra Leone during the 1990s. Ibrahim Kamara, Sierra Leone's UN ambassador, said in July 2000, "We have always maintained that the conflict is not about ideology, tribal or regional difference.... The root of the conflict is and remains diamonds, diamonds and diamonds."[22]

Even prior to the 1990s, corruption, cronyism, and illegal mining had squandered the country's diamond riches, to the point where few government services were functioning and educational and economic opportunities were scarce. Pressure from international lenders for financial austerity and cuts in

---

[21] UNESCO, "The World Heritage List," <www.unesco.org/whc/heritage.htm#debut>, and "World Heritage List in Danger," <www.unesco.org/whc/danglist.htm>, both viewed 11 August 2001; United Nations, op. cit. note 50, pp. 10–12; "Miners' Rush for Coltan Threatens Rare Gorilla," *Environment News Service,* 13 April 2001; "One Minute to Midnight for Great Apes," *The Ecologist*, July/August 2001, p. 15; Blaine Harden, "The Dirt in the New Machine," *New York Times Magazine*, 12 August 2001.

[22] Mamara quote in Barbara Crossette, "Singling Out Sierra Leone, UN Council Sets Gem Ban," *New York Times*, 6 July 2000.

the government workforce only worsened the situation. The International Rescue Committee has reported that one-third of all babies in the diamond-rich Kenema District die before age one. UNDP ranked Sierra Leone last on its Human Development Index in 2001.[23]

Throughout the 1990s, Sierra Leone suffered from rebellion, banditry, coups and coup attempts, and seesawing battle fortunes. In March 1991, the Revolutionary United Front (RUF) invaded Sierra Leone from Liberia and seized control of the Kono diamond fields. The RUF had strong backing from Liberian warlord (now president) Charles Taylor. The ranks of the RUF contained disaffected young men from slum areas, illicit diamond miners, Liberian and Burkinabe mercenaries, and others who welcomed the opportunity for pillage and violence. But many others (including a large number of children) were forcibly recruited. Though the RUF professed to act on unresolved grievances, its principal aim was to gain control over the country's mineral wealth. Characterized by banditry and brutality, the rebellion claimed more than 75,000 lives, turned a half-million Sierra Leoneans into refugees, and displaced half of the country's 4.5 million people.[24]

Faced with the RUF rebellion, the government expanded its armed forces from 3,000 to 14,000. This undisciplined, ineffective, ragtag army brought together ill-trained soldiers, militiamen from neighboring Liberia, urban toughs, and street children involved in petty theft. Mary Kaldor of the University of Sussex, in commenting about the latter, noted that, "they were given an AK47 and a chance to engage in theft on a larger scale." Government soldiers often supplemented their meager pay through looting and illegal mining.[25]

Rebel forces and parts of the government army actually collaborated at times. Government soldiers by day sometimes became rebels by night. This cooperation between supposed adversaries culminated in May 1997 when disgruntled

---

[23] Smillie et al., op. cit. note 2, pp. 8, 14; David Keen, "Going to War: How Rational Is It?" <www.onwar.org/warand money/index.html>; Reno, op. cit. note 10, p. 48; International Rescue Committee from Arms Trade Resource Center, "March Update," distributed by e-mail, 7 March 2000; United Nations Development Programme (UNDP), *Human Development Report 2001* (New York: Oxford University Press, 2001), Table 1.

[24] Events in Sierra Leone are documented and analyzed by Africa Confidential, "Special Reports. Chronology of Sierra Leone: How Diamonds Fuelled the Conflict," <www.africa-confidential.com/special.htm>, viewed 9 September 2001; by Human Rights Watch, "Sierra Leone: Priorities for the International Community," June 2000, at Global Policy Forum, <www.globalpolicy.org/security/issues/diamond/hrw2.htm>; by Smillie et al., op. cit. note 2, pp. 8, 14, 15; by UN Security Council, "Tenth Report of the Secretary-General on the United Nations Mission in Sierra Leone," 25 June 2001; and by United Nations, Security Council, "Report of the Panel of Experts Appointed Pursuant to Security Council Resolution 1395 (2002), Paragraph 4, in Relation to Liberia," S/2002/2470, 19 April 2002, p. 15.

[25] Character of government forces from Keen, op. cit. note 23, and from Kaldor, op. cit. note 3, p. 94.

government soldiers staged a coup against a government that had been elected just a few months earlier, and invited the RUF to join the new junta.[26]

Sierra Leone is a comparatively small diamond producer, but a large share of its gemstones is of very high quality and, therefore, sought after. The RUF purchased arms and sustained itself through control of the diamond fields, but diamond wealth has also been a constant source of internal friction. At first, RUF fighters did the mining, but later, the group relied more on forced labor, including that of children. The group's annual income has been estimated at $25 to $125 million. RUF diamonds have entered the world market disguised as Liberian, Guinean, and Gambian diamonds.[27]

Charles Taylor's Liberia has played a pivotal role. An investigative UN panel found evidence that "Liberia has been actively supporting the RUF at all levels, in providing training, weapons and related matériel, logistical support, a staging ground for attacks and a safe haven for retreat and recuperation, and for public relations activities." Under Taylor, Liberia became a major center for diamond smuggling, arms and drug trafficking, and money laundering.[28]

Following the near-failure of a 1999 peace accord, the war in Sierra Leone finally wound down in 2000 and 2001. Aside from the need to demobilize and reintegrate tens of thousands of combatants, the challenge of dealing with ongoing illegal diamond mining – and the possibility of continued conflict over the issue – remains.

Democratic Republic of the Congo

Resource pillage has also been a key factor in the two conflicts that have engulfed the former Zaïre, now called the Democratic Republic of the Congo, in devastating violence since 1996. During the 1996–1997 conflict in Zaïre, which ended with the overthrow of the Mobutu dictatorship, international investors aggressively negotiated lucrative resource deals with Laurent-Desire Kabila's rebel Alliance of Democratic Forces for the Liberation of Congo-Zaïre (ADFL). Among the investing companies were such major corporations as De Beers, Anglo-American Corporation, Barrick Gold Corporation, Banro American Resources, American Mineral Fields, and Bechtel Corporation. In April 1997, when it was clear that Kabila's forces had the upper hand, the ADFL signed an $885 million contract with American Mining Fields, a U.S. firm craving the copper, cobalt, and zinc deposits that had fallen under rebel control.[29]

[26] Keen, op. cit. note 23; Keen, op. cit. note 6, pp. 35–36; William Reno, "War and the Failure of Peacekeeping in Sierra Leone," in Stockholm International Peace Research Institute (SIPRI), *SIPRI Yearbook 2001: Armaments, Disarmament and International Security* (New York: Oxford University Press, 2001), p. 151.

[27] United Nations Security Council, op. cit. note 2; Rachel Stohl, "UN Imposes Diamond Ban on Sierra Leone," *Weekly Defense Monitor*, 14 July 2000.

[28] United Nations Security Council, op. cit. note 2; Smillie et al., op. cit. note 2, pp. 11, 47.

[29] Dena Montague, "Stolen Goods: Coltan and Conflict in the Democratic Republic of Congo," *SAIS Review*, Winter-Spring 2002, pp. 106–110; Robert Block, "As Zaire's War Wages, Foreign Businesses Scramble for Inroads," *Wall Street Journal*, 14 April 1997.

Resource looting, death, and suffering were far more widespread during the second war, which began in 1998. It claimed an estimated 2–3 million lives, and displaced at least another two million people. In August 1998, Ugandan and Rwandan troops invaded, assisting rebel groups seeking to overthrow the Kabila government. Angola, Zimbabwe, Namibia, and Chad dispatched troops in support of Kabila. According to one estimate, more than 100,000 foreign troops were, at one point, involved in Congo. Political and military factors played an important role in triggering the conflict. Several of the intervening forces wanted to thwart their own rebel groups operating from Congolese soil. Rwanda, in particular, was concerned that remnants of the Hutu Interahamwe militias that had carried out a campaign of genocide in 1994 were using Congo as a staging ground for ongoing hit-and-run attacks.[30]

But whereas the initial motivation was primarily related to security concerns, the opportunity to plunder the enormous resource wealth of Congo, in the context of lawlessness and a weak central authority, soon became the primary incentive. Congo is extremely rich in minerals and gemstones such as diamonds, gold, coltan, niobium, cassiterite, copper, cobalt, zinc, and manganese.[31]

During the first year of their invasion, the foreign forces and their rebel allies resorted to outright plunder of stockpiled raw materials. Once the stockpiles were exhausted, they organized a variety of methods to extract additional resources. For instance, the Ugandan army has directly carried out gold mining activities; local Congolese have been put to work by Rwandan and Ugandan forces; child labor has been used in gold and diamond mining; local, artisanal miners were made to relinquish some of their finds, or were taxed; and companies of questionable reputation were given concessions to help exploit Congo's resources. A number of Belgian, Dutch, German, and Swiss companies, for example, have been involved in the illegal coltan trade.[32]

---

[30] Estimates of deaths reported in Norimitsu Onishi, "African Numbers, Problems and Number Problems," *New York Times*, 18 August 2002; displacements from Taylor B. Seybolt, "Major Armed Conflicts," in SIPRI, op. cit. note 26, p. 26; foreign troops from "Peace Here Means War Elsewhere," *The Economist*, 23 June 2001, p. 44; Colette Braeckman, "Congo: A War Without Victors," *Le Monde Diplomatique*, English language edition, April 2001. Uganda and Rwanda initially supported the Rassemblement Congolais pour la Democratie (RCD), formed in August 1998. But as tensions increased between the two countries and led to fighting in 1999 and 2000, they backed different factions, respectively RCD-ML and RCD-Goma. As the Congo conflict has worn on, the different rebel forces have increasingly splintered, with shifting alliances forming. See Christian Dietrich, *Hard Currency: The Criminalized Diamond Economy of the Democratic Republic of the Congo and Its Neighbors*, The Diamonds and Human Security Project, Occasional Paper No. 4, Ottawa, Canada, June 2002, pp. 39, 40.

[31] United Nations, Security Council, "Report of the Panel of Experts on the Illegal Exploitation of Natural Resources and Other Forms of Wealth of the Democratic Republic of the Congo" (New York: 12 April 2001), pp. 41–42.

[32] Ibid., pp. 11, 14; United Nations, Security Council, "Addendum to the Report of the Panel of Experts on the Illegal Exploitation of Natural Resources and Other Forms of

The conflict has enabled Rwanda and Uganda to become major exporters of raw materials that they do not possess or only possess in limited quantities. Looted resources have become a major source of their foreign exchange, and enabled both countries to finance their military presence in the Congo.[33]

To stave off defeat and continue the war, the cash-strapped Congolese government used the country's resources to purchase weapons and secure allied support. But the state-owned mining companies, Gécamines (copper and cobalt) and Miba (diamonds), had become dysfunctional through corruption, mismanagement, theft of equipment and spare parts, and lack of investment during the Mobutu years. Copper production was at one-fifteenth its peak, cobalt output at one-sixth, and diamond production had fallen by almost half. The only way to obtain revenues quickly was to grant concessions and enter into joint ventures with foreign firms in return for up-front payments.[34]

Kabila also appealed to Zimbabwe, Angola, and Namibia for military assistance. Although the governments of these nations had strong political and strategic motivations for dispatching troops, they also demanded compensation, and Kinshasa used resource wealth as an incentive for its allies to stay involved.[35]

Angola's and Namibia's commercial pursuits are modest; those of Zimbabwe are far more extensive. The leadership of the Zimbabwean military formed Osleg (Operation Sovereign Legitimacy), a company that was supposed to pay for the military presence in Congo. Among other ventures, Osleg become involved in running the Sengamines diamond concession (including alluvial deposits near Mbuji Mayi and kimberlite deposits in Tshibua). Ridgepoint Overseas Developments, a Zimbabwean firm whose officials include Zimbabwe's justice minister and a nephew of President Mugabe, was awarded management of three Gécamines copper- and cobalt mines.[36]

As a result of intense diplomatic efforts and growing international pressure, the foreign troops were largely withdrawn by late-2002. However, as a UN expert group noted, even though the Congo war has diminished in intensity, Rwanda, Uganda, and Zimbabwe (in collaboration with Congolese government

---

(*continued*)
Wealth of the Democratic Republic of the Congo," 13 November 2001; Harden, op. cit. note 21, pp. 37–38; Musifiky Mwanalasi, "The View from Below," in Berdal and Malone, op. cit. note 3, p. 142; International Peace Information Service, *Supporting the War Economy in the DRC: European Companies and the Coltan Trade*, Antwerp, Belgium, January 2002.

33 United Nations Security Council, op. cit. note 31, pp. 3, 7, 14–19, 29–31; United Nations Security Council, op. cit. note 32, p. 20.

34 Emasculation of state mining companies from Wrong, op. cit. note 10, pp. 113–118, 122–125; United Nations Security Council, op. cit. note 32, pp. 5, 8, 9.

35 United Nations Security Council, op. cit. note 31, pp. 29–36.

36 United Nations, Security Council, op. cit. note 32, pp. 10–11, 16–19, 22; Michael Nest, "Ambitions, Profits and Loss: Zimbabwean Economic Involvement in the Democratic Republic of the Congo," *African Affairs*, No. 400 (2001), pp. 469–490. Ridgepoint from Reno, op. cit. note 10, pp. 57–58.

officials) left behind elaborate networks run by unscrupulous military, political, and business leaders that continue to pillage the Congo's mineral resources.[37]

Angola

Angola's involvement in the Congo war was but the most recent episode in its own history of seemingly interminable conflict. Angola had been almost continuously at war since its struggle for independence from Portugal (1961–1975). At first, it was superpower support (and Cuban and South African intervention) that sustained fighting between the MPLA government and UNITA rebels. But when the outside powers phased out their assistance in the late-1980s, both sides turned to the country's ample natural resources.[38]

Angola's oil and diamond wealth fueled arms purchases and enriched a small elite on both sides. Angola is the world's fifth-largest producer of non-industrial diamonds and the second-largest oil producer in sub-Saharan Africa. While the offshore oil wells have remained in government hands, control of the diamond mines has shifted back and forth. Early in the 1990s, UNITA controlled about 90 percent of Angola's diamond exports, but after a string of defeats, its share declined to about two-thirds in 1996 and 1997. After 1998, its revenues further declined due to additional territorial losses, depletion of some deposits, and the (limited) impact of UN sanctions. As a result, it is believed that UNITA's diamond income declined to between $80 million and $150 million per year, down from as much as $600 million annually a decade ago.[39]

UNITA had some of its own soldiers involved in diamond mining, but much of the work has been carried out by an estimated 100,000 bonded laborers, and semi-enslaved diggers deprived of even basic rights and working under

---

[37] United Nations Security Council, "Final Report of the Panel of Experts on the Illegal Exploitation of Natural Resources and Other Forms of Wealth in the Democratic Republic of the Congo," S/2002/1146, New York, 16 October 2002.

[38] Global Witness, *A Rough Trade: The Role of Companies and Governments in the Angolan Conflict* (London: 1998). MPLA is the acronym for the Popular Movement for the Liberation of Angola; UNITA stands for National Union for the Total Independence of Angola.

[39] Virginia Gamba and Richard Cornwell, "Arms, Elites, and Resources in the Angolan Civil War," in Berdal and Malone, op. cit. note 3, pp. 165–167; diamond production from Smillie, op. cit. note 2; oil production from BP Amoco, *2001 BP Amoco Statistical Review of World Energy* (London: Group Media & Publications, 2001). Shifting diamond control from Global Witness, op. cit. note 38, from Harden, op. cit. note 21, and from Michael Ross, "Natural Resources and Civil Conflict: Evidence from Case Studies," University of Michigan, Department of Political Science, 11 May 2001; use of diamond income from United Nations Security Council, "Final Report of the UN Panel of Experts on Violations of Security Council Sanctions Against Unita" (New York: 10 March 2000).

dangerous conditions. The rebel group also received "commissions" from diamond buyers operating in its realm.[40]

Until 1999, when De Beers decided to stop buying Angolan diamonds, UNITA had little difficulty selling its gemstones. For several years, De Beers pursued a no-questions-asked diamond-purchasing policy, being more interested in maintaining its market control than in the suffering that "blood diamonds" perpetuated. Diverse smuggling routes enabled UNITA to largely circumvent a 1998 UN embargo on its diamonds. UNITA was able to smuggle diamonds through the Central African Republic, Côte d'Ivoire, Morocco, Namibia, South Africa, and Zambia, with or without the knowledge of the governments of these countries. Burkina Faso, Mobutu's Zaïre, and Rwanda have served as safe havens for illicit transactions. UNITA was similarly able to evade a UN arms embargo by relying on a variety of arms brokers and delivery routes, and by securing the complicity of several governments that provided false end-user certificates for weapons. Weapons came primarily from Bulgaria and other Eastern European countries.[41]

Ordinary Angolans suffered enormously during the long war. Almost 30 percent of all children die before they reach the age of six, and nearly one half of all Angolan children are underweight. Two-thirds of the country's population survives on less than a dollar a day, and 42 percent of adults are illiterate. Unsafe drinking water and a pervasive lack of health services have combined with food shortages to limit life expectancy to just 47 years.[42]

The country now has perhaps the best chance to emerge from its long ordeal. Following Jonas Savimbi's death and a considerable weakening of his forces, the two warring sides signed a cease-fire in April 2002, committing the rebels to undergo disarmament, demobilization, and reintegration.[43]

HOW RESOURCE EXTRACTION CAN TRIGGER CONFLICT

In many instances, resource extraction is itself the source of conflict. Around the world, the operations of oil, mining, and logging companies are causing severe tensions with local populations – often indigenous communities. In Ecuador and Peru, in Nigeria and Cameroon, and in Indonesia and Papua New Guinea, broadly similar scenarios of environmental destruction, economic inequity, and social alienation are unfolding.

---

[40] United Nations Security Council, op. cit. note 39; Global Witness, op. cit. note 38. Number of 100,000 diggers from Andrew Cockburn, "Diamonds: The Real Story," *National Geographic*, March 2002, p. 23.

[41] United Nations Security Council, op. cit. note 39; United Nations Security Council, op. cit. note 2; Global Witness, op. cit. note 38.

[42] Global Witness, *A Crude Awakening* (London: 1999), p. 4; UNDP, *Human Development Report 2002* (New York: Oxford University Press, 2002).

[43] Cease-fire from Rachel L. Swarns, "Angolans Cheer the Peace and Hope It Will Stay Awhile," *New York Times*, 5 April 2002.

Typically, these operations confiscate land from local people without proper compensation. They cause an array of environmental problems by poisoning drinking water, destroying arable land, clear-cutting forests, and disrupting hunting and fishing grounds. Moreover, they introduce social disruptions and communal tensions: roads etched into previously inaccessible areas bring about heavy influx of construction workers, miners, and loggers and, sometimes, migrant populations. While the burdens and disruptions are all too real, the economic benefits from resource extraction mostly accrue to outsiders: the central government, multinational corporations, and assorted foreign investors. When the affected communities resist, they are often met with severe government repression.

WEST PAPUA (INDONESIA)

Indonesia features some of the most intense resource-triggered struggles, which, to a large extent, are the product of policies pursued during the long years of the Suharto dictatorship (1966–1998). Under Suharto, licenses were awarded to domestic and foreign businesses that were closely linked to, or broadly supportive of, the regime. In part, because this practice was heavily tinged by corruption and favoritism, it brought about precisely the kind of imbalance of benefits and burdens described above, and, with it, the seeds of conflict. The Suharto-era policy of "transmigration" – encouraging the movement of people from the most densely-populated parts of the country to outlying provinces – has added fuel to an already combustible situation. Since 1998, the rapid growth in illegal resource extraction has complicated the picture and added new strains of conflict. The Indonesian military and police are involved in both legal ventures, and illegal logging and mining conducted through front companies and joint ventures with private timber barons. These activities, along with protection rackets under which illegal operators pay for protection against prosecution, raise half or more of their operational budgets.[44]

After West Papua was forcibly incorporated into Indonesia in 1961, a rebel movement known as OPM (Organisasi Papua Merdeka, the Papuan Freedom Organization) arose in the mid-1960s and advocated the establishment of a separate state. But OPM did not gain much support from the local population until the 1970s, when it harnessed grievances against a large-scale mining operation.[45]

U.S.-based Freeport-McMoRan Copper & Gold Inc. is operating Grasberg, the world's largest open-pit gold mine. Profits from the operation have been the single biggest source of tax revenue for Indonesia. Land owned by the indigenous peoples, the Amungme and Kamoro, including a mountain sacred to

[44] International Crisis Group, "Indonesia: Natural Resources and Law Enforcement," ICG Asia Report No. 29, Jakarta and Brussels, 20 December 2001, pp. 10, 18.

[45] Ross, op. cit. note 39, p. 27; "No Flags for Papua," *The Economist*, 12 October 2000; Human Rights Watch, *Violence and Political Impasse in Papua* (New York: July 2001), p. 2.

them, was taken over without their consent by a 1967 agreement between Freeport and the Suharto regime. Not only have many villages been displaced, but mine wastes have also been dumped on downstream tribal lands. In 1998, for example, some 200,000 tons of ore were dumped into the Ajkwa river system. These mine tailings have turned 230 square kilometers of the river delta into a lifeless wasteland.[46]

From the beginning, the local tribes opposed Freeport's presence, but this opposition was not linked to OPM's armed separatism until 1977. Indonesian security forces retaliated by bombing and burning villages. Freeport has maintained close ties with the armed forces, providing transportation, accommodation, and funding to the troops in return for military protection. Military reprisals not withstanding, land rights conflicts, compensation demands, human rights violations, and environmental damage keep triggering violent and nonviolent protests.[47]

The Papuan striving for independence has grown far beyond OPM to a broad, civilian-based movement. But Jakarta dispatched thousands of additional troops after the Papuan Congress declared independence in June 2000. Civilians were attacked, peaceful protests banned, key Papuan leaders arrested, and access by journalists and human rights observers severely restricted. Papuan militants, in turn, have attacked military forces and non-Papuan migrants. Although the violence is currently at a relatively low level of intensity, the death toll since 1961 may be as high as 100,000. Under a special autonomy package adopted by the Indonesian parliament in November 2001, West Papua is to receive a larger share of the revenues derived from its natural resources. However, its implementation is difficult, promises of autonomy have failed to satisfy Papuans, and Jakarta considers the province too valuable to grant it full independence.[48]

Bougainville (Papua New Guinea)

The world's largest open-pit copper mine, owned jointly by mining giant RTZ (80 percent) and the central government (20 percent), began operating at

---

46 "Risky Business: The Grasberg Gold Mine," Project Underground Reports, <www.moles.org/index.htm>, viewed 9 July 2001; "The Strains on Indonesia," *The Economist*, 3 December 2000; "Provocation," *The Economist*, 30 November 2000; Abrash, op. cit. note 15, pp. 38–39; Michael Shari, "Freeport-McMoRan – A Pit of Trouble," *Business Week*, 31 July 2000.

47 Ross, op. cit. note 39, p. 28; *Rape and Other Human Rights Abuses by the Indonesian Military in Irian Jaya (West Papua), Indonesia* (Washington, DC: Robert F. Kennedy Memorial Center for Human Rights, May 1999); Abrash, op. cit. note 15, p. 40.

48 Human Rights Watch, op. cit. note 45, pp. 2, 3, 10, 11; International Crisis Group, "Indonesia: Resources and Conflict in Papua," ICG Asia Report No. 39, Jakarta and Brussels, 13 September 2002; Jim Lobe, "Indonesia's Hard Line Strengthens Secessionists in West Papua," *Foreign Policy in Focus*, 1 July 2001. Autonomy package from Jim Lobe, "Indonesia: Aceh Arrests Could Portend Increased Polarization, Violence," *Foreign Policy in Focus*, 1 July 2001.

Panguna in 1972. But the severe social and environmental impacts of the mine reinforced demands for secession (Bougainville was governed by Australia from 1920 until 1975, when it was subsumed into newly independent Papua New Guinea, despite local protests).[49]

Papua New Guinea's (PNG's) constitution declared that mineral rights belonged to the state, violating Bougainville traditions of land ownership and reinforcing the alienation of rule by a different ethnic group. Copper revenues of US$500 million per year went to the central government and foreign investors, but the local population saw relatively few benefits. The presence of an affluent expatriate mining community and the influx of a large number of workers from other parts of PNG (locals were paid considerably less than other workers) intensified Bougainvilleans' resentment of the mine. The mine also led to major social disruptions, including an unraveling of the island society's matriarchal structure. Mine tailings and chemical pollutants damaged approximately one-fifth the total land area, forcing village relocations, decimating food and cash crops like cocoa and bananas, contaminating rivers, and depleting fish stocks.[50]

Bougainvilleans' complaints and demands for adequate compensation were ignored. In 1988, they launched a campaign of sabotage that, spurred by human rights violations of government forces, quickly developed into guerrilla war. The mine fell to rebels and was closed down in May 1989. The government reacted to this major loss of revenue by launching a series of ultimately futile military campaigns and a blockade of the island that led to the death of thousands of civilians. Government forces committed atrocities and burned thousands of homes, but failed to recapture the mine.[51]

Following defeat in one of the army's many attempts to recapture the island, in 1997 a desperate Julius Chan – PNG's prime minister at the time – offered US$36 million in World Bank funds to the British mercenary firm Sandline International in a last-ditch effort to dislodge the Bougainville rebels. However, senior army officers – incensed that their own budget was being cut – forced Chan to cancel the Sandline contract and resign.[52]

Pressure mounted on both sides to end the war, with growing realization of the horrendous costs of an unwinnable conflict. In 1998, a cease-fire was signed and a small international peacekeeping force deployed. The conflict was

---

[49] Michael T. Klare, *Resource Wars: The New Landscape of Global Conflict* (New York: Metropolitan Books, 2001), p. 196; "Chronology," *Accord*, No. 12/2002, special issue on "Weaving Consensus: The Papua New Guinea-Bougainville Peace Process," <www.c-r.org/accord/accord12/chron.htm>.

[50] Mary-Louise O'Callaghan, "The Origins of the Conflict," *Accord*, op. cit. note 49. Annual revenues from Klare, op. cit. note 49. Environmental damage from Volker Böge, "Bougainville: A 'Classical' Environmental Conflict?" Occasional Paper No. 3, Environment and Conflicts Project (ENCOP), Bern, Switzerland, October 1992.

[51] O'Callaghan, op. cit. note 50.

[52] "Chronology," *Accord*, op. cit. note 49; $36 million figure from Klare, op. cit. note 49, p. 198.

formally ended in August 2001, following intense negotiations facilitated by New Zealand and Australia. The PNG parliament approved autonomy status for Bougainville in March 2002, with the prospect that a referendum on independence could be held in 10–15 years.[53]

RESPONSES

Confronted with severe conflicts in Sierra Leone, Angola, and Congo, the UN Security Council has increasingly examined the role of resources in perpetuating these wars. It imposed a number of embargoes on the illicit diamond trade and the purchases of arms, equipment, and fuel paid for with diamond money (See Table 6.1). The diamond embargo on Liberia, for example, has had some success, contributing to a dramatic decline in Liberian-based diamond smuggling. However, it has become painfully obvious that existing sanctions are being violated by unscrupulous producers, traders, bankers, and governments. There is an urgent need to step up international efforts to monitor compliance with sanctions, and to improve the capacity to enforce embargoes and investigate violations so that traffickers can no longer operate with impunity.[54]

Growing energy is being directed toward efforts to make it more difficult for resources gained through conflict to be sold on world markets. By far, the most attention has gone to the diamond industry. A growing number of governments are backing schemes under which only diamonds with proper certificates of origin are considered legal, although such a system can be undermined by poor enforcement and circumvented by intricate international smuggling networks. Hence, support has been growing for establishing a standardized global certification scheme, and, from 2000 to 2002, representatives from about 40 nations, the diamond industry, and a number of NGOs conducted negotiations toward developing an international system.[55]

Although the Kimberley Process scheme (named after the town in South Africa where the first meeting was held) is an important step forward, it has a number of critical shortcomings. It relies primarily on voluntary participation and adherence by governments and industry, and lacks the international

[53] "Chronology," *Accord*, op. cit. note 49; Robert Tapi, "From Burnham to Buin: Sowing the Seeds of Peace in the Land of the Snow-Capped Mountains," *Accord*, op. cit. note 49; "Papua New Guinea: Security Council Members Back Peace Plan," *UN Wire*, 15 August 2001; "Autonomy Approved for Bougainville," *UN Wire*, 28 March 2002.

[54] Human Rights Watch, "Neglected Arms Embargo on Sierra Leone Rebels," Briefing Paper, 15 May 2000, as posted on Global Policy Forum, www.globalpolicy.org/security/issues/ sierra/00-05sl4.htm, and from United Nations, op. cit. note 2.

[55] Alan Cowell, "New 'Labels' for Diamonds Sold by Sierra Leone," *New York Times*, 28 October 2000; United Nations, Security Council, op. cit. note 24, pp. 26, 28–29. "Diamonds: Kimberley Process Reaches Breakthrough on Certification," *UN Wire*, 22 March 2002; Alan Cowell, "40 Nations in Accord on 'Conflict Diamonds'," *New York Times*, 6 November 2002.

Table 6.1 Resource Conflicts and United Nations Sanctions.

Date (Number of resolution)	Security Council Action
November 1992 (788)	Arms embargo against Liberia.
September 1993 (864)	Embargo on deliveries of arms, military equipment, and fuel to Angola's UNITA rebels after their rejection of the 1992 election results.
August 1997 (1127)	Additional sanctions against UNITA (freezing of bank accounts; prohibiting foreign travel by senior UNITA personnel; closing of UNITA offices abroad).
October 1997 (1132)	Embargo on arms and oil supplies to Sierra Leone; travel ban on members of military junta (oil embargo terminated in March 1998).
June 1998 (1171)	Arms embargo and travel ban on anti-government forces in Sierra Leone.
June 1998 (1173)	Embargo on direct and indirect import of Angolan diamonds not approved under an Angolan government Certificate of Origin regime.
May 1999 (1237)	Established a panel to investigate violations of sanctions against UNITA.
July 2000 (1306)	Embargo on direct and indirect import of all rough diamonds from Sierra Leone. Following the establishment of a new monitoring regime, the embargo was narrowed to non-official exports in October 2000.
March 2001 (1343)	Demands that Liberia cease financial and military support for RUF, and cease imports of Sierra Leonean rough diamonds that do not have an official certificate of origin; embargo on arms deliveries to Liberia and travel ban against its political and military leaders (extended for another year in May 2002); embargo against Liberian diamond exports threatened unless Liberia can show that it is not supporting RUF.

Source: Compiled from United Nations, Security Council Documents Full Search, <www.un.org/Docs/sc>, from United Nations, op. cit. note 12, and from United Nations, op. cit. note 50, pp. 41–45.

authority to monitor and enforce rules. Participation in the "chain of warranties" that follows the initial export (as diamonds are sold and resold, polished, and incorporated into jewelry) is to be voluntary, and monitoring and enforcement are left to self-regulation. Other weaknesses concern data collection, standardization, and analysis.[56]

It is clear that a number of businesses – mining companies, trading firms, airlines and shipping companies, manufacturers, and banks – carry a degree of responsibility for the events that have triggered campaigns against blood diamonds and other conflict resources. This responsibility ranges from an active

[56] U.S. Government Accounting Office, op. cit. note 2, pp. 17–21.

role (in which companies are directly and knowingly involved in illicit resource exploitation), to a silent complicity (in which firms do business with repressive regimes because of lucrative contracts), to a passive enabling role (in which few questions are asked by companies down the supply chain about the origin of raw materials or about money being laundered).

International embargoes and UN reports have begun to create greater transparency. NGO campaigns have tugged at the cloak of complicity through investigative reports and by "naming and shaming" specific corporations, in an effort to compel them to do business more ethically or to terminate their operations in certain locations. Such campaigns have been most potent in the case of companies that sell highly visible consumer products, or whose corporate logos and slogans are familiar to millions.[57]

At the end of the 1990s, the diamond industry was hit by a wave of bad publicity and faced the threat of consumer boycotts. De Beers, the industry's monopolist, was sufficiently embarrassed by the London-based NGO Global Witness – which revealed that the company had knowingly purchased diamonds from Angola's UNITA rebels – that it decided to adopt a more responsible policy, urging the rest of the industry to follow suit. Similarly, when the role of coltan in the Congo war become more widely known, consumer electronics companies scrambled to avoid the kind of negative publicity that the diamond industry had endured. Companies like Ericsson, Nokia, Motorola, Compaq, and Intel suddenly scrutinized their supply chains and put pressure on mineral processing firms to stop purchasing illegally mined coltan. Kemet and Cabot, U.S. companies processing coltan, cancelled orders for ores originating from the Democratic Republic of the Congo, and Belgium's Sogem (a subsidiary of Umicore) terminated its partnership with a Congolese coltan supplier. The Belgian airline Sabena stopped its coltan shipments from Rwanda to Europe.[58]

Where development aid and private investments continue to flow into the extractive sector, they should, in principle, go only to governments that are accountable to their own citizens. A new initiative (simply called "Publish What You Pay") by philanthropist George Soros and a large coalition of NGOs from developed and developing countries proposes that natural resource companies be required, as a condition of being listed on leading stock exchanges and financial markets, to disclose all taxes, fees, royalties, and other payments they make to host governments. Such a step would shed some light on often opaque financial transfers, and increase accountability of how such payments

[57] In the mid-1990s, for example, human rights and environmental organizations launched campaigns aimed at Shell (for its role in Nigeria) and at Amoco, Texaco, ARCO, and Petro-Canada (for their roles in Myanmar); Marina Ottaway, "Reluctant Missionaries," *Foreign Policy*, July/August 2001, pp. 47–48.

[58] Harden, op. cit. note 21; Nicole Gaouette, "Israel's Diamond Dealers Tremble," *Christian Science Monitor*, 1 June 2001. Electronics companies' reaction from Harden, op. cit. note 21, p. 38, and from Kristi Essick, "Guns, Money and Cell Phones," *The Standard: Intelligence for the Internet Economy*, 11 June 2001. Kemet and Cabot from United Nations, Security Council, op. cit. note 32, p. 6.

are used. The initiative states that "mining, gas, and oil companies cannot control how governments spend taxes, royalties, and fees. But they do have a responsibility to disclose the payments they make so citizens can hold their governments accountable. Companies that fail to do so are complicit in the disempowerment of the people of the countries to which the resources belong."[59]

There is growing awareness that natural resources will continue to fuel deadly conflicts as long as consumer societies import and use materials irrespective of where they originate and under what conditions they were produced. Support is growing for the idea that companies need to adopt more ethical ways of doing business. Shareholder activism and campaigns for ethical investing can help achieve these goals but it is clear that activities to date are only a beginning. Governments and international organizations will need to work hard to create greater corporate transparency. So far, western nations have been all too ready to turn a blind eye in order to protect the interests of their own corporations.

Action will also be necessary on a number of other fronts. Countering the massive proliferation of small arms, something that can be achieved only with broad international cooperation and sustained effort, is essential. So is a substantial improvement in the world's peacekeeping and peace-making capabilities – overcoming the severe handicaps that such efforts now face. Promoting democratization, justice, and greater respect for human rights are also key tasks, along with efforts to reduce the impunity with which some governments and rebel groups engage in extreme violence. Another challenge is to facilitate the diversification of economies away from a strong dependence on primary commodities to a broader mix of activities. A more diversified economy would provide better economic balance, reduce vulnerability to the "resource curse," and lessen the likelihood that natural resources become pawns in a struggle among ruthless contenders for wealth and power.

Investing in human development, improving health and education services, and providing adequate jobs and opportunities for social and economic advancement will go a long way toward reducing the risk that a country's natural resource endowment will become its undoing. This is an investment that needs to be made not only by the governments concerned but also by the World Bank and other multilateral development agencies that have generously funded oil, mining, and logging projects. It must also be a priority for the rich nations that have long benefited from cheap raw material supplies, while turning a blind eye to the destruction at their source.

[59] The coalition includes, among others, Amnesty International, Christian Aid, Friends of the Earth, Global Witness, Oxfam, Save the Children, and Transparency International. George Soros, "Transparent Corruption," Financial Times, 13 June 2002; Transparency International, "Press Release: George Soros and NGOs Call for Rules to Require Corporations to Disclose Payments," <http://www.transparency.org/pressreleases_archive/2002/2002.06.13.publish_pay.html>, viewed 25 June 2002.

Part II

Artisanal and Small-Scale Mining, Labour and the Community

7

Introduction Part II

Gavin M. Hilson

Many social scientists undertaking research in rural regions of the developing world would argue that, the livelihoods and needs of grassroots communities must not be overlooked when drafting policies and legislation at the national level. In the case of artisanal and small-scale mining, it is evident that many governments regularly ignore such socio-cultural criteria as literacy, training, education, and cultural traditions, in key regulatory procedures. In response, several artisanal and small-scale miners have begun voicing complaints over the demands of newly-introduced legislation, shortages of technical and financial support, and inappropriate policies. Many more have elected to ignore regulations outright by operating illegally in locations well out of the reach of authoritative bodies.

Confronted with the challenge of having to control rampant illegal artisanal mining activities, accompanying illicit mineral marketing, and disease and crime brought on by a rapid intensification of operations, many governments are searching for ways to facilitate improved regularization in the sector. One key is to first improve the level of understanding of artisanal and small-scale mining at the national level, which requires, *inter alia*, in-depth analysis of the conditions of artisanal and small-scale communities, the lifestyles of its inhabitants, and industry needs. These findings can then be used to further refine draft legislation, as well as existing laws, codes and policies in place for artisanal and small-scale mining.

Through many recent projects, international agencies such as the International Labour Organization (ILO) and the United Nations Development Program (UNDP) have illustrated the importance of undertaking community-based research *before* implementing policies and regulations for resident artisanal and small-scale mining operations; there is now considerable evidence purporting such approaches. Many point to how the ineffectiveness of the majority of activities undertaken to regularize, and deliver support to, artisanal and small-scale miners have been a direct result of both a lack of foresight on the part of authorities, and the failure to collect pertinent cultural and socio-economic data. Problems have been exacerbated by the cavalier approaches taken to address priority issues in the industry. In the majority of cases, emphasis has been almost exclusively placed on finding technical solutions to mining and processing problems (Jennings, 1999), when, in fact, "the implementation of technical changes, modifications and improvements require in almost any case detailed knowledge of the cultural, social, economic and organizational context of the miners" (Hentschel et al., 2001, p. 40). Governments must therefore draw more systematically on local knowledge when framing potential solutions to problems in the artisanal and small-scale mining sector. The "grassroots" approaches required for research of this scope have long been advocated by influential NGOS and development departments/agencies such as GTZ and DfID.

The chapters presented in this section of the book seek to address a number of the important issues that governments must take into account when implementing policies and regulations for artisanal and small-scale mining. Specifically, the chapters provide an overview of conditions on the ground, describing the labour and gender characteristics of artisanal and small-scale mining, the conditions of mining communities at the grassroots, and the effectiveness of many of the community-based industry initiatives in place.

In the next chapter, Gavin Hilson attempts to put into perspective the importance of taking into account community-side concerns when strategizing to improve working conditions in artisanal and small-scale mining regions. He explains how most of the top-down management approaches adopted by governments to date to address pressing problems in the sector have failed to take into account many of the important socio-cultural elements of the mining communities themselves. The author details the schematics and weaknesses of conventional small-scale mining developmental approaches using a series of case studies. It is against the background of this analysis that measures and strategies are then proposed for improving small-scale mining developmental strategy.

In Chapter 9, Beatrice Labonne of UNDP provides details of a seminar on artisanal and small-scale mining held in Yaoundé, Cameroon, November 19–22 2002. The seminar, which was entitled *Artisanal and Small-Scale Mining in Africa: Identifying best Practices and Building the Sustainable Livelihoods of the Communities*, aimed at providing a forum for debate and partnerships. The seminar successfully brought together some 70 participants, including many highly-qualified senior policy-makers from several African countries, and experts from UNAIDS, the African Development Bank, the World Bank and DifD. Although the seminar focused exclusively on the current state-of-affairs in the African artisanal and small-scale mining industry, the reported findings are directly relevant to all developing countries with flourishing artisanal and small-scale mining sectors. The author details many of the issues discussed by the delegates at the conference, including their views on pressing industry problems, the efficacy of industry support services, and the state of small-scale mining in selected countries.

In Chapter 10, Norman Jennings of the International Labour Organization (ILO) examines the important labour-related challenges facing the small-scale mining industry today; special emphasis is placed on female and child labour. After providing a statistical breakdown of global employment in the sector, the author briefly discusses many of the discriminatory-related issues confronting the industry's female contingent. A link is also drawn between poverty and excessive child labour in small-scale mining, after which measures are proposed to improve working conditions at sites.

Chapters 11 and 12 focus on the roles of women in artisanal and small-scale mining. In Chapter 11, Jennifer Hinton, Marcello Veiga and Christian Beinhoff profile the many roles played by women in artisanal mining communities. The authors further argue that additional research emphasizing the enhancement of women in artisanal mining is a key to bridging the gap between many of the well-articulated industry-specific technical and socio-economic changes prescribed by policy-makers. In Chapter 12, Suzanne Tallichet, Meredith Redlin and Rosalind Harris, focus more on the gender inequality issues omnipresent in small-scale mining. Specifically, the authors provide an overview of the many socio-economic and cultural barriers inhibiting women's participation in the sector, making frequent reference to recent feminist frameworks of women in development. The chapter concludes by proposing possible measures that, if adopted by governments and international agencies, could help to alleviate gender inequality in the industry.

The final chapter of this section, by Malebabo Sakoane, provides a brief review of many of the diseases affecting artisanal and small-scale mining communities. Apart from distinguishing the ailments associated with mining activity and mineral processing, the author makes a concerted effort to outline how diseases in small-scale mining regions affect women, children and miners themselves. Many of the diseases affecting the large-scale miners and farmers operating in developing countries are also described.

This section of the book is intended to acquaint government officers and policy-makers with conditions on the ground.

REFERENCES

Hentschel, T., Hruschka, F. & Priester, M. (2001). *Global report on artisanal and small-scale mining (ASM)*. Draft Report prepared for Mining, Minerals and Sustainable Development Project, London.

Jennings, N.S. (Ed.) (1999). *Small-scale gold mining: Examples from Bolivia, Philippines and Zimbabwe*. International Labour Office, Sectoral Activities Programme, Industrial Activities Branch, ILO.

8

Accounting for Community-Side Issues in the Artisanal and Small-Scale Mining Developmental Agenda

GAVIN M. HILSON

In the developing world, social scientists are increasingly neglecting the methodological implications of social and cross-cultural research. Without a proper understanding of the socio-cultural context of indigenous communities and associated study sites, frames of reference can be entirely at a variance. Such has been the case with small-scale mining research, which, to date, has mainly emphasized the technical, geological and chemical aspects of the industry. Minimal effort has been made to understand the needs of the artisanal and small-scale mining communities themselves.

The purpose of this chapter is to describe the importance of taking into account community-side concerns when strategizing to improve working conditions in artisanal and small-scale mining regions. Many of the top-down approaches adopted by governments to date to address pressing problems in the sector have failed to take into account the important socio-cultural elements of small-scale mining communities. Consequently, the industry is now "bedevilled with too many regulations that are mostly designed to constrain it", and its support-related projects, most of which have been left to stand on their own, have often "gently wound down due to a lack of continued government support or supervision" (Jennings, 1999).

The discussion begins by examining the link between poverty and small-scale mining. The section that follows describes the conventional small-scale mining developmental approaches that have been adopted by governments. The chapter then presents a series of case studies in an attempt to illustrate the weaknesses of conventional small-scale mining developmental strategies. The section that follows describes a series of initiatives for improving small-scale mining developmental strategy. Concluding remarks are then provided.

POVERTY AND, ARTISANAL AND SMALL-SCALE MINING

In the majority of cases, artisanal and small-scale mining are poverty-driven activities that employ redundant large-scale mine labourers, seasonal farmers, and nomadic rural dwellers. Most of these individuals are low-skilled and poorly educated, and, as a result, earn subsistence wages. Moreover, the majority are subjected to working conditions that are by no means ideal, as most mine sites are surrounded by unsanitary waters and contaminated soils that contain a plethora of disease-carrying pathogens. Poverty of this nature is rampant throughout the rural regions of the developing world, but continues to be overshadowed by burgeoning discussions on urban population growth, expanding city slums and shantytowns.

The relationship between small-scale mining and poverty has become more apparent in recent years (World View, 2001; Hilson, 2002a). In Burkina Faso, for example, which ranks 171st among 174 countries on the World Bank Development Index, approximately 60,000 individuals are engaged in the mining of gold on a small scale. Here, an estimated 45% of the country's population is classified as "poor", which is why an increasing number of individuals are seeking to exploit the mineral riches of resident gold veins. Similarly, in Peru, where at least 20,000 people are involved in small-scale mining, some 37% of the population is classified as "poor", and another 16% as "extremely poor". Illegal mining is now widespread throughout the country, largely because of the shortage of economic opportunities.

In the majority of countries where it is practised, artisanal and small-scale mining has precipitated mass urban-rural migration. At present, approximately 50% of the global population (2.8 billion) live on a daily wage of less than US$2 (Table 8.1). As a result, many of the families that had originally migrated to urban centres in search of employment have since relocated to remote rural regions where artisanal and small-scale mining is practised (in search of job opportunities).

The dynamics of small-scale mining culture and associated poverty, however, are poorly-understood issues. This is evidenced in governments' continuous attempts to both deliver support services and introduce formalities to the sector; most have proven ineffective because key social and cultural factors

Table 8.1 Distribution of the global poor.

	Poor on US$1 per day (millions)	Poor on US$2 per day (millions)
East Asia and Pacific	54	252
China	213	633
Latin America and Caribbean	61	159
Middle East and North Africa	6	85
South Asia	522	1095
Sub-Saharan Africa	302	489
Total	1157	2714

Source: World Bank, 2001.

have not been taken into account. As Downs (2000, p. 601) explains, "complex relationships exist between human nature and needs, cultural evolution and ecological dynamics", and "it is at the local/community level that the connections among ecosystem health, the health of human communities, and individual livelihoods are most apparent and crucial" (King & Hood, 1999, p. 49). Yet, in spite of growing consensus that "cultural ecology stresses the importance of local knowledge" (Batterbury et al., 1997, p. 127), and that, "the rapidly rising importance of social and cultural issues and the need to deal with them effectively, both domestically and internationally, represents a clear and new reality for the mining industry" (Clark & Clark, 1999, p. 190–1), most artisanal and small-scale mining policies and support services continue to neglect such socio-cultural issues as miners' relationships with land, rural livelihoods, and community needs.

The International Labour Organization (ILO) estimates that, between 11 and 13 million people are employed directly by the industry, and that overall, between 80 and 100 million depend on its existence. Over the past 10–15 years, substantial effort has been made to differentiate between the industry's illegal (artisanal) and legal (licensed) operators, but surprisingly little has been done to categorize miners according to key social and cultural criteria. More specifically, without a proper understanding of *why* people mine on a small scale, employment data could easily be misconstrued as meaning that the industry provides economic opportunity, when, in fact, it mainly provides *economic relief*.

The preceding discussion has dealt almost exclusively with a growing segment of the industry – that comprised of individuals driven to artisanal and small-scale mining regions in search of employment. The other, and equally significant segment of the industry, consists of individuals engaged in mining for socio-cultural reasons. Many of Ghana's *galamsey* gold miners, for example, continue to mine concessions awarded to large-scale mining companies, contending that they have cultural ties to the land their ancestors had mined for centuries (Hilson, 2002b). In another case, the Ndjuka people of Suriname have long engaged in alluvial gold mining, repeatedly expressing "vested interests in conserving the forest that is the home and source of subsistence for their families" (Heemskerk, 2002, p. 331).

Before undertaking initiatives to provide support services and assistance to artisanal and small-scale miners, governments must first study the sociology of small-scale mining. The task requires going beyond the conventional examination of illegality and regularisation, and involves a detailed analysis of the individual needs of miners and their working conditions.

CONVENTIONAL ARTISANAL AND SMALL-SCALE MINING DEVELOPMENTAL STRATEGY

As Bulkeley (2000, p. 290) explains, "the 'local' has more recently gained recognition as a key arena for the pursuit of sustainable development", the "rationale being that as the level of governance closest to the people [local authorities] play a vital role in educating, mobilizing and responding to the public". However, typical developmental strategy has failed to "reflect both the demand and supply side of the equation", and has mainly emphasized "institutions rather than the strategies and choices of the rural poor" (Redclift, 1992, p. 255). This problem is perhaps best illustrated by the technology transfer process to developing countries, which often lacks the scientific and technological infrastructures, along with the requisite training facilities, to effectively absorb a system imported from a developed country. Thus, in many cases, equipment goes unused in the recipient developing country, or, over the long term, proves ineffective because of underlying educational, cultural and sociological factors. Effective grassroots developmental strategy, therefore, calls for the examination and integration of *both* institutional and community-side elements.

The foregoing discussion established that conventional small-scale mining developmental strategy has been top-down in nature, featuring a series of undertakings that have only managed to facilitate marginal improvements. Few governments have addressed the important community component – *i.e.* key behavioural-side issues – which requires preliminary analysis of the needs, demands and opinions of recipient artisanal and small-scale mining populations (see Table 8.2 for examples of important areas that could be investigated

Table 8.2 Examples of areas requiring investigation prior to the implementation of artisanal and small-scale mining initiatives.

Area of research	Importance
Gender and age characteristics of the industry	It is a well known fact that certain tasks are confined to different genders and age groups. If particular areas of operation are to be improved, each group must be specifically targeted and approached. For example, if the grinding of stones is a task carried out by middle-aged women, it would be futile to introduce heavy, complex milling machinery as a substitution strategy.
Levels of education and training	The industry has long been characterized by predominantly manual operation methods. Introducing technical machinery, therefore, calls for first determining how adept the populace would be at operating such equipment. This, in turn, enables a government to determine the degree of training required.
Cultural beliefs and values	Often, miners contend that they have cultural ties to areas in which they mine. As many small-scale miners are tribal, these beliefs can extend back centuries. Thus, there is merit to determining *what* miners themselves value before implementing a solution that potentially strains such relationships.
Indigenous knowledge	Tapping into indigenous knowledge could prove invaluable in many ways. The miners themselves are well acquainted with the field, and could provide advice about neighbouring tribes, prospective mining areas and existing mine-related problems.
Production capabilities	The production capacity of small-scale mining regions should be determined if assistance is to be provided. This prevents over-demand and potential breakdown of introduced equipment.

during the course of conducting such preliminary research). As a result, there has been a tendency to implement technical "quick-fix" solutions in the sector. Although most artisanal and small-scale mining problems require some degree of technical input, the belief that they can be solved using exclusively technique-orientated approaches is a common misconception.

More specifically, the "implementation of technical changes, modifications and improvements" requires "in almost any case detailed knowledge of the culture, social, economic and organizational context of the miners" (Hentschel et al., 2001, p. 40). For example, during a personal interview with a Ghana-based small-scale mining consultant who regularly conducts fieldwork in Burkina Faso, it was explained that,

> People continue to propose technology as an all-out solution to the industry's environmental problems but what researchers are now doing in Burkina is carrying out sociological studies to gain an understanding of target mining populations and the ways in which they think... One cannot tackle the industry's problems from one dimension, and must therefore have doctors, sociologists, scientists, and social scientists analyze each before solutions are devised.

To help further illustrate how conventional small-scale mining developmental approaches have failed to yield marked improvements at sites, the next section of the chapter presents a series of short case studies, which collectively put into perspective the weaknesses of conventional small-scale mining developmental strategy. As is shown, overlooking the crucial behavioural (community) side of the industry can lead to a plethora of problems, including mismanaged funds, unanticipated equipment expenditures, staff shortages, and educational challenges.

CASE STUDIES

Four case studies are presented in the discussion that follows. Each reflects a different area of operation commonly overlooked by governments, international agencies and industry organizations in their pre-assessments of target artisanal and small-scale mining populations. The first two case studies emphasize general areas of the industry, and the final two cases focus on more specific issues.

Case study #1: support-related initiatives for small-scale miners

Artisanal and small-scale mining support services are needed if operations are to progress along an efficient course. Most of the attempts made to deliver both financial and technical assistance to small-scale miners, however, have been ad hoc, thereby constraining "efforts to promote better organization and work practices; increase the productivity of small-scale mines; and lessen the adverse labour and social effects" (Jennings, 1999). A shortage of support simply hinders opportunity for improvement, and can perpetuate a number of problems such as illegal mining and environmental degradation.

It has been suggested that NGOs may be the most suitable agencies for providing assistance to small-scale miners, since they do not require timely repayments or direct returns on investment to continue operating. Numerous examples of successful NGO-sponsored assistance have been described in the United Nations' report, *Recent Developments in Small-Scale Mining* (UN, 1996a). In Peru, for example, the Proyecto de Investigacio'n y Aplicacio'n de Technolgi'a Apropriada has aided the Santiago Ananea mining co-operative in improving mining techniques, enabling miners to increase production. Similarly, in Chile, the Cenda Foundation has had considerable success in providing technical assistance that has emphasized both environmental management and community development.

Other economic support-related initiatives have gradually folded because governments have failed to involve NGOs, and have complacently relied on resident banks to provide financial assistance. The involvement of state banks and lending institutions has often been characterized by disappointing results, largely because borrowers fail to repay their loans (UN, 1996b). In China, for example, there is now "very little in the way of support services for artisanal (coal) miners" (Gunson, 2001) because most Chinese lending institutions demand substantial collateral for loans. Similarly, in Peru and Bolivia, poor credit-worthiness in the small-scale mining sector has led to decreased access to formal credit markets, as resident commercial lenders are now reluctant to finance anything outside of well-established large-scale mining projects; failed repayment of loans awarded to small-scale miners was a main reason why the Banco Minero of Bolivia and Peru was forced to close.

Additional problems have occurred in the process of administering technical assistance to miners. Specifically, many governments have exhibited a lack of technical awareness whilst undertaking such support-related initiatives. In Tanzania, for example, although the Ministry of Minerals and Energy has pledged tens of millions of dollars to improve its small-scale mining segment, a study by Mutagwaba et al. (1997) confirms, inter alia, that: (1) mines inspections are still rare and too casual; (2) there continues to be a lack of technical advice being provided to miners; and (3) most working tools are *still* fabricated at sites. In another example, the Ghanaian government, in the early-1990s, attempted to improve the efficiency of resident operations by introducing equipment to miners. Using funds granted by the World Bank, a wide range of extraction and processing equipment was purchased, despite the fact that no preliminary investigation was carried out to determine *which* equipment would be most appropriate. The units purchased proved to be more suitable for large-scale operations, and as a result, the government was forced to auction all of the equipment at discounted prices; the entire exercise thus proved futile.

Case study #2: mercury retorts

Mercury pollution has proven to be one of the most serious environmental and health-related concerns in the artisanal and small-scale mining industry. It is used to assist in the extraction of gold from alluvial deposits and primary ores but is often applied carelessly and discarded freely into local waterways after use. In the natural environment, metallic mercury undergoes a change in speciation, and acute exposure to the resulting methylmercury can cause motor control deficits, muscle ataxia, and visual impairments, with convulsions preceding death. The challenge for governments keen on reducing the adverse impacts of mercury at small-scale gold mines lies in effectively minimizing emissions at the source. This, however, can only be achieved by distilling the mercury amalgam in an enclosed circuit. To date, the most efficient environmental technology that has been devised for such a task is the mercury retort, which is a simple system assembled with a closed crucible connected to a condenser, designed so that mercury from gold amalgams evaporates when heated (Fig. 8.1).

Although standard retorts reduce mercury emissions by 90%, and the most advanced of set-ups have proven to be 99% effective, their use is far from widespread in artisanal and small-scale gold mining regions. In fact, there are a wide range of implementation problems associated with retort usage, few of which governments have effectively addressed. First, and foremost, many authorities have overlooked the importance of providing training to artisans. There is some preparedness required *prior* to using retorting technology, which, in turn, has deterred numerous miners from adopting these and associated apparatuses. For example, to prevent explosions, a wet rag should be applied to the outlet, and not placed directly into the water (Hollaway, 1993). Moreover, the retort must be sufficiently heated in order to ensure efficient operation, which requires it to be clamped – not screwed – to the condenser using asbestos string. As most artisanal and small-scale miners exhibit low levels of technical expertise, the aforementioned preparations commonly work to discourage the adoption of retorts.

The second, and perhaps most significant problem, however, is that of cost. Governments continue to overlook the earning capacities of miners, offering retorts at prices that are clearly beyond their budgetary means. In Zimbabwe, for example, where GNI per capita is US$460, retorts sold through the channels

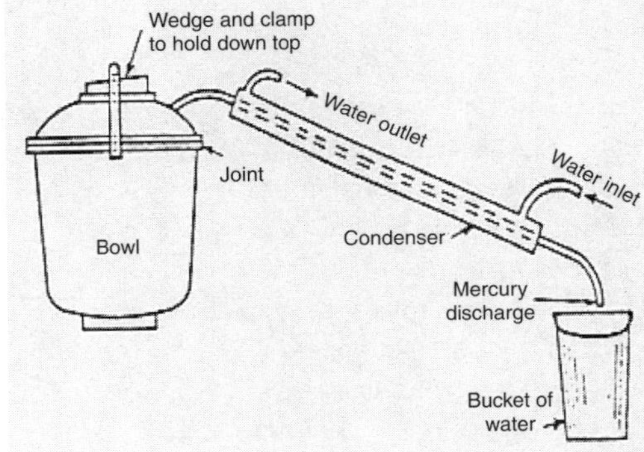

Figure 8.1 Standard mercury distillation retort.
Source: Preister and Hentschel, 1992.

of the Intermediate Technology Development Group (ITDG) are in the range of US$150. In another example, Ghana, where GNI per capita is even less at US$340, the original retorts designed by the German non-profit GTZ were in the range of US$250. Furthermore, retorted gold can, on occasion, feature discoloured surfaces, typically perceived as "impure" by local merchants, and thus rarely fetches market prices.

Finally, in the process of introducing mercury retorts, many governments have failed to take into account the level of understanding in target mining communities. As already explained, many miners work the same lands as their ancestors before them, and utilize the same processing strategies. The introduction of retorts, therefore, potentially "alters" existing mining processes. In particular, miners have expressed a concern over the opaqueness of many of the metal retorting apparatuses, maintaining that gold is potentially "lost" from the enclosed circuit. More specifically, there is emerging belief among miners in such countries as Ghana, Zimbabwe, Bolivia, PNG and the Philippines that gold is lost altogether from the system; these groups contend that there is no evidence to prove otherwise, given that the processing itself cannot be seen. One noteworthy attempt to rectify this problem was made by UNIDO. The organization designed and introduced to certain mining communities in Asia and Africa its transparent Thermex retort (Fig. 8.2). However, despite enabling miners to observe mercury being released from the amalgam and its condensation (Babut et al., 2003), the design, at a value of US$500, is highly unaffordable. Moreover, the system is fragile and potentially breakable, as it is constructed from glass, which poses an additional financial risk to cash-strapped miners keen on making the purchase.

In summary, retorts have not been widely adopted in small-scale gold mining communities because of a series of cost, training and culture-related problems. Governments, in their attempts to introduce retorting systems to target communities, have failed to adequately address these, and related, issues.

Figure 8.2 Thermex transparent retort designed by UNIDO.
Source: Babut et al., 2003.

Case study #3: small-scale mining licensing systems

It is recognized that, "to facilitate formalization" in the small-scale mining industry, "registration procedures and regulatory compliance can be simplified for subsistence miners in order for them to operate in a favourable institutional and legislative regime" (UN, 1996b). However, in introducing legal formalities to informal channels of the industry, governments have failed to effectively bridge critical information gaps, and acquire feedback from miners on newly-introduced regulations. As a result, small-scale mining is now "bedevilled with too many regulations that are mostly designed to constrain it" (Jennings, 1999).

Up until the mid-1980s, migratory small-scale miners were largely ignored and unmonitored by government authorities, but over the past 10–15 years, an increasing emphasis has been placed on formalizing operations. It is now argued that licensing for small-scale mining is an important step towards improving the efficiency of the industry. It is a primary example of how governments, in recent years, have attempted to introduce formalities to what has long been viewed and thus treated as an informal sector of industry. However, the delays, exorbitant paperwork and costs inherent with licensing procedures continue to be major disincentives to registration. Again, in the designing of small-scale mining licensing procedures, governments have overlooked the perceptions and capabilities of miners.

In Brazil, for example, though the need for effective licensing and regulatory procedures is recognized, "the great majority of mining operations are unauthorized and unregulated and the environmental and social consequences are serious" (Bezerra et al., 1996). More specifically, the series of laws in place are intended to promote mining co-operatives by creating small-scale mining reserves. Although many are functioning, most small-scale miners prefer individual panning licenses because they better associate themselves as a civil association rather than a co-operative forced to comply with co-operative requirements (Bugnosen, 1998). A major flaw with the existing licensing procedure in Ghana is that the registration process, overall, is largely voluntary, requiring significant initiative to be taken by the miner. Registered miners are required to provide *voluntary* feedback to the government on a wide range of issues, including environmental impact assessment, technology usage, production and sales. It is therefore not surprising that an overwhelming majority of small-scale miners elect to operate illegally (there are nearly six times as many illegal small-scale mine operators in the country). Similarly, in the Philippines, small miners are required to undergo a stringent evaluation and registration process to obtain a small-scale mining permit (PD 1899) to mine artisanal works. It is highly likely that these challenges and grievances are also causing a number of operators to mine illegally.

Case study #4: the Shamva Mining Centre, Zimbabwe

In Zimbabwe, which has the second largest mineral wealth in Africa, there are approximately 350,000 individual gold panners. It has been conservatively estimated that, in addition to riverside activity, the country has 5000 small-scale gold mines in operation, many of which are located within the Zambezi Basin and Central Plateau. In an attempt to provide a wide range of support services to resident small-scale gold miners, the Zimbabwe Small-Scale Miners' Association, under the guidance of the London-based Intermediate Technology Development Group, constructed a contract gold mill and training centre in the north-eastern Shamva mining area in 1989. Heralded as the most significant support-related initiative undertaken in the industry to date, the services of the "Shamva Mining Centre" have proven to be insufficient to meet miners' needs. The undertaking clearly shows a lack of foresight on the part of its designers, who clearly miscalculated miners' demands for its services.

Initially, the Centre was successful because it enabled miners to improve process efficiencies by providing them access to reliable technologies. The milling services employed at the Centre resulted in income increases of as much as 30 percent for certain miners. Furthermore, the fees paid by miners for gold processing are directly related to the final prices at which gold is purchased by the Reserve Bank, and are therefore set at affordable rates (Hentschel et al., 2001; Hilson, 2002a). In 1996 alone, the Reserve Bank purchased four tons of gold from small-scale miners (Zava, 1997), much of it through the Centre.

In recent years, however, the Centre has experienced its share of problems. Most of the discussions on the Centre focus on its positives, indicating that although it was originally built to serve the needs of approximately 40 local gold miners, it now processes the ore of over 500 small-scale operations (currently more than 4,000 tonnes per annum) and provides advice on many mining-related topics. Few, however, have attempted to explain *why* the Centre has expanded in size. As already noted, the Centre's problems have largely been a result of poor planning, the most significant of which stem from vast miscalculations of user demand. By the early-1990s, it became evident that the capacity of the ball mill installed was insufficient to meet the growing needs of miners. The pre-existing mill, which had the ability to produce only one tonne ore/hr, forced many miners to wait between three to six weeks to have their ore processed. This undoubtedly led many to engage in fast, unmonitored mining processes alongside riverbanks, and to sell product through illegal channels.

Additional problems have resulted from poor business decisions. Not only does Shamva now lack an experienced and competent manager, but it also continues to experience serious financial problems, due in part to the committee's declaration, in January 1999, that it had sufficient capacity to "run (it) without external assistance" (Mugova, 2001). As a result, the Centre is operating well below its staffing capacity (Hentschel et al., 2001), which, in turn, has further crippled its ability to provide support services to resident miners.

To summarize, each of the above-mentioned case studies demonstrates how governments have overlooked key community side issues in the process of strategizing to improve the conditions of resident small-scale mines. It has been shown that, if the perspectives of miners and the characteristics of their operations are not taken into account before implementing policies and support-related measures, a number of production, staffing, and financial problems can occur. The next section of the chapter describes a series of initiatives for improving artisanal and small-scale mining developmental strategy.

IMPROVING SMALL-SCALE MINING DEVELOPMENTAL STRATEGY

It is clear from the discussion thus far that, in their attempts to improve the efficiency of small-scale mining operations, many governments have overlooked key community-side elements. The failure to integrate both institutional and behavioural-side solutions into decision-making processes is largely a result of the adoption of "top-down" management approaches, which mainly emphasize the state providing services to people living within a defined area. Each begins from a basic policy decision, and then "examines the extent to which its objectives are realized over time" (Wallis & Dollery, 2001, p. 249). In such approaches, "the initiative of setting the agenda and the onus of developing resultant plans of action falls upon the international or national institution" (Iyer-Raniga et al., 2000, p. 231). In the context of artisanal and small-scale mining, although the community plays a pivotal role in determining the success or failure of a support-related undertaking, "top-down" approaches fail to take into account the "voice" of target populations; as a result, unanticipated problems can arise (Fig. 8.3). The key

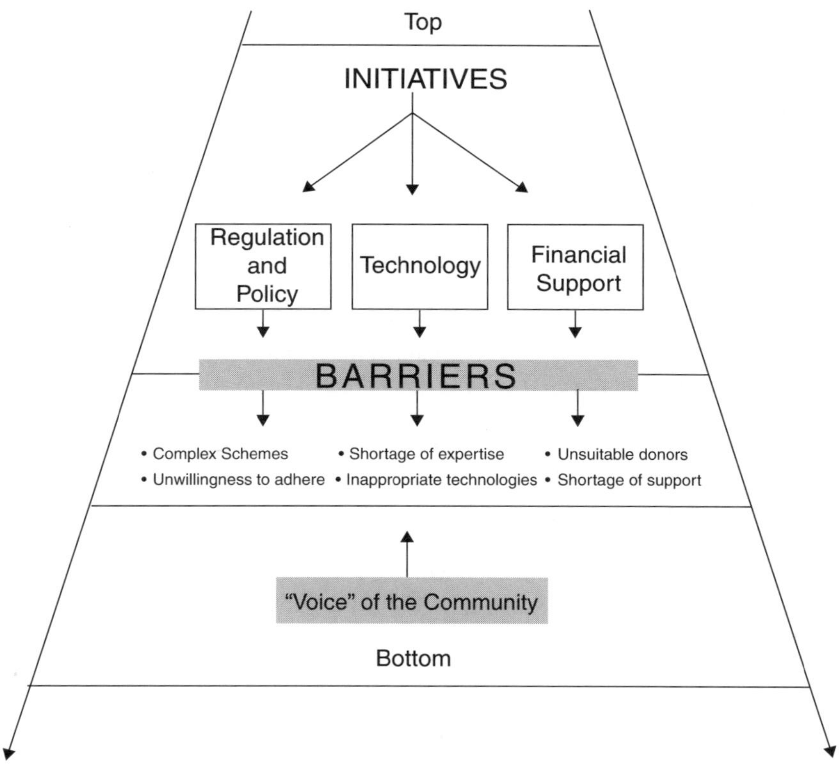

Figure 8.3 Schematic representation of "top-down" management approaches adopted for artisanal and small-scale mining. In the process of implementing support-related measures (i.e. policies, technologies, and financial assistance schemes) from the "top-down", various problems can occur, largely because community-side issues have not been properly taken into account.

to improving the delivery of artisanal and small-scale mining support, therefore, is progressive adoption of "bottom-up" management approaches, which target problems more effectively at the grassroots. Thus, the issue at hand is not necessarily the initiatives themselves, but rather the ways in which they are implemented.

Step #1: construction of small-scale mining district support centres

An unwillingness to adopt decentralization strategies has proven, time and time again, to be a major impediment to effectively delivering support to artisanal and small-scale miners. For example, a number of countries in Africa have highly-centralized political systems, which have had commanding influences on day-to-day operations in resident small-scale mining industries. In Ghana, despite the existence of local small-scale mining district centres, industry policies and licensing procedures are implemented in the country capital of Accra. This, in turn, has led to a number of problems. Poor networking, communication and road infrastructure have caused delays in licensing, and have prevented authorities from effectively monitoring support services that have been implemented for small-scale mining. In fact, such is the case in most African countries, where the majority of artisanal and small-scale mining regions are well out of reach of authorities.

A necessary first step, therefore, is to construct regional small-scale mining support centres and empower each with administrative-related tasks, as they are capable of providing badly needed guidance and support to miners. The construction of these centres enables government staff to be in contact with miners on a daily basis, and thus provides an opportunity to tackle pressing problems at a local, rather than national, scale. Units cannot only be equipped with "hardware" – namely, equipment and technologies capable of facilitating improved production – but could also be staffed with engineers, technicians and geologists capable of providing hands-on assistance to miners. The establishment of regional service centres also creates opportunities for effective education, providing a platform to host training seminars in the areas of environmental protection, health and safety and financial management.

Attempts have already been made by certain governments world wide to construct regional offices and support centres for small-scale miners. These provide a number of services, including financial support, technological assistance, and registration and licensing inquires. In Ghana, for example, shortly after the legalization of small-scale gold mining in 1989, the government established eight district centres in the southern part of the country (although only seven remain) to help register small-scale gold miners, supervise and monitor their operations, and provide training facilities and assistance (Iddirisu & Tsikata, 1998). Each is staffed with a mine engineer and a mines inspector (provided by the Department of Mines) to register claims, provide technical advice, and encourage the safe and productive operation of local mines (Davidson, 1993).

In Venezuela, UNECA (UNit of gold Extraction and Controlled Amalgamation) Centres (financed jointly by the Venezuelan government and UNIDO) were established to provide a number of environmental and technical services to artisanal gold miners. As Veiga and Beinhoff (1997) explain, the first such centre, constructed in Playa Blanca, employed four technicians and one engineer who oversaw the amalgamation of concentrates from over 70 barges. More importantly, however, UNECA Centres provide miners with the following benefits:

- Improved gold recovery from gravity concentrate
- Better prices for gold
- No mercury vapour exposure
- No need to buy mercury illegally
- Access to information pertaining to legal mineral titles
- Access to information pertaining to financial support

The Centres also provide the public with information related to mercury use and exposure, and educational material about mining and the environment.

These, and examples of other small-scale mining educational/training centres are described fully in Table 8.3.

Step #2: makeshift changes in research approaches

As explained at length throughout this discussion, preliminary research must be undertaken before implementing support-related measures for artisanal and small-scale mining. In the developing world, there has been an emphasis on "technical solutions for 'basic needs'" (Schulpen & Gibbon, 2002, p. 1). Science and technology, however, only offer "a glimmer of hope in dealing with

Table 8.3 Examples of small-scale mining support centres.

Country	Name/type	Summary
Ghana	District Support Centres	A total of seven support centres exist in the southern part of Ghana, each of which provides both basic support and purchasing services to miners.
Zimbabwe	Shamva Mining Centre	A centrally-located services and support facility for artisanal miners.
Venezuela	UNECA Centres	Provide a wide range of services to artisanal gold miners, particularly in the area of mercury management.
Chile	Empressa Nacional de Mineria (ENAMI)	Operates four regional concentrators, treating ores purchased from numerous small and medium copper miners. Also maintains 16 mineral purchasing centres dispersed throughout the country.
Morocco	CADETAF, a governmental agency	Offers a wide range of services to artisanal miners extracting high-grade lead and zinc ore in the Atlas Mountains. Also operates seven regional ore purchasing centres that provide basic materials and essential services to miners.

problems of developing countries" (Ryan & Mothibi, 2000, p. 375). Such is the case in the small-scale mining industry.

It is imperative that, before industry initiatives are undertaken, pertinent background data is first collected, which can be a challenge of monument proportions, given that most developing countries have acute shortages of information collections on important developmental topics. As Bulmer (1983) explains, "development officials often lack adequate data on basic population parameters and indicators of development such as agricultural production or infant mortality", and in many situations, the position is more a case of poor quality data than of no data. For example, in many developing countries, one of the primary sources of injury data are newspaper reports (Ghaffar et al., 2001). Moreover, "researchers in developing countries find that some of the most frustrating problems they face are in the library, not the laboratory", as they "have less access to the traditional channel of (paper) publications" (Samik-Ibrahim, 2000, p. 2). For example, in a study by Majid (2001), it was discovered that only 50% of surveyed Pakistani, Bengali and Sri Lankan research institutions used CD-ROM technology – long heralded to be idealistic information technology for centres based in the isolated areas of the developing world. In short, "the official statistical collection system and infrastructure in many developing countries is much more fragile than in developed ones, and in social statistics, which have relied primarily on administrative sources, the collection system is particularly weak" (Bulmer, 1983, p. 5). In the case of small-scale mining, there is often a shortage of data in the areas of equipment usage, registration, and production, and, in the most extreme of situations, an absence of tabulated censuses for miners, in which case, researchers are forced to rely on estimates put forth by experts.

Once whatever available information is secured, small-scale mining researchers must undertake socioeconomic research in target communities. Specifically, an attempt should be made to determine, *inter alia*, the composition (*e.g.* age, gender, etc.) of target populations, education and literacy levels of communities, and technological expertise. This research is imperative given that there are often major differences between the rural and urban regions of the developing world; there are major geographical, cultural and regional variations throughout developing countries, many of which are extreme. For example, different types of agriculture may be found in the same society, ranging from nomadic practices through subsistence farming, to commercial agriculture. Moreover, marked ethnic, language and racial differences are common (Bulmer, 1983). Collection of this information requires listening "to the desires of those he/she (the researcher) is attempting to serve" (Parsons, 1996).

Papadopoulos et al. (1998) first proposed the model for "culturally competent researchers", which is based on four concepts (summarized from Papadopoulos & Lees, 2002). The first, *cultural awareness*, begins with the researcher examining and challenging his/her personal value base and understanding how values are socially accepted. The second element, *cultural knowledge*, is achieved through interacting with people from different indigenous groups. This involves drawing expert feedback from many disciplines, including biomedicine, anthropology, sociology and psychology. The third element, *cultural sensitivity*, is achieved by considering participants in research as true partners, the outcomes of which involve a process of facilitation, advocacy and negotiation that can only be achieved on a foundation of trust, respect and empathy. The final element is *cultural competence*, which requires synthesis and application of previously gained awareness, knowledge and sensitivity. It is crucial that the research undertaken is sensitive to the ethnic and cultural backgrounds of the communities involved.

There is merit using this model as guidance when undertaking community research in the small-scale mining industry. In many cases, without access to primary information sources, small-scale mining researchers will be forced to incorporate baseline study components in their research approaches. The following methods are recommended (Majid, 2001): (1) beneficiary assessments; (2) epidemiological & anthropometric surveys/censuses; (3) ethnographic investigations; (4) household & health surveys; (5) longitudinal village studies; and (6) participatory assessments. As Table 8.4 explains, each approach can help a small-scale mining researcher acquire important community information that could factor prominently into key decision-making processes at the government level.

Step #3: simplification of small-scale mining licensing schemes and implementation of sector-specific regulations

According to Jennings (1999), "if small-scale mining is to be encouraged to operate legally, legislation must be (at least) even-handed in allowing small-scale miners access to suitable land for prospecting and permits that provide clear security of tenure for a reasonable period so that small-scale mining can become established". It is therefore crucial that newly-introduced licensing schemes and legislation is kept simple, bearing in mind that artisanal miners have long operated without regulations, and have the tendency to perceive the implementation of complex legislation as an infringement of their mining rights.

The argument put forth by governments is that legislation is key to improving the efficiency of resident small-scale mines, as it helps to set much-needed performance benchmarks for the industry. In a study by Bugnosen (1998), in which the small-scale mining legislation of 18 countries was analyzed, it was concluded that "almost all of the small-scale mining legislation examined aims to assist small-scale miners who are nationals of the given country". It was further noted that, there are "indications in some countries that preferential access to certain mineral lands are provided to citizen small-scale miners", accomplished "by designating specific areas" for activity (Bugnosen, 1998, p. 8).

In a number of other countries, the approaches taken to implement regulations and codes for small-scale mining have been largely ad hoc. For example, many governments have responded to the challenge of regulation in the industry by accommodating and promoting the small- and large-scale mining segments using the same umbrella of legislation. This, however, has led to a number of problems, the most important of which relates to competition for land plots

Table 8.4 Recommended data collection methods for small-scale mining researchers.

Data collection method	Description	Use in small-scale mining research
Beneficiary Assessments	Systematic data collection methods such as interviewing over a limited time	Can be used to determine basic needs of the mining population such as health, environmental and safety awareness
Epidemiological & Anthropometric Surveys/Census	Bio-medical surveys of populations to determine patterns of morbidity, mortality and nutrition	These can be used to determine the quantity of illegal and legal miners in a given mining community, and the health of miners, which can be used to determine individuals' economic stability
Ethnographic Investigations	Anthropological research techniques, especially direct observations	Much can be determined through direct observation, including the quality of work conditions, the level of mechanization and the nature of mining practices
Household & Health Surveys	Structured interviews of a representative household sample	Excellent means of obtaining opinionated feedback from miners on a number of key issues, including views of proposed technological undertakings, financial schemes and licensing schemes
Longitudinal Village Studies	Wide variety of methods ranging from direct observation and recording, periodic semi-structured interviews with key informants and the village population, to survey interviews in several different observation periods	Excellent means of obtaining opinionated feedback from miners on a number of key issues, including views of proposed technological undertakings, financial schemes and licensing schemes
Participatory Assessments	Ranking, mapping, diagramming and scoring methods	Once research has been undertaken, findings can be categorized and perhaps, charted, which could help authorities make key decisions

Source: Modified from Majid, 2001.

and mineral resource (Davidson, 1993). In short, the question of transferable mining rights – a principal issue for the artisanal miner – still needs to be answered in many developing countries (UN, 1996b).

Even many of the regulations that have been implemented *specifically* for small-scale mining have proven ineffective. As Kumar and Amaratunga (1994, p. 17) explain, "such schemes do not have the stringent conditions imposed on large-scale mining operations". Reconnaissance, exploration and exploitation issues "are not envisaged in small-scale mining operations which are more 'open' where a statutory right to prospect and mine is acquired". Since the conditions imposed on small-scale miners are minimal, their minerals rights and titles are less secure than their large-scale counterparts.

Governments are therefore encouraged to shy away from regulating small-scale mines with general mining legislation. Not only does this strategy fail to regulate the industry appropriately, but the adoption of such a lackadaisical regulatory approach is also clear indication that improved industry efficiency is not a national goal. The key is to keep legislation simple, and, at the same time, maintain even and effective enforcement, which helps to ensure that small-scale miners are operating within the law.

The most significant improvement that could be made legislatively is simplification of small-scale licensing schemes. As noted earlier, many of the governments that have sought to regularize small-scale mining operations have implemented registration systems, whereby prospective miners are required to apply for a license to mine on a small scale. However, the complexity of many such schemes, in combination with the length of time it can take to secure a license, have proven to be major deterrents for many miners contemplating operating within the legal channels of the industry. For example, in Ghana, in addition to the registration process being largely voluntary, in many cases, it has reportedly taken months to approve a license. In another example, Brazil, it has been noted that whilst the mines themselves are not illegal, because of the complexity of the environmental licensing and the ineffective administration at the local level, many are not complying fully with the law (Hennies et al., 1996).

In short, licensing schemes should not be designed to discourage miners from registering with the government, but rather encourage them to operate legally. They can also be used more positively if seen as a policy tool rather than a form of revenue raising (Burke, 1997).

Step #4: discourage illegal mining

Perhaps the most significant socio-economic problem facing governments is illegal mining – the largest share of the small-scale mining sector, its growth a direct result of the implementation of inappropriate legislation and an uneven enforcement of laws. Though it is frequently argued that illegal mining may be a result of devalued currencies and an unavailability of foreign exchange resources, it is most commonly practiced in nations where state enterprises have ownership and control of operations. Kambani (1995) provides estimates of illegal small-scale mining activity in a number developing countries, including:

- Colombia: 80% of emeralds, valued at US$800 million per year;
- Zambia: 50% of emeralds, valued between US$200 and US$300 million per year;
- Zaire: an estimated US$400 million worth of gold and diamonds are smuggled illegally each year;

- Zimbabwe: 50% of emeralds, valued at US$3 million, are mined and sold illegally each year; and
- Brazil: most of the US$2 billion worth of gold mined by small-scale miners.

In order to tackle the illegality issue with improved strategy, "governments must be prepared to move beyond the establishment of legal frameworks, to identify deposits and areas amenable to small-scale development, including the preliminary evaluation of their technical and economic viability at different areas of operation" (Davidson, 1993, p. 319).

Realistic market and buying prices must first be set for miners. In fact, it has been proven that, where governments have been prepared to pay competitive international prices in foreign exchange or local currency tied to free market exchange rates, clandestine sales have reduced (Davidson, 1993). In Tanzania, for example, the implementation of a mineral trade liberalization policy in the late-1980s, facilitated a rise in the quantity of legally traded gold, which increased from US$550,000 in 1985 to US$38.78 million in 1992 (Hentschel et al., 2001). In another example, Ghana, a series of regularization activities undertaken in the late-1980s has resulted in the collection of over US$140 million in revenues from mineral sales that would have otherwise been lost (Labonne, 1996), and by the early-1990s, had contributed to well over US$70 million in foreign exchange earnings for the government (Davidson, 1993). It has been shown through experience that to prevent rampant illegal mining, the difference between government and global prices paid for minerals must not exceed a margin of 5% (Noetstaller, 1994).

However, detailed analysis is needed to determine *why* miners are operating outside of the legal bracket; this is undoubtedly a key to further eradicating illegal mining activity. The following examples help to underscore the importance of doing so:

1) Many miners remain unregistered with governments because they have little knowledge of legal requirements. In such cases, implementation of licenses and state mineral purchasing services are futile exercises.
2) A number of other small-scale miners are choosing not to register with government authorities because they claim that there are few incentives for doing so (Hentschel et al., 2001).
3) In other cases where governments have established marketing boards, such as in Guyana, small-scale miners may prefer to deal with unofficial traders to get better prices, and even in situations where the official exchange rate closely reflects the black market rate, small miners may still prefer dealing with unofficial traders to avoid paying royalties.
4) Finally, it is crucial that miners' perceptions of government-organized mineral purchasing services are determined. As Davidson (1993) explains, many miners prefer complementary services with medium- and large-scale counterparts, as they are not at the mercy of local and international mineral traders, who do not always deal with smaller producers on an equitable basis.

These, and related, elements must be taken into account when tackling the problem of illegal mining, and before deciding on a strategy for mineral purchasing and licensing.

CONCLUSION

By following each of the aforementioned guidelines, governments put themselves in an improved position to both tackle pressing problems in the small-scale mining sector, and implement relevant support-related measures. Many of the support-related initiatives undertaken to date have only managed to facilitate marginal industry improvements. Part of the blame can be placed on governments' increasing dependence upon imported western ideologies and technical solutions. The main reason, however, is their inherent neglect of key cultural issues. Specifically, without having predetermined the needs of small-scale mining communities, the education and literacy levels of target populations, and the technological expertise of resident artisans, how can an effective small-scale mining developmental strategy be devised?

This chapter has brought many of these concerns to the forefront. It has underscored the importance of factoring the "voice" of the small-scale mining community into industry decision making processes, and has presented a series of illustrative case studies that collectively put into perspective the potential problems that can arise when important community-side issues are overlooked.

REFERENCES

Babut, M., Sekyi, R., Rambaud, A., Potin-Gautier, M., Tellier, S., Bannerman, W. & Beinhoff, C. (2003). Improving the environmental management of small-scale gold mining in Ghana: a case study of Dumasi. *Journal of Cleaner Production*, 11(2), 215–221.

Batterbury, S., Forsyth, T. & Thomson, K. (1997). Environmental transformations in developing countries: hybrid research and democratic policy. *The Geographical Journal*, 163(3), 126–132.

Bezerra, O., Verissimo, A. & Uhl, C. (1996). The regional impacts of small-scale gold mining in Amazonia. *MERN Bulletin* 9.

Bugnosen, E. (1998). *A preliminary assessment of small-scale mining legislation and regulatory frameworks*. Intermediate Technology Group, London.

Bulkeley, H. (2000). Down to earth: local government and greenhouse policy in Australia. *Australian Geographer*, 31(3), 289–308.

Bulmer, M. (1983). General Introduction, p. 3–26, in *Social research in developing countries*. New York: John Wiley & Sons.

Burke, G. (1997). Policies for small-scale mining: the need for integration. *Raw Materials Report*, 12(3), 11–14.

Clark, A.L. & Clark, J.C. (1999). The new reality of mineral development: social and cultural issues in Asia and Pacific nations. *Resources Policy*, 25, 189–195.

Davidson, J. (1993). The transformation and successful development of small-scale mining enterprises in developing countries. *Natural Resources Forum*, 17(4), 315–326.

Downs, T.J. (2000). Changing the culture of underdevelopment and unsustainability. *Journal of Environmental Planning & Management*, 43(5), 601–621.

Ghaffar, A., Hyder, A.A. & Bishai, D. (2001). Newspaper reports as a source for injury data in developing countries. *Health Policy and Planning*, 16(3), 322–325.

Gunson, A.J. (2001). *Hunan case study: the Changsha training center for small-scale coal mines.* MMSD Workshop on Artisanal Small-scale Mining, London.

Heemskerk, M. (2002). Livelihood decision-making and environmental degradation: Small-Scale Gold Mining in the Suriname Amazon. *Society & Natural Resources*, 15, 327–344.

Hennies, W.T., da Silva, L.A. & Chaves, A.P. (1996). The small mining enterprises and their environmental effects in Jaguariuna, Sao Paulo State, Brazil (pp. 685–689). In *Proceedings of the 5th International Symposium on Mine Planning and Equipment Selection, Sao Paulo, Brazil*. Rotterdam: A.A. Balkema.

Hentschel, T., Hruschka, F. & Priester, M. (2001). *Global report on artisanal and small-scale mining (ASM)*. Draft Report prepared for Mining, Minerals and Sustainable Development Project, London.

Hilson, G. (2002a). Delivering aid to grassroots industries: A critical evaluation of small-scale mining support services. *Raw Materials Report*, 17(1), 11–17.

Hilson, G. (2002b). The environmental impact of small-scale gold mining in Ghana: identifying problems and possible solutions. *The Geographical Journal*, 165(1), 57–72.

Hollaway, J. (1993). Review of technology for the successful development of small scale mining. *Chamber of Mines Journal*, 35(3), 19–25.

Iddirisu, A.Y. & Tsikata, F.S. (1998). *Mining sector development and environment project.* Regulatory Framework Study to Assist Small Scale Miners, prepared for the Minerals Commission.

Iyer-Raniga, U. & Treloar, G. (2000). A Context for Participation in Sustainable Development. *Environmental Management*, 26(4), 349–361.

Jennings, N.S. (Ed.) (1999). *Small-scale gold mining: examples from Bolivia, Philippines and Zimbabwe*. International Labour Office, Sectoral Activities Programme, Industrial Activities Branch, ILO.

Kambani, S.M. (1995). The illegal trading of high unit value minerals in developing countries. *Natural Resources Forum*, 19(2), 107–112.

King, L.A. & Hood, V. (1999). Ecosystem health and sustainable communities: North and South. *Ecosystem Health*, 5(1), 49–57.

Kumar, R. & Amaratunga, D. (1994). Government policies towards small-scale mining. *Resources Policy*, 20(1), 15–22.

Labonne, B. (1996). Artisanal mining: an economic stepping stone for women. *Natural Resources Forum*, 20(2), 117–122.

Majid, S. (2001). Trends in using CD-ROM in academic libraries of three South Asian countries – Pakistan, Bangladesh and Sri Lanka. *Aslib Proceedings*, 53(2), 68–76.

Mugova, A. (2001). *Presentation: The Shamva mining centre project.* MMSD Workshop on Artisanal Small-Scale Mining, November 19–20, London.

Mutagwaba, W., Mwaipopo-Ako, R. & Mlaki, A. (1997). *The impact of technology on poverty alleviation: The case of artisanal mining in Tanzania*. Research Report 97.2, Research on Poverty Alleviation. Dar es Salaam: Inter Press Tanzania Ltd.

Noetstaller, R. (1994). Small-scale mining: practices, policies and perspectives. In A.K. Ghose, (Ed.), *Small-scale mining: A global overview*. Rotterdam: A.A. Balkema.

Parsons, L.B. (1996). Engineering in context: Engineering in developing countries. *Journal of Professional Issues in Engineering Education and Practice*, 122(4), 170–176.

Popadopoulos, I., Tilki, M. & Taylor, G. (1998). *Transcultural care: A guide for health care professionals*. Wiltshire: Quay Books.

Popadopoulos, I. & Lees, S. (2002). Developing culturally competent researchers. *Journal of Advanced Nursing*, 37(3), 258–264.

Priester, M. & T. Hentschel. (1992). *Small-scale gold-mining: Processing techniques in developing countries*. Germany: Bertelsmann Publishing Group International.

Redclift, M. (1992). A framework for improving environmental management: beyond the market mechanism. *World Development*, 20(2), 255–259.

Ryan, T.B. & Mothibi, J. (2000). Towards a systemic framework for understanding science and technology policy formulation problems for developing countries. *Systems Research and Behavioural Science*, 17, 375–381.

Samik-Ibrahim, R.M. (2000). Grounded methodology as the research strategy for a developing country. *Forum: Qualitative Social Research*, 1(1), 1–5. http://qualitative-research.net/fqs

Schulpen, L. & Gibbon, P. (2002). Private sector development: Policies, practices and problems. *World Development*, 30(1), 1–15.

UN (United Nations) (1996a). *Developments in small-scale mining*. Economic and Social Council, United Nations, New York.

UN (United Nations) (1996b). Recent developments in small-scale mining: A report of the Secretary-General of the United Nations. *Natural Resources Forum*, 20(3), 215–225.

Veiga, M.M. & Beinhoff, C. (1997). UNECA Centres: A way to reduce mercury emissions from artisanal gold mining and provide badly needed training. *Industry and Environment*, 20(4), 49–51.

Wallis, J. & Dollery, B. (2001). Government failure, social capital and the appropriateness of the New Zealand model for public sector reform in developing countries. *World Development*, 29(2), 245–263.

World Bank (2001). *Global economic prospects and the developing countries*. World Bank, Washington.

World View (2001). *World development indicators*. World Bank, Washington.

Zava, J. (1997). Panning power. *Orbit* issue 66 http://www.oneworld.org/vso/pubs/orbit/66/pan.htm

9[1]

Seminar on "Artisanal and Small-Scale Mining in Africa: Identifying Best Practices and Building the Sustainable Livelihoods of Communities"

BEATRICE LABONNE

Because artisanal mining is largely driven by poverty, it has grown as an economic activity, complementing more traditional forms of rural subsistence earnings. The Mining, Minerals and Sustainable Development (MMSD) Southern Africa report (2002) acknowledges that Artisanal and Small-Scale Mining (ASM) is "typically practised in the poorest and most remote rural areas by a largely itinerant, poorly educated populace, men, and women with few employment alternatives" (MMSD, 2002). However, the sector has the potential to empower economically disadvantaged groups and enrich nations by virtue of its low investment costs and short lag-time, from discovery to production.

In spite of the growing regional economic importance of ASM, there is a lack of reliable regional and national socio-economic and poverty statistics to accurately assess the *real* economic significance of the sector. For many policy makers, this absence of data hinders the formulation of growth-based rural poverty reduction policies, and their subsequent implementation. It also limits the impact of sectoral strategies and programmes. Over the years, entities responsible for the ASM sector, such as mining ministries, have been challenged to develop and implement policies capable of mitigating the industry's adverse socio-economic and environmental impacts. Globally, the dynamism of ASM is still not well understood – most notably, the economics behind the marketing of gold, diamonds and gemstones. As a consequence of the local inflationary

[1] This chapter reports the findings from a plenary forum on ASM held in Yaoundé, Cameroon, 19–22 November 2002.

tendencies of ASM, one still has to demonstrate whether short-term income gains at the local level can translate into medium and long-term poverty reduction. It is not clear whether ASM gains outweigh its often-severe, socio-economic and environmental drawbacks. Experience shows that ASM stakeholders have divergent interests, and that it can become a viable economic activity that contributes positively to poverty reduction, provided that these diverse interests can be adequately addressed and reconciled.

As it is highly unlikely that a single "best-practice" model for ASM can be applied to all countries, there is a need to devise strategies that are realistic, well founded in communal realities, and supported by the communities themselves. This requires efforts both in dialogue and partnership at all levels of decision-making, from national government to the community level.

The seminar on *ASM in Africa: Identifying best Practices and Building the Sustainable Livelihoods of the Communities* aimed at providing a forum for debate and partnerships. The Yaoundé seminar brought together some 70 participants, including highly qualified senior policy-makers from several African countries, notably, Burkina Faso, Cameroon, Central African Republic, Congo Ivory Coast, Ethiopia, Gabon, Guinea, Mali, Mozambique, Nigeria, Sierra Leone, South Africa, Tanzania, Togo, and Zimbabwe. Participants from international organizations included experts from UNAIDS, the African Development Bank, the World Bank, DfID, and the French bilateral cooperation. Individual ASM miners from the host country, Cameroon, also attended.

By bringing together key stakeholders, the organizers' goal was to encourage a fresh exchange of views and experiences concerning the main issues that illustrate the challenge of balancing the social, economic, and environmental concerns of ASM. The specific objectives of the Seminar were as follows:

a) To identify, analyse and propose a set of novel policy options for governments, IGOs, donors and NGOs, on the basis of case studies and "best practices". These policy options encompass the macro, meso and micro (sectoral) dimensions of ASM to enable full mainstreaming of this activity into local development planning and national poverty reduction efforts (PRSP process). The focus was on approaches and actions consistent with community empowerment initiatives.
b) To offer a forum for debate between ASM experts and practitioners on poverty reduction issues to promote novel policy and realistic implementation mechanisms.
c) To draw upon lessons and recommendations for implementation in other African countries and elsewhere through existing networks such as the Communities and Small-Scale Mining (CASM) knowledge bank.
d) To re-energize the interest of governments, donor and NGO communities, and the private sector in this important but overlooked sector.

The seminar was co-organized by UNDESA and UNECA. It was hosted by the Ministry of Mines, Water and Energy of the Republic of Cameroon, and for the UN by the Sub-regional Development Centre of UNECA for Central Africa.

The Seminar was opened by Mr. M. Abbo, Secretary-General of the Ministry of Mines, Water Resources and Energy; Ms. Patricia de Mowbray, the UN Resident Coordinator in Cameroon; Mr. Halidou Ouedraogo, Chief Sub-Regional Development Centres, Coordination Unit, UNECA; and Ms. Béatrice Labonne, Senior Adviser, UNDESA. The opening remarks underscored the major challenge for African poverty, and a commitment to reduce it in line with the Millennium Development Goals. The importance and potential contribution of the ASM sub-sector as a means to alleviate poverty was highlighted. It was pointed out, however, that major efforts are required to solve the ASM development conundrum in order for ASM communities to benefit economically and socially from their access to natural resources. In spite of its economic vitality, the sector does not seem able to become an engine for growth and lasting poverty reduction. Moreover, it is often associated with deadly diseases, and socially and environmentally unsustainable practices. Examination of efforts to disseminate and promote ASM "best practices", as well as the experiences of countries to formalize the sector, was seen as a suitable starting point for discussion.

It was stated that policies should aim at empowering ASM communities and commit a broad-based partnership for well-defined targets. On the issue of improving the livelihoods of people whose incomes are derived directly or indirectly from ASM, participants were invited to propose "doable" policies, and shift from a "what to do" to a "how to" frame of action. It was emphasized that ASM plans should be grounded in economic, social and cultural realities, and that strategies and policies must reflect the true delivery capacity of government; bridge the gaps between various levels of decision-making; and reflect the various interests of ASM stakeholder parties. As minerals are finite, it was judged important to measure the impacts of these policies in terms of sustainable poverty reduction, inclusive of additional and alternative opportunities for income generation. Finally, it was reiterated that a broad-based partnership between the various ASM national and international stakeholders was needed to concretely tackle the daunting task of improving livelihoods in ASM communities.

This chapter reports the findings of the conference, and details the recommendations put forward at discussions for improving conditions in ASM communities, with special emphasis on Africa.

SESSION 1: SETTING THE STAGE

The keynote presentation by ASM specialist Mr. Kevin D'Souza, a mining expert at Wardell Armstrong, offered a very insightful and comprehensive review of ASM in Africa, particularly its constraints and challenges. Mr. D'Souza began by describing the broad spectrum of issues that must be considered in relation to ASM, including legislation, institutional capacity, child labour, gender inequality, health and safety, the environment, marketing and credit. He went on to explain that the fundamental problem is that both the ASM sector and governments are caught in a "negative cycle of cause and effect". A poverty trap results from a denial of choices and opportunities, whilst living in a marginal and vulnerable environment. He indicated that in the past, most efforts made by donor agencies endeavored to solve one aspect of the ASM puzzle alone, whilst ignoring others. This shortsightedness has often been compounded by donor and government reluctance for long-term engagement in the sector. Under such conditions, the original objectives behind millions of dollars in aid have not been met.

During his presentation, Mr. D'Souza highlighted some of the challenges and constraints involved in changing policy. He warned that progress towards the economic viability of the sector, with the ultimate goal of sustained poverty reduction, depended on a drastic change in the official perception of – and attitude towards – the ASM sector. Only then can strategies, methodologies and implementation mechanisms bring about the necessary incentives for the sector to fully contribute to resilient rural development. The need for all ASM stakeholders to be prepared for the long haul to overcome the many challenges that plague this vulnerable sector was underscored. The aim should be to wholly integrate ASM into the momentum of national social and economic development.

Mr. D'Souza stressed that a holistic approach must be adopted in order to increase the human, financial, physical and social capital available to ASM communities. All issues must be addressed together, which calls for a radical rethinking of current ASM policies. Stakeholders must work to mainstream ASM issues with the current national process of the Poverty Reduction Strategy Papers (PRSPs). The sector must also align itself with the wider social, economic and environmental principles of sustainable development to mitigate threats to social, economic and biophysical systems throughout Africa. Such a change in approach may facilitate the leverage of donor funding towards ASM communities.

The fact that ASM exploits finite and non-renewable resources was highlighted. For Africa, the challenge now is to capitalize on livelihood opportunities – namely, ensuring that ASM plays a part in fostering other economic pursuits within the goal of lasting poverty reduction and sustainable development.

Finally, Mr. D'Souza called on participants to think realistically in order to move forward and away from what he called "dead policies". He encouraged seminar participants to build on the momentum of the activities of the UN agencies, ILO, DfID, CASM and the MMSD in order to mobilize support for the sector in a holistic manner.

The second presenter, Mr. Jean Le Nay, an Inter-regional Adviser on Development Planning at UNDESA, provided an overview of the current thinking pertaining to poverty reduction within the broader development agenda. With regard to African countries, and Least Developed Countries (LDCs), in particular, Mr. Le Nay outlined the various stages that have led to the current focus on poverty reduction as the declared priority, and the central goal of development. He began his presentation by illustrating the progressive evolution of policy-thinking and prescriptions of the Bretton Woods Institutions (BWIs). Mr. Le Nay indicated that this evolution, away from the strict macro-economic based structural adjustment (SAP) of the 1980s, to the design of a poverty reduction strategy in late-1999, resulted from a mix of pragmatism and peer pressure. He noted, in a precise manner, the advocacy role played by UN agencies, as well as the specific people-centered approaches endorsed by UNDP. According to Mr. Le Nay, the outcome of the World Summit for Social Development (Copenhagen (1995)) was a "watershed". The Summit drew global attention to the urgency of eradicating absolute poverty, and the need to redress social ills. More precisely, it recommended the inclusion of social development goals within SAPs.

Mr. Le Nay explained the process, characteristics and objectives of the BWIs-backed Poverty Reduction Strategy Papers. He made clear that if the main objective of the PRSPs is to reduce poverty, then the process is furthered, as it encompasses the modalities for achieving this goal; economic growth underpins poverty reduction, and a stable and pro-business macro-economic environment should stimulate growth. Good governance, transparency and an increased participation of civil society are regarded as a required social assurance. The PRSP approach differs from the previous BWI orthodoxy, as it provides an economic voice and role to the poor themselves. The caveat is that economic growth alone is insufficient to reduce poverty in a lasting manner.

In the second part of his presentation, Mr. Le Nay highlighted the impediments to the implementation of PRSPs. One is the unlikelihood of a steady and strong rate of economic growth in terms of GDP, particularly in Sub-Saharan LDCs. Many LDCs are unable to generate growth that can meaningfully pull people out of poverty. In these countries, the private sector is typically very weak, with limited growth potential, as it is dominated by the informal sector, notorious for being motivated by subsistence and survival.

Finally, Mr. Le Nay stressed that, in spite of the mixed messages from the international community, poverty reduction has become the yardstick by which development is assessed. The BWIs promote a partnership approach at both national and international levels. It is expected that this partnership should stimulate greater solidarity between the North and the South.

The third presenter, Mr. James Carnegie, a UNDESA consultant on sustainable livelihoods, explained that the Sustainable Livelihoods Approach (SLA) was not a new development fad still to be tested on the ground. Very much to the contrary, it was developed in response to dissatisfaction with results from past development efforts. Mr. Carnegie stressed the evolutionary dimension of the SLA, as it is built on practices that are working. As a result, the SLA is now used by many international agencies, including UNDP, the World Bank and DfID (UK Department of International Development).

The SLA focuses on peoples' livelihoods, how people can improve their coping strategies, and how their livelihoods can be improved and sustained. The speaker illustrated the seven key SLA principles, which are deeply focused on poverty eradication; the central issue may not be mining per se, but is nonetheless poverty-related. Therefore, much needs to be done to improve the livelihoods of ASM communities, as well as the people who are affected by ASM activities.

Mr. Carnegie emphasized that the process should be participatory and transparent. Specifically, the views and livelihood priorities of artisanal miners must be factored in: the people must be in the driver's seat, and empowered. Furthermore, the approach relies on partnerships among stakeholders. It is also important to address the issues that are affecting ASM communities. Strength, assets and opportunities should be clearly identified in order to provide a foundation for development efforts. Linkages between the community (micro), service providers, and government at the local (meso) and national and international levels (macro) must also be well understood. Actions taken at the community level must be coherent with those planned at the meso and macro levels. Finally, the learning process is continuous. The ASM environment is notably dynamic, and there is a need for governance and constant adaptation by service providers to meet the challenge of a characteristically fluid milieu. The sustainability of ASM livelihoods hinges upon the integration of social, economic and environmental dimensions. An important consideration is that ASM is exploiting non-renewable and finite natural resources.

The presenter explained the objectives, process and expected outcomes of the following four country case studies: Guinea, Mali, Ethiopia and Ghana. He particularly illustrated the added value of the Participatory Rural Appraisal methodology used to elicit the views of the people themselves. Each study was completed by a national validation workshop, whereby stakeholders reviewed, finalized and fine-tuned individual country findings and recommendations. Mr. Carnegie concluded by stressing that local people should be empowered to become more involved in managing their own development, and in articulating their demands; they should no longer be content with having a role as passive recipients of services. The greater self-reliance of the ASM sector implies that service providers can reconcile bottom-up aspirations with top down efficiency requirements.

The presentation of Mr. Antonio Pedro, a Senior Economic Officer and Minerals and Energy Programme Leader at UNECA in Addis Ababa, Ethiopia, concluded the session on "Setting the Scene". He called for more concrete action and fewer grand declarations. Over the last decade, concerns over ASM had been highlighted in a number of regional and sub-regional meetings, and conferences. Mr. Pedro listed the resulting declarations, plans of action and protocols. He lamented that these were merely "paper policy," as the majority of the identified actions and measures were not being implemented on the ground.

Although the ASM sub-sector is regarded as a high priority, notably in the fight against poverty, little progress has been reported. While Mr. Pedro regarded all of these stated objectives as very relevant, he noted the lack of political will to embark on any meaningful action. The slow or absent progress of implementation was of concern. He listed nine major impediments to the application of these policies. In addition to identifying a lack of resources – human, financial, institutional and technical – as a major hurdle, Mr. Pedro also commented on the lack of impact and sustainability of many ad hoc programmes. The poor definition of roles and responsibilities, along with the lack of benchmarking, deadlines and indicators of achievement were identified as serious shortcomings. The absence of a means of verification and monitoring instruments called into question the seriousness of these policies.

Mr. Pedro indicated that African leaders had designed the New Partnership for Africa, NEPAD, as a genuine commitment to improving the welfare of their people. NEPAD aims to promote good governance; to cure the continent of its persistent economic malaise; and to sustainably reduce poverty by attracting billions of dollars of private and public investment. NEPAD holds the mining sector as an attractive and conducive environment for investment. Efforts should be directed to create a conducive and transparent regulatory and management framework, improving the quality and availability of information on mineral resources, and establishing "best practices" to ensure that extraction is environmentally sound.

Finally, Mr. Pedro clarified The Economic Commission for Africa's area of intervention in the mineral sector. He stated that the focus of UNECA is on policy analysis, dissemination of "best practices", advocacy and awareness raising, monitoring policy implementation, and the organisation of stakeholder dialogues.

SESSION II: COUNTRY CASE STUDIES

As previously noted, four country case studies were discussed at the seminar. Their manner of presentation followed the same format as before – namely, a general overview of ASM communities, the ASM community situation assessed through the SLA, strategies for poverty reduction in the ASM sector, and the specific expectations of national teams.

Guinea[2]

In Guinea, the following two areas were surveyed: Kouroussa/Kineiro (gold), and Kerouane/Banankoro (diamonds). Mr. Diaby stressed the important economic role played by ASM; in the case of gold, ASM follows timeless traditions. Some 100,000 people are directly involved in the sector. Six tonnes of gold were produced in 1997, and diamond production totalled two million carats between 1996 and 1999. Mining is mostly a seasonal activity in Guinea, carried out during the dry season. During the rainy season, agriculture-based income-generating activities take precedence. Mr. Diaby went on to describe the organizational and social structure of the resident ASM sector, the support it receives from government, as well as the role of NGOs and community-based organizations (CBOs). He also pointed out the differences between the diamond and gold

[2]The Guinea Case Study was presented by Mr. Fodé Diaby, Coordinator of the ASM project in Upper Guinea, Ministry of Mines, Geology and Environment, Conakry, Guinea, and Mrs Aissatou Toure, Division Chief, Division for the Promotion of Women, Ministry of Social Affairs, Conakry, Guinea.

mining segments. He elaborated on the role of "masters" in diamond mining, and pointed to higher soil degradation in diamond mining areas.

Mr. Diaby indicated his satisfaction with the results obtained from the SLA. In particular, he was convinced that the survey results will help develop better-targeted policies. It was timely to take advantage of the existing economic dynamism of ASM regions in order to further income diversification and the sustainability of the ever-increasing population. Because ASM is associated with social and environmental degradation, these policies should be multi-sectoral, participatory, and strengthen the link between various levels of government. The strengthening of agricultural CBOs is particularly needed in order to optimise farming incomes.

A lively discussion followed. Many speakers were of the view that the role of the "masters" was controversial, and did not contribute to poverty reduction in the ranks of the miners. It was noted, however, that the "master" phenomena is not a problem *specific* to Guinea. In such a system, miners were not benefiting from their work, and had little chance for self-reliance. Many participants recommended that, if governments are serious about reducing poverty in a lasting manner, then the "master" practice must be discouraged. Other participants pointed out that sectoral policies such as those concerned with mining, have little understanding of the tenets of PRSPs. At the same time, the majority of PRSPs give lip service to the contribution of the ASM sector. Efforts should be made to better understand its rapport with the other rural sectors. Some participants felt that the findings of the SLA approach should be better utilized to formulate practical responses to the rural development challenge. Its goal should be the social and economic empowerment of miners and their communities. It was felt that it was necessary to improve the livelihood option of ASM communities, as many participants felt that these conditions were often close to slavery. The reality behind the child labour phenomena was debated, as was the scourge of HIV/AIDS in mining regions. It was suggested that "best practices" for ASM, as well as feedback on failures, should be disseminated at the regional level. The role of large mining companies was also discussed. Many felt that a good neighbourly relationship should be promoted.

Ethiopia[3]

Ethiopia is a country of 65 million inhabitants, approximately half a million of who are small-scale artisanal miners. The project focused on gold mining in the Southern Ethiopian Hayadima region, 600 km from Addis Ababa, and dimension stones in Senkelle in the Oromia region, 130 km west of the capital. As is the case in Guinea, artisanal mining is recognized by law. Mr. Gesit described the major problems affecting artisanal mining, focusing particularly

[3]The Ethiopia case study was presented by a team consisting of Mr. Gesit Tilahun of the Mineral Operation Department, Ministry of Mines, Ethiopia, and Mr. Demile Yismaw of the Ministry of Finance & Economic Development, Addis Ababa, Ethiopia.

on gold mining. The problems listed are similar to those in other ASM countries. However, the ASM sector attracts little visibility in the country.

The study recommendations were two-pronged: improving mining methods and diversifying activities away from mining. Referring to the case of the Senkelle region, it was deemed important to increase the quality of dimension stones in order to improve added value. The speaker felt that there was enough information for project definition, and that concrete actions were now needed. He was of the view that the study could have targeted regions even poorer than Senkelle. New demonstration projects should be launched using the SLA, with the communities themselves given more responsibility; moreover, governments and donors have to be sensitized for funding activities in this area.

The discussion focused on the inability of the government to assist the ASM sector, largely because of its lack of resources. In response, some participants indicated that there was a need to identify programmes, which would break bonds of dependency on financiers. Moreover, communities should be better empowered to ensure that projects are not short-lived. Although the PRSP includes the ASM sector, it was felt that actions were not very well articulated.

Mali[4]

This presentation did not adhere to the set format, but nevertheless brought very interesting issues to discussion. Mali is famous for its gold mining tradition and its large number of mine sites. Large mining companies are also operating in the country side-by-side artisanal miners, known as traditional and migrant *orpailleurs*. It is estimated that 200,000–300,000 people are operating in the country's gold fields, with women constituting up to 60–70% of on-site labourers, depending on the location. The gold production from the ASM sector is in the range of six tonnes per year. In Mali, as in other West African countries, gold mining is a seasonal occupation, although it has become the main income-generating activity.

Women are active at every stage of gold mining, from production to retail; yet, there is no gender-specific task. Unfortunately, in Mali, as in many other mining countries, strength does not come in numbers. More specifically, opportunities to work are not synonymous with status and empowerment. Thus, at sites, women not only have fewer traditional rights, but they also receive less money for their work, and are disproportionately affected by social ills. A direct consequence of the existence of women in the country's mining industry is the presence of numerous children at sites. Mr. Keita indicated that there are also child-specific tasks, such as baby-sitting, and the sorting and transporting of ore.

According to Mr. Keita, the poor economic performance of miners was primarily a result of organization deficit. Unable to assess the size and

[4]Mr. Seydou Keita, Expert, PAMPE/DNGM, Ministry of Mines, Energy and Water, Bamako, Mali, presented the Mali case study. He was assisted by Ms. Maimouna Traore, Expert, Poverty Reduction Strategy Coordination Group, CSLP, Bamako, Mali.

exploitability of mineral reserves, the miner is unable to devise a practical business plan. He therefore has few savings, and no access to finance for capital investment to increase production. This hand-to-mouth way of life compels the miner to take loans from traders and licensed buyers. It is not unusual for a miner to lose 40–60% of his production in exchange for the rental of equipment. Mr. Keita indicated that, contrary to previous thinking, the artisanal miner no longer discovers deposits. He is therefore forced to return to prior work sites. These sites are usually quickly depleted because of population pressures.

Mr. Keita listed strategic objectives and related actions, which have been identified by the national validation workshop. These include capacity building, maintaining social cohesion, increasing land farming, developing access to micro-credit to improve technology and investment, and stimulating rural capital formation. However, no organisation was identified to implement the identified actions.

The speaker also provided an update on the community support programme of the Sadiola Mine (Anglo Gold). The programme has several elements, one being to support the local *orpailleurs*. The objectives are to improve gold miners' livelihoods, develop entrepreneurial skills, reclaim degraded lands, and better integrate the sector into overall rural income generation by creating links.

The Sadiola Programme elicited many questions. It was felt that it was a very interesting initiative that should become the starting point for long-term poverty reduction. To maximize sustainability, it was felt that mining companies should integrate their social activities within government sectoral programmes, particularly, education and health. Participants were of the view that the resulting income-generating activities should stimulate self-reliance in rural areas, and lessen economic dependency on mines. The discussions also focused on the perceived shortcomings of the traditional *orpaillage* organization. Many members of the audience believed that it was keeping the *orpailleurs* in their poverty trap; it was noted that women and children were the first victims of the system. Finally, some participants felt that governments, NGOs and community-based organisations should strengthen their cooperation.

Ghana

In the absence – for logistical reasons – of the Ghana team, Ms Beatrice Labonne read to the audience the major findings of the Ghana case study and national workshop recommendations. She noted that the workshop had been officially opened by Mr. Benjamin Aryee, Chief Executive of the Minerals Commission of Ghana. In his address, Mr. Aryee had emphasized that the key goal of the country was to make indigenous small-scale mining more sustainable, and the Commission had been pushing for the establishment of alternative livelihoods or local economic development projects. Efforts to reduce land use conflicts were also made possible by providing more land for ASM activities. The goal of the 1989 regularization of the ASM sector was not to increase production but rather to reduce its ill effects, in particular, environmental impacts.

The Report stated that, in spite of its economic and fiscal importance, the formal and industrial mining sector had yet to make an impact on rural transformation, and that Ghana remains an agriculture-based economy. The Report also revealed that the Ghana Poverty Reduction Strategy had not identified the ASM sector among the vulnerable groups. Both ASM miners and *galamsey* (*i.e.* those working without mining titles) had not been properly studied as a group, nor were they considered "vulnerable".

The Report emphasized that ASM regions have a relatively underdeveloped infrastructure, and that agriculture had become less attractive to younger people due to the depressed prices of farm products. The high cost of living in ASM regions has led to a hand-to-mouth lifestyle. Moreover, the inadequate marketing facilities were hurting both miners and other community inhabitants. The community members were unhappy with the level of support provided by local governments. Although agriculture and mining complement each other, and should be equally supported, it was reported that community members were unable to take advantage of this. The community urged the government to take the necessary actions. According to the Report, the community has a very limited vision of market opportunities with respect to alternative livelihoods. This limited vision is regarded to be the most negative factor for increasing community self-reliance. For Ms. Labonne, the Ghana Report and workshop findings were important, as they made clear that ASM could not be addressed in an isolated manner. Ghanaian decision makers and stakeholders well understand that efforts should be increasingly directed towards building synergy between the various rural sectors. Non-ASM activities should become equally attractive economic opportunities to stimulate development and economic growth. Diversified rural livelihoods are the goal.

The need to improve community assertiveness is well understood by the Ghanaian Government. The Minerals Commission and large mining companies are increasingly using participatory approaches to empower communities, and help them to better assess their vulnerability, capacity and needs.

The following five key actions were identified by the national workshop: obtaining baseline information on poverty in ASM regions; reviewing options to address the issue of poverty/sustainable livelihoods; identifying institutions and their roles at the meso level; providing sets of institutional, technical, financial and other support measures/guidelines necessary to identify the viability of alternative income-generating activities, as well as identifying funding schemes; and establishing a steering committee with members from several ministries at the macro and meso levels. As the floor was opened for questions, the land use issue attracted the attention of participants because in Ghana, there are currently land conflicts between large- and small-scale miners. To resolve disputes of this nature, many governments have had to make additional efforts to reconcile modern land titles with traditional rights. The marketing channels need to become more transparent, and miners should be given more choices to avoid the creation of de facto buying cartels. The ASM sector should interface with other rural sectors, and policies should be practical, taking into consideration the implementation weaknesses of the majority of government agencies.

As mining is finite, some participants supported the creation of a social capital fund. Miners should be sensitized to the limitation of the ASM sector, and look for alternative and complementary options.

SESSION II: PRESENTATIONS ON CASM, CDE AND ON THE HIV/AIDS CHALLENGE IN THE ASM SECTOR

CASM

Mr. Jeffrey Davidson, a Senior Mining Specialist at the World Bank in Washington DC, made a presentation on Community and Artisanal and Small-Scale Mining (CASM). His presentation was appropriately titled "Putting CASM to Work in Africa". He explained that the CASM mission was to "reduce poverty by supporting the integrated sustainable development of communities affected by, or involved in, artisanal and small-scale mining in developing countries". CASM was established with four main goals in mind: (1) mitigating or eliminating the negative environmental, social and cultural effects of ASM on affected communities; (2) reducing occupational health and safety risks to miners; (3) improving the policy environment and institutional arrangement governing ASM; and (4) increasing productivity and improving the livelihoods of miners.

Mr. Davidson echoed Mr. D'Souza's presentation, as he spelled out some of the ASM realities in socio and economic terms. It is estimated that 80 million people worldwide are directly or indirectly dependent upon ASM activities for their livelihoods. As their numbers are growing, however, miners are finding it increasingly difficulty to maintain even a subsistence lifestyle.

Mr. Davidson gave details of the CASM global initiative, its structure, activities, initial budget, and current sponsors, namely, DfID, the World Bank, and UN agencies. CASM is fulfilling a critical coordination function, first, and foremost, in terms of knowledge-sharing to improve the design and delivery of assistance. CASM is also a stakeholder network – a forum that exchanges information on past and emerging experiences – and a facilitator to link projects with funding.

The presenter went on to detail the recent activities and achievements of CASM, including the initiation of the small grants programme. This programme aims at supporting networking, knowledge-sharing activities, and the organisational efforts of miners themselves at the grassroots level. Reference was made to the knowledge centre on the CASM website (www.casmsite.org), which includes contact and bibliographic databases. Very shortly, a database on ASM technical assistance will be posted on the site.

During the subsequent discussions, some participants lamented that CASM was not in a position to provide funding for operational projects. Some felt that the "small grants" were financially meaningless. Other speakers questioned the focus on community, rather than the miners themselves. There was a debate on the evolution of donor policy – specifically, moving away from direct assistance and toward more capital-intensive projects that focus on policy support and monitoring. There were also questions posed on the added-value of CASM creating a special African network collaborating with other networking initiatives.

CDE

The next speaker, Mr. Meite Vaflahi, Coordinator of the Sectoral Coordination Unit, Centre for the Development of Enterprise (CDE) in Brussels, presented information regarding African/Caribbean/Pacific countries (ACP), which contain some 30% of the Earth's proven mineral reserves. Using charts, Mr. Meite provided an overview of the economic significance of the large-scale mining sector in ACP countries. He stated that the CDE focuses on the development of small and medium sized enterprises (SMEs), and that it is currently assisting some 160 enterprises, providing project evaluation, marketing studies, pre-feasibility and feasibility studies, environmental impact assessment, facilitation, and financial assistance. These programmes are grounded in the concept and principles of sustainable development.

Mr. Meite indicated that CDE supports SMEs, the turnover from each of which ranges between 60,000 and 200,000 euros, provided that they agree to advance the goals of sustainable development. He also indicated that CDE assistance yielded some very positive and sustainable results, as demonstrated by several case studies, and that the organization also supports business associations such as SSM organizations.

Many participants were unaware of the CDE programmes, whilst others argued that ASM did not have the capacity to attract the attention, and support, of the CDE.

HIV/AIDS

The presentation of Dr. Pierre Mpele, Team Leader of the inter-country team of West and Central Africa UNAIDS, focused on the HIV/AIDS *problematique*. He outlined the issues as they related to ASM communities. Dr. Mpele was very direct, warning the audience that serious efforts must be made, as HIV/AIDS is a potential time bomb in these communities. He began his presentation with some very dark figures on the spread of the pandemic in Africa. With 28.5 million reported cases of HIV/AIDS, Africa has 70% of the world's recorded cases of the disease. The incidence of the infection is rising in the majority of African countries; one in every four Africans is directly related to the disease. A new report indicates that the spread of HIV/AIDS was widening, and is fuelling the famine in Southern Africa, as many infected farmers are unable to work in their fields. HIV/AIDS is the leading cause of death in Africa, and is reversing the development clock. In Botswana, the average life expectancy has decreased from a peak of 63 years in 1990 to a present-level of 48.

Dr. Mpele stressed that the epidemic was no longer simply a health issue, but a development crisis. Women have also been seriously affected by the disease, and now comprise close to 50% of the global population living with HIV/AIDS. However, the resulting economic crisis is driving an increasing number of women into unsafe sexual liaisons. The presenter described the vicious circle of HIV/AIDS and poverty. Simply put, poverty, mobility and migration, combined with socio-cultural beliefs, are driving the spread of HIV/AIDS in Africa.

Mining regions have been particularly affected. Mines commonly attract single males from poor regions. These workers live in mining camps away from their families during long periods, and, because of their expendable incomes, sex trades inevitably flourish. Dr. Mpele believes that the infection rate can be reduced if a community-based strategy, featuring approaches that favor local development responses, is implemented. Partnerships must also be stimulated through dialogue and exchanges, with the aim of transforming the artisanal miner from a target into an empowered actor.

Whilst acknowledging the colossal dimension of the problem, Mr. Mpele provided several practical recommendations to be addresses at both the national and international levels. First, and foremost, a national multi-disciplinary team should be created to monitor the impact of HIV/AIDS at ASM sites. He also recommended establishing a taskforce among UN agencies such as UNDESA, UNAIDS, ILO, ADB, and UNDP. Such a group could assist governments with facilitating the forging of partnerships, assessing national situations, mobilizing resources, the support of programme formulation, the monitoring and assessment of results, drawing lessons from failures, and disseminating "good practices".

Specific actions were recommended to be undertaken at mine sites. The first recommended task was to assess, and understand, the social, cultural and economic drivers of the epidemic. It was maintained that prevention programmes should be implemented among mining populations, and should incorporate initiatives emphasizing the provision of condoms, voluntary testing, and medical assistance to the sick. Finally, Dr. Mpele stressed that HIV/AIDS sensitization needs to be reinforced with poverty reduction.

The participants were taken aback by the frankness of the presentation. After some hesitation, the discussion rolled on. Many participants described what their country was doing to fight the AIDS epidemic. One government appears to be tackling the problem head-on. Many additional questions put to Dr. Mpele pointed to building broad partnerships among ministries, donors, communities and the private sector.

SESSION III: PRESENTATION AND DISCUSSION OF THE DRAFT COMPENDIUM OF BEST PRACTICES FOR ARTISANAL AND SMALL-SCALE MINING, ASM IN AFRICA

The Compendium was commissioned by UNECA. Because of a number of factors, the small-scale mining sector continues to perform poorly, thus trapping most of its participants in a vicious cycle of poverty. However, UNECA's objectives were to disseminate and discuss some of the encouraging efforts that are now being deployed by a number of African governments, including

Ghana, South Africa, Tanzania and Zimbabwe. The Compendium was designed to capture these practices for dissemination across the continent. It should be considered a work in progress that requires continuous updating in tandem with the evolution and dynamics of the sector.

This presentation was made by Mr. Wilson Mutagwaba,[5] who explained that the ASM compendium on "best practices" in Africa was divided into three major topics: background to the project; a general overview of small-scale mining; and selected "best practices" in small-scale mining. While the first two chapters helped to provide a general overview of small-scale mining in Africa, the third chapter (*i.e.* "best practices") featured the following topics:

- Criteria for the Selection of Best Practices;
- Mining Policy;
- Mining Legislation;
- Technology;
- Health and Safety;
- Minerals Marketing;
- Institutional Capacity;
- Access to Credit and Finance;
- Technical Assistance Programmes;
- Women in Mining;
- Child Labour;
- Research and Development; and
- Co-operation between Small and Large-scale Miners.

The presentation on "best practices" was made under the following topics: Policy Formulation and Legislation; Minerals Marketing; Financing; Environmental Management; Health and Safety; Technology Development and Utilisation; Capacity Building; Technical Assistance; Gender Mainstreaming and Reduction of Child Labour; and Promoting Co-operation between Large- and Small-Scale Mining. Overall, participants appreciated the efforts made to put together such a comprehensive list of "best practices". However, some participants – especially those from French speaking Africa – were concerned that the compendium had only been produced in English, thus limiting its full utilization. UNECA, therefore, was called upon to ensure that the Compendium was translated in French. Moreover, most participants wanted to know the criteria used to evaluate and measure the success of the selected practices. More specifically, it was noted that the listing of "good policies" and legislation was a substantially different exercise to putting measures into practice. It was therefore judged important to detail the selection processes of the presented "best practices".

Based on this request, participants were informed that UNECA was prepared and would incorporate a section on "success factors" into the Compendium. Other participants expressed concerns that some practices that have been

[5]A UNECA consultant and Managing Director of MTL in Tanzania, who provided the text on the Compendium.

successfully implemented in one country may, in fact, fail in another; there was consensus regarding the need to discuss these differences. It was explained by the presenters that some failures of the practices were not because the practices were "bad" but rather ineffective because of other prevailing socio-economic influences within the recipient country. The case of the Shamva Mining Centre in Zimbabwe was given as an example. Specifically, it was explained that the Centre, the operational failures of which are now well-documented, is a "good practice", replicated by a number of other countries. However, it has proven to be minimally ineffective in the Zimbabwean context strictly because of national economic difficulties. It was also explained that the inability to translate "good practices" on paper into workable initiatives is a result of many factors, including political will, and the existing economic and fiscal policies (in the country). Most participants identified the inclusion of ASM policies in countries' rural development programmes as a means of ensuring the integration of ASM activities into the rural economy. As an example, the case of the Sadiola Mining Project was given, which has been able to generate alternative employment opportunities for communities by integrating small-scale mining and rural development. Participants also cited the approach being taken in South Africa by Mintek. With financing from various state grants, Mintek assists developing rural communities by putting much emphasis on beneficiation and added value activities.

In conclusion, and in accordance with the demands made by many participants of the Compendium, a special section is being added that describes the success factors that have been used to select "best practices". In short, specific measures were used to determine each. For example, the assessment of "best practice mining policy" was evaluated using the following indicators:

- The simplicity and transparency of licensing systems for small-scale mining;
- The applications used for licenses handled by district and regional offices;
- Licensing systems for minerals trading handled at the regional level; and
- The simplicity of environmental management regulations specific to small-scale mining.

In accordance with these indicators, the mining policy and legislation of Tanzania, for example, was determined to be a "best practice".

Participants called on governments to put in place supportive mechanisms that would enable such "best practices" to be made a reality on the ground. Participants agreed that ASM could be made to contribute positively towards viable community life and economies, particularly in rural areas.

SESSION IV: BREAK-OUT GROUPS, VIEWS, CONCLUSIONS AND THE WAY FORWARD

In order to draw conclusions and propose novel, yet, doable, policy options, the participants were divided into two linguistic groups. For each of the two groups, the main objective was to make recommendations, and articulate actions and tasks to reduce poverty and improve the livelihoods of ASM communities. Each group worked independently on the basis of a common "Group Task" sheet, which was provided as a guideline.

The recommendations of the Yaoundé Seminar are embodied in the following vision statement: "Contribute to sustainably reduce poverty and improve livelihoods in African artisanal and small-scale mining communities by the year 2015 in line with the Millennium Development Goals."

It was admitted that ASM continues to operate poorly, holding miners and their communities in a poverty trap. A consensus developed among the seminar participants that poverty reduction should become the top priority. Poverty and its social corollaries, such as abject working conditions for children and the spread of diseases, should be addressed head on. This implies a genuine paradigm shift for the ministries in charge of the management of the mining sector and the ASM sub-sector. If a minority of African countries have already embarked on this challenging path, the multi-dimensional tenets of poverty reduction are not being well-understood by the majority.

The seminar provided a frank forum for debating these new realities. The recommendations put forward point to a new and improved direction. It was agreed that more emphasis should be placed on people-centered and participatory projects, whereby the *real* needs of the community are addressed in policy mindful of the community's strengths, assets, absorption capacity, and internal and external vulnerability. Conversely, the traditional ASM staple projects – i.e. sectoral- and capital-intensive programmes – should only be funded if they fall within, and strengthen the overall, poverty reduction actions of the country. The discussions clearly indicated that although they had high expectations, these costly projects have yielded either insignificant or unsustainable results.

Mining authorities may be compelled to revisit their contribution and commitment to poverty reduction in order to become an active partner in the Poverty Reduction Strategy Paper (PRSP) process. This commitment goes beyond the provision of fiscal revenues, foreign exchange, and the provision of mining employment. Poverty reduction and its multi-dimensional characteristics and implications should be mainstreamed into sectoral policy. Conversely, the national strategy for poverty reduction should not reduce the mining sector to a "cash cow". Multifaceted mining development should be fully integrated into the governance structures and its poverty reduction platforms. Governance structure should be adapted to facilitate this integration.

SUMMARY

The Yaoundé seminar on ASM, the aim of which was to "Contribute to sustainably reduce poverty and improve livelihood in African artisanal and small-scale mining (ASM) communities by the year 2015 in line with the Millennium Development Goals", provided consensus on many of problems plaguing ASM today, and, perhaps more importantly, provided recommendation for possible directions forward. This chapter has described many of the dialogues at the seminar. It is the intention of this concluding section of the chapter to summarize, mainly in bullet-point format, the chief activities and themes of the conference, the recommendations made by its delegates, and proposed undertakings of the upcoming years.

Goals for institutions

- To acknowledge and reflect upon ASM sectoral issues in national legislation and codes.
- To mainstream poverty reduction strategies into mining policy inclusive of ASM policies.
- The integration of ASM policy into the Poverty Reduction Strategy Paper process, with linkages to other rural sectors, and to develop a strategic framework for PRSPs.
- Revisit the existing thinking on ASM legislation (namely, traditional land rights, and the modern land use legislation nexus) and role of the central government.
- Strengthen Institutions:
  - Improve the availability of appropriate technologies
  - Develop appropriate analytical and business skills
- Undertake necessary reforms for the ASM sector: improve policies; institutions; processes and livelihoods; reduce child labour; ensure gender equality; improve health and safety; forge partnerships; promote the sustainable use of natural resources; increase infrastructure development; and improve land use management.

Challenges facing the ASM sector

Delegates identified the following as the main challenges facing the ASM sector:

- Dwindling rural livelihood choices in a marginal environment in remote regions.
- Increasing numbers of people seeking a livelihood in ASM.
- Limited public budgets and competing needs.
- Increasing poverty exacerbated by HIV/AIDS/STDs, natural disasters, etc.
- Increasing occurrence of child labour.
- Severe gender inequality.
- Increasing pressure on available resources (institutions, land, mineral resources, etc.).

What to do? Governments and development partners

- Formalize government commitment to ASM issues.
- Revisit mining policies in order to assess their capacity as an engine for poverty alleviation (link to the PRSP process).
- Increase the profile of ASM in International Financial Organizations (IFIs) and donor agencies.

- Undertake necessary reforms:
  - Implement appropriate legislation for ASM;
  - Revisit existing ASM policies and legislation with regard to traditional land rights on modern land use legislation, and the role of central government: update rules, regulations, and legislation;
  - Adopt appropriate and enforceable health and safety guidelines;
  - Adopt appropriate and enforceable environmental guidelines;
  - Establish partnerships with NGOs;
  - Ensure gender equality;
  - Launch child labour reduction programmes;
  - Provide credit facilities, and co-operative saving schemes by making available credit and loans schemes, micro credit, and credit co-operatives; and
  - Ensure free and equitable markets.
- Improve the availability of appropriate technologies.
- Facilitate institutional strengthening and community training.
- Ascertain areas suitable for ASM activities (improved knowledge of mineral resources) and improve methods of exploration, extraction, processing and marketing in order to maximize the efficiency and effectiveness of ASM as a business venture.
- Community health issues: establish HIV/AIDS and STD community health awareness programmes.
- Stimulate the formation of stakeholder partnerships (government at all levels, NGOs, banking organizations, professional organizations, mining companies, etc.).
- Identify alternative livelihoods strategies:
  - Integrate the ASM sector into rural community development programmes;
  - Stimulate capacity-building, and technical and organizational development;
  - Facilitate access to, and the development of, basic social services and transport infrastructure; and
  - Streamline marketing channels.
- Facilitate community-led activities:
  - Execute a sensitization and empowerment campaign to promote community organization and micro-business development;
  - Develop analytical and business skills;
  - Establish ASM co-operatives and associations;
  - Implement community-based saving plans for productive investment;
  - Remove gender-based constraints and appoint women leaders to stimulate alternative income generating activities;
  - Implement health cooperatives for the prevention and care of sick people, particularly those living with HIV/AIDS;
  - Increase awareness, and have more stringent law enforcement and monitoring coupled with the generation of alternative income opportunities;
  - Increase family support services and the provision of affordable education to reduce child labour at ASM sites; and
  - Build community-based partnerships with local authorities, and the local private sector and opinion leaders.

Tasks for international stakeholders (private sector, IFI, donors and NGOs)

- Identify and disseminate best/good practices regulations;
- Present the recommendations/vision statement of the Yaoundé Seminar to the World Bank/EIR Regional consultative workshop in Maputo, Mozambique, 13 January 2003 (UNECA–UNDESA);
- Establish a Yaoundé communication network through CASM and encourage other countries to join (March 2003);
- Identify available resources for ASM support (CASM);
- Review existing baseline surveys to assess relevance to the "Yaoundé vision statement" in selected countries (CASM and UNDESA, September 2003);
- CASM AGM and learning event in Africa, September 2003;
- Identify key stakeholders (affected, interested, beneficiaries, providers, developers, donors) to build the Yaoundé Network by August 2003;
- Establish an inter-agency (UNAIDS, UNDESA, ECA, etc.) working group on HIV/AIDS in mining by August 2003.

National level tasks for stakeholders, governments, the private sector, NGOs and CBOs

- Lobby and increase the profile of ASM issues within governments, the private sector and ASM organizations.
- Convene national workshops:
  - Build partnerships with government and the private sector; and
  - Promote the collection of baseline survey data to help identify key issues, both positive and negative, and establish common benchmarks.
- Collate and consolidate existing information (Liaise with global ASM networks such as CASM).

REFERENCE

Mining, Minerals and Sustainable Development (MMSD) (2002). Report of the Regional MMSD Process, University of the Witwatersrand, South Africa.

10

Addressing Labour and Social Issues in Small-Scale Mining

NORMAN S. JENNINGS

Small-scale mining – labour-intensive, non-mechanized, artisanal mining – is estimated to employ between 11.5 and 13 million people world-wide, mostly in developing countries and often at a subsistence level. This number is, if anything, likely to increase over time. Moreover, up to 100 million people are estimated to depend on small-scale mining for their livelihoods.

The number employed in small-scale and large, formal mining is about the same but there are important qualitative differences. Jobs in the small-scale mining industry are often precarious and far from conforming with international labour standards; small-scale mining is seldom a source of what the ILO calls "decent work".

Like most economic activities, small-scale mining has positive and negative aspects. It is closely linked to economic development, particularly in rural areas of developing countries where it is prevalent; it can help stem rural-urban migration; it can make a major contribution to foreign exchange earnings; it can enable the exploitation of what might otherwise be uneconomic resources; and it can be a precursor to large-scale mining. However, small-scale mining often exhibits many traits that are of concern to the ILO and others who are working to assist it. These include: safety and health (at the workplace and in the community), working conditions, child labour, gender and development issues, labour inspection, training, workers' rights (particularly in terms of the lack of employment contracts or guaranteed wages), environmental concerns, and community impoverishment.

While many authorities seek to discourage small-scale mining, labelling it as dirty, dangerous, and damaging, efforts to eliminate it risk foundering on the bedrock of economic necessity that drives people to take up, and continue to participate in, small-scale mining altogether. For impoverished communities, it holds the promise of cash earnings, but it tends to reinforce a vicious circle of appalling working conditions, significant environmental damage, and poverty in the very communities whose survival depends on sustainable small-scale mining.

This chapter examines briefly the extent to which the different labour and social issues that affect small-scale mining are being, or should be, addressed, and how action has evolved over the past few years so that small-scale mining can contribute to mineral production in ways that provide decent work and lasting socio-economic benefits to those involved and their communities.

WHAT IS PRODUCTED BY SMALL-SCALE MINES?

Small-scale mining is best known in relation to the production of gold and other high-value metals and precious stones; it is a "gold-rush" type of activity that is newsworthy. Furthermore, the more valuable the product, the better the prospects for a windfall, and the more likely that miners will be willing to work far from home, even in cases without work contracts or regular wages. Often, in exchange for hard labour, food and shelter are provided, as well as promise of a percentage of the value of the material found; debt and poverty are often the result.

Most minerals are mined on a small scale, albeit to varying degrees. Moreover, most quarrying for building and road-working materials, especially in developing countries, is carried out at small-scale mines. At such operations, people tend to work for wages, generally based on production, for the quarry owner or the contractor who manages the concession. In some countries – mainly in Asia – hundreds of thousands work at small-scale coal mines. These workers also tend to be paid a wage, but there is often at least one intermediary between the miner and the mine owner. In such cases, wages can be heavily discounted. Responsibility is also diluted – for *e.g.*, as far as working conditions, and occupational safety and health are concerned – effectively to almost nothing.

As far as mineral production from small-scale mines is concerned, the long-standing estimate of 15–20 per cent of global non-fuel minerals appears to be of the right order. This output, which excludes coal, is produced by a workforce of about the same size as that of the large-scale sector (which produces five to six times as much). In terms of high value products, approximately one-sixth of the world's diamonds are produced at artisanal and small-scale mines, accounting for some US$1.2 billion a year; gold valued at over US$200 million a year is produced by small-scale mines in at least six countries; and, in certain countries, such as Zambia, some 15 per cent of emeralds are produced at small-scale mines. In some countries, small-scale mine production exceeds that of large mines, generally because of the nature of the deposits, or as a result of falling prices. In the Democratic Republic of Congo, for example, even after the price of coltan dropped dramatically, the "very low" labour cost at small-scale mines kept it profitable (UN report, 2002, p. 21).

EMPLOYMENT AT SMALL-SCALE MINES

The difficulty in estimating the number of people engaged in small-scale mining arises from its informal nature in many locations, the lack of official data,

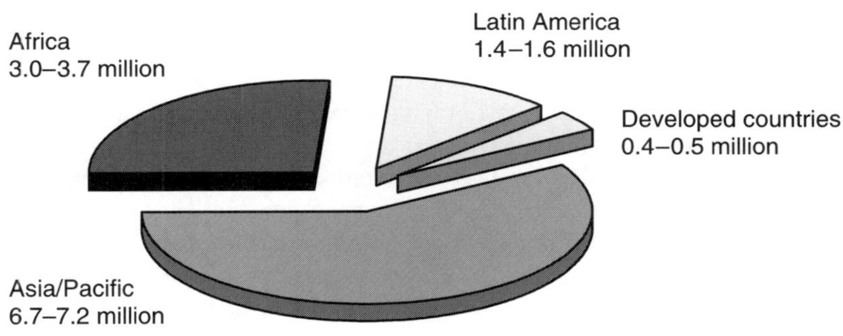

Figure 10.1 Employment in small-scale mining.

the transient and seasonal nature of its activities, and problems of definition. Estimates by ILO in 1999 of somewhere between 11.5 and 13 million people directly engaged in small-scale mining (Fig. 10.1), with 80–100 million people directly depending on it, have not been seriously challenged (ILO, 1999, p. 6).

The sporadic data that have been produced subsequently show that it is a reasonable estimate, and one which is likely to increase rather than fall; in any event, it is significant, even when compared to the workforce of formal mining operations, where over 3.5 million jobs were lost during the 1995–2000 period alone (ILO, 2002, p. 3). The small-scale mining workforce often comprises women and children (the former sometimes in the majority), is rarely organized by trade unions, and often has no recognizable contract or guaranteed wage.

AN ILO PERSPECTIVE

In 1999, through a report and a tripartite meeting, the ILO examined a variety of social and labour issues at small-scale mines. The following elements of the conclusions from the meeting highlight the action that a representative group of stakeholders considered is needed to bring small-scale mining into the mainstream as a source of decent work and sustainable livelihoods. The conclusions also show the extent to which assistance is required – from governments, employers' and workers' organizations, donors, NGOs, development agencies and financial institutions – to achieve sustained improvement (ILO, 1999 May, p. 60), Some of this assistance is now being provided.

Safety and health

Health and safety are clearly important for both small-scale mineworkers and their communities. While it is impossible to say how many deaths and injuries occur in small-scale mining, due to under-reporting and the clandestine nature of much of the work, the risks of fatal and disabling accidents are high. The risk of mining accidents, however real, is not as acute as the health hazards and sickness found in many mining communities. Silicosis and mercury poisoning are occupational hazards of many small-scale miners, and extend to the entire community, including wives and children. The widespread lack of access to health care makes it impossible to gauge the extent of these diseases. Other infectious diseases are abound (See Chapter 13 for an overview of diseases in artisanal and small-scale mining communities).

The lack of reliable data limits the development of effective assistance programmes and the improvement of Occupational Safety and Health (OSH) performance; the major OSH problems are well-known. Governments should be encouraged to establish a regime of effective reporting on OSH in small-scale mining, placing emphasis on the prevention of accidents and disease. Large mining companies and workers' organizations should share their methods and experiences with governments dealing with small-scale mining. They should also co-operate with small-scale miners and their communities to increase awareness of the benefits of safe and healthy mining, and set appropriate examples. Many improvements in safety and health in small-scale mining require minimal investment, which, in any case, is quickly paid back via higher productivity. Changes for the better are frequently achieved merely by modifying work practices and work organization.

Women and small-scale mining

Although women constitute as much as 50 percent of the small-scale mining workforce – compared with typically 5–10% of the formal mining workforce (including clerical and administrative jobs) – they do not receive 50 percent of the rewards. There are many barriers and constraints to women's participation in the industry, not all of which are gender-based. However, a female-specific approach may be necessary to overcome existing imbalances between men and women in some aspects of small-scale mining.

Measures to improve women's participation include the provision of schooling for their children, and education for women themselves. Governments have the responsibility of providing these services but in doing so, they may need to enlist the support of social partners and other relevant bodies. In acting to increase women's participation in small-scale mining, governments should ensure that they do not inadvertently introduce other forms of discrimination. Providing easier access to finance might quickly enable women's participation to increase.

Child labour and small-scale mining

Child labour, which is widespread in much of small-scale mining, is closely linked to poverty. The more remote or informal the activity, the more likely children are to be involved. Reliable data are scarce but this does not obscure the significance of the problem. Make no mistake; children undertake *all* mining activities, often for little or no pay, and to the detriment of their growth, intellect and health. They face a bleak future unless steps are taken to reduce poverty, increase educational opportunities, and remove them definitively from this hazardous work. Most child labour in small-scale mining can be classified among the worst forms of child labour in terms of the ILO Convention (No. 182).[1] Such child labour should be eliminated, and the ILO has programmes in all regions to tackle child labour at small-scale mines.

Employers' and workers' organizations can help small-scale mining to become more efficient, productive and prosperous, which would eliminate the "need" for child labour. Governments must improve data collection using, where necessary, the resources and expertise of IGOs and NGOs. The immediate removal of children from the most hazardous work should be accompanied by measures to ensure that family income does not suffer. This includes improving mine productivity, safety and health. For example, if an adult is injured or becomes sick, children may well be brought back into the workforce, despite having been removed under the terms of the Convention.

Legislation

The legal limbo of much small-scale mining prevents its recognition and leads to large economic losses. The legalization of small-scale mining is a key step toward achieving sustainability. However, to be effective, legislation must be even-handed, pragmatic and user-friendly. The transparent issue and transfer of permits, and the minimizing of bureaucracy, are important areas in need of attention. When regulations are being drafted, those most concerned should be consulted and relevant examples from elsewhere taken into account. Most importantly, regulations should be applied fully and without distinction. The UN report on the Democratic Republic of the Congo cited earlier shows the extent to which illegal activity, much of it in small-scale mining, can debilitate a country's resources.

LINKS BETWEEN LARGE AND SMALL MINES

There is often tension and a clash of interest between large and small mining parties. There are also many challenges facing companies that open their doors to small-scale mining. These include: avoiding an influx of small-scale miners and people looking for work at the large mine who might turn to small-scale mining; ensuring that sufficient resources exist to enable small-scale mining to continue for an agreed period; and getting government agencies to accept and meet their responsibilities towards the community. While private mining companies often play a role in community development, they cannot be expected to replace the government. The expanding search for exploitable mineral resources, particularly in developing countries, means that there is potential for increased tension over mineral resources. Therefore, care is needed to establish the right relationship between small-scale and large miners.

---

[1] The Worst Forms of Child Labour Convention (No. 182), 1999. By November 2002, 132 of the ILO's 175 member States had ratified this Convention. The text of the Convention is at: http://www.ilo.org/ilolex/english/convdisp1.htm.

The technical information, best practices, OSH data and information that exist throughout the mining industry should also be made available to those involved in small-scale mining. Employee and workers' organizations, particularly at the local level, should assist small-scale miners improve their operations and the social conditions of their communities. Donor assistance to mobilize efforts on behalf of small-scale mining should be sought. Those concerned should see how they could act as mentors to small-scale mines.

THE EVOLUTION OF ASSISTANCE TO SMALL-SCALE MINING

Although small-scale mining activities have remained pretty much the same for decades, waxing and waning with, for example, the availability of resources, price and the health of the large-scale formal mining sector, approaches to assisting it have evolved over the last 20 years or so. Essentially futile attempts to define and compartmentalize small-scale mining in the 1970s were followed by a technical, productivity-linked approach in the 1980s, and by a broader socio-economic approach that involved environmental concerns in the early-1990s. A focus on poverty, gender and child labour issues, culminating in sustainable livelihood and community issues, was the approach at the turn of the century, stimulated by the ILO report and meeting.

A hitherto largely ad hoc approach has constrained efforts to promote better organization and work practices, increase the productivity of small-scale mines, and lessen adverse labour and social effects. Erratic policy and decision-making has led to confusion among administrators and managers of large and small mines, and has sometimes caused conflict at mining locations. A lack of co-ordination in the provision of assistance has not helped. Most projects to assist small-scale mining have failed or have not led to lasting improvements because they treated small-scale mining as a sub-set of large, formal mining. Most of the emphasis has been on finding technical solutions to mining and processing problems, with little attention being paid to the underlying economic, social and labour issues. Another factor in their relatively short-lived success has been the low priority given by a number of governments to this sector.

The ILO meeting pointed out that any assistance to small-scale mining must take into account the immediate needs of those involved and their communities. At the national level, government leadership is central, particularly as far as permits, exploration, credit and marketing are concerned. Governments should consider setting up national or local consultative groups that could develop assistance and direct it to where it is needed most. Governments should also be aware of the effects of their fiscal and other policies on small-scale mining. Employers' and workers' organizations should use their considerable expertise to assist small-scale mining – especially in developing countries – by providing information, resources, training and technical assistance, such as the adaptation of technology or equipment for use at small-scale mines, and marketing. The ILO and other agencies should ensure that information on successful small-scale mining projects is collected and disseminated. Technical approaches to increase productivity can pay long-lasting dividends. Productivity growth is the major source of sustained improvements in real incomes which, in turn, raise the demand for goods and services in the community and beyond. It will also counterbalance any income effects of removing children from the small-scale mining workforce.

The evolving focus on community issues was further stimulated by the establishment of the Communities and Small-Scale Mining network (CASM)[2] in 2001. CASM was launched as a multi-donor networking and co-ordination facility. Its mission is to reduce poverty by supporting integrated sustainable development of communities affected by, or involved in, artisanal mining in developing countries. CASM functions as a knowledge-based community. With limited funds at its disposal, it supports and promotes the development of projects and approaches, by individuals, communities, and institutions that will directly or indirectly contribute to the reduction of poverty and the construction of more viable livelihoods in rural areas where small-scale mining is a significant activity.

In 2001–02, the Mining, Minerals and Sustainable Development (MMSD)[3] project spent considerable time discussing artisanal mining and raising its profile within the larger, formal mining sector. One of the important outcomes of the Global Mining Initiative (GMI) Conference in Toronto in May 2002, which focused on the results from the MMSD project, was the "Toronto Declaration" of the newly formed International Council on Mining and Metals.[4] As far as small-scale mining is concerned, an important element of the ICMM Declaration, which reflects the view of much of the mining industry, was the following statement: "ICMM recognizes that: … Artisanal, small-scale mining [and orphan site legacy issues] are important and complex. However, they are beyond the capacity of ICMM to resolve. Governments and international agencies should assume the lead role in addressing them." Thus, when seeking to address small-scale mining issues, the corporate view is that the initiative should come from elsewhere – a change of emphasis from the outcome of the ILO meeting some three years earlier.

This institutional reluctance of the mining industry to get involved in what are perceived as development issues is understandable. However, at the practical level, including under the umbrella of CASM, individual companies are making a considerable effort to provide a sustainable foundation to small-scale mining activity on, or near, their operations. Nonetheless, assistance to small-scale mining is still sporadic and, apart from the activities of CASM, poorly co-ordinated.

Until the link between small-scale mining and poverty was highlighted in the late-1990s, neither major donors nor governments showed much interest in assisting it. This gave little incentive for banks and other organs at the local

---

[2] www.casmsite.org.
[3] www.iied.org/mmsd/finalreport/index.html.
[4] www.icmm.com.

level to participate. Moreover, the assistance that was provided by donors and development agencies tended to go to relatively few countries, and had mixed results. Small-scale mining was just not high enough on the agendas of governments and international donors to attract widespread, and sustained, attention. However, the growing importance of addressing sustainable development issues and reducing poverty across the board has continued to raise the profile of small-scale mining as a sector that, with assistance, could provide minerals and a basis for sustainable rural livelihoods.

In September 2002, the *Plan of Implementation* adopted at the World Summit on Sustainable Development noted that enhancing the contribution of minerals and metals to sustainable development includes action at all levels. This involves fostering sustainable mining practices through the provision of financial, technical and capacity-building support to developing countries and countries with economies in transition for the mining and processing of minerals, *including small-scale mining* (emphasis added), and, where possible and appropriate, improving value-added processing, upgrading scientific and technological information, and reclaiming and rehabilitating degraded sites. This *Plan* is directed at all those involved, including large mining companies.

TOWARDS BEST PRACTICE

If small-scale mining is to be a sustainable source of minerals and mineral-based products, a conscious effort must be made to ensure that it is both an anchor for, and a springboard to, other productive activities. If this is not feasible, efforts should focus on finding other activities for those involved, rather than allowing them to become marginal, counterproductive, and a major social and economic burden. There are plenty of examples that illustrate that as soon as there is an alternative to small-scale mining, most people will take it.

A sustainable livelihoods approach to small-scale mining that is being developed and tested in a number of African countries includes the following elements:

1) Mainstreaming poverty alleviation into national policy making in all sectors, including minerals.
2) Promoting small-scale mining as a catalyst and an anchor for other productive activities to stimulate the development of complementary and alternative productive ventures necessary for sustainable poverty alleviation.
3) Placing people first through pro-poor strategies and participatory methodologies aimed at strengthening the organizational capability of grassroots communities, therefore favouring a bottom-up approach.
4) Reversing the focus from "hands-on state intervention" to the creation of private enterprises, including services, and especially micro-enterprises and co-operatives (Labonne, 1999).

Approaches such as this, which emphasize the need to improve the productivity and profitability of small-scale mining and the health and safety of its workers, will be instrumental in linking it to rural development and ensuring that the industry can coexist productively with other economic activity, including large mining operations. These and related issues were examined at a seminar in Yaounde in November 2002 with the objective of identifying best practices and building the sustainable livelihoods of small-scale mining communities (Chapter 9).[5]

CONCLUSIONS

Policies and programmes are needed to put small-scale mining on a stable footing so that it can provide decent work for the millions of workers and entrepreneurs involved, and thereby generate lasting community benefits. Small-scale mining needs to be brought into the mainstream by: ensuring title and property rights; enabling access to finance; addressing labour and social issues; improving working and living conditions; minimizing environmental impact; enabling access to technical and business skills; and linking it to the broader economy of the community concerned.

The social, labour, environment and economic issues that underpin small-scale mining are interrelated. Thus, when assistance on these issues is being developed and implemented, due consideration should be given to the other issues – namely, legal, financial, technical, cultural and political – that affect it. Conversely, other assistance should anticipate and take heed of any social and labour implications for small-scale mining. For assistance to succeed in a lasting way, there must be immediate, tangible benefits for target groups. It is therefore vital that all the stakeholders play an active, practical and co-ordinated role.

It is only through co-ordination, commitment and co-operation at all levels, and by increasing the resources available to assist small-scale mining, that assistance can be delivered to where it is needed in an efficient and effective way. The respect for basic rights stimulates productivity and thus job growth and development. Perhaps this could be a good place for all those concerned to start when setting out to assist small-scale mining.

REFERENCES

IIED/MMSD (2002). *Breaking new ground: Mining, minerals and sustainable development*. Report of the Mining, Minerals and Sustainable Development Project, London, Earthscan.

ILO (1999). *Social and labour issues in small-scale mines*. Geneva.

——(May, 1999). *Note on the Proceedings*, Tripartite Meeting on Social and Labour Issues in Small-scale Mines, Geneva, 17–21 May 1999, 60 pp.

[5] Economic Commission for Africa: *Draft compendium on best practices in small-scale mining in Africa*, Doc. ECA/RCID/003/002 (Addis Ababa, 2002), 83 pp. (This paper was written before the seminar took place.)

——(2002). *The evolution of employment, working time and training in the mining industry*, 83 pp.

Jennings, N.S. (2000). Small-scale mining: a sector in need of support. In *Mining Environmental Management*, Jan 2000 (London, Mining Journal), pp. 17–18.

United Nations (2002). *Final report of the panel of experts on the illegal exploitation of natural resources and other forms of wealth in the Democratic Republic of the Congo*, Doc. S/2002/1146 (New York, 16 Oct 2002), 59 pp.

UNECA (2002). *Draft Compendium of best practices in small-scale mining in Africa*. Document ECA/RCID/003/002 (Addis Ababa, 2002). 85 pp.

World Summit on Sustainable Development (2002). *Plan of implementation*. Advance unedited text (Johannesburg, 5 Sep 2002).

11

Women and Artisanal Mining: Gender Roles and the Road Ahead

JENNIFER J. HINTON, MARCELLO M. VEIGA AND CHRISTIAN BEINHOFF

A hot, dry wind envelopes a statuesque woman as she kneels over wind-sifted trays of tin-bearing pulverized ore in Uis, Namibia. In Bolivia, a nine-year old girl scrambles down a steep pit wall yet again to refill her bucket of metal-rich sand. And yet another woman stokes the fire in her wood burning stove in the Philippines, releasing the mercury from doré in a poorly ventilated kitchen; the thick black soot coating the kitchen wall contains more than 15% mercury. Up to her knees in muddy water, a woman pans for gold to supplement the meagre family income in a Malian "orpaillage". The faces are as varied as these scenarios but there is one commonality – artisanal mining represents an opportunity. To some, participation is driven by the allure of riches; however, for many women, artisanal mining signifies an opportunity to relieve the strains of poverty.[1]

Artisanal miners[2] employ rudimentary techniques for mineral extraction and often operate under hazardous, labour-intensive, highly disorganized and illegal conditions. Despite these factors, artisanal mining is an essential activity in many developing countries, particularly in regions where economic alternatives are critically limited. The International Labour Organization (1999) estimates that the number of artisanal miners is currently around 13 million in 55 countries, which is roughly equivalent to the global workforce of large-scale mining. From this, it has been extrapolated that 80 to 100 million people

---

[1] The term *poverty* is used in a broad sense in this chapter. As described by Çagatay (2001), poverty refers to "human poverty", which includes lack of assets, dignity, autonomy and time in addition to income poverty.

[2] *Artisanal mining* is used to denote all small-scale as well as medium and large-scale mining that may be illegal or legal, formal or informal. Artisanal mining may be better characterized by a lack of long-term mine planning and use of rudimentary techniques (Hinton et al., 2003).

worldwide are directly and indirectly dependent on this activity for their livelihood. ILO further estimates that artisanal mining activities have increased by up to 20% in the past decade.

Approximately 30% of the world's artisanal miners are women who occupy a number of roles ranging from labour-intensive mining methods to the processing aspect of artisanal mining, including amalgamation with mercury in the case of gold extraction. As processing activities are often conducted in the home, women and their families can be at great risk from mercury poisoning and silicosis. In many cases, the roles of women in artisanal mining communities differ significantly from those of men, and extend well beyond direct participation in mining activities – this added facet brings with it different contributions and a completely unique set of risks and opportunities.

This chapter intends to explore existing and evolving gender[3] roles of women in artisanal mining communities, and provides a rationale and strategy for women to maximize potential benefits from participation in the sector. Women are often overlooked by initiatives and development programmes directed at catalyzing the transformation of artisanal mining. Due to their critical role, not only in mineral production, but also in the development of sustainable communities, combined with their susceptibility to poverty, enhancing the role of women in artisanal mining may be a means to "bridge the gap" between the well-conceived technical and socio-economic changes often prescribed for artisanal mining, and the actual facilitation of positive transformation of the artisanal mining sector. This may be accomplished in a number of ways, including:

- Gender-sensitive technology assistance initiatives;
- Enhancement of other skills, including managerial and accounting;
- Financial support through the establishment of credit lines and micro-lending programmes;
- Support for the acquisition of mineral titles;
- Consideration of women in the development of regulations and policies;
- The awareness of health and safety issues, with consideration of children who may accompany their mothers or take part in artisanal mining activities; and
- The challenge of social norms, which prevent women from benefiting from these activities.

It has been well documented that inequities in political power, distribution of income, capital assets, and access to education and information have resulted in the increased susceptibility of women to chronic poverty. In some cultures, this is exacerbated by the fact that women do not always have control of their earned income or they occupy positions in the unpaid economy (*e.g.* subsistence agriculture, domestic work). Ultimately, it is crucial that women be empowered to

[3] *Gender*, as applied herein, refers to the behaviours, attitudes, values, beliefs, etc. that a particular socio-cultural group considers appropriate for males and females. The authors adhere to the belief that gender roles are fluid and can shift over time, space and in different contexts (Butler, 1990).

transform their skills and capabilities into well-being. Artisanal mining is only one micro-industry; however, as Labonne (1996) articulated "(artisanal) mining... may become a stepping stone towards economic fulfillment, contributing to a better future for women and men in many developing countries".

PARTICIPATION OF WOMEN IN ARTISANAL MINING COMMUNITIES

Artisanal mining communities around the world are diverse, dynamic and distinct – they vary from culture-to-culture, region-to-region and mine-to-mine, and change over the course of time. The women within these communities are also heterogeneous and unique; however, they tend to be engaged in specific roles throughout the world. Typically, they are labourers (*e.g.* panners, ore carriers and processors), providers of goods and services (*e.g.* cooks, shopkeepers) and are often solely responsible for domestic chores. Women's responsibilities in mineral processing activities range from crushing, grinding, sieving, washing and panning, to amalgamation and amalgam decomposition in the case of gold mining. Less commonly, women are concession owners, mine operators, dealers and buying agents, and equipment owners. In many locales, women function in multiple capacities. For instance, a woman working as a panner may also obtain income as a sex trade worker and a cook. Despite the diverse and important roles undertaken by women in artisanal mining, limited reliable information is available on this topic. The numbers of women involved have been estimated in several countries; however, there are very few accounts telling the stories of individual women and a paucity of information concerning the age, ethnicity, status, wealth, and health of both women and men in these communities. Since the participation of women in non-mining activities is often overshadowed by their involvement in mining, their direct and indirect roles in artisanal mining communities are discussed independently below.

Women as miners

Women's direct participation in artisanal mining varies throughout the world. In Asia, generally less than 10% of miners are women, whereas in Latin America, the proportion tends to be higher, at approximately 10–20%. The percentage of female artisanal miners is the highest in Africa, ranging between 40 and 50%. In some regions, the artisanal mining workforce is comprised of 60 to 100% women (ILO, 1999; Amutabi & Lutta-Mukhebi, 2001; Onuh, 2002). Women typically play a much larger role in artisanal mining than in the large-scale mining sector (WMMF, 2000). Statistics on the participation of women derived from country studies commissioned through the Mining, Minerals and Sustainable Development initiative are shown in Table 11.1.

As women often work part-time at informal mining operations, and occupy "ancillary roles" (*e.g.* cooks, service providers), there may be significant discrepancies between the *estimated* and *actual* numbers of women involved in

Table 11.1 Women in artisanal mining in selected countries.

Country	Number of Women	Proportion of Women
Africa		
Burkina Faso	45,000–85,000	45
Ghana	89,500	45
Malawi	4,000	10
Mali	100,000	50
Mozambique	18,000	30
South Africa	500	5
Tanzania	137,500	25
Zambia	9,000	30
Zimbabwe	153,000	50
Asia		
India	33,500	7
Indonesia	10,900	10
Philippines	46,400	25
Papua New Guinea	12,000	20
Latin America		
Bolivia	15,500	22
Ecuador	6,200	10
Total	596,000	

Source: (*after* Hentschel et al., 2002).

artisanal mining (Wasserman, 1999). Further to this, as women are more frequently associated with transporting and processing materials, as opposed to digging, they are not always identified as "miners" (Susapu & Crispin, 2001).

Africa

The involvement of women in Ghanaian small-scale mining has been well documented by Hilson (2001, 2002). Acting as licensed buyers (6%), concession holders (10%), and work group sponsors or participants (15–20%), women comprise approximately 15% of the legal small-scale metal mining labour force in Ghana. Hilson (2001) estimated that illegal involvement in the *galamsey* industry is up to 50%. The involvement of women in industrial minerals (*e.g.* clay, stone quarries, salt) is much greater, with the proportion of women in salt mining as high as 75%. In the Tarkwa Mining Region, Akabzaa and Darimani (2001) observed women working in all aspects of mining, processing, and marketing. As is found in other countries, women predominantly participate in sieving, sorting, the transport of ore and water, and washing, although involvement differs depending on whether activities are legal and a cooperative is present. For example, in the Akoon mining cooperative, due to risks associated with underground mining, women are not directly engaged in mining but are employed as bookkeepers and security guards. At the illegal Cocoase Camp, women pound rocks, and carry ore and water for wages that are 60% lower than those of men involved in ore digging and washing.

In Burkina Faso, approximately 90% of mineral processing activities are conducted by women (Gueye, 2001). Between 45,000 and 85,000 women work in gold mining alone and as many as 45% of all artisanal miners are women. At many alluvial gold mine sites (*orpillages*), women spend countless hours meticulously hand picking out nuggets. Women in Burkina Faso also take on an important role in the industrial minerals sector (*e.g.* sand, gravel, quartz), although there is a possibility that their full participation has not been quantified because of the intermittent nature of their involvement.

Approximately 10% of the Malian population depends on gold mining (~1 million people) (Labonne, 1998). Like Burkina Faso, Mali has a high percentage of women active (~50%) in artisanal mining with an emphasis on mineral processing (90% female) (Keita, 2001). High levels of female participation in this region are thought to be related to strong traditional associations with gold mining in some cultural groups, and droughts, which have driven "non-miners" into artisanal mining.

Women tend to dominate salt mining in many West African countries. In the village of Keana, Nigeria, women are exclusively involved in salt mining (Onuh, 2002). This process involves placing salt-rich sandy soil in pots. Water poured into the pots dissolves the salt and drains through a small hole in the bottom of each container. The salt-rich solution (brine) is collected in a hollowed-out log and heated on a fire in another pot until it crystallizes. Due to revenues generated from salt mining, the women of Keana are renowned for sponsoring their children to attend school.

In the Sudan, it is estimated that 35% and 10% of the gold miners are women and children in the Southern Blue Nile and Eastern Bayuda Desert regions, respectively (UNIDO, 2002). Women's participation in mining in the Sudan has not been ascertained, although it is anticipated that they play similar roles as in adjacent countries.

In a detailed report on artisanal mining in Southern African countries (Malawi, Mozambique, South Africa, Tanzania, Zambia and Zimbabwe), Dreschler (2001) observed that although female participation in artisanal mining is, on average, relatively high (~25%), women are consigned to subordinate or subsistence work. Dreschler attributes this disparity to the fact that, first, women do not participate in large-scale, formal mining, and therefore, do not acquire the same skills as men who more readily find employment at large-scale mines, and, second, women are tied to their households through familial obligations.

Approximately 40,000 people are active in small-scale mining in Malawi (Dreschler, 2001). Although the numbers of women participating in different industry activities is not well-documented, it is known that ~300 women are involved in lime production via a European Union sponsored association,[4] three women are participating in gemstone mining, and a presumably large number are involved in salt mining and illegal stone aggregate "handnapping". The

[4] The Lirangwe (women) Limemakers Association of Malawi has received funding from the European Union to purchase mills (Dreschler, 2001).

Table 11.2 Women's participation in small-scale mining in Tanzania.

Commodity	Direct	Indirect	Total
Gold	8,400	41,810	50,216
Diamond	523	505	1,028
Gemstone	17,866	56,430	74,296
Salt	9,876	7,585	17,464
Aggregates	14	37	62
Dimension stones	9,920	7,699	17,619
Grand total	46,599	114,066	160,685

Source: (*after* Dreschler, 2001).

Malawi Association of Women Miners includes women active in limestone, gemstone, and salt mining.

The majority of the 60,000 artisanal and small-scale miners in Mozambique are men (Dreschler, 2001). As in other locations, female involvement is mainly limited to ore and water transport, washing and panning, in addition to numerous domestic responsibilities (childcare, food preparation, etc.). A survey of mining productivity in Manica found that women produce roughly half as much gold as men (10–15 g per miner per month). This is believed to be mainly due to a lack of tools and knowledge about processing methods (Dreschler, 2001).

Approximately 25% of Tanzania's 550,000 artisanal and small-scale miners are women (Dreschler, 2001). As outlined in Table 11.2, gemstone mining is one of the most important sectors for women, but the gold, dimension stone and salt segments are also significant.

As in other countries, women in South Africa are driven to mining by poverty. Little information exists about women in the small-scale mining sector of South Africa, with the exception of the kaolin miners of Kwa Zulu Natal, where women and children are predominantly responsible for digging and selling clay bricks.

Women in Zambia act as mine owners and workers. An activity of particular importance in the Lusaka region involves crushing and selling marble for use in construction (Dreschler, 2001). Gemstone mining continues to be important throughout the country, although documentation of women's participation is not available.

In Zimbabwe, approximately 1200 of 20,000 legal mining claims have been registered by women (Mugedeza, 1996). In the formal (legal) sector, approximately 10% (or 3000) of miners are women, 70% of whom are full or part mine owners, with the remainder engaged in ore grinding using a mortar and pestle (Dreschler, 2001). Increasing numbers of women are registering claims in the formal sector as the process is responsive to various investors, including women. This contrasts drastically with the informal sector, where at least 50% of the more than 300,000 artisanal miners in Zimbabwe are women (Dreschler, 2001). Most of these women are involved in gold digging and panning (UNIDO, 2002). Women work part-time in the mining sector far more frequently than men.

A study of gender issues in mining in the Mukibira district indicates that women in Kenya play a central role in artisanal mining activities (Amutabi & Lutta-Mukhebi, 2001). The majority (80%) of the 2000 panners in the region are women, with proceeds supporting a population of approximately 10,000. The average gold panning family income is around 5,000 to 10,000 Kenyan shillings per month (US $60–120), which is high compared to those working in agriculture or the civil service. Local and indigenous women have been engaged in panning in Mukibiri for hundreds of years, passing down their expertise to younger generations. Because women are typically regarded as being more honest than men, they are frequently engaged in selling at the markets. Although women's groups have recently formed to market products themselves, most men and women involved in intermediate roles (e.g. buyers) are predominantly of European or Asian descent. Five tonnes of gold are produced annually in Kenya by panning.

Asia

Due to the informal nature and limited number of surveys conducted in Asia, in conjunction with significant variations in involvement from location-to-location and over time, women's participation in artisanal mining has been difficult to assess. In Papua New Guinea (PNG), Susapu and Crispin (2001) infrequently observed women working in artisanal mines, with the exception of family-based mines where multiple generations are involved. In nearby Indonesia, it is only in recent years that women's direct participation in mining activities has been documented. Currently, Indonesian women take part in ore crushing and act as mine owners and operators (Aspinall, 2001). Unlike many other countries, amalgam decomposition is almost exclusively the responsibility of men in both PNG and Indonesia. In Malaysia, licenses to reprocess tin tailings are issued only to women, who use wooden pans (*dulangs*) to recover the low value metal (Cope, 2000). With the exception of some involvement in business aspects of small-scale mining, women in China typically do not directly participate in mining or mineral processing (Gunson & Yue, 2001).

According to Chakravorty (2001), approximately 6–7% (~20,400) of artisanal miners in India are women. Due to their reliability, lower wage rates, and the infrequency with which they indulge in alcohol, women are sought after in mines. Generally, women are engaged in hand sorting and blending, particularly when it is preferred to mechanical means. Women are banned from working underground due to the *Mines Act* of 1952. In a detailed study of the socio-economic profile of women working in small and medium mines in selected regions in Eastern India, a more significant proportion of miners were found to be tribal women – entrance into the sector was primarily attributed to loss of agricultural land due to land use pressures from population growth (Nandi & Aich, 1996). Lower class women, or higher class women living below the poverty line are also involved in mining in India, particularly with loading bauxite ores.

The participation of women in Lao PDR has been well documented by Beinhoff (1998). Upstream of Luang Prabang on the Mekong and north of Houei Koa, Nam Ou, women are involved in excavating up to two metre deep pits in gold-containing sediments. Sediments are transported to the river, where stones are separated by hand and the clayey components eliminated by washing the material in conical pans. More than 80% of the panners are women, who typically handle 100 charges weighing 10 kg each, which equates to a daily throughput of one tonne per day per miner. Gold is difficult to isolate from the heavy metal rich concentrate, and recoveries of 50% are not uncommon. Mercury amalgamation is conducted using barehands and decomposition takes place in open steel pans – children often visit the site of mercury burning. Miners in this region spend 2,500 Kip on mercury to produce a quantity of gold valued at 20,000 Kip. In the District of Pak Ou at Nam Ou River, women are engaged in highly labour-intensive work. Excavation of gold containing sediments from a riverbed is conducted using 10 m long bamboo rafts tied to bamboo pillars in order to withstand the strong current. Miners use 3 m long digging booms with attached 50 cm long, point-nosed shovels, pulled along by a cable from a wooden capstan – this is powered mostly by women at the far end of the raft. North of Hatkhe in Nam Ou, female workers predominate in gold mining and more than 80% of the workforce (approximately 100 people) consist of girls aged between five and 12, working under extremely harsh conditions, typically under adult supervision (Fig. 11.1).

In the Philippines, women work shallow gold deposits with small groups or family units ("pocket miners"). Typically, women are involved in amalgam decomposition, sorting and milling (Murao et al., 2002). At the Kias Gold Mine, all members of the mining association are men, but women are often contracted by miners to conduct panning in exchange for payment in tailings that contain residual gold ("*linang*") (Bugnosen, 1998). Some of the highest rates of female participation (25%) in Asia have been estimated for the Philippines (Bugnosen, 2001). In Luzon and Mindanao, UNIDO (1998) has indicated that a significant proportion of miners are women.

Figure 11.1 Young girl engaged in mining in Lao PDR.

Latin America

Women in Bolivia occupy a number of roles ranging from labourers conducting the most labour-intensive and informal jobs at the mine site (*e.g.* digging, washing), to owners and operators, to women who work tailings ("*palliris*").[5] Women also work as intermediaries ("*las rescatiris*") who sell products to processing plants for *palliris* and mine operators. Most *rescatiris* are believed to be unlicensed. Women work as part of cooperatives (there are more than 100 mining cooperatives in Bolivia where approximately 7500 of 60,000 members are female) or conduct shift work, where remuneration is based on a percentage of material produced (typically 20–30%) during a shift (Jerez, 2001).

In Brazil, as in many other Latin American countries, women work at very small operations panning for gold along riverbanks to supplement family needs (Veiga, 1997). The entrance of women into small-scale production related activities at *garimpos* (artisanal mining sites) is often carried out by intermediaries, such as their spouses or relatives (Sena, 2001). Women at *garimpos* may be involved in panning, amalgam decomposition or refining (*i.e.* in gold shops), or may own and rent machinery. Most women occupy ancillary roles at the *garimpo*, such as cooks or sex trade workers. At medium-size operations, women are more likely to have secondary-level education and own their own mining property (Veiga, 1997). This scenario is, however, more typical of rock quarries, and sand and gravel operations than gold mines. In the context of medium-scale mining, Brazilian women have been quite successful in commanding teams of men and adopting novel methods.

In the emeralds mines of Campos Verdes, in Goias State, Brazil, women are in charge of the concentration step. Most *garimpeiros* (*i.e.* artisanal miners) in the region extract ore from underground workings. After manually crushing and screening, the friable schist is visually inspected by men in order to select the good stones. Women and children usually spend the day looking for smaller, less valuable stones in the tailings. Since 1981, Campos Verdes has been the primary producer of local quality emeralds. Despite this, its 12,000 inhabitants are among the poorest in Brazil. In the 1990s, an automated mine was constructed in Campo Verdes by a mining company from the state capital. Ore is transported by truck to a beneficiation plant, where it is crushed and screened. The screened material is conveyed through an encased glove box, where six women on each side of the conveyor hand pick the emeralds still aggregated to rock fragments and place them within a bucket in the enclosed system. Liberation of emeralds in the buckets takes place at the end of the day in an adjacent shed. After hand-picking, fines (<1 inch) and tailings are sold to local small-scale miners, who hire women to carefully collect any remaining emeralds (Fig. 11.2).

---

[5] *Palliris* are typically women (often widows or wives of miners) who supplement their incomes by extracting small amounts of metal from mine tailings (Jerez, 2001). Typically, *palliris* make less than 25% of the Bolivian minimum wage, although some women who inhabit mine sites also act as security guards, who make approximately 55–65% of minimum wage.

In Colombia, women involved in extraction are most commonly associated with subsistence activities, using pans and sluices to work rivers or rework tailings. The involvement of women exceeds men in subsistence practices in the Pacific region (Veiga, 1997). In some regions, Colombian women have been recognized as entrepreneurs with demonstrated success as owners-operators of mining operations and processing shops. In these roles, women have been known to employ alternative techniques, such as the use of nitric acid to dissolve mercury from the amalgams and the use of cyanide in *lieu* of mercury. Women have also been known to operate equipment (*e.g.* crushers) or act as gold dealers (Rodriguez, 2003). Unless the woman becomes head of the household due to death or abandonment of a spouse, men typically maintain control of finances in family mining operations.

In 1997, it was estimated that 15,000 to 30,000 artisanal miners were producing five to 10 tonnes of gold annually in Suriname (Veiga, 1997). A large

Figure 11.2 Women handpicking emeralds in Campos Verdes, Brazil.

number of these miners (~8000) are Brazilian garimpeiros (Veiga, 1997), although indigenous forest people, the Maroons, are also active. In the Sella Creek region of Suriname, hydraulic miners typically work in teams consisting of six labourers, a boss and overseer (Heemskerk, 2000). Both men and women work as shop owners, merchants, camp bosses and cooks; however, only men are directly engaged in hydraulic monitoring and gold processing.

In 1995, the number of artisanal miners in Venezuela was estimated at between 30,000 and 40,000, approximately 25% being illegal (Veiga, 1997). The proportion of miners that are women has not been determined, although it has been observed that women are often relegated to the most dangerous roles, based on the reasoning that they are more meticulous and thus better suited to "more refined" work, such as panning, amalgamation and amalgam decomposition. Despite the added chemical dangers associated with this, women are usually unaware of the hazards and are typically paid less then men.

Indirect roles of women

Despite their significance, women occupying roles not directly related to mineral production have received minimal attention by researchers, development programmes and governments. In addition to their contribution to productivity, women in artisanal mining communities are critical to community stability, cohesiveness, morale and general well being, and act as primary agents in facilitating positive change.

The lives of women at the *garimpos* – or mine sites – of Brazil, may be the best depicted of all women in artisanal mining (Fig. 11.3). Despite the often deplorable conditions and poor familial ties in hastily created *garimpos*, women

Figure 11.3 Girls in an established Brazilian mining community – high hopes but limited opportunities.

are attracted to *garimpos* by salaries often considerably higher than those obtained through similar work in rural or urban areas. Women who act as "cooks" are particularly significant, not only in terms of food preparation, but also with respect to managing food stocks and related financial resources. They also frequently provide administrative assistance to mine owners and represent a stabilizing factor in *garimpos* – this is achieved through maintenance of regular schedules, and provision of emotional and (occasionally) medical support (Rodrigues, 1993; Veiga, 1997). Women also act as goods and service providers, including owner-operators of bars and equipment owners. Lured by the promise of riches and opportunity (and often an initial lump sum of money), many young girls are brought to remote regions to work in "night clubs" as prostitutes. These bars are regarded as a center of conflict in the community, a place where men, alcohol and guns signify a dangerous mixture for women, particularly those involved in the sex trade. In later stages, as communities become more established, women act as community leaders, as observed in the well-established artisanal mining regions of Cachoeira do Piriá (Hinton, 2002) and Itaituba, Brazil (Rodrigues, 1993).

Interestingly, women's lives at *garimpos* are not unlike those observed in many communities throughout the world (Amutabi & Lutta-Mukhebi, 2001; Dreschler, 2001; Veiga, 1997). Within the community, women predominantly provide goods and services, in addition to a host of domestic chores. Women engaged in mining also tend to take on other jobs to supplement their incomes. In Guinea, although women undertake the same labour as men, inequities in pay (men are paid four times more for the same quantity of gold) often leads to a "troc", or trade of sex for additional money or gold (USAID, 2000). Although there is considerable variability from community to community, gender-based divisions of labour typically result in women working more hours per day than men.

Both direct and indirect involvement of women in artisanal mining is believed to be on the rise. This can be attributed to a number of factors, including: escalation of rural poverty from droughts and/or structural adjustment programmes resulting in a greater need to supplement incomes; outward migration of skilled male miners from artisanal mining areas due to increased large-scale mining development in other regions *or* in pursuit of other opportunities in urban areas; evolving cultural norms with respect to gender roles; lack of employment in other sectors; and high birth rates and growth of extended families.[6]

HEALTH AND SAFETY ISSUES

The World Health Organization (WHO) defines health as "a state of complete physical, mental and social well-being and not merely the absence of disease

[6] With escalating mortality rates in many countries due to AIDS and other causes, families are increasingly inheriting their relatives' children, resulting in larger families and a greater economic strain (Wente, 2002).

and infirmity." Health is not "an objective for living but a resource of everyday life." Poor health generates a vicious cycle – when spouses or family members are infirmed and their capacity to work is diminished, a "healthy" family member must work harder to pay for normal living expenses in addition to health costs. Ill health of a family member may initially drive women and children into mining. Arduous work, combined with inexperience in mining and lack of knowledge about chemical exposures, can further exacerbate the potential for injury or illness; thus, the cycle of ill health and poverty is perpetuated.

The health and safety issues that plague artisanal mining can primarily be attributed to: the informal and often illegal nature of artisanal mining,[7] economic demands that result in inadequate equipment and neglect of safety measures, and a frequent lack of expertise and insufficient training. Although the chemical dangers, in particular, those associated with mercury and cyanide misuse, first come to mind, most occupational hazards are a consequence of poor physical conditions, such as ground failure, shaft collapses and machinery accidents. Hydraulic monitoring[8] of secondary deposits can also be extremely unsafe, as there is potential to undercut hill slopes and induce landslides (Hinton et al., 2003). Methane or coal dust explosions are also significant in small coal mines, such as those found in China, where at least 6000 miners die annually (ILO, 1999). Poor lighting and ventilation, electrocution and explosives misuse are other frequent causes of accidents. Although accidents are underreported due to the illegality of artisanal mining, ILO (1999) states that non-fatal accidents in artisanal mining are still six to seven times greater than in formal, large-scale operations.

Women, men and children who work in artisanal mines face additional illness, injury and stress from dust and noise pollution, as well as extreme exertion from highly labour-intensive jobs. For instance, several hours of digging, carrying large weights great distances, and bending over in awkward positions (*e.g.* during panning or scavenging for gems) can result in painful, chronic injuries (*e.g.* lower back pain) and fatigue (Fig. 11.4).

Inhalation of fine, crystalline silica dust, which is generated from breaking and crushing rock, can result in silicosis. Silicosis is an incurable lung disease that kills thousands annually (WHO, 2000). Conditions resulting from silicosis include emphysema, lung fibrosis and silica-tuberculosis. Advanced stages of silicosis have been documented among women and children as young as 14 in Ghana (ILO, 1999). Primarily due to widespread silicosis, life expectancy in the Bolivian Altiplano is barely 48 years (Quiroga, 2000), an overwhelming difference from the national average of 63 years (World Bank, 2002). In Lusaka, Zambia, women are illegally involved in the manual crushing of marble, and face a high potential for developing pneumoconiosis (Dreschler, 2001). Crushing and grinding can be accomplished using a number of mechanical means (*e.g.* hammer

[7] ILO (1999) estimated that approximately 80% of artisanal mining activities take place outside of a legal framework.

[8] *Hydraulic monitoring* involves the high pressure application of water to "fluidize" metal-bearing highly weathered soils (Hinton et al., 2003).

Figure 11.4 Children engaged in panning in Laos.

mills) or manual techniques, such as a mortar and pestle. When the degree of mechanization is low, women are more likely to predominantly be engaged in this activity.

As women are most commonly involved in the processing aspects of artisanal mining, they are also highly susceptible to chemical dangers, particularly in association with mercury misuse in gold mining. Mercury is widely recognized as one of the most toxic metals known to man. Mercury vapour released during amalgam decomposition poses a serious hazard to women and others in close proximity to gold shops and amalgam decomposition sites. In many countries, gold decomposition takes place in the home (using the kitchen stove) or in small sheds adjacent to processing sites. In a study by Murao et al. (2002), women who conducted in-house amalgam decomposition in pocket mines of Luzon Island in the Philippines frequently had elevated levels of mercury accumulation in hair and exhibited symptoms of poisoning, including kidney pain, respiratory problems, and dizziness. Chronic exposure to mercury vapour can also result in gingivitis and muscular tremors. Mild cases of mercury poisoning have many psycho-pathological symptoms, such as depression and exaggerated emotional responses, which can be mistaken for alcoholism, fever, malaria or other tropical diseases. Exposure to acute levels can produce dysfunction of kidneys and urinary tract, vomiting, and, potentially, death.

The situation in El Callao, Venezuela exemplifies the need for mercury education programs for artisanal miners. In El Callao, where thousands of artisanal miners actively excavate ore from hillsides, the amalgamation process is performed by "moliners", who typically own moderately mechanized plants consisting of three to five lines of jaw crushers, small hammer crushers (capacity of five tonnes/h) and amalgamation tables (mercury-covered copper plates (Beinhoff, 2003). Approximately three to 20 grams of gold is recovered from each of the 50 kg bags of ore processed through comminution, concentration and amalgamation. During this operation, the *molineros* direct the pulp with their bare hands over the mercury-covered plates. By performing this operation, all people working in the ore treatment plant are constantly exposed to mercury vapors – many miners have evident symptoms of mercury toxicity. Amalgam decomposition is conducted by burning. Gold production of 650 *molineros* amount to approximately 45 kg gold per month (Beinhoff, 2003). On a visit to a *molinero*, Veiga (1996) documented the following:

> We have talked to a *molinero*, Mr. David Mejias who recently lost his brother with mercurialism symptoms. According to Mr. Mejias, his brother, who used to take care of the amalgamation work, died due to kidney problems, breathing deficiency and swollen heart. As Mr. Melijas was telling this story, his helper, now a woman, was burning amalgam in a shovel. At this point, he said: "from now on I will be inside of my office when she burns the amalgam". Mr. Melias has never seen a retort and no environmental or mining inspector has ever given him technical advice. He prefers to hire unaware women for the dirty work.

In addition to educational campaigns targeting women involved in amalgamation, spouses of miners can also play a key role in advocating better amalgamation practices to their partners. In addition to posing numerous other health risks, mercury vapours can cause irritability and depression; thus, communicating the potential for diminished sexual performance may be means of capturing miners' attention. A similar approach was actually explored by the Secretary of Mining of Goias State, Brazil, who in 1985 started a campaign promoting retort use. This included a brochure illustrating the effects of mercurialism. Impotence was stressed as one of the initial symptoms, which is somewhat questionable from an ethical standpoint but was nevertheless extremely effective in intriguing miners (Hinton et al., 2003). In 1995, in a talk to artisanal miners in the interior of Venezuela, M. Veiga explored this aspect by asking the women in the audience to point out impotent miners in the room. Although women did not act on this request, miners quickly became very interested in using retorts.

Women involved in reworking tailings may simultaneously be exposed to multiple pollutants. For instance, *palliris* in Bolivia may spend several hours per day working in tailings saturated with heavy, metal-rich acidic drainage and cyanide residues (Jerez, 2001). Cyanide, which is used as an alternative to mercury in recovering gold, is an effective asphyxiant and hydrogen cyanide gas (HCN), and can be fatal to humans at concentrations around 250 ppm in air. Chronic exposure to low concentrations of cyanide has been linked to neuropathological lesions and optical degeneration (Potter et al., 2001). In addition to cyanide and mercury, women reworking tailings may be exposed to highly toxic metals, including lead, cadmium and arsenic.

Women and the natural environment

Although mercury vapour can pose a serious health risk if inhaled, organic forms of mercury – specifically, methylmercury – is the greatest concern in terms of exposure from food. Metallic mercury discharged into the environment (air, water, tailings) from gold mining practices can be transformed into methylmercury, a readily bioavailable form of mercury. Due to its tendency to increase in concentration upward through aquatic food chains (*i.e.* it is biomagnified), individuals reliant on fish in mercury impacted areas may be at risk. Chronic exposure to moderate levels of methylmercury results in symptoms including: visual constriction; numbness of the extremities; impairment of hearing; impairment of speech; and impairment of gait. In cases of acute intoxication, muscular atrophy, seizures and mental disturbance are prominent. Women of childbearing age and their children are particularly susceptible, as methylmercury readily crosses placental barriers and is considered to be a developmental toxicant (Grandjean, 1999). Depending on the frequency and degree of exposure, effects can range from sterility, and spontaneous abortion, to mild to severe neurological symptoms.

As women are predominantly responsible for provision of food, and children and pregnant women are vulnerable to methylmercury, strategies to reduce methylmercury exposure should target women. This is the case in Bolivar State, Venezuela, where measures were taken to limit ingestion of carnivorous fish (Veiga, 1997). Efforts primarily focused on consumption advisories that used clear signage indicating safe and unsafe species, and the establishment of 'safe' quantities for consumption. An interesting component of the program involved the provision of recipes to dilute fish with vegetables, thereby reducing the amount of fish, and, as a result, the amount of methylmercury, ingested. This innovative approach presents an interesting opportunity to involve women in the development of educational programmes. Educational campaigns have employed measures, including the distribution of information packages, such as pamphlets, comics and videos, while others have relied on verbal communication through community meetings or door-to-door campaigns (Hinton, 2002).

In addition to mercury, exposure to other potentially toxic metals (*e.g.* cadmium, lead) can also occur through inhalation (*i.e.* of dust) or consumption of contaminated drinking water and food. This can be facilitated by metals mobilization, in association with acid rock drainage, or direct uptake from tailings into crops or grazing animals. Heavy metals can result in a host of negative health effects. Chronic exposure to cadmium, for example, can have effects that include kidney stones to osteomalacia, a form of rickets (WHO, 1996).

In informal economies in particular, the natural environment is critical to women's abilities to generate income and satisfy household needs. In addition to crop production, women are typically the main providers of water and biomass fuels, and further rely on the natural environment for medicinal plants and resins. Due to the direct link between women, family health and the natural environment, women can be highly effective in land management and particularly influential in advocating practices that prevent environmental damage and related human health effects. Women and other groups are less likely to invest time and resources into more sustainable practices on land they do not own (Sass, 2002).

In addition to chemical contamination, artisanal mining can detrimentally impact ecosystems through deforestation and the modifications of hydrologic systems, for example, through silt accumulation in rivers or construction of water reservoirs (Akagi & Naganuma, 2000). Siltation of rivers caused by discharge of tailings into waterways reduces light penetration and dissolved oxygen levels, thereby jeopardizing fisheries, and may result in flooding (Hinton, 2002). As flooding increases the net area of standing water, it also contributes to malaria and other mosquito-transmitted diseases. Deforestation can significantly impact women and families, due to the importance of forests for fuelwood and, sometimes, food and medicine.

Despite the evidence that indicates the strong link between mental and physical health (WHO, 2001), indicators of mental health or stress have not been studied in the context of artisanal mining. Factors such as heavy workload, poverty, and family illness are expected to negatively impact the mental health of men and women. The effects of migration may also weigh heavily on the psychological well-being of women. Migration of women to artisanal mining regions, as observed in the *garimpos* of Brazil (Sena, 2001; Veiga, 1997), is often related to poverty and economic crises in their homelands. In some cases, women migrate to mining areas for periods of time, leaving their children in the care of relatives (Heemskerk, 2000). Factors such as change in diet and stress associated with leaving traditional lands, often breaking social ties in the process, have been linked to negative impacts on women. In Mashonaland, Zimbabwe, a study of effects of rural economic development on women's blood pressure revealed notably elevated levels in mining areas in comparison to areas where large-scale agriculture and traditional economic activities took place (Hunter et al., 2000).

Women and children

Children are typically the responsibility of their mothers. At many mine sites, women work with young babies tied to their backs and toddlers at their side (Fig. 11.5). In cases where alternative childcare or schooling is unavailable or additional supplementation of income is needed, older children accompany women in artisanal mining activities, and often participate. In accordance with this, the health and safety risks faced by children depend on the type of mineral exploited, the techniques employed, and whether these activities take place in the home (*e.g.* in the kitchen) or at a mine site.

The participation of girls and boys in artisanal mining around the world has been well-documented (ILO, 1999; Hentschel et al., 2002) and continues to persist in response to economic pressures faced by impoverished families. Malnutrition, thermal injuries, and skeletal damage in young children resulting from child labour in artisanal mining has also been extensively documented (Wasserman, 1999). Although generally to a lesser extent than boys, girls are likely to be involved in activities such as digging, grinding, crushing and transporting materials, typically using tools such as picks, shovels and hammers and without any protective equipment (ILO, 1999). Girls are somewhat more

Figure 11.5 A miner and her child in Ghana.

susceptible to sexual exploitation than boys, and child prostitution is common, in part due to the status given to virginity, but also as young girls are deemed unlikely to carry HIV-AIDS or other sexually transmitted diseases.

Violence and sexual abuse

A major study conducted on artisanal mining in Latin America states that the living conditions of girls and women in these communities are "usually on the boundary between poverty and misery" (Veiga, 1997). Many girls are enticed from extreme poverty-stricken situations to work in "night clubs". Initially, they are loaned money to escape their poverty but very few can repay their debts, particularly given the exorbitant cost of food and accommodation imposed on them. Controlled by violence, most women work seven days a week and often suffer from malnutrition and sexually transmitted diseases. Often, sex trade workers are living in a veritable prison and attempts at escape are commonly punished by death (Veiga, 1997).

Incidences of violence towards women in artisanal mining communities have been documented throughout the world. In Guyana, rape of Amerindian girls by foreign miners has been reported (Anon, 2001). In addition, increases in acts of aggression by Amerindian men have also been attributed to the influence of miners and the "mining culture". Martha Bitwale of the Tanzania Women Miners Association has described the fear of sexual abuse associated with women conducting mining and exploration in remote areas (Machipisa, 1997). Maroon women admittedly battle sexual harassment and other hardships (*e.g.* malaria) only because of the absence of viable alternatives (Heemskerk, 2000). In the mining camp Huaypetuhe in the Madre de Dios gold mining region of Peru, high crime rates and incidences of rape and violence are, in part, attributed to the absence of police and lawlessness common in many artisanal mining communities (Kuramoto, 2001). Escalating violence in Ecuadorian communities has been attributed to rising poverty levels and inequity between men and women (World Bank, 2000).

Given the potential for acts of violence, combined with the sex trade typically present in artisanal mining communities, it is not surprising that many artisanal mining regions are plagued by high incidences of HIV-AIDS and other sexually transmitted diseases. At least 70% of women interviewed in a major mining area of Kenya reported at least one incidence of venereal or sexually transmitted disease (Amutabi & Lutta-Mukhebi, 2001). Educational campaigns by organizations such as USAID have had some success in promoting condom use and safer practices, thereby reducing transmission rates, but these efforts are challenging given taboos associated with their use, as well as cost.[9]

The pervasiveness of violence and sexual exploitation tends to depend, in part, on the characteristics of the community. For example, unlawful activities (*e.g.* drugs, violence and prostitution) may be more prevalent in *ad hoc* communities created in response to a "rush", than in well-established communities where government presence may be more significant, familial ties are stronger, and social cohesion is more evident. The influence of the various factors on artisanal mining communities and women in particular are described in greater detail below.

DIFFERENT FACES, DIFFERENT LIVES

The experience of women in artisanal mining communities is uniquely dependent on circumstance. For instance, one woman in a Peruvian mining village may undertake arduous tasks under virtually slave-like conditions, whereas another woman in the same community may independently own and operate a gold shop, using revenues to finance the education of her children. Every woman's story is unique; however, identifying in a more general sense the factors that shape women's realities in artisanal mining communities is also critical to better understanding how women affect, and are affected by, this sector. Key factors are presented in Figure 11.6 and described further below.

These factors are some of the critical drivers that define gender roles for women and men in artisanal mining communities and affect how mining activities impact the natural environment, the health of individuals and communities, and socio-economic conditions. In order to better understand how gender roles influence, and are influenced by, artisanal mining and other environment-impacting activities, issues beyond divisions of labour must be explored. Key topics include: women's and men's access to, and control of, resources; their

[9]In some artisanal mining regions, condoms are provided at subsidized costs. High condom costs relative to incomes has been documented in many areas. In Sierra Leone, for example, three condoms cost the equivalent of one woman's weekly earnings in the sex trade (Wente, 2002).

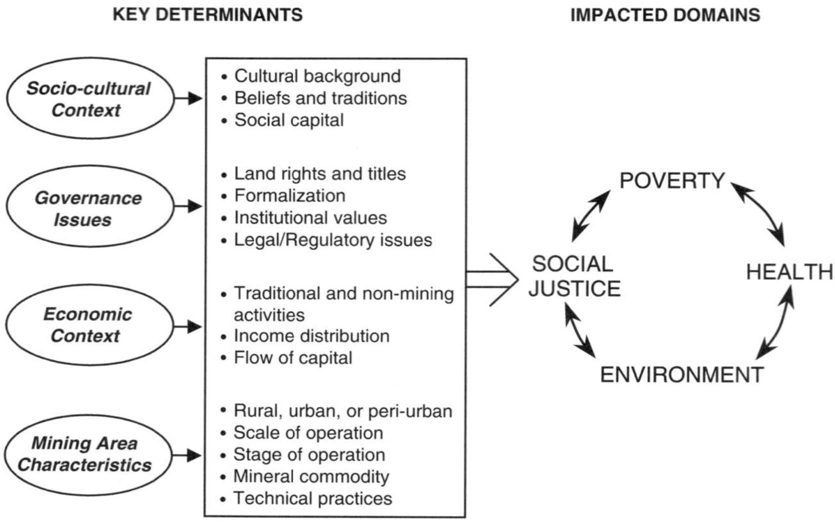

Figure 11.6 Key factors shaping women's realities in artisanal mining communities.

ability to attain knowledge of resources, their decision-making capacity or political power; and beliefs or attitudes that support or impede the transformation of gender roles. These issues are described further in the context of the determinants identified in Figure 11.5.

Socio-cultural context

Cultural beliefs and traditions strongly influence interactions between individuals and groups, the nature of community organization and societal rules or norms. Although these factors are critical, not only in terms of gender roles, but also with respect to the extent to which artisanal mining impacts and benefits communities, they have been poorly studied in the context of artisanal mining.

Likely, the most comprehensive analysis of gender roles in artisanal mining has been conducted by Heemskerk (2000). While studying the Ndjuka Maroons in the forests of Suriname, Heemskerk observed that the participation of women in gold mining in the Sella Creek region is primarily constrained by a lack of resources and limited mobility due to domestic and agricultural responsibilities. Ndjuka women are more likely to be engaged in mining if they have embraced urban values with respect to gender roles. This contrasts with other cultures, where centuries of mining tradition has determined women's involvement, as observed in mining regions of Mali (Keita, 2001).

In Suriname, Heemskerk (2000) determined that exclusively male pit-workers are paid based on gold recovery (~US$ 360/mo) and other workers (*e.g.* male and female cooks) are paid a fixed wage (~US$ 540/mo). Women, who are predominantly employed as travelling merchants (~US$ 90/mo), supplement their earnings by providing additional services (domestic and sexual). Even the lowest paid workers earn wages considerably higher than those found in other sectors. Despite the potential economic benefits and extensive social networks in the Maroon-dominated mining areas, Ndjuka women infrequently participate in this sector, presumably due to traditional gender roles.

In an assessment of women's participation in tin mining in the Jos region of Nigeria, Ogbe (2001) ascertained that the role of a woman is largely determined by her husband – in this case, mining is predominantly conducted by women. Although many differences exist between the 374 different cultural groups in Nigeria, women are generally subordinate to their husbands and in-laws, with little autonomy.

Education can be critical in transforming gender stereotypes and empowering women to derive greater benefits from artisanal mining. In many countries, girls are systematically discriminated against in terms of education – this is mainly attributed to family expectations concerning domestic responsibilities, in addition to social and cultural barriers. Amutabi and Lutta-Mukhebi (2001) determined that families in Kenya are more likely to send boys to school than girls, as has been observed in many other African and South East Asian countries (UNESCO, 2000). In some cases, girls are removed from school at the onset of puberty, due to mistrust of male teachers and students and fears of unwanted pregnancies (UNESCO, 2000). When girls do enter the school system, they tend to be responsible for considerably more extracurricular chores than boys, thereby resulting in longer hours worked daily and less attention paid to homework. Gender inequities may be further exacerbated by educational practices and beliefs that reinforce gender stereotypes. For instance, educators may associate men with public life and corresponding lifestyles (work, politics, power), and link women with domestic responsibilities.

Social beliefs may, in some cases, lead to poorer health for women compared to their male counterparts. In Eastern India, for example, disparities between women and men's health have been attributed to a lack of rest houses for women at mine sites, despite government regulations, and the accepted norm that men receive the larger portion of the food, a better part of the lodging, and contribute less to the upkeep of the household (Chakravorty, 2001). Ill health inevitably translates to losses in working days, and thus, wages.

A significant barrier to women's equitable participation in mining involves cultural taboos and superstitions. In Zambia, for instance, Namakau Kaingu, mine owner and Chairperson of the SADC Women in Mining Trust, described challenges in gemstone mining due to the belief that women should not approach a gemstone mine as the spirits of the stones would drive the gems deeper into the earth (Synergy Africa, 2001) or would disappear altogether (Kaingu, 2003). Some believe this is particularly significant during menses. Disappearance of the stones can be averted with the slaughter of a goat or cow, and the calling of the spirits of ancestors (Kaingu, 2003).

In N'tulo, Mozambique, women are believed to attract bad spirits, and therefore, are banned from working in the mines (Dreschler, 2001). They are, however, permitted to sell food and beer. In nearby Manica, women are prohibited from digging trenches but can transport and wash the gold-bearing ore.

Due to the belief that women bring bad luck, they have been banned or deterred from working underground in countries throughout the world (Robinson, 1998). Up until the mid-1980s, organized mining companies adhered to this superstition in Brazil, and, in the 1970s, coal miners in Eastern US states adhered to the belief that the presence of women underground would cause explosions (Bryson Hodel, 2000).

Menstrual taboos in the Maroon culture, which prohibit women from engaging in activities, including sex, cooking for men, or touching items used by men, essentially result in the exile of women to a menstrual hut for a portion of each month (Heemskerk, 2000). As a consequence, they are reportedly bored, isolated from the community, and their productivity is impeded. The perceived repercussions of violating the taboos include illness, death and other misfortune to people in the community. Some women working at mine sites take oral contraceptives to prevent menstruation (Heemskerk, 2000).

In some instances, women's participation in mining is a consequence of prejudice, often under the guise of concerns for women's safety. In the gold rush in Serra Pelada,[10] Brazil, women were strictly prohibited from entering the artisanal mining reserve (Veiga, 1997). In Zambia and Zimbabwe, regulations restrict women from working underground. In Zambia, Rita Mittal of the Association of Zambian Women Miners (AZWM) has described the difficulty women have with initializing mining in some areas due to the hostility of chiefs threatened by their presence (Machipisa, 1997). The justifications for discrimination in each of these cases may differ; however, the end result is ultimately a system that undermines a woman's ability to participate fully in mining.

Since the individual who has access to, and control of, land is likely to derive the most benefit from this finite commodity, issues concerning land rights and ownership are crucial to women's lives; it may even be the single most important factor in women's struggle for gender equality and empowerment (Mushunje, 2001). In some cultures, women maintain control of lands in association with matrilineal ties. This is the case on the phosphate-rich island of Nauru in the South Pacific, where mining has taken place over the past 100 years (Pollock, 1996). The Ndjuka Maroons of Suriname and French Guyana are also matrilineal, and women frequently own land, houses and canoes (Heemskerk, 2000). Despite this, rural Ndjuka women infrequently hold positions of religious or political significance, and have less access to capital, education, and the outside world than their male counterparts. In pre-colonial Zimbabwe,[11] women also

[10] *Serra Pelada* was a phenomenon wherein government intervention into the discovery of a major gold deposit resulted in the formation of the first artisanal mining reserve (Veiga & Hinton, 2002). Approximately 80,000 men were drawn to the region, wherein ~90 tonnes of gold was produced from a single open pit.

[11] Traditionally, Zimbabwean women were given land upon marriage and birth of a first child (Mushunje, 2001). With the institution of taxation systems, which relegated men to working in money-generating jobs (*i.e.* working in colonial mines or farms), combined with the *Land Husbandry Act* of 1951, women's land rights degraded, and their primarily role was transformed into the protector of food security.

PROFILES OF SUCCESS:

Namakau Kaingu, Mine Owner-Operator and Chairperson of the SADC Women in Mining Trust

Namakau Kaingu is a remarkable individual. She has overcome significant hurdles to achieve considerable success in gemstone mining in Zambia, and generously shares her time and expertise with other women. Namakau arrived on foot to a major gemstone mining area where women had not ventured before and were forbidden from participating in mining.* Initially, she was continually requested by men to leave the area, but Namakau persevered. She learned the requirements of the mining authorities and used her knowledge and determination to excel beyond the men working in the region. Namakau now owns her own well-developed mine that supports a community of 18 male workers and their wives and children. As the government school is 25 kilometres from the community, Namakau has built a school for local children and is committed to sustainability, environmental responsibility, and the well being of workers and their families.

In addition to her responsibilities at the mine, Namakau continues to advance women's involvement in mining through her volunteer activities. She is Chairperson of the SADC Women in Mining Trust, one of the most recognized women's mining associations in the world, and is currently in the process of assisting native women in forming their own mining association. Namakau also offers her time and expertise by conducting workshops for women in topics ranging from gemmology, mine development, basic geology, environmental preservation, and management and safety issues.

Given her impressive accomplishments and contributions, Namakau Kaingu has broken new ground as one of the most important role models for women in mining.

---

* As previously discussed in the chapter, numerous beliefs in Zambia persist concerning the disappearance of gemstones when women enter the mine; they are, therefore, prohibited from entering the mine.

had rights of ownership, but land allocation was based on the birth of a child, was often limited to a vegetable garden sufficient to provide for the family, and could be revoked by a husband (Mushunje, 2001). Clearly, ownership does not necessarily equate to access and control of land.

Due to its importance in terms of poverty reduction and community well-being (World Bank, 1998), the contribution of women to social capital, which features elements of social organization, networks, norms, and trust, is a vital issue in artisanal mining communities. In Kangaba, Mali, Beatrice Labonne described the commitment of Malinke women to community health (Labonne,

1998). Normally, Malinke women are required to turn over all gold recovered to their spouses. However, during the inter-season when agricultural responsibilities have concluded and full scale "orpaillage" activities had not yet resumed, men rest and women commence mining activities early. In this "moonlighting" period, women are allowed to retain the profits from gold mining in order to purchase food and spices, children's clothing and medicine. Labonne (1998) documented women assisting elderly women with the cultivation of gardens, and with revenue planning behind hiring a midwife for the community.

The capacity and tendency to organize is a significant element of social capital. There are numerous examples of women's cooperatives throughout the world. In Peru, in the adjacent mining settlements of La Rinconada and Cerro Lunar, 800 women belong to a cooperative of women *pallaqueras*, who rework tailings to extract gold (Kumamoto, 2001). When united, women are more likely to broach topics of rights and safety, may be more amenable to adopting alternative methods, and may be in a better position to access credit.

Governance issues

Governance refers to "the process whereby elements in society wield power and authority, and influence and enact policies and decisions concerning public life, and economic and social development" (Anon, 1996). Governance includes the conventional responsibilities of government but also concerns interactions between formal agencies and institutions and civil society. In the context of women and artisanal mining, key governance issues pertain to the effectiveness of policy in advancing equality, particularly in terms of land rights, representation of women in decision-making processes, and an institutional environment that is conducive to participation by women.

For rural households in particular, access to land is a key determinant of poverty, not only due to its relevance in terms of ownership of products, but also in terms of collateral (World Bank, 2000). Thus, policies that restrict or deter women from obtaining concessions or land rights further contributes to the feminization of poverty. In Kenya, female miners have access to land, but do not control land, and therefore, mining activities (Amutabi & Lutta-Mukhebi, 2001). Because men own the land, they also tend to dictate women's roles in production; as a result, they are usually consigned to transporting, washing and panning, and turning over the profits. In Zimbabwe, Burkina Faso and Cameroon, women have equal rights to own land, but it is almost solely controlled by men (Sass, 2002). In some countries, women may have legal access to land, and may not be aware of these rights, or these rights may be lost once upon the death of a husband or a divorce.

These barriers on women's land rights hinder their ability to access other resources. Often unable to use land as collateral to obtain loans, women have difficulty adopting alternative technologies and hiring labour when needed. In addition, women may not be able to access other supportive services, such as extension programs and training on innovative land management approaches. In certain countries, inequities are further exacerbated by laws that stipulate that women cannot take out a loan without the consent of her husband or father, as is the case in Botswana and Lesotho (Carr, 1993). Difficulty obtaining financing has also been documented by Hilson (2001).

In many developing countries, in rural communities in particular, women's voices are largely absent from political decision-making. The effects of this inadequacy can be felt at both a national (*i.e.* policy-driven) and local level, and results in women's perspectives, needs, knowledge and proposed solutions being largely ignored (Sass, 2002). Exclusion of women in decision making can result in policies that criminalize and further marginalize their activities. Sass (2002) documented a case in El Salvador where efforts to protect a mangrove ecosystem involved restrictions on fishing in estuaries and shoreline timber collection. Men, who fish in the open seas, were predominantly responsible for the policy but were largely unaffected by it. Women, who fish and collect fuelwood along the shoreline for subsistence purposes, were not consulted in policy development and continue to conduct these activities illegally. Incorporation of women into decision-making is a vital factor in the development of gender-sensitive policies and successful implementation of any programme that modifies the use of resources, such as artisanal mining.

Government agencies can play a key role in supporting women in artisanal mining. The National Council of Women in Ecuador, CONAMU, has been particularly successful in responding to the public sector, mainstreaming gender in various programmes and implementing a number of initiatives (World Bank, 2000). The Mineral Resources Department and other government agencies in Tanzania are also conducting workshops and seminars to create awareness and encourage female participation in mining (Dreschler, 2001).

One of the most significant barriers to women deriving benefits from mining relates to regulations that are predominantly oriented towards men. Women such as South Africa's Minister of Mining, Phumizile Mlambo Ngcuka, and Tanzania's Women's Minister, Halima Hatibu, are working to develop more gender appropriate mining regulations (Synergy Africa, 2001). As mining ministries may be difficult to approach for many women, particularly as those employed there tend to predominantly be male (Machipisa, 1997), these agencies could better assist women by diversifying their workforce and sensitizing their services. One strategy adopted by several countries, including India, Uganda, Brazil, and the Philippines, involves formal allocation of a percentage of seats to women on national and local bodies (Sass, 2002).

Economic context

The significance of non-mining activities (*e.g.* agriculture, forestry), and the control and flow of capital, speak directly to socio-cultural values and issues of equity within mining communities. As it is often poverty that drives many to artisanal mining, the economic context of mining plays a critical role in defining gender roles.

Certain jobs are specifically the domain of women, who are accordingly afforded lower pay. For instance, in many parts of Africa, gold panning is deemed

PROFILES OF SUCCESS:

Maria Rodriguez,* Mineral Processor and Smelter, Shop Owner

Maria Rodriguez has demonstrated how skill, diligence and a commitment to quality can earn the respect and trust of clients, and result in the establishment of a successful business. Since 1985, Maria has owned and operated a mineral processing and smelting shop that services artisanal and small-scale miners in a major gold mining region of a Latin American country*. Upon the passing of her husband eight years ago, Maria took on the management and operation of the shop independently, but the customers who had grown to appreciate the quality service she provided continued to frequent the shop. As a result, Maria's shop outlasted a decline in mining activities that saw 20 of 24 gold refining shops in the city disappear. Working up to 70 hours per week, Maria services between five and 10 clients daily who are mainly engaged in gold refining, although she also provides consulting services for processing a variety of minerals.

As in many Latin American cities, safety issues plague the operation of many businesses. In Maria's community, police are largely ineffectual, and generally cannot be trusted. Maria, her assistants and clients face the risk of violent confrontation with criminals as they enter and leave the shop. Maria faces additional hazards working with mercury and nitric acid; despite her awareness of health risks, she often foregoes precautions (*i.e.* use of a retort), as they are comparatively loud, uncomfortable and inconvenient. Maria's situation is further exacerbated by a policy environment that opts for apathy or threats of shop closure in *lieu* of provision of support. Despite these challenges, and an inability to generate capital, Maria has been able to make a decent living. She has put five children through university, and has been able to support other members of her family in a country hard hit by recession. Maria Rodriguez is certainly a remarkable example of perseverance, dedication, and excellence, and is a valuable role model for women throughout the world.

*Due to risks to Maria's personal safety, her name has been changed and the location of her operation has been omitted upon her request.

the responsibility of women, as it requires a gentle, sensitive touch (ILO, 1999). In other locations, women are relegated to unskilled, menial work, such as transporting ore or water. Even when women undertake the same work as men, they are often paid considerably less. In Siguiri, Guinea, for example, women and men work side-by-side, washing gold from the lateritic soil (USAID, 2000). For every five *calabashes* (a large carrying container) of ore that the women wash, male intermediaries receive the profits from four and the woman retains only one.

Increases in household income do not necessarily equate with improved well-being of family members, particularly in households where men continue to control finances. Quite often, women's work in the small mining sector is unpaid, and conducted to enhance the earnings of their husbands. In Kangaba, Mali, for instance, although women and men work side-by-side with men gold digging, Malinke tradition dictates that women turn over all gold to their husbands (Labonne, 1998). The same tradition typically applies to women in Kenya, although there are a number of examples where women manage family finances and give their husbands an allowance (Amutabi & Lutta-Mukhebi, 2001). Despite the expectation that women will turn over all profits to their spouses, in some Latin American countries, they have been known to withhold products and revenues from their husbands, often through informal agreements with local buyers (Sandoval, 2002).

Many studies indicate that the revenue generated by women in artisanal mining contributes more directly to the well-being of households then that of men (UNDP, 1999; ILO, 1999, Veiga, 1997). Specifically, the income generated by women is more likely to be directed towards improving the quality-of-life in the family – *i.e.* through education, food, agriculture, etc. – whereas men tend to spend revenue on gambling, prostitution and alcohol (Hentschel et al., 2001). In addition, when women receive and manage earnings, their economic dependence on men may decline, thereby testing existing gender roles.

In addition to different spending habits, increased female participation in artisanal mining may result in decreased attention to agricultural crops, resulting in poorer nutrition. Alternately, participation in other traditional occupations (*e.g.* crop production) may continue, resulting in a substantial increase in the work burden of women (Çagatay, 2001). In fact, one of the most significant gender inequities relates to time. Women are often exclusively responsible for domestic responsibilities even when they participate in the labour market, whereas men tend to be responsible for predominantly economic activities (Veiga, 1997). This reality, coupled with differing spending habits, generally means that women's participation in activities that generate revenue can be equated with increased labour hours. In the Mukibiri district of Kenya, it is estimated that ~90% of all mining work is performed by women, in addition to domestic chores.

Most artisanal miners are rural and poor (Hentschel et al., 2002). Due to migration of men to urban areas or large mining districts, women are often *de facto* heads of households; these households are often the poorest in rural communities. Male migration and return to households has been shown to modify values, which weakens women's positions in certain circumstances. In the Sierra region of Ecuador, for instance, it has been documented that men return home with a greater sense of "machismo" than is traditionally observed in indigenous Sierrans (World Bank, 2000). Alternately, in the absence of men, women's position may actually be enhanced, mainly because they become solely responsible for management of the household and agriculture during this period. In Suriname, men are often encouraged by their wives to migrate to mining areas, as they are ultimately better off economically than women with spouses who are not miners (Heemskerk, 2000). This phenomenon has also been observed in Brazil where men from the Northeast leave their families behind to become *garimpeiros* (*i.e.* miners) in the Amazon. *Garimpeiros* may send money home for years, but when their luck runs out, they establish new families in mining villages, and women are left to support the family independently.

Characteristics of the mining region or operation

Unlike the majority of the aforementioned factors shaping women's lives, most characteristics of a given mining operation are fixed (*e.g.* location, reserves, mineral resource, etc.). These features do, however, also provide insight into gender roles in artisanal mining communities, and a basis for addressing gender issues under specific circumstances. The influence of various characteristics of mine sites, specifically the location, mineral mined, and the scale and stage of an operation, are discussed in greater detail below.

Most artisanal mining communities are located in sparsely populated, rural, and often, remote, regions. Some exceptions are large mining districts, such as Ashanti in Ghana, or towns that form around discoveries and happen to develop into major centres. In comparison to their urban counterparts, women in rural communities typically have less access to services, such as health care and education, a reduced dependence on the cash economy for food and non-food purchases, and fewer health problems due to contamination (*e.g.* air, water pollution) from overcrowding (Ruel et al., 1999). In hastily-established artisanal mining communities, residents may experience the worst of both urban and rural environments, being able to access exclusively poor services, becoming increasingly dependent upon the cash economy, and exposing themselves to pollutants. A major review of gender issues in Ecuador (World Bank, 2000), where income inequity has worsened in recent years, found that women living in rural areas also have higher fertility levels and lower education levels than women in urban areas (4.1 versus 7.1 yrs).

Perhaps one of the most interesting determinants of the roles and responsibilities of women involves the type of commodity mined. In the extraction of high value products, such as gold, men more commonly maintain control of the mine site and are involved in digging, whereas women are typically relegated to washing, panning and transport activities. The basis for this division of labour is presumably related to the limited physical capacity of women for pit labour. Interestingly, women are engaged in digging and equally laborious activities for agricultural purposes, or in the extraction of high volume, low value commodities (*e.g.* clay, limestone, dimension stones). Women are more likely to participate in greater numbers and play a more significant role in control of the land and decisions surrounding its use for low value commodities, as observed in Nigeria (salt), South Africa (kaolinite) and Brazil (sand and gravel). Differences in participation in high versus low value commodities can be viewed either as a startling indicator of gender inequity in artisanal mining (*i.e.* women are in control of land of lesser value) or a reflection of women's entrepreneurial capacity; nevertheless, these minerals represent an important source of livelihood for women. In the case of gemstones, women are beginning to challenge norms – as observed in Zambia – and are taking on a greater role in this sector. Namakau Kaingu has stated that many opportunities exist for women in gemstone mining (Machipisa, 1997); the case of Zambia may therefore become an important exception to the rule.

Differences in methods employed for mineral extraction and processing also affect women. As observed in gold mining, women are predominantly involved in processing aspects and so are at risk of exposure to dust and mercury. In some cases, men are somewhat aware of the dangers associated with mercury use and consign women to undertake processing responsibilities.

It has been found that women's participation in artisanal mining generally increases with the decreasing scale of the operation, particularly in the case of small family operations, where mining takes place to supplement other income (*e.g.* from subsistence agriculture). This may also be related to disparities in education and training in small-scale mining techniques. Typically, women's direct participation decreases with increased scale of operation and prevalence of mechanization.

The stage of an operation strongly influences the nature of artisanal mining communities, and the lives of residents. Clark and Cook-Clark (1996) described two phases in the case of artisanal gold mining in Asia: a "rush" stage, which typically lasts between one and three years, and a post-rush stabilization period, wherein the number of miners tends to decrease by 30 to 50%. In many communities in the Amazon, the artisanal mining cycle is comprised of phases of discovery, migration, and relative economic prosperity, followed by periods of resource depletion, outward migration and economic destitution (Veiga & Hinton, 2002). Either scenario could apply to communities in the artisanal mining regions of the world. In general, hastily-created "rush" communities lack infrastructure, organization, the presence of government or law enforcement, and are often pervaded by violence, prostitution, drug and alcohol abuse, and gambling – elements that make for difficult conditions for community residents. During the Fourth World Conference on Women in Beijing (1995), Noeleen Heyzer stated:

> In conditions of rapid change – including environmental deterioration, the outmigration of men, changing economic activities and aspirations, and government interventions – women play an even more crucial role in the maintenance of livelihoods, cultural continuity, and community cohesiveness.

Artisanal mining communities formed as a consequence of "rushes" seem to experience all of the conditions of rapid change, generally with the exception of government intervention. It seems apparent that support of women and men during the resource depletion phase can be critical to the development of more stable, resilient communities.

CHALLENGING GENDER INEQUITIES, AND THE ROAD AHEAD

At the World Bank Roundtable on Regularizing Artisanal Mining (1995), Mrs. Ofei-Aboagye of the Ghana Institute of Management and Public Administration stated that primary constraints to the effective participation of women include "legal and social taboos; widespread illiteracy; and organizational, technical, and financial constraints". These issues are inextricably linked to gender inequity. Inequities have been evidenced by limited political power, access to resources (capital, information, education, etc.), mobility, and basic human rights. Given this reality, it is not surprising that "while the positive impacts (of artisanal mining) are hardly felt by women, they are hard-hit by the negative impacts" (Traore, 1997). Dreschler (2001) further articulates that "women (in Tanzania) make up 50% of the work force but do not receive 50% of the rewards."

It is apparent that social equality must be achieved before women and men will fully benefit from artisanal mining. As gender roles are socially and culturally constructed, social change induced by economic transformation, incentives and regulatory reform can contribute to the transformation of these roles (World Bank, 1994). Specific courses of action to facilitate this change may also include the strengthening of networks, gender mainstreaming in government agencies and assistance programmes, the adoption of strategies that are inclusive and accessible to both women and men, additional research into gender roles in this sector, and the advancement of sustainable livelihoods in artisanal mining communities.

Strengthening of networks

Connecting women with Non-Governmental Organizations (NGOs) and strengthening linkages between individuals or groups of women can markedly improve women's access to resources. Factors that affect access to information include the availability of time, literacy, control over household media, access to written material, and the ability to travel (Appleton et al., 1995). Within a formal network or organization, women are better equipped to respond to negative impacts of artisanal mining and take advantage of opportunities in this sector (Wiego, 2002).

Organizing women – for instance through creation of Miners' Association or business coalitions – has been recognized as a key step to obtaining funding. Women are increasingly being financed through microcredit programmes,[12] which have shown tremendous success in revolutionizing women's lives. The formation of organizations or cooperatives may put women in an even better position to receive financial support from international charities and religious organizations, or formal lending institutions.

Strong women's civil society organizations can make considerable strides in advancing women's issues in political agendas. The National Council of Women (Consejo Nacional de la Mujer, CONAMU) in Ecuador, which was formed largely as a result of the efforts of women's NGOs, is one such example (World

[12] Additional information on microcredit programmes can be found through Women in Informal Employment Globalizing and Organizing (WIEGO), a worldwide coalition of institutions and individuals concerned with improving the status of women in the economy's informal sector (www.wiego.org). A key example of a microcredit programme is the UN sponsored Microcredit Summit Council, whose goal is to reach 100 million of the poorest families, especially women, with micro-finance for self-employment.

Bank, 2000). In part due to the efforts of CONAMU, gender seems to be better mainstreamed into public programs, and the private sector has adopted a number of creative gender initiatives.

Women and assistance programmes

A number of technology assistance programmes have been carried out to facilitate the use of more environmentally sound techniques, advance the commercial benefits derived from this activity, and formalize the artisanal mining sector. Organizations that have made significant contributions in these endeavours include the UN Industrial Development Organization (UNIDO), Intermediate Technology Development Group (ITDG), Projekt-Consult GmbH, Project GAMA and the Brazilian Centre for Mineral Technology (CETEM).

The assumption that "what is good for men is good for their families" has been a major shortcoming of some technical assistance programmes. Most development programmes do not recognize the gender-specific nature of development, and are therefore, oriented towards men's tasks, interests and needs. A detailed study of systemic deficiencies in development programmes identified a shortage of gender-sensitive scientists, inadequate support systems for women, and an inequitable allocation of resources (Muntemba & Chimedza, 1995). Another common failing is an emphasis on commercial versus domestic benefits (Everts, 1998). In detailed assessments of the effects of development programmes on the lives of women involved in agriculture and forestry (Wightman, 2001; Wakhungu & Cecelski, 1995), it was found that the quality of their lives did not, in fact, improve, and even declined as a consequence of certain development initiatives. As women's direct involvement frequently decreases with increased scale and mechanization of a mining operation, it is easy to speculate how technical assistance programmes could negatively impact women. Incorporation of women in programme development and implementation represents an important opportunity for technical assistance programmes to build upon the demonstrated successes of past efforts.

One programme that has specifically targeted women was conducted by the Intermediate Technology Development Group in Zimbabwe (ITDG, 2002). Upon completion of courses on a number of mining-related topics, women applied this knowledge to obtain financing to initiate an environmentally sound, economically viable, small-scale mining operation. As a result of the success of the school, it has been extended to men and women in other regions.

In a recently implemented project in Mindanao, Philippines, UNIDO made special efforts to ensure that women equally participated in mining, and benefited from the introduction of new equipment and processing techniques (UNIDO, 1998). The Federation of Small-scale Miners in Mindanao played a critical role in encouraging and monitoring the incorporation of women into project educational and training programmes. A similar UNIDO project in Zimbabwe, Sudan, Tanzania, Lao PDR, Indonesia, and Brazil, entitled "Removal of Barriers to the Introduction of Cleaner Artisanal Gold Mining and Extraction Technologies", will implement specific measures to support the direct entry of women into artisanal mining, and will conduct special training and awareness programs for women that address health issues specific to them (UNIDO, 2002).

As women have little knowledge of the hazards associated with mercury, and are often involved in the processing of gold, it is logical that they be targeted by campaigns to reduce mercury use. They are also predominantly responsible for food preparation and, as women of childbearing age and children are particularly susceptible to methylmercury exposure from fish, gender-specific educational programs could prove highly effective at reducing exposure through ingestion (Veiga, 1997). Due to the importance of amalgamation in gold recovery, combined with generally poor health conditions related to a lack of sanitation, widespread disease (malaria, cholera, STDs, etc.) and generally limited access to health care providers, initiatives directed at reducing the comparatively invisible health impacts from mercury will have limited success if the programme does not also comprehensively address these broader community issues (Hinton et al., 2003).

Insufficiently broad participation in artisanal mining technical assistance programmes not only has ramifications for women, but for the success of the programmes as well (Gender Working Group, 1995). When technological innovations are derived from traditional, local knowledge and comprehensively consider the impacts on users and other stakeholders, communities are more likely to benefit from their development. In order to be truly comprehensive, technical assistance programs may need to disrupt local norms with respect to participation and authority, in order to promote inclusiveness and maximize community benefits.

Women and appropriate technology development

In the past, technology assistance programs have typically involved the transfer of modern, science based technologies as opposed to building upon traditional technologies that have evolved within communities (Appleton et al., 1995). A technology development programme would be more likely sustained if it built upon people's strengths, inherent skills and knowledge. Marcelle and Jacob (1995) further identified the need to make "historically and culturally specific recommendations in any policy intervention and to identify policy alternatives in a fully participatory mode." The ITDG, an organization involved in the development of technologies that support the alleviation of poverty, has stated that women's technical knowledge tends to be less prestigious and is of lower profile because they tend to focus on processes and organization of production, rather than equipment or hardware (Appleton et al., 1995). Incorporation of this process-oriented local knowledge into more traditional hardware-focused technologies could significantly improve artisanal mining technologies as a whole.

In addition to building on local knowledge, there are also opportunities to explore the application of innovative, locally developed technologies in some countries and transfer or modify these to other locales. For instance, an inexpensive method to decompose the mercury-gold amalgam and recover volatilized mercury has been developed in Papua New Guinea and China (Hinton et al.,

2003). Both systems essentially consist of a small metallic bowl contained with a larger bowl filled with sand or water. A larger bucket is placed over the smaller bowl, such that when the amalgam in the smaller container is heated, mercury is volatilized, condenses on the larger bucket and is collected in the underlying sand or water. Something so simple could result in considerable health benefits, particularly for the women and men involved in amalgam decomposition.

In a detailed study of gender and technology (Evert, 1998), it was found that interventions did not benefit women when: the "improvements" were not more convenient and accessible than traditional sources or activities (*e.g.* clean water wells), modifications were directed towards commercial uses (*e.g.* development of forests for resale when fodder needs were not being met), and technologies were generally inappropriate (*e.g.* "improved" stoves that did not consider the cultural value of traditional food preparation techniques). Projects directed at commercialization can also undermine the position of women by disrupting links to agriculture, forest and livestock sectors (*i.e.* severing links with food, fodder and fuel). In the context of artisanal mining, increasing the scale of an operation could impact resource flows by: altering the quantity and quality of water; increasing pressure on women to spend more time mining, thereby reducing their attention to other responsibilities; and causing more widespread environmental damage (*e.g.* deforestation, damage to fisheries). In many instances, a greater focus on commercialization has resulted in the loss of women's control of production and access to resources, transforming their role more and more into that of labourers (Appleton et al., 1995).

As time is one of the most important commodities for many women, it should be a critical consideration in appropriate technology development. More time can enable women to pursue more productive tasks, can reduce stress levels, and may ultimately improve quality-of-life. As many woman miners are responsible for carrying ore or water, the increased use of donkeys and/or carts may be an effective means to decrease transport time and labour requirements. In one credit scheme described by Everts (1998), a participatory approach was employed with Maasai women in Kenya to develop appropriate panniers for donkey transport. The project involved the development of a cooperative of 18 women divided into six groups. Each group of women raised half of the purchase price of a donkey and the creditors loaned the remaining amount. In order to overcome traditional concerns with the use of donkeys, Maasai men and women visited communities where donkeys were used. Women who participated in donkey time-sharing reduced their carrying time by approximately 25 hours per week, time that was spent on other tasks, leisure, and participation in community work. In a similar non-gender sensitive programme in Tanzania, the introduction of donkeys exacerbated gender inequities, as women had little disposable income and were unable to purchase donkeys (Everts, 1998). Men employ their donkeys to carry crop harvests.

In cases where processing activities take place closer to home, the construction of on-site mills could markedly reduce time and effort dedicated to carrying ore. A modification such as this, however, may be less attractive to women simultaneously conducting child care or when security issues are of

Figure 11.7 Venezuelan woman and a wooden "Batea".

greater concern. The state of existing machinery may also influence women's burden of work. In the Makate district of Tanzania, improved functioning of grinding mills considerably benefited women by reducing the tonne/km load by more than 90% (Everts, 1998). Although the mill was applied in an agricultural context, this lesson is readily transferable to artisanal mining.

One of the better known examples of appropriate technologies for women involved in artisanal mining is the wooden pan (Fig. 11.7). As this pan is lighter and equally effective as metallic pans, it has been well received by women throughout the world. This underscores the importance of developing technologies specifically for users.

Research on gender and artisanal mining

The paucity of reliable information concerning women's involvement in artisanal mining represents a major knowledge gap. In most reports on artisanal mining, a discussion of a women's role is altogether absent, or is grouped together with child labour, often in the same paragraph. This is particularly disturbing, as child labour is widely condemned; thus, it is implied, by association, that women's involvement should also be limited. In most instances, artisanal mining represents an opportunity for both women and men to relieve the burdens of poverty.

An additional shortcoming of most research conducted to date involves the use of terms like "ancillary", "secondary" and even "subordinate" in documentation of women's indirect participation in artisanal mining. As spouses of miners, cooks, shopkeepers and other service providers, women not directly involved in artisanal mining activities (in addition to those who are) play a pivotal role in the health and well-being of households, as well as in the community. Terms that imply women are of secondary importance place their participation on the periphery, and marginalize their achievements in the sector. This is further exacerbated by the tendency to omit part-time workers from statistical reports of mining activities; consensus is needed as to how part-time direct involvement should be enumerated.

The differential impact of both current practices and technical change on the lives of women and men also requires more in-depth research. This knowledge can significantly contribute to the development of gender sensitive strategies to advance positive change in artisanal mining communities and increase the likelihood that initiatives will be sustained.

Women and sustainable livelihoods

In a given region, artisanal mining can provide a viable means of employment for a limited number of people until depletion of the resource. Artisanal mining can also contribute to sustainable development of the surrounding community by supporting auxiliary enterprises (*e.g.* jewellery production) and agricultural development (Hinton et al., 2003). This sort of diversification may be well suited to women in artisanal mining communities for a number of reasons. The majority of women involved in small-scale enterprises are sole proprietors who make little use of hired labour, or participate in income generating groups on a part-time basis (Carr, 1993); thus, they may be easily adapted to small-scale ventures. In addition, women are frequently responsible for food security and have demonstrated competencies in agriculture.

Due to unequal access to information, credit and training, in addition to time-consuming domestic responsibilities, particularly in comparison to their male counterparts, women face significant challenges in establishing new ventures. They have difficulty obtaining capital and training because they do not know it is available (if it is) and how to access it. In Tanzania, for example, only 3% of the National Bank of Commerce clients are women (Carr, 1993). In addition, women may lack the skills to market products and manage business ventures. These challenges have been described by Marcelle and Jacob (1995) as the "double bind", which is a consequence of an unsupportive policy environment, combined with uncertain markets for products.

The Mining Sector Diversification Programme (MSDP) of Zambia made strides in conquering the "double bind" by sending five mining associations, including one Women in Mining association, to a major gemstone fair in Tucson, Arizona in 2003 (Anon, 2002). This endeavour will link the Zambian gemstone market with foreign buyers, and build the capacity of small miners to add value to their stones. Efforts to develop value-added processes in the production and marketing chain of gemstones, particularly through lapidaries and jewellery manufacturing, have been recognized as an important opportunity for women in Zambia (Kaingu, 2003) and many other countries (Hinton et al., 2003).

With the exception of the UN Development Programme (UNDP), which has outlined a strategy for the development of sustainable livelihoods in artisanal mining communities (Labonne & Gilman, 1999), very few examples of sustainable livelihoods initiatives have been documented. In Mali, the Sadiola Gold Mining Project was created by Anglo Gold to provide technical assistance to artisanal miners and support alternative revenue generating activities, such as agriculture, jewellery production, and fabrication of dyes and soaps (Keita, 2001). In the Tapajós region of Brazil, governments and NGOs are implementing jewellery schools in mining areas to add value to the miners' product and, in regions where the resource has been depleted, entrepreneurs have invested in cattle farming, and palm tree and "açaí-nut" production (Hinton et al., 2003). In KwaZulu Natal, South Africa, brickmaking provides a means to add value to kaolin mines by women, and has been shown to support sustainability in the community of Ozizweni (Dreschler, 2001). In Alta Floresta, Brazil, ex-miners have been working with local experts and the government to turn flooded mine pits into aquaculture operations (Otchere et al., unpubl).

Artisanal mining has the potential to be an important catalyst for entrepreneurial activities, and the formation of strong, resilient communities, but progress will be severely limited until all stakeholders have the necessary tools to affect positive change and challenge existing inequities. Women can play a vital role in this transformation, in part, through the development of sustainable livelihoods. Providing women with access to information, credit, and training combined with an amenable policy framework are key steps to facilitating this change.

CONCLUSIONS

An estimated 70% of the world's 1.3 billion poor are women and girls (UNIFEM, 2000). The feminization of poverty, combined with other factors (*i.e.* the outward migration of men, evolving cultural norms with respect to gender roles, lack of employment in other sectors, and high fertility rates) have led to the escalation of women's direct and indirect involvement in artisanal mining. Currently, approximately 30% of the world's artisanal miners are women, although, as women often occupy multiple roles in artisanal mining communities, their direct participation may be significantly higher when part-time work is considered.

This review has integrated information on women's direct and indirect participation in artisanal mining, in order to provide a basis for: identifying major gaps in the knowledge base, determining significant factors shaping gender roles in this sector, and recommending a course of action for policy makers, artisanal mining researchers, assistance programme officers, and other actors. Key findings are summarized as follows:

- Women working in mines are most commonly involved in transporting ore and water, washing, or seemingly "delicate" work such as panning, amalgamation and amalgam decomposition. Women also function as small mine owners and operators, but this tends to be more of an exception than the rule.

- In extraction of high value products, such as gold, men more commonly maintain control of the mine site. Women are more likely to participate in greater numbers and play a more significant role in control of the land, and decisions surrounding its use for low value commodities, such as industrial minerals.
- Women's direct participation in artisanal mining generally increases with the decreasing scale of the operation; thus, they are most prevalent in small family operations where mining takes place to supplement other income (*e.g.* from subsistence agriculture).
- Outside of mining, women often assume key roles in the provision of goods and services (*e.g.* cooks, sex trade workers, shop owners, housekeepers). Moreover, within the community, women are of fundamental importance in terms of food security, are critical to community stability, cohesiveness, and morale, and act as primary agents in facilitating change.

Significant barriers exist that prevent women from fully benefiting from this sector. In order to respond to inequities in political power, access to resources (capital, information, education and training, etc.), mobility and basic human rights, combined with the presence of legal, socio-cultural, and financial constraints, the following are recommended:

- Commitment to gender mainstreaming in government agencies, and appropriate recognition of women in policy frameworks, particularly in relation to land ownership rights.
- Adoption of strategies inclusive of, and accessible to, both women and men, and which support women's participation in political decision-making.
- Elimination of discrimination from educational systems, and provision of support for families sending children to school.
- Formal incorporation of gender issues, and the adoption of holistic approaches to artisanal mining communities though technical assistance and community development programmes.
- Promotion of micro-credit and other programmes that provide financing for women.
- Execution of educational campaigns that target women in order to mitigate specific health risks, and which provide the means for them to access other resources.
- Implementation of programs to train women in various aspects of mining, as well as in marketing, management and bookkeeping.
- In-depth research on women's involvement in artisanal mining communities, and the differential impact of current practices and technical change on the lives of both women and men.

The extent to which women are positively or negatively impacted by artisanal mining largely depends on issues of equity. As inequities are challenged, gender roles will inevitably evolve, and artisanal mining will be better equipped to support sustainability in the communities where it takes place. Ultimately, any work undertaken in artisanal mining must adopt a community-centric approach. Women have a demonstrated capacity to drive positive change. The next steps taken will be critical in determining whether women will be empowered to apply this capability in order to advance their lives and the artisanal mining sector as a whole.

REFERENCES

Akabsaa, T. & Darimani, A. (2001). Impact of mining sector investment in Ghana: A study of the Tarkwa mining region. *Draft Report Prepared for SAPRI.*

Akagi, H. & Naganuma, A. (2000). Methylmercury accumulation in Amazonian inhabitants associated with mercury pollution. *Proceedings of US-Japan Workshop on Human Health Effects of Low Dose Methylmercury Exposure, National Institute of Minamata Disease, Minamata, Japan, November 28–30, 2000* (pp. 133–145).

Amutabi, M. & Lutta-Mukhebi, M. (2001). Gender and mining in Kenya: The case of the Mukibira mines in the Vihiga district. *Jenda: A Journal of Culture and African Women Studies*, 1(2), 23.

Anon (1996). *Governance: A working definition.* Governance Working Group of the International Institute of Administrative Sciences, http://www.gdrc.org/u-gov/work-def.html

Anon (2001). *Gold mining: An industry on the verge of crises.* Report to Conservation International, http://www.conservation.org/xp/CIWEB/programs/awards/2001

Anon (2002). Gemstone Miners US Fair Bound, *Times of Zambia*, posted at AllAfrica.com, http://allafrica.com/stories/200212060252.html

Appleton, H., Fernandez, M.E., Hill, C.L.M. & Quiros, C. (1995). Claiming and using indigenous knowledge, Chapter 3. In G. Oldham, (Ed.) *Missing links.* UN Commission on Science and Technology, Gender Working Group book on Science and Technology Policy, IDRC, Canada, August 1995.

Aspinall, C. (2001). *Small-scale mining in Indonesia.* Country Study commissioned by MMSD, 30 pp., http://www.iied.org/mmsd/

Beinhoff, C. (1998). *Artisanal mining in Luang Prabang.* Lao PDR, Mission Report to UNIDO.

Beinhoff, C. (2003). Personal communication: Artisanal mining in El Callao, Venezuela, January 2003.

Bryson Hodel, M. (2000).Women miners a vanishing breed, *Courier-Journal*, Louisville, Kentucky, March 19, 2000.

Bugnosen, E. (1998). The Kias Gold Mine, Philippines. In *Small-scale gold mining: Examples from Bolivia, Philippines, and Zimbabwe.* ILO, www.ilo.org/public/english/dialogue/sector/papers/goldmine

Bugnosen, E. (2001). *Artisanal and small-scale mining: Philippines.* Country Study commissioned by MMSD, 8 pp., http://www.iied.org/mmsd/

Butler, J. (1990). *Gender trouble: Feminism and the subversion of identity.* Routledge publ., L.J. Nicholson, (Ed.)

Çagatay, N. (2001). *Trade, Gender and Poverty.* United Nations Development Programme, 45 pp.

Carr, M., (1993). Women in Small-scale Industries – Some Lessons from Africa, Ch. 7. In I.S.A. Baud & G.A. de Bruijne (Eds.) *Gender, Small-scale Industry and Development Policy.* London: IT Publications.

Carr, M. Chen, M. & Tate, J. (2000). Globalization and home-based workers. *Feminist Economics,* 6(3), 123–42.

Chakravorty, S.L. (2001). *Country paper on artisanal and small-scale mining in India.* Country Study commissioned by MMSD, 81 pp., http://www.iied.org/mmsd/

Cope, L.W. (2000). Malaysian tin. *Engineering and Mining Journal,* December 1, 2000.

Dreschler, B. (2001). *Small-scale mining and sustainable development within the SADC region.* Country Study commissioned by MMSD, 165 pp., http://www.iied.org/mmsd/

Everts, S. (1998). *Gender and Technology: Empowering Women, Engendering Development.* London: TOOLConsult.

Gender Working Group, (1995). Taking action: Conclusions and recommendations of the gender working group, Ch. 1. In G. Oldham, (Ed.) *Missing links.* UN Commission on Science and Technology, Gender Working Group book on Science and Technology Policy, IDRC, Canada, August 1995.

Grandjean, P. (1999). Methylmercury as a developmental neurotoxicant. In *Mercury as a global pollutant* – 5th Intl. Conf. Rio de Janiero, Brazil, (abstracts).

Gueye, D. (2001). *Small-scale mining in Burkina Faso.* Country Study commissioned by MMSD, 35 pp., http://www.iied.org/mmsd/

Gunson, A.J. & Yue, J. (2001). *Small-scale mining in the People's Republic of China.* Country Study commissioned by MMSD, 19 pp., http://www.iied.org/mmsd/

Heemskerk, M. (2000). *Gender and gold mining: The case of the Maroons of Suriname, working papers on women in international development 269.* A. Ferguson, (Ed.). Ann Arbour: Michigan State University: 48 pp., http://www.isp.msu.edu/WID/papers/abstracts.html

Hentschel, T., Hrushka, F. & Priester, M. (2002). *Global report on artisanal and small-scale mining.* Summary Report commissioned by MMSD, 67 pp., http://www.iied.org/mmsd/

Hilson, G. (2001). *A contextual review of the Ghanaian small-scale mining industry.* Country Study commissioned by MMSD, 29 pp., http://www.iied.org/mmsd/

Hilson, G. (2002). Small-scale mining and its socio-economic impact in developing countries. *Natural Resources Forum,* 26(1), 3–13.

Hinton, J.J. (2002). *Earthworms as a Bioindicator of Mercury Pollution in an Artisanal Mining Community, Cachoeira do Piriá, Brazil,* MASc

Thesis, Dept. of Mining Engineering, University of British Columbia, September 2002.

Hinton, J.J., Veiga, M.M. & Veiga, A.T.C. (2003). Clean artisanal gold mining: A utopian approach? *Journal of Cleaner Production*, 11(2), 99–115.

Hunter, J.M., Sparks, B.T., Mufunda, J., Musabayane, C.T., Sparks, H.V. & Mahomed, K. (2000). Economic development and women's blood pressure: Field evidence from rural Mashonaland, Zimbabwe. *Social Science and Medicine*, 50, 773–795.

Intermediate Technology Development Group (ITDG) (2002). Small-scale Miners: Women in Mining, http://www.itdg.org

International Development Research Centre (IDRC) (1995). *Missing links: Gender equity in science and technology*. Ottawa, Canada.

International Labour Organization (ILO) (1999). *Social and labour issues in small-scale mines*. Report for discussion at the Tripartite Meeting on Social and Labour Issues in Small-scale Mines, ILO, Geneva.

Jerez, D.B. (2001). *Regional/national study on small-scale and artisanal mining in Bolivia*. Country Study commissioned by MMSD, In Spanish, 41 pp., http://www.iied.org/mmsd/

Kaingu, N. (2003). *Personal communication: Zambian beliefs about women in mining*.

Keita, S. (2001). *The contribution of the Sadiola gold mining project to poverty reduction and the development of local mining communities*. Paper presented at MMSD, November 19–20, 2001, London.

Kuramoto, J. (2001). *Artisanal and informal mining in Peru*. Country Study commissioned by MMSD, 53 pp., http://www.iied.org/mmsd/

Labonne, B. (1996). Artisanal mining: An economic stepping stone for women, *Natural Resources Forum*, 20(2), 117–122.

Labonne, B. (1998). Of gold and women. *DESA News*, Vol. 2, Issue 1, February 1998.

Labonne, B. & Gilman, J. (1999). Towards building sustainable livelihoods in artisanal mining communities. In Proceedings from the Tripartite Meeting on Social and Labour Issues in Small-scale Mines, Geneva, Switzerland, May 17–21, 1999.

Machipisa, L. (1997). Rocky path for women miners. Interpress Service, http://www.moles.org/projectunderground/motherlode/gold/women5.html

Marcelle, G. & Jacob, M. Unlocking the double-bind facing women – science and technology policy for women in small and medium-sized enterprises. Chapter 11: In G. Oldham, (Ed.) *Missing links*. UN Commission on Science and Technology, Gender Working Group book on Science and Technology Policy, IDRC, Canada, August 1995.

Mugedeza, C. (1996). *Southern Africa-economy: Mining's unseen hands*. ANC Daily News Briefing, Friday 22 November, http://www.anc.org.za/anc/newsbrief/1996.

Muntemba, S. & Chimedza, R. (1995). Women spearhead food security. Ch. 4: In G. Oldham, (Ed.) *Missing links*. UN Commission on Science and

Technology, Gender Working Group book on Science and Technology Policy, IDRC, Canada, August 1995.

Murao, S., Daisa, E., Sera, K., Maglambayan, V.B. & Futatsugawa, S. (2002). PIXE measurement of human hairs from a small-scale mining site of the Philippines. *Nuclear Instruments and Methods in Physics Research B*, 189: 168–173.

Mushunje, M.T. (2001). *Women's land rights in Zimbabwe*. Report for the BASIS Management Entity, June 25, 2001, 19 pp., http://www.wisc.edu/ltc/live/bassaf0106a.pdf

Nandi, A. & Aich, P. (1996). Changing socio-economic profile of women working in small/medium mines, Eastern India. In *Proceedings of the Global Conference on Small/Medium Scale Mining, Calcutta, India, December 1996.*

Ofei-Aboagye (1995). Second session: Organizational, social, and women's participation issues. In *Proceeding from the World Bank Roundtable on Regularizing Informal Mining, Washington, DC, May 17–19, 1995.*

Ogbe, F.M.D. (2001). *Success stories of women in natural resources management in Africa: South West Nigeria*. UNU/INRA, ACCRA, 51 pp., http://www.unu.edu/inra/success_stories/nigeria.pdf

Onuh, B. (2002). *Salt women of Keana*. From *Newswatch* (Lagos) and reprinted at http://allafrica.com/stories/200211190755.html

Otchere, F.A., Veiga, M.M., Hinton, J.J., Farias, R.A. & Hamaguchi, R. (unpublished). *Transforming mining open pits into fish farms: A sustainable venture.*

Pollock, N.J. (1996). Impact of mining on Nauruan women. *Natural Resources Forum*, 20(2), 123–134.

Potter, J., Smith, R.L. & Api, A.M. (2001). An assessment of the release of inorganic cyanide from the fragrance materials benzyl cyanide, geranyl nitrile and citronellyl nitrile applied dermally to the rat. *Food and Chemical Toxicology*, 39, 147–151.

Quiroga, E.R. (2000). *The case of artisanal mining in Bolivia: Local participatory development and mining investment opportunities* (unpublished).

Robinson, K. (1998). A bitter harvest. *New Internationalist*, Issue 299, March 1998, http://www.newint.org/issue299/women.html

Rodriguez, R. (1993). *Women in the Garimpo, consequences of artisanal mining on society and the environment in the Amazon*, A. Mathis & R. Rehaag (Eds.) Cejup Graficentro, Belém, Brazil, *in portuguese*, pp. 149–155.

Rodriguez, M. (2003). *Personal communication: Interview and discussion of women in artisanal and small-scale mining in Columbia*. (*M. Rodriguez is a pseudonym*) January 2003.

Ruel, M.T., Haddad, L. & Garret, J.L. (1999). Some urban facts of life: Implications for research and policy. *World Development*, 27(11), 1917–1938.

Sandoval, M.C. (2002). *Personal communication*: *Women in artisanal mining in Latin America*. November 2002.

Sass, J. (2002). *Women, men, and environmental change: The gender dimensions of environmental policies and programs*. Population Reference Bureau, http://www.measurecommunication.org/

Sena do Nascimento, J.A. (2001). *Project database: Documentation of information about small and artisanal miners, Brazilian report*. Country Report for Mining Minerals and Sustainable Development Initiative, http://www.iied.org/mmsd/

Susapu, B. & Crispin, G. (2001). *Country study report on small-scale mining in Papua New Guinea*. Country Study commissioned by MMSD, 30 pp., http://www.iied.org/mmsd/

Synergy Africa (2001). Small-scale gemstone miners cry for help, on AllAfrica.com, http://allafrica.com/stories/200212060252.html

Traore, A.P. (1997). *Socio-economic aspects and gender issues related to artisanal gold mining*. In UNIDO Expert Group Meeting on "Introducing New Technologies for Abatement of Global Mercury Pollution Derived from Artisanal Gold Mining", UNIDO Headquarters, Vienna, July 1–3, 1997.

UNESCO (2000). Women and girls: Education not discrimination. *World Education Forum,* Dakar, Senegal, April 26–28, 2000, http://www2.UNESCO.org/wef/en-press/press-kit_wome.shtm

UNIDO (1998). *Assistance in reducing mercury emissions in highly contaminated gold mining areas in Mindanao – Phase I*. UNIDO Project Document DP/PHI/98/OO5.

UNIDO (2002). *Final proposal: Removal of barriers to the introduction of cleaner artisanal gold mining and extraction technologies*. Project Number GLO/01/G34.

UNIFEM (2000). *Progress of the world's women 2000: A new biennial report*, http://www.unifem.undp.org/progressww/2000/

USAID (2000). *Mining for gold in Siguiri: A close look at a high-risk population*. Report on Guinea, US Agency for International Development, http://www.usaid.gov/gn/health/news/001226_mines/index.htm

Veiga, M.M. (1996). *Advisory assistance on avoidance of mercury pollution from artisanal gold mining operations in state of Bolivar, Venezuela*. Report of the contract SI/VEN/94/801/11–51.

Veiga, M.M. (1997). *Introducing new technologies for abatement of global mercury pollution in Latin America*. Rio de Janeiro UNIDO/UBC/CETEM/ CNPq, Rio de Janiero, Brazil.

Veiga, M.M. & Hinton, J.J. (2002). Abandoned artisanal gold mines in the Brazilian Amazon: A legacy of mercury pollution. *Natural Resources Forum*, JNRF: 26(1).

Wakhungu, J.W. & Cecelski, E. (1995). A crises in power: Energy planning for development, Ch. 5. In G. Oldham, (Ed.) *Missing links*. UN Commission on Science and Technology, Gender Working Group book on Science and Technology Policy, IDRC, Canada, August 1995.

Wasserman, E. (1999). Environment, health and gender in Latin America: Trends and research issues. *Latin American Research Issues, Environmental Research Section A*, 80: 253–273.

Wente, B. (2002). It's the women who'll save Africa. *Globe and Mail*, December 14, 2002, A21.

Wiego (2002). *Women in informal employment globalizing and organizing*, www.wiego.org
Wightman, M. (2002). *Creating connections with women in India: Researcher profile*. Brenda Cranney; *IDRC Reports,* October 18, 2002, http://www.idrc.ca/reports/
WMMF (2000). Mining communities workshop: Artisanal mining, In: *Proceedings of the 1st WMMF, Toronto, Canada, March 8–11, 2000.*
World Bank (1998). *Social Capital Initiative Working Paper No.1*, April 1998.
World Bank (2000). *Ecuador gender review: Issues and recommendations*. A World Bank Country Study, World Bank Group Publ., Washington, DC.
World Bank (2000). *Ecuador gender review: Issues and recommendations*. World Bank Country Study, World Bank Publ., Washington, DC.
World Bank (2002). Bolivia at a Glance, Country Data, http://www.worldbank.org/data/countrydata/aag/bol_aag.pdf
World Health Organization (WHO) (1996). *Cadmium, in guidelines for drinking-water quality*. (2nd ed. Vol. 2.) Geneva, pp. 195–201.
World Health Organization (WHO) (2000). Silicosis, Fact Sheet No. 238, May 2000, http://www.who.int/inf-fs/en/fact238.html
World Health Organization (WHO) (2001). Advances in Behavioural Medicine, *The World Health Report – Mental Health: New Understanding, New Hope.* http://www.who.int/whr2001/2001/main/en/chapter1/001b2.htm

12

What's a Woman to Do? Globalized Gender Inequality in Small-Scale Mining

SUZANNE E. TALLICHET, MEREDITH M. REDLIN, AND ROSALIND P. HARRIS

Globalization represents an economic restructuring that exacerbates women's impoverishment world wide. It involves a transnationalization of capital, ascendance of finance capital, decentralization of industry, and reorganization of labour (Sachs, 1996, p. 141). The resulting changes, which include the impacts of structural adjustment programs, low commodity prices, declines in public and private employment (including the closure of larger-scale mines owned by corporations or the state), modified patterns of trading and farming, and inflation, have drawn rural workers – particularly women – away from their more traditional roles in subsistence agriculture, and toward more wage-dependent types of employment (ILO, 1999, p. 29). However, as rural women have entered the industries of the cash economy, they have become even more marginalized than when they engaged in subsistence economies (Third World Network, 1998). At the same time, many more women are also becoming household heads upon whom their children are solely dependent. Thus, rural women are more susceptible to poverty unless able to secure paid employment in other industries, such as small-scale mining. In the rural areas of developing countries, small-scale mining is often the best-paying pursuit available to women, especially those providing for their families while remaining near to their children and husbands, and, in certain cases, when working alongside both.

Small-scale mining was first recognized as an industry by the United Nations in 1972 (UN, 1972). Its operations are localized in predominantly the remote rural areas of Latin America, Africa and Asia. It involves the extraction of either highly valuable minerals such as gold, silver, and precious stones, or industrial minerals and related construction materials at, or near, the earth's surface (Hilson, 2002). Estimates show that small-scale mining produces between 15 and 20 percent of the world's non-fuel mineral production (Jennings, 1993). Initial capital investment is minimal, and productivity per worker is high. Moreover, it is believed that as much as 80 percent of small-scale mining world wide is performed illegally or without regulation (ILO, 1999). The work is arranged either informally, requiring mostly, if not exclusively, manual labor, or involves small organization of semi-skilled entrepreneurs utilizing semi-mechanized and slightly advanced processing techniques. But generally, the work is rudimentary, employing low-to-intermediate technology, accompanied by labor-intensive practices (Peiter et al., 2000; Hilson, 2002). Most small-scale miners lack the training and are unaware of the deleterious effects to their own health, safety and to the environment.

Because of its generally informal and low tech nature, low start up costs, and high labour intensity, small-scale mining activity has increased in recent years. Despite problems of environmental degradation, health, and safety, its prospects include boosting the economies of developing nations by relieving their debt via foreign exchange earnings, and providing lucrative employment to approximately 30 million male and female rural workers either directly or indirectly (ILO, 1999). Typically, the wages paid in small-scale mining pursuits are much higher than in other sectors of the economies of developing nations. Men who have been laid off from large-scale mining operations often readily find employment in the industry. More importantly, women miners have gained financial security as well as social independence from this pursuit. According to a recent study by the ILO (1999), women constitute an estimated 3.5 to 4 million of the world's 11.5 to 13 million small-scale miners, many of whom work part-time; another 1.5 to 2 million are estimated to be indirectly involved. However, these estimates also vary between five and 60 percent across continents. Female participation in small-scale mining is less than 10 percent in parts of Asia, and peaks around 60 percent in certain African countries. In Latin America, females constitute between 10 and 30 percent of the small-scale mining workforce. In certain countries, some women play a very prominent role in small-scale mining. For example, it is estimated that 75 percent of small-scale miners in Guinea are women; 50 percent in Madagascar, Mali and Zimbabwe; and 40 percent in Bolivia (Hilson, 2002).

The extent of women's involvement in small-scale mining has been steadily increasing, and is inversely related to the size of the mining operation. Countries where cultural convention has been defied, women play a dominant role as owners who hold permits and hire miners. In the majority of cases, they are members of cooperatives and other less formalized groups, and are employed as either paid laborers, or work as unpaid laborers with their husbands and children. Indirectly, women also provide support services to small-scale mining operations. Those actually involved in mining work in all aspects of the industry, but mainly carry out panning, carrying and processing tasks (ILO, 1999; Labonne, 1996). Although women sometimes earn the same salaries as men, they comparatively earn less overall, because most work in lesser-skilled menial jobs or perform more delicate tasks for which they are deemed to be better suited.

Clearly, the work of most female small-scale miners has been restricted by socio-economic and cultural barriers (Hilson, 2002). Since the vast majority of rural women in small-scale mining are poor, they lack formal education, training in mining techniques, and access to bank credit. In many developing countries, culturally, women are considered inferior to men both at work and home. At work, their status is generally lower, and, in cases when they supervise men, they are often disobeyed. Due to their subordinate status in the family, women often require their husband's consent before seeking assistance to finance their own mining operations.

The objective of this chapter is to examine gender inequality in the small-scale mining industry. It begins with a brief review of the socio-economic and cultural barriers to women's full participation in the sector, followed by a theoretical discussion that draws upon recent feminist frameworks about women in development. The chapter concludes with a discussion of what governments and international agencies can do to alleviate gender inequality via policies and programs aimed at increasing women's integration into the industry.

WOMEN IN SMALL-SCALE MINING

As developing world economies have restructured in response to globalization, small-scale mining has grown in 35 countries in Africa, Latin America and Asia by an estimated 20 percent since the late-1990s. Many of its miners are women seeking a temporary financial stop gap for themselves and their families. However, commensurate with their secondary status in their respective societies, women bear more health and safety burdens with fewer rewards than their male counterparts.

Global restructuring does not uniformly affect the lives of rural women world wide (Sachs, 1996, p. 147). Rather, the effects of these trends are mediated locally by existing social, economic, and cultural (ideological) forces. But generally speaking, they do serve to restrict the extent to which women can become fully integrated into small-scale mining. More specifically, across regions and types of mining, women face a common set of challenges to becoming full participants at all levels of small-scale mining. The following scenarios illustrate this reality.

Africa

In Africa, rural women have historically engaged in informal economic activities in order to supplement the subsistence agricultural production that takes place in and around their households. Understandably, then, there are more women engaged in small-scale mining in southern and eastern Africa than anywhere else in the world. As estimated by the Economic Commission for Africa, there are 1.8 million female miners. According to the ILO (ILO, 1999), African women participate at all levels of the industry. Some even own mines and have formed regional associations exclusively for the women participating in the industry. In locales where women are evenly distributed at different levels of operations, they are more likely to be treated more equitably, although this is not always the case. Women occupying positions commanding greater responsibility and authority have regularly encountered gender bias, which dampens their successes relative to men's. For example, in Tanzania, over a decade ago, small-scale mining legislation was enacted, granting more than 500 mining licenses to individual women, and to groups, cooperatives and companies headed by women. Nonetheless, as is the case in Ghana, Guinea, and Mali, Tanzanian men are reluctant to obey women who own and/or operate mines, largely because of tradition, and possibly because men commonly believe that they possess superior experience and knowledge about mining (ILO, 1999). In Ghana, women traders have resorted to using male agents to assist them in order to avoid gender discrimination, but doing so has created additional problems, as many have been either cheated or robbed outright by these middlemen.

When it comes to mining, the extent of female involvement varies between African countries, depending on the materials being mined (and the processing techniques used) and the accompanying socio-cultural proscriptions regarding women's societal roles. Although women predominate in non-metal mines and quarrying operations, in gold mining, for example, they constitute about five percent of the workforce in Gabon, 15 percent in Ghana, and almost a third in Guinea and Mali (ILO, 1999). Very few women work underground, although some remove gold close to the earth's surface using rakes and hoes, or through panning alongside streams. Rather, the gold that is processed (through crushing, grinding, washing, sieving and carrying) by women is typically carried out in the home. Much of this work is performed manually without the benefits of machinery. It would appear, however, that overall, women are less involved in the mining of high value minerals, and, when they are, they are restricted to gender-appropriate work, such as washing and other preparatory tasks.

Until recently, the laws of some African countries forbade women from mining underground. Additionally, according to the customs of some communities, women working at the mines are considered to be bad luck, and, in other communities, are forbidden to mine during menstruation. For example, in Zambia, people believe that if a woman goes near a gemstone mine, she will disturb the spirits of the stones and they will burrow more deeply into the earth (AllAfrica.com, 2001). Pioneering mine owner Namakau Kaingu, head of the Southern African Development Community (SADC) Women in Mining Trust, was shot at while sleeping in her hut one night by men trying to scare her off her claim. "That only gave me more courage," she said (Machipisa, 1997). "We face a lot of rejection and we are not taken seriously by people in the field". According to Association of Zambian Women Miners' secretary Rita Mittal, "there are a lot of traditional obstacles along the way. Chiefs feel undermined when they see women coming to mine their areas. They are hostile" (Machipisa, 1997).

Women are even more severely restricted when their work is carried out within the context of the family. Traditionally, African women are solely responsible for child care and other domestic duties. Their family responsibilities often tie them to a specific locale, therefore restricting their work hours. When women move to different mining areas, the strain of being separated from their families and isolated from their communities have proven particularly severe.

Moreover, because males are considered to be the sole heads of the household, women are subordinated to their authority. They often work with them for no pay and need their husbands' consent before becoming a permit holder or obtaining credit for personal entrepreneurial purposes.

Finally, a study undertaken for the United Nations Development Fund for Women (UNIFEM) found that few governments have undertaken studies to assess women's problems and the extent to which they benefit from engaging in small-scale mining. There are also very few policies or programs that directly address gender-based issues (ILO, 1999). For example, relative to men, women mine owners and miners alike have difficulty obtaining the technical advice and training appropriate to their level of involvement. Rather, they rely on family connections and the experience they have gained while operating in the industry. Moreover, they often have more difficulty getting sufficient financial support to run their mining operations. Banks and governments are generally unwilling to finance small-scale mining operations, the same UNIFEM study confirming that only six percent of Zimbabwean female miners were successful in securing loans for their operations, due in part to a lack of collateral but also because of bankers' negative attitudes toward women in business (ILO, 1999). Despite its positive attitude toward female-orientated economic activities, the government provides little support to women involved in small-scale mining. Thus, operations run by women are undercapitalized, lack technology, and are less efficient than those run by men.

Latin America

According to ILO (1999) estimates, about 20 percent of small-scale miners in Latin American countries are female, although estimates vary by country. They are represented at all levels of mining, but only between one and two percent assume leadership roles (Veiga, 1997). Although Latin American women generally engage in mining informally, most are organized into cooperative associations. In Bolivia, for example, of the 60,000 members in mining cooperatives, 7,500 are women (ILO, 1999). From the top to the bottom of the small-scale mining work hierarchy, women are involved as full cooperative members; working members of cooperatives; partially paid shift workers; traders; workers performing unorganized, clandestine tasks (*baranquilleros*); and as scavengers (*palliris*) at mine sites (ILO, 1999). Women are also employed as guards at these sites, some working double shifts as both guards and *palliris*.

Both men and women work together in cooperatives for equal remuneration at operations engaged in the extraction of gold, gemstones and tin ore. Women are generally responsible for breaking large stones with hammers, while men are responsible for grinding, a task requiring the use of heavy and cumbersome tools. Women are heavily involved in processing tasks such as sieving, washing, and using mercury to separate and amalgamate gold. Work at gold mines is especially arduous for women working informally at the bottom of the hierarchy. For example, *baranquilleros* endure long hours with low pay in hazardous working conditions, often standing in polluted water at high altitudes with little chance for advancement (ILO, 1999). *Palliris* gather and sort through the mine tailings of processing plants, which they either sell themselves or through intermediaries.

According to Women's Mining Organization member Janeth Barzola, in Peru "the women have to pay 60 percent of what they earn to middle men. Otherwise, the middle men refuse to supply them with water which they look after as if it were gold" (ICEM, 1998). Similarly, in Nicaragua, women have begun working by processing the waste from gold mines, which is then sold to middle men. According to Dalia Indiana, a female miner with the Federation of Mineworkers' Unions of Nicaragua, these individuals profit from the women's work because, "all the machinery comes from one owner. The women have to sell the gold to the middle men in order to get the machine" (ICEM, 1998). Moreover, as is the case in African countries, these women are triply burdened with their domestic duties, caring for their children, and carrying out their mining work that supports them. Frequently, they are criticized for neglecting their children when they are forced to leave them in order to earn their wages. Those who work with their husbands do so to enhance his wages, with little or no pay for themselves.

Asia

In Asian countries, where fewer women are found working in small-scale mining than in other countries worldwide, work is generally confined to sorting, packing, and loading minerals onto conveyances (ILO, 1999). In Thailand and Vietnam, both women and men are paid the same wage for the same work. However, women in India are paid a minimum wage or, for extra work, slightly more, while men are paid more for performing the same tasks – namely the carrying of minerals from one point of the mining operation to another. Women also scavenge for materials that may have fallen off trucks or were spilled during extraction. Tribal women work alongside their husbands while children are tended by elderly family members. Similar to women in other countries, their wages are rarely their own, and are used to provide for their families. Companies' increasing use of trucks is now beginning to threaten their job security (labourfile.org, 2000).

Similarly, in the Philippines, where the government is trying to stimulate developments in the mining sector, TNCs are moving in, forcing traditional miners, half of whom are women, out of their jobs, denying them of their livelihoods. Seventy-two year old Angela Bisitan, who has been a small-scale miner for most of her life, said she never had any problems buying items like rice. However, since being pushed out of her mining job, she can now only afford the low-quality variety (Malanes, 1997). According to Maura Almoza, a traditional miner from a small town 30 kilometers from Itogon, "the good times are only memories" because before the Benguet Corporation moved in "many of us could send our children to private schools and to college. We could also build comfortable homes and could afford amenities" (Malanes, 1997). At the same time, the mining company has instituted an unofficial policy banning women from open pit mining based on the superstition that they are bad luck at the mines. In fact, not only do male small-scale miners from the Itogon area (known as *camota* miners) forbid women from entering their pocket mines, but they also do not have sexual intercourse for at least one week before entering the mine tunnels to "ensure" their own safety underground (Cimatu, 2001).

FEMINIST THEORY ABOUT WOMEN IN DEVELOPMENT

Since the early-1970s, feminist scholars have examined the ways in which global economic processes have contributed to global gender inequality. Three distinct, but related, theoretical frameworks have contributed to recent research on women in development – more specifically, underdevelopment. The first, post-colonial approaches, as proposed by Spivak (1994), highlight the ambiguities and contradictions in traditional cultural positions held by women, and their often convenient and limited integration into the new larger economic structures. Secondly, discursive critiques of development bring forward constitutive gaps in global planning, and local and indigenous planning outcomes (Escobar, 1995; Sachs, 1993; Ferguson, 1994). Finally, indigenous and eco-feminist perspectives, widely known through the work of Shiva (1993, 2000), help to bring forward frameworks by which to bridge the gaps and omissions arising from current development policies.

This discussion on small-scale mining and the cultural limitations faced by women reflects the cultural double bind and contradictory (re)presentation of women arising from globalization. As noted in the introduction, women increasingly find themselves in socio-economic positions of self-dependence, even as the opportunities to pursue economic independence are culturally limited, or, in some cases, prohibited. The end consequence of this contradiction is a newly constituted female "object", who is both key to future economic development and retention of cultural integrity and identity through the recreation of gender "subject" positions and roles. This difficult position is described by Spivak (1994, p. 102): "Between patriarchy and imperialism, subject-constitution and object-formation, the figure of the woman disappears, not into a pristine nothingness, but into a violent shuttling with the displaced figuration of the 'third-world woman' caught between tradition and modernization".

This "violent shuttling" is seen in the disparate and limited integration of women into the small-scale mining industry, and the socio-economic position that drives this contradictory integration. On the one hand, women under globalization regimes are key to facilitating a transition to globalized economies. Moreover, their economic subsistence, as noted previously in this chapter, is increasingly dependent on their inclusion into the formal economy. On the other hand, the cultural roles of the feminine subject under patriarchy are integral to cultural integrity, linking the home and the workplace – *i.e.* the areas of private and public activities that comprise both discursive and structural coherence in a culture. Women's structural transitions in this interplay, in turn, not only raise issues of gender politics, but also multiple issues of cultural survival.

In this shuttling, the end effects for women are limitations that extend to both subject and object positions. Discursive reductions due to their displacement in culture determine their level and compensation for their work in the industry, thereby marginalizing them.

Anti-development theorists have consistently demonstrated that the contradictions inherent in past and current regimes of Western-driven development and planning extend to deliberate action to create this very contradiction under the auspices of "progress". Women's role in small-scale mining emerges at the axis of discursive constructions of the market, as well as discursive constructions of natural resources. This axis represents a narrowing of understanding of both market and resources. The market, for example, has now a discursive position that (re)places this former multi-faceted social institution into the form of a dispassionate and fair actor.

The market return on small-scale mining drives its growth in these nations' economies. As an actor, this discursive market is no longer a social institution to be regulated or managed for its service to the larger society, but is rather society's ruling force. As noted by Berthoud (1993, p. 71), "it has become increasingly self-evident that the market should no longer be viewed as an institution which must be regulated by external social forces, but, on the contrary, that it should be used to regulate society as a whole". A transition in the role and presence of the market in the global economy of necessity predominates, and all development efforts then focus on "freeing" the market to increase the benefits of development. In this way, even determined efforts to assist women into market economic participation establishes them in limited roles as worker-objects, above, and beyond, their family and cultural responsibilities.

Berthoud (1993, p. 71) further notes that, "development is held to be possible only for those who are ready to rid themselves entirely of their traditions, and devote themselves to making economic profit, at the expense of the whole gamut of social and moral obligations". Given women's crucial cultural and economic roles, the burden for this transition is placed upon them, even as they are displaced from the possibility of effective action as either a worker or a cultural protector.

The discursive formation of natural resources has also been altered because of globalization. Shiva (1993, 2001) stresses that natural resources constitute one of the boundaries of appropriate human activity – that is, the relation between natural resources and human society is one of reciprocity, not exploitation for the end goals of the market. This discursive transformation evolved not in opposition to, but in conjunction with, development and market discourse, and prescribed women's positions in both:

> Nature, whose real nature it is to rise again, was transformed by the originally Western worldview into dead and manipulable matter. Its capacity to renew and grow had been denied. It had become dependent on people. The development of people was thus essential for the development of nature (Shiva, 1993, p. 206).

The process of globalization in relation to small-scale mining does not have an unexpected outcome of women's concentration in lower paid and culturally restrictive employment. Rather, the primary importance of the market as an actor and the (re)formulation of resources, including female labor, as objects, requires the dismantling of culture and people to fit.

Discourse theory has been generally criticized for its analytical penchant for undermining and/or deconstructing without suggesting an alternative path. In relation to women's roles in small-scale mining, we are still left with one simple question: What, then, should we do?

Following Foucault (1994), most development discourse scholars are reticent to offer social or political advice or formulations of policy, even when approached. The nature of the development discourse and its social ramifications is such that any advice offered places one firmly back into the development discourse itself, despite the careful critique that preceded it. Ferguson (1994, p. 284) explains these limitations in his conclusion of his study of Lesotho: "These agencies seem hungry for good advice, and ready to act on it. Why not give it?... [because they] seek only the kind of advice they can take". Attempts to "fix" or ameliorate the effects of globalization must be (re)presented within the discourse in order to be given, and to be taken and put, into effect. Therefore, policy initiatives or programming objectives normally reflect the problems that already exist, and may tend to exacerbate their effects. Such initiatives and programs most often focus on, as stated by Shiva (1993), "the development of people" to make them more adept and amenable to the global market system. In this manner, attempts to address women's precarious economic and cultural contradictions under global development often result in merely "developing" them into more appropriate actors for their global role, rather than addressing the contradiction itself.

Beyond the mere difficulties in identifying points of productive action for activists and practitioners, development discourse also privileges and encourages activities that misdirect productive action. The process of globalization is not an insulated, unchanging process. In order to be effective, it must be self-examining and self-critical of its multiple practices. However, the outcome of the examinations of the globalization process remain limited due to the fact that the process is purely self-referential. As Ferguson (1994, p. 285) explains:

> In "development" as in criminology, "problems" and calls for reform are necessary to the functioning of the machine. Pointing out errors and suggesting improvements is an integral part of the process of justifying and legitimating "development" interventions. Such an activity may indeed have some beneficial or mitigating effects, but it does not change the fundamental character of those interventions.

The critique itself serves to validate, rather than undermine, processes of global development.

Yet, something must be done. Women's economic involvement in small-scale mining is growing, and the economic advantages (despite health and welfare disadvantages) will continue to propel their participation. The opportunity for policy, participatory and social action is located outside of the goals and objectives found in globalization and international market forces. Ferguson

(1994, p. 286) suggests that either intellectual or other outside engagement is appropriate under very restrictive conditions: "1) where it is possible to identify interests, organizations, and groupings that clearly represent movements of empowerment and 2) when a demand exists on the side of those working for their own empowerment for the specific skills and expertise that the specialist possesses". The emphasis here is not on creating empowerment for its own sake, or for the purpose of furthering already existent goals. Rather, engagement action is directed by those in the position, and driven by their desire and drive to alter the contradictions they face.

In sum, the economies of mining-dependent regions are characterized by boom and bust cycles accompanied by relatively high levels of segregation and poverty. During the past few decades, feminist scholars have maintained that Western development of underdeveloped countries has increased the amount of women's work, decreased their access to resources, and has added to the "feminization of poverty" (Sachs, 1997). They further maintain that women, as well as resources, have been exploited by male-biased development strategies.

While the foregoing section examined women's lives as affected by their participation in small-scale mining on three continents, the present section has offered a feminist interpretation of those issues that both directly and indirectly affect women in this industry. In the final section of this chapter, we turn to policy evaluation to complete a picture of "what should be done."

POLICIES AND PROGRAMS FOR THE FUTURE

Globally, women employed in the artisanal and small-scale mining industry continue to receive less recognition and remuneration largely because of the forces of economic globalization. As suggested by the foregoing feminist frameworks, policies and their corresponding programs should reflect the understanding that the market needs to be regulated so that it acquires a more managed status, rather than be allowed to become a force acting on its own. Such programs need to operate under the assumption that action must be taken to develop the small-scale mining industry, so that human needs and environmental standards are met, and not the reverse. Furthermore, it is imperative that the goals of these policies and programs increase the roles that women play in acquiring their own empowerment as actors determining their own positions in the industry, in the process, minimizing Western corporate influences. It is in keeping with these formulations that the following initiatives are identified below.

First, as suggested by Hilson (2002), there is a need to introduce cultural education programs in countries where small-scale mining is currently developing. Clearly, women are affected by the degree of patriarchal control in their respective societies; this determines their economic dependency on men, as well as the prevalence of a gendered ideology within their respective cultures. The latter involves the extent to which mining is seen as "men's work", based on superstitions and beliefs about female inferiority – namely, their perceived inability to perform strenuous mining tasks. This further maintains their subordinate status in the family, where they are more appropriately seen in terms of their domestic and child care responsibilities. In turn, all of this is reflected in their subordinate status in society, which influences their relationships with male miners, company personnel, and government officials, whereby they are denied the training, credit and other assistance that might enable them to secure mine ownership or other positions of authority over men in the industry. Members of the private, public and NGO sectors could do much to improve the work environment for women by sensitizing men to women's present and potential contributions to the industry.

Secondly, the illiteracy rate among rural women in these countries is cause for great concern. According to the 1990 Human Development Report, female illiteracy rate was approximately 17 percent in Latin America and 19 percent in Southeast Asia. Within various groups in south Asia, and in northern and Sub-Saharan Africa, the female illiteracy rate is at least 25 percent higher than that of the men's (Beneria & Bisnath, 2000). Thus, more educational programs are needed to improve female literacy so that they are able to participate in the training and technical assistance programs designed to improve knowledge of mining technology, safety, and environmental protection, as well as managerial skills and access to credit. If such changes are not made on the educational front, women will continue to be relegated to more menial roles in the small-scale mining industry.

Thirdly, as mentioned previously, women require training in the techniques and technology of small-scale mining to enable them to expand their roles in the work force, and eventually assume leadership positions. Female miners also require greater knowledge about the safety hazards associated with mining, and to be educated about safer, and more environmentally sound, mining practices. More often than not, processing is labeled as "woman's work", and, when processing minerals such as gold, women typically handle hazardous materials in their homes, thus jeopardizing the health of their children. Silicosis and mercury poisoning are not uncommon among female miners and their children, which has prompted the launching of various training programs. For example, in Zimbabwe, the United Nations Development Fund for Women (UNIFEM), with technical assistance from the Intermediate Technology Group (ITDG), has established a school to train female miners help start businesses of their own (UN, 1994). More such programs are urgently needed in countries where none presently exist.

Fourthly, women lack representation and political support from their own governments in their quests to attain equal opportunity in the small-scale mining industry. A 1994 United Nations (34–35) document (UN, 1994) declares the objectives for national governments by 2000 making necessary changes "to eliminate constitutional, legal, administrative cultural behavioral, social and economic obstacles to women's full participation in sustainable development and in public life. This strategy will, *inter alia*, ensure women's access to vocational training, all forms of credit, particularly in the informal sector, and property rights. Programmes to (a) support and strengthen equal employment opportunities and equitable remuneration for women in the formal and informal sectors,

(b) eliminate persistent negative images and prejudices against women, and (c) enhance the legal capacity of women to gain access to entitlement to natural resources, technology, banking facilities and the control of pollution at the workplace ..." Governments need to legalize small-scale mining where it is considered illegal. They could also remove legal constraints on licensing and permit holding, or pass legislation permitting women the right to legally own property and to declare their own income in order to obtain credit. For example, the Ethiopian government has recently enacted legislation promoting mining as a sustainable long term industry that provides for the training of female staff members to manage the same projects offered to men (UN, 1994). Ghana and Zimbabwe have passed neutral legislation allowing women and men equal access to mining titles (Labonne, 1996).

Fifthly, more women's support groups, such as SADC Women in Mining Trust, are needed. Established in 1997, this group consists of eight affiliated national associations from eight African countries: Angola, Botswana, Namibia, Mozambique, Swaziland, United Republic of Tanzania, Zambia, and Zimbabwe. Their mission is the recognition and advancement of women in small-scale mining, by lobbying, identifying and obtaining training and technical assistance, financial assistance, facilitating marketing and setting up a library and database (ILO, 1999). In Latin America, a three year project sponsored by a Dutch trade union confederation was launched to train women trainers to strengthen their mining unions (ICEM, 1998). Beginning in 1998, workshops for female small-scale miners were held in Nicaragua and Peru, and alliances have been formed between women in mining, textiles and other factory work in several Latin American countries. More of these initiatives are needed where they currently do not exist. There is also need for networking among mining women and affected communities, and alliance-building within economic sectors where oppressed and marginalized workers are found. These organizations can begin to put pressure on governments.

Finally, more research is needed to gain a thorough understanding of the specific situations women employed in the small-scale mining industry are experiencing in different continents, countries, and regions. These data could be used to justify the sponsorship of development projects that focus on both improving the livelihoods of the industry's women, and focus more on the development of human resources in general (UN, 1996).

What are these needs? To enlist the cooperation of governments, banks, miners groups and miners, data must be collected at the local level, and then shared with organizations involved with small-scale mining. The role universities could play is almost limitless. As Victoria Tauli-Corpuz, Director of the Tebtebba Foundation (Indigenous Peoples' International Center for Policy Research and Education), has stated, "we need a better understanding of globalization trends as they affect industries such as small-scale mining so that a counter globalization strategy can be developed that promotes basic human rights, particularly the rights of women and their children" (Third World Network, 1998). From here, education campaigns to increase community awareness of the global restructuring of mining could travel through established networks of female miners.

Clearly, the sustainability of the small-scale mining sector is directly related to the economic well-being of women and their families. But miners cannot do this alone. They need governments and international agencies, such as the United Nations and the World Bank, along with universities, to help address educational, technological, economic and environmental initiatives. There are important long-term benefits to such a plan that serve to boost both the economies of these developing nations and raise the standard of living of its impoverished citizens, especially women. When women are involved in small-scale mining as entrepreneurs or permit holders, as opposed to miners and laborers, there is much that they can do to improve their own economic status. They can provide employment opportunities for other women and serve as role models in mining or in other industries that are financed via entrepreneurial involvement in mining. This could also lift the limits that the traditions of their cultures have imposed; this would allow many more women the chance to lead lives that are more economically secure, if not advantageous, and would serve to elevate the status of women in other areas of their lives as well.

REFERENCES

AllAfrica.com. (2001). Zambia: Small scale gemstone miners cry for help. 30 July. <wysiwyg://52/http://allafrica.com/stories/200107300021.html>

Beneria, L. & Bisnath, S. (2000). Gender and poverty: An analysis for action. In F. Letcher & J. Boli (Eds.), *The globalization reader.* Oxford: Blackwell Publishers.

Berthoud, G. Market. In W. Sachs (Ed.), *The development dictionary: A guide to knowledge and power.* Witwatersrand University Press: Johannesburg, South Africa, pp. 70–87.

Cimatu, F. (2001). Benguet mine tunnels off limits to women. Philippine Daily Inquirer, 26 September. <wysiwyg://103/http://www.geocities.com/ferdibee/womines.htm>

Escobar, A. (1995). *Encountering development: The making and unmaking of the Third World.* Princeton, NJ: Princeton University Press.

Ferguson, J. (1994). *The anti-politics machine: Development, depoliticization, and bureaucratic power in Lesotho.* University of Minnesota Press: Minneapolis, MN.

Foucault, M. (1994). *The order of things: An archaeology of the human sciences.* New York, NY: Vintage Books Edition.

Hilson, G. (2002). Small-scale mining and its socio-economic impact in developing countries. *Natural Resources Forum*, 26, 3–13.

International Federation of Chemical, Energy, Mine and General Workers' Union (ICEM) (1998). Training women trainers: New ICEMLatin America Project. ICEM Info., Vol. 3 (1). <http://www.icem.org/info/no1-98/women.html>

International Labour Organization (ILO) (1999). *Social and labour issues in small-scale mines.* Report for discussion at the Tripartite Meeting on Social and Labour Issues in Small-scale Mines. International Labour Organisation, Sectoral Activities Programme, International Labour Office, Geneva.

Jennings, N.S. (1993). Small-scale mining in developing countries: Addressing labour and social issues. *Guidelines for the development of small/medium-scale mining.* United Nations. New York.

Labonne, B. (1996). Artisanal mining: an economic stepping stone for women. *Natural Resources Forum*, 20(2), 117–122.

labourfile.org (2000). Andra Pradesh: Tribal movements stop illegal mining. On website: *Mining in India: Movements, multinationals and malaise.* <http://www.labourfile.org/Mining/Report.htm>

Machipisa, L. (1997). Rocky path for women miners. 8 December. On website: *Women speak out on mining: Africa.* <http://www.moles.org/Project Underground/motherlode/gold/women5.html>

Malanes, M. (1997). Good times are gone in the mining industry. 8 December. On website: *Women speak out on mining: Philippines.* <http://www.moles.org/ProjectUnderground/motherlode/gold/women2.html >

Peiter, C., Villas-Boas, R.C. & Shinya W. (2000). The stone forum: implementing a consensus building methodology to address impacts associated with small mining and quarry operations. *Natural Resources Forum*, 24(1), 1–9.

Sachs, C. (1996). *Gendered fields: Rural women, agriculture, and the environment.* Boulder, Colorado: Westview Press.

Sachs, C. (1997). *Women working in the environment.* Washington, D.C.: Taylor & Francis.

Sachs, W. (1993). Introduction. In W. Sachs (Ed.), *The development dictionary: A guide to knowledge and power* (pp. 1–5). Johannesburg, South Africa: Witwatersrand University Press.

Shiva, V. (2000). *Stolen harvest: The hijacking of the global food supply.* Cambridge, MA: South End Press.

Shiva, V. (1993). Resources. In W. Sachs (Ed.), *The development dictionary: A guide to knowledge and power* (pp. 206–218). Johannesburg, South Africa: Witwatersrand University Press.

Spivak, Gayatri C. (1994). "Can the Subaltern Speak?". In P. Williams and L. Chrisman (Eds.), *Colonial discourse and post-colonial theory: A reader* (pp. 66–111). New York: Columbia University Press.

Third World Network (1998). *The globalization of mining and it's impact and challenges for women.* Paper delivered by Victoria Tauli-Corpuz, director of the Tebtebba Foundation (Indigenous Peoples' International Center for Policy Research and Education), at the International Conference on Women in Mining, Baguio City, January 1997. <http://www.twnside.org.sg/title/mining-cn.htm>

United Nations (UN) (1996). *Developments in small-scale mining.* Report of the Secretary-General. Economic and Social Council. Committee on Natural Resources. United Nations, New York. <http://www.un.org/documents/ecosoc/c7/1996/ec71996-9.htm>

UN (1994). *Economic and social development needs in the mineral sector: Small-scale mining activities in developing countries and economies in transition.* Report from the Secretary-General. Economic and Social Council. Committee on Natural Resources. United Nations, New York. <http://www.un.org/documents/ecosoc/c7/1994/ec71994-9.htm>

UN (1972). *Small-scale mining in developing countries.* Department of Economic and Social Affairs, United Nations, New York.

Veiga, M.M. (1997). *Mercury in artisanal gold mining in Latin America: Facts, fantasies and solutions.* Paper prepared for the Expert Group Meeting on UNIDO High Impact Programme "Introducing New Technologies for Abatement of Global Mercury Pollution Deriving from Artisanal Gold Mining," Vienna, 1–3 July.

13

Occupational and Other Diseases in the Small-Scale Mining Sector

MALEBABO SAKOANE

Small-scale mining is a low revenue, labour-intensive, subsistence activity that provides employment to a significant number of people in the developing world, most of whom inhabit remote areas where well-paid job opportunities are scarce. Although its operators are generally unskilled with limited formal education, average incomes in the industry are usually higher than those paid in other small-scale sectors of the economy. In certain regions, such as West Africa, small-scale mining is mainly a dry season activity; in such places, inhabitants turn to other income-generating activities such as farming during the rainy season. There are, however, many areas where it is performed year-round, and overall, a significant portion of the rural communities in Africa, Latin America and Asia are heavily dependant on small-scale mining for their survival.

Many people have no alternative but to practice artisanal and small-scale mining under harsh working conditions. In the rare cases that occupational safety and health regulations exist, they are seldomly enforced. Because of a lack of reliable data, occurrences of disease in small-scale mining regions have been poorly documented. Moreover, a lack of health care makes it impossible to gauge the extent of rampant diseases. Most miners are unwilling to use their hard-earned incomes to improve health practices, particularly in cases where benefits cannot be realised in the short-term. They therefore rely on self-help, traditional health services, and primary health care services, in the event of illness or injury.

This objective of this chapter is to outline the health hazards and diseases associated with this sector of industry.

ENVIRONMENTAL INFLUENCES

Over the long term, mineworkers are subjected and exposed to high levels of noise, toxic gases, dust, chemicals, vibration, insufficient lighting, extreme heat and cold, direct sun, repeated strain and contaminated water supplies (i.e. water containing silt, mercury and organic material). Underground workers are exposed to post-explosion irritations, and commonly experience pulmonary oedema (the build-up of fluid in the spaces outside the blood vessels of the lungs) and asphyxiation (lack of oxygen which interferes with the oxygenation of the blood). Such risks are aggravated by a lack of adequate ventilation, and the fact that many miners elect not to use protective respiratory gear. Furthermore, the combination of unbalanced meals and strenuous labour reduces resistance to disease.

COMMON DISEASES IN MINING REGIONS

The diseases of high prevalence in the mining regions of the developing world include malaria, Sexually Transmitted Diseases (STDs), diarrhoea, and tuberculosis (TB). Occurrences of malaria, diarrhoea and TB are mainly induced by deteriorating mining conditions. For example, extensive digging precipitates the development of pools of water, which form following episodes of rainfall; they then become breeding grounds for malaria-infected mosquitoes. In fact, malaria has had a serious impact in mining regions because many mine workers lack the finances to purchase protective medicines, and, therefore, expose themselves to the disease. There were as many as 578,000 cases of malaria reported in Brazilian mining regions during the period 1970–1989[1] alone.

Diarrhoea results from the consumption of contaminated water sources – namely, water from local rivers and unprotected wells. STDs, on the other hand, are widespread throughout many mining regions because of thriving prostitution, which attracts many men from their homes and families. High incidences of TB at sites are indication of the prevalence of HIV/AIDS in many mine camps and surrounding communities.

The often-harsh living conditions, along with a lack of information and education on disease prevention, contribute to high incidences of HIV and other communicable diseases among miners and their families. In Latin America, for example, the locations of small-scale and artisanal mining operations and incidences of infectious diseases appear to be highly correlated.

Airborne contaminants, such as rock particles and dust, are produced as a result of drilling, blasting, loading, and crushing. The inhalation of silica-rich dust, produced as a result of crushing ore with high silica content, can lead to silico-tuberculosis. Cases of work-related lung cancers are also common. People exposed to excessive dust for prolonged periods also put themselves at risk of developing permanent lung diseases such as silicosis.

Fumes produced during shot-firing operations contain toxic gases (such as sulphur dioxide, nitrous oxide, and nitric oxide). If inhaled, sulphur dioxide can

[1] Graduate School of the Environment, Macquarie University. *Gold Mining/Mercury*, http://www.gse.mq.edu.au, downloaded on 11/10/02.

cause diseases of the lung and respiratory disorders such as wheezing and shortness of breath. Even moderate concentrations of sulphur dioxide can stimulate lung deterioration in asthmatics. Higher sulphur dioxide levels may result in tightness in the chest and excessive coughing. Sulphur dioxide pollution is considered more harmful when particulate and other pollutant concentrations are high – something known as the "cocktail effect". Nitrogen dioxide can cause a variety of environmental and health impacts, and is a persistent respiratory irritant that can exacerbate asthma, and possibly increase susceptibility to infections.

Repeated or prolonged exposure to excessive noise levels leads to hearing impairment. Workers operating hand-held machinery, particularly pneumatic rock drills and pick hammers, can suffer from the effects of vibration in their hands and arms. Vibration White Finger (VWF) or "dead finger" starts when the fingers become numb, and can lead to gangrene (death and decay of body tissue) in the limbs.

For example, in Mererani, Tanzania, where the major hazards of mine-related work include heat, noise, vibration, and a lack of ventilation underground, complaints of tiredness, headaches and other stress-related symptoms are common. When pneumatic drills are used, the effects of vibration and dust (neither water nor masks are used to suppress or guard against dust) are inflicted on young people, but do not appear until years later. The carrying of heavy loads in awkward, crouching positions, has shown to be a precursor to the lower-back problems that frequently appear later in life. The complete lack of medical and health facilities in the area (other than certain private dispensaries manned by medical assistants and charity organizations) means that there is no screening and no indication of the effect of these working conditions on any of the workers. Some of the region's employers only contribute to medical expenses if they feel it is in their interest to have the worker return to work rather than find replacements.

DISEASES ASSOCIATED WITH MINERAL PROCESSING

The processing of non-metallic minerals from small-scale mines, such as coal, limestone and gemstones, is largely a dry, physical process that relies on classification of particle size and observation. Various stages of production include transport, size reduction, sorting, and, depending on the final product, packing product into sacks or other containers before loading it onto trucks or loading it in bulk, for final dispatch. Crushing and milling processes are notoriously hazardous because of the presence of machinery, noise and vibration. Workers participating in wet processes, such as sluicing and collecting concentrate, commonly have their hands and feet permanently wet, which, as a result, induces a high degree of skin maceration and cuts; waterproof boots, eye goggles and gloves are rarely used. Moreover, as many small-scale mineral processing plants operate 12–24 hours a day all year round, workers commonly complain of irritability, sleeping disorders and a loss of appetite.

The major hazards associated with such activities are linked to the use of tools and machinery (such as crushing equipment, mechanical sieves and conveyor belts), and common injuries include strains and sprains from over-exertion, and those resulting from slips and falls. Unguarded machinery and practices, such as hand sorting while sitting adjacent to inclined and elevated conveyor belts, further increase the likelihood of accidents. In plants where women and men fill sacks or other containers with powdered products, load them onto trolleys, and then transport them into waiting trucks, or carry the sacks on their backs, dust is pervasive. Only occasionally are scarves or dust masks worn.

In some countries, such as India, it is common for women, and, sometimes, teenage girls, to carry on their heads baskets of clay or stones that weigh as much as 25–30 kg. In many instances, these baskets are carried up slippery inclines, where the contents are deposited into crushers, hoppers, or onto a stockpile. The repeated lifting and carrying of sacks and baskets of material are precursors to lower back problems, which are further aggravated by poor posture during mining in confined spaces, panning and amalgamation. More specifically, the existence of narrow adits forces miners to work underground without standing; moreover, most miners are forced to slouch during sluicing and panning.

Once liquids, chemicals and high temperatures become part of mineral processing, hazards and the risk of accidents increase markedly, particularly burns (from chemicals or heat) and scalds (from hot liquids).

During recovery/processing of gold, the pre-concentrate may be mixed with mercury in a process known as amalgamation. Amalgamation recovery is often conducted in open spaces. The mixture obtained is then boiled in the open air and the gold is recovered. During the process of amalgamation recovery, a portion of the mercury is commonly released directly into the river system, and the remainder is released as gas into the atmosphere, some of which is inhaled by operators. In Peru, where amalgam is known to be processed everywhere, mercury poisoning tends to be higher among women and children than men, who tend to spend more time in the only uncontaminated area – *i.e.* the mine itself. Elemental mercury vapour is released during the heating of the mercury/gold mixture. The vapour released during amalgamation is absorbed into the lungs, where it is converted into methylmercury, which is then absorbed into the bloodstream. The effects of this on the body include colic, vomiting, gastro-enteritis, kidney/urinary tract infection, acute enteritis, gum ulcers and extreme sensitivity to light. Moreover, when absorbed into the central nervous system, elevated levels of methylmercury can cause tremors, speech disturbances, poor concentration and mood swings, insomnia, brain damage, and death (See Chapters 31 and 37 for more detailed discussions on the impacts of mercury in artisanal and small-scale mining communities).

Extended and repeated exposure to mercury vapour can eventually lead to chronic mercury poisoning. For example, an investigation around two riverside villages downstream from gold mining activities in Brazil revealed that 97% of the villagers had high concentrations of mercury in their hair (on average

15 milligrams per gram). Although these levels of mercury contamination were under the World Health Organisation's threshold of 50,[2] people still exhibited declines in manual dexterity and certain visual functions, such as the ability to distinguish contrasting lines.

It is possible that the high exposure of miners to metallic mercury increases susceptibility to malaria. More specifically, it is suspected that mercury exposure suppresses the immune system, making people more susceptible to malaria infection.[3]

INVESTIGATING DISEASES IN THE MINING COMMUNITY

Diseases associated with child labour

Working long hours under arduous conditions is difficult enough for adults. It takes an even more serious toll on the soft bones and growing bodies of children. In Tanzania, the eagerness to be the first to reach newly blasted areas and, hopefully, find tanzanite, means that little time elapses between underground blasting and workers re-entering of the mine. Some are believed to hide in the mine to ensure they will be first at the work face after blasting, as there is no reliable record of who is in the mine or not. With little or no pay and no choice of work, child workers face a bleak future, hoping to be taken onto the payroll as they grow older, and hoping to find some gemstones through scavenging.

Non-fatal asphyxiation due to lack of oxygen is a common occurrence among children. Additional medical complaints among child workers include septic wounds, malaria, gastroenteritis, general weakness, and aches and pains. In a study carried out in the Philippines, it was found that child mine workers were suffering from carrying heavy loads, and from mercury contamination. Some eighty percent of the sample population had suffered from respiratory ailments, and two-thirds from muscle-skeletal disorders.[4] They reported having had various symptoms, particularly joint and muscle pains. When examined, almost all of the children exhibited skin, ear, nose, throat and neck abnormalities, while 60 per cent had restricted lung function.

Women's health

Although the health risks of mining are similar for both sexes, the industry can be especially hazardous for women, especially if they come into contact with chemicals that present a health risk to foetuses or breast-feeding infants.

---

[2] International Development Research Centre (1997/98). *1997/98 Annual Report: Strategies and Policies for Healthy Societies Theme*, http//www. idrc.ca, downloaded on 11/09/02.

[3] World Resources Council (1999). *Malaria in the Brazilian Amazon*, http://www.wri.org, downloaded on 11/10/02.

[4] International Labour Organisation (1999). *Social and Labour Issues in Small-Scale Mines*, http://www.ilo.org, downloaded on 10/09/02.

Traditionally, in the developing world, men have been provided with a larger share of food, a better part of the lodging, and undertake less housekeeping work. Many female labourers, therefore, are forced to assume all of the domestic responsibilities, in addition to their hard mine labour. All of these factors have led to poor health conditions. Common ailments suffered by women include headaches, muscular and skeletal pains, and occasional fevers.

Community health

Insufficient supplies of clean water and inadequate sanitation, and the sickness and diseases which follow, underpin much of what is unsatisfactory about small-scale mining, particularly in areas that have seen sudden influxes of people in search of a living through mining. Where small-scale miners live at the mine site, or where processing is carried out in nearby communities, health and sanitation conditions are generally poor. Most people use pit latrines or the open bush; as a result, cholera, typhoid, dysentery, tuberculosis, bilharzia and enteric infections are common. Even long-term miners are unwilling or unable to afford the construction of hygienic sanitation facilities. Transient miners, therefore, have absolutely no incentive to do so.

Where water for domestic use is collected from open wells, there are often serious health implications. Only boreholes and protected wells can be considered to provide *safe* water. However, the drilling of boreholes and creation of protected wells requires money, which, in most cases, is not readily available. The continued use of river water for domestic purposes is of particular concern since it is also used for panning and bathing. The transient nature of many small-scale mining communities means that even if local authorities are aware of operations, there is little or no incentive to build long term infrastructure such as drainage, sanitation, wells, boreholes, and clinics. Directly related to this is the lack of educational facilities, which inevitably results in a high level of illiteracy. A lack of education and the resulting low employability is also mainly responsible for inducing excessive child labour, crime, prostitution and the spread of sexually transmitted diseases (including AIDS); there is an acute shortage of information on the prevalence of these diseases.

In areas where mercury is used to recover gold, it is possible that local people may consume contaminated fish. Mercury is first ingested by the fish, and then converted to methylmercury in the body of the fish; as already explained, methylmercury is more readily absorbed into the blood stream than elemental mercury. Mercury contamination can also lead to the poisoning of fish-eating communities downstream that may not even be involved in mining or the use of mercury. A World Bank study carried out in 1997 showed that 50% of schoolchildren had 100 times more mercury in their bodies in mining areas as compared to non-mining regions.[5]

---

[5] World Bank (1997). *Environmental assessments of Mining Projects,* www.worldbank.org, downloaded on 11/07/02.

DRAWING COMPARISONS WITH DISEASES AFFECTING THE LARGE-SCALE MINING SEGMENT AND SMALL-SCALE AGRICULTURAL OPERATIONS

Disease and large-scale mining

In both small and large-scale mining, the degree of work-related health risks depends on the type of mineral resource being mined. For example, emphysema, a lung disease involving damage to the air sacs/alveoli, is thought to be caused by the chemicals within coal dust, which are released into in the lungs, where they damage the walls of the air sacs. Chronic lung disease is common among migrant underground mine workers in South Africa.[6] Another example is chronic bronchitis, which is an inflammation of the bronchi, the main passage to the lungs. The severity of the disease relates to the amount and duration of exposure to coal dust. In the US, for example, an estimated 2000 miners die every year from lung diseases caused by exposure to coal mine dust.[7]

Tarkwa, a Ghanaian town with a long history of mining, (gold, bauxite and manganese), has the highest incidence of malaria in Ghana.[8] Here, there are also high incidences of pulmonary and silico-tuberculosis.

In South African gold mines, the rate of TB infection is high. However, the contribution of infections imported into mining communities vis-à-vis transmission within them has been poorly documented. It is believed that most incidences of TB in the mining community are a result of ongoing transmission, and that up to one third of TB cases is a result of previously failed treatment.

In South Africa, up to 20% of men currently engaged in gold mining have radiological silicosis. Silicosis and overcrowded living conditions have been shown to be risk factors for TB (the risk of developing active TB is increased three-fold). In Southern Africa, mine workers have been found to have a much higher incidence of TB than the general population (1000–2000 TB cases per 100,000 population per year compared to about 240 per 100,000 in the general population).[9]

In both small-scale and large-scale mining, HIV/AIDS infections are exacerbated by the migrant labour system and unaccompanied miners. In small-scale mining, the lack of information and education about HIV/AIDS and how it is transmitted, encourages infection. In large-scale mining, miners are relatively well informed about HIV/AIDS but often elect not to prevent or minimise the rate of infection. Often, they are in denial and feel powerless in the

---

[6] World Bank (2001). *Background Paper/Extractive Industries Review*, www.worldbank.org.mining, downloaded on 23/08/02.

[7] http://www.altindia.net, *Uranium Mining in Jaduguda, Bihar, Living in Death's Shadow,* downloaded on 12/10/02.

[8] The Centers for Disease Control and Prevention, *Mine Safety & Health*, http://www.cdc.gov, downloaded on 12/10/02.

[9] Whiteside, A. (Ed.). *Aids Brief for Sectoral Planners and Managers: Mining Sector.* http://www.und.ac.za/und/heard, downloaded on 17/10/02.

face of yet another disease. Health education programmes and free condom distribution in a gold mining community in Carletonville (South Africa) have not stopped commercial sex workers from having unprotected sex. This is due to the fact that clients refuse to use condoms. It was found that 69% the region's women are HIV-positive,[10] and infection rates of approximately 50% appear standard for Southern African mines.[11] In some developing countries, however, it is difficult to confirm the relationship between mining and the spread of already prevalent diseases such as malaria and HIV/AIDS.

The establishment of a large mine is often associated with an increased availability of health services. Moreover, employment and increased living standards bring about nutritional and psychological benefits, and hence, better health standards. However, this may not necessarily translate into improved community health if the facilities are not made available to the broader community.

A study of Navajo Indians in the Western United States found that an unusually high number of birth defects, including hydrocephaly, microcephaly, Downs Syndrome, cleft lip, and epilepsy, were common among more than 500 babies born between 1967 and 1974. The area is characterized by more than 350 abandoned open-cut uranium mines. There was also a significant increase in acute leukaemia in Grand Junction, Colorado, where uranium tailings were used in the construction of 6000 homes, schools, shopping centres and footpaths.[12] A study carried out in Jaduguda, India, the location of yet another uranium mine, shows that people from villages near the tailings ponds suffer ailments such as skin diseases, cancer, TB, fertility loss, bone and brain damage, kidney damage, hypertension, disorders of the central nervous system, congenital deformities, insomnia, nausea, dizziness, and joint pains. Here, one third of women exhibit fertility problems.[13]

It has been known since 1920 that uranium miners suffer high mortality from lung cancer caused by exposure to radio-active radon gas during their work. When inhaled, radon gas and its decay products easily lodge in the human lung, emitting energetic alpha particles that adversely affect the vulnerable layer of cells lining the lungs. Mining of radio-active minerals is not typical of small-scale operations, however.

Some of the detrimental health effects of mining on communities may surface after mining has ceased altogether. A case in point is the current legal case

[10] International Planned Parenthood Federation (2002). *Majority of Sex Workers in South African Gold Mining Community are HIV Positive*, http://ippfnet.ippf.org, downloaded on 12/10/02.

[11] Kunanayagam, R. et al. (2001). *Mining and Poverty Reduction*, http://www.worldbank.org, downloaded on 23/07/02.

[12] http://www.greenleft.org.au (1996). *Dollars for Death: why Uranium Mining Should End,* downloaded on 12/10/02.

[13] http://www.altindia.net, *Uranium Mining in Jaduguda, Bihar, Living in Death's Shadow,* downloaded on 12/10/02.

involving a South African asbestos mine that closed in 1968, and surrounding communities, which only recently have begun exhibiting higher incidences of various lung diseases.

Diseases in small-scale agriculture

The small-scale agricultural sector is also characterized by high incidences of TB, malaria and HIV/AIDS. In small-scale agriculture, more women are involved in production. Thus, in small-scale farming, infected women pose a three-fold labour loss, a threat to reduced care provision at home, not to mention the plethora of problems related to rearing children.

Some diseases are caused by exposure to pesticides and other toxic chemicals used in response to increased pressure for fertilizer to be used with shifts in crop production, hybrid seeds, and increased yields to match input costs. Agricultural workers involved in subsistence farming in the developing world also face hazards from imported pesticides and fertilizers, which have been banned for a number of health reasons in other producing countries. These fertilizers and pesticides are supposedly labelled with adequate warning, but by the time they reach farmers, often, labels have worn off or are written in a language that farmers do not understand. These health problems are compounded when importing countries themselves have insufficient knowledge of the hazards or toxicity of the fertilizers and pesticides they import.

WHAT NEEDS TO BE DONE?

The challenges facing the small-scale mining sector need to be addressed through appropriate legislation. Most current legislation only takes into account the needs of the large-scale mining sector. In South Africa, for example, the *Mine Health and Safety Act* of 1996 stipulates that "every mine with more than 20 employees must have a health and safety representative for each shift at each designated working place at a mine". This, however, does not apply to mines employing less than 20 employees, even though the risks may be similar.

In addition, mining communities must be formalized. This enables authorities to train individuals more effectively, and to make them better aware of appropriate mining methods, and their impacts on human health and the environment. Partnerships with large mining companies can go a long way towards educating small-scale miners on best practices and methods. More importantly, more research is needed to improve monitoring, and to design and implement solutions to problems. For example, most of the relevant information available on occupational health is qualitative. Given the large volume of employment in this sector, however, a better data base is needed on exposures and practices, and their impact on health.

Governments have to begin supporting the small-scale mining sector in this regard. They must begin to appreciate the role that the sector plays socially, and then act accordingly.

CONCLUSION

Small-scale mining alleviates poverty in rural areas where job opportunities are few. However, this source of employment is associated with a number of occupational diseases, including water-borne diseases, respiratory diseases, mercury poisoning, muscle-skeletal disorders, malnutrition, and STDs such as HIV/AIDS. The causes of these diseases range from ignorance and a lack of health facilities, to direct contact with dangerous chemicals and machinery. However, a shortage of research and information makes it is impossible to quantify these problems in many respects.

Similarly, in large-scale mining, miners suffer from numerous ailments, including respiratory diseases, skin and hearing disorders, and symptoms resulting from overexposure to radioactive minerals. The small-scale agricultural sector is also characterized by high incidences of TB, malaria and HIV/AIDS. Moreover, it experiences diseases caused by exposure to pesticides and fertilizers.

The health problems of small-scale mining communities can only be tackled if significant research is undertaken and the appropriate information is made available; this is a starting point for the design of relevant legislation, after which, governments can better tackle the rampant health problems in the sector. Governments must also involve target communities in research programs and data collection activities.

REFERENCES

UN Economic & Social Council (1996). *Developments in Small Scale Mining*, http://www.un.org, downloaded on 23/08/02, p. 5.

National Directorate of Mines (2000). Artisanal Mining Baseline Survey, Mozambique, Maputo, pp.

Third World Network, *Mined Out,* http://twnafrica.org.org, downloaded on 11/10/02.

Part III

African Case Studies of Artisanal and Small-Scale Mining

14

Introduction Part III

GAVIN M. HILSON

Africa is endowed with rich and diverse mineral resources. Its countries collectively produce some 60 metal and mineral products in mass quantities. Overall, the continent hosts approximately 30% of the Earth's mineral resources, including 60% of cobalt, 70% of platinum and 35% of gold (Table 14.1).

For thousands of years, numerous minerals – particularly gold, diamonds and gemstones – have been extracted and processed on an artisanal and small

Table 14.1 Proportion of world mineral reserves in selected African countries.

Commodity	Country	% of global reserves
Gold	South Africa	35.0
Coal	South Africa	10.9
Uranium	Namibia	6.7
	South Africa	9.4
Cobalt	Zambia	5.6
	DRC	26.0
	South Africa	0.2
Copper	Zambia	5.2
	South Africa	2.0
	DRC	4.6
Nickel	South Africa	8.4
	Botswana	0.6
	Zimbabwe	0.2
Titanium	South Africa	20.4
Chromium	South Africa	68.3
	Zimbabwe	20.5
Iron ore	South Africa	0.9
Manganese	South Africa	80.0
Vanadium	South Africa	44.5

Source: Sinkala, 2002, p. 15.

scale in Africa. These outputs formed the basis of wealth for many empires and kingdoms during colonial and pre-colonial periods. The true pioneers of African artisanal mining were the San ("Bushman") hunter-gatherers, who, as early as 40,000 years ago, began exploiting obsidian and chalcedony rock for stone implements and weapons.

In the 9th century AD, Arab merchants began travelling vast stretches of the Sahara Desert to engage in trade with West African civilizations whose kingdoms were based on artisanal gold mining activity. At approximately the same time, small-scale miners began intensively working various portions of the greenstone belts in Zimbabwe, Botswana, Tanzania, Mozambique and South Africa.

African small-scale gold and diamond mining experienced unprecedented growth in the 1800s, following an extended period of trans-Atlantic slavery, which had put continental mineral interests on the back burner for nearly 200 years. By the turn of the twentieth century, operations had agglomerated within some of the continent's most prominent mining regions of the present, including the Zambian Copperbelt, the many Lake Victoria gold fields of Tanzania, and the Obuasi region of the former Gold Coast. Today, artisanal and small-scale mining operations can be found continent-wide within a diverse set of environments (Fig. 14.1 highlights some important small-scale mining regions in Africa).

In their recent efforts to define "artisanal" and "small-scale mining" in national mineral policies, codes and legislation, African governments have relied on a wide range of criteria, including production, levels of mechanization, and concession size (see Table 14.2). Although the industry is far from achieving its full potential (UN, 2002), the continent's operations employ at least three million men, women and children (See Table 14.3 for artisanal and small-scale mining employment estimates for selected African countries) and make notable contributions to continental mineral output. For example, in sub-Saharan Africa, gold and gemstones valued at US$1 billion are produced by artisanal and small-scale miners each year (ILO, 1999). In Ghana, the government invested a modest US$1.4 million to construct regional buying stations that pay world prices to small-scale miners for their gold, a move that has resulted in the collection of well over US$140 million in revenues that would have otherwise been lost (Labonne, 1996). In other African countries, such as Guinea and the Central African Republic, mineral production is almost entirely confined to small-scale mining. In the Central African Republic, 90% of diamond and 100% of gold production is carried out by small-scale miners. In Guinea, the share of small-scale mining in national mineral production increased from 66% in 1990 to almost 100% in 1993 (UNECA, 1993; Bocoum & Samba, 1995; UN, 1996; Hilson, 2002).

This section of the book presents a series of case studies on African artisanal and small-scale mining. The cases provide a varied set of perspectives on the socioeconomic aspects of Africa's artisanal and small-scale mining industry, particularly, its precious minerals segment.

The opening chapter, by Esther Kazilimani, Florence Bukali and Gary McMahon, presents the findings of a recent baseline study conducted on

(1) Western Ghana's Birimian and Tarkwaian gold belts and Oda diamond fields.
(2) Diamond fields in Angola and Botswana, which continue to be a major cause of violent regional conflicts.
(3) Lake Victoria gold fields in Tanzania.
(4) Gemstone mining regions in Zambia, much of which occurs near the Copperbelt.
(5) Gold panning in Harare and Bulawayo, Zimbabwe
(6) Gold panning in Manica, Tete, Niassa and Nampula, Mozambique

Figure 14.1 Important small-scale mining regions in Africa.

artisanal gold mining in Mozambique. It initially provides background on the industry, and then present the results of the survey and complementary discussions conducted with artisanal miners/local community members from four important mining regions in the country. The chapter that follows, by S. Mondlane and D.S.M. Shoko, builds upon the discussion presented in the previous chapter, describing the legal framework in place in Mozambique for artisanal mining, the industry's employment characteristics, and the chief socioeconomic problems it faces.

Chapter 17, by Winnie V. Mitullah, Jason S. Ogola, and Monica A. Omulo, examines the state of small-scale gold mining in Kenya. Using a case study of the Migori District, the chapter provides background information on Kenya's

Table 14.2 Different criteria used in the definition of small-scale mining in selected African countries.

Country	Criteria used
Ethiopia	Annual Production, level of mechanization
Ghana	Capital investment, number of participants, concession size
Guinea	Type of mineral exploited
Ivory Coast	Level of mechanization
Senegal	Depth of working, crude production levels
South Africa	Capital investment
Tanzania	Annual production capacity
Zambia	Size of concession area
Zimbabwe	Size of concession area, capital investment

Source: UN, 2002

Table 14.3 Small-scale mine employment in selected African countries.

Country	Employment
Burkina Faso	60,000–70,000
Burundi	10,000
Central African Republic	45,000
Chad	10,000–15,000
Congo DR	150,000
Ivory Coast	10,000–15,000
Ethiopia	>100,000
Ghana	50,000–300,000
Guinea	40,000
Kenya	30,000–40,000
Madagascar	5000–20,000
Mali	100,000
Morocco	5000–10,000
Mozambique	700–100,000
Namibia	5000–10,000
Nigeria	10,000–20,000
Niger	440,000
Rwanda	5000–15,000
Senegal	3000
Sierra Leone	30,000–40,000
South Africa	10,000
Tanzania	450,000–600,000
Uganda	5000–10,000
Zambia	20,000–30,000
Zimbabwe	50,000–350,000

Source: ILO, 1999

mine workers, their employment and incomes, and the effects of small-scale gold mining activities on education and agriculture.

In the first of his two chapters on artisanal and small-scale mining in Tanzania, Crispin Kinabo examines the industry's regulatory and chief policy-related elements. Moreover, the author examines working conditions at sites, and the environmental problems operations are causing in many of the country's rural areas. In his second chapter, Crispin Kinabo focuses on the importance of the female contingent in the Tanzanian artisanal and small-scale mining industry. The author draws many comparisons to the global situation.

Two case studies of artisanal and small-scale mining in Zambia are then presented. The first case, by Munyindei Masialeti and Crispin Kinabo, focuses on the organizational aspects of the Zambian small-scale mining industry, which is dominated by gemstone extraction activity. Stephens Kambani builds on Masialeti and Kinabo's work, detailing the legislative and policy environment in place for small-scale mining in Zambia. The author also briefly describes the different characteristics of the gemstone and industrial mineral sub-sectors of the industry.

In Chapter 22, Oliver Maponga and Maideyi Meck examines the state of the alluvial gold panning industry in Zimbabwe. Specifically, the authors review the panning activities along the Mazowe River and its tributaries in Mashonaland Central Province some 40 km outside Harare, in an effort to provide insight into the dynamics of a typical illegal panning site in Zimbabwe. The chapter seeks to further the understanding of conditions that drive communities into illegal gold panning in the first place.

The concluding chapter of *African Case Studies of Artisanal and Small-Scale Mining*, by Benjamin Aryee, provides a comprehensive overview of artisanal and small-scale mining activity in Ghana. The chapter begins by defining small-scale mining in the Ghanaian context, and traces the development of the sector. After reviewing the industry's problems, constraints and challenges faced, the author proposes a framework within which the rationale of the sub-sector's activities can be analyzed.

REFERENCES

Bocoum, A. & Samba, A. (1995). *Le secteur artisanal en Republique Centrafricaine: aspects socio-economiques et organisationnels.* International Roundtable on Artisanal Mining, World Bank, Washington.

Hilson, G. (2002). Small-scale mining and its socioeconomic impact in developing countries. *Natural Resources Forum*, 26(1), 3–13.

International Labour Organization (ILO). (1999). *Social and labour issues in small-scale mines.* Report for discussion at the Tripartite Meeting on Social and Labour Issues in Small-scale Mines, International Labour Organization, Sectoral Activities Programme, International Labour Office, Geneva.

Labonne, B. (1996). Artisanal mining: an economic stepping stone for women. *Natural Resources Forum*, 20(2), 117–122.

United Nations (UN). (1996). Recent developments in small-scale mining: a report of the Secretary-General of the United Nations. *Natural Resources Forum*, 20(3), 215–225.

United Nations (UN). (2002). *The Compendium on Best Practices in Small-Scale Mining in Africa.* United Nations Economic Commission for Africa (UNECA), Draft Report, Addis Ababa.

United Nations Commission for Africa (UNECA). (1993). *Development and utilization of mineral resources in Africa.* United Nations Economic Commission for Africa (UNECA), Report of the Fifth Regional Conference, Ethiopia.

15

Artisanal Mining Baseline Survey of Mozambique

Esther Kazilimani, Florence Bukali da Graça, and Gary McMahon

In Mozambique, over 90% of all mining activities are classified as "small-scale" or "artisanal". They operate using prospecting licenses; marketing licenses issued by the Ministry of Mineral Resources and Energy (MIREME); and mining certificates issued by the provincial directorates of MIREME. Activities are concentrated in pegmatite areas situated between the provinces of Nampala and Zambezia. However, many other people are reported to be involved in the mining of semi-precious stones such as emerald, tourmaline morganite, and aquamarine (as well as gold) in Niassa, Tete, Manica and Cabo Delgado.

Mozambican artisanal and small-scale mining is characterized by intensive labour technology with low productivity, a lack of financing, a lack of skilled labour, poor safety measures, and limited geological information of prospective mineral areas. Operations are often associated with having an adverse impact on the environment, largely because of the extraction and mineral processing methods employed. These practices have also led to increases in various health problems, such as malaria, cholera, and bilharzia. Furthermore, as numerous miners are operating in remote locations, few have access to social services such as health, sanitation, or education.

This chapter will begin by describing the general situation of artisanal miners in Mozambique, and its most important mining areas. It then presents the results of a recent survey and discussions conducted with artisanal miners/local community members in four important mining regions in the country. The study emphasizes changes in socio-economic, environmental, and health conditions. Finally, recommendations are made that, if adopted, could improve the efficiency and sustainability of artisanal and small-scale mining in Mozambique.

CONTEXT

In recent years, the Directorate of Mines (DNM) of MIREME has undertaken mining sector policy reforms, which have culminated in improvements to the Mozambican mining regulatory framework; it clearly defines the Government as a regulator, promoter, facilitator and supervisor of resident mining activities, and the private sector as the industry's chief operator and implementing agent. DNM is in the process of preparing a Management Project, the main objective of which is to promote private investment in the sector through institutional strengthening. It is expected that, the strategy of sustainable exploitation of the sector's resources – namely, extraction in an environmentally sound manner – will contribute to the Government of Mozambique's objectives of reducing poverty and improving the quality-of-life of its rural people.

In 1999, MIREME, with the support of the World Bank, held a seminar on artisanal mining. At the time, a decision was made to strengthen DNM's capacity to manage the sector. Thus, a study was sponsored, the aim of which was to establish baseline information on the country's artisanal mining sector. The specific objectives of the study were to:

- Undertake a national survey of the artisanal activities in gemstone and gold mining in Mozambique;
- Describe and assess the organisation of artisanal mining in relation to production, processing and marketing;
- Highlight the constraints and problems affecting the sector's development;
- Assess the impacts of artisanal mining on the livelihood of local peoples, local economy, environment and human health; and
- Make recommendations on how to better manage, develop and regulate the sector.

The study was conducted in the following four provinces of Mozambique, each of which contains gold and gemstone mining operations: Manica, Tete, Niassa and Nampula.

STUDY METHODOLOGY

The following research methods were used in the study:

- Questionnaire;
- Observation;
- Focus Group Discussions;
- In-depth personal interviews; and
- Reviews of Health Records.

Teachers from nearby schools, health staff, and cultural and religious leaders were among the key informants that participated in the survey. The target groups were gold and gemstone miners, community members, community leaders, mineral traders, store owners and informal sellers (owners of tuck shops or "*barracas*"). Focus group discussions were conducted with men, women and children at mine sites and in communities. Because of logistical problems and time constraints, focus groups discussions were not carried out in all of the provinces.

Short interviews were conducted with members of the community, the feedback from which was used as complementary information. The issues discussed included local authority structures, infrastructure, ethnicity, proximity and identity, work and gender, cultural and religious practices, land ownership and use, settlement patterns (post war, mining, etc.), food security, access to natural resources, and economic activities.

Some 10% of the country's miners were interviewed. The approximate population of the mining sites visited in Manica Province was about 600 (500 in Munhena, 50 in Chitewe, and 50 in Sussundenga); in Niassa Province, N'tulo, the population was between 300 and 400; in Tete Province, Cassoca, the population consisted of about 60 miners; and in Nampula Province, it was estimated that there were 100 gemstone miners in Namuhara, and between 500 and 1000 gold miners in Murrupula. However, no interviews were conducted in Murrupula, as its population had already been surveyed in the recent past (Valoi et al., 2000). Nevertheless, observation of the on-going activities in Murrupula enriched the survey data and information collected. Random interviews were conducted with mineral traders, storeowners, community members, and community leaders at mine camps and in villages adjacent to the mine sites.

GEOLOGY AND MINERALIZATION OF THE SURVEYED AREAS

In Mozambique, artisanal mining operations are concentrated in areas containing pegmatite deposits – namely, within the provinces of Nampula and Zambezia, and certain geological environments in Niassa, Tete, Manica and Cabo Delgado. These areas contain precious and semi-precious stones such as emerald, tourmaline, morganite and aquamarine. Gold is mainly extracted in the provinces of Tete, Manica, and Niassa and, more recently, in Zambezia.

Manica Province

The artisanal mining sites in Manica Province are located between 12 and 25 km north of the town of Manica, along the greenstone belt. Here, gold is disseminated in veins within a silicified zone associated with a NW-SE oriented shear zone in a greenstone environment; the zone is approximately 100 meters wide. Red lateritic soil several metres deep covers most of the area. The weathered veins and shear zones closer to the bedrock, which are conserved within the laterite, are very much sought after by gold panners.

Previous exploration carried out in Manica has shown that the area contains low-grade, high tonnage gold deposits of disseminated mineralization in decomposed gneiss rock. Exploration work undertaken by the *Ashanti Gold Fields* determined the gold ore content of rock chips from re-opened trenches (3.69 g/t) to be the highest in the area. The gold content of shallow trenches (from highly decomposed rock) was determined to be between 0.11 and 0.69 g/t; these findings are consistent with those of old reports, which point to the region's average gold content being less than 1 g/t, increasing with depth (from 0.65 g/t at the surface and 3.11 g/t at a pit depth of 8 m).

Tete Province

In Tete Province, artisanal gold miners are found in Chifumbazi near to the Zambian border, as well as along the Luenha and Mazoe Rivers in Changara district. New gold deposits have been reported along the Tete-Songo Road in Marara near Chirodzi, and in Cassoca; a significant number of artisanal miners are now operating in these areas.

Gold and gemstone mining is also taking place in Marilongoe in Maravia, and also in Angónia and Tsangano districts. In Zumbo district, which borders Zambia and Zimbabwe, some pegmatite-related minerals are reportedly being extracted by artisanal miners from both neighbouring countries.

Niassa Province

Since 1992, Niassa has become one of the most important artisanal gold mining areas in Mozambique, featuring some thirty thousand (30,000) miners, the majority of whom originate from neighbouring Tanzania. The Lago and Sanga districts are the main regions of importance for artisanal gold mining activities, where considerable amounts of alluvial gold are produced and smuggled out of the country. The main mining sites in Niassa are M'papa, Mianzini, Chiwindi, M'peia, and, more recently, N'tulo, which is located along the Kalandon River some 20 km from the administrative post of Lupilichi.

Niassa Province is situated at the extreme northwest of Mozambique. Here, important pegmatitic mineralisations of tourmalines, beryl, colombo-tantalite, mica and lithium are processed. There is also a favourable metallogeny for titanomagnetites, beryllium, gold, sulphide and copper carbonates. Gold is extracted from both riverbeds and quartz veins.

Nampula Province

The alluvial gold mined at Murrupula originates from a primary deposit in Toconhu located approximately 20 km southwest of Murrupula Town, and 33 km northeast of the Muiane Mine. The veins are partially covered by weathered debris but appear fairly continuous and extend up to 1,000 metres in length; thickness varies up to six metres. These deposits have reportedly been worked previously, though there are no production figures. However, it appears that most work was conducted on this mineralised zone. It is probable that existing excavations represent exploratory works rather than extraction of ore.

Gemstone mining areas surveyed (Tete and Nampula)

One of the gemstone mining areas visited was Monte Kapapa in Tete Province, some 30 km from Angónia Town. No interviews were conducted with the miners, as operations at Mt. Kapapa were said to be conducted at night. However, a visit was made to the gemstone site to observe various trenches and pits. There was evidence of on-going works in the area, thus indicating that mining activity was, in fact, taking place at this site.

Other sites visited in Nampula Province were located in Mavuco, Chalaua, and Nkule-Mukurussi-Namuhara. Currently, most of the artisanal miners are concentrated in Mavuco, Chalaua, and Nkule-Mukurussi-Namuhara, where mainly aquamarine is being recovered using simple mining methods.

THE SOCIO-ECONOMIC AND ENVIRONMENTAL IMPACTS OF ARTISANAL MINING IN MOZAMBIQUE

Infrastructure and social services

Apart from Murrupula, which is located about 85 km from the provincial capital city of Nampula, and 21 km from the town of Murrupula along a good quality tarmac road, most of Mozambique's artisanal mining sites are located in remote and highly inaccessible areas (Fig. 15.1).

Namuhara is located about 145 km from Nampula. There is no feeder road to Namuhara, apart from wide paths and trails that have long been deeply cut and eroded. N'tulo in Niassa Province is approximately 285 km from the capital city of Lichinga. Cassoca in Tete Province is located around 80 km from the capital city, Tete, and Monte Kapapa in the same province is some 35 km from the town of Angónia.

During the rainy season, most of the roads are in bad condition, which impedes travel, communication and trade. Minimal maintenance is conducted

Figure 15.1 Access roads and bridges in Lupilichi, Niassa province.

on roads and bridges. In Murrupula, because of fairly good quality access roads, the mobility of traders from Nampula and Zambézia Province is high. Local traders have built shops and stalls, selling a wide range of commercial goods and buying gold throughout the year. In gemstone mining areas, the mobility of traders is not influenced by the condition of the access routes to these sites. Rather, the value of the products and the high buying power of the traders play an important role.

Health and educational services are limited in mining communities, both in terms of availability and overall capacity to respond to the needs of the population. Many Mozambican mining communities do not have access to potable water, educational facilities, and health services. In several mining areas, local rivers represent the main source of water used for drinking, cooking and mining. In places where basic infrastructure exists, facilities have not been upgraded since the end of the civil war in 1992. Moreover, most services are located far distances from inhabitants. The conditions of health units are inadequate in terms of facilities, medical supplies, equipment and health staff. Schools only go up to grade four and have serious shortages of teachers, desks, books, and classrooms.

Economic activities

In the majority of the mining areas visited, there were very few formal job opportunities available. This situation has come about mainly because of the closure of a number of mining companies in Manica and Nampula, as well as agricultural companies (sugar, copra, timber) in Zambézia Province, which had employed a significant number of people. As a result, artisanal mining has become one of the main income earning professions in the country. Numerous markets and stalls have surfaced in many Mozambican mining regions as a result of artisanal mining activity; some are vibrant and operate daily, selling such items as vegetables, fish, beer, soaps, batteries, and canned goods.

Authority structures in mining communities

In Mozambique, the local authority structures that have a relationship with the artisanal mining sector include: (1) the local government; (2) state administration; (3) and traditional and religious authority (*curandeiros, mwenes or barozos*) structures, such as the police, mining department, traditional chiefs (or *regulos*[1] in some cases), village secretaries, and mining camp leaders.

Although Christianity is the main religion among Mozambican miners, in Manica and Nampula, traditional religious worship also has an integral cultural

[1] The *regulo* is a remnant authority of the Portuguese colonial administration, whose role was to collect tax from the rural people, among other local administration duties. In some rural areas the regulo is still an important local leader.

role. This was evidenced through observing the activities of the *curandeiros* and traditional leaders, who frequently performed religious worship. It was reported that, ancestral worship in the form of rituals and prayers was usually performed at the grave sites of ancestors, and at sites of major importance; such rituals involve "asking" the dead to play a positive role in the lives of the living.

The *curandeiros* work in collaboration with traditional local chiefs. For example, in Manica, the traditional chief of *Munhena* area has appointed a representative at the mine site. His main duties are to demarcate mining plots, and to ensure through the *curandeiros* that rituals and prayers are conducted accordingly. Furthermore, the chief's representative is responsible for declaring rest and workdays for miners, and must resolve social disputes in the camp. In the event that mediation through the chief fails, incidents of crime at the camp and villages are usually referred to the local police.

In return for work, miners "pay" the chief's representative a sack of gold ore. Locals believe that this gesture and associated rituals will lead to good fortune and benefits – namely, reduced incidences of sickness, fewer accidents at mine sites, increases in production, and the discovery of rich pockets of gemstones. The relevance of traditional and religious authority to artisanal mining lies in the relationship that it establishes between miners and the land. To the Mozambican people, mines represent powerful symbols of belonging and ownership.

Although rituals and prayers are important phenomena in Nampula, the role of the chiefs or *mwenes* is mainly that of tribute collectors. There is a strong presence of *mwenes* and their subordinates, who await gifts (in the forms of food or money) to reinforce their authority over the land and its people.

Three local power groups discernible at the N'tulo campsite in Niassa were the village secretaries, miners' leaders, and traditional chiefs. The role of tribute collector is played by the village secretary, although he is often seen to consult and co-ordinate with the traditional chief of the area. The miners appear to be highly organised and dynamic in N'tulo camp; leaders play an active role in maintaining order at work sites. It also appears that they organize women's work at the site. The police and administration officials were said to regularly visit and "tax" the miners or impose fines on them, thereby frequently creating feelings of antagonism between the different parties. Feelings of antagonism between miners and mining department inspectors are also rumoured to exist in Namuhara in Nampula Province. This deteriorating relationship has resulted from the attempts made in the past by the Department of Mining to control the illegal trade of gemstones, which led to a confiscation of products, and, occasionally, the imprisonment of gold miners or traders.

In Tete, the secretary, as representative of the state and local administration, is recognized as an important leader in charge of *all* artisanal mining affairs, and who oversees the activity and directs entrants into the mining area. Village secretaries remain important authority structures, although their role is increasingly being questioned in the present multi-party government of Mozambique, largely because of their previous political affinity.

Settlement patterns

Although a number of the communities visited have been established for two to three generations, some had become quite depopulated during the civil war. Certain communities had acutely suffered the effects of the war, either directly or indirectly; one such community was Cassoca in Tete. At the time of the study, a number of its residents had just returned, and many of its other original inhabitants had not yet returned from neighbouring Zambia and Zimbabwe, where they had sought refuge during the war years.

In terms of settlement patterns, many people had migrated to ancestral lands during the post-war period, thus precipitating the formation of a series of resettled communal villages. Communal villages are remnants of *FRELIMO's* post independence policy (1977) of villagization, or planned settlement of rural villages to be formally administered by the state through the secretaries, reporting to the district administrator through the sub-district administrator or "*chefe do posto*".

The following two mining community types have emerged in the country: (1) a " village", or an existing rural settlement; and (2) a "camp", which is established on the mining site. The village-type consists of "invisible gold miners", mainly local people from existing settlements who usually exploit alluvial sites. Building structures are permanent, and were in existence long before gold was discovered. They consist of communities organised along ancestral lines. The camp-type of mining community settlement is usually semi-permanent. Both local people and individuals originating from distant communities, provinces and neighbouring countries reside in camp-type settlements. The houses in these communities are made of precarious construction materials using poles, thatch and mud. A number of these sites have appeared and disappeared with the exploitation of local gold deposits. For example, certain mine sites in Manica and Niassa have been in existence for over four years. There are indications that many more of these settlements will remain and become permanent communities even after gold deposits have depleted. Some miners have already diversified their activities, and are now cultivating backyard gardens and/or are engaged in trade. In Niassa, for example, the majority of the artisanal miners who have emigrated from Tanzania have now permanently settled in the area, employ local peoples, and have married into Mozambican communities. However, these artisanal miners, along with their children, are still perceived as "illegal" (foreign) inhabitants by local government authorities and are therefore occasionally subjected to "taxation", which is viewed by authorities as a permit for "legal" residence in the country. At the time of the interviews, these miners expressed a desire to be recognized by Mozambican local authorities; to secure legal rights; and for security of tenure of the land they have worked.

The distinction between types of mining settlements is important. There is a differentiation in terms of infrastructure and sanitation requirements, as well as in terms of understanding and planning support to improve the livelihood of the residents. In camp-type settlements, populations occupy peri-urban environments, and are subjected to a wide diversity of social risks such as reduced food security for the family, social disintegration, homelessness, landlessness, marginalization, increased morbidity and mortality. Moreover, the residents of camp-type settlements are more prone to contracting disease or epidemics, due largely in part to a high concentration of people in small spaces, poor sanitation, an unavailability or inadequacy of basic services, and the poor working conditions of artisanal mining overall. The village-type of settlement has the advantage of having more space, and thus provides a greater opportunity for ensuring family survival through agriculture.

Ethnicity, proximity and identity

In Niassa, many of the interviewees identified themselves as belonging to the Nhanjas, Jauas, and Swahilis ethnic groups, which extend into neighbouring communities, provinces, and countries such as Malawi, Tanzania and Zambia. Similar situations occurred in other provinces. Apart from ethnic relations, these miners also have relations of kin and friendship that extend across borders. These factors constitute strong senses of belonging and identities among the population, which stretch into communities far beyond mining sites.

The relevance of these relations to this study comes into play when exploring the organisational, environmental, health, legal and regulatory issues, and support structures in artisanal mining. For example, whereas these relationships explain the level of social cohesion in mining communities, they may also explain factors such as the level (routes) of migration following gold or gemstone discoveries, duration of stay (or settlements), resulting social organisation, interaction, and impact on the environment.

Work and gender

As with most of rural Africa, work in artisanal mining communities follows gender lines. More specifically, cultural barriers continue to hinder women's involvement, and the valuation of their roles in artisanal mining reinforces Mozambican public perception of a woman's role as mother, food provider, or housewife. In the case of agriculture, for example, which was the predominant activity in most of the areas visited, the role of men and women are markedly different. Men are seen as the income-earners, and are therefore either cash crop growers or are in-charge of the proceeds or income earned from cash crops. Although women take part in cash crop production as well, their roles as mothers, housewives and caretakers predominate. Activities such as long distance trade (across neighbouring countries, trade in fish), gold digging, mineral trading, and fishing are performed by men.

The existing social differentiation and valuation of tasks performed by men and women has been reproduced in artisanal mining (Figs. 15.2 and 15.3). For example, in N'tulo, Niassa, women were "not allowed" to work at the mining sites. The role of women in N'tulo was restricted to selling food and beer and owning what was termed as "hotels" – restaurants that provide meals to miners

Figure 15.2 Miner digging a quartz vein in Munhena, Manica province.

Figure 15.3 Women transporting gold ore to the primary processing stage sites in Muhena, Manica province.

and other people at the mine sites. Those who own "hotels" are generally miners' wives, while beer brewers and sellers are usually unmarried women. In Manica and Cassoca, women are not allowed to dig trenches for gold. Their main tasks include transporting ore, processing non-alluvium deposits, and selling food and beer. However, many women in Manica pan gold from deposits previously exploited by men; in Mount Kapapa, gemstone mining takes place at night and is only conducted by men.

GENERAL PROFILE OF ARTISANAL MINERS

In Mozambique, the majority of artisanal miners are men, and, as already explained, women are active in transportation, washing, panning and downstream activities such as the selling of food and beverages. Children between the ages of six and 10 are also present at mine sites, and are involved in activities similar to those undertaken by women. As indicated earlier, the origin of artisanal miners was traced from communities within the same district, other districts in the provinces surveyed, and from neighbouring provinces and countries (or at border posts) such as Tanzania, Malawi, Zambia and Zimbabwe, which share similar ethnic identities.

Mozambican artisanal miners are highly mobile. They move from one site to another, as new discoveries are reported, or as deposits are fully exploited. However, some artisanal miners still linger on old sites, especially women working old tailings. Artisanal miners are aged between 20 and 50 years old, although the majority are between the ages of 25 and 35. There was no formal organization observed among those artisanal miners encountered; groups are highly unstructured, comprised mainly of "next of kin" and friends.

Only 3.4% of the 153 artisanal miners interviewed had mining titles. Only 27 artisanal miners reported having received training in either geology, mining or mineral processing. Some 22 of these respondents were from Manica, and were trained by the personnel of mining companies established in the province during the colonial period.

Extraction methods

The mining methods and equipment used to extract and process gold and gemstones were similar in all four provinces. For both gold and gemstones, pits and trenches are first dug, whereby more than one metre of overburden is removed using shovels. When the ore-bearing horizon containing gold is reached, the ore is then extracted using picks and shovels and heaped on the ground. The transporting of the ore from the pits to the surface is conducted using ropes and buckets. Once enough ore is accumulated, it is separated from coarse and unwanted material.

Pits are often close together – *i.e.* between five and 10 feet apart; some are as deep as 10 meters. Rudimentary ladders made from logs are used in pits. Occasionally, "steps" are dug into the sides of a pit to facilitate the handling of ore and equipment. Groups of four or five men typically work both alluvial and non-alluvium gold and gemstone deposits, and are usually self-employed. Individual miners usually work in small pits and trenches of about four metres deep. Safety and other important protective measures were not observed. Respondents were, however, fully aware of the potential for accidents and hazardous working conditions at their mines, and requested for support in the form of protective boots, caps, and masks.

In Chiwindi, Niassa Province, and Munhena in Manica Province, where gold is being mined from quartz veins, ore is ground using hand mortars, screened using sieves, washed in sluice boxes and pans, and then amalgamated (Fig. 15.4).

For gemstones, diggings take place using similar tools to those being deployed in gold mining areas. After removal of overburden, gem-bearing material is manually sorted from gangue. Gemstone particles are isolated and separated from ore.

In the case of alluvium deposits, various technologies are first used to identify mineralized portions of riverbeds. These are then excavated to depths of two metres, using shovels and picks. In general, the top beds (between 25 and 50 cm and, sometimes, as much as 1.5 m) are characterized by detric materials such as sand, boulders, and quartz. The mineralized bed contains abundant boulders, pebbles, and sand, argillaceous sand, and heavy minerals. Mining and processing contributes to soil erosion, sedimentation and water contamination. Children aged between 10 and 16 years were observed helping their parents in Tete; on their own in Manica; and as hired hands for transporting ore in Manica and Niassa. In Cassoca, babies were observed lying nearby stagnated pools while their mothers worked. Another group of children, all of whom were under five years old, was found at a mine site close to their mothers; the mine site is seen as the only playground while all members of the family are busy at work. In Manica, Murrupula, and Tete, most children attend school in the morning. In the afternoon, they join their parents at the mine site, where they then assist with the transporting and processing of gold. The mine site in N'tulo-Niassa is located a significant distance from the only school in the region. In fact, no children from this area attend school. In Manica, there were reports of children missing classes or leaving school altogether, for the sole purpose of working at local mines.

Gold production is an estimated 10–15 grams per month per miner. In Manica, production levels were stable and slightly higher. Because of a shortage of tools

Figure 15.4 Women and children panning old tailing accompanied by babies in Murrupula, Nampula province.

and mass inefficiencies in mining methods, women and children produce significantly less gold.

Processing of gold and gemstones

The main processing methods used in all of the artisanal mining areas surveyed were sluice boxes, and wooden pans for washing and mercury for amalgamation.

Sluice boxes and wooden pans are made by local craftsmen. Although constructed locally, sluicing is still generally an expensive means of processing because of the high cost of the 200 litre petrol drum, which is used as a screen and feeder to the sluice box itself. In N'tulo in Niassa, Munhena in Manica, and Murrupula in Nampula, an intensive use and hiring of more sophisticated sluice boxes was observed (Figs. 15.5 and 15.6). The sluices used in Sussundenga appeared less sophisticated and were crafted from tree barks. The process of washing (gravity separation) begins with the ore being transported – mainly by women and children – and then fed into the feeding drum. The feeding drum is made from a 200-litre drum longitudinally cut with small holes punched at the bottom.

A large screen is placed atop the feeding drum and is used to separate ore from all undesired materials. A 20-litre gallon drum is used as a water "feeder"; water is drawn from a hole punched into one of its sides. This, in turn, is used to wash, transport and separate fine gold particles from gangue. The material is then washed down into a series of elongated riffled boxes fitted with a piece of blanket or sack, which captures the gold as tailings are discharged into the river. The blanket is then washed into a pan where the gold pre-concentrate is

Figure 15.5 Grinding and screening process using mortals and sieves in Chiwindi, Niassa province.

Figure 15.6 A locally built sluice box used to recover gold, N'tulo, Niassa province.

separated from the remaining particles. In Nampula and Manica, sluice boxes are actually placed near or along riverbanks. To accumulate sufficient water for washing, rivers are diverted to form pools; as previously indicated, tailings are normally discharged directly into river systems.

In Sussundenga, mining and processing takes place almost everywhere – *i.e.* in the riverbed, along riverbanks, and in trenches and pits dug near to watercourses (Fig. 15.7). The gravel in Sussundenga appears to be more productive than those of other mining areas in Manica, but the gold recovered is very fine and therefore difficult to capture using the conventional rudimentary sluice boxes crafted from local tree bark. Water is obtained by diverting, collecting and channelling rivers through mounted metal pipelines. All miners benefit from such a system and it is common to see groups or individuals washing in queue.

In most of the gold-producing areas, gold particles are very fine, and recovery rates (using the locally built sluice boxes and pans) vary between 20 and 35%, as most of the finest gold particles are lost during washing; clearly, the processing methods used to recover gold are very inefficient, and most miners are not fully aware of other technologies. The fact that women manage to recover notable quantities of gold by reworking tailings is an indication of exactly how much gold is lost during the primary gold production phases. The amount recovered by women during this exercise amounts to close to half of the quantity obtained by the men engaged in primary extraction.

In Manica, where mercury amalgamation is most common, the pre-concentrate obtained during processing is mixed with mercury. Amalgamation is often performed near rivers and in open places; the pre-concentrate is mixed with mercury to recover fine gold particles. The mixture obtained is then burned in the open air, thus producing the final purified gold product. During the amalgamation process, notable quantities of mercury are released directly

Figure 15.7 Pools of water near the river bank used for washing gold in Munhena, Manica province.

into the river system; inhaled by workers, most of whom wear no protective masks; and released into the atmosphere. Mercury was observed to be used intensively by gold miners in Manica and Niassa. However, in a number of the other areas visited, miners reported not being fully aware of the technology.

Gemstone processing methods are very simple. After the mining of gem-bearing rocks, miners simply manually sort ore, using the obvious visual-optical characteristics of gem particles – namely fluorescence and shine – as the primary criteria for separation; no equipment is used at this stage. The next step in the processing of gemstones is "cobbing", which is performed during the marketing stage. This consists of the removal of non-gem material attached to gemstones. Pieces are cleaned using hammers, chisels, pliers, pincers, and, occasionally, grinding wheels. These tasks must be performed extremely carefully in order to prevent cracking, which drastically reduces their value. During the final stage of processing, a hammer is used to remove any traces of other materials.

Marketing and distribution of gold gemstones

In Mozambique, trade in both gold and gemstones is dominated by private dealers (See Table 15.1). However, other important players include the Fundo de Fomento Mineiro (*FFM*) – a mining development fund, the main objective of which is to promote and develop the resident small-scale sector – and licence holders. It is estimated that there are 250 gold and gemstone traders active in the areas surveyed.

In Tete, both the Mining Development Fund (FFM) and private dealers are major gold buyers. With the exception of Changara District, few private dealers

Table 15.1 Different buyers of artisanal mining production.

To whom do you sell your production	Niassa	Tete	Nampula	Manica
Dealers or Traders	14.7%(10)	81.8%(9)	64.7%(11)	51.7%(31)
Government (FFM)	82.4%(56)	18.2%(2)	35.3%(6)	43.3%(26)
Licence Holders	–	–	–	–
Others	2.9%(2)	–	–	5.0%(3)
Total number	68	11	17	60

are active at other mining sites in the province, largely because of remoteness and inaccessibility. The majority of miners travel to the province's capital city or neighbouring countries to sell gold and gemstones.

The gold mining regions of Niassa are highly accessible as well. In fact, most gold buyers come from Tanzania, where most of Niassa's gold is sold. Niassa miners also travel and sell gold directly in Tanzania, where, after selling product, they purchase consumables and other goods for the local market, and bring many back to Mozambican mining areas for resale and/or for their own consumption. In Manica and Nampula, gold marketing is dynamic because of easy access to mining regions. Dealers from the provinces' capital cities control the local gold and gemstone trades. Dealers later sell gold in the cities, where a number of gold craftsmen and lapidarians work the raw product into jewellery. Most of the jewellery is sold locally and later exported illegally.

The relations between mineral traders and miners are strictly commercial; there are no repeated buying arrangements or credit. However, in Manica, where mercury is used intensively in gold processing activities, buyers provide mercury in advance to the miners to help "guarantee" a subsequent direct sale of the product. Although this act does not seem to influence the price of gold, it has nevertheless had a tendency of creating a dependency and indebtedness among miners, who feel obligated to produce and sell the gold to the same traders.

Prices are usually set by traders without prior consultation with miners. In Murrupula and Namuhara, traditional chiefs, who regularly consult with mineral traders, also play a role in setting local market prices for gold and gemstones. Gold prices of all of the surveyed sites were in the range of US$7 per gram, the highest prices being reported in Manica and Niassa. In Tete, gold was slightly cheaper, selling at a price of US$6.60 per gram (See Table 15.2).

In Namuhara, gem quality aquamarine can normally fetch up to US$25 per gram. An average of 600 to 900 grams is sold in the region per month, as buyers or intermediaries travel from other cities in Mozambique, as well as from neighbouring Zimbabwe, Zambia and Malawi. Other notable buyers come from Germany, Taiwan, the US, and South Africa.

The Mining Development Fund (FFM) plays a significant role in supporting the marketing of gold in Tete, Manica, and Niassa. FFM helps direct production to official market channels, thus establishing official production records and helping miners to market their products in areas where there is little or no competition.

Table 15.2 Amount received per gram of gold or gemstone.

Amount received for sale of gold production/ gram/week	Niassa	Tete	Nampula	Manica
<5USD	32.2%(19)	40.0%(2)	75.0%(9)	35%(17)
5–7	11.9%(7)	20%(1)	–	60.4%(29)
7–9	32.2(19)	40.0%(2)	16.7%(2)	–
10–12	10.2%(6)	–	–	4.2%(2)
>12 USD	13.6%(8)	–	8.3%(1)	–
Total number	59	5	12	48

Table 15.3 Summary of socio-economic impacts.

Actual impacts	On-going impacts (critical issues)
Positive impacts • Increased employment opportunity • Improved income • Enhanced local economy *Negative impacts* • Influx of people into mining sites, and related social and health problems • Loss of agricultural land • Loss of natural resources (vegetation, water contamination)	• Insufficient employment in artisanal mining communities areas • Resettlement of miners/families in new areas • Migrants to mining areas (seeking jobs, business, war returnees) • Poor infrastructure • Poor basic social services such as health, education and water supply • Small local business development (stalls, informal market selling) • Loss of natural resources and effects on shortages of firewood supply, reduced forests and bushes and its effect on ethno-botany (traditional herbalists) • Loss of scenic view

SOCIO-ECONOMIC IMPACTS OF ARTISANAL MINING IN MOZAMBIQUE

As artisanal mining is new to many areas of Mozambique, the socio-economic impacts of its activities are not always readily discernible. What *is* eminent from field visits is the fact that resident artisanal mining activities have increased local employment, enhanced incomes, stimulated migration to mining areas, and precipitated the development of new settlements. At the same time, however, small-scale mining has had an adverse effect on natural resources, the environment and human health (See Table 15.3 for a summary of these impacts).

Increased employment opportunities

When asked to provide reasons for why they are involved in mining today, most miners indicated that a lack of regional employment opportunities prompted their involvement. In Manica, Nampula, and Zambézia, recent closures of mining and agricultural companies have increased the number of unemployed. A wide-range of income-earning opportunities are provided through agriculture, fishing from Lake Niassa/Malawi, cross-border trade (provides income to many inhabitants in Niassa), and livestock production (considered an important source of income and social security in Tette), but the revenues accrued from artisanal mining supplement many individual incomes. Although during the rainy season, the rural population works the fields, artisanal mining activities often intensify after harvest, at which time certain produce and incomes obtained from the sale of agricultural products are used to support various mining activities (for *e.g.* in Manica, Ntulo and Murrupula). However, in Cassoca and Namuhara, where water is scarce, mining intensifies during the rainy season. An abundance of water is necessary for effective gemstone excavation, as ore is naturally washed away, thus making it easier to identify prospective gem material. Moreover, the losses of gemstone particles resulting from hand sorting activities during the dry season are easily recovered, and production thus increases considerably.

Some 90% of the active population of mining communities engages in diverse activities such as extraction, transporting, processing, trade, and businesses established at markets. However, for most people, artisanal mining is but a seasonal and/or part-time activity.

Improvements in income

A significant number of respondents reported improvements in income due to artisanal mining. The majority of people surveyed in Niassa (83.3%), Nampula (68.8%) and Manica (49.1%) indicated that artisanal mining was their main source of income. As artisanal mining has only recently begun in Tete, it was not surprising that 91.7% of the respondents indicated agriculture to be the most important source of income (See Table 15.4).

The percentage contribution of income from artisanal mining for those directly involved in mining is estimated to be between 20 and 40%. In Nampula, the production of a large range of precious and semi-precious stones contributes considerably to the total income generated by thousands of artisanal miners. Women also benefit directly from increased income obtained from miners, informal trading, and transporting goods. Local chiefs and their representatives

Table 15.4 Reported main source of income.

What is your main source of income?	Niassa	Tete	Nampula	Manica
Agriculture	10.6%(7)	91.7%(11)	31.3%(5)	43.9%(25)
Fishing	4.5%(3)	8.3%(1)	–	3.5%(2)
Hunting	1.5%(1)	–	–	3.5%(2)
Mining	83.3%(55)	–	68.8%(11)	49.1%(28)
Total number	61	12	16	57

benefit directly from income, gifts and tributes received as a result of their roles as traditional and religious leaders in mining communities.

As can be seen in Table 15.5, the extra income obtained from artisanal mining contributes to the survival of the household. It is used to purchase food and other household consumables such as soap and clothing, and to hire additional agriculture labour. In Niassa, reinvestment in mining (through the purchase of working implements) proved significant (40%) among artisanal gold miners. Miners' family members not living in the camps also benefit from remittances.

Improvements to the local economy

Members of rural communities reported that mining activities have led to many improvements to the local economy, including increases in income, upgrades in agriculture, increased educational support, increased medication, and the establishment of small businesses (shops and informal trading operations) in mining areas (See Table 15.6). Subsequent agricultural improvements (especially in Niassa and Nampula), and the establishment of small businesses, have further helped to enhance the vibrancy of artisanal mining economies countrywide.

The range of products observed to be sold at markets surrounding Mozambican mine camps include, agricultural products, dried fish, locally-brewed or imported beers, wines and sprits, various clothing, and other household consumables, such as paraffin and matches. In Manica, recreational

Table 15.5 Use of mining income as reported by respondents.

How is income from mining used?	Niassa	Tete	Nampula	Manica
Remittances home to my family	46.9%(30)	83.3%(10)	88.2%(15)	73.3%(44)
Banking	1.6%(1)	–	–	5.0%(3)
Buy working tools	40.6%(26)	8.3%(1)	5.9%(1)	18.3%(11)
Investment in other activities	10.9%(7)	8.3%(1)	5.9%(1)	3.3%(2)
Total number	64	12	17	60

Table 15.6 Reported improvements as a direct result of mining.

Improvements resulting directly from mining	Niassa	Tete	Nampula	Manica
Increase in Income	21.4% (3)	50% (7)	40% (2)	84.6% (11)
Improved agriculture	35.7% (5)	14.3% (2)	40% (2)	7.7% (1)
Able to send children to school	14.3% (2)	14.3% (2)	20% (1)	7.7% (1)
Able to buy medicine	7.1% (1)	14.3% (2)	–	–
More shops	21.4% (3)	7.1% (1)	–	–
Total number	14	14	5	12

activities, such as video viewing, were also observed. Stall owners include residents of respective mining communities, and people from other provinces or countries who have established themselves in the area. Women are among the beneficiaries of the increased income generation brought by artisanal mining activities, as camps provide a staple market for agricultural and food products. At the same time, however, numerous women are found to be prostituting in certain areas, and are often given tedious mine-related tasks such as ore pounding and washing.

ENVIRONMENTAL IMPACTS

The environmental impacts from mining activities identified by those individuals interviewed include, water contamination, river sedimentation, soil erosion, the destruction of vegetation, and various health-related risks. Other environmental problems reported were associated with the use and exposure to mercury. In Manica, where mercury is used intensively by artisanal miners, some 93% of respondents reported some degree of awareness related to the risks of using mercury in gold processing. The type and level of magnitude of these impacts varied according to mine location and deposit type.

As can be seen in Table 15.7, the most widespread environmental impacts associated with the extraction of non-alluvial deposits are erosion and the destruction of vegetation. In communities where alluvial gold mining is occurring, water contamination, sedimentation and soil erosion are rampant. In Manica and Nampula, rivers are commonly diverted, the water from which is used to fill open pits for washing. Tailings are directly disposed into rivers that are normally used by the local population for irrigation and domestic consumption. The worst cases have been reported in Manica, Nampula and Niassa, where mining has been practised for more than five years. In Manica, the problem has been exacerbated by the intensive use of mercury in gold processing. Women and children are seen working on riverbeds and beaches using mercury to recover

Table 15.7 Environmental problems as reported by communities.

Environmental Problem	None (%)				Significant (%)				Very Significant (%)			
	N	T	Nam	M	N	T	Nam	M	N	T	Nam	M
Water contamination	100	71.4	66.7	72.7	–	–	33.3	27.3	–	28.6	–	–
River sedimentation	–	83.3	–	50	100	–	100	50	–	16.7	–	–
Soil Erosion		83.3	100	66.7	–	–	–	33.3	100	16.7	–	–
Destruction of access roads	100	91	100	66.7	–	–	–	33.3		9	–	–
Destruction of religious grounds	–	91	100	75	–	9	–	25	–	–	–	–

N = Niassa, T = Tete; Nam = Nampula; M = Manica.

alluvial gold. The wide availability of mercury from neighbouring Zimbabwe, coupled with its low cost, makes amalgamation a highly acceptable and widely disseminated practice in Mozambican artisanal gold mining regions. No mercury usage was observed at the other alluvial sites visited, largely because few people outside of Manica appeared to have any knowledge of amalgamation.

In N'tulo, although gold recovery is performed without the use of mercury, respondents exhibited symptoms of mercury poisoning. In all likelihood, this was a result of the mining activities concentrated upstream in Chiwindi, where a primary gold deposit is currently being exploited, the particles from which are being amalgamated.

HEALTH ISSUES

This section of the chapter documents the major health problems confronting the inhabitants of Mozambican artisanal mining communities, the means by which many are coping with these impacts, and the causes and possible solutions to problems. Malaria was reported to be most widespread disease in the mining areas visited. This was also confirmed by a review of health records (conducted only in Manica Province), which showed an increase in the number of incidents of malaria reported at health posts (See Tables 15.8 and 15.9).

Table 15.8 Reported common health problems.

Common Health problems	Niassa	Tete	Nampula	Manica
Malaria	20% (14)	16.1% (14)	14.6% (6)	25.9% (15)
Cholera	5.7% (4)	12.6% (11)	12.2% (5)	8.6% (5)
STD	14.3% (10)	10.3% (9)	9.8% (4)	10.3% (6)
Pulmonary problems	7.1% (5)	13.8% (12)	14.6% (6)	6.9% (4)
Diarrhoea (due to water quality)	14.3% (10)	13.8%(12)	14.6% (6)	15.5% (9)
Bilharzia	14.3% (10)	12.6% (11)	14.6% (6)	17.2% (10)
Conjunctivitis	17.1% (12)	12.6% (11)	12.2% (5)	15.5% (9)
Other	7.1% (5)	8% (7)	7.3% (3)	–
Total responses	80	87	128	58

Table 15.9 Reported increase of disease over the past two years.

Disease increased over past 2 years	Niassa	Tete	Nampula	Manica
Malaria	28.6% (26)	30% (3)	21.4% (3)	48.4% (45)
Cholera/diarrhoea disease	5.5% (5)	40% (4)	28.6% (4)	8.6% (8)
STDs	36.3% (33)	20% (2)	14.3% (2)	24.7% (23)
Pulmonary problems	17.6% (16)	10% (1)	21.4% (3)	18.3% (17)
Other	12% (11)	–	14.3% (2)	–
Total responses	91	10	14	93

Malaria was also reported to be single greatest cause of death among miners and community members. The possible cause of the increase in malaria was the accumulation of rainwater in the pits dug by miners. As most are not reclaimed, the stagnated waters have become breeding grounds for infected mosquitoes.

STDs were also reported to be on the increase, as well as chest-pains and cholera. It is probable that the increase in STDs is due to there being a lot of miners who have left their families behind. Mining camps have proven to be active grounds for prostitution; the STD problem, in turn, has been exacerbated by an overall lack of condom use. During focus group discussions conducted in Manica and Niassa, it was discovered that both men and women do not like to use condoms. Some women – namely, widows and divorcees – reported having no choice but to agree to have unprotected sex, as the incomes derived from the prostitution is used to supplement their incomes. However, some respondents noted that they would use condoms if they were accessible and affordable, a case in point being an incident in Manica Province, where numerous men collected free condoms being distributed. Some artisanal gold miners contacted the health consultant to ask if condoms could be washed and reused, which is clear indication of there being a lack of information on their use.

The health posts visited in both Manica and Niassa confirmed that there were increased numbers of STDs cases. In Manica, the increase began about the same time that the nearby mine had commenced operation; there was almost a seven-fold increase (*i.e.* from about two patients a week to between three and four patients a day). Nurses reported that they sometimes had condoms on hand but that overall, supplies depleted very quickly, and that supply is insufficient to meet the demand. Officials from all provinces, with the exception of Tete, reported that STDs were widespread in mining regions. Manica had the greatest abundance of information on STD prevention. Store owners reported that they do not sell basic medicines. Niassa was the only province where shops sell condoms, which are imported from Tanzania.

In all likelihood, the dust within the work environment is the main cause of respiratory deterioration – namely, lung and pulmonary problems. As explained earlier, most miners have no protection from dust while working during the dry season.

An estimated 20% of respondents reported having had diarrhoea within two weeks of the time in which this fieldwork was undertaken. This is an indication of poor water quality, crowded living conditions, and unhygienic and unsanitary work conditions. Incidents of cholera and diarrhoea were probably caused by poor water quality, as over 60% of all respondents reported using local river water and unprotected wells as a source of drink. Another probable cause could be the reported lack of latrines. In Manica and Niassa, few respondents reported having latrines, and of those that do exist, few are maintained.

Although health units and pharmacies are said to be more than 10 kilometres distance from mine sites in nearly three quarters of the provinces, respondents reported having visited health posts or pharmacies whenever ill. After further investigation, it was discovered that respondents, do, in fact, visit health posts, traditional healers (*curandeiro*) and/or administer medication themselves.

During focus group discussions in Niassa, some respondents indicated having visited a Tanzanian hospital near the border as opposed to the facility located in Lupilichi, which, at the time of their visiting, was virtually devoid of medical stock.

In Manica, where fine gold ore is mixed with mercury by hand and then burned to recover gold, acute mercury exposure was observed. More specifically, exposure to mercury occurred via direct skin contact during amalgamation, or as a result of inhaling mercury vapour during the firing of the amalgam. Handling or the eating of food with hands contaminated by mercury has proven to be a major pathway of direct mercury ingestion. During panning, miners typically use plastic dishes or wooden pans for amalgamation. Domestic utensils, therefore, become easily contaminated, as the same utensils are used at home for food preparation. Storage and handling of mercury in residential compounds further increases the risk of mercury exposure. In Manica Province, mercury poisoning was substantiated by recent studies conducted in the Sussundenga and Munhena areas, which determined that high levels of mercury exist in rivers in close proximity to mining areas, as well as in areas downstream from operations. University researchers have taken urine, hair and nail samples for analysis, and preliminary results indicate that there is strong possibility of mercury contamination; these samples have been sent abroad for analysis and confirmation.[2]

Additional accidents have been reported during extraction processes. Focus group discussions revealed that cave-ins occurred because of unskilled digging, an absence of material to secure the walls, and an overall lack of methodological work procedures.

RECOMMENDATIONS AND CONCLUSIONS

In order to facilitate improved sustainability in Mozambique's artisanal mining sector, increased emphasis should be placed on improving the organization of operations.[3] Given the lack of communication and co-operation among miners, it seems likely that an outside organization such as DNM will have to play a lead role in facilitating improved organization within the sector. Such an approach would entail DNM collaboration and co-ordination with a number of important bodies, including:

1) Institutions, projects and programmes that emphasise alternative livelihood development, education, training, infrastructure development and participation in the sustainable utilisation of natural resources. One existing example of such a setup is the co-management project of the Small Scale Fisheries

[2] Unpublished papers on mercury contamination from the University of Eduardo Mondlane, Department of Geology.

[3] Sustainability in this context means more environmentally friendly production methods, as well as reinvestment of earnings from mining into alternative livelihoods.

Development Institute (IDPPE) in Nampula and Zambézia Province, funded by IFAD/OPEC and the Mozambican government; and
2) The Agricultural Department and relevant NGOs, for the purpose of providing extension support and to develop alternative income activities for artisanal mining communities; programs that target women and children should be a priority.

It is recommended that the following mechanisms be used to achieve these goals:

- Workshops, for feedback and dissemination of survey results;
- Workshops, for training authority community leaders and other community groups;
- Workshops, to sensitise potential partners for collaboration and co-ordination in the sector; and
- Permanent dissemination of relevant information and educational programmes related to mining, environment, social and health issues.

The following *specific* initiatives should also be considered:

- Training courses for small-scale miners on environmental issues, health and safety;
- Awareness campaigns on occupational health risk and the environment, addressing aspects specifically related to the use of inadequate small-scale mining methods;
- Dissemination of appropriate and cheaper processing technologies such as those of the GTZ-sponsored Riverbed Gold Projects in Harare and Bulawayo;
- Environment assessment studies in highly potentially contaminated mining areas (soil and water sampling);
- Technical support from research and finance institutions dealing with environmental and social issues;
- Support alternative income generating activities; and
- The monitoring of social, environmental and health effects of artisanal mining.

Most governments do not have the resources to effectively control artisanal mining operations, as many are located in remote areas, and are widely dispersed. Only by slowly building trust between the miners' groups and the government can DNM and MIREME fulfil its role as a manager, developer and regulator. DNM's role could be enhanced through increased collaboration with existing local authorities, who have legitimate authority over the people and the land to resolve such issues as regulating entrants into mining areas; controlling water pollution; mobilizing contributions in the form of work, time and money for activities that contribute to improved health; and improving and enhancing social cohesion among the artisanal mining communities. These include, but are not limited to, the construction of wells, building and encouraging the use of latrines at mine sites, and arbitrating disputes between miners and mineral traders. DNM could also provide training through workshops, as well as on-the-job training to empower and strengthen existing authorities to manage and control artisanal mining activities at the local level.

To improve production, DNM must address the equipment requirements of miners' working groups. Individuals must be provided with hand-operated pumps, screens, concentrators, pans, as well as wheelbarrows for the transportion of ore. Dissemination of such equipment could be achieved via the promotion of tool rental and credit schemes. DNM should also put in place extension services capable of facilitating the transfer of simple technologies for environmental protection. For example, back filling the overburden and tailings from washing processes would enable the rehabilitation and reclamation of mined areas. The construction of simple drainage ditches and catchment systems could further facilitate the re-circulation and conservation of water during screening, as well as reduce sediment loading in nearby rivers. To provide additional support for the marketing of gold and gemstones, DNM must establish regional purchasing centres. This is imperative because it provides facilities for the marketing of mineral products in remote locations.

Additional research is needed to determine the *extent* of mercury poisoning in Mozambican artisanal gold mining communities – a task that should also be conducted by DNM. As a starting point, DNM must facilitate an exchange of experiences and lessons learned from other SADC countries, as well as draw from valuable lessons learned in other parts of the world. To make communities aware of the impact of artisanal mining on the environment, IEC materials should be developed and aimed at changing work habits that have adverse effects on the environment.

DNM could collaborate with the National Council for the Fight Against AIDS to develop and disseminate information on the prevention of AIDS to miners. As a first step, a strategy on STD/HIV/AIDS should be elaborated on. Certain NGOs could be engaged to work and advise DNM. Through MIREME, DNM should pursue possible collaboration and co-ordination with MOH for the sole purpose of addressing the lack of basic medicines such as aspirin and chloroquin. One avenue possibly worth exploring is the hiring of training community health agents (APEs).

Information Education and Communication (IEC) programs need to be developed for mine workers and community members on various matters, including the prevention and treatment of STDs, malaria, diarrhoea, and occupational health risks, particularly with respect to mercury poisoning.

Further research must be carried out to identify and determine the various occupational health risks related to the use of mercury, and ways in which DNM could increase its capacity to manage and develop artisanal mining. DNM must implement programs to monitor the social, environmental and health effects of resident artisanal mining activities. Environment assessment studies (soil and water sampling) should be conducted in potentially highly contaminated mining areas. A gender assessment of artisanal mining could also be conducted to assist female artisanal miners.

In summary, this study shows that artisanal gold mining contributes to the economic and social well being of Mozambique's rural people. The most significant of these impacts relates to employment generation. The incomes generated from the industry's activities not only supplement family earnings, but also enable seasonal farmers to invest funds to improve agricultural outputs.

However, the study also shows that artisanal gold mining is associated with various social, environmental and health-related risks. The inefficient technologies commonly featured in gold extraction and processing techniques, coupled with the constraints associated with economic support infrastructure, contribute to low productivity. The remoteness of the activity and the absence of basic social services such as medical care and education are further impeding its development. Moreover, the transient, informal, and disorganised nature of the activity makes it difficult to monitor, regulate and control operations.

Many of the negative impacts could be strongly mitigated with small changes in technology and production practices. Moreover, higher yields from slightly more sophisticated technologies should easily make their introduction profitable. However, as is clear from the surveys and interviews undertaken in this study, artisanal mining in Mozambique is almost completely disorganized beyond the small production unit. Accordingly, for significant change to take place, it will be necessary for some organization to take a lead role with respect to information and communication.

REFERENCE

Valoi, G., Kazilimani, E. & Bukali, F. (2000). Artisanal baseline survey: Mozambique, Final Report. Ministry, Mineral resources and energy, Maputo, 124 pp.

16

The Socio-Economic and Environmental Impacts of Artisanal and Small-Scale Mining in Mozambique

S. MONDLANE AND D.S.M. SHOKO

Gold panning has been widespread in the Archaean part of Mozambique since the Monomotapa Empire *C.* 1500 AD (Manuel et al., 1999). Presently, artisanal and small-scale miners produce 100% of the country's gold; collective industry output is in the range of 400 kg per year. Not only does the sector contribute positively to national GDP, but it is also an important source of livelihood for many residents.

The definition of an artisanal and small-scale miner has been the subject of lively debate among researchers. In Mozambique, artisanal and small-scale miners are those who use rudimentary instruments, and who produce no more than five kg of mineral output per month. Their activities are also highly environmentally disruptive, and are often carried out illegally. However, since the closure of 90% of mining and agricultural companies in Manica and Zambézia provinces in the 1980s and 1990s, artisanal and small-scale mining (ASM) has become one of the main sources of income for many Mozambican communities.

Resident artisanal and small-scale miners are often organised in small groups (up to five) – usually consisting of members of the same family (Manica) or friends[1] – although self-employed workers (49.3% in Niassa, 83% in Manica and 100% in Tete) are most common. According to a Baseline Survey conducted in Mozambique in 2000 (see previous chapter), local authoritative structures such as government, state administration, traditional and religious groups (*curandeiros, mwenes or barozos*), play a central role in determining the social and power relationships within artisanal mining communities. The police, mining departments, traditional chiefs (or *regulos*[2] in some cases), village secretaries and mine camp leaders reportedly exercise their authority in these communities.

---

[1] Matos 2001, personal communication.

[2] The *regulo* is a remnant authority of the Portuguese colonial administration, whose role was to collect tax from the rural people, among other local administration duties. In some rural areas, the *regulo* is still an important local leader.

The purpose of this chapter is to provide a general overview of Mozambique's ASM industry, and the socio-economic impacts of its activities.

LEGAL AND INSTITUTIONAL FRAMEWORK

Mozambique's *Mining Law* was last revised in 1986 as Statutory Instrument Number 2/86. After 14 years of application, it had become apparent that the Law was in need of amendment. An ongoing review is therefore striving to readjust the Law to meet the new mining activity challenges posed within the recently approved "Geological and Mining Policy", guided by Statutory Instrument 4/98. It is also aiming to harmonize (the Law) with recently implemented legislation, such as the *Land Law* (19/97), the *Environmental Law* (20/97), the *Fiscal Law*, and the *Arbitrage, Conciliation and Mediation Law* (5/99 and 11/99 respectively). The fiscal regime aspects of the *Mining Law* have already been amended in conformance with the new *Fiscal Law* (5/99) and the new *Land Law*.

Currently, *the Mining Law* allows for the securing of the following:

- A prospecting and exploration licence;
- A mine concession title (for exploitation of mineral resources);
- A quarry licence; and
- A mining certificate.

The revised *Mining Law* will feature a number of major changes, such as the inclusion of the mineral medical waters; detailed requirements for a Reconnaissance License during the preliminary stages of mineral exploration; and legislation for a mining "Senha",[3] which authorises ASM activities in pre-designated areas of less than 1000 hectares. It is expected that the new Law will be revised in a fashion to encourage private investment in the country's mining sector, and will designate the government full responsibility for regulating, facilitating, promoting, and monitoring mining activity. The State, however, will continue to provide the necessary geological survey services, and compile the national mineral inventory.

The *Mining Law* will also be amended to include a specific section on the environment (as per the *Environmental Law*); a section emphasizing the importance of prioritizing mining as an economic activity; and a section governing the marketing of mineral resources. Presently, the marketing of mineral resources by both individuals and companies is regulated by Decree 31/95. Operators must apply for a licence, which entitles them to buy and sell mineral resources.

There are three basic types of licenses overall, which are awarded to the following groups of people: (1) those buying minerals for their own use (including jewellery); (2) those buying for the purposes of selling, with or without added value, to other operators (nationally); and (3) those buying and selling with or without added value, at the national and international levels. Each is issued at a fixed tax rate of 500,000,000 MZM (US$250), 1,000,000,000 MZM (US$500),

[3] License for small-scale/artisanal miners in designated areas.

Table 16.1 Percentage of Mozambican artisanal miners with mining titles.

Do you have any mining title?	Niassa	Tete	Manica
Yes	3%	9.1%	3.4%
No	97.0%	90.9%	96.6%
Total (interviewed)	66	11	59

Source: Valoi et al., 2000.

and 2,000,000,000 MZM (US$1000) respectively. However, a survey by Valoi et al. (2000) has shown that most small-scale miners do not have such mining titles as required by Law (Table 16.1).

SMALL-SCALE MINING IN MOZAMBIQUE

In terms of composition, the small-scale mining sector in Mozambique is not unique in the slightest, and therefore, the problems it faces are strikingly similar in scope to those faced by most comparative industries world wide. Mozambican ASM operators are highly mobile, and, in many cases, are seasonal. The sector features an array of basic tools, and produces a limited quantity of mineral. Operators usually sell their product to local dealers, who generally dictate the buying price.

As already noted, the gold and gemstones extracted in the country are confined to a small scale (Fig. 16.1 shows the location of important gold and gemstone ASM operations in Mozambique), as there are no large-scale mines currently in operation. However, the products mined by Mozambican artisanal and small-scale miners extend beyond gold and gemstones, and include semi-precious stones, gravel, aggregates, coal, limestone, guano, silica sand, bentonite, bauxite and ornamental stones.

The introduction of the Economic Structural Adjustment Program (ESAP) in 1986 led to a mass displacement of workers. This, along with an extended period of drought between 1990 and 1993, prompted many retrenched workers and peasants to pan gold in certain provinces of Mozambique. By 1999, the number of gold panners had reached 20,000 in Manica Province alone, and, in the entire country, over 60,000 people were reportedly involved in ASM (Mondlane, 2001). In provinces such as Tete, ASM is entirely seasonal, practised only during the dry season. In Niassa and Manica, 30% of miners practise the activity seasonally in order to complement earnings from agriculture, which is mainly practised in the rainy season.

The *exact* number of people involved in economic activities such as agriculture, fishing, hunting and mining, is not known. However, the Baseline Survey conducted in 2000 suggests that the largest proportion of income for artisanal miners comes from mining activities, the notable exception being Tete, where some 91.7% of respondents (Valoi et al., 2000) identified agriculture as a primary source of employment (see Table 16.2).

Source: Mondle, 2001.

Figure 16.1 Small scale mining activity in Mozambique.

Table 16.2 Main sources of income of artisanal miners.

Sources of income	Niassa	Tete	Manica
Agriculture	10.6%	91.7%	43.9%
Fishing	4.5%	8.3%	3.5%
Hunting	1.5%	0%	3.5%
Mining	83.3%	0%	49.1%
Total (interviewed)	66	12	57

SOCIAL, ECONOMIC AND ENVIRONMENTAL ISSUES

As has been explained throughout this book, the general lack of financial and technical resources within the ASM sector has often led to haphazard and wasteful mining, inefficient mineral processing, illegal trading, precarious living and working conditions, as well as severe environmental and social problems. This situation has often been exacerbated by the absence of favourable legal, regulatory and institutional support frameworks. The result has been a reduction in potential incomes for artisanal and small-scale miners, and governments. More specifically, the interplay of the above constraints often gives rise to negative cycles of *cause* and *effect*, in which decreasing miners' incomes result in fewer revenues for both local and central governments. This narrows the State's revenue base, resulting in inadequate operational resources, which, in turn, limit a government's ability to effectively administer and supervise mining activities. However, it is pertinent to note that while ASM methods can lead to wastage of non-renewable mineral resources, and can be hazardous to human and environmental health, they can also enrich nations and economically empower disadvantaged groups by virtue of their low investment costs and short lead time required from discovery to production. Artisanal and small-scale miners can also work deposits that are either too small or of much lower grades than those economically viable for large-scale mines. Several authors have examined the socio-economic and environmental implications of gold panning in Southern Africa (*e.g.* Hollaway & Associates, 1992; SARDC et al., 1994; Manuel et al., 1999). Each is in general agreement that ASM constitutes a major source of income for poor families in rural areas (Bezerra et al., 1996; Manuel et al., 1999), and also helps to curb urban-rural drift.

Socio-cultural issues

ASM is almost always associated with some degree of poverty. Although the sector is frequently credited with providing employment to largely unskilled labourers, requiring small initial capital and infrastructure for start up, and producing minerals at depressed prices, the obvious trade-offs have been, at times, low-income levels and widespread poverty among so-called self-employed workers. As artisanal and small-scale miners begin with very little start-up capital, and often go into business during periods of harsh economic depression induced by drought and economic structural adjustment, most earn wages below the poverty datum line (Chiwawa, 1993). In most SADC countries, including Mozambique, with the exception of agriculture, mining is the only known alternative source of employment in rural areas. Employment figures in the sector have increased considerably during times of drought (Shoko, 2002).

Provision of infrastructure and social services are keys to improving the sustainability of ASM activities. It is important because most, if not all, sites are located in remote, highly inaccessible areas (Fig. 16.2). There are some exceptions in Mozambique, such as a number of sites in Manica and Nampula Provinces, which are linked by tar and dust roads of good standards. However,

Figure 16.2 Precarious conditions of access roads and bridges in Lupilichi/Niassa Provinces.

generally, access roads to, and from, service centres are in a poor state, especially during the rainy season; this, in turn, inhibits the effective marketing of gold produced by artisanal and small-scale miners. There is minimal maintenance of tertiary and rural roads by the government, although some initiatives by non-governmental organisations, promote projects of "work-for-food", which maintain some of the tertiary and rural roads in Manica, Nampula, Zambézia and Niassa Provinces.

Health and educational services are limited within Mozambican mining communities, both in terms of numbers, and capacity to respond to the needs of the population (Valoi et al., 2000). Generally, basic infrastructure is located significant distances from mining communities. Moreover, facilities were not upgraded at the conclusion of the civil war in 1992 and after recent flooding events (during the 2000 and 2001 rainy seasons). Many mining communities do not have access to potable water, in which case, rivers and pits are used as the primary sources of water for drinking, cooking and mining activities.

Most of Mozambique's artisanal and small-scale miners are nomadic, temporarily settling in mining camps (Fig. 16.3) surrounding mineral-rich areas, and then migrating once a new site with better ore grades is discovered. These miners compete with the local population for water and land. Since 1975 (post-independence war), settlement patterns have generally been complex, largely because of a resettlement of refugees during the civil war (1976–1992); most never returned to their original homes.

In communities where mining activities have been established, two mining community types have emerged: the village type and the camp type. The "camp" is similar to a peri-urban environment, in which people are more subjected to social risks. People in "camp" settlements have reduced food security, increased mortality, poor sanitation, are typically homeless and landless, and suffer from social disintegration and marginalization. The "village" type of settlement,

Figure 16.3 Small scale mining settlement at Munhene deposit, along Revue river in Manica Province.

however, has the advantage of having more space and therefore provides greater opportunity for ensuring family survival through agriculture. Illegal immigrants from neighbouring countries are reportedly settling and bringing families to camps. These artisanal miners and their children are subjected to "taxation", which serves as a means of compensating for their residence.

For a community to function effectively, it requires an authority structure to organise, regulate and resolve disputes and prevent crimes. Mining communities are no exception; most have authority structures based on a combination of indigenous beliefs and local governmental frameworks. In Mozambique, the authority structures that interplay in ASM mining communities include local government (police), state administration (mining department), village secretaries, traditional leaders (*traditional chiefs* or *regulos*), religious figures (church/spiritual authority – *curandeiros, mwenes* or *barozos*), and mine camp leaders. The traditional chief's representative declares rest and workdays for miners, and is also responsible for settling social disputes at camps. If mediation through the chief fails, cases of crime at the camp or in the villages are reported to the local police. In return for his work, miners "pay" to the traditional chief's representative a bag (c. 50 kg) of gold ore to honour his role and power. Miners firmly believe that to succeed in their work, they require protection from their ancestors. Ancestral worship, led by traditional chiefs, includes rituals and prayers, and is performed at gravesites.

Valoi et al. (2000) reported that *curandeiros* work in collaboration with local chiefs. For example, in Manica, the traditional chief of *Munhena* area appointed a representative at the mine site. Its main duty is to allocate mining plots and ensure, through the *curandeiros*, that rituals and prayers are conducted at sites.

Administrative authorities are normally involved in regulating and maintaining law and order at mine sites. The police and administration officials are said to regularly visit and "tax" the miners, or impose fines upon them in instances where safety and environmental issues are not adhered to. This often creates a feeling of antagonism between miners and government authorities (mining department inspectors), a problem that dates back before 1994, when authorities attempted to control the illegal trade of gemstones, which often result in the confiscation of products, and occasional imprisonment of miners or traders.

At many of the mine sites surveyed by Valoi et al. (2000), the following ethnic groups were identified; Nhanjas, Jauas, Swahilis (also present in Malawi, Tanzania and Zambia), Macuas (Cabo Delgado Province), Sena (Sofala Province), and Nhungues (Zambia, Zimbabwe and Malawi). Among these, and most other groups, the majority were men. Women are predominantly involved in transporting, washing, and the panning of ore. Other supplementary services performed by women in support of artisanal mining include childcare, various family duties, household chores, and food and beverage sales. Children between the ages of six and 10 are often involved in activities similar to those performed by women. Overall, women and children constitute an estimated one third of country's mining population of 20,000 (Mondlane, 2001; MMSD, 2002).

There are a number of cultural barriers and superstitious beliefs that hinder women's involvement in ASM activities. These range from the indigenous belief that the woman's primary roles are as mother, food provider, and housewife (Valoi et al., 2000). In recent years, however, female participation has increased, with some women reportedly running more efficient mining enterprises than many of their male counterparts; however, most frequently find it challenging to secure financial, legal and technical aid. In some parts of the country, such as N'tulo, Niassa, women are not permitted to work at the mine site because they are believed to attract "bad spirits". In such cases, they are only involved in food and beverage sales. In Manica, women are not permitted to dig trenches, and are confined to ore transportation and processing duties.

Children aged between 10 and 16 years (Fig. 16.4) commonly work alongside their parents in Tete, operate on their own in Manica, and are hired for transporting ore in both Manica and Niassa. In Manica, Zambézia and Tete, most children attend school during mornings, and rendezvous with parents at mine sites in the afternoons. In Manica, there are reports of children skipping classes or leaving school entirely in order to work at the mines (Valoi et al., 2000). Although child labour is illegal in most countries, it is fairly rampant in informal business sectors in developing countries, particularly in the ASM sector (see Chapter 10).

Economic factors

The generally poor economic performance of ASM is largely a result of miners not operating within legal frameworks – perpetuated by the fact that respective governments rarely assume the required facilitation role (in inspection, market provision, increasing access to finance, and establishing infrastructures). The

Figure 16.4 Children and women reworking old damps and washing for gold in Murrupula – Nampula province, Mozambique.

Ministry of Mineral Resources and Energy (DPMIREME) in Mozambique has helped to convince miners to adopt safe mining methods and appropriate waste management systems. To help prevent the direct discharge of tailings into rivers, the DPMIREME has instructed miners to divert waterways, construct dams, and wash and pan within basins outside of the main riverbed. These efforts have yielded positive results in Manica Province, the immediate impact being a reduction in both siltation and mercury contamination in rivers. However, to ensure that miners adhere to the aforementioned techniques, DPMIREME staff has had to inspect sites regularly but efforts, overall, have been crippled by insufficient funding. The FFM in Manica Province also offers technical assistance to miners, and, at one stage, even attempted to organize miners into co-operatives and supply them with machinery such as Knelson diesel concentrators.

Up until only recently, there were major land use conflicts between small- and large-scale mining parties in Mozambican gold mining regions. However, large-scale mineral exploration companies ceased activities altogether in the late-1990s (see Table 16.3), mainly because of shortages of economic reserves. Initially, conflicts had arisen because small-scale miners were encroaching on large-scale concessions (occasionally after nightfall). A case in point was the discovery of a rich quartz vein within the exploration concession of North Rand Company by small-scale miners in 1998; eventually, police were called upon to guard the area during the day and night (Mondlane, 2001). Occasionally, however, small-scale miners have worked areas rehabilitated by large-scale companies, such as the Chua River near Manica, where artisanal miners have mined regions rehabilitated by the ALMA/BENICON Company.

Table 16.3 Exploration and mining companies in Manica Province.

Exploration Company	Mining Company	End of Activity	Remark
Ashanti Goldfields		1999	Insufficient reserves
Trillion Co.		1998	Insufficient reserves
North Rand		1998	Insufficient reserves
	Mincor (Monarch)	1997	Financial problems
	Alma/Benicon	1995	Insufficient reserves
ZIMOZ		1998	Reason not known
AUSMOZ		Active (2001)	

Source: Mondlane, 2001.

Table 16.4 Different buyers of artisanal/small-scale mining products in Mozambique.

To whom do you sell your production?	Niassa	Tete	Manica
Dealers or Traders	14.7%	81.8%	51.7%
Government (FFM)	82.4%	18.2%	43.3%
Commercialisation Licence Holders	–	–	–
Others	2.9%	–	5.0%
Total (interviewed)	68	11	60

Source: Valoi et al., 2000.

There are an estimated 250 private gold traders in Mozambique. However, the country's complex gold market not only features legalized buying agents, but also the "Fundo de Fomento Mineiro (FFM)" mining development fund, and a plethora of unregistered buying agents. The complexity of the market structure is compounded by widespread illegal purchasing and production. The legal framework mandates that every gold producer sells mined product to the FFM, which has been established by the Ministry of Mineral Resources and Energy for the purpose of:

- Contributing to better knowledge of Mozambican mineral resources;
- Funding policy implementation for the country's mining industry;
- Funding and assisting the promotion of resident ASM activities; and
- Ensuring that the investment and the output from ASM activity is used to further develop the country's mining sector.

The FFM has also been involved in purchasing gold produced by artisanal and small-scale miners in Manica, Niassa and Tete provinces (Mondlane, 2001). Valoi et al. (2000) identified these provinces as the locations where most of the country's gold is sold to private buyers. The FFM is the second largest buyer of gold, mainly because private dealers offer higher prices, Niassa Province being the only known place where more gold is sold to the FFM than dealers (Table 16.4). However, the FFM plays an important role in *supporting* the marketing of gold in Tete, Manica, and Niassa, where it channels the gold produced by small-scale miners to official markets, and helps to establish production records (see *e.g.* Table 16.5). It also determines, using the Reserve Bank's

Table 16.5 Estimate of gold commercialized by the formal sector in Manica Province.

Year	1997	1998	1999	2000	2001		
Months	Aug.–Dec.	All	All	All	Jan.–Feb.	March	April
Au (g)	2067	9527	11535	15837	740	5086	5438

Note: In January and February, the formal sector bought very little gold because of the rainy season in Manica. In March and April, the roads to Maputo and Beira were closed due to floods in the central part of the country, hence the limited presence of buyers from other provinces and capital cities.
Source: Mondlane, 2001.

policy as reference, an appropriate price for gold, which it then offers to the industry's producers. Yet, dealers modify the gold buying prices set by the FFM, in turn, offering exploitative prices, much to the detriment of desperate small-scale miners. In Murrupula and Namuhara, traditional chiefs, in consultation with mineral traders, also play a role in price fixing for gold (Valoi et al., 2000). Consequently, different prices are in place throughout the country, varying from US$6.6 to $7 per gram; the highest prices paid are in Manica and Niassa, while the lowest prices have been reported in Tete.

Some producers travel to provincial capital cities to sell their gold, whereas others travel to the country's capital city of Maputo, or even neighbouring countries (Zimbabwe, Zambia and Malawi). Most dealers and traders do not hold the requisite licences to trade in minerals. As the government has not strictly enforced their adoption, it has enabled these merchants to avoid paying high taxes. In Manica and Nampula, ready access to mining areas has made local gold marketing dynamic. Dealers from the provincial capitals of Nampula and Zambézia Provinces – Nampula and Quelimane respectively – control the gold business. These merchants sell gold in cities, where a number of craftsmen work the products into jewellery, which is then sold locally in these cities and in Maputo. In terms of illegal smuggling, illicit marketing is particularly pronounced in Niassa. Tanzania is readily accessible from Niassa's gold mining areas, and most of Mozambique's gold buyers are, in fact, Tanzanian, to whom most indigenous miners sell non-quantifiable amounts of gold directly, often within the Tanzanian borders.

The relationship between buyers and miners is strictly commercial, involving cash payment and few credit arrangements. However, in Manica, buyers reportedly provide mercury in advance to miners to help assure their purchase of the gold product. Although this action does not seem to influence the price fixing for gold, it does have a tendency to create a dependency and indebtedness among miners, who feel obligated to produce and sell the gold to the same traders.

Small-scale miners are eligible for funding from the Commercial Banks, provided that they have at least 20% of the required resources to co-fund a proposed project. The FFM is the only source of direct credit for small-scale miners, although the figures and the number of beneficiaries of this credit scheme are not available. However, the Fund has provided several miners with economic assistance, especially those operating within the gemstone field of Alto Ligonha.

Table 16.6 Support schemes to artisanal miners (adapted from Valoi et al., 2000).

Do you get any kind of support?	Niassa No	Tete No	Nampula No	Manica No
Financial	21.4%	100%	100%	91.1%
Equipment	7.1%	100%	100%	55.4%
Other	21.4%	100%	100%	97.1%

Source: Valoi et al., 2000.

To qualify for funds from the FFM, candidates must file a request to its president. In cases where funds are needed for the purchase of equipment, he/she is required to append the appropriate financial quotations. Applicants must also include the following documents within the application package: (a) copy of their mining licence; (b) a credible feasibility study of the project; (c) proof of availability of 20% of finances for the proposed project; (d) guarantees in property equivalent for the requested loan amount (*i.e.* a mortgage); (e) a plan for pay back of the credit; and (f) proof of market for the commodity to be produced. As the foregoing requirements are prohibitive to most ordinary artisanal/small-scale miner, most have not qualified for funding (see *e.g.* Table 16.6), and, of those who have managed to secure credit, most have not honoured the pay-back plan. Some do not even invest the funds in mining, opting to invest awarded moneys outright or acquire luxury commodities (*e.g.* vehicles). There simply exists no adequate means of monitoring the progress of the implementation of proposed projects and the disbursement of awarded funds. In recognition of this problem, the government has proposed the implementation of a new regulatory scheme with recommendations from the World Bank. What has made the assessment particularly challenging is the fact that there are few reports available on non-governmental financial support to small-scale miners in Mozambique.

Environmental and health-related impacts

As already explained, most ASM activity in Mozambique is seasonal, occurring during the non-agricultural season, or when agricultural yields are poor as a result of rampant drought or flooding. The number of artisanal and small-scale miners has increased significantly in Mozambique since 2000, largely because of two successive and devastating floods during the 2000–2001 and 2001–2002 rainy seasons, which were accompanied by short rainy seasons. It can be concluded from field visitations that ASM in Mozambique has been an important source of employment, has intensified immigration, and facilitated local economic improvements. It has also precipitated the formation of new settlements; caused a mass degradation of natural resources (loss of vegetation, land, etc.); and has created a wealth of social problems, including conflicts, and health- and sanitation-related complications.

In Mozambique, the environmental impacts of ASM activity are quite visible. Most result from deteriorated soils (siltation and water quality), deforestation, and the use of mercury in gold recovery. Deforestation has occurred largely

Table 16.7 Amount of mercury used to obtain a gram of gold.

Quantity Hg per g Au	Number	Manica
<1 g	17	53.1%
1–5	11	34.4%
5–10	1	3.1%
>15 g	3	9.4%

because of the sudden sprawling of "gold rush camps". The construction of surrounding mining villages has also required an extensive cutting of trees to create space, and for building materials and fuel. In Manica and Nampula, rivers are diverted and water is collected in open pools for gold washing.

Mercury is used intensively in gold processing activities in Manica and Niassa Provinces, where primary gold quartz veins are worked. The amount of mercury that is used to process one gram of gold (Table 16.7) is clear indication of how much mercury is consumed during gold recovery – hence, potentially released into the environment. However, water and river contamination has not only resulted from mining activity in Mozambique. Gold mining activity is rampant upstream along the banks of Mazoe, Luenha, Revue and Zambezi rivers, which drain eight of the fourteen SADAC countries – namely, Angola, Botswana, Malawi, Namibia, Tanzania, Zambia and Zimbabwe (Mondlane, 2001).[4] The communities in the vicinity of mine sites, in turn, are exposed to mercury through inhalation and/or consumption of contaminated water. Non-gold mining communities, particularly those situated downstream, are at major risk of exposure to methylmercury through consumption of mercury-contaminated fish. Vegetables grown on the riverbank are irrigated using the contaminated river water.

Miners themselves are at a high risk of mercury exposure due to direct skin contact during amalgamation and via inhalation of mercury vapour during the roasting of the amalgam. Handling or eating food with hands coated with mercury leads to additional contamination. The storage and handling of mercury in artisanal gold mining areas further increases exposure. The process of amalgamation itself frees up to 60% of mercury into the atmosphere. Previous studies in Brazil have shown that about 60% of the mercury used in gold ore processing is lost into the atmosphere during amalgamation, while 30% enters river systems through tailings.

The transfer of appropriate mercury technology to Mozambique has been hindered by a number of factors, including costs, a lack of training, and suspicion on the part of the miner. In Manica, ASM operators do not use mercury retorts and TermEx technology because they are too expensive to purchase and

[4]SADC: Project AAA.4.1 – Environmental Impact of mining and related industries on the water quality of the rivers of the Zambezi Basin, and Project AAA.4.3 – Investigation of the pollution, river bank degradation and siltation caused by small scale mining and the use of mercury and cyanide.

Table 16.8 Environmental problems, as reported by communities.

Environmental problem	None (%)				Significant (%)				Very significant (%)			
	N	T	Nam	M	N	T	Nam	M	N	T	Nam	M
Water contamination	100	71.4	66.7	72.7	–	–	33.3	27.3	–	28.6	–	–
River Siltation	–	83.3	–	50	100	–	100	50	–	16.7	–	–
Soil Erosion		83.3	100	66.7	–	–	–	33.3	100	16.7	–	–
Destruction of access roads	100	91	100	66.7	–	–	–	33.3	–	9	–	–
Destruction of religious sites	–	91	100	75	–	9	–	25	–	–	–	–

N = Niassa, T = Tete; Nam = Nampula; M = Manica.
Source: Valoi et al., 2000.

maintain. Some miners cited the unavailability and low efficiency of various retorting systems as additional reasons for not adopting them altogether (Valoi et al., 2000).

In Manica, where mercury is intensively used in gold processing, some 93% (53) of respondents reported being aware of the risks associated its use but as already explained, opt not to deploy alternative techniques because they are untrained and cannot afford to pay the prohibitive purchasing and maintenance costs.

Table 16.8 shows people's perceptions of the environmental problems resulting from ASM in Manica, Tete, and Nampula provinces. Water contamination, river siltation (gold mining), soil erosion, destruction of vegetation and health-related risks were reported as the main environmental problems.

The most common health-related problems in ASM communities result from rampant diseases, including malaria, STDs, diarrhoea, and tuberculosis. In fact, ASM communities serve as a vector for the spreading of disease. Most Mozambican ASM communities do not have access to the appropriate health services, largely because of the remoteness of sites and the high mobility of mining communities. When efforts have been made to establish health and educational facilities, such as the case of Nampula Province, where a community of 10,000 people was assisted by government in constructing a school and hospital, they are typically abandoned once deposits are exhausted, as miners vacate the area in search of other prospective locations.

Malaria is the most common disease in Mozambican ASM communities, induced by the highly unsanitary conditions exacerbated during the rainy seasons, when the pits and pools dug by miners become breeding grounds for disease-carrying mosquitoes. Malaria was reported to be the most significant cause of death among miners and community members (Valoi et al., 2000). Another common ailment, diarrhoea, is also caused by unsanitary working conditions. The sickness is brought on by the fact that most miners are forced to consume poor quality water; hygienic problems are worsened by the complete absence of toilets.

Like most mining communities, Mozambican sites contain a heterogeneous mixture of people from different places, most of whom are without their families. This encourages sexual relations with prostitutes. Various STDs, including HIV/AIDS, and therefore prevalent, accounting for the third most common group of diseases in the industry.

CONCLUSION

In Mozambique, ASM is plagued with a number of problems, each of which must be resolved in order for the sector to contribute to growth and development in a more positive fashion. These stumbling blocks are both socio-economic and environmental in nature, and are directly and indirectly related to poverty, population pressures exerted upon natural resource stocks, as well as a general lack of knowledge.

The alleviation or elimination of poverty is believed to be the panacea for the ASM sector's plethora of problems. Poverty is endemic, particularly in the remote rural areas of Mozambique, where ASM is, in fact, proving to be a means of poverty alleviation. The limited success with poverty alleviation in the sector in Mozambique, as is generally the case in most of the southern African region, has deep roots, fuelled largely by a lack of interest on the part of the central government. This cavalier stance is, in part, rooted in the remote and scattered nature of the economic activity, which makes the registration and formalisation of these activities prohibitively expensive. As long as the sector continues to operate outside of the State's legal and institutional framework, the central government will continue to lose out on the full benefits accrued from mineral sales, royalties and taxes. Without such clearly defined benefit streams, the government will continue to be reluctant to play its usual facilitation role of providing infrastructure (*e.g.* schools, hospitals/clinics and roads). Moreover, if effective registration and formalisation procedures are not in place, artisanal and small-scale miners cannot access micro-finance for development capital, which forces them to make use of middlemen for marketing their minerals (who very often exploit their ignorance and desperation) – hence, perpetuating further the vicious cycle of poverty that characterizes the industry.

REFERENCES

Bezerra, O., Veríssimo, A. & Uhl, C. (1996). The regional impact of small scale gold mining in Amazonia. *Mining and Environmental Research Network*, 10, 50–53.

Chiwawa, H. (1993). *Environmental impacts of cooperative mining in Zimbabwe*. 4th Congress, Organisation for Social Science Research in Eastern and Southern Africa, Debra Zeit, Ethiopia, 25 pp.

Hollaway, J. & Associates (1992). *Environmental effects of mining in Southern Africa*, A presentation to the Mining and Environment Workshop for the SADC.

Manuel, I., Muacanhia, T., Zacarias, R. & Vicente (1999). Exploração artesanal do ouro no Distrito de Manica: Degradação ambiental versus desenvolvimento; Congresso de Geoquimica dos PALOPs.

MMSD – Final report (2002). Breaking New Ground (*Mining, Minerals, and Sustainable Development*); Earthscan, London and Sterling, VA; 441 pp.

Mondlane, S. (2001). Small Scale Mining and Sustainable Development in Southern Africa: A baseline survey for Mozambique. In B. Dreschler (Ed.), *Small-scale mining & sustainable development within the SADC region, MMSD Southern Africa Report on Research Topic 1; RT1 ITDG, Harare,* pp. 27–58.

SARDC, IUCN and SADAC (1994). *The state of the environment in Southern Africa, Gaberone.*

Shoko, D.S.M. (2002). *Small scale mining and alluvial gold panning within the Zambezi Basin: an ecological time bomb and tinderbox for future conflicts among riparian states*. London: Weaver Publishers (in press).

Valoi, G., Kazilimani, E. & Bukali, F. (2000). Artisanal Mining Baseline Survey: Mozambique; Final Report. Minist. Min. Res. And Energy; Maputo; 124 pp.

17

The Socio-Economic Aspects of Artisanal Gold Mining in Migori District, Kenya

WINNIE V. MITULLAH, JASON S. OGOLA, AND MONICA A. OMULO

Migori District is located in Nyanza Province of Western Kenya (Fig. 17.1). It lies in the southern part of the Winam Gulf in Lake Victoria. Here, the gold that occurs in the quartz veins of the Archeaean rocks of the Nyanzian Group was first discovered in 1920. Although episodes of large-scale mining occurred in Nyanza until the mid-1960s, artisanal mining has persisted to date. Currently, mining takes place in the entire district of Rongo in the north, around Migori town in the south, and at Masara and Macalder in the west. Once auriferous quartz veins have been extracted from host rocks, ore is crushed and panned, after which mercury is added to the tiny gold particulates. The processed gold is then sold to local buyers, who are either agents or self-employed in the business. Finally, gold is sold to dealers in Nairobi, although some product lands in the hands of smugglers, who, in turn, transport it out of the country through illegal channels.

In Kenya, artisanal gold mining takes place on both farmland and homesteads in the western part of the country; in the Migori, Siaya and Kakamega districts; and in the Turkana district in the north-west. The number of people involved in mining activities countrywide fluctuates from 30,000 during peak periods, to 10,000 during low periods. Despite being renowned for featuring periodic gold rushes, most of the country's sites are ill-equipped to endure the pressures of rapid population growth, particularly during periods when newly-discovered gold reefs are intensively worked. Workers are quite mobile, frequently migrating from one mine site to another, depending on the availability of gold. Once a site has been exhausted, water logged, or is considered too dangerous to work, miners simply move to new prospective mining regions. This, however, causes substantial environmental damages. Frequent movement also disrupts family life, particularly children's education and farming activities.

The chapter examines the socio-economic aspects of gold mining and related activities in Migori District, Kenya. It describes the background of mine workers,

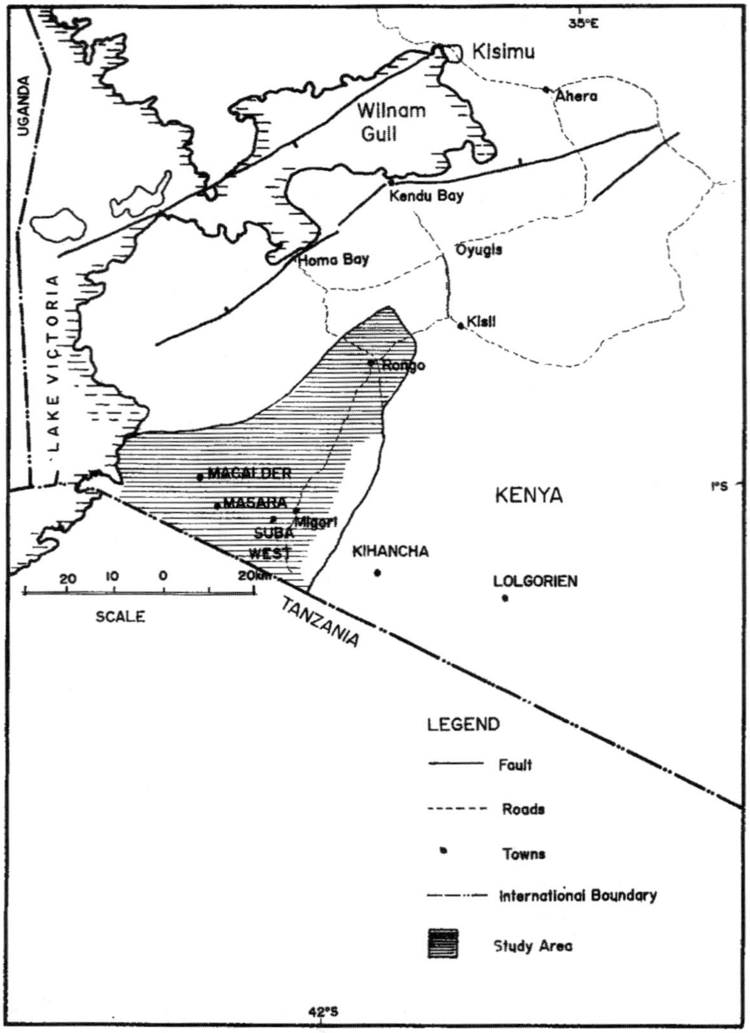

Figure 17.1 Location map of the mine sites in Migori District .
Source: Ogola et al., 2002.

their employment and incomes, and the effects of mining activities on education and agriculture.

MATERIALS AND METHODS

The data used in the study reported in this chapter are drawn from a previous study of artisanal gold mining in the Migori District (Ogola et al., 1999). The study covered three areas: technical aspects related to artisanal gold mining; the impacts of artisanal mining on the environment and human health; and the socio-economic aspects of activities. The chapter focuses heavily upon the socio-economic aspects of the study. Data were gathered from 51 entrepreneurs working in eleven gold prospects within the district, two focus group discussions, and a number of key informants.

MINE WORKERS

The majority of individuals employed in the Kenyan gold mining industry are young. The present study revealed that most were born between 1961 and 1970 (51%), and between 1971 and 1980 (27.5%); only one of the people surveyed was born before 1950 (Table 17.1). The average age of surveyed mine workers was 31 years, and there was little difference between the average ages of (surveyed) males (32 years) and females (30 years). The country's Micro and Small Enterprises (MSE) sector has been known to attract the young, economically-active segment of the labour force. Indeed, many people, and, in certain cases, entire homesteads, were found to be wholly dependent upon mining and related activities. At some mine sites, individuals who had left urban and other formal sectors of employment had joined artisanal gold mining forces.

The age distribution shows that the majority of those involved in mining activities were born when gold mining within the Migori District was at its peak. Some of the gold reefs that are currently being exploited are leftover from past large-scale mining undertakings; in other cases, new reefs have been discovered altogether as a result of artisanal mineral prospecting.

Although the majority of surveyed workers (56.9%) were married, there was a high percentage (21.5%) of divorcees/widowers. However, more male respondents (66%) were married than female respondents (50%); 13% of surveyed females were single; and 37% were divorced, separated or widowed. Compared to similar studies, the percentage of married women in this study was comparatively low. This is partly explained by the high numbers of single, divorced, separated and widowed female migrants who have been pushed into opportunities far from their homes and/or in residential areas.

Gender analysis indicates that most gold miners are men, with only a few women involved in the industry as workers. Some women are traders, buying ore, and either processing or hiring other people – mostly men – to process for them. Most women are also engaged in ore crushing and panning, while others provide services to the industry.

Table 17.1 Age distribution pattern of surveyed mine workers.

Year of birth	Number of respondents	Percentage
Before 1950	1	1.9
1951–1960	10	19.6
1961–1970	26	51.0
1971–1980	14	27.5
Total	51	100

Table 17.2 Education levels of surveyed mine workers.

Education level	Frequency	Percentage
No formal education	6	12
Lower primary	7	14
Upper primary	25	49
Secondary	13	25
Total	51	100

Those involved in the Kenyan gold mining industry originate from both within and outside mining districts. However, a significant percentage of outsiders who had moved with their families now face accommodation and related problems of service provision. Rapid population growth exposes miners to harsh conditions, especially during the early stages of mining. However, as confidence in activities grows, miners begin constructing shelters with the hope of staying longer.

In Kenya, educational level has important implications in the employment market. It determines upward mobility, access to economic opportunities, and self-actualisation. The educational status of surveyed mine workers was comparatively low, as only 25% had secondary education; some 63% had primary schooling and 12% were illiterate (Table 17.2). Other studies conducted in the country (McCormick et al., 2001; Kyahesi, 2001; Kinyanjui, 2001; Wegulo et al., 2001; Njeru et al., 2001) have shown comparatively higher levels of education among SME operators. For example, in a study by McCormick et al. (2001), it was reported that of 304 women entrepreneurs sampled, 118 had attained secondary education (39%); two had post-secondary qualifications; and four, a university education.

EMPLOYMENT AND INCOME

Gold mining provides employment to inhabitants of the rural communities in Migori Gold District and beyond. Activities include underground ore extraction, transportation of ore to the surface, ore crushing and panning, gold recovery from gold/mercury amalgam, and gold trading (Table 17.3). The majority of mine workers surveyed did not have any prior formal employment. Specifically, only 11% were engaged in formal employment, and approximately 36% were involved in a variety of activities in the informal sector (Table 17.4). Those who did not hold an occupation before becoming involved in mining were either been unemployed or attended school. Out of those who were engaged in the informal sector before joining the mining segment, 41.2% were female, while 58.8% were male. The majority of respondents indicated that they began working at the District's mines after 1990.

The majority of mine workers surveyed considered the occupation to be permanent, whereas only a small proportion viewed mining as temporary employment. Some (49%) described mining as merely a part-time engagement,

Table 17.3 Mining and related activities.

Type of activity	Frequency	Percentage
Ore extraction	13	25.5
Transport	7	13.7
Ore crushing	9	17.6
Panning	8	15.7
Gold Trading	14	27.5
Total	51	100

Table 17.4 Engagement of workers before joining artisanal mining.

Type of engagement	Frequency	Percentage
Unemployed/in school	25	53.2
Informal sector	17	36.2
Formal sector	5	10.6
No response	4	–
Total	51	100

Table 17.5 Mean net-income per month by gender (US$).

Peak/low mining period	Mean male	Mean female
Peak	250	96
Low	40	22

the incomes from which are used to supplement that received from other sources of employment they are reportedly involved in.

The funds invested in artisanal mining originate from a number of individual sources – mainly personal savings or revenues obtained through the sale of property (*e.g.* land, cattle or severance from previous formal employment, etc.). Although initially, one must raise about US$50 in capital before being able to mine, realistically, one must raise at least US$1,000 to reap any reasonable interest from the trade.

An analysis of educational levels *vis-à-vis* incomes revealed that individuals having secondary school diplomas or higher qualifications were the industry's top money makers. In fact, those with higher education were involved in the more lucrative sub-sectors of the industry, which require greater investment. Such individuals also have effective gold trading networks. Some had been previously employed, and had saved funds for investment, thus enabling them to acquire mine pits; employ people to mine and process gold; and, by bypassing middlemen, sell product at higher prices in Nairobi.

Income levels varied between male and female workers, and during peak and lag periods (Table 17.5). Generally, men had higher incomes than their female counterparts. An analysis of incomes between male and female workers revealed a difference in value of 50%, thus supporting the contention often advanced by many researchers (*e.g.* Parker & Torres, 1994; McCormick & Mitullah, 1995; Graham et al., 1998) that women generally earn less than men in SME activities. Graham et al. (1998) observed that women often concentrate on one economic activity alone, while men were engaged in a variety of tasks. This phenomenon is best explained by the fact that Kenyan women have numerous domestic chores and are charged with caring for children. Although these chores are not economic activities per se, they are nevertheless major reasons why many women close their businesses early, often during busy lunch hours. Conversely, men open their businesses early, and can continue to attend to their activities uninterrupted throughout the day.

EFFECT OF MINING ON EDUCATION IN KENYA

In Kenya, gold booms are often blamed for irregular school attendance and eventual school dropouts. A survey in Migori District confirmed this position. A number of school children were found to be actively engaged in gold mining activities. At most sites, children appeared to provide different forms of labour, although most were engaged in ore crushing and panning, with only a small number involved in actual extraction of gold from reefs. Most of the children interviewed provided a number of reasons for being engaged in mining activities. The main reasons given were that parents had not bought them the uniforms and the necessary books, or had failed to pay school fees. Others simply stated that they were helping their parents by earning wages at the mines. A small group of youths was brave enough to point out that they themselves had opted to undertake mining operations. Some children have not abandoned school completely, but do occasionally skip classes in order to earn money working as miners. Other children work overnight, and, as result, are often too tired to attend school during the day. In short, the high premium placed on gold mining and related activities disrupts learning, with a number of children not recognising the difference education *could* make to their lives.

When children combine schooling and gold mining activities, they are often economically better off than their teachers. Some students are even able to lend teachers money, an aspect that is ethically problematic for both students and teachers. It makes students place less value on education compared to their gold mining activities.

The attitude of most parents toward schooling has also been responsible for the poor educational response from Kenya's rural youths. Many parents do not appear to understand the value of education, preferring that their children work at the mines. Others argue that their children only work at the mines during weekends and school holidays. However, one mother remarked that once a school pupil has access to "gold money", he or she no longer adheres to serious education. Thus, any youth attending school should not be involved in mining activities at all. Our survey confirmed this position. It was established that those who had completed secondary education and proceeded to colleges or universities had never worked in the mines during their youth.

The teachers and education officers within the mining district further confirmed that gold mining had disastrous effects on education. One officer observed that in one of the divisions where gold mining has been ongoing for several decades, educational standards had deteriorated. Cases of purchasing forged teaching certificates with the "gold money" earned in the area are not uncommon, particularly during times when gold production is low or has ceased altogether. In certain cases, the culprits have been unable to read or write, and have had to be dismissed from the teaching profession.

MINING AND AGRICULTURE

An overwhelming majority of Kenyans are engaged in agriculture and related activities – the principal source of livelihood and income in the rural regions of the country. Mining and agriculture are somewhat interrelated since those who move into mining villages require food. However, mining prevents villagers from farming; in the end, most miners purchase large quantities of food that they would otherwise have produced on their own farms. In short, gold mining diverts people's attention from agriculture because farming is regarded as a lower income activity.

Apart from the lack of interest during periods of lucrative gold activity, poor mining methods, which have been responsible for causing environmental degradation, have also adversely affected farming. Many mine sites essentially become wasteland (Fig. 17.2), and post-mining attempts to farm such land have

Figure 17.2 Moulds of abandoned pits and trenches in the Kamagambo area, Migori District.

resulted in stunted crops and poor yields; as a result of mining, waste rock has replaced previously rich soils.

Deforestation, haphazard disposal of waste rock, tailings dumps, unfilled mine pits, soil erosion, and the siltation of rivers, have also occurred throughout Kenyan mining regions. The removal of trees for use in mine workings has caused notable deforestation and soil erosion. Occasionally, mining has precipitated deforestation *outside* of mine regions, as logs have been removed in distant forested areas and then transported to mine sites for use. A great number of other trees have also been threatened by miners that have dug for ore at their bases.

Whereas the *Mining Act of Kenya of 1976* stipulates that mined sites must be rehabilitated once mining has ceased, resident artisanal miners have not adhered to this legislation. However, legislatively, artisanal miners cannot be blamed. More specifically, this category of mining is not legally recognised by the Government, as no mining law covers its activities. Moreover, *actual* miners are not the landowners, which is why once mining has ceased, most simply migrate to new sites. Thus a landowner is often left with gaping holes and collapsed mine workings on his land.

Although gold mining has had an effect on agriculture, and government officers recognise that it inhibits farming, some Kenyans engaged in gold mining, do, in fact, combine it with agriculture. However, in cases where soil is infertile, farming is not possible, and therefore, gold mining is the only means of survival.

POVERTY IN THE MIDST OF WEALTH

Our findings show that gold miners and those working in related activities in Kenya earn decent incomes in a rural context. However, the wages they earn do not accurately reflect their living standards and health. Most work and live in poor conditions, devoid of sanitation facilities and basic services. The worst hit are those who either work underground, or are involved in ore crushing and the burning of gold-mercury amalgam. These individuals earn comparatively low amounts, and receive no health benefits or coverage, despite continually being exposed to deleterious health-related entities.

In Kenya, the greatest beneficiaries of artisanal gold mining are landowners, who simply lease land at exorbitant costs, and collect money on an hourly or daily basis without any investment. Apart from "renting" deposits, landowners are also entitled to a mineral share of up to 50%. During peak mining periods, they collect up to Kshs. 500,000.00 (approximately US$6500) per week. This type of money in a rural setting results in excessive lifestyles, characterized by perpetual drunkenness and reckless marriages involving several wives.

Alcoholism, divorce and remarriage are common among artisanal miners. However, poverty commonly occurs within the first few months following the completion of mining. Even the landowners, many of who became millionaires overnight, do not fully escape the vicious cycle of poverty.

To justify this vicious cycle, miners commonly associate the wealth derived from gold with a number of myths. For example, they believe that gold money is "haunted", and, if invested in any business, will inflict bankruptcy and eventual poverty among investors. However, in isolated cases, individuals have managed to save and invest "gold money" successfully. A more effective summation of the situation is as follows: because artisanal miners generally do not plan and lack budgets, most simply spend profits recklessly.

DISCUSSION

Gold mining and related activities in Migori District bring both wealth and problems, depending on how acquired money is spent. Artisanal gold mining is generally a positive undertaking in most rural areas of Kenya, as it is an important source of employment and income. The earned income has improved people's lives, enabling many to invest in good quality housing, businesses and livestock. At the same time, however, resident gold mining activities have had a number of negative effects on surrounding communities. They have disrupted general rural lifestyle, stimulating immoral practices such as prostitution, divorce and remarriages. They have also interfered with youth education, agriculture, and have been responsible for notable environmental degradation.

The rapid urbanisation that takes place wherever rich gold reefs are discovered creates social and environmental explosion. This, in turn, disorientates the local population by creating a lifestyle alien to villagers. To some extent, this situation prevails because of the Government's lack of recognition of artisanal mining. Mining activities are generally unplanned, and operations are conducted without any regulation and guidance. Recognition of artisanal mining could facilitate improved planning, environmental management and monitoring, and would also create some sense of responsibility among the miners themselves, helping to persuade them to operate in a more sustainable manner.

The educational and agricultural sectors within gold mining areas need to be monitored by the appropriate government bodies. Both pupils and their parents do appear to place a high premium on education, seemingly distracted by the short-term benefits accrued from gold mining. An assessment of children working at mines revealed that occasional work and exposure to money literally removes children from school; this applies to both girls and boys. Boys leave school to work at the mines, whereas girls are lured out of school to marry young men who have become temporarily wealthy from gold mining. This is quite problematic since anecdotal evidence and findings from this study show that gold mining activities are, in fact, short-term ventures. Once mining ceases, a number of youngsters are lured into criminal activities, since they have no alternative employment, and consequently, no source of income.

The assertion made by some parents that school children can work at mines both during weekends and holidays is unjust. These time periods must be seen as times for independent studying or for "homework" sessions. In any case, Kenya is signatory to the convention barring children from hard labour, and any parent who allows a child to work at the mines is in violation of this international law.

Finally, labour intensive mining prevents villagers from undertaking subsistence farming activity. Once mining activities cease, famine generally occurs, as there is no food security in the area. Some villagers even find it hard to revert back to farming, having grown accustomed to the large sums of money acquired from gold mining. Furthermore, a number of landowners lose their land as a result of mining activities, which transform it into wasteland with gaping holes and tailings dumps; as previously explained, in the process, the land becomes unsuitable for agriculture.

REFERENCES

Graham, A., Mitullah, W.V. & Kopiyo, G. (1998). *Baseline survey of women street vendors in Kenya*. MATRIX Development Consultants and DFID, Nairobi.

Khayesi, M. (2001). Matatu workers in Nairobi, Thika and Ruiru towns in Kenya: Career Patterns and Conditions of Work. In P. Alila & P.O. Pedersen (Eds.), *Negotiating Social Space: East African Micro Enterprises*. Africa World Press, Inc., Trenton, Paper No. 528.

Kinyanjui, M. (2001). Employees in small enterprises in Nairobi: Job search and career patterns. In P. Alila & P.O. Pedersen (Eds.), *Negotiating Social Space: East African Micro Enterprises*. Africa World Press, Inc., Trenton, Paper No. 528.

McCormick, D. & Mitullah, W.V. (1995). Policy experiences of women in small scale enterprises. Paper presented at *UNESCO Meeting*, Nairobi.

McCormick, D., Mitullah, W.V. & Kinyanjui, M. (2001). *Enhancing Institutional Capacity for Policy Development, Dialogue and Advocacy: Role of Associations and Other Community Based Organisations*. University of Nairobi, Institute for Development Studies, Nairobi.

Njeru, H.N. & Njoka, J.M. (2001). Women entrepreneurs in Nairobi: The socio-cultural factors influencing their investment patterns. In P. Alila & P.O. Pedersen (Eds.), *Negotiating Social Space: East African Micro Enterprises*. Africa World Press, Inc., Trenton, Paper No. 528.

Ogola, J.S., Mitullah, W.V. & Omulo, M.A. (1999). *An integrated artisan gold mining management in Migori District, Kenya*. A Research Project Report submitted to the Royal Netherlands Embassy and Kenya National Academy of Sciences, Nairobi, p. 8.

Ogola, J.S., Mitullah, W.V. & Omulo, M.A. (2002). Impact of gold mining on the environment and human health: A case study in the Migori Gold Belt, Kenya. *Environmental Geochemistry and Health*, 24, 141–158.

Parker, J.C. & Torres, T.R. (1994). *Micro and Small Enterprises in Kenya: Results of the 1993 National Baseline*. Gemini Project, Final Report.

Wegulo, F.N & Obulinji, H.W. (2001). The interface between farm and nonfarm activities among the Mumias sugar cane growers. In P. Alila & P.O. Pedersen (Eds.), *Negotiating Social Space: East African Micro Enterprises*. Africa World Press, Inc., Trenton, Paper No. 528.

18

A Socio-Economic Study of Small-Scale Mining in Tanzania

CRISPIN KINABO

Tanzania's economic policy is based on three main strategies: (1) inflation/ poverty reduction; (2) achieving consistent GDP growth; and (3) improving the country's balance of payments (see Appendix A for a breakdown of the country's chief economic indicators). Since the implementation of its 1992 legislation, inflation has reduced substantively (Government of Tanzania, 1997a; Government of Tanzania, 1997b; Government of Tanzania, 1999). There has also been an increased participation of private companies (large and small) in the mineral exploration, mining, mineral processing and marketing sectors (Hestor, 1998; Tesha, 2000).

Since the enactment of the *Tanzania Investment Act* in 1997 and the modified *Mining Act* in 1998, the country has been active in promoting resident mineral exploration and investment activity. It also had a "gold rush" in the greenstone belts at the southern end of Lake Victoria. Today, exploration companies from Australia, Canada, South Africa, Sweden, and the UK are very active in the country, in many cases, working together with local small-scale miners. The future of Tanzanian mining appears promising but currently, a lack of adequate support continues to pose major problems (Financial Times, 1999). However, recently, with the support of international development funds, some effort has been made to improve local infrastructure.

Tanzania is cradled on an Archaean craton. Younger crystalline rocks rim this granitic nucleus with sediments and volcanic material originating from rifted grabens, the coastal plain and inland basins. Rocks belonging to the Archaean, Proterozoic, Palaeozoic, Mesozoic and Cenozoic are well represented and mineralized, and contain notable quantities of economic minerals (See Fig. 18.1). These deposits are being developed and mined by artisanal and small-scale miners, and large mining groups.

In Tanzania, the significance of artisanal and small-scale mining activities, along with the number of people involved in this sub-sector of industry, is frequently underestimated. To provide an overview of global employment estimates, according to the May 1999 issue of *Earthworks Magazine*, the UN calculated that, in 1996, six million people were involved in small-scale mining activities world wide. A more recent report by the International Labour Organization (ILO, 1999) suggests that the artisanal and small-scale mining workforce had increased to 13 million people by 1999. The report on a baseline study conducted by Tan Discovery (Tan Discovery, 1996) estimated that in Tanzania, the sub-sector employs 550,000 people, of which between 30 and 50 percent are women (ILO, 1999). These data indicate that Tanzania employs a substantial percentage of the global artisanal and small-scale mine workforce. The average Tanzanian mine worker has a family dependence of eight to nine (*i.e.* a father/ mother, four or five children, and two relatives) people, which translates into between 4.4 million and five million country-wide. In effect, artisanal and small-scale mining supports about 12.6 to 14 percent of the population of the country (Kinabo, 2000a).

The objective of this chapter is to provide an overview of the country's small-scale mining sector, the legislative framework in place for operations, and a breakdown of the benefits and problems associated with the activity.

Artisanal and small-scale mining activities in Tanzania are labour intensive, and feature poor processing skills and make use of rudimentary tools. Most operations perform poorly environmentally, characterized by:

- A poor handling of toxic chemicals;
- A wastage of mineral resources as a result of the deployment of inefficient recovery techniques;
- A wide scattering of shallow and deep pits; and
- Substantial mine waste and tailings.

However, there are some organized small-scale gold mines in the country, such as those operated by the Blue Reef Mining Company in the Geita District of the Mwanza Region. The company has managed to mechanize its mining and processing plant, and adheres to a Code of Practice supplied by the Ministry of Energy and Minerals. The company also maintains production and sales records, which are furnished to the relevant authorities, and uses appropriate technologies such as retort systems, amalgamation tanks, and special amalgamation ponds in gold mines, all of which meet Government regulations.

AN OVERVIEW OF SMALL-SCALE MINE PRODUCTION AND MINING TECHNIQUES IN TANZANIA

Basic statistics on artisanal and small-scale mining

In Tanzania, artisanal and small-scale mine operators are mainly engaged in the extraction of gold and gemstones, and, to a lesser degree, salts, industrial minerals, gypsum, dimension stones, coal, limestone, diamonds, sand and aggregates (See Fig. 18.1 for an overview of the country's geology). Between 1987 and 1996, the number of artisanal and small-scale miners increased from 150,000 to 550,000 (Tan Discovery, 1996). Table 18.1 provides a breakdown of the number of individuals involved in each segment of the industry.

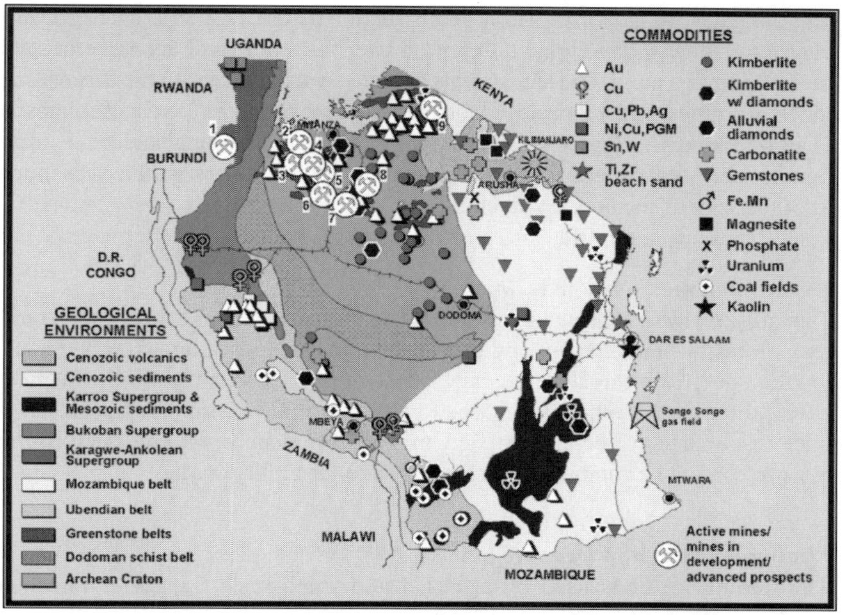

Figure 18.1 An overview of Tanzania's geological environments.
Source: Tesha, 2000.

Table 18.1 Employment totals in selected segments of the Tanzanian artisanal and small-scale mining sector.

Mineral commodity	Region	Men involved	Women involved	Total
Coloured gemstones	Arusha, Tanga, Rukwa Morogoro, Ruvuma Mtwara	226,430	72,842	299,272
Diamonds	Shinyanga Kwanza	1480	1028	2508
Gold	Shinyanga, Musoma, Rukwa, Singida, Mwanza, Mbeya, Ruvuma	171,113	51,553	222,666
Sand	Main towns and cities	872	136	1008
Salt, lime, & aggregates	Dar es Salaam, Tanga, Mtwara	8499	17,448	26,056
Gypsum	Kilimanjaro	590	0	590
Dimension stones	Kilimanjaro	225	45	270
Total	Tanzania	409,209	143,161	552,370

Mineral production

The production of mineral commodities in Tanzania during the period 1994–1998 is presented in Table 18.2 (Hestor, 1998); the current contribution of the mining sector to national GDP is 2.3 percent. However, many of the details regarding the production of various construction materials such as sand, gravels, clays, and stones, have gone unreported over the years, and are therefore not included in this table. These materials are chiefly used for building purposes in towns and cities. Diamond excavation in the country is carried out by Williamson Diamond Mines – a large-scale mining company – and artisanal and small-scale miners. The artisanal and small-scale segment of the industry produces as much as 50 percent of the country's gemstones.

Exploration methods and technology

Typically, resident artisanal and small-scale miners determine the mineral content of ores by panning soils and taking sediment samples along river beds, following specific mineral veins exposed at the surface of the Earth. In most cases, there is no transition phase from exploration to mining. Specifically, once resources have been identified, mining commences without establishing reserve content or conducting the appropriate feasibility studies.

Mining methods and technologies

The mining methods used by artisanal and small-scale miners in Tanzania can be grouped into one of two categories. The first covers open cast mining methods, which are used during shallow pit excavation. They are principally used to mine alluvial gold, rubies, rhodolites, alexandrites, garnet, sapphire, chrysoberly, gypsum, sands and gravel. The second category of methods is those which are used exclusively to mine underground. Such techniques are commonly used to mine gemstones (tanzanite and rhodolite), coal, and to extract gold from reefs.

In both cases, little attention is paid to the disposal of waste products. In fact, the industry, overall, is responsible for substantial land degradation, pitting and air pollution. During rainy seasons, waterbodies are polluted from excessive silt and surface runoff from sites.

Table 18.2 Mineral production from 1994–1998.

Commodity	1994	1995	1996	1997	1998
Salt (t)	84298	105000	86700	119710	97000
Phosphate (t)					
Apatite/P_2O_5	N/A	6700/2080	28020/8686	3000/930	1935/600
Gypsum (t)	7536	1052	8765	8800	8800
Silica sand (t)	4200	4200	4200	4200	4200
Kaolin (t)	541	596	1332	1300	1300
Limestone (t)	648474	1062081	120000	120000	120000
Coloured gemstones	48509	111403	142160	142000	48518
Soda ash	300	300	300	300	300
Gold (kg)	2861	320	318	8082	6870

Beneficiation methods
Artisanal and small-scale miners depend on low-level technologies for processing their ores. Certain minerals, such as sand, clay, gypsum, and oryx, do not have to undergo processing, and are therefore shipped to market once extracted. However, a second group of minerals, which includes gold, diamond, gemstones, dimension stones, kaolin, lime, and construction aggregates, must be processed and refined before being marketed.

GOLD MINING

Artisanal and small-scale gold mining involves the extensive excavation of pits using simple tools such as hand hammers, picks, hoes, and shovels (Figs. 18.2 and 18.3). After prospective ore has been extracted, gold is recovered through repeated crushing, grinding, gravity concentration, and amalgamation. The mercury- gold amalgam is then fired in the open air or in a distillation system (*i.e.* a retort) to produce mercury vapour and elemental gold. For centuries, such procedures and technologies have been used to produce gold in Tanzania (University of Dar-es-Salaam, 1996).

During the amalgamation process, mercury is introduced into the environment through the disposal of process water and tailings, while open air firing

Figure 18.2 A gold panner in Tanzania.

Figure 18.3 Small-scale gold miners in the Lake Victoria area.

of gold-mercury amalgam releases mercury vapour into the atmosphere. Mercury emissions from informal gold mining operations pose a serious environmental concern in Tanzania (University of Dar-es-Salaam, 1994).

PROCESSING OF GEMSTONES

There is a wide variety of gemstones mined and processed in Tanzania, including rubies, sapphires, tanzanite, green grossular garnet (tsavorite), tourmaline, emeralds, aquamarine, alexandrite, amethyst, scapolite, iolite, spinel, apatite, chrysoprase, garnets (rose, rhodolite, almandite), zircon, and malaya garnet (spessartine).

Due to the unique geological nature of resident gemstone deposits – more specifically, the propensity for mineralization to occur in small pockets – the processing of mined gemstone differs from one gem type to another. Generally, underground gemstone mining and processing involves rock breaking using low-energy explosives, followed by material cobbing and hand sorting.

The consequences of using low-level technologies to process ores include:

- wastage of minerals due to poor ore and mineral recovery during mining and mineral processing, respectively;
- direct discharge of process water and tailings into rivers;
- spillage of tailing dams during heavy rains;
- erosion and desertification;
- damages to river banks in alluvial mining regions;
- a scattering of mine waste products; and
- mercury pollution.

However, certain mines have attained high levels of mineral recoveries due to repetitive scavenging of processed tailings, a task normally performed by women and children.

LEGISLATIVE FRAMEWORK

In 1990, the Tanzanian Government introduced a series of policies aimed at liberalizing the economy. As a result, the mining industry experienced a 51 percent growth in 1991, and an additional 24 percent growth in 1992. The Government also introduced measures aimed at controlling illegal gold mining activities which, at the time, were common within small-scale mining regions. The Bank of Tanzania benefited enormously from the legislation, reportedly purchasing a total of 13.84 tons of gold between April 1990 and April 1994. The artisanal and small-scale gold mining sub-sector was the main contributor to the country's mineral export earnings during this period. In 1992 alone, 76 percent of the country's mineral export earnings resulted from gold mined by artisanal miners.

Mineral policy of 1997

The Government of Tanzania expects the resident mining sector to contribute at least 10 percent of the country's GDP over the next three decades. To help achieve this goal, it issued the *Mineral Policy of 1997*, the aim of which is to promote foreign investment and persuade the private sector into assuming a leadership role in local mineral development and marketing. With regard to mineral development, the role of the Government is well defined within the Policy, described as:

- Regulating, promoting and facilitating private investment.
- Acting as a service provider in the case of extension services for artisanal and small-scale miners.

The Government is particularly keen on stimulating a transformation from artisanal to small-scale mining; integrating mining into the national economy; and addressing social and environmental problems related to mining (Government of Tanzania, 1997b).

The objectives of the *Mineral Policy of 1997* with regard to small-scale mining are:

- To promote small-scale mining and encourage investment;
- To assist in the identification of small-scale ore deposits and making them accessible;
- To provide education and training services in the area of marketing, and to finance skill development; and
- To further encourage and emphasize the transformation of artisanal miners to formal small-scale miners in ways that do not deprive (miners) of their livelihoods.

Strategies for achieving the policy objectives for small-scale mining

The Government has emphasized the need to both improve small-scale miners' access to credit, and mainstream small-scale mining loans administered through formal financial institutions. The strategies that the Government is now following include:

1) Supporting the formation of formal enterprise groups, such as miner associations;
2) Formalizing traditional funding systems by promoting, inter alia, hire-purchase systems, forward sales, and mutual group savings schemes;
3) Encouraging banks to develop and finance expertise, and to establish mobile banking systems as well as commercial banks in mining areas;
4) Encouraging financial institutions to support small-scale mining by formulating affordable credit schemes for the sub-sector and adjusting start-up capital requirements;
5) Promoting the use of third party guarantees to enable other institutions assist miners in obtaining loans;
6) Facilitating the creation of mineral property markets to enable prospectors to sell their properties to developers at competitive prices;
7) Working, in conjunction with NGOs, toward the establishment of mining co-operative banks, and informal financial arrangements such as rotating savings and credit associations;
8) Advocating efforts to increase miners' incomes through value-added activities;
9) Putting in place a mining trust fund to finance simple mining equipment and other inputs, and devising effective mechanisms for the replenishment of such a fund; and
10) Conducting awareness programmes to promote savings culture among small-scale miners.

Rationalizing artisanal and small-scale mining

The Government has recognised the positive economic contributions made by artisanal and small-scale mining (Government of Tanzania – The Mineral Policy, 1997; Ministry of Energy and Minerals, 1998; Nyelo, 2000). These include the discovery of mineral occurrences, mineral production, and creation of employment in the rural communities. The Government is committed to supporting the small-scale mining sub-sector. As already explained, it is working toward transforming artisanal mining activities into more organised and modernized small-scale mining units, and is also promoting modalities of mineral marketing, which, in turn, helps encourage transparent business transactions and discourage smuggling. It has identified, and is pursuing, the following steps to facilitate these changes:

1) Transforming and upgrading artisanal mining into organised and modernised mining;
2) Making available appropriate and affordable mining tools, equipment and consumables, and encouraging their local manufacture and supply;
3) Promoting partnerships between local small-scale miners and large-scale investors to facilitate technology transfer and optimise mineral resources exploitation;

4) Providing supportive extension services for mining, mineral processing and marketing;
5) Streamlining and simplifying the licensing of artisanal miners and mineral dealers;
6) Preparing, disseminating and enforcing a code of conduct in mining and mineral processing; and
7) Promoting marketing arrangements that are responsive to the requirements of the artisanal and small-scale mining sub-sector.

The Mining Act 1998

Specific provisions for support to small-scale mining were introduced in the *Mining Act 1998* under Division D (Government of Tanzania, 1999). These provisions aim at:

1) curbing illegal mining, or the illegal trading and smuggling of mineral products onto the black market;
2) addressing environment effects;
3) assisting small-scale miners to operate in a more organised manner;
4) providing technical support to small-scale miners;
5) creating employment opportunities; and
6) promoting viable small-scale mining activities.

The *Mining Act 1998* states that primary (small-scale) mining licenses can only be acquired by individual Tanzanians or by companies whose management is Tanzanian. However, it further states that a primary mining license can easily be converted into a mining license in the event of a small-scale miner wanting to involve foreign investors in his small-scale mining operation.

The Mining Act in relation to artisanal and small-scale mining

The country's new mining policy and supporting legislation recognizes the significance of the artisanal and small-scale mining sector. As noted earlier, this has been in response to the Government's recognition of the sector's contributions to employment and thus, its ability to alleviate poverty. To reiterate, it is estimated to date that more than 600,000 people directly and indirectly depend on this sector in Tanzania.

The small-scale mining sector continues to dominate mineral production, especially gold and gemstones, thus contributing significantly to national export earnings and foreign exchange. Other advantages include its ability to occur in remote areas and on marginal reserves that would otherwise be uneconomical, as well as aiding in the discovery of large ore bodies, as exemplified by the case of Kahama Mining Corporation's Bulyanhulu Mine, which is now under mine development.

The experience of Tanzania after the policy changes

The revised *Mining Act* actually became law on 1st July 1999, and the regulations came into effect on 2nd July 1999. From this point onward, special consideration has been given to small-scale mining. For example, the duration of primary mining licenses was increased from one to five years. In addition, the primary mining license was incorporated into national mineral rights, thus providing the holder with the same incentives (such as reductions on import duties, equipment and fuel) provided to the management of large/medium scale miners. The Act further empowers the Minister of Energy and Minerals to declare special areas for small-scale mining. Apart from encouraging artisanal and small-scale miners to acquire primary mining licenses to protect their workings, they are advised to join regional miner associations and other economic groups, such as credit societies, in their areas.

Institutional framework

Tanzanian Government institutions are responsible for establishing laws and regulations, and overseeing their enforcement. The institution in charge of the small-scale mining sub-sector is the Ministry of Energy and Minerals (MEM) through the Mineral Resources Department (MRD), which is responsible for administering the activities of mining on behalf of the Ministry of Energy and Minerals. MRD has the following four departments: (1) Geology; (2) Mines; (3) Laboratory Services; and (4) Mineral Trading. The Department of Mines is responsible for granting prospecting rights and mining claims, administering law and regulations, offering technical assistance, monitoring production and sales records, and safety. It is important to mention that, since trade liberalization began in Tanzania in 1987, the state has changed from being an implementing agency, to a promoter and facilitator; mining associations have assumed the implementation role. These include the following:

- REMA – Countrywide, there are more than 17 registered associations, its active and inactive members rising from 150,000 in 1987 to 550,000 in 1996. The main objectives of REMA are:
  - Uniting small-scale miners at the regional level;
  - Holding meetings where miners can express their opinions and problems;
  - Assuring stable mineral markets by working together through consultation with FEMATA (FEMATA represents REMA at the national level and its members are elected from REMA);
  - Bridging miners' interests with Government through dialogues; and
  - Arranging training programs for miners.
- TAMIDA – Started in early-1990, and had more than 64 members in 1996. Its main tasks are to bring together miners and brokers involved in gemstone trading activities. Specific responsibilities include:
  - Providing legal advice (to dealers) on mineral rights according to existing mining laws and regulations;
  - Working with FEMATA and REMA in order to create conducive mineral markets between foreign buyers and dealers; and
  - Arranging mineral auctions in collaboration with MEM.

- The Tanzanian Chamber of Mines – Formed in 1994. Its main purpose is to protect the interests of both small and large-scale miners. Specifically, the Chamber is responsible for:
  - Promoting safety in mining areas;
  - Protecting the environment in areas that contain mines;
  - Reconciling mining disputes; and
  - Continual improvement in the mining industry through wages, labour, taxation, laws and legislation.

The major handicaps of these associations include a lack of operational funds, uncommitted leadership, weak/poor planning and management, poor technical facilities to reach their members, and a lack of alternative sources of funding to finance association activities.

Marketing

According to a recent report (Tan Discovery, 1996), local trading of mineral products mined by small-scale miners in Tanzania can be divided into the following three categories:

- Small brokers: operate by linking market services, either from a miner to a mineral broker, or a dealer. The number of such brokers varies from 1000 to 5000, depending on the type and volume of the mineral commodity being produced, the season of production, and quality of the stones extracted.
- Mineral brokers that are traders or miners: buy mineral commodities at mine sites, and then sell them to official dealers or mineral smugglers. In March 1996, the Government legalized the activities of mineral brokers; most claimholders in gemstone mines acquire brokers' licenses.
- Mineral dealers: include licensed and unlicensed operators dealing with gold and gemstones in mine centres. This third group of mineral trading contributes the most to the Government's revenues.

A World Bank-funded study conducted in Tanzania in 1995/1996 found that 60 percent of gemstones, and between 70 and 85 percent of gold produced by artisanal and small-scale miners, were being smuggled out of the country (Tan Discovery, 1997). Generally, gemstone buyers use different approaches to purchase a commodity as inexpensively as possible. Among the common strategies include:

- Booking production by providing credit finance to miners; and
- Providing incentives to miners that enhance business relationships.

In spite of the economic changes made during 1987, mineral smuggling has continued in the country because of the following reasons:

- The finance of better trade in foreign countries (due to fluctuations in the national currency);
- It is a means of avoiding paying taxes and Government revenues;

– The long bureaucracy associated with acquiring export licenses for mineral commodities;
– Better prices in foreign market channels.

However, the changes that have been made by the Government since implementation of the new *Mining Act 1998* have created an environment more attractive to domestic mineral marketers.

ENVIRONMENTAL FACTORS

While protecting the environment through environmental impact assessment (EIA) continues to be the national policy, the majority of artisanal and small-scale miners are not aware of the necessary procedures. In fact, only a small group of Tanzanian gold miners appear aware of the environmental consequences of activities, and have adopted environmental sound technologies to address pressing problems (Kinabo, 2000). Specifically, notable reductions in mercury pollution have been achieved as a result of using environmental technology assessment (EnTA) applications. Developed by the United Nations Environmental Programme (UNEP) to assist with EIA and in the evaluation of alternative technologies, EnTA techniques have enabled miners to considerably reduce the amount of mercury spillage and seepage into the environment, especially in the Lake Victoria Goldfields region, where the processing of gold is performed alongside river beds draining into Lake Victoria (Fig. 18.4).

To date, both artisanal and small-scale gold miners in Tanzania have used a variety of amalgamation methods. However, in the process, significant quantities of mercury have been dispensed into the air, water, and soils, and are now

Figure 18.4. Environmental damages in Tanzanian small-scale gold mining regions.

found in trace quantities in local foods. Field data indicate that, miners typically use one (non-recoverable) gram of mercury to recover approximately one gram of gold (UNIDO, 1999; SANTREN, 2000). In Tanzania, gold burners, women and children are the main victims of poisoning because most lack information about its impact on human health.

In an attempt to address this problem, the Small-Scale Mining Team of Southern African Network for Training and Research on the Environment (SANTREN, 2000) began running a series of educational training programs, and introduced alternative technologies it believed to be more environmentally sound. The Team also emphasizes the use of collective processing centres for amalgamation, retort systems, and the application of gravimetric methods. Since many miners have agreed to use gravity concentration equipment like sluice boxes and shaking tables (as observed in Geita and Matinje mining sites south of the Lake Victoria Gold Fields in Tanzania), the Network is convinced that gold mining can be conducted safely with zero mercury emissions to the environment.

SANTREN has also been active in demonstrating simple technologies that are frequently ignored by miners, or which have yet to be introduced to them. Examples include the use of amalgamation drums or pans instead of using bare hands for mixing mercury with gold bearing slurry; the application of home made retorts (built with standard plumbing water pipes) instead of open air burning (of gold mercury amalgam); and the use of protective gears while amalgamating and distilling gold from amalgam.

In its attempts to introduce alternative technologies to mine centres, SANTREN has used EnTA applications, which have enabled it to critically examine the wide range of mercury technology available. The application of EnTA to the various amalgamation technologies used by artisanal and small-scale miners in the Lake Victoria Region in Tanzania has proven to be very successful. A detailed EnTA technical manual for the abatement of mercury at gold mines can also be obtained from the Co-ordinator of the Small-Scale Mining Project of SANTREN, P.O. Box 35052 Dar-es-Salaam, Tanzania (kinabo@hotmail.com), or from SANTREN Head Office, Mount Pleasant, Box 167, Harare Zimbabwe.

SOCIO-ECONOMIC IMPACTS

Occupational health and safety

The mining activities conducted in Tanzania pose several occupational health and safety hazards to miners (Tan Discovery, 1996). Gold miners frequently inhale silica dust during milling operations, while individuals extracting coloured gemstones commonly inhale graphite dusts or asbestos dust during the excavation, transportation and processing of ore materials. Another problem in underground operations is a lack of ventilation, widespread unsanitary air and excessive heat. Poor ventilation was a major cause of the Merelani Tanzanite Mines disaster, in the northern town of Arusha, which claimed more than 39 lives in 2002 (see Table 18.3). As miners rarely use protective gears, most are affected, over the long term, by the high noise and vibration at the workplace. Obsolete and poorly maintained equipment such as ball mills and pumps, pose an additional safety risk to miners, and poor sanitation and working conditions in underground mines results in waterborne diseases such as diarrhoea and dysentery. Although the Ministry of Energy and Minerals has issued a code of practise for small-scale miners, most continue to work in unsafe conditions. Drugs and alcohol are also widespread throughout region.

Another potential health problem relates to the excessive use of mercury. Recent analysis shows that food grown near small-scale gold mining regions has been contaminated by the mercury used at the mines. Table 18.4 presents results of analysis of food stuffs grown in one mining villages in Tanzania.

Child labour

The Tanzanian Government prohibits children under the age of 14 from working in the formal industry sector in both urban and rural areas; however, youth employment continues to be widespread. The ILO estimates that 3.4 million out of 12.1 million children in the country (under the age of 18) work on a regular basis, and that one out of every three children in rural areas is economically active as compared to one of 10 in urban (ILO, 2003). Children employed

Table 18.3 Fatal accidents at the Mererani Mine, Tanzania (1996–2002).

Causes of accident	No. of deaths	Percent of deaths
Suffocation	86	40
Explosion	13	6
Collapse of hanging wall	17	8
Rock fragments flying from blasting	10	5
Falling into shaft	7	3
Break of a hoisting cable	6	3
Flooding (El Nino floods)	71	33
Mine waste heap failure or slide	3	1
Falling objects	1	0.4
Total	214	99.4

Table 18.4 A comparison of mercury contents in food stuffs grown in a mining and non-mining area.

Plant sample collected	Artisanal farming village, max value in ppb	Artisanal gold mining village Maximum value (Hg, ppb)
Sweet potatoes *(Iponea batata);*	20	462
Yams *(Calocasia esculcuta).*	2	105
Beans *(Phaseolus fulgaris);*	Nil	412
Groundnuts *(Arachis hypogaea);*	Nil	0.01

as miners generally receive lower wages than their adult counterparts, despite working comparable jobs. Between 1,500 and 3,000 children work at unregulated tanzanite mines in Arusha alone.

The Government has worked with the ILO's *International Program on the Elimination of Child Labour* to develop a national plan of action to address the issue, and, in 2000, implemented a program for the elimination of child labour. The Government ratified ILO Convention 182 during the same year.

Although child labour at mines is comparatively small to other sectors of industry, the frequency of injuries and work-related illnesses among youths is high. ILO surveys show that mining and quarrying is one of the most dangerous sectors for working children. In the small-scale mining industry, there are rarely limits on the length of the working day. Occupational safety and health regulations, when they exist, are rarely enforced. Children often work without adequate protective equipment and clothing for up to 12 hours a day, with only one 30–60 minute break. The health and development of child miners is further threatened by deep and poorly reinforced pits, poor ventilation, excessive noise, intense vibrations from machines, excessive heat or cold, high humidity levels, awkward working positions, and extremely arduous work. Deaths from explosions or cave-ins are common among children, and there is a constant threat of respiratory illnesses, induced by dust and gas poisoning. First-aid and medical facilities are rarely available at the workplace. Workers typically have no access to rehabilitation or social security schemes, schooling, and vocational training.

Children in the Tanzanian artisanal and small-scale mining industry are involved in the following areas of operation:

- Up to 6,000 children toil in the country's quarries, either alone or with their mothers, to produce heaps of gravel that are later sold to building contractors. Sand quarrying and small-scale aggregate and chip making, which also attracts children as young as 4–5 years, is a common practice in most cities and towns in Tanzania. They are also engaged in gem sorting and grading, as observed in some gemstone mining areas of Kalalani, Mwarazi, Merelani, Tunduru and Luangwa. In Uchira Village, Moshi, children are hired to carry volcanic blocks from the mine to the sizing site, and to load rubble into transport trucks (Tan Discovery, 1996).
- In the processing of ore tailings and mercury amalgamation in the gold processing centres of Lake Victoria Goldfields; the nature of the work poses serious risks to children's health, safety and welfare. Naturally, those exposed to mercury are susceptible to acute mercury poisoning (UNIDO, 1999).
- The underground tanzanite mines in Arusha, NE Tanzania, have more than 3000 children called "Snake Boys" who work in deep narrow mines. These operations, in most cases, are more than 100 m deep, with poor ventilation systems, lighting and safety.
- Service providers (*e.g.* security surveillance, etc.) and grocery suppliers (food, drink, water, clothes, etc.).
- Mining gold from narrow workings.

As a result, many children are prematurely exposed to mine-related diseases, sexual abuse, drugs, alcohol, and crime. They are also generally deprived of social welfare (*e.g.* health, education, etc.).

Women and mining

In Tanzania, the total number of women engaged both directly in artisanal and small-scale mining activities, and indirectly as service providers, ranges between 140,000 and 300,000 (Tan Discovery, 1996). Although women constitute between 25 and 50 percent of the country's artisanal and small-scale mine labour force, their remuneration is far less than their male counterparts (ILO, 1999). Their participation is discriminated by legal, social, economic and environmental factors. Clearly, the provision of schooling to children, and education for miners is the foremost step towards liberating women and eradicating child labour in the sector.

In 1999, the ILO estimated that there were between five and six million female miners in developing countries. After Zimbabwe, Tanzania has the world's largest female mining contingent. Field census statistics conducted in 1996 indicate that there were some 552,000 miners operating in Tanzania, of which at least 143,000 were women (Tan Discovery, 1996). Women are engaged in the mining and processing of gold, gemstones, industrial minerals, salt, building and construction materials. Most, however, are involved in the extraction of gemstones and gold.

As in many places in the developing world, the driving force behind female participation in the Tanzanian mining industry is poverty. The main causes of poverty in Tanzania, particularly in its rural areas, include deteriorating world market prices of crop products (*e.g.* cotton, coffee and sisal); successive long term droughts in sub-Saharan Africa, where people depend mainly on livestock grazing and subsistence farming; economic adjustment and restructuring programs; and high inflation rates. As a result, women are forced to pursue additional income generating activities.

In Tanzania, as in most Africa countries, women play a key role in the development of the family, particularly in rural areas, where artisanal and small-scale mining is widespread. They are expected to secure food for their families, and provide water and energy sources; women also engage in informal activities such as gem and stone quarrying.

The main hardships faced by Tanzanian women are as follows:

- Low wages and a shortage of credit – apart from being underpaid, Tanzanian women generally have low access to credit, and are therefore trapped at a subsistence level of mining, extracting what is easiest, and confining themselves to a precarious existence.
- Poor education and mining skills – provision of education and field skills to female miners, and schooling to children, are two major steps toward liberating women and eradicating child labour in the sector.

These cultural barriers impose a heavy family burden that limit their independence and mobility; women are not considered to be the head or joint head of a family in Tanzania, and, therefore, do not have property or incomes in their name. Moreover, women keen on managing a mining business often face resistance from their husbands and families.

OVERCOMING CONSTRAINTS

These gender-specific obstacles can be overcome by integrating several social and economic factors. Sample strategies include the following:

- Capacity building: There is a high level of illiteracy among the female contingent of the Tanzanian population. Thus, there is an urgent need to increase school enrolment among young girls. Like every citizen, a miner has the need for, and right to, information, training, and consultation on social, economic and environmental issues; there is a need to deliver these services on-site.
- Entrepreneurial and micro-finance schemes: In Tanzania, women at mines are engaged in a variety of activities. Promotion of small-scale industries downstream would therefore reduce pressure and help to supplement artisanal mining. Provision of easier access to finances – through micro-finance programs – would help to facilitate increased female participation in the mining industry. All miners should be encouraged to acquire mining licenses, a procedure that should be grounds for being eligible for such credits; miners with legal documents could also be granted the flexibility of being able to enter joint ventures with other individuals, groups and large companies, and the permission to sell their mineral rights outright.
- Alleviating poverty in rural areas: Surveys have shown that 50 percent of all Tanzanians are poor, while 36 percent of the poor live in "very poor" conditions. Almost 60 percent of the rural population is poor compared to 39 percent of the urban population. Priority, therefore, should be given to improving rural artisanal and small-scale mining regions.
- Formation of women associations: The existence of women miners associations and financial clubs can be a powerful tool in uniting female miners. One notable example of an existing association is the Tanzania Women Miners Association (TAWOMA), which aims at improving women's participation in mining through awareness; lobbies for support and the advancement of women miners though identification of training and technical needs; conducts training; develops database collateral to guarantee funds; and facilitates the marketing and establishment of mining databases and libraries. Such associations help to improve access to loans by forging partnerships with existing large-scale mining companies. The associations also help to facilitate the rapid development of informal financial institutions, such as rotating savings and credit associations.

DISCUSSIONS AND CONCLUSIONS

To improve operating conditions in the Tanzanian artisanal and small-scale mining sector, changes must be made on numerous fronts. First, and foremost, partnerships must be forged between artisanal and small-scale miners, and large-scale operators. Secondly, the Government must begin addressing a number of aspects in national policies. Finally, some responsibility must be assumed by NGOs and the miners themselves.

Forging partnerships between artisanal/small-scale mining parties and large-scale miners

In Tanzania, many foreign companies have entered into agreement with small-scale miners, after converting their primary mining licenses into mining licenses. One notable example is at the Tembo Mine, located in Geita District. Here, a foreign company has entered into an agreement with a small-scale miner who has been granted permission to participate fully in mining and mineral processing on site, and in the marketing of the final products. Other companies provide equipment and working capital, as in the case of Meremeta, where, in exchange for equipment and support, small-scale miners are required to sell their gold to the (local) Meremeta dealer.

While the working partnerships between artisanal and small-scale mining parties, and large-scale miners has improved in Tanzania's gold mining areas overall, there have been clashes reported in gemstone areas. Such has been the case at Merelani Tanzanite Mines, where there is a land/claim dispute precipitated largely by local political intervention.

Overall, however, the relationship between the parties remains quite strong. A number of factors have made such cohabitation possible, including:

- The fact that the country is relatively stable politically in comparison with neighbouring countries.
- The existence of a new legal and regulatory framework that emphasizes the streamlining of licenses with guaranteed security of tenure, stability of fiscal regime, freedom of commercial operation, the right to trade mineral rights, and access to foreign exchange at market rates.
- A favourable geology amenable to both small and large-scale mining.
- The policies of various large-scale mining projects, including the Ashanti Goldfields Company, Golden Prode Mine, and Kahama Mining Company. These companies have created direct and indirect employment for people.
- The implementation of community development programs (by various large-scale miners), which emphasize improvements to rural infrastructure. Major projects (*e.g.* roads, water supply, schools and health centres) can now be observed in gold mining areas south of Lake Victoria.
- The willingness of large-scale mining companies to both provide technical assistance to small-scale miners, and improve the infrastructure of rural communities.

The role of the Government

The Tanzanian Government, as a policy making body, is aware of the great potential of mining to reduce rural poverty. Although the country's mining policies and frameworks are well developed, policy makers have paid little attention to devising regulations and laws for liberating women and children in the sector. Additional priorities from a legislative and policy-making standpoint include the following:

- Improving the organization of artisanal mines: The State must simplify licensing procedures so that miners are able to readily acquire mining licenses. Moreover, the Ministry of Energy and Minerals must ensure that artisanal miners acquire and use explosives properly, which will help to reduce the occurrence of accidents at local gemstone mines.
- Facilitate and implement poverty alleviation programs at mines: Generate alternative economic activities, and provide education to artisanal and small-scale miners on key social and labour issues such as health, education and the environment.
- Regulating mining and downstream economic activities: The Government must provide supportive extension services to ensure that women can participate fully in all levels of the industry.
- Using the available media and public structures to promote mine safety: This includes prohibiting all forms of work (to women and children) that are physically and/or morally harmful, and developing lobbying skills so that women can petition for laws to be changed if they are unsuitable for their needs.
- Develop negotiating skills for women so that they can liaise directly with city council officers, police, small contractors and middle-men, and representatives from civic and political organizations, through their own representatives or mining associations. Moreover, developing leadership skills among women who work outside of the formal sector, and promoting partnerships between artisanal and small-scale miners, and large-scale operators.
- Taking appropriate measures to promote increased literacy and education – particularly among girls. The establishment of vocational training programmes for the unemployed at mine centres could help to build capacity among miners, and increase the employment rate at mines.
- Facilitate and promote the eradication of child labour and STDs at mines through reproductive health education, child care/nursery centres, and schooling facilities. The Government must enforce a minimum age of work at mines, such that children – particularly school children – are prohibited from working at mine camps.
- Ensure that both men and women working under the same conditions are remunerated equally.
- Take appropriate measures to ensure that women are working in a healthy environment with proper management plans for the acquisition, storage, transport, use and disposal of toxic wastes such as mercury, and mercury-contaminated water and tailings. The Government must enforce existing laws and regulations, and establish a regime for effective reporting on safety and health, so that it can invest more practically in the prevention of accidents and diseases.
- Ensure that miners abide by the *Code of Practice for Small Scale Miners* prepared by the Ministry of Energy and Minerals.

Role of local mining communities

Local Government authorities, small-scale miners and farmers must work together and implement collaborative programs that:

- Ensure that food supplies are cultivated in approved contamination-free areas, free from heavy metals and cyanide complexes;
- Emphasize the conducting of field inspections during planting seasons, to better ensure that crops are grown in safer places;
- Ensure that contaminated sites are properly cordoned off and cannot be accessed by domestic animals;
- Assess water resources available and their potential to become contaminated by mining activities;
- Ensure that drinking water wells are periodically inspected to determine whether or not they are located sufficient distances from sources of potential contamination;
- In consultation with the zonal mine offices, to ensure that potential prospecting and mining claims are issued in locations sufficient distances away from water runways;
- Evaluate and approve sources of drinking water separated from process water before issuing building site approvals; and
- Ensure that small-scale mining is conducted safely and efficiently.

Additional functions for large mining companies

Large-scale mining houses have to draw programs that emphasize:

- The protection of groundwater through the monitoring of bore holes, continuous data collection, interpretation, education, and provision of technical support to local jurisdictions and the Water Departments;
- Preventing waste water from contaminating groundwater and surface runoff that drains into Lake Victoria;
- The control of dust emissions from operations; and
- The prevention of acid mine drainage (AMD) formation.

This chapter has provided a general overview of the Tanzanian small-scale mining industry. The chapter that follows (Chapter 19) examines more in detail the role of women in the (Tanzanian small-scale mining) industry.

REFERENCES

Financial Times (1999). Country survey: Tanzania key sectors – mining: another time, another gold rush. *The Financial Times*, March 31, 1999.

Government of Tanzania (1997a). *The Mineral Policy of Tanzania 1997*. Dar-es-Salaam, Tanzania: Government Press.

Government of Tanzania (1997b). *The Tanzania Investment Act 1997*. Dar-es-Salaam, Tanzania: Government Press.

Government of Tanzania (1999). *The Mining Act 1998*. Dar-es-Salaam, Tanzania: Government Press.

Hestor, B.W. (1998). *Tanzania – Opportunities for mineral resources development*, 3rd Edition, Ministry of Energy and minerals, United Republic of Tanzania.

ILO (2003). *ILO Declaration on fundamental principles and rights at work.* 86th Session, Geneva, June 1998, http://www.ilo.org/public/english/standards/decl/declaration/text/

ILO (1999). *Social and labour issues in small-scale mines.* Geneva, 17–21 May 1999.

Kinabo, C. (2000a). A paper on *Community education for the small-scale gold miners on the health hazards of mercury in Lake Victoria Zone.* SANTREN Publications, pp. 1–6.

SADC Mining Sector Coordinating Unit (1997). Made in the SADC: SADC *Regional Mining Review.*

SANTREN (2000). *Workshop report, community education for artisanal and small-scale gold miners on the health hazards of mercury.* Southern African Network for Training and Research on the Environment (SANTREN) held in Kwanza, Tanzania 25–28th September 2000 p. 39. http://www.ilo.org/public/english/standards/ipec/simpoc/tanzania/ra/mining.pdf

Tan Discovery (1996). *Report on baseline survey and preparation of development strategy for small-scale and artisanal mining program in Tanzania.* Ministry of Energy and Minerals/Tan Discovery Consulting Company. World Bank Project. p. 135.

Tesha, A.L. (2000). *Growth and diversification in mineral economies.* Regional Workshop for Mineral Economies in Africa, Cape Town, South Africa, November, 15 pp.

UNIDO (1999). *Introducing new technologies for abating mercury pollution caused by informal gold mining operations in Tanzania.* UNIDO/Ministry of Energy and Minerals, United Republic of Tanzania. 1998–1999, p. 25.

University of Dar-es-Salaam (1994). *Report, effects of mercury pollution caused by small-scale mining activities in Lake Victoria Region.* Department of Geology, University of Dar-es-Salaam/National Environmental Management Council (NEMC) 1993/94 p. 20.

University of Dar-es-Salaam (1996). *Report on environmental aspects of mining and industrialization in Tanzania.* A SIDA/SAREC Project, Department of Geology, University of Dar-es-Salaam, 1994–1998, p. 95.

APPENDIX A: BASIC BACKGROUND COUNTRY STATISTICS (SADC, 1997)

Area:	945,090 km^2
Population:	31.271 m (July 1999 est.)
Population growth rate:	2.14% (1999 est.)
Population density:	33/km^2 (1999 est.)
Capital:	Dar-es-Salaam, some Government offices have been transferred to Dodoma which is planned as the new national capital
Currency:	Tanzanian shilling (TSh)
Exchange rate:	US$1 = 803TSh (Feb 2000)
Inflation:	13.5% (1998 est.)
Languages:	Kiswahili (official), Kiunguju (name for Swahili in Zanzibar), English (official), Arabic (widely spoken in Zanzibar), many local languages
Value of mineral production:	Diamonds (US$16.64 m), semi-precious gemstones (US$7.95 m), gold (US$2.0 m) (export figures for 1997)
Value of exports:	US$0.952 billion (f.o.b. 1998 est.)
Value of imports:	US$1.46 billion (f.o.b. 1998 est.)
Trade balance:	US$0.508 billion (1998 est.)
Main exports:	Coffee, manufactured goods, cotton, cashew nuts, minerals, tobacco, sisal
GDP:	US$22.1 billion (1998 est.)
GDP per capita:	US$730 (1998 est.)
Total labour force:	13.495 m (1995 est.)
Literacy rate:	67.8 percent (1995 est.)
Bordering countries:	Burundi, Kenya, Malawi, Mozambique, Rwanda, Uganda, Zambia (Map 1)

19

Women and Small-Scale Mining in Tanzania

CRISPIN KINABO

Women comprise approximately 51.6% of Tanzania's population, and constitute some 54% of the economically-active population in the country's rural areas (National Reports, 1995), where they engage in farming, livestock management and mining. Moreover, the majority of entrepreneurs involved in the country's informal industry sector are female, where they are notorious for performing laborious activity. Traditional cultural barriers are preventing many women from accessing and controlling the economy, largely because men continue to dominate household decision-making in Tanzanian society.

One Tanzanian industry in which women are playing an increasingly important role is artisanal and small-scale mining. The International Labour Organization (ILO) estimates that approximately 13 million people in the developing world are currently engaged in artisanal and small-scale mining, some 40–50% of whom are women (ILO News, 1999). However, additional female participation is impeded by a series of legal, social, economic and environmental factors (ILO Press Release, 1999).

Provision of schooling to children and education is the foremost step towards empowering female miners. With enhanced skilling through education, women are better able to alleviate poverty on their own. Moreover, over the long-term, improved education for women reduces child labour (as more educated women are able to earn lucrative wages and can thus send their children to school). The resulting enhanced literacy, in turn, puts women in a better position to: (1) participate in local micro-finance schemes and programs; (2) improve production practices through the use of appropriate technologies; (3) access markets; and (4) combat problems with environmental pollution.

DEFINITION OF ARTISANAL AND SMALL-SCALE MINING

The definition of "artisanal and small-scale mining" varies from country-to-country according to a wide range of criteria (MMSD Southern Africa

Draft Report, 2002), including:

- Social factors – *i.e.* level of inflation, poverty, number of people involved;
- Economic factors – *i.e.* level of technology, capital investment, equipment and tools, raw ore produced per mine; and
- Environmental factors – *i.e.* aerial coverage, level of emission of contamination and pollutants, volume of material excavated.

Whereas much of the developed world community views artisanal and small-scale mining as a dirty, dangerous, and disruptive engagement, most developing world governments see it as a means of survival in peri-urban and rural areas, and as an industry that is potentially profitable (ILO Sectoral Activities, 1999).

In Tanzania, small-scale mining is characterized by:

- Minimum investment, and an unlimited number of workers – *i.e.* the fact that any individual in the vicinity of a mining region can quite conceivably engage in operations with minimal capital (all that is needed is a few weeks of food stock, a pick and a shovel);
- Poor mechanization – *i.e.* activities featuring rudimentary tools such as a mine pick, and which exhibit poor ventilation and lighting; and
- Quick cash returns, and inefficient extraction and processing methods prone to causing environmental degradation.

Artisanal mining activities in Tanzania provide a variety of socio-economic benefits, not the least of which is employment to thousands of rural inhabitants. This chapter, however, exclusively examines the role of women in artisanal and small-scale mining activities in the country.

THE GLOBAL SITUATION REVISITED

In 1999, the ILO provided a regional breakdown of small-scale and artisanal mine employment worldwide. It was as follows: Asia, 7.2 million; Africa, 3.7 million; Latin America, 1.6 million; and the developed world, 0.7 million (ILO Sectoral Activities, 1999). As already noted, between 40 and 50% of the industry's workforce is female and statistics prove that the sub-sector provides employment – and therefore, a means of existence – to many rural populations (see Table 19.1). In terms of the number of people employed, Tanzania's mining workforce ranks second to only Zimbabwe's, and features the most women of any workforce in Africa.

Naturally, the number, and perhaps roles of women and the extent of their participation in artisanal and small-scale mining varies from country-to-country. However, what appears to be commonplace throughout Africa is the fact that female entrepreneurial artisanal miners face a number of traditional cultural barriers; in many African countries, only men are entitled to mining rights, or can be mine title owners. In Tanzania, few women in gold and gemstone mining centers are permit holders. More specifically, there are few existing systems under female supervision. Although female participation in the industry is well recognized at both the local and national levels, the majority is unskilled, and,

Table 19.1 Global participation of men and women in mining.

Country	Gender: % Female workforce	% Male workforce	Total people in mining
Asia excluding India & China	~10	~90	1,800,000
Latin America	20	80	1,600,000
Africa excluding Zimbabwe and Tanzania			2,550,000
India	30	70	1,100,000
Tanzania	40	60	600,000
Zimbabwe	50	50	350,000
China	No data	No data	4,300,000
Developed countries			700,000

Table 19.2 Artisanal and small-scale mining in Tanzania in 1996.

Mineral commodity	Men involved	Women involved	Total
Coloured gemstones	226,430	72,842	299,272
Diamonds	1,480	1,028	2,508
Gold	171,113	51,553	222,666
Sand	872	136	1,008
Salt, lime & aggregates	8,499	17,557	26,056
Gypsum	590	0	590
Dimension stones	225	45	270
Grand total	409,209	143,161	552,370

is therefore forced to engage in tedious and lower-paying work such as ore crushing and grinding. Many women also perform various gold panning and amalgamation activities (Kinabo et al., 2000).

WOMEN AND ARTISANAL MINING IN TANZANIA

Many Tanzanian women are involved in the artisanal mining and processing of gold, diamonds, various gemstones (*e.g.* tanzanite, ruby, sapphire, garnets, alexandrite, amethyst, etc.), industrial minerals, building and construction materials, and salt. The number of people employed in the county's artisanal mining industry fluctuates annually; the highest recorded census of miners occurred in 1996 (552,000 people), of which 25% were women (Table 19.2) (Tan Discovery, 1996).

Why have women turned to artisanal mining in Tanzania?

The increased involvement of women in the Tanzanian artisanal and small-scale mining sector has not come about by accident. There are several factors that have precipitated their expanded involvement (SANTREN, 2000). Some of these factors are highlighted below.

Dwindling cashcrop prices

The majority of Tanzania's rural population relies on subsistence farming for their existence. Despite regular fluctuations in market prices, the value of Tanzanian agricultural commodities has, in fact, decreased over the past three decades. The dwindling international selling prices of cash crops such as cotton, coffee, and sisal, prompted many female farmers to begin engaging in artisanal and small-scale mining. These women were unable to produce sufficient crops for survival, which forced them to search for a new means of survival at low cost. Artisanal and small-scale mining proved to be an adequate substitute.

Long-term drought

In recent decades, successive long-term droughts of varying severity and duration have affected many countries south of the Sahara, Tanzania being one of them (Wolgin, 2001). The country's existing agro-pastoral practices are predominantly reliant on natural rainfall – something which is clearly beyond the control of small-scale farmers. The variations in climate that have occurred in the country in recent years have seriously threatened rain-fed agriculture in certain areas. Frequent and extensive drought has led to acute food shortages and famine, prompting many more individuals based in the country's rainfall-dependent economy to switch to mining as a profession.

Changes in socio-economic structure

In the 1980s, many of Tanzania's former state-owned mines (*e.g.* Williamson Diamonds Ltd, Nyanza Salt Mines, Coastal Salt Works Co., Kiwira Coal Mines, Buck Reef Gold Mining Co., Minjinku Phosphate Co., Ltd. Pugu Kaolin Mines, and Lupa Gold Mining Company) underwent closure or restructuring. At the time, many experienced male and female mineworkers began undertaking small-scale mining activities. Many *other* women not involved in the industry at the time were also forced to join their family members already engaged in mining business ventures. Although rare in the SADC Region, greater economic independence is a main reason why many women migrate to artisanal and small-scale mining centres in the first place. Such has been the case in many of the country's gemstone mining regions (*e.g.* at Merelani's tanzanite sites, and in Arusha in northeastern Tanzania), and in various gold mining centres in the Lake Victoria Goldfield and Mpanda Mineral Field. Here, many women own mining claims, and employ both men and women for surface and underground works on a "cooperative" basis (Kinabo, 2000).

High inflation rates

Much of Tanzania's population is in absolute poverty; between 1980 and 1994, it was an estimated 50% of the population. Per capita food production growth between 1990 and 1997 was a mere 2.7% (Tanzania 1999). The country's high levels of inflation reflect a volatile economy in which money does not hold its value for very long. Workers require higher wages to cover rising costs, and are disinclined to save whatever they earn. Producers, in turn, are seen to raise their selling prices to cover these increases, scale back production to check their costs (resulting in lay-offs), or avoid investing in future production altogether. High inflation rates have forced many peri-urban women to work in sand and aggregate quarries in order to support and maintain their families. Primary examples of secondary mining activities taken up by women include aggregate crushing in major cities and towns, lime burning, and sand collection in many parts of the coastal zone of the Indian Ocean.

Poverty alleviation

Women migrate to artisanal and small-scale mining regions with hopes of improving their poverty-striken states, and although considerable achievements have been made, poverty continues to be a serious problem in the country's rural areas. Tanzania is the seventh poorest country in the world (Elections, 2000), and many of its inhabitants are demoralized by their low living standards. The average income for a Tanzanian is inadequate to meet basic needs. Malnutrition, particularly among children, along with poor education and disease, are serious problems affecting many Tanzanians. Moreover, there are insufficient numbers of extension workers in place at the grassroots level. More specifically, the existing party of agricultural workers, health workers, and primary school teachers is barely noticeable. With problems compounding daily, more women are forced to endure harsh conditions at mines.

PROBLEMS FACED BY FEMALE MINERS

There are a number of problems affecting both male and female artisanal and small-scale miners in Tanzania. First, and foremost, each lacks human and material resources, and the assistance needed to improve work and living standards. They use poor rudimentary tools, which results in a heavy dependence on manual labour; unplanned work that occurs in an inefficient environment with high risk of health hazards and low safety standards; and uncontrolled, illegal activities due to an ignorance of governing mining laws. In addition, African woman play a key role in the development of the family, especially in rural areas where artisanal and small-scale mining mainly takes place. Here, the woman typically undertakes several activities such as securing food for the family, the provision of domestic water and energy, and informal commercial tasks such as stone quarrying or gold mining. Recent interviews conducted with women engaged in the quarrying of limestone for construction purposes (Kinabo & Lema, 2000) indicate that each typically works more than 18 hours per day. Their daily activities can be broken down as follows:

- Eight hours for quarrying;
- Six hours searching for alternative family support;

- Four hours gathering domestic water and energy supplies, and for breast feeding; and
- Six hours for sleep.

These tasks and constraints, in turn, prevent many women from improving their participation in artisanal and small-scale mining. In most cases, the main reason for female participation in the industry altogether is the desirability of working in a low-cost profession, combined with the need for relief from high inflation rates. However, regardless of area, ore type, or mining technique, there are specific problems that *exclusively* beset women working in artisanal and small-scale mining. Some of these hardships are highlighted in the discussion that follows.

Economic hardships

Although women constitute up to 50% of the global artisanal and small-scale mining workforce, they do not receive anywhere close to 50% of the reaped rewards (Kinabo, 2001a). Unlike men, most women in Tanzania lack access to credit and finance, which prevents them from participating fully in artisanal and small-scale mining. More specifically, without adequate credit, women are trapped at a subsistence level of mining, extracting what is easiest, and thus confining themselves to a precarious existence.

Lack of technical know-how

Limited technical know-how in relation to their male counterparts, compounded by rampant illiteracy, continues to be a major disadvantage for prospective Tanzanian female miners. The anarchic nature of artisanal and small-scale mining, along with the poor working and living conditions at mine sites, are sometimes sufficient reasons in themselves to prevent women from getting involved in the first place.

Social commitments

Social productive issues such as a lack of representation and support, a lack of management and administrative skills, perceptions about a female's status in society, and other cultural barriers, impose a heavy family burden that limits female independence and mobility. In Tanzania, women are not considered to be the head or joint head of a family and therefore have no property or income in their names. Married women face opposition from their husbands and families to operate a mining business, which sometime takes them to remote areas for long periods of time, thus making it difficult to combine work with family and other household responsibilities.

OVERCOMING CONSTRAINTS

Clearly, female employment in artisanal and small-scale mining is constrained by a number of factors, particularly legal and financial obstacles, and socio-cultural taboos, which collectively undermine a woman's capacity in Tanzanian society. To overcome these constraints, a greater awareness of their potential contribution to mining is needed.

Capacity building

One major indicator of a country's poverty is the high level of illiteracy. In Tanzania, the literacy level has diminished. Since the 1990s, it has decreased from 90% to its current level of 68% (EFA, 2000). Moreover, the net enrollment rate for primary school pupils is decreasing, having gone from 90% in the early-1990s to 77.8% in the late-1990s; school attendance rate for basic education is also on the decline. The resulting high unemployment rates in villages has led to substantial poverty because the economy is unable to generate enough employment opportunities to meet the needs of its labour force. Rural areas have also been unable to create gainful employment opportunities for the youths that have managed to complete their primary school education.

It is imperative that female miners are provided key information, training and consultation regarding the social, economic and environmental activities at sites. Increased education and on-site training is a key to improving planning and management. The dissemination of appropriate technologies would lead to increased mineral production, which, in turn, would improve the well-being of miners. Provision of schooling for their children is a key to reducing child labour at mines. Mobilizing and sensitizing women will enhance economic performance, help to eradicate child labour, and reduce incidences of HIV-AIDS in mining regions (Kinabo, 2001b). Government institutions and the private sector should train extension service workers to provide hands-on assistance to female miners.

The involvement of women in local government programs, in conjunction with increased assistance, would inevitably enhance female participation in rural Tanzanian society. A variety of government programs must therefore be implemented, particularly those that sensitize the community on issues of hunger, malnutrition, and the improvement of reproductive health. Such programs would help to reduce maternal deaths and infant mortality. Family planning clinics must be improved and extended to many parts of the country, particularly mining regions. Reproductive health education must also be made available to all age categories at all educational institutions.

Entrepreneur and micro-finance schemes

Stakeholders must promote industries that supplement artisanal mining. This can be accomplished by improving productivity, profitability and working conditions.

Providing more readily accessible finance enhances and enables female participation in many "downstream" economic activities. Microfinance schemes should emphasize miners' access to simple and appropriate mining equipment, as well as facility upgrades. The financing institution should assume monitoring responsibilities to ensure that borrowers use credit funds for the purposes intended, have the capacity and discipline to repay borrowed money, and that funds are managed appropriately overall. Credit programs featuring effective, informal lending practices, such as small, short-term loans that increase gradually in size, should be made available to poor women; the mode of application for credit should be simple and quick. For the management of non-payment risks, women should be required to form assurance groups or a representative association. The traditional funding system of savings and credit should be improved to include high purchase systems, forward sales and mutual group savings (Linda, 1997). Female miners should be encouraged to acquire legal mining licenses, participate in joint ventures, and sell their mineral rights when they desire.

Alleviating poverty in rural areas

Poverty is not uniformly distributed geographically in Tanzania. Distinctions can be made in both rural and urban poverty situations, as well as within gender and agro-ecological zones. Surveys have shown that 50% of all Tanzanians are poor, and that 36% of the poor live in "very poor" conditions (National Reports, 2001). There is also evidence indicating that poverty is more widespread in the rural areas of the country. Almost 60% of the rural population is poor, compared to 39% of the urban population. The Government should therefore prioritize improving existing infrastructure and creating new infrastructure in rural regions.

Formation of women associations

The formation of female mining associations and financial clubs can be a powerful tool to unite resident women miners. A major example of an existing organization doing just that is the Tanzanian Women Miners Association (TAWOMA), which is currently aiming to improve female participation in mining through awareness; lobbying for support and the advancement of female miners though identification of training and technical needs; training; establishing database collateral to guarantee funds; and facilitating marketing and establishing mining databases and libraries. To further improve access to loans, the association should form enterprise groups with existing large mining companies such as Geita Gold Mine, and Kahama Gold Mines, the management from which have already expressed a willingness to provide support to artisanal miners. Associations should also promote the formation of informal financial institutions such as rotating savings and credit associations.

Improvements to infrastructure

In Tanzania, living standards are deteriorating – particularly in rural areas – because of basic infrastructure inadequacies and accompanying rapid population growth. Upgrading infrastructure is therefore an important task, not only because it leads to improved living standards but also because it leads to the upgrading of mineral industrial bases. Local governmental authorities are confronted with the task of having to implement rural development programs in the areas of transportation, communications, electric power, and water and sewage. They also face the challenge of reducing the rural-urban migration of women, which can be accomplished by improving economic growth, increasing infrastructure development, upgrading social services, and ensuring that non-renewable natural resources are utilized efficiently.

RECOMMENDATIONS AND CONCLUSIONS

Before any positive action can be taken to improve female involvement in the sub-sector, it is necessary to understand the root cause of their involvement. This can be determined by examining the economic, social, technical and environmental characteristics of impoverished artisanal mining communities. Both the Government and the private sector must work to improve the livelihoods of artisanal women miners in communities. Where there is a large number of miners and a limited quantity of mined resource, stakeholders are forced to promote alternative livelihood opportunities for female miners, such that the remaining few can develop resources economically. Good governance through enforcement of the exiting legal and regulatory framework will undoubtedly lead to reductions in environmental problems and high crime, and improve the social and health conditions of miners. This will, in turn, facilitate increased formalization in the sub-sector, and improvements to the livelihoods of female artisanal miners by assuring them access to education and credit institutions, mining rights and investment opportunities.

The role of government

As a policy maker, the Government is aware of how mining contributes to poverty reduction. To a large extent, the country's mining policies and frameworks are well developed, and, if enforced properly, can have a positive influence on the development artisanal and small-scale mining. A primary example is the *Mineral Policy of 1997*, which stipulates the Government's desire to promote small-scale mining through education, training and increased access to credit. Another example is the *Mining Act of 1998* under Division D, which outlines specific provisions for small-scale miners. Both the Act and Regulations of 1999 give special considerations to small-scale mining, including encouraging miners to acquire mining licenses; increasing the period of primary mining licenses from one year to five years; and providing equal mineral rights to small-, medium- and large-scale miners. The majority of the contents of the framework have been well elaborated in the local Swahili language, and are also summarized in the "Code of Practice for Small-scale Miners", issued initially in 2000 by the Ministry of Energy and Minerals, United Republic of Tanzania. The Government, therefore, appears to recognize the importance of enhancing technical incentives to small-scale mining.

As stated in the Mining Policy, the role of the Government is to regulate, promote and facilitate development in the sector. However, it needs to assess both the assets and strengths of female miners, as well as their ability to access and manage the available mineral resources, and facilitate strategic plans for the community. The issues pertaining to the liberation of women and children in the sector that need to be discussed among policy makers and community representatives, and, eventually, addressed, include:

1) Transforming and upgrading artisanal mines into formal, organized mines. The State has to begin simplifying licensing procedures so that miners can acquire mining licenses and increase their chances to trade available mineral rights in their respected claims.
2) Regulating the mining and resulting downstream economic activities to increase female participation.
3) Sensitizing women on awareness of their rights, using available media and other public structures, including the promotion of female participation in social, economic, cultural and political activities, as well as encouraging the equal participation of women from the grassroots to the national and international arena. This will entail prohibiting all activities physically and/or morally harmful to women and children at mines, minimizing prostitution, remediating unhealthy and unsafe work environments, and empowering women to organize and lobby to have laws changed if they prove unsuitable to their needs.
4) Developing negotiating skills among women so that they can negotiate directly – through their own representatives or mining associations – with local government bodies.
5) Enhancing the existing partnerships between small-scale and large-scale miners, and to develop leadership and entrepreneurial skills for women involved in the informal industry sector.
6) Taking appropriate measures increase literacy among girls, and to minimize the discriminatory actions taken towards women in education. The establishment of vocational training programs for the unemployed at mine centers would help to build capacity among miners and thereby reduce employment at mines.
7) Fighting child labour and preventing the spread of pandemic HIV-AIDS and other sexually transmitted diseases at mines, through reproductive health education, child care and nursery centers, and schooling facilities. The Government must enforce a minimum age of work at mines, so that school children are prohibited from working at mine camps.
8) Involve women in socio-economic and environmental policy-making processes.
9) Taking appropriate measures so that women are working in a healthy environment with proper management plans for the acquisition, storage, transport, use and disposal of toxic wastes such as mercury and mercury contaminated processed water and tailings. All miners, regardless of their social and cultural dissimilarities, should adhere to the *Code of Practice for Small-scale Miners* prepared by the Ministry of Energy and Minerals of the United Republic of Tanzania.

The role of the private sector

Similar to most SADC countries, Tanzania faces major challenges with regard to the empowerment of women. There are several discriminatory laws and practices that indirectly or directly prevent women from actively participating at different levels of development. The private sector has the task of changing the situation and increasing female participation by:

1) Working to provide women with choices;
2) Initiating the formation of women's associations, financial clubs and networks. Such unions build unity among women whose work is not recognized, which increases chances of securing legal advice and assistance for formalizing work;
3) Providing women access to other organizations that offer facilities such as skills training, credit and loan facilities, legal assistance, health advice and counseling;
4) Supporting small and medium business enterprise development at mines, the formation of cooperatives, and credit schemes among women cooperative members;
5) Sensitizing and educating women and the public in general on gender issues, and ensure that women are equally represented at all levels of decision-making. In parliament, for example, 20% of seats are currently reserved for women, and, in local governments, 25% of positions are reserved for women. The private sector has the task of lobbying and stimulating debate among the public and members of parliament such that women are equally represented in parliament and are encouraged to participate in local and general elections; and
6) Carrying out research on women, traditions and gender-based discrimination practices, to better understand how to promote gender equity in society.

REFERENCES

Elections (2000). Invest at home, Mkapa advises people from Pemba Saturday, October 21, 2000 By the Guardian Election Team, 2000.

ILO News (1999). http://www.us.ilo.org/news/ilowatch/9906.html

ILO Press Release (1999). Small-scale mining on the increase in developing countries. Health and safety risks imperil miners, women and children Monday 17 May 1999 ILO/99/10. http://www.ilo.org/public/english/bureau/inf/pr/1999/10.

ILO Sectoral Activities Programme (1999). Tripartite Meeting on Social and Labour Issues in Small-scale Mines, Geneva, 17–21 May 1999, International Labour Office Geneva Copyright ©1999 International Labour Organization ILO.

Kinabo, C. & Lema, E. (2000). Women Participation in Mining. SANTREN's National Training Workshop on Women in Mining and Environmental Protection in Tanzania, held in Mwanza Tanzania, 21–24 August 2000.

Kinabo, C. (2000). Mercury, Health and Environment. A paper on an International Workshop on the Environmental Management of Artisanal and Small Scale Mining Activities, SANTREN Publication, September 2000, Mwanza, Tanzania, p. 34.

Kinabo, C. (2001a). SANTREN Training Manual on "Child Labour in Mines", Regional Workshop on Women in Mining and Environmental Control, SANTREN Publication, Dar-es- Salaam, Tanzania 9 –11 February 2001, 21 pp.

Kinabo, C. (2001b). SANTREN Training manual on "Occupational Health and Safety in Mines", Regional Workshop on Women in Mining and Environmental Control, SANTREN Publication, Dar es Salaam, Tanzania 9–11 February 2001, 54 pp.

MMSD Southern Africa Draft Report (2002). Prepared for a meeting of the MMSD Southern Africa Steering Committee held in Johannesburg, 11–12 February, 2002.

National Reports (2001). Tanzania, English version. http://www.socwatch.org.uy/2001/eng/national%20reports/tanzania2001_eng.htm

SANTREN (2000). Community Education for Artisanal and Small-Scale Gold Miners on the Health Hazards of Mercury. Southern African Network for Training and Research on the Environment SANTREN held in Mwanza, Tanzania 25th to 28th September 2000, p. 39.

Tan Discovery (1996). Baseline Survey and Preparation of Development Strategy for Small-Scale and Artisanal Mining Program in Tanzania. Ministry of Energy and Minerals/Tan Discovery Consulting Company, World Bank Project 1996, p. 135.

Tanzania (1999). Economic Trends and Outlook: Chapter II – Economic Trends and Outlook. U. S. Department of Commerce – National Trade Data Bank, September 3, 1999.

Wolgin, J.M. (2001). A STRATEGY FOR CUTTING HUNGER IN AFRICA Commissioned By: Technical Committee of the Partnership to Cut Hunger in Africa. Version March 26, 2001 http://www.aec.msu.edu/agecon/fs2/africanhunger/wolgin_eng1.htm

20

The Socio-Economic Impacts of Small-Scale Mining: The Case of Zambia

MUNYINDEI MASIALETI AND CRISPIN KINABO

Mining has been the economic and social backbone of Zambia since the commissioning of the first large-scale exploitation of the copper-cobalt deposits in the Copperbelt Province in the 1930s. Its contribution to national employment, GDP, export earnings, and state revenue has been estimated to be 10%, 15%, 80% and 13% respectively (Hoadley et al., 2002; Simukanga, 1998). The country's main mineral exports are copper, cobalt, their by-products (gold, silver, selenium), gemstones (precious and semi-precious stones), and dimension stones (marble, granites and slates). Apart from dimension stones, production and reserves of the other minerals mined in Zambia are significant both within the Southern African Development Community (SADC) region (Fig. 20.1) and on a global scale (Table 20.1).

However, mineral production of both metallic ores and gemstones has decreased, after reaching peak production levels in the 1970s and 1980s. To date, copper and cobalt have been exploited on a large scale in the Copperbelt Province, while gemstones and dimension stones have mainly been produced at small-scale mining operations. However, the overwhelming majority of resident small-scale miners produce gemstones – namely, emeralds, amethysts, aquamarines and tourmalines. Production is concentrated within the following five areas: (1) Ndola rural (emeralds) in the Copperbelt Province; (2) Lundazi (aquamarines and tourmalines) in the Eastern Province; (3) Mkushi (aquamarines and tourmalines) in the Central Province; (4) Itezhi-tezhi near Mumbwa (aquamarines and tourmalines) in the Southern Province (see Fig. 20.2); and (5) in the Kalomo area (amethyst) in the Southern Province. Figure 20.3 provides an accurate overview of the geographical occurrences of gemstones in Zambia, and Table 20.2 provides a breakdown of small-scale mine production in the SADC Region.

According to an undated leaflet on "Gemstones in Zambia", produced by the Ministry of Mines and Minerals Development, beryl was the first gemstone

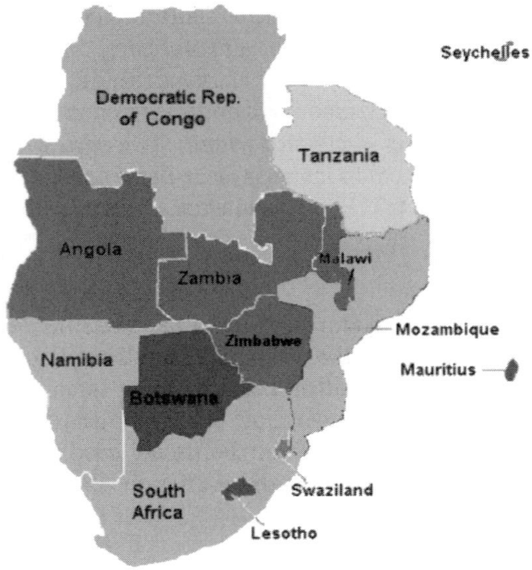

Fig. 20.1 The SADC region.

Table 20.1 Zambia's contribution to world mineral production and proportion of world mineral reserves.

	World production (1998)			Global reserves		
Mineral	Percentage contribution by SADC	SADC countries with >5%	Other SADC producers	Country	Percent	SADC total, %
Cobalt	42	DRC Zambia	Botswana RSA Zimbabwe	DRC Zambia RSA	26 5.6 0.2	31.8
Copper	4		Zambia RSA Botswana Namibia Zimbabwe DRC	Zambia DRC RSA	5.2 4.6 2.0	11.8
Semi-precious stones##	Not available	Tanzania Zambia	RSA Botswana Zimbabwe	Zambia Tanzania	Not available	

DRC – Democratic republic of Congo, RSA – Republic of South Africa, ## – small-scale mining.
Source: Modified after Hoadley, M. et al., 2002.

discovered in the country. The discovery, which was made in 1928, occurred west of Luanshya in Ndola rural, where the existence of emeralds was later confirmed some three years later in 1931. At the same time, aquamarine was discovered in Lundazi District in 1931; the presence of amethyst in the Mwakambiko Hills in Kalomo was established in the 1950s.

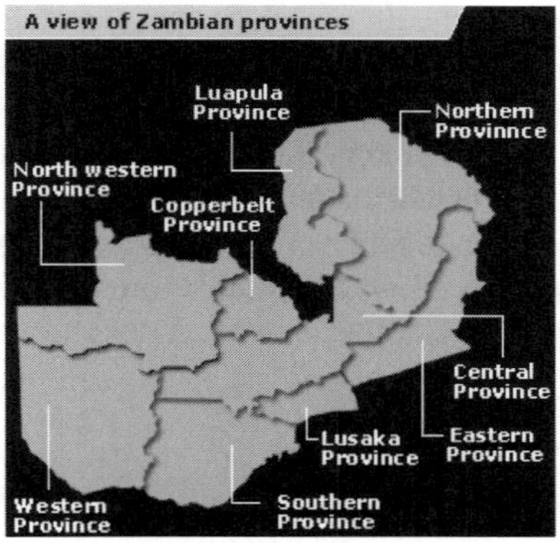

Figure 20.2 Provincial map of Zambia.

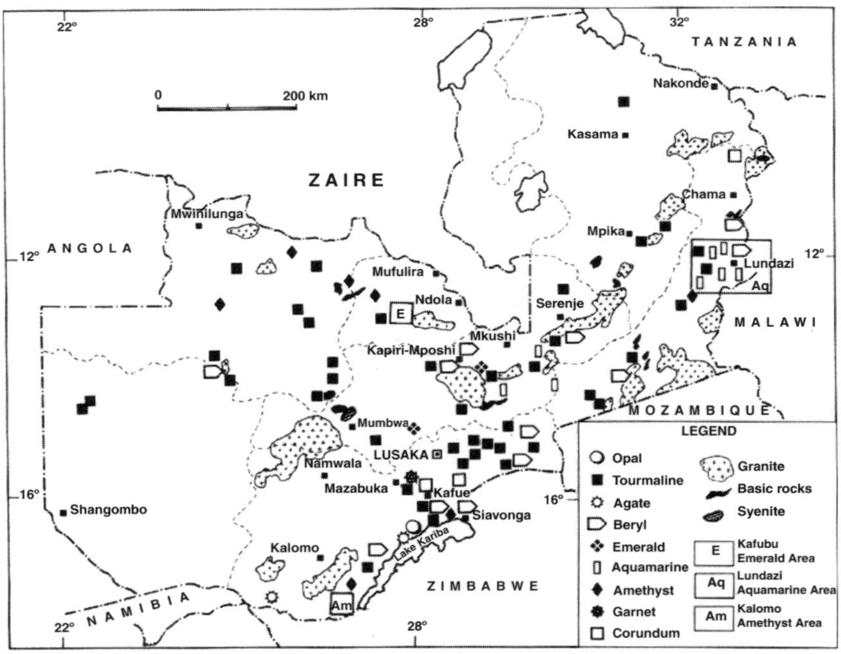

Figure 20.3 Gemstone occurrences in Zambia.
Note: Zaire is the former name of DRC.

It has been estimated that gemstones alone constitute 25% of Zambia's total mineral output (Taupitz, 1991). In terms of value, outputs of emeralds account for approximately 80% of the earnings generated in the gemstone mining sector, and, in terms of total production, amethyst accounts for the largest share of output (Mwenechanya, 2000). On the international market, annual emerald sales from Zambia have been estimated at US$200 million (Kambani, 1995), accounting for approximately 20% of global production (Sikatali & Mambwe, 1994). Unfortunately, only an estimated US$20 million (10%) is recorded by the State, as the majority of emeralds are sold through illegal channels.

This is the first of two chapters that examine the socio-economic characteristics of the Zambian small-scale mining industry. While the second of these chapters aims to provide an overview of the policy changes made in the industry, this chapter focuses more on the production, employment and locational characteristics of Zambian small-scale mining operations.

Table 20.2 Minerals mined by small-scale in SADC region.

	Malawi	Mozambique	Tanzania	South Africa	Zambia	Zimbabwe
Main Minerals for small-scale mining	Lime, Building-materials, Gemstones	Gold, Gemstones	Gold	Gold, Gemstones	Gemstones	Gold, Tantalite
Minerals mined						
Baryte						X
Bauxite		X				
Chromium						X
Clay	X	X			X	X
Coal	X	X		X		X
Cobalt					X	X
Copper		X			X	X
Diamantes		X	X	X	X	
Feldspar					X	X
Gemstones	X	X	X	X	X	X
Gold	X	X		X	X	X
Graphite				X		X
Gypsum			X			
Iron						X
Kaolin				X		X
Lead					X	
Limestone	X	X	X	X	X	X
Lithium						X
Magnesite						X
Nickel						X
Ornamental & Dimension Stones	X	X	X	X	X	X
Phosphate				X		X
Platinum						X
Quartz					X	X
Salt	X			X		X
Sand	X	X	X	X	X	X
Silver				X	X	X
Sulfur					X	X
Talc	X	X		X	X	X
Tin						X
Tungsten						X
Zinc					X	

Source: after Svotwa, 2001.

SMALL-SCALE MINING AND ITS CHARACTERISTICS

As is repeatedly mentioned throughout this book, there is no universally accepted definition of "small-scale mining"; each is based on a single criterion or a set of criteria. Primary examples include: number of persons employed; size of concession lease; size of reserves; a mine's productive capacity; degree of capitalization; and requirements in terms of mine safety (Berger, 1982; ILO, 1999; Hentschel et al., 2001). The Mineral Resources Forum (http://www.natural-resources.org/minerals/smscalemining/news.htm), managed by the United Nations Conference on Trade and Development (UNCTAD), lists a host of definitions currently being used by different countries. The mining community in Zambia has been divided into "large-scale" and "small-scale", according to investment capacity and capability. This system of classification has been incorporated into national mining legislation, and is used to classify mining rights according to mineral type and area (Table 20.3).

The number and distribution of various small-scale mining rights – excluding an artisan's mining right – are listed in Table 20.4. In terms of area, any land plot exceeding these specifications is considered "large-scale". However, certain mining rights are known to expire as new ones are being granted. The greatest concentration of license holders is found in Ndola rural (283 emerald/beryl), followed by Lundazi (66 aquamarine), and Kalomo (65 amethyst). Of the estimated 1095 "small-scale mining rights" that have been granted in Zambia to date, 542 have been awarded to artisans.

From a structural and technical point of view, small-scale mining practices in Zambia can be placed into the following three categories (Tembo et al., 2000): "artisanal", "traditional small-scale", and "advanced small-scale". "Artisanal" refers to the smallest and simplest of operations. They feature simple tools and are informal business enterprises. They can take the form of spontaneous practices without legal title to property, as well as activities with registered claim to land plots. Artisanal mining is predominantly perpetuated by illegal miners (phantom soldiers*) who move swiftly to a reported new find. A number of

Table 20.3 Classification of small-scale mining rights in Zambia, according to the *Mines and Minerals Act.*

Mining right	Mineral type	Area	Period of tenure
Artisan's mining right*	All minerals other than gemstones	Up to 5 Ha (0.05 km^2)	2 years
Prospecting permit*	All minerals other than gemstones	Up to 10 Ha (0.10 km^2)	2 years
Small-scale mining licence	All minerals other than gemstones	Up to 400 Ha (4 km^2)	10 years
Gemstone licence	Gemstones	Up to 400 Ha (4 km^2)	10 years

* Non-renewable.

* "Phantom soldiers" is a label used for illegal miners involved in the gemstone mining areas of Zambia, who are fond of recovering quality stones either through exploration and extraction, or by scavenging discarded mine waste from developed mines *in cognito*.

Table 20.4 Distribution of small-scale mining rights in Zambia.

Number of licences		Minerals												
Province	District/Place	a	am	b	c	d	e	g	l	q	ra (rc)	s	sa	t
Central (53)	Chibombo (2)		✓	✓		✓								
	Kabwe (2)										✓			
	Kapirimposhi (3)	✓								✓				✓
	Mkushi (16)	✓					✓							✓
	Mpika (5)	✓		✓						✓				
	Mumbwa (11)	✓	✓											
	Serenje (14)	✓		✓						✓				
Copperbelt (283)	Luanshya (6)			✓					✓	✓				✓
	Ndola rural (283)			✓			✓			✓	✓			
	Mpongwe (1)	✓												
Eastern (85)	Chama (8)	✓						✓		✓				✓
	Katete (1)									✓				✓
	Luangwa (1)													✓
	Lundazi (66)	✓						✓		✓				✓
	Nyimba (9)		✓	✓								✓		✓
	Petauke (1)		✓											
Lusaka (12)	Chinyunyu (1)			✓										
	Chongwe (7)	✓		✓			✓			✓				✓
	Kafue (1)		✓											
	Lusaka (3)							✓						
Luapula (1)	Mansa (1)												✓	
Northern (3)	Chinsali (3)						✓							✓
North-Western (26)	Mwinilunga (10)		✓			✓				✓				✓
	Solwezi (16)		✓	✓	✓					✓				
Southern (89)	Choma (2)		✓						✓					
	Gweembe (3)		✓											✓
	Itezhi-tezhi (5)	✓												
	Kalomo (65)		✓							✓				
	Mazabuka (3)						✓	✓						
	Mboroma (1)		✓											
	Monze (2)		✓							✓				
	Namwala (7)	✓	✓	✓						✓				
	Sianzovu (1)		✓							✓				
	Sinazeze (1)													
Western (1)	Sesheke (1)							✓						
Total = 553														

a – aquamarine; am – amethyst; b – beryl; c – citrine; d – diamond; e – emerald; g – garnet; l – limestone; ra (rc) – rock aggregate (crystal); s – spessatite; sa – sapphire; t – tourmaline.
Source: Ministry of mines and minerals development, 2002.

such workings are found within the gemstone mining segment of the industry. The "traditional small-scale mining" category is comprised of the registered and licensed non-mechanised or semi-mechanised operations run by society members or entrepreneurs with use of hired labour; these operations have basic management structure, and lack financial resources, as well as appropriate management and technical skills. The majority of small-scale mines operating in Zambia are considered "traditional small-scale". The third category, "advanced small-scale mining", comprises the legally constituted, highly mechanised, operations, the management of which undertakes reasonable geological investigations and mine planning; because of their advanced nature, most are highly productive. These mines tend to employ qualified staff, including geologists and mining engineers.

However, over the past few years, women and children in peri-urban and urban areas, especially in Lusaka, have carried out stone crushing operations as a source of livelihood. Although such activity is illegal, the efforts taken by the Ministry of Mines and Minerals Development, the local council, and the Environmental Council of Zambia, to stop the practice, have been futile.

SOCIO-ECONOMIC IMPACTS

Economic impacts

Small-scale mining contributes to Zambia's economic development in a number of ways. Being a labour-intensive and rural-based activity, it creates employment in mining, mineral processing, lapidary and other associated downstream activities, and provides a source of livelihood for rural communities. From an employment angle, urban communities mainly benefit from the existence of processing and lapidaries (a few), which are typically based in citified areas. The total number of people employed at Zambian small-scale mines has been estimated to be 30,000, of which 9,000 (30%) are women. Table 20.5 provides small-scale mining employment estimates for selected developing countries, showing where Zambia stands in relation to other developing countries.

Resident small-scale mine products – particularly gemstones – are mainly exported to Israel, India, Hong Kong and, of late, South Africa. As previously explained, gemstones provide a major portion of the country's foreign exchange earnings. In 1997, out of the US$822 million contribution made by mining to Zambia's total export earnings of US$1050 million, gemstones contributed some US$225 million in foreign exchange. Apart from contributing to foreign exchange, licensed small-scale mine operators pay all forms of taxes and royalties, thereby providing the government with important sources of revenue.

However, on the negative side, the illegal selling of small-scale mine products – particularly gemstones – results in the loss of vast amounts of potential foreign exchange earnings for the government. This is manifested by the disparity between the revenue (US$20 million) recorded by the State and sales in the international market (US$200 million) for emeralds. Apart from the illegal mining and marketing of gemstones, stone crushing in Lusaka and other

Table 20.5 Employment in small-scale mining in selected developing countries.

	Employees				
		Women		Children	
Country	Total number ('000)	Number	% of total	Number	% of total
Bolivia	72	15,500	22	–	–
Brazil	67 small-scale miners 300–400 garimpeiros	–	–	–	–
Burkina Faso	100–200	45,000–85,000	45	–	–
China	3–15	–	–	–	–
Ecuador	92	6,200	7	4,600	5
Ghana	200	89,500	45	–	–
India	500	335,00	7	–	–
Indonesia	109	10,900	10	2,180	2
Malawi	40	4,000	10	–	–
Mozambique	60	18,000	30	–	–
Papua New Guinea	50–60	12,000	20	18,000	30
Peru	30	–	–	–	–
Philippines	185	46,400	25	9,300	5
South Africa	10	500	5	–	–
Tanzania	550	143,000	26	>3,000	>0.5
Zambia	**30**	**9,000**	**30**	**–**	**–**
Zimbabwe	350	1,53,000	<50	–	–

Source: Modified after UN 1996; IIED 2002.

peri-urban areas results in a loss of revenue to both the local councils and government.

Social impacts

Small-scale mining affects Zambian society in a number of negative ways. First, operations have led to minor, serious, and fatal accidents, most of which have resulted from pit highwall failures, unstable steep highwalls, and undercutting. Most of the accidents involving both legal and illegal miners are not reported to Ministry of Mines partly because of a fear of consequences on the part of the mine owners, and the remoteness of the operations in relation to Ministry offices. For instance, only six fatal accidents involving seven people were reported to government authorities between 1984 and 1995. A lack of safe drinking water and social services such as education, health clinics (the exception being the Kalomo amethyst area), and protective clothing, combined with unsuitable sanitary conditions and low wages, have continued to daunt the communities based in gemstone mining areas. Respiratory diseases are now prevalent among the stone crushers in Lusaka, who commonly carry out manual crushing in a dry environment without the use of protective devices such as respirators.

Unprotected mining excavations in both urban areas and rural mining regions continue to pose safety hazards. In urban areas, these excavations have not only become havens for criminals, but also breeding grounds for disease-carrying mosquitoes during the rainy season. Further, cases of children drowning and inadvertently falling in these excavations have been reported in urban areas. In one case, a miner drowned in one of the flooded pits.

Finally, illegal mining and marketing activities have rendered the security situation in most gemstone areas very unsafe; in many cases, deaths have occurred as a result of disputes.

The next chapter (Chapter 21), by Stephens Kambani, provides an overview of the policy changes made in Zambia that have had an effect on indigenous small-scale mine development. It both expands upon many of the ideas presented in the current chapter, and provides a concise overview of pertinent small-scale mining legislation in place in the country.

REFERENCES

Berger, A.R. (1982). The importance of small-scale mining: a general overview in proceedings of an AGID workshop on strategies for small-scale mining and mineral industries, Mombasa, Kenya, April 14–25, 1980, AGID report N0.8, 1982, compiled and edited by Nielson, James, M., pp. 1–12.

Hentschel, T., Hruschka, F. & Priester, M. (2001). Draft Global report on Artisanal and Small-scale mining, MMSD project Report. Prepared for MMSD meeting, London, November, 2001.

Hoadley, M., Limptlaw, D. & Weaver, A. (2002). Mining, minerals and sustainable Development, The report of the regional MMSD process, Volume 1, MMSD Southern Africa, School of Mining Engineering, University of the Witswatersrand, Private Bag 3, Wits 2050.

International Institute for Environment and Development (IIED) (2002). Artisanal and small-scale mining in Breaking new ground: the report of mining minerals and sustainable development (MMSD) Project, Published by Earthscan for IIED and WBCSD, ISBN: 1 85383 907 8 paperback/1 85383 942 6 hardback, pp. 315–334.

International Labour Organisation (ILO) (1999). Report for discussion at the tripartite meeting on Social and Labour issues in small-scale mining, Geneva, 17–21 May 1999, International Labour Organisation, Sectoral Activities programme.

Kambani, S.M. (1995). The illegal trading of high unit value minerals in developing countries. *Natural Resources Forum*, 19(2), 107–112.

Mwenechanya, S. (2000). A concept/draft paper on *a framework for the accelerated development and growth of small scale mining in the SADC region,* UNECA, SADC MCU.

Natural Resources (2002). (http://www.natural-resources.org/minerals/smscalemining/news.htm).

Sikatali C. & Mambwe, S.H. (1994). Mineralisation and potential of the gemstone industry in Zambia, Industrial Minerals in Developing countries,

Edited by Mathers S.J. and Notholt, A.J.G., AGID report series Geosciences I International Development No. 18 pp. 263–272.

Simukanga, S. (1998). Mineral Resources: Issues Associated with Their Management for Sustainable Development in Zambia in Summary Proceedings of the UNU/INRA Regional Workshop, Accra, Ghana – March 1998 Africa's Natural Resources Conservation and Management Surveys, Edited by J.J. Baidu-Forson, pp. 101–104.

Svotwa, R. (2001). Small-scale mining and sustainable development – Zimbabwe, in Small-scale mining and sustainable development in the SADC region, MMSD Research topic 1, Final report compiled by Bernd Drechsler, Harare, August 2001, pp. 145–187.

Taupitz, K.C. (1991). Technical introduction, Small Scale Mining Technology Workshop (Abstracts edition), 18–20 February 1991, Livingstone, Zambia.

Tembo, F., Kambani, S., Katongo, C. & Simasiku, S. (2000). Mining and Geology of Ndola Emerald Area.

UN (United Nations) (1996). Developments in Small-scale mining, Report of the Secretary General, United Nations Economic and Social Council, Committee on Natural resources, third session, 6–17 May 1996, E/C.7/1996/9.

21

The Impact of Policy Changes on Small-Scale Mining in Zambia

STEPHENS M. KAMBANI

Zambia is richly endowed with a wide variety of mineral resources, as it rests firmly astride the Central African plateau, a great depository of mineral wealth stretching from Shaba Province in Congo DR in the north, to the gold and diamond fields of South Africa. A substantial proportion of the world's known reserves of copper and cobalt are found in Zambia. Other important mineralizations include gold, zinc, tin, gemstones, coal, and a variety of industrial minerals. More recently, diamond-bearing mineralizations have been detected in various places with potential to become future diamond mines.

Of all the minerals mined in Zambia, gemstones and building materials (sand and stone aggregates) are the prime targets for small-scale miners. However, in terms of value of output and number of people involved, gemstones are the most important.

Although there are many definitions used to define small-scale mining (SSM) worldwide, in the Zambian context, it is defined according to the *Mines and Minerals Act* of 1995. Four types of mineral rights are issued for SSM, as indicated in Table 21.1 (*Mines and Minerals Act*, 1995). In the Zambian case, types of minerals and size of area, as shown in Table 21.1, fixes the definition of SSM.

This chapter presents an overview of SSM in Zambia. Initially, the key features of Zambia's legal and policy framework for mining are discussed, as

Table 21.1 Defining parameters for SSM in Zambia.

Type of mining right	Minerals	Area
Prospecting license	All minerals except gemstones	Less than or equal to 10 km^2
Artisanal license	All minerals except gemstones	Not exceeding 5 ha
Small-scale mining license	All minerals except gemstones	Not exceeding 400 ha
Gemstone license	Gemstones	Not exceeding 400 ha

well as how they have impacted mineral resources development and the SSM sector. As there are different economic characteristics between gemstones and industrial minerals, each is discussed separately in different sections.

POLICY AND MINING LEGAL FRAMEWORK

Long before the current mining policy was launched in 1995, the mining sector did not have a policy. The "unwritten" policy was tailored around large-scale copper mining, which has dominated the Zambian economy for more than six decades. As can be expected, the medium and SSM sub-sectors of the Zambian mining industry were neglected at the expense of large-scale mining. This is reflected by the existing large gap between their levels of development in terms of skills, technology and productivity. SSM remains largely an artisanal activity that features low levels of technology and management skills.

The launching of Zambia's *Mining Policy* in 1995 signified an important step towards recognizing that mineral resources development require a blueprint to ensure that government can maximize benefits from the mining sector, and, at the same time, safeguard the environment. The Policy recognized SSM as a key player in the economy. However, the development of the *Mining Policy* should not be seen as an isolated case, as it was part of a major economic reform program that the new government embarked on in 1991.

Economic restructuring involved the introduction of free market reforms. Thus, markets hitherto controlled by the government were liberalized. The controlling of prices and foreign exchange rates was soon abolished, and the engine for propelling economic growth was placed on the private sector. The government eventually established the Zambia Privatization Agency, which was commissioned the task of assisting government with the privatization of state-owned companies, which, at the time, had dominated the economic landscape of the country. The government's new role was that of passing legislation and monitoring.

Other important landmarks of the reforms included the promulgation of new pieces of legislation such as:

- The *Securities Act* No. 38 of 1993;
- The *Investment Act* No. 39 of 1993;
- The *Companies Act* No. 26 of 1994;
- The *Mines and Minerals Act* No. 31 of 1995; and
- The *Income Tax Act*.

The *Securities Act* regulates the securities industry, established the Securities Exchange Commission, and defines its objectives and functions. Following its implementation, the Lusaka Stock Exchange (LSE) was established, which became fully operational on February 21st 1994; it provides an important facility for the mobilization of investment capital, including that for mining (Securities Act, 1993).

Unfortunately, the LSE has not been overly successful in raising capital to finance mining, particularly high-risk activities such as mineral exploration and SSM. Given the poor state of the economy, the teething problems of the LSE, and the concept of a stock exchange as a vehicle for investment being a relatively new initiative to Zambians, it is hoped that its role will improve in the future.

The *Investment Act* of 1993 was passed to revise the law relating to investment in Zambia, so as to provide a comprehensive legal framework for investment, and to repeal the *Investment Act* of 1991. As a result of implementing the Act, the Investment Centre, which was formed in 1991, was allowed to continue to perform its pivotal role of promoting and coordinating government policies, and facilitating investment in Zambia; the Centre was to provide a one-stop support facility to investors (Investment Act, 1993).

With regard to the promotion of investment in the mining sector, the Centre appears to have been effective. This is reflected by the increased presence of both local and foreign investors involved in mineral exploration and mining activities, including the SSM of gemstones and gold compared to the period prior to 1991.

The new *Mining Policy* was passed to make provision with respect to the prospecting and mining of minerals in Zambia. Its objectives are as follows (MMMD, 1995):

- **The Privatization of Mining Companies:** To systematically and expeditiously privatize state mining and marketing companies, and to make the private sector the principal producer and exporter of mineral products.
- **Development of New Mines:** To promote private sector initiative in the development of new mines by providing a stable legal and fiscal regime to attract private investment in exploration and development. The aim is to increase and diversify mineral and mineral-based production and exports.
- **Small-Scale Mining:** To promote the development of SSM by providing an appropriate legal and fiscal regime.
- **Gemstone Mining and Marketing:** To promote the development of gemstone mining, and to facilitate the liberalization of marketing arrangements.
- **Industrial Minerals:** To promote the exploration and exploitation of industrial minerals by providing information on the availability and uses of local industrial minerals.
- **Energy Minerals:** To promote private sector initiatives involving the exploration and exploitation of energy minerals.
- **Ferrous Minerals:** To encourage the privatization of the ferrous industry – recognized as a pre-requisite to industrial growth.
- **Environmental Protection for Mining:** To ensure that the ecosystems in the vicinity of mining operations are protected, and to ensure that a high level of health and safety is maintained among workers and the inhabitants of affected areas. In recognition of the damaging nature of mining operations, any mining permit issued must be accompanied by an approved Environmental Impact Assessment (EIA) and Management Plan. In addition, the legislation provides for the establishment of an Environmental Protection Fund.

- **Down-stream Processing:** To promote the local processing of raw mineral materials into finished products for added value by providing an appropriate fiscal regime. In addition, the enactment of higher royalties to the exporters of unprocessed materials *vis-à-vis* those selling finished goods requiring no additional processing.
- **Infrastructure:** To develop infrastructure in areas with potentially extractable mineral deposits. The *Development Agreement* embodied in the legislation provides for putting in place a mechanism for the development of infrastructure in areas of mineral exploitation.
- **Labor:** To create employment and training opportunities for Zambians in the mining industry. To this effect, the government has put in place legislation that ensures employment and the training of Zambians by mining right holders.
- **Marketing:** To develop a liberalized and free marketing environment for mineral products. In the case of gemstones, the establishment of a Gemstone Exchange, and the provision of trading in gemstones through a Gemstone Sales Certificate are designed to facilitate movement towards a liberalized and free market environment.
- **Institutional Support:** To provide efficient and effective support services to the mining industry via:
  1) The decentralization of the Ministry's services;
  2) The establishment of an Investment Promotion Unit;
  3) The establishment of a Mining Advisory Committee;
  4) The removal of bureaucratic licensing procedures;
  5) The maximization of the utilization of national research and development institutions to enhance development in the minerals industry; and
  6) The provision of computerized information systems and databases.

Like all policies, the foregoing requires continuous review in order to assess achievements, failures, and relevance, and to allow for the incorporation of new ideas.

Table 21.2 summarizes the performance of the policy thus far, with special reference to SSM.

GEMSTONE MINING

Zambia has a rich endowment of high-quality gemstones, and is one of the world's leading producers. Its primary gemstones include emeralds, amethyst, tourmalines, aquamarine, garnets, beryl, citrine and rock crystal; of these, emeralds are the most important. The country is currently the world's second largest supplier of emeralds (after Colombia), its US$200 million annual output accounting for some 20% of global supply (Simwanza, 2002). The deposits of opal and tanzanite that have recently been discovered also have potential for commercial production. Figure 21.1 provides an overview of gemstone localities in Zambia.

Table 21.2 Summary of policy reviews.

New Policy Objectives	Comment
Privatization of state-owned mining Companies	The privatization of large-scale mining was successfully concluded on March 31st 2000. With regard to SSM, only Mindeco Small Mines Limited was privatized. The disposal of the remaining interests in Kagem Emerald Mining Ltd. and Kariba Minerals Ltd. has been clouded by the lack of transparency and not yet completed.
Development of new mines	Since 1991, there has been an unprecedented increase in mining activities, especially at the SSM level. There are now over 465 gemstone licenses, compared to less than 10 issued before 1991. Close to 30 of these are fully operational mines.
Promotion of small-scale mining	The new *Mines and Minerals Act* incorporates small and artisanal mining as special cases. A simplified licensing, reporting and taxation system was introduced. Unfortunately, a lack of resources hinders the Ministry of Mines from effective supervision and provision of technical assistance essential to the promotion of SSM.
Promotion of gemstone mining, and the facilitation of liberalized marketing	Prospecting and mining licenses have been combined as a result of the passing of the Act. Gemstone marketing has been liberalized, although a review of policy is necessary to reduce the illegal exporting of gemstones. The Gemstone Exchange has not yet been formed but will soon be operational with the assistance of the European Union-funded five-year Mining Diversification Project (2002–2004). Arbitrary income taxes not based on declared production are of major concern to gemstone miners, most of whom are not productive.
Promote exploration and exploitation of industrial minerals	Industrial minerals remain largely unexploited, mainly due to low industrial activities in the country. However, illegal quarrying (stone crushing activities) around major cities has grown, and is causing serious concerns in terms of environmental degradation.
Ferrous minerals	Despite the presence of iron ore deposits, not much has been done to exploit reserves.
Environmental protection for mining	Statutory Instrument No. 29 – *The Mines and Minerals (Environmental) Regulations* – was passed in 1997. Before a mining license is issued, EIAs are required, although at present, only large-scale miners are required to comply. With the expansion of SSM activities, however, there is an increasing need to enforce existing regulations.
Down-stream processing	To date, there has been little progress made to process minerals locally. As most gemstones are exported in raw form, there is a need for a review of policy to help facilitate a reverse in trend.
Infrastructure development	Efforts involving the provision of mine infrastructure have not been overly successful because of resource shortages. Most small mines are located in remote areas lacking appropriate communication facilities, all-weather roads, power, schools and hospitals.
Develop a liberalized and free marketing environment for mineral products.	The liberalization of marketing has been successful. The government has since dissolved state marketing companies for all minerals, including gemstones.
Provide efficient and effective institutional support services to the mining industry	A lack of funding has limited the Ministry of Mines' capacity to deliver quality services to the sector, including the maintenance of an up-to-date geological database, the provision of technical services, and the implementation of monitoring services. A lack of affordable finance (credit) has further hindered development.

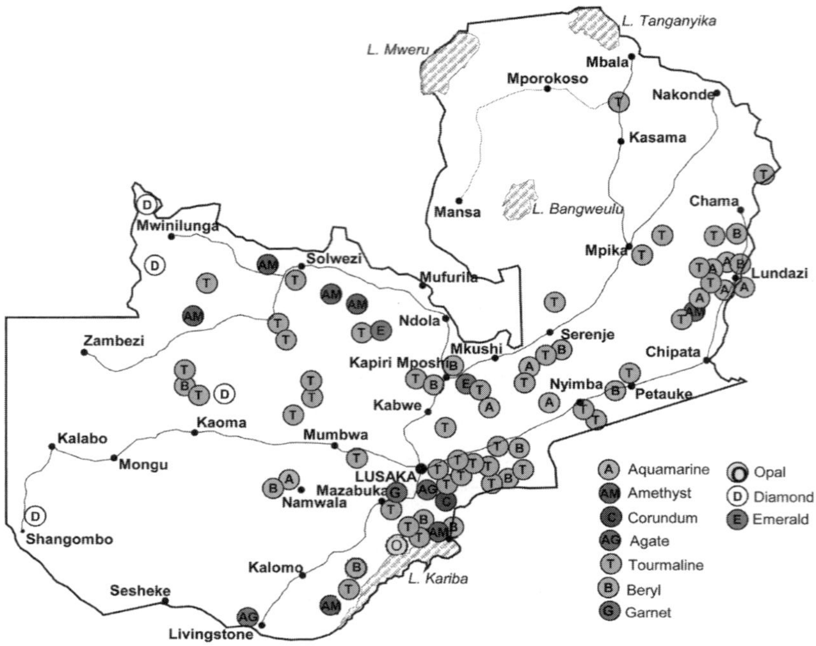

Figure 21.1 Gemstone localities in Zambia.

Table 21.3 shows the declared production figures for gemstones in Zambia during the period 1991–1999. However, these figures are only indicative because most production is undeclared (Kambani, 1995), although as Figure 21.2 indicates, it is apparent that there was a remarkable increase in the amount of declared production following the introduction of the new mining policy in 1995, which, among other things, liberalized the marketing of gemstones.

Despite the wide distribution of gemstones, there are four major gemstone-mining areas in Zambia. They are:

- Ndola rural, for emeralds;
- The Chipata and Lundazi area, for aquamarines and beryls;
- Lake Kariba, for amethyst; and
- The Nyimba and Petauke area, for garnets and tourmalines.

Structurally, the gemstone sector comprises small companies, syndicates, cooperatives, individual owners, as well as various associations. The latest records of gemstone licenses issued by the Ministry of Mines are summarized in Table 21.4. There are also numerous unlicensed operations mining various kinds of gemstones.

The following associations represent the sector:

- Association of Zambian Women in Mining;
- Emerald and Semi-Precious Stones Mining Association of Zambia (ESMAZ);

Table 21.3 Declared gemstone production and sales figures 1991–1999.

Year	Amethyst		Aquamarine		Beryl		Emerald		Garnet		Tourmaline		Total	
	Quantity (kg)	Value (US$)	Quantity (kg)	Value (US$)	Quantity (kg)	Value (US$)	Quantity (kg)	Value (US$)	Quantity (kg)	Value (US$)	Quantity (kg)	Value (US$)	Quantity (kg)	Value (US$)
1991	139,957	1,049,388	0	0	110	40,130	618	10,687,779	4	652	46	21,911	140,735	11,799,860
1992	512,274	1,868,112	65	37,197	1,466	242,946	2,647	2,838,883	91,827	1,943	34	23,665	608,313	5,012,745
1993	650,258	2,377,135	113	40,737	202	66,547	666	1,556,168	1,685	32,771	994	59,023	653,919	4,132,382
1994	188,611	1,501,430	188	105,030	911	200,142	438	1,821,524	26,192	54,624	537	146,848	216,877	3,829,597
1995	746,411	2,401,140	365	298,151	510	380,793	639	2,874,823	177,043	665,990	1,454	59,869	926,422	6,680,766
1996	1,359,792	3,200,830	866,243	213,712	2,106	722,739	1,181	6,477,167	26,624	169,465	7,175	293,671	2,263,120	11,077,584
1997	1,149,040	2,139,459	574,540	465,383	3,261	760,758	1,047	9,201,207	16,100	33,775	5,764	606,502	1,749,752	13,207,084
1998	717,064	1,439,943	199,844	782,854	4,055	281,233	1,933	7,263,096	2,212	76,270	6,253	481,695	931,362	10,325,090
1999	794,947	2,071,858	5,766	835,146	20,285	464,703	1,831	9,567,500	4,016	32,011	17,299	488,412	844,144	13,459,629
Total	6,258,353	18,049,296	1,647,124	2,778,209	32,906	3,159,992	11,002	52,288,147	345,703	1,067,499	39,555	2,181,595	8,334,645	79,524,738
AVG	695,373	2,005,477	183,014	308,690	3,656	351,110	1,222	5,809,794	38,411	118,611	4,395	242,399	926,072	8,836,082

AVG = Average
Source: Ministry of Mines, 2000.

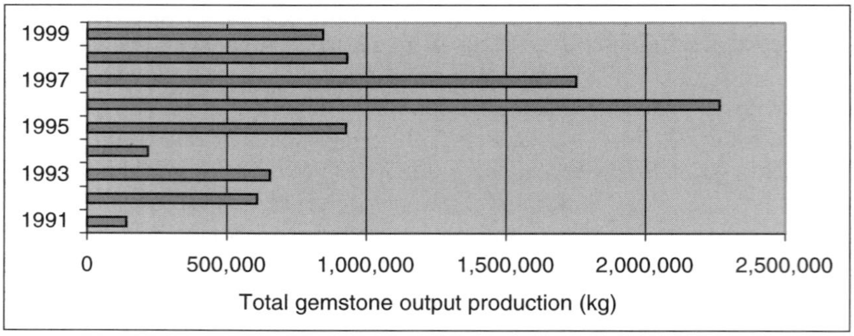

Figure 21.2 Gemstone production 1991–1999.

Table 21.4 Number of gemstone licenses issued and distribution.

Gemstone type	Number of licenses issued	Location and distribution
Emeralds	350	Ndola Rural (343), Mwinilunga (2), Mkushi (2), Serenje (1), Chinsali (1), Mpika (1)
Amethyst	72	Kalomo (55), Mumbwa (4), Kafue (1), Mpika (1), Namwala (2), Solwezi (6), Petauke (1), Gwembe (2)
Aquamarine	10	Lundazi (5), Namwala (3), Mumbwa (1), Solwezi (1)
Tourmaline	22	Lundazi (9), Nyimba (4), Mumbwa (2), Mazabuka (2), Serenje (2), Mkushi (3)
Garnet	7	Chama (3), Sesheke (1), Kapiri-Mposhi (1), Nyimba (2)
Diamond	1	Mwinilunga
Total	462	

Source: Ministry of Mines, Zambia.

- Kalomo Miners Association;
- Zambia Gemstones and Precious Metals Association;
- Zambia Emeralds Association;
- Zambia Gemstone Corporation;
- Zambia Integrated Mine and Mineral Traders Association;
- Zambia Small-scale Miners Association Fund;
- Masansa Gemstone Miners Association; and
- Chibale Small-scale Miners Association.

Of these associations, the most active are the initial four.

Gemstone mining is conducted at all levels. At the lowest end are artisanal miners, most of whom are operating illegally. They typically mine on other people's licensed prospects, as well as newly discovered deposits. Basic tools, such as picks and shovels, are employed in this level of mining, with hammers and chisels used in the recovery process.

Figure 21.3 shows a dangerous pit at a typical amethyst mine in Kalomo district, while Figure 21.4 shows some of the basic tools, such as wheel barrows, shovels and sieves, used in the recovery of amethyst at another mine in the same area.

Figure 21.3 A dangerous pit at an amethyst mine in Kalomo district.

Figure 21.4 Basic tools used in the recovery of amethyst, Kalomo district.

Open pit mining is the most preferred, and widely used method by Zambian artisanal miners. Underground mining is also practiced but mainly when individuals seek to conceal their operations. This is a prevalent practice in emerald mining.

The next level of operation consists of traditional small-scale miners. These are licensed mine owners whose operations feature some type of management structure and hired labor. Unfortunately, most of these miners are not functioning efficiently because of a lack of financial resources. Furthermore, technical, business and management skills are usually lacking among this category of miners. Consequently, the majority operate at an artisanal level using antiquated equipment.

At the highest level are the highly mechanized, and well-organized, small-scale miners. A typical miner in this category has access to mining equipment and machinery such as excavators, bulldozers, air compressors and dewatering pumps; some of this equipment is hired or leased.

There is also a group of large-scale gemstone mines such as Kagem Mines Ltd, Kuber Minerals Ltd., and Kariba Minerals Ltd. Kagem is a joint venture between the government (55% shares) and the Hagura Group – an Israeli and Indian partnership (45%) – producing emeralds (Fig. 21.5). Kariba Minerals is also a joint venture between government and the private sector, and mines amethyst, while Kuber Minerals is a private emerald mining company, which also has interests in amethyst mining.

The major constraints faced by the sector can be summarized as follows, many of which are interrelated:

- Policy related constraints;
- Financial constraints;

Figure 21.5 Emerald mining at Kagem Mines Limited.

- Technology constraints;
- Lack of technical and management skills; and
- Marketing constraints.

The effects of policies, and how they have impacted the development of the gemstone-mining sector, have already been discussed in detail.

Of the above-listed constraints, a lack of finance takes the centre stage. It has had far-reaching ramifications in the development of SSM in Zambia. Without finances, mine owners are unable to procure essential mining equipment and machinery. Furthermore, a lack of finances limits miners' ability to acquire much-needed technical and management skills. The acquisition of technology is also constrained without finances.

The majority of small miners in Zambia also experience marketing constraints. There are two important factors at play in this respect. First, miners lack basic knowledge about gemstone valuation. This is important for miners keen on securing fair prices for their products. Secondly, because the local market is highly undeveloped, the lack of access to international markets and marketing information has proven to be a formidable constraint to miners. In an effort to address these marketing constraints, there have been plans put forward to establish a local Gemstone Exchange. However, a lack of funds has impeded this initiative.

Recognizing the potential economic importance of gemstone mining to the economy, the Zambian government, under the auspices of the European Union, initiated a five-year project (2002–2004) to further develop the sector. Initially, it was named the "Sysmin Project" but has since changed to the "Mine Diversification Program"; it has a cost of 30 million Euros. Its principle objectives are:

- To provide funds (mostly grants) for mineral exploration;
- To provide loans for mining equipment and machinery;
- To provide expertise and funds for the establishment of the Gemstone Exchange;
- To provide training for miners; and
- To offer training in practical gemology.

INDUSTRIAL AND BUILDING MATERIALS

This sub-sector of SSM comprises numerous illegal artisanal miners, and a limited number of registered operations. The latest records at the Ministry of Mines indicate that there are 96 licenses issued to mine various types of industrial minerals and building materials; they are distributed throughout the country. However, this is dominated by minerals related to the construction industry, such as lime and aggregates, which are located within, or in close proximity to, markets in the cities or towns.

With regard to building materials, the lack of employment opportunities in other sectors of the national economy has made the artisanal mining of stone

Figure 21.6 Part of the heat-treated rock being extracted.

Figure 21.7 Stone aggregates and building sand in the background for sale along a major road in Lusaka city.

aggregates and building sand an important source of livelihood for many unemployed peri-urban city dwellers. The mushrooming urban populations have created a high demand for housing, and, in effect, a huge demand for a cheap source of building materials, which are supplied by these illegal miners.

Simple methods are used by artisanal operators mining stone aggregates. The process involves heating an exposed rock with used car tires. Once heated, the rock is quenched with water to induce cracks. The rock is then reduced (crushed) to appropriate sizes with hand-held hammers (Fig. 21.6). Women, with the assistance of children or family members, typically crush stone to desired sizes.

Since these illegal operations are conducted near homesteads, they are a major source of environmental degradation in urban areas, especially in the capital city of Lusaka (Fig. 21.7). Vast tracks of prime land near the City Airport, west of the heavy industrial area, and in many places within the city, have been severely degraded. These abandoned trenches have ruined the scenic view of the city and pose a danger to humans. Furthermore, during the rainy

Figure 21.8 Concrete block making at a quarry in Lusaka.

Figure 21.9 Excavation left by a quarry in Chilenje South, Lusaka.

season, they become breeding places for mosquitoes infected with malaria. The burning of used car tires for heat treatment of rocks causes air pollution that is a nuisance. Since the product is sold along motorways, the activity is also hazardous to motorists.

Licensed quarries are usually well organized small-scale activities. Such mining operations involve the drilling, blasting and transporting of rock to a crusher. Products normally include stone aggregates, quarry dust and concrete blocks (Fig. 21.8). More recently, the production of dimension stone from marble, pink quartz and granites has assumed greater importance. These are largely exported.

Due to the cost sensitivity of transporting low-value bulk materials like aggregates, quarries are located near, or within, Zambia's cities. Since there is no form of land reclamation being performed, they contribute to land degradation (Fig. 21.9). These operations are also a source of dust and noise pollution. Dust emanates from crushing operations and transportation by dump trucks, whereas noise originates from crushing, transportation and blasting; both are a major concern to residents near the mine sites. Blasting vibrations have, in some cases, damaged nearby structures.

REFERENCES

Kambani, S. (1995). The illegal trading of high unit value minerals in developing countries. *Natural Resources Forum*, 19(2), 107–112.

MMMD (1995). *Zambia's mining policy.* Ministry of Mines and Minerals Development.

Simwanza, Z. (2002). *An assessment of the lapidary industry in Zambia.* Master of Mineral Sciences degree dissertation, University of Zambia.

The Mines and Minerals Act, No. 31 of 1995.

The Investment Act, No. 39 of 1993.

The Securities Act, No. 38 of 1993.

22

Illegal Artisanal Gold Panning in Zimbabwe – A Study of Challenges to Sustainability along the Mazowe River

OLIVER MAPONGA AND MAIDEYI MECK

Zimbabwe has a long history of small-scale mining mainly involving gold, iron ore, copper, diamonds and tin. The complex Zimbabwean geology – in particular, its gold geology – has led to an exceptionally large number of artisanal and small-scale mining operations. Mining of the artisanal[1] nature is recorded to have taken place well before colonisation in the 1890s, as indigenous populations produced hunting tools from iron and used gold as a medium of exchange (Maponga & Ngorima, 2002). As early as the second century AD, locals mined copper and iron on an artisanal scale for domestic use.

Zimbabwe's artisanal and small-scale mining sector has gained great prominence since the early-1980s, following realisation of its importance in regional economic development. The 1993 Harare United Nations Conference on Guidelines for Development of Small and Medium Scale Mining (Harare Declaration) represented a major turning point in the development of the sector. The Conference helped to elevate the status of small-scale mining by putting into perspective its economic importance. As explained numerous times in this book, the sector supports the livelihoods of between 11.5 and 13 million people around the world (Veiga & Hinton, 2002; Heenskerk, 2002; ILO, 1999). The International Labour Organisation (ILO) estimates that a combined 60 million people depend on small and artisanal mining for their livelihood, both as a primary source of living or as a noteworthy complement. Local nomenclature such as *galamseys* (Ghana), *makorokoza* (Zimbabwe) and *garimpeiros* (Brazil) has evolved around the world, clear indication of the universality of artisanal and small-scale mining.

[1] Artisanal mining is characterised by the absence or low degree of mechanisation, low safety standards, poorly trained personnel, large influx of migrant workers, low remuneration, low productivity, chronic shortage of capital, illegality, unknown mineral reserves and poor environmental management systems.

The growth of small-scale mining activities around the world has since received an additional indirect boost from pressures exerted on developing nations by the competition for international capital. Specifically, the opening of new geologically prospective areas in Africa, Asia, Eastern Europe and Latin America has heightened competition for capital worldwide. Many developing nations have looked inward to small-scale miners to explore and exploit scores of small mineral deposits, and to fill the gap in local investment in minerals development. As this chapter will show, Zimbabwe has attempted to develop systems to promote small-scale mining during the past decade with limited success. Although its systems are still fraught with weaknesses, Zimbabwe has become one of the leading nations in recognising the potential of the sector as a potent force in regional economic development, and has created a framework that has facilitated the growth of formal gold panning. On a world scale, promotion of the sector has continued on an ever-increasing scale since the Harare Declaration, with the World Bank, UN and the UK Department for International Development (DfID) playing leading roles.

Small-scale mines are defined in the literature in terms of size of operation; numbers of employees, their qualifications and skills; scale of production; amount of investment needed *vis-à-vis* turnover; the type and quality of equipment used; managerial/organisational structure of operations; and/or the size of target mineral reserves being extracted (Ghose, 1993). By all accounts, artisanal mining represents the lower end of the mineral industry's production spectrum, in the majority of cases, undertaken as a means of survival. At a slightly higher level are small-scale miners, who are generally trained professionals drawn into the sector as entrepreneurs; they tend to use more sophisticated mining and mineral processing methods. The Zimbabwean Ministry of Mines and Mining Development defines "small-scale gold producers" as those producing less than 15 kg of gold per year.[2] However, it is not the intention of this chapter to define who, or what, a small-scale miner is; nor does the discussion seek to dwell on the debate. Rather, the various definitions used in the literature – and the weaknesses and strengths inherent of each – are accepted. Given the plethora of definitions, the term is used loosely throughout this chapter.

In Zimbabwe, artisanal and small-scale mining revolves around seven major minerals – namely, chromite, dimension stones, diamonds, gold, tantalite, tin and other gemstones. The artisanal sector is synonymous with black indigenous mining, as very few white Zimbabweans are engaged in small-scale mining, at least at the artisanal level. Various authors have used an array of generic schemes to classify small-scale mines in Zimbabwe (Maponga, 1993). Yet, the group exhibits diverse characteristics, which can generally be described as organised/formal/ legal or unorganised/informal/illegal. Simply put, both artisanal and small-scale mines differ from a technical, as well as structural, perspective. Formal small-scale mining production systems in Zimbabwe include families, co-operatives, syndicates, and registered individual miners. Also

[2] In Zimbabwe, mines producing 15 to 150 kg of gold per year are considered "medium-scale", and those producing more than 150 kg of gold are considered "large scale".

included are panners registered according to both the 1991 Regulations and the *Mines and Minerals Act* (Chapter 165). A complete description of the various forms of formal small-scale mines in Zimbabwe is provided by Maponga (1993).

Recent writings (Maponga & Ngorima, 2002; Maponga, 1993; Hollaway, 2000) on the small-scale and artisanal mining sectors in Zimbabwe have discussed its importance in terms of employment generation and as a source of income, the challenges it faces, and how it improves the livelihoods of people residing in the impoverished rural regions of the country. For example, analysts note that the small-scale mining sector contributes an average of 20% to total mineral production (Maponga, 1993). Anecdotal evidence also suggests that small-scale miners account for approximately 25% of total gold output in Zimbabwe, although official records show current contribution at only 1%. This discrepancy is due to the fact that most of the gold produced by small-scale miners (particularly those residing in the artisanal sector), which is estimated to be 10 tonnes annually, is lost to the parallel market (Maponga & Ngorima, 2002).

As argued by Maponga and Ngorima (2002), the sector, if properly integrated into rural economic development programmes, could become an important springboard for growth and vibrancy, as well as a source of primary and secondary income, and employment linkages. In areas where artisanal and small-scale miners are active, the sector still exhibits the characteristics of enclave development. More specifically, very few linkages to the local regional economy are evident, apart from the commercial activities that service the consumption requirements of miners and their families.

The *Panning Regulations*, promulgated in 1991, were designed to bring panning into the mainstream. By this time, illegal gold panning had developed into an economic base for many communities. Moreover, the government realized that the sector was no longer a police issue: it could not be "wished away", hence, the enactment of regulations. Gold panning has since become a mixture of both legal and illegal activities. On the one hand are formally registered panners operating in accordance with the 1991 regulations and the *Mines and Minerals Act*, and on the other hand are illegal panners who are scattered along the country's rivers and also scavenge old mine workings and dumps. Although panning has a history of over 3,000 years, a significant upsurge in illegal activity has occurred along major rivers in the last 20 years, as many panners have begun exploiting known, and proven, occurrences of alluvial gold for their survival. In the process, an estimated eight million tonnes of material is moved by panners each year (Maponga & Ngorima, 2002).

Currently, an estimated 5,000 km of the country's major rivers are being panned for gold both legally and illegally. Panning also occurs on auriferous reefs, and in abandoned mines, old workings and dumps. In some areas, panning densities – i.e. the number of panners found along a stretch of a river – can vary from less than 50 people/km to more than 100 people/km (George-Kumirai, 2002). Panning densities of as many as 300 panners per kilometre have been reported along the Mazowe and Angwa Rivers in Mashonaland West. The banks of such rivers as the Tokwe, Mazowe and Mzingwane, which are believed to be rich in alluvial gold, are now littered with abandoned pits, trenches, and temporary cluster and linear settlements. Table 22.1 lists the most panned rivers in Zimbabwe.

Table 22.1 Major panning sites in Zimbabwe.

Province	Names of major panning sites (rivers)
Mashonaland Central	Mazowe, Mudzi, Nyadire, Rwenya, Inyangombe
Mashonaland West	Angwa, Sanyati, Mupfure, Munyati
Matebeleland North and South	Insiza, Mzingwane, Gwayi, Bubi
Midlands	Shangani, Gweru
Masvingo/Manicaland	Runde, Nyanyadzi, Odzi, Save

Gold panning, both legal and illegal, is an arm of the small-scale mining sector that has received significant attention in recent years, mainly because of its environmental impacts. In fact, as far as gold is concerned, panning has become synonymous with illegal mining because of the huge numbers of illegal gold panners now found working along the country's rivers. Illegal gold panning constitutes a segment of informal[3] small-scale mining systems in Zimbabwe. The usually higher and quicker returns offered by gold panning have resulted in a larger group of illegal small-scale miners compared to other minerals. However, the boom in the tantalite market between 1999 and late-2001 introduced a new dimension to illegal mining in Zimbabwe. During this period, tantalite panning replaced gold as the most attractive mining activity in some areas of Zimbabwe because of greater returns accrued from high international prices (Maponga, 2000); the analysis of the role of international prices in commodity booms put forward by Heemskerk (2001) applies here.

In Zimbabwe, panning has provided an alternative economic base for many communities orphaned by economic contraction in recent years, retrenched workers, and those ravaged by devastating droughts in the last decade; it is simply a social safety net for many Zimbabweans. As noted by Maponga and Ngorima (2002), the sector has grown from a "get rich quick" and temporary means of economic survival, to a primary means of livelihood for entire villages and families who have settled along river banks to engage in both legal and illegal panning. An estimated 350,000 people are directly involved in both legal and illegal panning in Zimbabwe (Maponga & Ngorima, 2002), although the industry now supports over 600,000 people both directly and indirectly.

By reviewing the activities along the Mazowe River and its tributaries in Mashonaland Central Province some 40 km outside Harare, the study presented in this chapter provides insight into the dynamics of a typical illegal panning site in Zimbabwe. The findings of the survey put into perspective the social, environmental, and economic challenges/opportunities associated with artisanal and small-scale mining. An outline of the challenges faced in dealing with an illegal, yet, socially and economically important, sector based in an impoverished economic setting is presented. The objective of the chapter is to advance the

[3] "Formal" and "legal", and "informal" and "illegal", are used interchangeably in this discussion.

understanding of conditions that drive communities into illegal gold panning. The chapter also seeks to examine opportunities for sustainability within the sector.

The analysis proceeds from an interrogation of the underlying causes of the growth of panning in Zimbabwe. An examination of the sector's socio-economic dynamics using information from this case study is provided in the third section, and is followed by an analysis of the challenges to sustainability. Concluding remarks are then offered.

EXPLORING THE "GOLD PANNING RUSH" SINCE 1980

During the mid-1980s, a phenomenon typifying a "gold rush" stimulated the growth of panning as a primary economic activity for many families in Zimbabwe (Maponga & Ngorima, 2002). Although alluvial panning has taken place since the 1800s, the sector has grown exponentially over the past decade, and indications are that it will continue to grow in the future because of the immense potential for small-scale gold mining in the country. The gold rush in Zimbabwe has its own peculiar dimensions, each of which requires separate treatment. The Amazon gold rush in Latin America, which was also fuelled by attractive gold prices during the early-1980s, continued into the 1990s (Veiga & Hinton, 2002; Boas, 2001).

In our view, the principal micro and macro economic and social factors that have precipitated the growth of panning in Zimbabwe in the past twenty years include the following:

1. High gold prices in the 1980s (opportunity);
2. Downsizing and closure of mines;[4]
3. Economic decline;
4. Poverty; and
5. Poor rainfall (intermittent droughts).

As a result of these, and other, ancillary factors, alluvial panning has evolved into a full-time economic activity supporting the livelihoods of over half a million Zimbabweans both directly and indirectly (Maponga & Ngorima, 2002). It is now a recognised mode of survival in many rural areas, especially during poor rainfall seasons. The push/pull framework shown in Figure 22.1, which incorporates movements in the growth of the Zimbabwean economy, price rises, poverty and drought, explains the growth that has occurred in this sector. In our view, panning is the real, and perceived, "prosperity box" in this dichotomy. The "opportunity effect" is also relevant in the alluvial panning rush.

To put the growth of the sector into proper perspective, the chapter examines each of these push and pull factors in turn. However, it is important to emphasise that this dichotomy excludes the social and environmental costs of the rush.

The spike in the price of gold during the early-1980s, as shown in Figure 22.2, precipitated a gold rush around the world. A boom in gold exploration occurred as a result of the rise in price, such that by 1983, gold accounted for

[4] 44 gold mines closed between 1996 and 2001, 14 of which were small-scale operations.

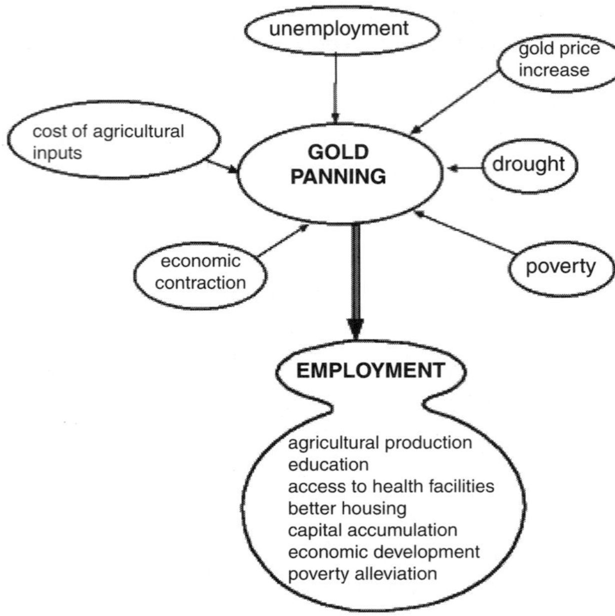

Figure 22.1 Push/pull framework in the growth of panning.

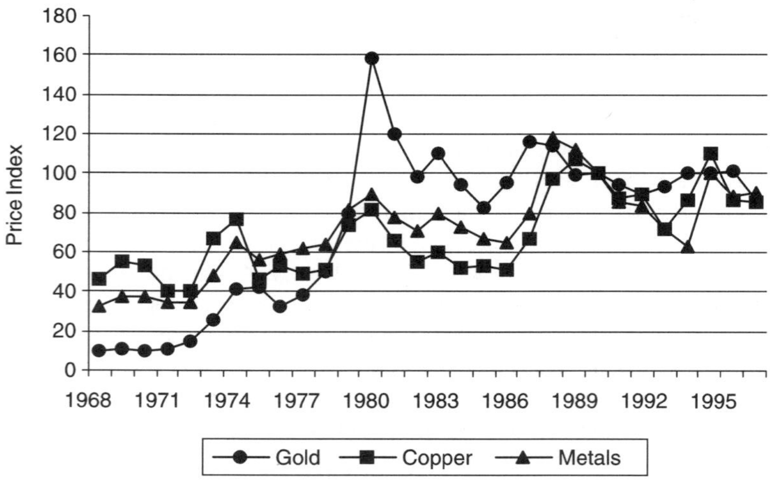

Figure 22.2 Indices of market prices for gold,[5] copper[6] and base metals[7] (1968 to 1997)[8].
Source: Maponga, 1999.

[5] US$ per fine ounce.
[6] US cents per pound.
[7] Includes copper, aluminium, iron ore, tin, nickel, zinc and lead.
[8] 1990 = 100.

44% of mineral exploration budgets in Australia and South Africa, 65% in Canada, and 55% in the United States. For Australia, the share rose from 11% in 1979/80 to 64% in 1996/97, purely because of price increases and technological developments. The international boom was reflected in Zimbabwe by the rise in the number of alluvial panners working the country's rivers during the early- to mid-1980s. As noted earlier, a rise in *garimpeiro* activities in the Amazon occurred at the same time. The boom popularised panning due to high returns.

The economic contraction experienced since the mid-1980s explains, to a great extent, the reasoning behind the growth of both legal and illegal artisanal and small-scale mining in Zimbabwe. High unemployment, a symptom of economic decline, has been one of the major causes of the rise in alluvial gold panning, as high school graduates, the jobless, underemployed, and retrenched workers seek a means of survival. The introduction of the World Bank-prescribed Economic Structural Adjustment Program (ESAP) in 1991 marked a major turning point in the fortunes of the economy. Despite all of its good intentions, targets set for the programme were not met, as liberalisation resulted in increased budget deficits, high interest rates, low economic growth rates, high inflation and an increasing domestic savings gap. Unemployment rose from 7.2% in 1988 to an estimated 64% in 2002. During the same period, overall economic growth declined from 5.9% to −8.5%. Inflation rose from 22.6% in 1995 to 87.5% in 2001 (it is currently estimated to be around 135%), during which time the government budget deficit increased from −Z$7.7 billion (US$140 million) to −Z$45.4 billion (US$810 million). The high level of government expenditure and low growth in income reduced the domestic savings gap from 21% of GDP in 1995 to 7.4% in 2001. These economic problems forced retrenched workers to turn to panning as a safety-net.

The removal of subsidies was one of the major prescriptions of ESAP, and this led to an increase in the cost of agricultural inputs – items that the Government had a deliberate policy to subsidise in order to promote small-scale agriculture. The rise in input costs resulted in communal and resettled farmers diversifying into panning in order to generate supplementary funds to finance crop farming. An economy based on illegal alluvial gold panning soon evolved in many rural areas, as subsistence farming dependent on natural rains suffered negatively from intermittent droughts and from the rise in costs of fertiliser and seed; illegal panning provides a supplementary source of income for rural communities. Farmers operating in poor marginal areas were the first to move into panning.

Zimbabwe's economic fortunes are intricately linked to the performance of the agricultural sector. This is amply demonstrated by its high contribution to GDP, foreign currency earnings, raw materials for the manufacturing sector, and employment. Many Zimbabweans rely on the sector for sustenance as farmers, farm workers, and processors of agricultural products. The sector directly and indirectly employs 60% of the total labour force, and accounts for 20% of GDP and 46% of total exports.[9] Boom and bust cycles in the sector are reflected

[9] Zimbabwe Background Paper, Paper Prepared for FANRPAN, Ministry of Lands, Agriculture and Rural Resettlement, October 2002.

in overall economic growth patterns. For example, the drought periods of 1991/92 and 1994/95, during which growth in the agricultural sector was −23.2% and −7.6% respectively, coincided with GDP growth rates of −5.4% and −1.0%. A GDP growth rate of −4.1% was also experienced during 1999 as a result of a slowdown in economic activities, and declining donor support since 1998. The devastating drought experienced during this period negatively affected communal resettlement and commercial agriculture.

As the push and pull framework in Figure 22.1 shows, communal and resettlement farmers, their employees and families have been forced into illegal gold panning for survival. Prior to the droughts of 1981/82, 1991/92 and 2001/02, panning was primarily a dry-season activity occupying farmers after the end of each agricultural season (Maponga & Ngorima, 2002; Hollaway, 2000). During rainy seasons, there is little time for panning, and the high rivers make panning dangerous. The 2001/2 agricultural season saw an increase in illegal panning due to poor rainfall in Sub-Saharan Africa, which caused total crop failure in many parts of the country. Furthermore, the devastating effects of cyclone *Eline* destroyed crops in many parts of Zimbabwe during 2001, forcing many to seek alternative forms of economic livelihood. The thesis is that, as returns from agriculture have declined due to poor rains, communal and resettled farmers, and farm employees have turned to panning as a survival strategy to reduce income instability.

Panning has influenced labour utilisation on surrounding farms. As reported by Maponga (1997), some commercial farm-owners have switched to seasonal employment, unable to maintain a full labour force during poor seasons. Farm workers pan during the off-season to support consumption patterns and smooth out peaks and troughs in labour utilisation on the farms. Proceeds from alluvial panning reduce insecurity from uneven incomes, and enable farm labourers to meet continuous consumption requirements. This study shows that with declining employment opportunities, the complementarities between panning and agriculture have made panning a major source of growth for communal and resettle-ment agriculture (Maponga & Ngorima, 2002). The survey showed that both illegal and legal panners had an agrarian background, further confirming that the poor agricultural seasons have a direct effect on increased panning activities. Yet, competition for labour between agriculture and mining becomes an issue, as panning negatively impacts on commercial and resettlement agriculture by drawing labour away from farms. Despite this, farm labourers are able to earn a living during poor seasons.

Communal agriculture in Zimbabwe benefits from farming inputs derived from formal employment in urban areas in the form of remittances. Left on its own, communal farming generally cannot generate enough surpluses in the form of seed and fertilizer to recycle into the next season (Maponga & Ngorima, 2002; Boas, 2001). It is truly a subsistence sector generating limited surplus revenue, and injections of financial resources from other sectors help to kick-start the new agricultural season. The reduction in government fertiliser and seed packs due to ESAP, coupled with reduced remittances from urban areas, have pressured rural communities into seeking other sources of inputs.

Low barriers to entry and exit result in huge changes in the absolute numbers of miners per period of time. Miners can easily move in and out of panning, depending on agricultural conditions and urban employment. The high density of illegal gold panners along rivers during periods of poor rains and the dry season, is a manifestation of the low barriers to entry. It is easier to mine illegally than to register a claim legally. As one panner put it,

> You do not need much to go into panning, capital investment is next to nothing, you just follow those already panning in order to find prospective areas and before you know it, you are an experienced panner and are making lots of money.

Fortunes of the "prosperity box" in Figure 22.1 have had a magnetic effect, and have been a significant pull factor for many people. In addition, the fortunes of front-runners have had significant positive demonstration effects on people from surrounding areas. A new and strong follow-on phenomenon resulting in an inflow of panners typifying a "gold rush" has emerged along many of Zimbabwe's rivers.

Another important pull factor into gold panning relates to improved knowledge of the existence of a wide variety of ores in diverse rocks. Communal and resettlement farmers have grasped knowledge on the richness of Zimbabwe's geological environment. Small gold deposits known to exist in Zimbabwe render themselves to relatively easy exploitation by artisanal and small-scale means. Moreover, some panners are former mine workers with knowledge of mineral occurrences and associations.

Despite its increasing social and economic importance, its potential for growth and the existence of a regulatory framework, the sector remains, by and large, illegal. It is characterised by an ever-growing parallel market trade in gold. The findings of the study of illegal panning activities along Mazowe River and its tributaries reveal the dynamics of the sector.

SOCIO-ECONOMIC AND ENVIRONMENTAL DYNAMICS OF ILLEGAL ARTISANAL GOLD PANNING ALONG MAZOWE RIVER

Area description

The study was undertaken along the Mazowe River, the location of one of the major panning sites in Zimbabwe. The river cuts through the Harare-Shamva Greenstone belt, which is comprised of andesitic and dacitic metavolcanis rocks of the Bulawayan Group, as well as metasediments and felsic metavolcanics of the Shamvaian group. The Mazowe River lies in the Mazowe subcatchment area, which is part of the Zambezi Catchment area that covers over half of Zimbabwe. Most of the subcatchment that feeds the river under study is underlain by the Zimbabwe Craton. The greater part of the area consists of greenstones that include the Dindi Greenstone Belt, Makaha Greenstone Belt, and Harare-Bindura-Shamva Greenstone Belt. Granitic terrain also covers a fair share of the subcatchment area, while the far north of the catchment area is underlain by Zambezi Mobile Belt gneisses. These rocks are known to produce the greater part of minerals in Zimbabwe, and gold, in particular.

Though most of the area's mining activities are concerned with gold production, nickel mining, dolerite quarrying and pegmatite mining are also prevalent. The alluvial gold mined in the Mazowe Valley and its tributaries is believed to derive from mainly Shamvaian rocks (Maponga & Ngorima, 2002). As noted in other reports, it is possible that the gold originates from the Bulawayan rocks (Maponga, 1997). Tributaries of the Mazowe also erode greenstones and are rich in alluvial gold; its major tributaries are the Nyadiri and Marodzi rivers. However, a number of other tributaries passing through the ancient Archean terrain washing away gold also feed into this river.

The location of the study area in relation to local geology is shown in Figure 22.3.

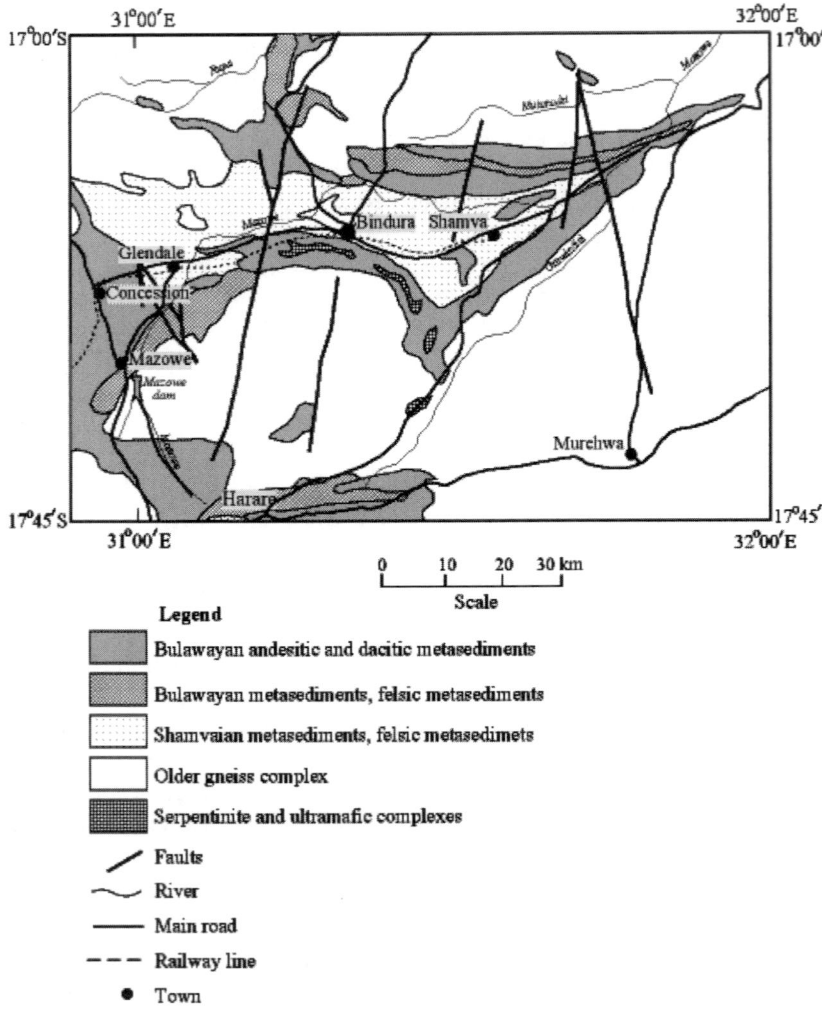

Figure 22.3 Location map of the Mazowe River Valley Area.

Thus, the panning activities along the Mazowe River attest to the extensive gold deposits derived from the greenstone belts in the river's catchment area. Gold discharged from the weathering process in gold-bearing rocks finds its way into water courses. The processes of eluviation and illuviation result in the deposition of gold in the river rubble, riverbanks and flood plains of Mazowe River areas, which are targeted by panners. The geology of the surrounding area provides fertile ground for both legal and illegal alluvial panning (Maponga, 1997).

Mining, mineral processing and marketing

As has been reported by other commentators (Maponga & Ngorima, 2002; Mahlangu, 1992; George-Kumirai, 2002), panners use unscientific methods for prospecting to identify mineable areas. Prospecting methods range from knowledge of spirit mediums, dreams, oral tradition, known previous mining areas, and trial and error. The growth of acacias, occurrence of dry soil and poor herbage, and the occurrence of small clover-shaped herbs are also used as indicators. Test panning along rivers is also used to identify potential areas.

The mining methods used are equally unscientific and depend on many factors, including the nature of the occurrence of gold (alluvial or vein), and the location of the mining area. Methods such as tunnelling for alluvial gold occurring underground, bank undercutting, and trenching, are common. Reef mining involves the selective working of auriferous veins, mainly underground. Twenty metre long trenches, and tunnels as deep as 15 m were observed along the Mazowe River.

Gold recovery methods are numerous and vary from place to place along the river, according to such factors as texture of gold and buyers' specifications. The study showed that 90% of panners used sluice boxes and James Tables of the nature shown in Figure 22.4 to recover free gold. While nuggets are hand picked, fine gold is trapped by napkins and old blankets, which are then washed into panning dishes to recover gold.

Figure 22.4 Appropriate technology: a sluice box and a James table.

For reef gold, ore is first pounded or ground through the rubbing of stones into fine powder, followed by panning using both wooden and metal dishes, and then amalgamation using mercury and heat, usually in an open flame, both indoors and outdoors (Fig. 22.5).

The survey[10]

Panning along the Mazowe River in Mashonaland Central, as is the case along many of Zimbabwe's rivers, has blossomed into a year-round activity, involving informal groups of miners, syndicates, "co-operatives" and families. The research reported in this chapter covered a 20-kilometre stretch of the Mazowe River and many mining settlements along the river's tributaries. A semi-permanent gold-reef mining and panning community located at Tafuna Hill adjacent to Mazowe River was also visited.

A total of 140 randomly selected respondents made up of permit holders, panners tributing claims, and illegal reef and alluvial gold panners were interviewed. All respondents at the Third World Syndicate, a huge panning settlement along the river, recovered alluvial gold from flood plains, river beds and banks of the Mazowe River and its tributaries, whereas 80% of respondents from Tafuna Hill worked on auriferous veins. The sample consisted of 75 male panners of whom 49 were aged between 18 and 35 years. The average ages for men and women were 29 and 33 respectively, and 64% of males were married, compared with 69% of females. The majority were young couples.

Identification of panning settlements along the river is easy as the pole dagga and grass structures stand out boldly. The structures depict poverty based on the quality of building materials used but they mask incomes earned daily as shown by the expensive furnishings, beds, wardrobes and radios found in some of the "shacks". These sub-standard structures exist in communities where miners claim to earn thousands of dollars daily, enough to afford decent houses. Panners argued that because they operate illegally and are migratory, there was little logic in building permanent structures.

Figure 22.5 Burning amalgam in an open area.

[10] A similar survey was undertaken by the same author in 1995.

It is instructive to review the socio-economic dimensions of illegal panning and its interaction with local economic activities in order to evaluate its overall sustainability.

ILLEGAL GOLD PANNING AND SUSTAINABILITY: A SURVEY OF ISSUES AND CHALLENGES

The growth of transitory or nomadic illegal alluvial gold panning has created various challenges and opportunities in regional growth and economic development. The sustainability of the sector has come under intense scrutiny from analysts in recent years due to its impact on the environment. Yet, its social and economic effects are often ignored in such criticisms.

The environmental problems associated with illegal gold panning as outlined in many studies include physical effects such as vegetation destruction, sterilisation of resources, river siltation, and chemical pollution from mercury (Maponga & Ngorima, 2002). Economic impacts include generation of employment leading to higher incomes, savings, investment and provision of adequate health services. However, the social challenges and opportunities emerging from illegal panning have, thus far, been dealt with rather superficially in most studies, despite the important role it plays in the economic emancipation of poor communities in rural areas.

Although the sector has been labelled wasteful of a depleting resource due to high grading (cherry picking), and hazardous to human and environmental health, it empowers economically disadvantaged communities, and has the capacity to enrich nations and regions (Maponga & Ngorima, 2002; Maponga 1997). The low capital requirements of artisanal and small-scale mining, shorter lead times, and the ability of the sector to exploit marginal deposits, make it appropriate for developing nations with large pools of surplus labour and highly prospective geology. Illegal alluvial gold panning presents many challenges and opportunities to social, economic and ecological systems. Figure 22.6

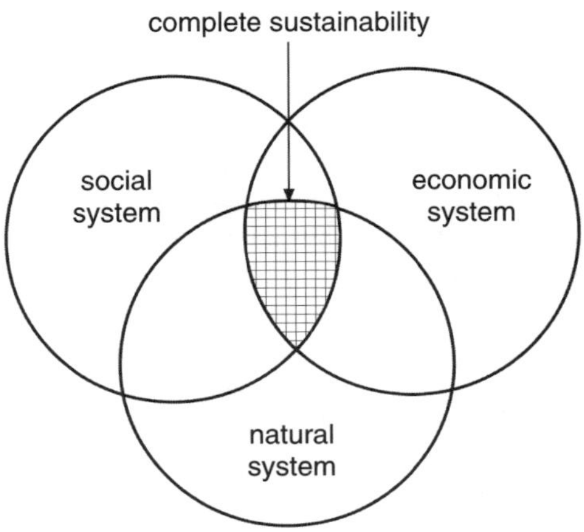

Figure 22.6 The sustainability framework.

adapted from the Mining, Minerals and Sustainable Development (Group) in Southern Africa (2002) report represents the widely accepted view of sustainability, which will be the basis of our evaluation of the sector.

Without dwelling on the sustainability debate, an activity can be described as completely sustainable if it simultaneously satisfies economic prosperity, ensures environmental quality, and promotes social equity. In other words, sustainability is about:

- Living within limits;
- Understanding the interactions among the economy, society and environment; and
- Equitable distribution of resources and opportunities.

However, despite the appreciation of the meaning of the concept among analysts, there is continuing debate over sustainability of a depleting (finite) resource. Although a wasting asset, mining, including panning, is clearly consistent with sustainable development and sustainability (Maponga & Munyanduri, 2001). First, as long as exploration continues to generate reserves to replenish old ones as new material uses evolve, mining is sustainable.[11] Secondly, the development of efficient technology helps to bring lower grade deposits into production (lower cut-off grades). Finally, environmental impact assessment and education, and the utilisation of mining proceeds for social and economic development, makes the sector sustainable (Maponga & Munyanduri, 2001). It is the final front where, in our view, illegal gold panning, though contravening Government laws, has an important social role to play, through resource utilisation at a local level and by providing employment and supporting agriculture. The challenge to sustainability, therefore, is to ensure that revenues from illegal activities benefit local communities.

The following section examines the sustainability of illegal gold panning based on the proposition in Figure 22.6.

SOCIAL SUSTAINABILITY AND ILLEGAL GOLD PANNING

An activity is socially sustainable if it results in the improvement of human health, raises income levels, and improves the living conditions of the poor majority (Traore, 1994). Using these measures of social sustainability as a point of reference, illegal alluvial panning is a significant contributor to this facet of overall sustainable development. Yet, as shown in Figure 22.6, social sustainability is just one facet towards achieving complete sustainability; it is inadequate on its own.

The sector's strongest contribution towards social sustainability is in poverty alleviation, job creation and capacity-building within communities through the acquisition of new survival skills. These are critical areas of sustainable development, particularly in marginalised areas like the Mazowe River valley, where the majority of inhabitants are communal and resettled farmers, whose harvests have been negatively affected by poor rainfall in recent years. Income from the sale of gold contributes significantly to poverty reduction within these communities, and inhabitants use earnings to provide for their families.

---

[11] See also Mikesell (1994) and Tilton (1996).

The creation of both primary and secondary employment opportunities has important wealth distribution effects. Panning is highly labour intensive, and, as such, incomes earned are distributed among many people within the community, which raises the standard of living and has many positive flow-on effects.

The survey showed that on average, panners produce between 0.2 and 1 gram of gold per day. Taking into account the prices paid in Mazowe, during the time of the survey, this translates into between Z$1,000 (US$18) and Z$5,000 (US$89) in income per day for alluvial gold. For gold panned from reefs, output is comparatively higher but prices paid are relatively lower, netting Z$4,500 (US$80) per gram in the same area. The sampled panners indicated that they worked, on average, six days per week, and produced as much as five grams of gold per week. This earns them Z$25,000 (US$446) per week (during "reasonably good weeks"), and is close to the estimated Z$35,000 (US$625) that the average Zimbabwean family required per month for a decent life, at the time of the survey. In an economy where the minimum wage for farm labourers is Z$7,500 per month (US$134), earnings from panning make a significant contribution to welfare. Although production is erratic, findings of this study reinforce recent studies in the Philippines and Indonesia that found that earnings in the sector were as much as 500% and 200% above earnings in other economic activities. As one panner put it,

> Panning is a gamble, you can spend as many as two days without recovering a gram or even a point (a tenth of a gram), yet on some days you can recover as much as five grams of gold.

The estimated national unemployment rate of about 64% has made the sector an attractive provider of sources of livelihood for many unemployed school leavers and retrenched workers. As already discussed, the sector provides both primary and secondary employment opportunities for rural communities and other urban dwellers unable to make a living in towns and cities. The emergence of other social reproductive activities to support the requirements of panners and their families is an additional aspect of the sector's important contribution to economic activities in the study area. However, other analysts have identified underemployment of skilled labour as a challenge within panning communities, as highly skilled workers marginalised by retrenchment are forced into panning.

As already explained, what makes the sector important in employment creation and the creation of sources of livelihood is the ease of entry and exit. The sector is capable of absorbing many people during crisis situations (such as drought) and releasing workers to engage in other economic activities (such as agriculture at the time when the seasonal rains come).

The increasing numbers of panners in recent years are a manifestation of poverty, and an indication of the positive benefits brought about by the sector. Gold panning is a survival strategy to avert hunger and poverty. Panners tell stories of being able to build up both social and physical capital using proceeds from panning. Women panners interviewed alluded to having been able to purchase herds of cattle for the first time, or restock and acquire furniture for their homes using the proceeds from gold sales. The accumulation of capital from panning activities has important implications for the promotion of sustainable rural livelihoods among communities along Mazowe River. Physical capital accumulated through savings from gold income can be converted into monetary wealth during tough times, thereby enabling families to survive. Positive multiplier effects are derived from building core farming capital such as implements (ploughs and cultivators) and stock, and from savings in a typical neo-classical economic fashion. The capital enables farmers to generate more income from crop farming.

The livelihood of communities along Mazowe River demonstrates that agriculture, panning and local economic activities are intricately linked. Incomes earned by panners and the skills learned have helped to "kick-start" community development in the communal areas around Mazowe. This has partially reduced the need for continued government support, improved infrastructure, and created opportunities for other small and medium-scale business opportunities.

Many respondents indicated that they were able to pay school and examination fees, and buy uniforms for children of school-going ages. This is an important contribution to human capital development, and further enhances capacity at a local level.

Families and informal groups work as mining "gangs" in many of the panning operations along Mazowe as shown in Figure 22.7.

An interesting and important gender dimension was evident within panning villages, as women play peculiar roles in these operations. A significant number of women are independent operators who are managing their own "mines". This has strengthened their economic disposition and some boast to earning more than their husbands working in urban areas or on nearby farms. One woman panner had this to say: "We no longer have to look to our husbands for all domestic requirements. We can provide some of them through earnings from panning". Financial empowerment of women has been shown to result in both improved nutritional status of children and care (Maponga & Ngorima,

Figure 22.7 A syndicate working a reef deposit.

2002). Some 46% of panners interviewed were women. In an industry traditionally dominated by men, this figure, which is close to the national average of 44%, shows that illegal panning results in a reasonably fair distribution of income among men and women. Table 22.2 adapted from an MMSD Report compares the number of women in artisanal and small-scale mining in different countries.

A form of division of labour and specialisation is evident in terms of roles played by members of each family in an operation (Fig. 22.8 shows a family

Table 22.2 Artisanal and small-scale mining employment in selected countries.

Country	Number of miners	Number of women in workforce	Share of women in workforce (%)
Bolivia	72,000	15,500	22
Burkina Faso	100,000–200,000	45,000–85,000	45
Ecuador	92,000	6,200	7
Ghana	200,000	89,500	45
India	500,000	33,500	7
Indonesia	109,000	10,900	10
Malawi	40,000	4,000	10
Mali	200,000	100,000	50
Mozambique	60,000	18,000	30
Papua New Guinea	50,000–60,000	12,000	20
Philippines	185,000	46,400	25
South Africa	10,000	500	5
Tanzania	550,000	143,000	26
Zambia	30,000	9,000	30
Zimbabwe	350,000	153,000	44 (46*)

*From survey for this study.
Source: Breaking New Ground, 2002.

Figure 22.8 A family operation along Mazowe.

after a typical day's work). Many factors, including physical disposition and the accessibility of gold, determine these roles. For selective vein mining (reef), men go underground to extract ore, manually hauling it to the surface; women and children pound the ore, fetch water and play a greater role in the amalgamation process. Children tend to assist parents, rather than earning money for themselves. Yet, issues of child labour in these family operations are a concern, as children often work exceedingly long hours (Maponga & Ngorima, 2002; Maponga, 1997). They either miss school, which is commonly located great distances away from panning sites, or, when they do attend school, are often too tired to concentrate fully.

Business opportunities for the benefit of women have emerged within panning villages in the form of water markets and mobile kitchens. Panning has created a market for water for miners working at distances away from rivers. Women and young girls fetch water and sell it to these panners. Mobile kitchens and "beer halls" operated by women were also observed along the river.

Increased income flows have enabled panners to access paid health services from clinics in Shamva and Bindura, and local district clinics. Three quarters of interviewed panners acknowledged an improvement in health since they began gold panning, their incomes enabling them to pay for treatment at clinics. They also attribute improved health, better standards of living, and the consumption of a more balanced diet, to increases in disposable income.

Thus, social sustainability defined to include poverty alleviation, employment creation, gender empowerment and the promotion of education is, by and large, being promoted by the activities of illegal gold panners along the Mazowe River. However, there continues to be a number of pressing social problems. For example, the HIV/AIDs pandemic is a major social challenge nationally and within panning communities. An estimated 2,500 people die of AIDS-related illnesses per week in Zimbabwe. Our survey showed that the temporary mining villages along Mazowe River are mainly inhabited by "migrant" miners, and have attracted a huge population of commercial sex workers. The spread of HIV/AIDS and other sexually transmitted diseases is certainly accelerated in such an environment, particularly because much of the panning population is young and sexually active. The lack of reasonably located and well-equipped hospitals and clinics, coupled with the unavailability of proper sanitary facilities, has created immense social challenges in terms of effective delivery of health services. More specifically, there are very few decent sanitary facilities at most panning sites. Shallow pit latrines are used by the majority of panners; use of the "bush system" is also common. The study reconfirmed findings of a 1995 survey in the same area, which observed that sanitary facilities were substandard [20]. According to this survey, 95% of illegal miners use shallow pit latrines as toilets (98% in 1995 study); 60% drink untreated water (32% in 1995 study), of which 35% drink river water (27% in 1995); and 25% consume water from open wells (21% in 1995). Disused mine shafts are also used as sources of drinking water.

The widespread use of mercury by the panners for final gold recovery is another major health hazard within the study area. As already explained, panners handle mercury without protective clothing and they burn amalgam in open areas; this is hazardous to human health (Boas, 2001; Mutagwaba & Hangi, 1995). The study observed that miners also roasted ore indoors, which results in inevitable inhalation of mercury fumes. Although no studies have been undertaken to determine potential mercury poisoning in the area, adverse long term health effects are a reality among panners. For example, studies of the Umzingwane River, where panning is also rife, report mercury levels of 525 μg/g in river water and 50 ng/g in soils away from the river. Water collected from near gold recovery sites along the Mazowe also exhibited unusually high levels of mercury (200 μg/g). The disposal of used mercury increases the risk of poisoning within the food chain, as panners draw water for cooking and drinking, catch fish for food, and domestic animals and aquatic life are left exposed.

Human health and safety dangers are aggravated by the dangerous working environment prevalent in panning areas. Protective clothing, including basic wear such as boots, was absent from the surveyed panners. Accidents due to the collapse of underground workings are a common occurrence and fatalities are high due to unplanned tunnelling, which induces the collapse of tunnels and pits. Disused mine shafts are a danger to other miners, surrounding communities and their livestock. In 2002, the local media has reported about fifteen deaths related to collapse of disused mine shafts. Poor ventilation in tunnels is another health problem faced by panners, who typically use candles for lighting purposes, which, in turn, causes respiratory problems.

In a manner consistent with self-employment, panners overwork themselves to enhance output. The study showed that on average, panners have a ten hour, six day working week. In the study, 33% of the panners worked between nine and twelve hours a day; this endangers health and contributes to fatigue-related accidents. Panners indicated that they work around the clock during lean times.

The human settlements that have mushroomed along the banks of the river have created social and environmental problems.

Pole, dagga and grass thatch houses are a common form of dwelling along the Mazowe River. Several issues challenge the sustainability of these settlements. First, the use of timber as a primary construction material, has resulted in mass deforestation. Secondly, the nomadic nature of activities means new settlements have to be constructed, as panners move to new panning sites. Thirdly, the location of homes is haphazard, as some dwellings are based along the banks of the Mazowe River. Makeshift homes offer very little protection from rain and wind; miners' accommodation is sub-human in the majority of cases.

ECONOMIC SUSTAINABILITY AND ILLEGAL PANNING

The ability of an activity to change society's consumption patterns and increase disposable income is one of the measures of economic sustainability. Other components include increased urbanisation of a community, and the request for urban-like services. These are some of the positive outcomes of increased incomes from illegal panning. The study noted increased income and expenditure linkages with nearby urban areas such as Shamva and Bindura. Tuck shops and commercial-like services offering different goods have also emerged around the panning villages. At one large panning village visited, 25 small shops and stalls selling various goods were counted. Also in existence were vegetable stalls, meat selling points, shoe repair shops, and second-hand clothes dealers. A "small shopping centre" has also evolved to service the needs of illegal panners and take advantage of increased incomes. Beer drinking is the most common form of recreation; beer selling points have therefore sprouted at most panning sites along the river. Sellers of various types of merchandise from as far as Bindura, Shamva and Harare, visit panning villages to sell their wares. Such economic linkages have significant ripple effects not only locally but also in the source areas of the vendors.

The increased income enhances the purchasing power of communities and enables them to change consumption patterns much more readily. Gold output as high as 20 grams per month from alluvial material was declared by some panners and at current prices paid by buyers on site; panners earn Z$100,000 (US$1785) per month. Such income levels are several times more than the current minimum wage of Z$8,000 (US$143) per month among industrial employees and close to salaries earned by University lecturers at State Universities in Zimbabwe, excluding other employee benefits. Although gold output is highly variable, as it depends on many factors, including luck and experience, some panners reported output as high as 50 grams per day on exceptionally good days. Because of its labour intensive nature, panning has higher employment multipliers compared to large-scale operations, which are more capital intensive. As a result, income distribution is fairer within panning sites.

The benefits of panning were amply demonstrated by one illegal panner, who noted:

> Panning seasons are much shorter than crop seasons. Even during the rainy season, when the river is high, we will pan, at selected points. Yet, with crop farming, it is a waiting period of four months to harvest whereas panning is instant cash, its like lotto.

The challenges to economic sustainability from illegal panning relate to local utilisation of financial benefits, which results in enclave development and "cherry picking" among miners. Although some panners confirmed to having been at the same site for more than four years, no visible local infrastructural development is occurring around their mining areas. In fact, some locals complain that resources are siphoned by "foreigners" providing very little benefit to the local community. Economic and social linkages are not particularly strong (except through retail trade). Panners often complain that owners of the small convenience shops are, in the majority of cases, non-locals. They argued that instead of benefiting them, the rise in incomes in panning areas contributes to a rise in prices of commodities to the disadvantage of locals not involved in panning altogether.

Conflicts between miners and communities are common along the river and within communities themselves. The conflicts revolve around regional self-determination and control over resources, land use, cultural survival, and pollution and land degradation.

Land disputes between illegal panners, claim owners and farmers are a major problem. Farmers along the river valley complained of illegal panners encroaching onto their farms, especially those working on narrow vein deposits, who continue following the narrow veins. Disputes also occur among panners on claim locations, especially where claims are sublet. These disputes provide illegal panners with the opportunity to "invade" panning sites and cause more damage, as they are not answerable to anyone. The study noted that for legal panners, claim owners did not monitor work by tenants; their interest was in output only. This created an opportunity for illegal activities.

Mining villages have their own cultures, values and norms, which very often conflict with those of the local communities; this can also be a source of serious conflict. The nomadic nature of panning activities results in a temporary disruption of normal activities in an area. In this study, representatives of villages around the panning sites complained about a rise in prostitution in the area. They blame it mainly on the sprouting of transitory villages. High levels of unemployment and the lack of opportunities in the villages results in teenage boys and girls moving into the villages, after which, prostitution becomes widespread. Disease soon becomes a major problem in the area. Internal conflicts are also fuelled by drunkenness, and are further aggravated by the fact that most panners are not locals, and, as a result, are viewed as benefiting from a resource that they are not "entitled" to.

The use of uneconomic mining and recovery methods is one of the major challenges to achieving sustainability in the sector. Reef mining is highly selective, and targets only areas visibly rich in gold; as a result, gold locked in pyrite may be lost in tailings or even sterilised in waste. Panning becomes wasteful when only free gold is recovered. The commonly used method, gravity concentration, is not the most effective of systems, as recoveries are low, which results in valuable material being lost in tailings. Sterilisation of the resource usually occurs, as future recovery from tailings may be more costly. Efficiency in extraction, therefore, impacts on income generation capacity, sustainability of the resource, and the environment. The causal relationship in most cases is that a rise in recoveries results in higher incomes, higher standards of living, and improved environmental management.

Although gold exports are an important source of foreign exchange for the nation, revenues lost to the parallel market have continued to grow, with the mushrooming of the underground gold economy (Maponga & Ngorima, 2002). The other challenge emanates from the market, which is exploitative and takes advantage of poverty. In most cases, a buyers' market is created because panners are desperate to sell gold.

ILLEGAL GOLD PANNING AND ENVIRONMENTAL SUSTAINABILITY

Notwithstanding its economic importance, gold panning has been the subject of intense debate among policy makers and academics, mainly because of its adverse environmental effects. In fact, environmental sustainability is the sector's biggest challenge. Environmental sustainability emphasises equitable and sustainable use of environmental and natural resources for the benefit of current and future generations. It encompasses the following five facets:

- Promotion of sustainable agriculture;
- Promotion of rural development;
- Combating deforestation;
- Conservation of biodiversity; and
- Protection of the atmosphere.

Illegal gold panning challenges sustainability from each of these perspectives; using the results of the survey, this section of the chapter qualitatively evaluates the sector against each of these measures.

The discussion on social sustainability dwelt, at length, on the positive impact of illegal panning on agriculture. This chapter has demonstrated that illegal panning is complimentary to communal and resettlement farming along the Mazowe River valley, providing incomes to lubricate crop farming. As an alternative economic base from which resources for agricultural inputs are derived, illegal panning enables farmers to survive from one season to the next, independent of the fate of the previous season. In a narrow sense, agriculture is being promoted. However, the issue of competition for labour between lucrative illegal gold panning and crop farming may also affect agricultural output, food security and sustainability.

The environmental issues related to illegal gold panning are the primary sore point in the sector. Both the primary and secondary activities of panners challenge environmental sustainability, especially with regard to the use of forest and water resources.

Prospecting and mining involves the clearing of vegetation, which results in damages to the banks and beds of rivers and tributaries. Undercut banks and trenches across river banks are visible along the drainage system in the study area. The loosened soil from the river banks becomes susceptible to erosion, and is discharged into the river. A typical sight along the Mazowe River valley is shown as Figure 22.9.

The widening of river channels reduces energy within the river and its tributaries, resulting in siltation. The loosening of material and vegetation removal in the river catchments are major contributors to erosion. Siltation and the lowering of the water table have resulted in the drying out of various portions of the rivers and their tributaries. For example, the Musambanyama and Zvitokwe rivers are almost extinct because of panning activities (Boas, 2001).

The pits, tunnels, and trenches created by panners are dangerous for both animals and human beings as they develop into gullies and dongas with time. Final recovery sites are littered with tailings dumps, which cause mudflows into the rivers during the rainy season; these destroy the underlying vegetation. Figure 22.10 shows remnants of selective mining of auriferous veins.

Another indirect environmental consequence of illegal panning is mass forest destruction; extensive cutting occurs to provide material for houses, fuel wood and material for making panning dishes. Even though metal dishes are

Figure 22.9 A panned river bank.

Figure 22.10 Remnants of selective mining of auriferous veins.

also used by panners, wooden panning dishes are an important tool in the recovery process. However, a single panning dish requires a tree of about 40–50 cm diameter, and thus selective tree cutting occurs to satisfy demand. Our survey showed that a panning dish has an average life span of three years, which has created a panning dish industry.

The clearance of vegetation exposes the surface to soil erosion, and excessive silt loads are discharged into the river. Stretches of the Mazowe River have been heavily silted. Yet, the adverse impacts of panning depend on the mine site's susceptibility to agents of erosion and the distance of panning activities from the river.

However, land degradation is immense within panning areas, but it is difficult to apportion accurately the extent of destruction, for poor farming and over stocking are also contributory factors.

IS ILLEGAL GOLD PANNING SUSTAINABLE?

Does illegal gold panning undermine environmental, economic and social systems? As argued in this chapter, none of the three facets is complete on its own. The answer, therefore, depends on priorities. If poverty alleviation is the priority, which it is in our case, then panning is indeed sustainable as both an income opportunity and a source of livelihood.

Although illegal panning exists outside the law, a lot can be done to improve the effectiveness of the current legislation to ensure that illegal activities are, at least, sustainable. The legal framework governing mining in Zimbabwe includes, among other relevant pieces of legislation: *Mines and Minerals Act* (Chapter 165), *Mining (Alluvial Gold) (Public Streams) Regulations* (1991), *Mining (Management and Safety) Regulations* (1990), and the *Natural Resources (protection) Regulations* (1975). The promulgation of the *Panning Regulations* removed major structural constraints to the growth of the sector, and afforded rural communities the opportunity to participate in formal economic activities through extraction of a local resource.

The drafting of these regulations was an attempt to empower rural councils to manage their own resources through decentralisation and the transfer of power and decision-making to Rural District Councils (RDCs). This was intended to enable RDCs to manage resources at a local level and enjoy resultant benefits. Theoretically, decentralisation allows for better planning of the use of physical capital, since the same community bears the costs of resource misuse and any associated negative externalities. The strategy was based on the premise that local commitment and ownership could be effective drivers of local resource utilisation and policing. The regulations are an attempt to introduce private property rights among panners and within the council, but the majority of panners still operate outside the law. As noted in this chapter, those that operate within the law and side by side with illegal miners, seldom observe regulations, as there is little motivation to do so.

Illegal panning has continued to grow outside, and alongside, these regulations, which have demonstrated that control is almost impossible unless secure and enforceable property rights are in place; this is what the sector needs.

Our view is that at a minimum, improving the capacity of RDCs to discharge their function as stipulated in the regulations, the gold marketing systems, and the provision of extension services, are areas in need of urgent attention. Each is examined in turn.

Capacity building in Rural District Councils (RDCs)

This study has demonstrated immense need for capacity building at a local level for the purposes of instilling a sense of ownership. Organisational, administrative, financial, and circumstantial capacity – keys to ensuring that regulations are followed and illegal panning is minimised – are lacking at RDCs. Moreover, RDCs lack the capacity to undertake the responsibilities stipulated in the regulations. For example, they have not been empowered to cancel previous permits when minimum conditions are not met. This observation raises the question of whether or not mining experts are required at RDCs, or if training could be part of capacity building in the councils to ensure that they are able to police conformity with the legal provisions of the *Panning Regulations*.

Other analysts have observed violation of regulations by RDCs. More specifically, it was discovered that permits were being issued to non-residents. This creates monitoring problems, as such panners are able to easily move to new areas before environmental rehabilitation conditions are met. The lack of a central databank or panning claims registry, means that panners violating regulations can easily register in other areas.

Although in theory, the legal environment provides entry points through which people are able to establish greater control over their lives and livelihoods, an effective system should link rights to obligations. The adage that communities can only become self-reliant when they control access to their resources, receive benefits and revenues, actively invest in their resources, and are involved in their management, can be fulfilled. In this case, self-reliance becomes the basis for true empowerment.

Gold marketing strategies

As noted elsewhere in this chapter, most of the gold produced by illegal miners is lost to the parallel market. This, in itself, is a threat to sustainability, as earnings in the parallel market deny the country of potential export earnings. A system that ensures that all gold produced is sold to Fidelity needs to be devised through a combination of attractive prices and conveniently located buying points. Artisanal miners are subsistence in nature; they require a market at the "source" because the small volumes of mineral they produce make it uneconomic to travel to Harare to sell gold. Thus, the parallel gold market has evolved naturally.

Our view is that the government stands to gain more from accessing all gold lost to the parallel market. If the estimated annual loss of about 10 tonnes of gold to the parallel market is correct, by plugging this hole, the economy could reap massive benefits. The parallel market surfaces largely because of the low prices offered officially. Another indirect loss of government revenues emanates from the non-payment of taxes by illegal miners and those who buy the gold. This, in turn, denies the government of a source of revenue to support national programmes for social development.

A gold support price, similar to that in place for larger formal producers, is also needed to collect all gold produced. Alternatively, RDCs could be allocated alluvial concessions, and be empowered to collect royalties or act as purchasing agents for Fidelity. Collected revenues or economic rents can be channelled into other productive activities for the benefits of the local economy. Artisanal mining could then become a catalyst and an anchor for other productive activities, as it would stimulate growth, including infrastructural development within regions.

Small-scale mining services centre

Several studies have identified financial and technical constraints as major problems faced by the sector; these could be effectively addressed through mining banks and extension services. Efficiency and productivity are keys to enhancing the sustainability of illegal panning. These factors impact on income generation capacity, the sustainability of the resource, and environmental management. Mobile small-scale mining service centres can assist miners increase throughput, and hence, help to overcome productivity constraints. Increasing mineral recoveries also minimises high grading, which impacts sustainability by reducing the income earned by miners and wasting important resources through sterilisation. The provision of extension services facilitates integration of national economic activities and the development of complimentary services.

As long as it remains subsistent, illegality will remain a common and permanent feature of gold panning. A services centre addressing constraints faced by the sector would enable panners to develop a legal and permanent source of livelihood. If those keen on mining become full-time miners, and leave farming to farmers, the total benefits would be immense.

The success of the current land reform exercise has important lessons for the growth and development of the small-scale and artisanal mining sectors. The programme has seen the government committing significant financial and manpower resources to assisting the newly resettled farmers. A similar approach could be used for both legal and illegal gold miners.

Environmental management

As argued in this chapter, the largest blemish to the sustainability of artisanal mining stems from its relation with the environment. Collection of royalties by RDCs presents opportunities for creating an environmental management fund, from which rehabilitation resources could be drawn. A community approach to environmental problems is a potential solution to mercury problems, for example. Output from individual producers may be too low to justify, economically, the acquisition of retorts but as a group, miners could invest in "community retorts". RDCs could, for example, set up these centres using royalties collected from panners. However, this system only works if panners are registered, which makes the legalisation of activities all the more crucial.

CONCLUSIONS

This chapter has demonstrated that, illegal gold panning in Zimbabwe is not a police issue and cannot be solved by periodically rounding up illegal panners and making them pay fines. It has grown to become a part of the social structure, playing an important role in rural development, and thus needs to be supported to strengthen its role as a source of sustainable livelihood. Arresting illegal panners succeeds in creating an underground panning sector where proceeds are lost to the parallel market; this inadvertently works to further disadvantage the nation.

The chapter has argued that, despite its largely illegal nature, gold panning, when undertaken in a socially, economically and environmentally responsible manner, presents unique opportunities for enhancing regional economic growth and development. Its proper integration into national development programmes is important for complete benefits to be realised. Legalising gold panning is part of the process to ensure that the activity proceeds in an orderly manner, the delegation of responsibilities to RDCs is perhaps the other important step. Yet, such delegation without commensurate financial and human resources to the councils, has proved ineffective thus far. To be effective, powers and accepted rights should be enforced by a flexible mechanism.

To achieve sustainable economic growth and improve the livelihoods of rural communities, a system that captures all income generated by the sector and channels it to important areas, is needed. The loss of gold to the parallel market needs to be averted through improved prices and flexible buying arrangements.

Low levels of technology result in poor recoveries and this compromises production efficiency, and hence, sustainability. The chapter has argued that the provision of mobile extension services to improve mining, processing technology, marketing and environmental management is required. Artisanal mining has to be transformed into an efficient industry.

The study showed that an unbalanced relationship exists between those who benefit, and those who bear the risks. Whereas the local community bears the full costs of environmental degradation, non-locals reap the positive social and economic benefits from panning. Such a relationship poses the biggest social threat to the activities of panners along the Mazowe River. Migrant miners upset the social structure, and the culture in panning villages "pollutes" local systems.

In conclusion, the illegal nature of panning in Zimbabwe demands that a social approach, which considers the conditions under which miners operate and the factors leading to the growth of the sector, be adopted. Contrary to the view that as a poverty-driven activity, the emergence of other economic forms of livelihood will result in panners abandoning their trade, panning has become an integral part of the Zimbabwean mining system. Although numbers may fluctuate, panning is a recognised source of livelihood for many Zimbabweans. A system that enables illegal panners to graduate into sustainable sources of livelihood is needed since the sector leverages development and generates wealth. The challenge, therefore, is as follows: help, entice or convince the participants of the sector to operate legally; become more productive; and to be less polluting and more socially beneficial.

REFERENCES

Boas, R.C.V. (2001). Mercury in Brazil as a result of Garimpo Operations. In R.C. Boas, et al. (Eds.), *Mercury in the Tapajos Basin* (pp. 9–30). CNPq/CYTED: Rio de Janeiro.

Breaking New Ground (2002). *Mining, Minerals and Sustainable Development (MMSD) Report*, International Institute for Environment and Development, London.

George-Kumirai, J. (2002). *An Assessment of Factors that constrain the Implementation of S.I. 275 of 1991 (Mining (Alluvial Gold) Public Streams) Regulations*, by Rural District Councils in Zimbabwe: The Case of Insiza Rural District Council. Unpublished Master of Policy Studies Dissertation, University of Zimbabwe, Harare, 58 pp.

Ghose, A.K. (1993). New configuration for small scale mining for developing countries. In A.K. Ghose (Ed.). *Small scale mining: a global overview* (pp. 29–42). New Delhi: Oxford.

Heemskerk, M. (2001). Do international commodity prices drive natural resource booms? An empirical analysis of small scale gold mining in Suriname. *Ecological Economics,* 39, 295–308.

Heemskerk, M. (2002). Livelihood decision making and environmental degradation: Small scale gold mining in the Suriname Amazon. *Society and Natural Resources*, 15, 327–344.

Hollaway, J. (2000). Lessons from Zimbabwe for best practice for small and medium scale mines. *Minerals and Energy Raw Materials Report*, 15(1), 16–22.

International Labour Office (ILO) (1999). Social and labour issues in small scale mining. *Report for the Tripartite Meeting on Social and Labour Issues in Small-Scale Mines*, Geneva, 17–22 May.

Mahlangu, T. (1992). *Gold panning in Zimbabwe*. Unpublished BSc Engineering Honours (Metallurgy) Project, University of Zimbabwe. 59 pp.

Maponga, O. (1993). *Small Scale Mining Operations in Zimbabwe, Ottawa, Canada:* International Research Centre Publishing (IDRC).

Maponga, O. (1995). *Gold Panning along the Mazowe River and Tributaries: A SAREC funded progress report.* IMR. Report No. C669, 12 pp.

Maponga, O. (1997). Small scale mining and the environment: The case of alluvial gold panning and chromite mining. In A.J. Ghose, (Ed.). *Mining on a small and medium scale: a global perspective* (pp. 185–211). London: Intermediate Technology Publications.

Maponga, O. (1999). *Offshore Mineral Exploration and Mining Investment: The Australian Experience*. Unpublished PhD Thesis, Curtin University of Technology, Perth, Western Australia, 297 pp.

Maponga, O. (2000). *Tantalite Boom: Prospects and Implications for Zimbabwean Miners.* IMR Report C789, 15 pp.

Maponga, O. & Munyanduri, N. (2001). Sustainability of the dimension stone industry in Zimbabwe – Challenges and Opportunities. *Natural Resources Forum*, 26(2), 113–126.

Maponga, O. & Ngorima, C. (2002). Overcoming environmental problems in the Gold Panning sector through legislation, and education: the Zimbabwean Experience. *Journal of Cleaner Production*, 11(2), 147–157.

Mikesell, R. (1994). Sustainable development and mineral resources. *Resources Policy*, 20(2), 83–86.

Mutagwaba, W. & Hangi, A. (1995). Environmentally sustainable gold mining in the Lake Victoria regions, Tanzania. *Africa Mining* 1995, Windhoek. London: IMM: 423–31.

Neisser, W.E. (1993). *Problems associated with the use of mercury by small-scale gold miners in developing countries*. Unpublished M.Eng. Thesis, McGill University, Montreal, Canada: 104 pp.

Tilton, J. (1996). Exhaustible resources and sustainable development: Two different paradigms. *Resources Policy*, 22(1/2), 91–97.

Traore, P.A. (1994). Constraints on small-scale mining in Africa. *Natural Resources Forum*, 18(3), 207–12.

Veiga, M.M. & Hinton, J.J. (2002). Abandoned artisanal mines in the Brazilian Amazon: A Legacy of Mercury Pollution. *Natural Resources Forum*, 26, 15–26.

23

Small-Scale Mining in Ghana as a Sustainable Development Activity: Its Development and a Review of the Contemporary Issues and Challenges

BENJAMIN N.A. ARYEE

The heightened international interest in the issues and stakeholders of, and the framework within which, small-scale mining has been carried out in recent times, can be traced to the focus created by the 1972 report of the UN Department of Economic and Social Affairs (UNDESA) on small-scale mining in developing countries (Davidson, 1993). The many conferences/workshops/meetings, largely sponsored by multilateral agencies such as the UN and the World Bank, as well as other activities at both the national and international levels (Davidson, 1993; World Bank, 2001), that have since taken place have aimed chiefly at developing and implementing policies for small-scale mining. However, as Davidson (1993) indicates, very little has been achieved in terms of dealing with the issues, and surmounting problems and constraints identified at the 1972 conference, largely because of the piecemeal, and sometimes rather benign, approaches adopted by countries.

Recognizing that little progress had been made globally with regard to small-scale mining issues, the World Bank hosted a seminal meeting on the subject in 1995. In attendance were representatives from multilateral agencies and 25 countries. One of the key conclusions reached at this conference was the need for integrated solutions to the sector's problems, and improved cooperation between various institutions.[1]

---

[1] At subsequent meetings convened by UNIDO and ILO, it was suggested that a coordinated approach is needed if significant progress is to be made in the sector. In response, the World Bank established the Consultative Group for Artisanal and Small Scale Mining (CASM) – a forum that provides a coordinated approach to assessing and addressing many of the problems in this sub-sector. CASM is responsible for developing policy guidelines, providing advice, disseminating best practices and experiences, raising funds and project/program implementation. It examines and funds proposals for assistance to the artisanal and small scale mining (ASM) sectors, according to pre-established criteria (World Bank, 2001).

Davidson (1993) admits that Ghana has committed to improving its small-scale mining activities and enhancing the sub-sector's contribution to the national economy. At the time of regularizing the hitherto illegal small-scale mining of gold in Ghana, it was estimated (Aryee, 1987) that in 1985 alone, over US$15 million in gold was being smuggled to neighbouring countries, and therefore, lost. However, by the end of 1993 – some four years after regularizing the county's small-scale mining sector – a total of 95,000 ounces of gold, valued at US$33 million, had been purchased by official agents.

By following up and cooperating with various international initiatives, and undertaking work on its own, Ghana has made tremendous progress in regularizing its small-scale mining sector, particularly the precious minerals segment. In 2001, output from small-scale mining accounted for 6% and 95% respectively of the total value of gold and diamonds produced and formally reported, compared to the 2% and 40% recorded in 1989 (the year in which small-scale mining was completely regularized in the country).

The main objectives of this chapter are as follows:

- To define small-scale mining in the Ghanaian context;
- To trace the development of the country's small-scale mining sector;
- To propose a framework within which the rationale of the sub-sector's activities can be analysed;
- To highlight the performance and achievements of the sub-sector;
- To review problems, constraints and challenges faced, and the progress made in dealing with each; and
- To review proposals aimed at addressing the issues and challenges facing the sub-sector.

WHAT IS SMALL-SCALE MINING?[2]

As Davidson (1993) and Jennings (ILO, 1999) acknowledge, it is difficult to find a universally-accepted definition of small-scale mining. Both small-scale and artisanal mining have been defined differently in different countries. However, as the UN and Intermediate Technology Development Group (ITDG) definitions quoted below show, small-scale mining is generally defined in terms of a combination of production factors, levels of sophistication with which minerals are exploited, and/or according to investment and employment criteria.

> Small-scale mining is any single unit mining operation having an annual production of unprocessed material of 50,000 tonnes, or less as measured at the entrance of the mine. (UN, 1971); and
>
> Small-scale miners are "poor people; individuals or small groups who depend upon mining for a living; who use rudimentary tools and

[2] Portions of this section of the chapter have been excerpted from a paper entitled Trends in Small-Scale Mining of Precious Minerals in Ghana – A Perspective on its Environmental Impact, published in the Journal of Cleaner Production.

techniques (*e.g.* picks, chisels, sluices and pans) to exploit their mineral deposits" (ITDG, 2001).

According to the World Bank Group, "small-scale mining is largely a poverty-driven activity, typically practiced in the poorest and most remote rural areas of a country by a largely itinerant, poorly educated populace with little other employment alternatives". The World Bank estimates that while some 13 million small-scale miners are operating under harsh and risky conditions with minimal incomes (in about 30 countries), as many as 100 million people could depend on the activity for their livelihoods world-wide (World Bank, 2001).

Mining in Ghana is primarily regulated by the *Minerals and Mining Law,* 1986 (PNDC Law 153), as amended by the *Minerals and Mining (Amendment) Act,* 1994, Act 475.[3] Sections 77 and 75 of Part X of the Law make provision for the special treatment of small-scale mining activities, stipulating that:

- it is to be limited to the citizens of Ghana;
- areas may be designated as small-scale mining areas for prescribed minerals by the Minister; and
- with respect to such areas, the application of any provision of the Law may be excluded or modified.

These provisions apply to all minerals mined on a small scale in the country.

The winning of precious minerals (gold and diamonds) has been the dominant sub-sector of Ghana's small-scale mining industry, its mines accounting for about 67% of the sector's operations (World Bank, 1995). While the winning of diamonds on a small scale was legal prior to 1989, the mining of gold was not. The government, therefore, enacted the *Small-Scale Gold Mining Law* (PNDC Law 218) in 1989, which effectively legalized small-scale mining completely.

According to PNDC Law 218, "small-scale gold mining" is defined as

> … mining of gold by any method not involving substantial expenditure by an individual or group of persons not exceeding nine in number or by a co-operative society made up of ten or more persons (Government of Ghana, 1989a).

The definition includes (i) the generally subsistence activity termed artisanal or "peasant" – using only rudimentary/artisanal implements – as well as (ii) the more sophisticated commercial mines in operation that have relatively low levels of production, and which generally involve limited capital investment. It should

[3] Mirroring clause 257(6) of Ghana's fourth republican Constitution of 1992, Section 1 of PNDC Law 153 vests all minerals in their natural state in the President and the Government on behalf of and in trust for the people of Ghana. The Minister responsible for mines is empowered to act on behalf of the President in matters relating to the Law. The Law also defines a mineral as "any substance in solid or liquid form occurring naturally in or on the earth, or on or under the seabed, formed by or subject to geological process including building and industrial minerals but does not include petroleum … or water" (Government of Ghana, 1986).

be noted, however, that the more sophisticated small-scale mining concession holders/owners usually employ others and may even get contractors, some of whom may be expatriates, to work for them.

The definition stipulated for gold has been widely applied to other minerals, albeit under different legal regimes; whereas gold has been included under the more elaborate provisions of PNDC Law 218, the provision for other minerals exist within the "largely discretionary" Section 77 of PNDC Law 153.

THE HISTORICAL DEVELOPMENT OF THE SMALL-SCALE MINING SECTOR IN GHANA

Gold

Gold mining has been practised in Ghana[4] for several centuries. It became a prominent activity in the late-fifteenth century, when the Ashanti Empire opened up trans-Saharan Arab trade routes to the North of Africa (Bogoso, 2002). The region's gold, which was mainly derived from alluvial materials or nuggets collected from river beds, was bartered for other metals such as iron and brass. Recorded trade with the Moors and the Phoenicians thus predates European incursions, which began in 1471 (Anin, 1990). In his 1689 work, Bossman (cited on page 4 of Anin, 1990) confirmed a predominance of alluvial gold workings in Ghana, making repeated reference to "women and boys in Elmina and Axim" collecting and washing beach sands and gravels with fresh water (to obtain the gold) "after violent rains". From these humble beginnings, the indigenous mining of gold progressed steadily through difficult times to become what it is today.

Writing in 1732, Barbott (cited on pages 4 and 5 of Anin, 1990) also pointed out that, alluvial gold in the form of grains, lumps, or fine dust was won using gravity techniques, and by washing gold-bearing alluvial materials collected from the bed of the Ankobra River. In 1812, Meredith (cited on page 5 of Anin, 1990) affirmed the existence of alluvial workings, and also documented attempts made by native miners to work hard rock gold deposits in local rain forests. He explained that such lode gold mining was performed by excavating "un-timbered" pits measuring up to two feet wide and 80 feet deep until "a dark coloured stone which is interspersed with gold" was reached. The gold was recovered using local implements – namely, hoes and calabashes – as well as imported hammers and chisels; the recovered stones were then ground into powder and washed (Anin, 1990 p. 5).

[4]The ancestors of present-day Ghana were said to have migrated, around the twelfth century, from the ancient Ghana Empire, which was located in some present-day areas of southwestern Mauritania, western Mali and parts of Senegal. Its foundation was laid by the north African nomads – the Berbers – in the late-fourth century, but was subsequently dominated by the Soninkes, a Mande-speaking people living in the region bordering the Sahara. The Ghana Empire's greatness emanated from its wealth generated through its serving as the southern terminal for the trans-Saharan trade route.

In his Annual Report to the British Colonial Office (cited on page 5 of Anin, 1990) in 1889, Governor Sir William Branford Griffiths indicated that "native gold workings" had developed to the point where, occasionally, timbered shafts (rather than pits) of about 200 ft square with windlasses and air shafts were being used at Prestea.

According to Anin (1990), since there was no effective way of de-watering pits and shafts, gold mining was generally limited by the level of the water table; it was therefore more widespread during the dry season, a time when groundwater is at its lowest level. Marked improvements were achieved in the industry in the late-nineteenth century, following the introduction of water pumps. In many cases, shafts in underground mines reached depths of 200–300 feet.

Up until this time, gold mining was generally conducted on a communal basis. The traditional chief on whose land the gold was being mined was generally entitled to one-third of the riches won, and therefore oversaw the organization of the activity (Anin, 1990). In fact, as gold performed a wide range of functions in Ashanti society, it was in the interest of chiefs to efficiently monitor activities. Not surprisingly, certain controls and regulations had been exercised for over two centuries (Mackay & Schnellman, 1987). Because of the vested local interest in resident activities, artisanal gold mining continued to be widespread in Ghana even after the Frenchman Pierre Bonnat, and other notable Europeans, introduced modern exploration and mining methods to the country in the late-1870s. Enterprising natives, who had neither the capital nor technical ability to participate in large-scale mining, were also compelled to continue to operate at the artisanal level. Almost every deposit subsequently worked using large-scale methods had been discovered through the "exploration" works of resident small-scale miners.

The 25 gold concessions that had been granted to European companies in 1884 had increased to 400 by 1888. However, this rampant granting of concessions reduced the availability of prospective land for indigenous miners, which, in turn, precipitated conflict. As Botchway (1990) explains, this was largely because of the desire of the British to control and regulate sources of gold known to be emanating from the 'Gold Coast' (now known as Ghana). Despite the attempts made between 1894 and 1897 to vest waste lands, forest lands, and minerals in the name of the British Queen, respective bills failed to be approved in parliament, in part because of protests from the indigenous population. One noteworthy change was that Part III of an 1897 Bill *did* become the *Concessions Ordinance of 1900*, Section 32 of which made provisions for the rights conferred by concession contracts – *i.e.* Certificates of Validity.

Various legislation that had prohibited or limited certain activities by small-scale gold miners, and had stunted the growth of the industry vis-à-vis the large-scale mining sector, was also implemented in the early-1900s. Notable examples include:

- Cap 93 of 1909 – *Gold Mining Products Protection Ordinance*, which limited indigenous mining to using "native" (artisanal) methods, and prohibited natives from dealing in gold obtained using "European" (modern) methods (labeled "illicit dealing"). This law also made it an offense to undertake goldsmith activities without a license.[5]

- Cap 184 of 1935 – *Mercury Ordinance*, which made it illegal to import, possess or deal mercury.[6]
- *Concession Ordinance* of 1936:.This replaced the 1900 Ordinance. Section 38 stipulated that native mining could not take place without a non-transferable, annual license. Such workings could not exceed 15 feet in depth, thereby rendering artisanal deep pit mining illegal.

Under these laws, the onus of proof of innocence largely fell on the accused. These laws therefore led to the domination of gold mining in Ghana by English-owned mining companies operating on a large-scale, as well as the virtual eradication of the small-scale sub-sector.

Following Ghana's independence in 1957, the *Minerals Act of 1962* was passed, vesting all minerals in the President for the people of Ghana. It chiefly addressed large-scale mining issues but also saved existing mining rights, "customary or otherwise." The preceding *Exchange Control Act of 1961* empowered the Minister of Finance to appoint banks, bodies or persons to deal in gold. However, to stem the smuggling of gold and other minerals, the *Minerals (Control of Smuggling) (Amendment) Act of 1965* was passed, making the buying, selling, or possession of gold illegal, unless under license. Traditional small-scale gold mining was thus branded illegal (see Aryee, 1987).

Notwithstanding their ban, (illegal) small-scale gold miners – known as "galamsey"[7] – flourished, and winnings from illicit operators were mainly smuggled for sale outside of the country through a well-orientated black market. This practice, while causing significant environmental damage and contributing little to national coffers, enriched neighbouring countries, which were exporting gold despite lacking operating mines (Aryee, 1987).

Diamonds

According to the Diamond Marketing Corporation (DMC, 1984), the first recorded diamond find in Ghana was made in February 1919 by the geologists A.B. Kitson and R.O. Teale in the Abomosu stream in the Eastern Region of Ghana, in the-then Gold Coast. Susequently, Kitson and Junner discovered what is today known as the Birim Diamond Field (1920). The Bonsa Diamond Field was later discovered in 1921. Production increased from about 7,000 carats in 1920 to nearly 1.5 million carats by the end of the 1930s, making Ghana the second largest diamond producer by quantity, and fourth by value,

---

[5] While the intention may have been to prevent smuggling from large-scale mines, this law, in restricting the adoption of improvements in mining and processing techniques, and also dealings in gold products by natives, stunted the development of native mining methods.

[6] Anin (1990) notes that licenses under Cap 93 and Cap 194 "were rarely, if ever, granted to natives of the Gold Coast".

[7] "galamsey", from a corruption of the phrase "gather them and sell".

worldwide (Government of Ghana, 1989b). While mechanized diamond winning has taken place in Ghana since the mid-1920s, small operators using simple hand tools and whose operations predate large-scale workings, have had a great impact on the diamond industry, their activities facilitating the discovery of various diamond fields.

Of the diamond fields that have been discovered, it is the Birim Field which is today still being extensively exploited by both the resident large-scale mining company (Ghana Consolidated Diamonds Ltd.) and several small-scale operators.

In January 1963, the government established the Diamond Marketing Board (DMB) as a State Corporation under Legislative Instrument (L.I.) 401, which empowered it with the responsibility of marketing all diamonds produced in Ghana. In 1972, as a result of the promulgation of the Diamonds Decree (NRCD 32) and L.I. 916, the DMB was transformed into the Diamond Marketing Corporation (DMC), and given expanded responsibilities for controlling and promoting the development of the resident diamond industry, as well as the marketing of diamonds in the country. As the Government's *1989 Investment Prospectus* notes, while the country's large-scale producers – *i.e.* Ghana Consolidated Diamonds Limited (formerly Consolidated African Selection Trust Limited) and Cayco (Ghana) Ltd. – continued to sell through DMC, most of the output from small mines was sold to smugglers (see also Aryee, 1987), who paid convertible currency against DMC's payments in overvalued Ghanaian cedis (Government of Ghana, 1989b).

DMC, operating under its mandate, sold the diamonds it purchased through a retail outlet it had established in Antwerp, Belgium up until 1986, when it was officially closed.

Other minerals mined on a small scale in Ghana

Historically, apart from gold and diamond extraction, small-scale mining activities in Ghana have encompassed the communal winning of salt for food and preservation; kaolin for cosmetic and local medicinal purposes; limestone for terrazzo chippings; sand and stone aggregates for the building industry; and other industrial minerals for various domestic purposes. For several centuries, salt was being mined along the coast in places like Ada, Elmina, Saltpond, Apam, Keta, Winneba, Nyanyano and Prampram. These, along with several other smaller locations, have traditionally had the necessary topography, soil types, and rainfall patterns needed for salt production. Ghana, along with Senegal, has therefore had a monopoly over solar salt production in West Africa.

During ancient times, most salt works were communally operated, with each family, gate or section of the local community being apportioned an area for winning. Such works centred on winning/harvesting the salt left behind after the saline water in ponds adjoining lagoons, or the lagoons themselves, had dried up during the dry season. Uniquely different operations occur in the Nyanyano area (located about 25 miles east of Accra), where miners source their brine from hand-dug wells/pits of 3–4 feet wide by 8–10 feet long, and 10–12 feet deep, into which concentrate oozes (Fig. 23.1). The brine is then transferred into

Figure 23.1 Hand-dug salt well containing very concentrated brine (salt solution).

Figure 23.2 Salt crystalizing pans with a hand-dug well in the foreground.

shallow evaporation/crystallizing pans (Fig. 23.2). The shallow pans and the near-saturated brine ensure crystallization of the salt within five days.

Because of the rainfall pattern along the coast, dryer locations towards the east like Keta and Ada have greater potential for salt production. Here, the dry season is particularly lengthy. For example, when the Keta basin area was flooded, the Ada Songhor Lagoon became the single most important source of natural solar salt winning along the coast of West Africa.

It is imperative to note that, while some improved technology has been introduced to gold miners over the years, salt winning has generally retained its traditional artisanal and informal form. This has resulted in a paucity of sector-specific production and employment data.

RECENT DEVELOPMENTS[8]

Regularization of small-scale gold mining

The increasing awareness that the continued marginalization of the small-scale gold mining sector was detrimental to the economy, prompted the World Bank to finance a study that investigated the phenomenon. Section 1.201 of the resultant report stated that, "…The purpose of the study is to recommend institutional and operational measures to improve the contribution of small-scale diamond and gold mining to Ghana's economy" (Mackay & Schnellman, 1987).

Following completion of this study, the *Small-Scale Gold Mining Law*, PNDC L 218, was enacted (May 1989), which effectively legalized and regularized small-scale mining in Ghana. As a result, the small-scale mining of precious minerals (mainly gold but to some extent, diamonds) began to flourish along the northeast/southwest gold belts of the country, and alongside placer deposits countrywide. Licensed miners have since been awarded the right to operate on concessions registered by the government. In a number of cases, such operators have access to extension services, and are generally well organized.

The Small-Scale Mining Project (SSMP) was established in 1989–90 by the government,[9] with assistance from the World Bank and GTZ;[10] the organizations helped to facilitate the regularization process and improve the efficiency of the sector's operations. As part of the Project, eight Small-Scale Mining District Centres (distinct from and generally spanning a number of the political administrative districts) were constructed in Tarkwa, Bibiani, Assin Fosu, Akim Oda, Dunkwa-on-Offin, Kibi, Konongo, and Enchi, each of which was identified as having intense small-scale mining activities (Fig. 23.3). Their functions have been to facilitate the registration and legalization of illicit workings, while educating operators and providing technical extension services. Each is equipped with a 4 × 4 pick up truck and a motorcycle, as well as radio communications and

[8] This section has excepts from a paper entitled "TRENDS IN SMALL-SCALE MINING OF PRECIOUS MINERALS IN GHANA – A PERSPECTIVE ON ITS ENVIRONMENTAL IMPACT" published in the JOURNAL OF CLEANER PRODUCTION; with the author as co-author.

[9] Under the supervision of the Minerals Commission, Mines Department, Geological Survey Department and Precious Minerals Marketing Company Ltd.

[10] Deutsche Gesellschaft für Technische Zusammernarbeit (GTZ) GmbH is the German Federal Agency for Technical Co-operation.

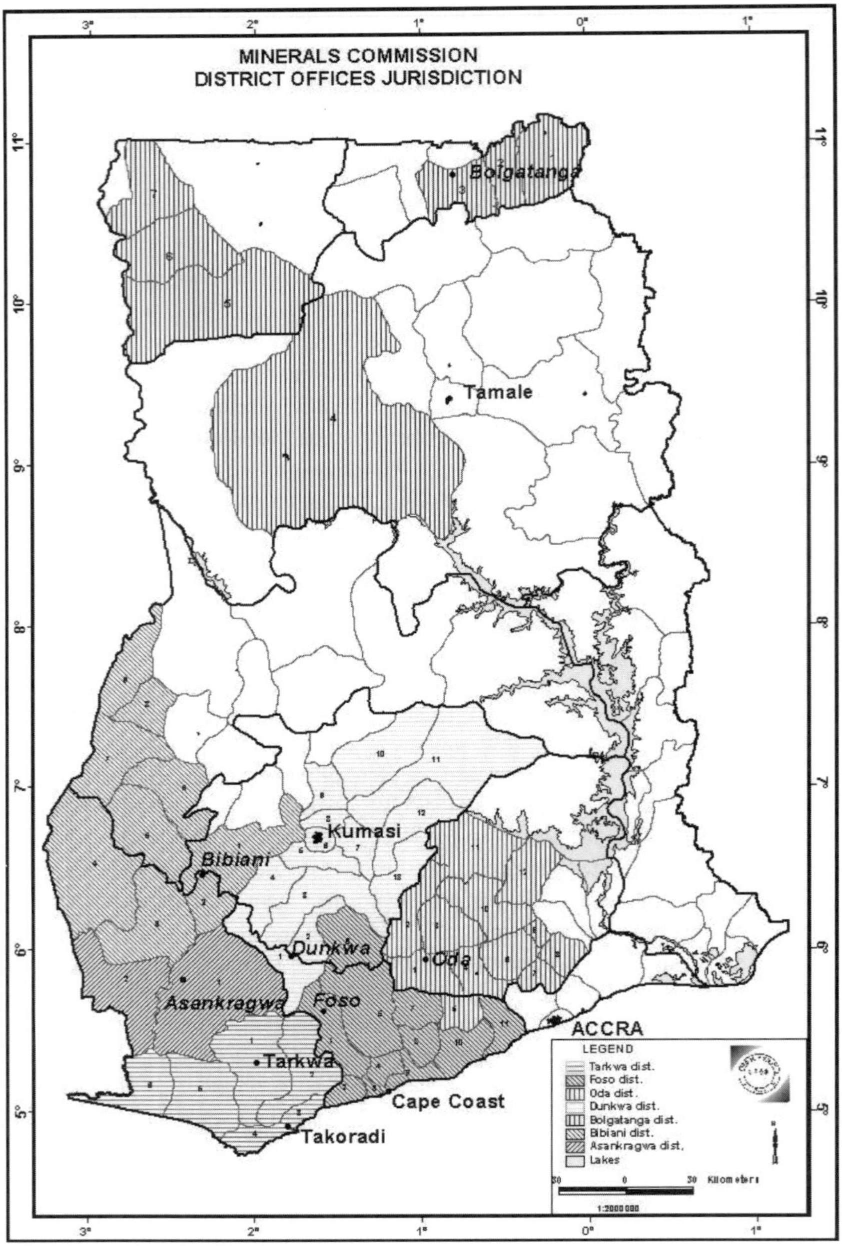

Figure 23.3 Locations of Ghana's seven small-scale mining districts.

other office equipment and supplies, and is manned by experienced mining engineers who tend to small-scale mining areas.

In 1991, after the Project had run for the initial two-year period, it was deemed appropriate to establish a corresponding department. Thus, the Small-Scale

Mining Department was created in the Minerals Commission,[11] and assigned the responsibility of assisting the Minister of Mines in the granting of mining rights, and with the monitoring of activities, in the small-scale mining sector.

The Konongo and Kibi Centres were closed down in late-1991 and early-1996, respectively, due to reduced small-scale mining activity; however, a new Centre was set up at Bolgatanga in 1996 in response to increased local activity. Furthermore, the Enchi Centre was relocated to Asankragwa in 1994, in response to a shift in the concentration of small-scale mining activities. As shown in Figure 23.3, these Centres continue to monitor fairly large areas and provide the following services:

1) the sharing of information, and
2) the provision of advice, assistance and training in best practices for mining, processing, marketing, environmental sustainability, health and safety.

The Centres also serve as a first point of contact for any individual or group of individuals wishing to acquire a license for small-scale mining.

In accordance with the *Small Scale Gold Mining Law* (PNDCL 218), small-scale gold mining licenses[12] may be awarded to Ghanaians aged 18 years and above, with respect of the following areas for the periods specified:

- 1.2 hectares in the case of one person or a group of persons not exceeding four in number (for up to three years, renewable);
- 2 hectares in the case of any group of persons not exceeding nine in number (for up to three years, renewable); and
- 10 hectares in the case of a grant to a co-operative society comprised of ten or more persons and registered companies (for up to five years, renewable).

Notwithstanding the promotional activities that have been implemented to facilitate regularization, there continues to be a large number of illegal workings. These miners, who work without licenses and have no concessions of their own,

[11] The Minerals Commission – organized in 1984 – is the government agency with "overall responsibility for recommending mineral policy, advising government on mineral matters, reviewing mining sector activity, and promoting development."

[12] The process of small-scale mining license acquisition involves, *inter alia*, the submission of a completed small-scale mining application form and ten copies of the site plan of the area being applied for to the Minerals Commission through its district centre. Following the collection of a field inspection report from the district mining centre, a report from the District Chief Executive of the political district in which mining is intended to be undertaken, an environmental permit from the Environmental Protection Agency (EPA), and the payment of the requisite fees, the application is then evaluated, and recommendations are made to the Minister responsible for the granting of a license. If the Minister approves, he signs an agreement between the applicant and the Government of Ghana. The applicant then takes the signed agreement (license) to the Chief Inspector of Mines to obtain an operating license before commencing work on the land. Small-scale mining licenses are subject to renewal after three or five years, depending on the size of the concession.

Figure 23.4 Windlass used to haul ore from underground workings.

operate illegally on concessions leased to large-scale mining companies and/or within the areas prohibited from being mined. They are poorly organized and operate in a haphazard, "hit and run" manner; operators regularly have confrontations with both the security personnel of large-scale mining companies and state law enforcement agencies.

Mining and processing methods

In general, the mining and processing methods currently employed by artisanal miners have not significantly changed over the years. For instance, operators rely solely on traditional/manual methods of mining, still inclined to use simple implements such as shovels, pick-axes, pans, chisels and hammers, as well as windlasses to haul ore from deep pit working (Fig. 23.4); moreover, they still use footholes (though sometimes iron ladders) to access workings underground (Fig. 23.5). However, operators at various stages of progression into mechanized small-scale mining can be found throughout the country. In many cases, companies using fairly sophisticated technologies have been permitted to work with, and support, small-scale miners.

Mining methods

The mining methods employed by small-scale miners of precious minerals vary according to the type of deposit being exploited and the location. In

Figure 23.5 Deep underground pit, showing footholes for accessing workings.

Ghana, the methods used in the small-scale mining, of particularly, gold, can be placed into one of the following three groups:

- shallow alluvial mining;
- deep alluvial mining; and
- hard rock (lode) mining.

Shallow alluvial mining techniques, which are popularly referred to as "dig and wash", are used to mine the shallow alluvial deposits typically uncovered in valleys or low-lying areas with little or no overburden. Such deposits are at depths not exceeding three metres. Vegetation is initially cleared, and soil is removed until the gold-rich layer is reached. The mineralized material is then removed and transported to nearby streams for sluicing. It should be noted that, in view of the relative ease of reaching such deposits and treating such ores, a significant proportion of the industry's operations are of this nature. For similar reasons, illegal workings are predominantly of this type.

Deep alluvial mining techniques are adopted in cases involving the extraction of deep alluvial deposits from the banks of major rivers such as the Ankobra, Tano, Pra, Birim and Offin, as well as along old river courses. Such techniques involve excavating a pit and digging until the gold-bearing gravel horizon, typically located at depths of seven to 12 metres, is reached. The sides of pits are stepped-in terraces or benches to ensure that they do not collapse. The gold-bearing gravel is then removed and sluiced to recover the gold.

Hard rock techniques are deployed when mining gold-bearing reefs, which can be located close to the surface or can be deep-seated. Holes are sunk to intercept the reefs, after which they are worked along the strike. Where such reefs are weathered, small-scale miners use chisels and hammers to break ore.

In cases where ore is hard, explosives are commonly used, despite being prohibited throughout Ghana.

Processing methods

Small-scale gold miners invariably prefer free milling ores to sulphidic ores. Therefore, gravity concentration using sluicing is the main method of processing. Current regulations do not permit the use of cyanidation or other leaching techniques. In any case, these are not likely to be widely adopted by the small-scale miners of Ghana should they be introduced, because of the longer duration needed to recover mineral from ores. In the case of alluvial ores, the traditional ore processing method, which usually yields a recovery rate of about 60%, involves the sluicing of mined material to obtain gold concentrate. Mercury is then added to the concentrate and mixed to form a gold amalgam, which is heated to separate the gold.

When processing hard rock ores, traditional or manual methods featuring artisanal implements are typically used. This largely stems from an unavailability of capital to purchase the requisite crushing and milling equipment to facilitate the process. The manual method of gold extraction from hard rock ore involves "pounding" (crushing and grinding) ores using locally-made metal mortars and pestles. The resultant powder is mixed with water and sluiced to obtain a gold concentrate, which is later amalgamated with mercury.

It is also worth mentioning that, an increasing number of Ghanaian small-scale miners have, in recent years, received financial and technical support from both foreign and local investors registered as "Mine Support Service Companies".[13] This has facilitated a transition from artisanal or traditional mining techniques, to semi-mechanized and mechanized mining methods in some mining districts, particularly, Tarkwa. In cases where the deposit is alluvial in nature, mechanized operations commonly feature sizing trommels and Knelson concentrators, as well as sluice boxes, which are used to process the material. In hard rock operations, on the other hand, jaw crushers, hammer mills, ball mills and modified corn mills are also used for communition.

In the case of diamonds, a mining method similar to the shallow alluvial mining technique described earlier is used. In short, mined material is transferred onto a screen where it is washed. The screened gravel (6 mm) is hand jigged into a concentrate, from which the diamonds are picked.

MARKETING SMALL-SCALE MINE OUTPUT

Provision and streamlining of marketing operations

One of the objectives of regularization is to increase the proportion of gold output from the informal sector being sold through official channels. In 1987, it

[13] Mine Support Service Companies are companies that have been accorded certain concessions in view of the services they provide exclusively to companies operating in the mining sector of Ghana.

was estimated that, prior to regularization, the proportion of production that had been lost via smuggling was of the order of US$15 million (Aryee, 1987). At the time of regularizing the small-scale gold mining industry, the Diamond Marketing Corporation (DMC) was expanded and renamed the Precious Minerals Marketing Corporation (PMMC) under PNDC Law 219. PMMC provides purchasing services for small-scale gold and diamond miners. The organization was also charged with promoting the development of precious minerals and jewelry industries in Ghana. To improve the efficiency of diamond sales, PMMC established regional offices and empowered buyers to purchase diamonds directly from small-scale miners and other individuals who had raw diamonds mineral for sale; it also provides storage and security, and transportation services to the airport for those exporting. For providing these services, PMMC is paid a flat rate of 2% of the value thereof.

Under the new PMMC, gold produced by small-scale miners is sold either at the purchasing centers that have been set up specifically for that purpose, or to Licensed Buying Agents (LBAs), who purchase precious minerals in the field on its behalf.

Competition

The government, mindful of the potential benefits of competition between marketing operations, has indicated its intention to subject the gold produced by small-scale miners to market competition, a strategy in line with the current approach to diamond purchasing. In 1994, therefore, a private company, Miramex Ghana Limited, was permitted to establish a one-year pilot project to purchase and market gold, mainly from small-scale miners. Although Miramex immediately started making purchases judged to be appreciable (see Table 23.1), the rather phenomenal change was with PMMC purchases. Over the last five years, PMMC has purchased, on average, 60,000 oz of gold each year from artisanal and small-scale miners, nearly three times the amount purchased (*i.e.* 18,614 oz) from these parties during the period 1989–1993.[14] However, as a

Table 23.1 Gold Purchases & Exports (oz).

Year	PMMC	Miramex
1994	56,937	35,407
1995	57,612	71,066
1996	54,637	40,574
1997	66,795	26,924
1998	70,287	54,008
1999	76,988	38,795
Total	383,256	266,774

Source: Minerals Commission.

[14] PMMC's purchases continue to increase (2000, 92,990 oz; and 2001, 147,531 oz), though Miramex's subsequent performance has been checked (2000, 52,672 oz; and 2001, 38,065 oz).

result of the licensing of Miramex, there have been marked increases in the volume of gold re-channeled from unofficial to official channels. More specifically, PMMC, mindful of the competition, established centres at Kumasi, Bolgatanga, Tarkwa, Wa, and a central facility in Accra, which better enabled it to increase its market share. Additionally, when monopolist, PMMC offered only 93% of the world market price of gold to its customers, but later increased the proportion to 98% in response to competition. In short, the existence of two competing agencies facilitated an increase in gold purchases, and the provision of higher prices to small-scale gold producers.

The increased purchases were laudable, as the gold was purchased by official agents, which reduced local mineral smuggling considerably. Competition has increased in recent times, although rather cautiously.[15]

Development and registration of mine support service companies

In the early-1990s, the government began to register and extend certain concessions to a number of companies. Specifically, the government's aim was to enable this group of businesses to provide certain value-added support services (*e.g.* geological and mining engineering, consultancy, mineral assaying, contract mining, drilling and blasting, and explosives manufacture) to mining companies – mainly, large-scale operators. Other companies arrived on the scene in the late-1990s to provide services specifically to the small-scale mining sub-sector. These companies provide, *inter alia*, technical (including equipment and other input fabrication), financial, and management services. Some are involved in contracting mining and beneficiation services, particularly for gold. While the ownership of small-scale mining concessions is limited to Ghanaians, companies providing mine support services to small-scale operators are, by law, permitted to be wholly foreign-owned. The infusion of such companies has facilitated increased mechanization in the small-scale mining sector, thus increasing its capacity to undertake hitherto previously near-impossible activities.

Bilateral/multilateral donor assistance

A number of donor-agencies have assisted the operators of Ghana's small-scale mining sector at various stages of its development. Both the World Bank (IDA) and GTZ (the German aid agency) provided technical, logistical and financial assistance between 1986 and 1990, during which time they helped to facilitate improved organization in the sector. Specifically, the World Bank provided funding for an initial study, which precipitated regularization, and also provided some logistics for the embryonic project. GTZ's role centred on technical assistance and training of both project staff and small-scale miners; following

[15] Examples include (i) the registration of other smaller buying companies; and (ii) allowing small-scale miners/Licensed Buying Agents to export their output/purchases through PMMC, if they can amass gold parcels of 50 kg.

organizational changes at GTZ, the provision of certain logistics were passed to BGR[16] in 1994. However, German bilateral assistance came to an end in 1998/99, although funding for an equipment-hire purchasing scheme,[17] which commenced in 1993 with a seed capital of US$87,000 (68% equipment and 32% working capital), stopped one year earlier. The scheme was unsustainable because of high inflation, low patronage and a rather low repayment rate (24%). Nevertheless, a revolving fund is being considered to finance the provision of mercury retorts to small-scale miners to abate mercury pollution.

In 1995, the World Bank, along with the Nordic Development Fund (NDF), co-financed the *Mining Sector Development and Environment Project.*[18] The Project provided funds to the government that were used to strengthen the capacity of mining sector institutions to help them carry out their functions of promoting and regulating investment in the mining sector in an environmentally sound manner. The funds were also used to enhance technical knowledge and promote the use of appropriate equipment and technologies at small-scale mining sites.

EMPLOYMENT AND EDUCATIONAL ISSUES: FEMALE AND CHILD LABOUR IN GHANA'S SMALL-SCALE MINING SECTOR

The various estimates (ILO, 1999; Appiah, 1998) that have been made suggest that employment in Ghana's small-scale mining sub-sector ranges between 50,000 and 300,000 people. The rather wide range of estimates bears testimony to the lack of readily available, accurate and reliable data – a direct result of the furtive nature of a large part of such workings, which are based in rather remote areas. The analysis presented in this section of the chapter is mainly based on literature review of indigenous works,[19] the 1999 ILO Report on Labour Issues (ILO, 1999), as well as the use of an approach akin to the "Delphi" method, which was used to survey the information base at District Mining Centres of the Minerals Commission during July–August of 2002.

Female labour in Ghana's small-scale mining industry

It has been noted by the ILO that the level of participation of women in small-scale mining is generally much higher than in the large-scale sector (See Chapters 11, 12 and 29 for a more thorough overview of the role of women in artisanal and small-scale mining). It is estimated that of the 11.5–13 million

[16] Bundesanstalt Für Geowissenschaften un Rohstoffe (BGR) – the German Federal Institute for Geoscience and Natural Resources.

[17] Operated in the Central, Western and Eastern Regions of Ghana.

[18] World Bank Credit 2743-Gh for US$12.3 million and NDF Credit No. 156 for US$5.5 million.

[19] Mainly two works that concern female participation in small-scale mining in Ghana; one was a study for GTZ, conducted on January 1992, and the other was a paper presented at the World Bank International Round Table on Artisanal Mining in May 1995.

small-scale miners worldwide, as much as 30% could be women, with an additional 1.5–2 million involved indirectly. The Economic Commission of Africa estimates that of the 3–3.7 million employed at small-scale mines continent-wide, between 40 and 45% are women (ILO, 1999).

Offei-Aboagye (1992) indicates that while women represent 51% of Ghana's population, they are less than proportionately represented in small-scale mining (15% of the small scale gold mining workforce and up to 20% of sponsors of work groups). However, the ILO work indicates that a greater percentage of women undertake sieving, washing and carrying duties at clay and stone quarries, and participate in the illegal mining of precious minerals (galamsey). Thus, as much as 50% of the industry is female, along with some 75% of individuals producing salt on a small scale.

Offei-Aboagye (1992) identified three areas of the (Ghanaian) small-scale mining industry in which women are especially active. They are as follows:

- Investors, as either concession holders or as sponsors of work gangs, though not in great abundance because of restricted access to capital, and limited knowledge of opportunities due to a lack of education.
- Employees as sorters, carriers, panners, and gang cooks.[20] Women constitute less than 10% of this group.
- Small business owners dealing in either general goods or cooked food.

Offei-Aboagye (1992) also found that: (i) 80% of the women involved in small-scale mining are below 40 years old; (ii) more than 50% of women involved in small-scale mining are married or are involved in a steady conjugal relationship with men; and (iii) a significant proportion had dependent children, with as much as 59% having children under six years of age.

The work also noted that educationally, most of the women involved in small-scale mining in Ghana do not have a high level of education; 93% do not have secondary education, while 33% are illiterate. Thus, their functional illiteracy level of some 77% is higher than that of men (58%). It was also found that more than 40% of women had to engage in additional economic activities – mainly subsistence farming and trading – in order to support their families (Offei-Aboagye, 1995; ILO, 1999).

Despite facing numerous difficulties and discrimination, more than 50% of female employees and 75% of small business owners found small-scale mining worthwhile, and intend to continue their activities provided that conditions persist. It is clear that while various reasons have been adduced for the participation of women in small-scale mining, the predominant factor is economic – *i.e.* it serving as a fairly quickly realizable means of providing or supplementing the family income. However, the participation of Ghanaian women in small-scale mining has been constrained by a number of factors, including a relative lack

[20] In the exploitation of other minerals on a small scale basis (including quarrying), women are also found engaged in crushing using a variety of implements, including hammers, grinders and sieves.

of, or limited access to, funding, particularly on the basis of credit; a relative lack or limited technical know-how; and discriminatory cultural and socio-economic practices endemic in the typically rural settings in which small-scale mining takes place. Little progress has been made to address these constraints, although affirmative action to address some of these factors, if not all, has been advocated.

Child labour[21] in Ghana's small-scale mining industry

In 1997, the ILO estimated that there were more than 250 million child labourers worldwide (ILO, 1997) but more recently, UNICEF estimated that child labour contingent consisted of as many as 400 million (UNICEF, 1999). However, child labour need not be harmful, and, in some cases, could be beneficial, provided that it does not occur at the expense of education (Blunch & Vener, 2001).

Child labour is found in many vocations, including mining, and is widespread throughout the poorer regions of the world. It should be noted that even where small-scale mining activities are legal, the participation of children may be illegal. In Ghana, only persons who are 18 years of age or older have voting rights, in accordance with Section 42 of the 1992 Fourth Republican Constitution of Ghana. However, as part of the skills imparting process, the Labour Decree of 1967 limits children under 15 years to nothing but light agricultural work or domestic duties. For youths aged between 15 and 18, the law permits some form of employment, but debars such children from working at any mine or underground work.

Child labour is found to some extent in both the legal and illegal small-scale segments of Ghana's mining industry. The small size of children makes them attractive labour for reaching and exploiting hard-to-reach deposits; their cheap (often family members) labour makes them attractive labour for transporting materials, assisting with food preparation/provision, and running errands. With respect to the rationale for child labour, "some children work to augment the family income – on their own volition or urged to do so by their parents; others are orphans or have left home and are working to survive" (ILO, 1999).

Child labour can only be reduced in the small-scale mining sub-sector if the activity can be put on a sounder, more sustainable economic footing, in a way that would positively impact the "family" income. Thus, as the ILO puts it, "merely removing children from (such) workplaces without providing the means for them to go to school and ensuring a compensatory contribution to the family income, will eventually result in them drifting back to the work they left, or perhaps, to worse occupations" (ILO, 1999), along with associated long-term impairment.

---

[21] The literature contains different definitions of "child labour". For instance, Cangarajah and Coulombe (1997) note that, "the literature distinguishes child labour and child work, where the latter is less harmful and probably healthy, and includes helping the household in various chores and household activities. These activities may take place after school hours or during holidays more intensively. ..." Apart from the specific reference to "harmful child labour" which is quoted from Blunch and Vener (2001) as referring to child labour which conflicts with schooling and therefore human capital accumulation, no such distinction is made in this write-up.

It is against this background that the ILO, through its International Programme on the Elimination of Child Labour (IPEC), is pursuing activities to:

- Prevent children from being put to work in small-scale mines;
- Withdraw children from dangerous activities;
- Improve children's working conditions as a first stage towards eliminating child labour;
- Gain a better idea of the living and working conditions of children involved in small-scale mining;
- Make children, parents, employers, private and public institutions and the public more aware of the dangers of putting children to work at small-scale mines; and
- Provide working children and their families with viable alternatives.

In Ghana, it has been established that there is a positive relationship between the distances to primary and secondary schools and harmful child labour. Moreover, though the government has tried to improve access to primary education and legislation to prevent child labour in a bid to address this problem, it has not been very successful since economic activities in Ghana still have active inputs from youths (Blunch & Vener, 2001).

Synthesis

Using the "Delphi" method, district centres were surveyed for estimates of how pervasive child and women labour is in Ghana's small-scale mining industry (see Table 23.2). Table 23.3 also tabulates data on women and child labour, as well as data on educational background.

Apart from Akim Oda, where diamonds are the dominant mineral won, the main mineral mined in all other districts is gold. In Asankragwa, child labour is much more pronounced in illegal activities than at other centres. However, despite the variation of the collected data (*e.g.* eight years at Assin Foso and 19 years at Akim Oda), the minimum age of child miners (13–14 years) was found to be close to the average (of the minimum). Because of the physical nature of the activity, the number of participants aged 55 and older is rather negligible.

Table 23.2 Summary Data on Women and Child Labour in the Vicinity of the Small Scale Mining District Centres of the Minerals Commission.

District	Female labour (%)	Child labour (%)	Minimum child age
Tarkwa	20	2	15
Bibiani	15	5	15
Dunkwa	14	6	12
Asakrangwa	31	10	12
Bolgatanga	25	15	10
Akim Oda	10	1	19
Assin Fosu	5	3	13

Source: Questionnaire/Delphi Exercise, August 2002.

Table 23.3 Data on Women & Child Labour and Education from use of Delphi Method.

A. BIBIANI

Age/gender matrix

Age group	Female	Male	Total
15–30	10%	65%	75%
31–50	5%	20%	25%
Total	**15%**	**85%**	**100%**

Average Age: 30 years.

Education matrix (highest level attained by group)

Male		Female	
Age group	Level	Age group	Level
15–30	SSS	15–40	JSS/MSLC
31–50	Tertiary		

B. ASANKRANGWA

Age/gender matrix

Age group	Female	Male	Total
12–18	0.75%	14.25%	15%
18–30	12%	48%	60%
30–55	3%	22%	25%
Total	**15.75%**	**84.25%**	**100%**

Average age: 35 years.

Education matrix (highest level attained by group)

Age group	Male	Female
12–18	JSS	JSS
18–30	JSS/SSS	JSS
30–55	MLSC/JSS	Illiterates

C. AKIM ODA

Gender/age matrix

Age group	Female	Male	Total
Below 25	3.5%	22.3%	26%
25–35	3.0%	35.6%	39%
35–45	2.0%	17.8%	20%

(*Continued*)

Table 23.3 (*Continued*)

Age group	Female	Male	Total
45–55	0.8%	10.7%	11%
Above 45	0.7%	2.7%	3%
Total	**10.00%**	**89.00%**	**99.00%**

1% Child labour not included; average age: 32 years.

Education matrix (highest level attained by group)

Gender	Male	Female
Highest level	SSS/6th form	Vocational/Comm. Sch.

But typically, MSLC/JSS; and 10% no formal education

D. DUNKWA

Gender/age matrix

Age group	Female	Male	Total
12–25	21%	62%	82%
25–45	2%	12%	14%
45–55	0.23%	4%	4%
Total	**23%**	**77%**	**100%**

Up to 7% Child labour included here.

Education/gender matrix

Education	Female	Male	Total
None	4.6%	7.7%	12%
Basic	17.3%	51.6%	69%
Secondary	1.2%	15.4%	17%
Tertiary	0.0%	2.3%	2%
Total	**23.1%**	**77.0%**	**100%**

E. BOLGATANGA

Age/gender matrix

Age Groug	Female	Male	Total
12–18	3.00%	0.12	15.00%
Above 18	25.00%	0.6	85.00%
Total	**28.00%**	**72.00%**	**100.00%**

Average age: 30 years.

(*Continued*)

Table 23.3 (*Continued*)

Education matrix (highest level attained by group)		
Gender	Male	Female
Highest level	Tertiary (polytechnic)	Secondary

But typically no formal education, with few educated largely to basic level.

F. TARKWA

Age/gender matrix

Age group	Female	Male	Total
Under 18	0.02%	1.98%	2%
18–45	15.00%	60%	75%
45–60	0.42%	20.58%	21%
60 and above	0.01%	1.99%	2%
Total	**15.45%**	**85%**	**100%**

Average age: 33 years.

Education matrix (highest level attained by group)

Age group	Female	Male
Under 18	JSS	JSS
18–45	MSLC/JSS	SSS
45–60	MSLC/JSS	SSS
60 and above	MSLC/JSS	MSLC

Education (typically MSLC/JSS); SSS leavers only 10% because of lack of alternative employment opportunities. About 40% of JSS leavers engage in small-scale mining to accumulate money to learn trade.

G. ASSIN FOSO

Gender/age matrix

Age group	Female	Male	Total
8–17	1.0%	2.0%	3%
18–45	4.0%	89.0%	93%
46–70	0.0%	9.0%	9%
Total	**5.0%**	**100.0%**	**105%**

Education matrix (highest level attained by group)

Age group	Female	Male
8–17	Basic	Basic
18–45	Basic drop out	SSS
46–70	Nil	MSLC/JSS

But typically illiterate (about 52%).

In most of the small-scale mining districts, at least 10% individuals are illiterate, with no formal education, although invariably, the majority (>50%) of miners – across all age groups – have only had educational training spanning basic to Junior Secondary (JSS) or Middle School (MSLC) levels. A large part (approximately 40%) of JSS participants claim to have begun participating in the sector with the intention of securing enough funds to enable them to learn a trade and then "move on". However, the numbers do not indicate much of "moving on". The data again reveal that the few miners who have progressed to Senior Secondary School are mainly male between the ages of 18 and 40. Most who have reached the tertiary level (largely polytechnics, although there are some graduates involved) are aged between 30 and 50 years. While the stimulus for the involvement of senior secondary school level participants is the shortage of alternative employment opportunities, tertiary educational level participants are typically retired, laid-off large-scale mine workers, or are individuals that provide expert services to mainly legalized small-scale miners. The data support the major finding of Offei-Aboagye (1992) – *i.e.* that women participating in small-scale mining activities in Ghana are at a significant disadvantage because of illiteracy.

Two of the major conclusions of a recent *Centre for Labour Market and Social Research Working Paper* (01–03) on child labour in Ghana (Blunch & Verner, 2001) were as follows:

1) That there is a positive relationship between poverty and child labour in the country; and
2) That girls are more likely to engage in harmful child labour.

The first conclusion is attested to by the data gleaned about small-scale mining in Ghana, with the activity being more predominant in the poorer areas – the rural areas *vis-à-vis* the urban centres. However, the second conclusion seems to run counter to the situation in Ghana's small-scale mining sector, as demonstrated in Table 23.3. In Ghana, men of all age groups dominate the activity.

Being largely a rural activity, a worrying aspect for both women and children participants in small-scale mining relates to the health implications of their activities. The ILO (1999) indicated that women and children as young as 14 years old have been diagnosed with having occupational diseases such as silicosis (in advanced stages) and mercury poisoning contracted through the grinding of hard-rock gold ores and the burning of gold amalgam. These health problems have been exacerbated by the fact that there are inadequate health care facilities in rural areas.

RATIONALE FOR SMALL-SCALE MINING IN GHANA

In a historical context, small-scale mining was undertaken as one of a plethora of traditional economic activities – all of them using fairly rudimentary, though historically appropriate, technology. However, despite the significant improvements in technology achieved in economic activities of all sorts, there is still a proliferation of rudimentary artisanal small-scale mining activities worldwide.

As indicated by the World Bank (2001) and Davidson (1993), small-scale mining is closely intertwined with elevated levels of poverty, and is proving to be a means of survival for large numbers of people in developing countries. Generally, people make use of resources having the lowest marginal costs, which are typically the most abundant and/or easily accessible. It is therefore no surprise that small-scale mining in Ghana is prevalent in areas which, in addition to exhibiting high levels of poverty, also feature:

- A fairly sizable mineral resource base that is fairly easily worked;
- Reasonably widespread expertise in mining and mineral processing through either a long history of traditional artisanal mining (*e.g.* subsistence farmers during the off-peak farming season); an association with industrial scale mining activities (*e.g.* laid off large scale mining workers); or an invasion of migrant miners from other communities; and/or
- A lack of any other reasonably affordable economic activities.

Other economic activities may or may not yield higher returns. As a result, most people inhabiting such rural areas tend to depend on either small-scale mining or subsistence agriculture for survival.

As already explained, small-scale mining activity is predominantly undertaken by the younger generation – *i.e.* those who do not have the means to start another venture or to support themselves over any lengthy gestation period. Without satisfying at least one of these requirements, they are left with a choice between the relatively quick return provided by small-scale mining, or migrating to urban areas for anticipated greener pastures. Individuals are more likely to reap quicker returns working mineral deposits than in other ventures, largely because the average income earned is often higher than the potential earnings from other subsistence activities or even formal sector employment available in areas in which small-scale mining, especially illegal workings, take place. As Hilson (2001) notes, with a GNP per capita of US$390, an average small-scale miner in Ghana could earn about US$ 7 per day – in other words, US$1,820 annually (Appiah, 1998).

An analysis of small-scale mining within the framework provided by Abraham Maslow's "hierarchy of needs" (Maslow, 1943) shows how this "attitude" may significantly explain the way small-scale mining (of mainly precious minerals) is conducted in Ghana. Maslow's "hierarchy of needs" is a theory explaining the motivation for human behaviour. The framework consists of the following five groups of "needs":[22] physiological; safety; love and affection and belongingness;

[22] Physiological needs" are the very basic needs, such as air, water, food and sleep; "safety needs" relate to the establishment of stability and consistency in a chaotic world; in terms of importance, "love and belongingness" is the next category of needs, since humans have a desire to belong to groups, to feel loved and accepted by others; "esteem needs" cover both self-esteem from competence or mastery of a task and the attention, recognition and admiration from others. The need for self-actualization is "the desire to become more and more what one is, to become everything that one is capable of becoming."

esteem; and self-actualization (Fig. 23.6). The theory stipulates that human beings are motivated by unsatisfied needs, and states that if a basic need is not satisfied, "higher" needs will not become motivators. More specifically, when the deficiency in a need is met, other (higher) needs emerge that dominate and dictate behaviour. Once these have been satisfied, new (and even higher) needs emerge. Thus, individuals first attempt to satisfy their basic physiological needs (food, clothing, shelter, etc., which tend to be economic in nature) to some appreciable extent, after which they attempt to meet security/safety needs, until they reach self actualization.

The World Bank noted that "small-scale mining is a largely poverty driven activity, typically practiced in the poorest and most remote rural areas of a country" (World Bank Group, 2001). Small-scale miners are therefore at the stage of Maslow's framework where they are pursuing "physiological needs" (this and the corollary effects are together referred to as "the Maslow Effect"). For such miners, the concept of a safe working environment would only become relevant if, and when, they are able to at least partially meet their basic economic needs. Socially acceptable, and ultimately, esteemable behaviour, is therefore at an even higher place for such miners. Using Maslow's framework as a basis for interpretation, small-scale miners who seem to throw caution into the wind and operate precariously, inefficiently and in an environmentally unsound manner, are only acting rationally – *i.e.* satisfying their physiological needs first, and foremost. In short, the Maslow Effect adequately explains *why* such miners are not likely to operate in an environmentally friendly manner.

Until poverty in Ghana can be reduced and controlled, it will be exceedingly challenging for small-scale mining and its rather environmentally degrading activities to be halted.

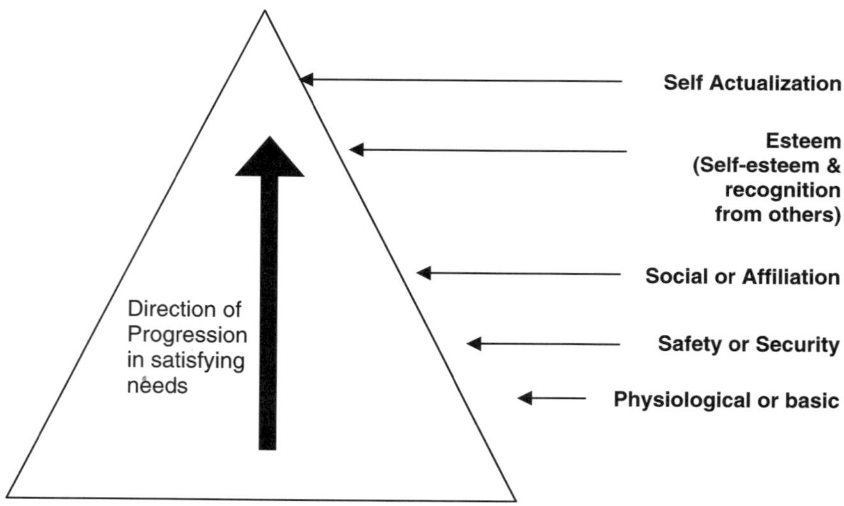

Figure 23.6 Conceptual Representation of Maslow's Hierarchy of Needs.

PERFORMANCE OF GHANA'S SMALL-SCALE MINING SECTOR

Production

In terms of value and quantum of economic activity generated, gold and diamonds are the major minerals exploited using small-scale mining methods in Ghana. Other minerals exploited using artisanal and small-scale methods, such as kaolin, limestone and sand (also stones through hand-breaking), are generally only significant on a localised basis.

Since 1989, the proportion of national gold output originating from artisanal and small-scale mines has increased to 8% (by volume) and 6% (by value); some 1.1 million ounces of gold valued at more than US$330 million were mined on a small scale between 1989 and 2001 (Table 23.4). Over the same period, 6.5 million carats of diamonds, valued at nearly US$140 million and accounting for 69% of Ghana's total (diamond) production, have been won by small-scale operators. The sub-sector's share of production rose from a mere 13.2% in 1988 to 83% in 2001. In terms of value, the proportion has increased from 4% to more than 95% over the same period (Table 23.5).

Other contributions

One of the reasons the government decided to regularize and streamline small-scale mining activities in 1989 was its employment generation potential.

Table 23.4 Comparative gold production and revenue figures (1989–2001).

	Production			Revenue		
Year	Small scale (oz)	Total Ghana (oz)	% Small scale to total Ghana	Small scale (US$Mill.)	Total Ghana (US$Mill.)	% Small scale to total Ghana
	a	b	c = a/b	d	e	f = d/e
1989	9,272	429,836	2.2	3.4	159.9	2.1
1990	17,234	534,630	3.2	6.3	201.6	3.1
1991	15,601	847,560	1.8	5.3	304.4	1.7
1992	17,297	999,950	1.7	6.1	343.4	1.8
1993	35,145	1,257,489	2.8	11.5	434.0	2.6
1994	89,520	1,428,011	6.3	34.7	548.6	6.3
1995	127,025	1,717,654	7.4	48.7	647.3	7.5
1996	112,349	1,597,575	7.0	36.0	612.4	5.9
1997	107,094	1,755,240	6.1	28.4	579.2	4.9
1998	128,334	2,382,339	5.4	36.6	687.8	5.3
1999	115,784	2,557,315	4.5	35.2	710.8	4.9
2000	145,662	2,447,591	6.0	40.9	702.0	5.8
2001	185,596	2,369,909	7.8	39.3	617.8	6.4

Source: Minerals Commission, based on Reports from PMMC and Miramex.

Table 23.5 Comparative diamonds production and revenue figures (1988–2001).

Year	Production			Revenue		
	Small scale (Carats)	Total Ghana (Carats)	% Small scale to total Ghana	Small scale (US$Mill.)	Total Ghana (US$Mill.)	% Small scale to total Ghana
	a	b	c = a/b	d	e	f = d/e
1988	34,158	259,358	13.2	0.2	3.5	4.4
1989	151,606	285,636	53.1	2.1	5.2	40.5
1990	484,876	636,503	76.2	14.3	16.5	86.5
1991	541,849	687,736	78.8	17.4	18.6	93.8
1992	442,266	656,421	67.4	13.0	19.3	67.5
1993	375,400	590,842	63.5	11.6	17.3	66.9
1994	405,830	757,992	53.5	11.2	20.4	54.9
1995	337,830	631,708	53.5	8.3	14.8	56.3
1996	443,244	714,738	62.0	8.9	13.4	66.1
1997	558,241	829,524	67.3	6.2	11.3	54.9
1998	570,186	805,742	70.8	6.1	10.6	57.0
1999	476,744	681,576	69.9	5.5	9.0	61.6
2000	686,551	989,851	69.4	11.8	**11.8**	99.7
2001	973,033	1,169,633	83.2	19.6	20.5	95.5

Source: Minerals Commission, based on Reports from PMMC.

Currently, the Minerals Commission estimates that over 80,000 Ghanaians are involved in the small-scale mining of gold and diamonds alone. The ILO estimates that between 50,000 and 300,000 people (ILO, 1999) are employed countrywide in the entire small-scale mining industry. With estimates of between four and nine dependents per small-scale mine worker, there could be anywhere between 200,000 and 2.7 million people relying on small-scale mining revenues in Ghana. As a result, the industry is frequently labeled locally as being *poverty alleviating*, and is said to have reduced rural-urban migration.

PROBLEMS, CHALLENGES AND CONSTRAINTS

Notwithstanding its positive contribution to the economy of Ghana over the years, small-scale mining has caused a wide range of problems. Furthermore, there are a number of barriers impeding its capacity to become a more sustainable activity. The major problems in the sector are as follows: environmental impacts; complications associated with illegal operations, popularly known as *galamsey*; inefficient operations; and socio-economic complications.

Environmental problems

Of all the problems caused by small-scale mining in Ghana, as Hilson (2001) has rightfully identified, environmental impacts and land-use conflicts are the

Figure 23.7 Typical post-small scale mining landscape in Ghana.

most significant. As with virtually all resource exploitation activities, some degree of environmental damage results from the operations of small-scale miners. This has largely manifested itself as impact on the physical environment – namely, deforestation and land degradation, as well as water and atmospheric pollution. It should be noted, however, that illegal miners, whose operations by their very furtive and clandestine nature are not amenable to being monitored by field officers, are the cause of the most severe forms of environmental damage, which results from the small-scale mining of precious minerals.

Deforestation and land degradation are common phenomena at many uncontrolled small-scale mining sites in Ghana. Large tracts of agricultural lands are destroyed as a result of the removal of vegetation and the disturbance of soil structure. Growth supporting topsoil is usually removed indiscriminately during mining, thus rendering land virtually incapable of supporting plant growth, and exposing surfaces and making them susceptible to erosion. Mining activities produce "moon-like" landscapes comprised of unstable piles of waste, abandoned excavations, and vast stretches of barren land. Excavated pits are rarely filled, and become receptacles for water; the resulting pools become breeding grounds for mosquitoes and thus pose serious dangers to both humans and animals. An example of such land degradation is depicted in Figure 23.7.

In many small-scale mining regions, drainage systems have been negatively affected by local operations. Discharges from sluicing, suspended solids, and mercury used in gold amalgamation activities, have seriously impacted the quality of many rivers and streams. Improperly disposed tailings have also found their way into streams and rivers during heavy rains, thereby creating sedimentation problems and rendering streams unusable for both domestic and industrial purposes. Removal of vegetation also induces soil erosion, which, in turn, increases the turbidity of surface run-off. Drainage of lubricants and other oils into streams lead to the de-oxygenation of water, which poses a threat to aquatic life.

Although gaseous pollutants are emitted from certain operations, the effect of small-scale mining on the atmosphere has not been considered significant as activities are carried out in ambient air. However, small-scale mining operations that generate dust not only pose a threat to the environment but also to human health, since the particles produced fall within the respirable dust range and are thus capable of causing dust-related diseases.

Another major problem results from uncontrolled mercury use. As already explained, at most small-scale gold mines operating in Ghana, mercury[23] is added to the concentrates derived from sluicing. The resulting amalgam is typically burned over an open flame, which releases mercury fumes into the atmosphere. In some cases, the burning of amalgam is conducted in poorly ventilated rooms; many small-scale miners have rejected the idea of using protective apparatuses – namely amalgamation retorts[24] – that help to separate gold from mercury within an enclosed circuit. The poor storage, handling and use of mercury commonly results in additional spillage, which poses a threat to agriculture and other land-dependent economic activities.

The seriousness of the mercury pollution problem in Ghana was realized during a recent study undertaken in Dumasi – a village where small-scale alluvial gold mining activities are widespread – near Prestea in the Western Region. The April 2000 study, which featured a follow-up validation a year later, was a joint investigation by the Minerals Commission and the United Nations Industrial Development Organization (UNIDO). Its purpose was to determine the extent of mercury pollution caused by galamsey operators in Dumasi. The results revealed elevated levels of mercury in both miners and non-miners. Of the 187 subjects studied, 31 had levels of mercury beyond the WHO thresholds.

Problems associated with illegal small-scale mining

In Ghana, illegal small-scale miners operate on the concessions of other licensees and within areas forbidden to mining. They are also known to carry out their operations in residential areas, farms, nature sanctuaries, parks, and alongside roads, railway lines and power transmission lines. The haste with which such miners carry out their "hit and run", furtive, and clandestine operations, make them more susceptible to accidents. They also waste mineral

[23] Mercury is a silvery-coloured liquid metal that is considered a neurotoxin. It is very volatile (easily changed to gas at low temperatures) and it is alleged that inhaling its vapours is by far more serious a threat than swallowing the liquid mercury itself.

[24] In addition to the relatively high cost of the retort, an apprehensive/skeptical attitude towards new technology has largely accounted for this rejection.

resources by "hand picking" deposits and discarding potentially economic mineralization.

The following factors have contributed to prolonged illegal small-scale mining in Ghana:

1) *Land use conflicts between large and small-scale mining parties*: Illegal small-scale mining has been attributed to the unavailability of land. Large-scale mining operations – both exploration and exploitation rights – have alienated large tracts of land. It is therefore argued that the only viable option remaining for small-scale operators is to work alienated lands or other restricted areas.
2) *Lack of information on the potential (mineral) yields of awarded concessions*: It has also been postulated that such miners have continued to operate illegally because there is little incentive for them to regularize/register. The argument made is that,
   a) Most of the small-scale mining licences awarded are not based on initial exploration work performed on the property; and
   b) Small-scale mining officials have not been able to effectively guide miners in their selection of "payable" concessions.

Thus, many miners feel that they are no better off operating legally.

3) *Inadequate law enforcement*: It has also been argued that the penalty for being caught while undertaking illegal small-scale mining is not disincentive enough to deter the activity. While the *Small Scale Mining Law* stipulates fines of up to two million cedis (US$250), actual fines have generally been far lower. For instance, in four cases involving 15 illegal operators arrested on the concession of a large-scale operator in the Ashanti Region during 2001–2002, the fines imposed ranged from only 200,000 to 500,000 cedis (US$25–62) or, in default, up to nine months imprisonment. Similarly, in the Bolga area of the Upper East Region, penalties for recorded illegal mining cases have ranged from 300,000 cedis to 700,000 cedis (US$38–88) or, in default, between three and six months in prison. Such fines are readily paid by either the illegal miner himself or his sponsor – generally, the buyer of his output. A main cause of this problem is the rather rapid deterioration the cedi (Ghanaian currency) has suffered. Thus, fines quoted in cedis in laws lose their value correspondingly over time, since legislation is rarely amended accordingly. Additionally, it has even been rumoured that some members of the law enforcement agencies are on the pay-roll of such miners, which, in turn, enables them to continue operating illegally.
4) *Downturn in the large-scale mining sector*: With the recent downturn in the prices of minerals, particularly gold, all mines went to work at streamlining their operations to become more cost-effective. In many cases, this led to mechanization, thereby inducing layoffs, redundancies and retrenchments. This, in turn, resulted in a pool of unemployed miners and eventually, precipitated increased illegal small-scale mining activity.

Problems with respect to inefficient operations
According to Mackay and Schnellman (1987):

> In hard rock gold mining only the highest grade material is taken while ground which may well be profitable to work is sterilised by the operations. During excavation 10–20% of the recoverable gold is often lost and a further 5–10% may be lost in crushing. Recovery on the sluices and at amalgamation is probably not more than 50–60% at the best. Actual recovery may therefore be of the order of only 30%.
>
> In alluvial working by pitting about half the area is often left unworked, as is the base of the gold or diamond bearing gravel below water level. Sluicing for gold recovers 50–60%, of the material raised which probably represents 25–30% of the original *in situ* resource. Washing for diamonds recovers only the larger stones, so that recovery is estimated to be of the order of 25–30% of the total *in situ* diamond content (Mackay & Schnellman, 1987).

While the estimates here may be said to be overestimating the losses, they nevertheless provide an indication of the inefficiency of small-scale mining and processing methods.

Socio-economic problems
Social problems, including drug abuse and sexual promiscuity, are rife among small-scale miners who have an air of apparent affluence. This results in cultural adulteration, and the proliferation of various social vices, as well as sexually transmitted diseases. Problems are particularly intense in illegal mining regions.

Akin to their large-scale counterparts but to a more serious degree, small-scale miners – especially illegal operators – are continuously struggling with other parties for land resources. However, small-scale miners are rarely in any position to compensate other potential users of land to enable them priority access.

RECENT AND/OR PROPOSED MEASURES FOR ACHIEVING SUSTAINABLE DEVELOPMENT IN THE GHANAIAN SMALL-MINING INDUSTRY

Policies pursued
Until only recently, most of the policies pursued and measures put in place for small-scale mining have emphasized increased legalization. For instance, in most cases, licensing fees have been kept at a minimum and small-scale miners have not been made to pay any taxes to Government, while technical support and extension services have been emphasized and provided free-of-charge.

More recently, however, policies have generally been re-focused, aiming to improve the efficiency of small-scale mining operations through:

- Encouraging the improvement of generally low-tech mining and mineral processing methods currently being employed by artisanal miners;

- Supporting and encouraging artisanal miners to adopt semi-mechanized and mechanized techniques; and
- Facilitating the adoption of more environmentally-friendly methods of operations.

Mitigative measures in place

The Ghanaian government, in its efforts to streamline the small-scale mining sub-sector and make it more sustainable, has implemented a number of measures to address pressing problems. Many of these are examined in the discussion that follows.

- *Environmental permitting*: In recent times, to ensure that environmental impacts are kept at a minimum, small-scale miners in Ghana are required to obtain an environmental permit from the Environmental Protection Agency (EPA) before obtaining a license to mine. To award an environmental permit, the EPA requires submission of an environmental overview of the intended operation, which includes a brief description of the operational methods to be used; a site plan of the area in which mining is to be undertaken; a summary of the anticipated environmental impacts; and proposed mitigation measures and reclamation proposals. The granting of an environmental permit places enormous responsibility on the small-scale miner, who must ensure that sound environmental practices are in place. The activities of field officers – in particular, site visitations and threats of sanctions – have been known to serve as stimuli for miners to eschew mining and ore processing practices that are detrimental to the environment.
- *Tributer schemes*: By implementing "tributer systems", large-scale mining companies have also played a constructive role in addressing the problem of illegal small-scale mining and the environmental impacts of associated activities. Such schemes involve the registration of all small-scale miners encroaching on a property and the subsequent allocation of suitable portions of the concession for small-scale mining; such portions are generally areas considered uneconomical to exploit using large-scale mining methods. Under tributer systems, small-scale operators are required to sell their winnings to the company through special pricing arrangements. The forging of these agreements enables large-scale companies to better monitor and control the small-scale miners operating within the boundaries of their properties, and, more importantly, to prevent them from causing extensive damage to the environment for which the concession owner is accountable. Though not very widespread, such a scheme was implemented by Abosso Goldfields Limited, which commissioned a similar scheme in the past.[25]

[25] Abosso subsequently employed most of these small-scale miners when the mine's operations expanded, and the scheme came to an end. Some incidences of illegal operations still occur, though.

- *Establishment of a legal framework for regularizing and streamlining small-scale mining*: A legal framework for registering and carrying out small-scale mining operations in Ghana has been established. The framework seeks to streamline and facilitate the licensing and provision of technical support to such workings.
- *Deterrence through penalties and law enforcement*: Illegal small-scale mining, being a multi-dimensional activity, can best be tackled through the collaborative effort of mining sector institutions, large-scale mining companies and law enforcement agencies. In this light, to deal with the issue of ineffective fines imposed by lax laws, "Penalty Units" have been introduced. According to the Fines (Penalty Units) Act 2000, Act 572, fines imposed by any law are to be expressed in penalty units. The pecuniary value of a penalty unit can then be specified and amended as appropriate by the Attorney General through a less cumbersome (as opposed to amending an Act, to change the levels of fines) legislative route. The levels of fines can now be made to reflect current values as much as possible, and therefore, hopefully serve as sufficient deterrent to illegal small-scale mining. Additionally, with the intensified government effort made to improve working conditions of law enforcement agencies in general, and, more specifically, their sensitization to the implications of the activities of illegal miners, it is expected that these agencies will begin to enforce laws prohibiting illegal small-scale mining activities in the country more proactively.
- *Cooperation to reduce conflict between large and small-scale mining activities*: Such conflicting situations have been resolved in a number of ways. Valuable information has been generated on the prospectivity of areas over which large-scale mining companies have explored since the mid-1980s. While some of these prospects may not be economical for large-scale exploitation, they may be amenable to small-scale operations. Armed with this knowledge, the government has sought to demarcate such areas for small-scale mining when large operators give up their land concessions, either wholly or partly. One notable example was the demarcation of patches of the 124.3 square kilometres original Konongo-Obenemase gold concession in July 2002. This occurred in consultation with a new investor applying for the property, when the previous large-scale operator surrendered it to government.

There have also been instances where, for good neighbourliness, large operators have deliberately ceded suitable portions of their concession to government for the purpose of demarcating land to small-scale miners. Such has been the case with Resolute Amansie Resources Limited, Ghana Consolidated Diamonds Limited, and Anmercosa (Jappa area). Another example was the Bonte Gold Mines Limited 2001–2002 case, where a conflict situation had developed as a result of small-scale miners being inadvertently issued licences which overlapped Bonte's concession. After negotiations, the management of Bonte agreed to cede 2.7 square kilometres of its 48.5 square kilometres alluvial concession to the small-scale miners.

While not being a legal requirement, as the examples have shown, a number of large-scale mining companies have, by adopting such a philosophy, largely reduced, if not completely eliminated, clashes between themselves and illegal small-scale miners. In the process, they have made more land available for legal small-scale mining.

- *Provision of alternative economic activities*: The provision of alternative sources of viable employment in areas having the potential for small-scale mining is a key to facilitating the abandonment of destructive activities, and inducing a switch to less damaging but equally profitable and sustainable vocations. Most large-scale mining companies are involved in the development/execution of one form of alternative livelihood project. These are expected to reduce the more hazardous illegal small-scale mining activities, encroachment on to large-scale mining concessions, and help to regularize and streamline activities.
- *Technical assistance and extension services*: The provision of technical assistance to small-scale miners has been identified as a priority. The government has attempted to deliver technical services to small-scale miners through its district offices, which are based in close proximity to operations. The assistance provided has been largely advisory in nature but has also involved monitoring to ensure compliance with regulations and other legal requirements.
- *Educational drive*: Another important line of action being pursued is education. Miners are regularly being advised to adopt more efficient mining and processing methods, and to practice environmental management. The need to create environmental awareness among small-scale miners has been recognized. Thus, along with intensive educational programmes aimed at improving technical competence, management and accounting/record keeping practices, and educational drives aimed at making miners aware of environmental regulations and guidelines, are being undertaken in all seven small-scale mining districts.

From the standpoint of environmental management, the educating and training of small-scale gold miners on mercury pollution abatement is an ongoing activity. It is expected that such programmes will help make miners aware of the benefits to adopting environmentally sound practices at sites. Through education, it is also hoped that miners will change their pre-regularization attitudes and begin perceiving small-scale mining activities as viable economic ventures that require planning, organization and execution to achieve desired results.

- *Technology transfer*: This has largely been a private-sector-led initiative, based on economic rationality. Technology imported from China is quite widespread, especially within the hard-rock gold producing areas around Tarkwa. Such equipment is generally inexpensive and affordable to small-scale miners, and has positively impacted efficiency and productivity. For similar reasons, technology from India is also being considered. This approach – *i.e.* technology transfer – coincides with the government's aim to mechanize artisanal operations.

Mitigative measures being considered for adoption
This section will highlight measures that have either been implemented on a pilot scheme basis or are currently being researched and therefore being considered for more general application in Ghana. A number of these measures were investigated using funding from a World Bank and Nordic Development Fund Credit (Mining Sector Development and Environment Project, Credit 2743 GH and NDF 156).

- *Pilot reclamation of lands mined by small-scale miners*: Under a pilot scheme initiated by Ghana's Minerals Commission, three agricultural plots (totaling 205 hectares) that had been devastated by small-scale mining activities, were rehabilitated and re-cultivated with economic trees and other indigenous plants. The three areas involved were Ablorman in the Greater Accra Region, which had been damaged by sand wining; the Nueng Forest Reserve in the Western Region that had been degraded by gold mining; and Buadua in the Eastern Region, which had been devastated by diamond winning. The plots in Ablorman and Buadua have been returned to the original landowners, and the Reserve has been handed over to the Forestry Commission.

The objective of this project was to demonstrate that, following intensive small-scale mining, land *could* be reclaimed and used for other economic activities. It also aimed to convince the residents of communities in close proximity to small-scale mining activities of the importance of demanding and subscribing to such rehabilitation.

- *Mercury pollution abatement*: As already explained, in 2000, a mercury abatement project was undertaken with the assistance of UNIDO and the French Government. Phase I of the study (US/GHA/99/128 – Assistance in Assessing and Reducing Mercury Pollution Emanating from Artisanal Gold Mining in Ghana) was conducted in a galamsey village in the Western Region (in the town of Dumasi), where hard-rock gold mining is widespread. As previously indicated, the results of the study revealed rather elevated levels of mercury in the galamsey operators themselves, other residents as well as their environment. A second phase of this study has recently been initiated in another Western Region village (Jappa).

Following the results of Phase I of the study, an intensive educational campaign on the toxicity of mercury was launched. Additionally, in the short term, mercury retorts – donated by UNIDO (Daily Graphic, June 13, 2001) – have been introduced; these systems enable small-scale gold miners to capture mercury fumes when they evaporate during amalgam burning. Though the adoption of the retort has not been as widespread as expected, a noticeable reduction in the misuse of mercury has been observed at Dumasi. Similar experiences have occurred in some small-scale gold mining areas of the Ashanti Region.

The government is aiming to make the possession of mercury retorts a requirement for securing small-scale gold mining licenses in the near-future. Over the long-term, the implementation of a five-year European union-sponsored Mining Sector Support Programme (formerly known as SYSMIN) from

2003/4 is expected to assist with identifying and adopting alternatives to mercury in gold processing. Alternative technologies from India and South Africa are also being considered for adoption and promotion.

- *Provision of improved geological information to small-scale miners*: This has been a stated objective of the government since the regularization of small-scale gold mining in 1989. However, due to the absence of an adequate funding scheme, few initiatives have materialized. Under the World Bank Credit, a programme aimed at providing better geological information to small-scale miners was initiated. Geologists worked in the field to identify suitable areas and delineate recoverable ore bodies for small-scale mining; these areas were then licensed to small-scale miners. Unfortunately, such activity has not been ongoing because of the limited funding available for such purposes.

However, increased provision of pre-licensing exploration data/information should help to entice illegal miners to regularize their activities and ultimately, improve the economic well-being of rural communities.

- *Promoting the use of improved processing techniques and equipment*: Under the World Bank Credit, a consortium from Ghana and South America was contracted to develop environmentally friendly processing techniques that will help to increase the productivity and yields of small-scale miners. Various mining and mineral processing equipment procured by the project was successfully tested by the consultants in major small-scale mining areas in the Western and Central regions of Ghana. The acceptability of equipment in the mining communities in which the tests were carried out was so high that inhabitants wanted to retain the equipment. While most concession owners could not afford the equipment directly, other small- scale miners and groups have applied for, and have been allowed the use of, various technologies on concessionary terms. It is expected that the adoption of the appropriate sizes (*i.e.* scaled-down versions) of these techniques and equipment will improve the efficiency of such small-scale mining operations and their environmental performance.
- *Educational drive*: The government has worked together with donor agencies to undertake studies that, from a policy standpoint, target ways to improve specific aspects of small-scale mining.
- *Local economic development or alternative livelihood projects*: If alternative economic activities to small-scale mining became available and were, at a minimum, equally profitable, small-scale mine workers could divert their attention to newer and more sustainable employment opportunities. Additionally, such alternative livelihood or local economic development (LED) projects would, by making available additional economic opportunities, help to address some of the socio-economic challenges of small-scale mining.

A study recently initiated by Ghana's Minerals Commission has identified a number of such LED projects, some of which are now being conducted by mining companies. The mine-initiated projects provide members of communities with re-skilling opportunities and entrepreneurship training. One of the conclusions of the study was that both large-scale mining companies and surrounding communities would be willing to support, participate in, and cooperate with, LED initiatives. A second phase of the study, the aim of which is to establish the feasibility and sustainability of such alternatives, has been completed. The results are being considered to assist in the development of a framework for implementation that will help improve socio-economic conditions within mining communities, as well as promote improved environmental management.

Following an exploratory mission of a UN Department of Economic and Social Affairs team to Ghana in June 2000 at the invitation of government and UNDP, a local consultant was appointed in April 2002 to undertake studies to identify a set of policy options and best practices for eradicating poverty in artisanal mining communities. This study, which is entitled *Poverty Eradication and Sustainable Livelihoods: Focusing on Artisanal Mining Communities*, is meant to be part of an African regional and multidisciplinary project, and has been completed and subjected to stakeholder review.

- *Proposal on Assistance to Small-Scale Miners and Mining Communities being considered for adoption under the Poverty Reduction Programme of Ghana's HIPC Initiative*: A proposal has been made and is being considered for adoption under the Ghana Poverty Reduction Programme. The Programme, if implemented, is expected to promote improved economic well being, the adoption of upgraded environmental practices, and generally enhance quality-of-life. The proposal entails identifying land areas with mineral deposits that are suitable for small-scale mining. It requires (i) a review of reports submitted by exploration companies to the government over the years; (ii) selecting areas with potential for small-scale mining; (iii) undertaking further exploration to better establish location as well as size and grade of reserves; and (iv) permitting small-scale mining in demarcated areas. The resultant improved assurance and ease of "making money" (compared to the current "trial and error, hit and run" mode of operation) should:
  1) Persuade illegal miners to regularize their operations;
  2) Enable a reasonably commensurate processing fee to be levied;
  3) Enable/facilitate monitoring and policing to ensure technically and economically efficient, as well as environmentally sustainable, operations.
- *Promotion and development of other minerals*: In recent times, Ghana's almost monopolistic position for producing solar salt in the West African Sub-region has been recognized as a means of generating revenue, foreign exchange, employment and economic development. The government has therefore mounted an intensive campaign to promote the development of this potential, as well as ancillary industries.

CONCLUSION

Provided that the measures that have been put in place or are being considered are streamlined, marked improvements in environmental management and operational efficiency can be achieved in the Ghanaian small-scale mining industry. This is a key to reducing poverty in Ghana's small-scale mining communities.

REFERENCES

Anin, T.E. (1990). *Gold in Ghana.* London, U.K.: Selwyn Publishers Ltd.

Appiah, H. (1998). Organization of small-scale mining activities in Ghana; *The Journal of the South African Institute of Mining and Metallurgy*, 98(7), 307–310.

Aryee, B.N.A. (1987). Legalisation of Small-Scale Gold Mining in Ghana: A Look at Some Economic Aspects; Appendix 'A2.3' to the Main Report, "M&S 8703A" on the World Bank sponsored study – "The Regularisation of Small-Scale Gold and Diamond Mining in Ghana" conducted by Mackay & Schnellmann Ltd. of England, as Consultants, and the Minerals Commission.

Aryee, B.N.A., Ntibery, B.K. & Atorkui, E. (2002). Trends in the small-scale mining of precious minerals in Ghana – A perspective on its environmental impact; In print – *Journal of Cleaner Production*, Elsevier Science Ltd.

Barbot, J. (1732). *A description of the coasts of North and South Guinea*; (cited on page 5 of Anin, 1990).

Blunch, N.-H., Verner, D. (2001). *Revisiting the link between poverty and child labour: The Ghanaian experience*; Centre for Labour Market and Social Research Working Paper 01–03 of April 2001.

Bogoso (2002). *An overview, Bogoso Gold Limited.*

Bossman, A. (1698). *A new and accurate description of the coast of Guinea*; (cited on page 4 of Anin, 1990).

Botchway, F. (1990). *The legal regime of small-scale gold mining in Ghana: A case study of the Tarkwa area*; a Dissertation presented to the Faculty of Law, University of Ghana, Legon, for the award of LLB degree.

Canagarajah, S. & Coulombe, H. (1997). *Child labour and schooling in Ghana*; Policy Research Working Paper No. 1844, World Bank, Washington, D.C.

Davidson, J. (1993). The transformation and successful development of small-scale mining enterprises in developing countries. *Natural Resources Forum*, 17(4), 315–326.

Government of Ghana (1986). *Minerals and Mining Law, P.N.D.C. Law 153.*

Government of Ghana (1989a). Small-Scale Gold Mining Law 1989, P.N.D.C. Law 218.

Government of Ghana (1989b). *Investment Prospectus, Gold and Diamonds in Ghana; Volume II – The Diamond Fields of Ghana.*

Hilson, G. (2001). *A Contextual Review of the Ghanaian Small-Scale Mining Industry*; A report prepared for the Mining, Minerals and Sustainable Development Project (MMSD), International Institute for Environment and Development (HED).

ILO (1997). *IPEC at a Glance;* International Labour Office, Geneva.

ILO (1999). *Social and Labour Issues in Small-Scale Mines.* Report for discussion at the Tripartite Meeting on Social and Labour Issues in Small-scale Mines, Sectoral Activities Programme, International Labour Office, Geneva.

ITDG (2001). @ *http://natural-resources.org/minerals/smscalemining/definitions.htm*

Mackay & Schnellmann (1987). Regularization of Small-Scale Gold and Diamond Mining in Ghana Main Report "M&S 8703A" on the World Bank sponsored study which led to legalizing small scale gold mining in Ghana, Accra.

Maslow, A.H. (1943). A theory of human motivation. *Psychological Review*, 50, 370–396.

Meredith, H. (1812). *Account of the gold coast of Africa;* (cited in Anin, 1990).

Offei-Aboagye, E.O. (1992). *A study of the participation of women in small scale mining*. Report No.1 for GTZ (GMBH) Small Scale Mining Project.

Offei-Aboagye, E.O. (1995). *Women's participation in artisanal mining in Ghana*. Paper prepared for the World Bank International Round Table on Artisanal Mining, Washington, D.C., May 1995.

UN (1971). @ *http://www.natural-resources.org/minerals/smscalemining/definitions.htm*

UNICEF (1999). *The State of the World's Children, Report*; UNICEF, New York.

World Bank (1995). *Staff Appraisal Report, Republic of Ghana Mining Sector Development and Environment Project.* World Bank Report No. 13881-GH, Industry and Energy Operations, West & Central Africa Department, African Region.

World Bank Group (2001). @ *http://www.worldbank.org/html/fpd/mining/key/artisanal/artisanal.html*

Part IV

Asian Case Studies of Artisanal and Small-Scale Mining

24

Introduction Part IV

Gavin M. Hilson

A review of the available literature will show that there is comparatively little material that examines the social, cultural, and economic aspects of artisanal and small-scale mining in Asia. The socio-cultural elements of Asian mining communities began receiving considerable attention in the early-1980s, when the activities of many international mining companies operating in Papua New Guinea, Indonesia and surrounding countries were profiled thoroughly for the first time. Global interests in Asian mining grew considerably in the late-1980s, following widespread reportage of the environment complications induced by BHP's Ok Tedi Mine in Papua New Guinea. Specifically, mass discharges of waste tailings to surrounding land, which, combined with the accidental release of hundreds of tonnes of cyanide into the Fly River System, displaced numerous *Wopkaimin* communities.

The multimillion dollar lawsuits that followed the Ok Tedi Disaster, along with the extensive documentation of complications of similar magnitude experienced at Bougainville, are issues that have long dominated the political environment of Asian mining. What continues to be overlooked is the fact that *most* Asian countries have highly-developed mining sectors with active artisanal and small-scale contingents (See Table 24.1). In India, for example, approximately 3000 small-scale and artisanal works allegedly produce 50% of the country's non-fuel output. In China, at any given time, as many as three million artisanal and small-scale operators could be involved in coal mining (ILO, 1999), and, in the Philippines, at Mount Diwalwal on the Island of Mindanao alone, 100,000 people are engaged in artisanal gold mining and associated processing activities (Hollaway, 1997). The Asian branch of the global artisanal and small-scale mining industry is not only extremely diverse in terms of the minerals it produces but also has the majority of the industry's participants, with China, India, Indonesia, and Papua New Guinea alone accounting for well over four million of the world's estimated 11.5–13 million artisanal and small-scale mine workers (Table 24.1).

This section of the book explores the Asian artisanal and small-scale mining industry through a series of detailed country cases studies. Chapters 25

Table 24.1 Small-scale mining employment estimates in selected Asian countries.

Country	Estimated employment
China	3,000,000
India	500,000
Indonesia	465,000
Pakistan	90,000–370,000
Papua New Guinea	50,000–60,000
Philippines	200,000
Viet Nam	35,000–40,000

Sources: United Nations (1996), ILO (1999), Ávila (2000); Hentschel et al. (2001).

through 27 focus on artisanal and small-scale mining in India. In Chapter 25, Kuntala Lahiri-Dutt builds from earlier chapters by Hinton et al. and Tallichet et al., examining women's participation in the Indian mining economy – specifically, within labour-intensive, small and surface bound operations. The author uses a case study of the stone quarries of Rajmahal region in Eastern India to explore gender issues in the artisanal and small-scale mining industry. In Chapter 26, Mrinal Ghose briefly examines the impacts of small-scale mining in India, and outlines possible measures capable of improving conditions in the sector. The final chapter on India, by Ajoy Kumar Ghose, provides a brief review of small-scale mining in the country in the past, reviews the current status of developments, and seeks to provide a configuration of the shape of things to come.

The next two chapters of this section examine the state of small-scale coal mining in China. In Chapter 28, Ziran Zhong provides a comprehensive overview of the policy environment in place for resident small-scale coal mines. Apart from describing major policy changes, the author examines pertinent legislation in detail, as well as the wide-ranging impacts of the country's recent coal mine closure campaign. In the next chapter, Philip Andrews-Speed, Guo Ma, Xunpeng Shi and Bingjia Shao use coal mines in China as a case for examining the impacts of small-scale mine closure. The authors initially review current approaches to planning for the closure of large-scale mines, and identify a number of features that distinguish small-scale mines in the context of mine closure. The authors then provide a detailed account of the impact of, and some responses to, the enforced closure of large numbers of small-scale coal mines in China between 1998 and 2002.

In Chapter 30, William Murray bridges a major gap in the small-scale mining literature by providing an overview of small-scale mining in Mongolia. Apart from a small group of World Bank reports that describe the general mining environment in the country, no study has examined the current state-of-affairs in the Mongolian small-scale mining industry. Emphasizing the gold segment, the author explains that small-scale mining is a relatively recent informal activity in Mongolia, barely acknowledged by government policy or the donor community. The chapter discusses the history and prospects of informal mining in relation to some of the major forces that have impacted Mongolian society since the beginning of its transition to a market economy.

In Chapter 31, Danilo C. Israel and Jasminda P. Asirot examine the state of small-scale gold mining in the Philippines, quantitatively assessing the economic cost of the mercury pollution emanating from its activities. The authors also discuss related environmental and developmental problems associated with the industry, and conclude by prescribing measures to both address these problems and promote better management in small-scale mining. Next, Hamid Etemad and Kamaleddin S. Salmasi, using a case study of Indonesia's "Contract of Work System", explain how emerging cooperative mining policies are superior to many of the conflictive practices of the past. The authors also review the relevant theoretical underpinnings of mining investments in general, as well as small-scale mining in terms of its operational risks and rational returns.

The concluding chapter of this section, by Geoff Crispin, provides a thorough account of small-scale gold mining activity in PNG. The author describes its main characteristics, its function in the rural economy, and the main assistance-related projects commissioned on its behalf.

REFERENCES

Ávila, E.C. (2000). *La llamada pequeña minería: un renovado enfoque empresarial*. División Recursos Naturales e Infraestructura, Santiago, Chile.

Hentschel, T., Nruschka, F. & Priester, F. (2001). *Global Report on Artisanal and Small-Scale Mining (ASM): Draft Version Only*. Prepared for Mining, Minerals and sustainable Development (MMSD) Project, International Institute for Environment and Development (IIED), London.

Hollaway, J. (1997). Small-scale mining: how to combine development with low environmental impact. *Industry and Environment*, December, 44–48.

International Labour Organisation (ILO) (1999). *Social and labour issues in small-scale mines*. Report for discussion at the Tripartite Meeting on Social and Labour Issues in Small-scale Mines. International Labour Organization, Sectoral Activities Programme, International Labour Office, Geneva.

UN (1996). Recent developments in small-scale mining: A report of the Secretary-General of the United Nations. *Natural Resources Forum*, 20(3), 215–225.

25

Not a Small Job: Stone Quarrying and Women Workers in the Rajmahal Traps in Eastern India

KUNTALA LAHIRI-DUTT

Women's participation in the mining economy has significantly diminished in recent years with the advent of mechanized technology. In the case of India, however, women miners are still concentrated in large numbers, employed at rudimentary, labour-intensive, small and surface bound operations. Using extensive fieldwork conducted over the course of a decade, this chapter examines the gender roles and associated issues in informal mines, with a focus on the *pathar khadans* (stone quarries) in the Rajmahal region of eastern India. As the Government of India regards "stone" or basalt as a minor mineral, it provides little official data on mining of this type or about its labour structure.

SCOPE

Mining has played an important role in the development of human societies and economies. It is an activity in which humans interface with the environment and development in complex and intertwining ways. In the past, the discovery of minerals, such as gold and silver, resulted in population shifts and economic growth in the New World. The extraction of minerals, gems and coal continues to provide a foundation for local economies in many parts of the world. In most places, local communities are involved in extraction processes, although mining is notorious for bringing unforeseen changes to a region's social fabric. However, when local communities participate in mining, does the mining economy reflect a shift in traditional gender roles, or are new identities formed and negotiated? Moreover, where are women involved in this sort of mining?

From a technical point of view, there are two basic types of mining and quarrying: surface and underground. In an underground operation, the participation of women workers is essentially restricted by international regulations that came into effect in the last century; thus, women tend to seek employment at surface operations. Quarrying is a form of mining distinguished by the fact that the excavated product is used for building or architectural purposes, rather than subjected to further processing, as in the case of an extraction of a metalliferous component of a rock, or the combustion of coal to obtain energy. Stone, such as marble, granite, limestone and sandstone, is quarried by splitting blocks of rock from a massive rock surface. The desired product can take the form of a dimension stone suitable as building blocks and tiles, or can be crushed for use as gravel or aggregate. In 1999, the gross world production of stones was in the range of 55 million tonnes (GDM, 2000), with the highest level of production recorded in China, followed by Italy, Germany, USA, Spain, Japan and India; these countries account for just over one half of global production. About 60% of quarried materials are considered useful or marketable.

India is the largest producer of dimensional stones (27% of world production), and is a major exporter of stone. The breakdown of stone production in India in 1998 was as follows: marble, 3.6 million tonnes; granite, five million tonnes; sandstone, 5.5 million tonnes; flaggy limestone, 2.1 million tonnes; and slate, three thousand tonnes. However, the informal nature of much of the quarrying activity in the developing world makes the reliability of these data questionable, particularly when the low profit levels of stones such as basalt are taken into account, which results in many transactions going unrecorded. Indian quarries are typically small operations that proliferate next to one another in a suitable geological location; they have large numbers of women workers (Chakravorty, 2001; Nandi, 1996; Aich, 1996), and are also renowned for having bonded and child labour (PRIA, 2001; Alfa, 1999).

As a background, it is perhaps useful to consider the scale of quarrying operations in the developing world, as well as their status in terms of formality or otherwise. From the point of view of the scale of operation, mining is commonly divided into two categories: large and small; the concept of scale is widely used in mining literature. However, such classification commonly obscures the unity or diversity across scales, giving rise to a false dichotomy. For example, many small quarries situated in close proximity to one another can cover an extensive area. A more accurate classificatory scheme, therefore, is level of formality (that is, informal and formal), which takes into consideration the economy of the mining operation, as well as the economic context in which it is placed. Against such classification, informal mining can be defined as the low-capital, labour-intensive extraction activity in developing countries in which local communities participate in significant proportions. Within such operations, the mining-community interface cannot be easily explained using exclusively assessments of geology and scale. There is a wide range of workers involved in informal mining, including traditional artisans, male wage workers, and women who participate either on their own or alongside their families and children (Carino, 2002; Burke, 1993). Although there are clearly many different types of labour performed at informal mines, the work typically undertaken by women frequently goes unnoticed and is little appreciated. Women tend to comprise an "invisible workforce", despite the fact that the jobs they perform are integral to the functionality of the industry.

SMALL AND INFORMAL MINES IN INDIA

How are small mines defined in India? The Government of India's *Mines and Minerals (Regulation and Development) Act* or MMRD Act of 1957 divides all minerals into two categories: major, and minor or small. The Act denotes, "building stones, gravel, ordinary clay, ordinary sand other than sand used for prescribed purposes (used for refractory, ceramics, metallurgical, optical and stowing in coal mines purposes, manufacture of silvicrete cement, sodium silicate, pottery and glass purposes), boulder, shingle, chalcedony pebbles (used for ball mill purposes only), limeshell, kankar, and limestone used in kilns for manufacture of lime used as building material, brick earth, fuller's earth, bentonite, road-metal, rehmatti, slate and shale used for building material, stone used for household utensils, marble, quartzite and sandstone when used for purposes of building or for making road-metal and household utensil and saltpeter" as India's "minor" minerals. This list clearly shows Indian's artisanal heritage of quarrying diverse minerals, and, at the same time, indirectly emphasizes the informal character of their exploitation. Sahu (1992, p. 3) described small mines as:

> Those whose production, or excavation quantity is limited in tonnage and not very large, mostly manually operated and sometimes employing machines to small capacity. Such mining activities are usually confined to deposits which are shallow in depth and small in extent.

Occasionally, a large number of small operations may be found occupying a significantly large area. The case study presented here is of one such region in eastern India, where there is extensive production of stone from several small quarries covering a rather large area.

The salient features in Sahu's definition are as follows: small production, labour intensiveness, the shallow nature of deposits, and low technology deployment. Interestingly, many so-called large collieries would qualify as "small", especially if compared to the mines of other important coal producing nations. Approximately 85 minerals are exploited in India at about 4,000 working mines. Of these, approximately 1,000 can be considered large mechanized opencast or underground operations. The remainder are small or "Class B" sites. In India, it has been estimated that the small-scale sector accounts for only 8% of national mineral output, and employs only 18% of the country's mine workforce (NISM, 1993, 1994). However, no data are available on production from informal mines, either on an individual or aggregate basis. Similarly, contributions from women miners, who, again, constitute a large segment of the workforce, are rarely mentioned in government documents. For example, the *National Mineral Policy*, 1993, states the following about small mines:

> Small and isolated deposits of minerals are scattered all over the country. These often lend themselves to economic exploitation through

> small-scale mining with modest demand on capital expenditure and short lead-time, they also provide employment opportunities for the local population. Efforts will be made to promote small-scale mining of small deposits in a scientific and efficient manner whilst safeguarding vital environmental and ecological imperatives. In grant of mineral concessions for small deposits in scheduled areas, preference shall be given to the scheduled tribes.

Details concerning the important functions of women in this type of mining are excluded from the Policy; in fact, details of workers and working conditions are not mentioned at all. Regardless of size, all mines in India fall under a plethora of government rules and regulations:[1] the *Mines and Minerals (Regulation and Development) (MMRD) Act*, *Mines Act*, *Forest Act*, and *Environment Act*. With their mostly inadequate technology and problems of mobilizing financial resources, small mines suffer innumerable handicaps, which lead to major inefficiencies. Among these is the excess worker problem, which has led to a large-scale retrenchment of women labourers in recent years. The country's small mines are typically associated with poorly-developed infrastructure; they are virtually nonexistent in the periphery of government vision. The immensity of the mining sector bureaucracy encourages informal, rather than formal, operations.

After describing the technology and work practices of the stone quarries of Rajmahal Traps, this chapter explores gender relations in Eastern India's informal mining industry. The study presented in this chapter asks the question: how do these informal mines reflect mainstream gender relations? More specifically:

- What work do women perform in the industry, and by what means?
- What are the socio-economic conditions of female workers, and their perceptions of gender relations in the quarries?
- To what degree of exploitation are women subjected to in these mines?

The work presented in this chapter constitutes a decade of intensive fieldwork that involved soliciting inputs from local indigenous people, particularly women miners, village *pradhans* (chiefs), other important community representatives, local Non-Government Organization (NGO) employees, district administrators, quarry owners, and labour contractors.

---

[1] According to a handbook on coal published by a local district administration, there are as many as 16 laws covering the entire gamut of India's mining activities. Important environmental legislation relevant to mining include: *The Water (Prevention and Control of Pollution) Act,* 1974; *The Air (Prevention and Control of Pollution) Act*, 1981; *The Forest (Conservation) Act*, 1981; *The Mines Act*, *Coal Mines Regulation and Circulars issued by the Director General of Mines Safety*; *The Environment (Protection) Act*, 1986; and *The Mines and Mineral Regulation and Development Act*, which was amended in 1988, making it obligatory on the part of mine operators to implement environmental safeguards, including reclamation measures for mined land.

THE RAJMAHAL TRAPS: HOME OF BASALT STONES

The basaltic Rajmahal Traps are located in Eastern India, some 250 km north-west of Calcutta within the bordering area of the states of West Bengal and Bihar. They occupy an area of approximately 1,000 sq km, and are estimated to be 600 metres thick, and to contain 30 metres of inter-trappean sedimentary beds measuring between 10 and 20 flows of basalt with thicknesses varying from 15 to 91 metres (Pascoe, 1959; Fox, 1931; Krishnan, 1958). Within the fine-grained Rajmahal traps, a columnar structure has developed in many places due to prismatic jointing (Wadia, 1975), which has facilitated *pathar khadans* – basalt stone quarrying in the region. The stone is fine grained, dark coloured, clearly hard ringing under a hammer, and fracturing with sharp edges like flint. The stone quarries occur in an area of transitional landscape orientated in a north-south direction, between the Rajmahal Hills to the west, and the Ganga Plain to the east. In this approximately 2,000 sq km area, the geomorphology sequentially changes from depositional plains, to erosional plains, to highly eroded plains with residual hills, uplands and mesa-like hills; the changes in the physical and cultural landscape are rather dramatic in this short distance. Moreover, the hill zone extending north-south from Sahibganj to the Brahmani River is traversed by two local streams, the Gumani and Bansloi.

The area's largest and most elongated block of hills occurs north of the Gumani, and averages 250 metres in height. Its highest points are Kochla Pahar, Chaugharia Pahar and Khuta Pahar. The second massive block extends from Basko Pahar to Chuparvita but at somewhat higher elevation. The third sizeable block is situated between the Gumani and the Bansloi at an elevation between 250 and 300 metres. The block between the Bansloi and the Brahmani is the fourth massive basaltic landmass.

Historically, the region's terrain has played a major role in the development and distribution of settlements, further influenced in recent times by the stone quarrying industry. Incisions in streambeds reveal knicks that expose the underlying basaltic material; at these junctions, quarrying becomes a feasible undertaking. Today, the inter-fluvial promontories of the erosional plains are heavily mined. Initially, the quarrying of basalt occurred around Sahebganj town, which was in close proximity to the railways. However, over time, the industry expanded to Pakur, where the low mounds of Malpahari, Kulapahari and Rajgram have provided ample opportunity for vigorous quarrying.

TERRA NULLIUS? INDIGENOUS COMMUNITIES AND LAND OCCUPATION

The social history of the region partly explains how women of indigenous communities and quarrying have interacted. The Rajmahal region has been inhabited by *adivasis* or tribal peoples but the land was always desired by the mainstream Hindu populations.

Rajmahal region is part of the *Damin-I-koh*, the core of which is inhabited by indigenous groups. Since ancient times, the groups that have occupied the region have depended on the natural resources provided by local dense jungles for their livelihoods. In all likelihood, the Pahariyas – both the Malers and Malpahariyas peoples – were the earliest occupants of the area; the Santhals are now its largest indigenous group (Kumar, 1986). The Malers were slash-and-burn cultivators, their material culture revolving around the forests, which provide a source of shelter, food and beverage; raw materials such as sabai grass, lac, silkworm, honey and bamboo; and a base for forest-dependent cultivation. The Malpahariyas have settled mainly in Dumka and Pakur, and have maintained a separate identity to the other indigenous groups of the region. They practice a *kurua* type of cultivation shifting, whereby dry crops such as millet, maize and beans are harvested using manure as fertilizer. They also hunt with bows and arrows, and fish using basic traps and nets, as well as spears. They also rely on local forests to earn a subsidiary livelihood.

The Santhals descended to these lands towards the end of the eighteenth century, at which time, other cultivating Hindu castes were migrating from the plains to the uplands of Chotanagpur Plateau (Dutta-Majumdar, 1955). Famines added to their desire for additional land, and they soon moved onward to the Rajmahal Uplands. The *zamindars* (local landlords) of Bengal hired Santhals to clear up new land for agriculture. The cultivating Hindu castes pushed the Santhals to the upper parts of the hill slope, whilst the Santhals, in turn, pushed the Pahariyas to the hilltops.

In places where the valleys of the east-flowing rivers open up to the Ganga plain, *Ghatwali* (caretakers of the "*ghats*" or hills) kings had been posted during the medieval period as standing militia to protect the plains' agriculturists from possible tribal invasions. Eventually, the government encouraged the Santhals to settle in the down-slope parts of the region, granting them land that was free of taxes and ordinary courses of law. By 1832, the colonial British government was trying to resettle the Pahariyas in a part of *Damin-I-Koh* to stop lowland *zamindars* from making inroads into the hills. However, they refused to vacate their upland locations. These fertile strips of land between the hills and the boundary were soon occupied by the Santhals (Hunter, 1868), who, with their strong tradition of group work in their hunting and food-gathering activities (Somers, 1977), were able to reclaim much of the jungle lands of Rajmahal region. This attribute of Santhal society would eventually enable their women to participate in quarrying activity in large numbers.

The Santhal people continued to harvest large crops from the virgin soils of *hal hashila* (recently reclaimed) lands (Wolfe, 1966) but gradually became incorporated in a market economy dominated by caste Hindus in which they were powerless to negotiate. There were minor internal conflicts between the Santhals and Pahariyas but the clashes between the caste Hindus and the *adivasi* groups over land ownership eventually precipitated a sharp division in modes of production, social relationships, behaviour patterns and attitudes toward life.

The introduction of the East Indian Railway Loop Line opened up new avenues for employment (Dutta-Majumdar, 1955). New land-owners began to buy land previously owned by Santhals, who, in turn, began to migrate away from Rajmahal to the newly-established plantations of Assam (Gausdal, 1960).

Some had even become indentured labourers in the rubber plantations of Mauritius (Hunter, 1877). In 1820, a tract spanning nearly 480 km in circumference, comprising portions of Rajmahal, Godda, Dumka and the Pakur subdivisions, was marked to protect the interests of the Pahariyas and to prevent the encroachment of lowland *zamindars* of Bhagalpur (Kumar, 1986); the Santhals were welcomed in the valleys and slopes of the demarcated areas. Following the rebellion of 1855, when the normally-timid Santhals and Pahariyas burst against the oppression of the *sahibs* (white colonial administrators), *dikus* (outsiders) and *bhatias* alike, the Santhals' land was returned, and *adivasi* lands were made non-transferable to non-tribal people (Somers, 1977).

However, there are major differences between titular and usufructuary control. By exerting usufructuary control, quarrying activities have begun on legally non-transferable *adivasi* lands. In many cases, the *khadan malik* or the quarry-owner persuades the *adivasi* landowner to part with his land under the false pretext that the quarry owner will improve its quality by removing restrictive stones. In return, owners are to receive a nominal sum per acre per year but the quarry-owners are only entitled to the surface rights of the land for a 10-year lease period without changing the configuration of the plot (Das, 1993). After economically-viable stones are exploited, the ruined land that is returned to tribal control is typically riddled with large hollows, overburden dumps and waste material. The law states that the lessee must provide compensation to the lessor in account payee cheques but since the *adivasi* lessor is often illiterate, the quarry owners take full advantage of the situation and get away without paying any compensation.

The *adivasis* have remained in the Rajmahal region, helping to develop its stone quarrying industry but only as poorly paid labourers with uncertain jobs and incomes. The expanding mineral extraction economy gradually introduced a kind of reluctance on their part to cultivate the lands that at one time, provided shelter and other important resources. The land was lost to them, as were the resources necessary for their survival; the *adivasis* thus became wage labourers on the very land that had once belonged to them. The dispossession of the *adivasis* meant that the land and its resources took on a different meaning – something that is to be exploited without any concern for the future – because it no longer belonged to those who had lived there for generations. Women, who, in *adivasi* societies, had a pride of place as workers and copartners in production, have taken on a much more insignificant role in the *pathar khadan* economy. More specifically, the new mining and quarrying economy has attributed a rather low status to women, using them as much as possible without due value for their labour.

STONE QUARRYING IN RAJMAHAL

Let us revisit the physical properties of Rajmahal Traps, or, in fact, its stones. The region's basalts are bluish-green in colour, hard and heavy, and of porphyritic texture without any ash or zeolitic beds, and are available in abundance. During ancient times, basalt was used in the construction of the now-ruined city of Gaur. It was also used as a supplement in doorposts and lintels in certain temples and buildings of Rajmahal town. Whilst constructing the railway bridge over the

Ganga at Mokama in Bihar, stones were supposedly tested by the Eastern Railways and proved to be the best quality of all of Asia's building and construction stones, largely because of their ability to withstand great pressures. This was the greatest impetus behind the large agglomeration of small stone quarrying operations or *khadans* in Rajmahal Traps – an occurrence unparalleled not only in eastern India but in the entire country.

The technologies used in these stone quarries are rudimentary, the only mechanized unit being the crusher. The physical conditions are also favourable for quarrying activity; low mounds underlined with good quality basalt are suitable surfaces for economic quarrying. In such topography, the amount of labour required to remove the overburden is minimal because of the proximity of hard rocks to the surface. Stone quarrying, a highly labour intensive activity, has benefited from the comparatively poor agricultural sector, and the availability of plentiful cheap indigenous local labour. The cost of labour is low, which enables high labour intensity. Many *adivasis* now undertake quarrying activity along with women as working partners; without the *adivasi* labour, the stone quarries would not have come into existence. Although women played an important role in assembling a necessary labour force, as we shall see, they have now ended up as subordinates to the *khadan* industry.

As per Regulation 106 of the *Mineral, Mining Rules (MMR)*, 1961, the *khadans* are required to adhere to particular requirements. Most notably, rock walls must not exceed a slope beyond 60 degrees; the height of each bench should be restricted to six metres (Das, 1985); the roads within the *khadan*s must be more than seven metres wide; and the slope of the road should be less than one in ten (Mukherjee, 1984). As in other forms of opencast mining, the removal and disposal of overburden requires vast quantities of land. However, as the processing of stones is almost always performed within the area acquired for the *khadan*, and the fact that the thin soils of upland Rajmahal do not support prosperous agriculture, the acquisition of large tracts of land for various activities (of the *khadans*) is not especially challenging.

The first quarry operators were very much in tune with the state of the minerals industry elsewhere in India at the time (Rothermund & Wadhwa, 1978), the-then *Maharaja* of Kashimbazaar and *Maharani* of Pakur investing the earnings of their *zamindari* estates. The colonial *sahibs*, however, initiated expansion of the industry. In 1870, an early entrepreneur, Atkinson, was attracted to the bluish-green compact basalt of Udhua *nala* (stream) situated close to the Ganga. Ambler followed him to set up his own unit in Maharajpur in 1877 (O'Malley, 1910). Initially, the waterways – more specifically, the accessibility to the Ganga – proved vital in the selection of quarrying sites. As a result, some good-quality basaltic deposits, located at a distance of 2–3 km from the river, could not be exploited. During summer, the stones were transported in carts from the quarry-site to the Ganga, whilst during the rainy season, the boats could enter the quarrying site directly through the *nala* or stream channel. Prior to the commencement of stone quarrying in Rajmahal, the Calcutta Municipal Corporation imported basalt. There are hardly any official data or accounts of quarrying from these time periods, and, apart from special circumstances, government agencies took little interest in developing quarries. Government departments such as the Indian Railways and the Public Works Department were the major customers of the stone produced by these quarries. By the time the railways were expanding in this part of India, and the need to build bridges such as the Sara Bridge at Mokama was being felt (toward the last quarter of the nineteenth century), references to Pakur quarries were beginning to be made. Rare personal histories and memoirs provide more information from which one can draw a rough sketch of the historical growth of the quarrying industry. For example, after 1909, quarrying activity began to increase at Malpahari near Pakur, and was soon complimented by the investments of local landlords and royalty. From Malpahari, quarrying extended to Rajgram within Bengal, under the initiative of the *Maharaja* of Kashimbazaar. Between 1930 and 1950, a great expansion of small and medium sized quarries took place along the railway line north of Pakur up to Maharajpur.

Following the partition of India, Sindhi and Punjabi businessmen who had crossed the India-Pakistan border and had brought wealth in the forms of gold and cash, settled in Pakur to invest in stone quarries (Das, 1993). The building boom that followed independence provided the first impetus to the quarries at Malpahari, and in the area between Pakur and Maharajpur. Strangely, not much of this has been recorded in the Government of India's documents. Only the Census Report of 1961 provides an inventory of the number and locations of stone crushing units in different parts of the districts of Santhal Parganas and Birbhum. The later censuses completely ignored this industry, a period that coincided with the most rapid growth of the *khadans*. The growth of this unorganized industry continues unabated, and remains largely unrecorded and disorganized, although most miners operate legally under an individual licensed ownership scheme. Numerous small quarries also emerged outside of the principal clusters of Malpahari and Rajgram, supplying boulders and ballasts to the crushers based in close proximity to railway loading sites at Pakur. Many of these quarries were ephemeral operations and were quickly abandoned, although some survived, in spite of the diseconomy of the poor transportation system of bullock carts.

Since the 1960s, sizeable industry expansion has occurred in Barharwa, Taljhari, Bakudih, and Borio villages, in connection with the construction of the Farakka Barrage across the Ganga. From this time onward, quarries began shifting away from railway track locations, depending on available road transport. In fact, the three phases of growth of Rajmahal's stone quarries are broadly associated with three different locations: riverside (1870–1900), railside (1900–1960) and roadside (1960 onwards). The 1961 Census recorded 65 crushers in Santhal Parganas of Bihar, and four in Birbhum of West Bengal. In Santhal Parganas, 63% of crushers were located in Pakur. Presently, no less than 260 quarries operate in the region stretching from Rampurhat in the south, to Sakrigali in the north (straddling Bengal-Bihar border). More than 65% of these quarries are located in the district of Sahebganj alone. There is, however, no recorded data since 1961 that effectively illustrates the magnitude of the period of extensive growth experienced in the region that began at the time.

Whereas quarries are all worked by local indigenous peoples – both women and men – Sindhis, who hail from the province of Sind in Punjab, mainly control the ownership of the *khadans*. This ethno-cultural group is now domiciled in Pakur but typically maintains an insular existence. Other owners are non-locals as well, and include Punjabis, Marwaris and Biharis; these groups have other business interests in adjoining areas such as retail and wholesale trade, transport and manufacturing. Many Biharis are related to the Maharajas of Pakur or were involved in various services at the *Raj* estates. Over the last one hundred years, when the *khadans* began to flourish, the ownership pattern evolved to reflect the overall pattern of growth in the Indian mining scene.

Presently, the Sindhis own more than 80% of the quarries, whilst the remaining operations are owned by Marwari, Punjabi, Bihari and Bengalee entrepreneurs. The quarries have an elaborately hierarchical social composition in which the *kamins* or women *adivasi* labourers belong to the lowest rank, forming the weakest section in the entire *khadan* economy.

PRODUCTION IN *KHADANS*

What is the cumulative production from these quarries? The entire production of stones of the Rajmahal Trap quarries is destined for markets outside of the region. No official records exist but some informed guesses could be made from personal observations and the data for stone goods exported by the Railways. The Ward section of Pakur Railway station provides relatively reliable information regarding stones carried in railway wagons but these data do not include the produce handled by road transport. Within the region, the quarries of Malpahari contribute a rather high share – approximately 90% – of the total volume of stone traffic. Stones produced by the crushers at Nagarnobi, Kulapahari and the roadside quarries west of the Railway track contribute no more than 10% of national output. In the last 30–40 years, stone production has increased sharply, keeping pace with increased demands for urban housing and other infra-structural constructions. Even these Railway Ward data tend to understate the figures, as Railway wagons are often loaded beyond their 600 cu ft capacity. Loading is performed by quarry owners, in collaboration with corrupt railway staff, in order to minimize freight charges and to disguise production/sales figures.

In India, no agency keeps tabs on the road traffic movement, records pertinent data, or maintains authentic information on stone production. At the exit point, all trucks passing through Pakur must pay a road tax that is supposedly recorded in logbooks but these figures are, in fact, more manipulated than those provided by the *Quarry Owners' Association*. However, quarry owners are inclined to keep the temporal increase of their production data a secret to avoid paying income tax.

Around Rajmahal, local consumption of stones is minimal. The railway data put total rail export at 52,894,200 cu ft but from personal observation, a more accurate – yet, conservative – estimate can be made based upon the number of trucks carrying stone goods in the Rajmahal Traps region. This road export may be taken as the equivalent of 250 trucks carrying loads of 300 cu ft for 300 days, (the monsoons having a few slack days), which amounts to 22,500,000 cu ft. Assuming a manipulation of the freight records to be about 100 cu ft per every 600 cu ft, the total annual production figure is in the range of 88 million cu ft or more than three million tonnes, based on informed guesswork. If these quantities are taken as saleable product, with a waste stream of 40%, the gross production of the Rajmahal region amounts to some five million tonnes, which is not consistent with the production data provided earlier. Indeed, basalt is not identified as a separate item in Indian stone data, and here, we are only dealing with one quarrying region.

The general trend in transport over the last 15–20 years has been a gradual decrease in rail compared to road. This is due to the increased availability of trucks compared to railway wagons, and the fact that trucks can deliver stones to the point of demand, thereby eliminating intermediary stages of loading and unloading at break of bulk points. The cost of transporting stones by road is slightly higher; a wagon of stone from Pakur to Calcutta costs approximately Rs 3,000 (US$60) by rail, whereas about Rs 4,000 is required to transport the same quantity of freight to the same destination via road. Again, the average annual production data hardly provides an accurate picture, as there is a distinct seasonality involved in the production of stone. During the months of February–June, prior to the monsoon decline, production is at its highest. *Khadans* are almost always closed during the rains, with the exception of certain roadside quarries situated away from the congested Malpahari cluster. These units also hire more tribal workers than the others and actually increase their production during the rains. The lower parts of larger *khadans* are typically submerged with water during these times, and although pumps are used to drain shallower quarries, in deeper excavations, the depth of water logging can be as much as 15 metres, which can cause a stoppage of work altogether. Moreover, because construction work also slows during this time, the demands for mined product are less. Managing a quarry efficiently becomes near-impossible during the time of the monsoons.

Quarrying activity picks up again from November onward, peaking in February because of the harvest-related activities in agriculture. The uneven topography, combined with the inadequate rainfall of Rajmahal region does not permit a thriving agriculture but during times of peak harvesting and sowing, labourers who would otherwise work in quarries, revert back to their traditional occupation. The harvesting and planting seasons are also festive seasons for local indigenous groups (*adivasis*) and some local/converted Muslims. As people are required to spend more money during these periods of festivity, they try to supplement their poor incomes by working more hours during the April–May and December–January periods. There is no season-based wage incentive for laborers, as many are marginal farmers, sharecroppers and landless labourers, who represent attached labourers to land-owning households. The bulk of the quarrying labour force is therefore partially bonded agricultural labourers who find it hard to leave their villages during peak agricultural seasons lest they are required by their landowners in exigencies. Women workers are usually tied to their family's movement, working as part of a "family unit". The labour contractors or the owners find the family labour system rather amenable to their needs: the payments are made to the family unit as a whole treating women as the equivalent of about half a man's work. Children are also part of the family unit, and often follow their mothers in their work.

TECHNOLOGY AND ECONOMICS OF *KHADANS*

A rudimentary extractive industry, stone quarrying has several interesting technological and economic aspects that are worth exploring. Almost all of the work performed in the *khadans* is manual and hence the largest component of cost is the wage bill. In fact, the labour intensive character of the industry has made it an easy form of capital investment, capable of providing a high rate of return; this labour intensive character is typical of informal activity. The patriarchal nature of the mining economy has determined a low status for all women involved in *khadan* work, and has assigned them the heaviest, and most monotonous, of jobs with the least security and most risks. Without the hard work put in by women workers, many of the quarries would not operate. However, there is little recognition of the fact that women are exploited at every level, whether at home or at work.

The work in the quarries, although manual, is elaborate and may involve several stages, each completed distinctively by both men and women. Being entirely informal in nature, the quarrying industry does not offer any benefits related to these jobs; diseases like tuberculosis and asthma are common and cause early deaths. The *adivasi* labourers are not tied to any particular quarry and tend to move from one to another. The quarry-owners do not offer any medical facility, and the nearest hospital is many kilometres away.

During the early days of quarrying, an intensely heated iron pot, such as a cauldron, was placed on massive stones for blasting, after which, the rock was cooled with water. Repeated heating and cooling caused the development of cracks in the rock, which was later hammered into pieces; this process was entirely manual. Today, the actual task of dislodging the stones is undertaken by men, who blast the explosives drilled into the rock either manually with a hammer or using a drilling machine. In Santhali, the blasting operation is referred to as "*hul*", meaning a stir or rebellion. Special caution is required for blasting the layer called "*kekri*", because the stone splinters may scatter a great distance. Blasting with gunpowder has become obsolete, and today, mainly gelatin is ignited with electricity through a remote control device. The removal of the overburden is performed manually using pick-axes, spades and crowbars, and then dumped into adjacent vacant lands by *adivasi* women, who carry waste rock in their baskets and spades. The blasted and loose boulders are then separated from the rock wall manually using pickaxes, crowbars, hand drills and hammers. Men perform this job, although at the next stage, the cutting of sized-boulders and loading of boulders onto trucks are both predominantly performed by women. The boulders are then hauled from the *khadan*s to the crushing sites for further processing. Here, women usually unload boulders from the trucks. After quarrying, the stones are shaped into different sizes to suit the requirements of customers.

The largest stones, *boulders*, are used to reinforce riverbanks and canal banks, and used in embankments. *Stone bricks* were once used widely for road construction and for paving tram tracks in Calcutta. With the advent of concrete, they have since become nearly obsolete. *Chellies* are intermediate products that require further processing before they can be used for various construction purposes. *Chips* are derived from crushed chellies, and *ballast* consists of pieces of stone meant for lining the railway tracks and for road building. Certain types of ballast used to be hand-made by women but that process is gradually changing. Chips are the ultimate finished product of the basalt stone quarrying industry. Processed into different sizes for different purposes, chips are used in concrete casting and in massive gravity structures like dams. Smaller chips are used in the construction of residential houses, and the smallest chips are used in the casting of roof and floors. Transport from the crusher to chip dumps is an activity performed by conveyor belts. The by-products of *khadans* include stone dusts that are used for land filling, and low quality mortar used for temporary structures. The basalts are covered on the top by a red earth, which is removed as overburden. This earth, when granular, is locally called "*morrum*", and is used in road construction because water can percolate through it swiftly. Gravels are by-products of inter-trappean beds, and are used for road building and lining river and canal banks. In the entire stone quarrying industry, crushers are the only mechanized units.

Mechanization of the operational processes in stone quarrying in the Rajmahal Traps region has remained low because of the inexpensiveness of labour and the lack of enforcement by the government of any labour rules and laws. Women workers bear a huge burden of this largely manual work at insignificant pay levels. As safety rules are non-existent, accidental deaths are usually not compensated for, and it remains up to the individual to take the necessary precautions for personal safety. A philosophy of "looking after the weaker sex" pervades the mines. Women are excluded from explosive usage and forbidden from using machines such as crushers when they are present. They are assigned mainly transport duties; they move head loads of baskets with masses as much as 50 kg. Local trucks carrying boulders from the *khadans* to the crushing site do much of the on-site transport jobs. For this purpose, temporary labourers are employed through *thikadars* or labour contractors. Women workers dominate the on-site truck transportation, as these jobs are even more uncertain than those in the *khadans*. Off-site transport costs are borne by the purchasers.

The cost of production varies from firm to firm, depending on their size, the quality of rocks, and the availability of labour. In Pakur quarries, a seasonal variation occurs during the rains as additional charges are levied for pumping rainwater that has accumulated in the *khadans*. The running costs of a quarry include wages for labour, electricity and royalty, whereas the investments made for infrastructure, management and machinery like crushers, compressors, generators and buildings are regarded as fixed costs. The cost of production also varies from product-to-product because they involve different sets of jobs and different rates of payments. An amount of 100 cu ft of stone chips requires about 150 cu ft of boulders to be crushed. Wage payments are made in three ways: piece rates, fixed daily rates, and contractual rates. In reality, labourers are paid mainly on a contractual basis for operations like blasting, drilling, truck loading and unloading. The crushing of stone chips from chelly generates about 40 cu ft of stone dust for every 100 cu ft of product. This dust is carried away at the rate of 10 *paise* per *tukri* or basket. The wages of machine cleaners, greasers, *tickli* (token)-distributors and water carriers are paid fixed daily rates that are low by even Indian standards, not exceeding Rs 1 per 100 cu ft of chips produced. However, the proportion of these costs varies according to the scale of production, rising in smaller firms and diminishing slightly in larger ones. The salaries of managerial and supervisory staff are usually higher but are also paid fixed rates. Their total ratio comes to about Rs 30 per cu ft of stone chips. The cost of electricity to produce 100 cu ft of stone chips is approximately Rs 10. The amount of royalty paid by the quarry owners is fixed at the rate of 80% of the pit-head value of stone. In Pakur, the pit-head value of 100 cu ft of stone is about Rs 45, leaving an end sum of only Rs 36. Fixed costs are as follows: Rs 200,000 for a crusher; Rs 50,000 for a compressor drilling machine; and Rs 50,000 for the construction of a small shed.

Coming back to the work performed by women labourers in the *khadan*s, we can sum up the observations as follows. Work is undoubtedly tedious and extremely labour-intensive, with little intrusion of capital or technology. If technology, such as crushers or explosives, is used, then these job areas immediately become male prerogatives; women perform all of the heavier burdensome jobs of transporting using head loads of baskets, and sorting stones according to size. They are pushed into types of work within *khadans* that are more uncertain or less secure. The quarrying economy of the *khadans* of Rajmahal is therefore not gender neutral. Despite a lack of mechanization and the informality of *khadan* work, women workers are still relegated to a lower level of production, performing tasks that are burdensome jobs and which are associated with great risks.

The next section of the chapter examines the physical and social changes that have been brought about by the stone quarrying industry. An effort is made to show how these changes have made it even more difficult for the women inhabiting poorer communities to survive.

KAMINS IN *KHADANS*

As noted, women workers are invaluable to the stone quarrying industry of Rajmahal Trap region. They are commonly referred to as "*kamin*", a feminine equivalent of the term "*coolie*". *Kamins* used to be common in Indian collieries and other mines (Roy Chaudhury, 1996), constituting some 40% of the total workers in *khadans* of various capacity, although this figure is based on informed guesswork. Today, there are approximately 40,000 labourers – both male and female – working in the entire region, including Bengal and Bihar; they comprise the *adivasi* communities and Muslim groups migrated from the nearby Murshidabad district. Local semi-*adivasi* groups such as bauri, bhumij and kurmi also work regularly in the *khadans*. This total figure refers to only those who work here on more or less a regular basis, or who have come through various labour contractors or *thikadars*. Besides these, there are also those who work in *khadans* as temporary workers for trucking companies. Workers usually come from an area of approximately 20 km around the *khadan* zone. The women workers came into quarrying work along with their families, and are often treated as a single unit with their male counterparts. Women also come with children of all ages, who often work in the *khadans* with their mothers.

The system of recruitment in *khadans* is interesting, and is usually the same for both male and female workers. Originally, there were two systems: *zamindari* and *nokrani* (service tenancy). Under these systems, the *zamindars* owning *khadans* would offer agricultural plots to workers on the condition that they would also work in quarries. It was clearly a semi-feudal arrangement with an industrial variation established to ensure a steady flow of labour. At the same time, it provided a degree of freedom to the quarry workers. This *zamindari* system of recruitment gradually changed, because of the lack of arable land around the quarries and the stronger attachment of labour to the land than the *khadan*. This, again, created a problem of unsure labour supply. Owners' concern over this uncertain labour supply situation resembles that in Indian collieries; in 1920, Mr Burt of the Coalfields Committee lamented:

> All our efforts should be directed towards obliterating all agricultural ideas out of the miners' heads. It is miners we want, until we have a mining class, our product chart will be like the peaks of the Himalayas. (Quoted in Roy Chaudhury, 1996).

Hence, an alternative system of recruitment had evolved based on contract, through contractors or *thikadars* solely responsible for the regular supply of workers in the quarries. Once brought to the *khadans*, the workers are paid by the quarry-owners; some *thikas* are offered payment on a per capita or commission basis. The agents of *thikadars* managed a sizeable supply of male and female labourers to the *khadans* through the payments of advance donations for local rituals, or the granting of various allowances. However, in recent years, there have been many incidences of voluntary employment of local *adivasis* without the intermediation of a *thikadar*.

Both the labour and care of *kamins* were needed for the flourishing of the stone quarrying industry in Rajmahal. From the beginning, *khadan* owners were eager to employ women workers as they found *kamins* to be more suitable for the tiring and endless jobs of sorting and carrying head loads. The male labourer usually came with his wife and children to work in *khadans*. Both parents worked along with their children, rested, slept and ate together. In fact, the quarry owners encouraged the *adivasis* to bring their female partners to minimize costs and to ensure a certain supply of labour. The wage levels for women were lower, and still are, in spite of their heavy manual workload. Initially, women were treated as part of family labour, and even when treated as a separate unit, the valuation of their labour was usually determined by quarry owners and labour contractors. *Kamins* remain important loaders in the *khadans*, working as partners to male cutters or as individual workers.

Although there are some nascent trade unions and activist groups working in the Rajmahal region, gender awareness has yet to grow among them. Living conditions close to the *khadans* are very poor. Dwellings have only one thatched room that acts as a kitchen-cum-bedroom, and lack toilet facilities. The heavy work, lack of medical facilities and poor living conditions are reflected in very high mortality rates and unsatisfactory health conditions among both women and men. However, women also bear the reproductive burden for their families, and are often the subjects of lust for local goons or leaders. Almost all *khadans* workers are illiterate and poor, which makes their exploitation all the more easier. Literacy rates are even lower among women. Often, *kamins* working individually are huddled together in tiny hutments, working hard to send their remittances to their villages to provide for the elderly in their families. Yet, it is clear that the patriarchal nature of Indian society has penetrated tribal values to ascribe a lower place for women. Stories of desertion and rape are not uncommon: Buti Mejhen of Pakur was raped several times by the *khadan* manager and when the news leaked, she was summarily thrown out of the *khadan* with the stamp of "immorality" put on her by her male colleagues. Her husband left her and took up another wife. Here, there is clearly a sense of solidarity, as males work more effectively than the class solidarity or ethnic camaraderie. In fact, *adivasi* communities in eastern India that never bothered much about a "woman's character", have changed in essentializing the values of mainstream society that go against the interests of women in quarries. Quarrying has led to a reversal of not only women's roles and statuses, but has also altered the way *adivasi* families look at their women and their work.

The chaotic nature of the *khadans* and extremely poor living conditions make certain that women are forced to accompany their families and be treated as part of a family labour unit. Thus, women are trapped in a vicious circle of subsistence in the *khadan* economy. Widows or single women tend to stick to their family relatives, and cases of sexual harassment are common. As low wages and the hard manual nature of work encourage drunkenness and alcoholism among men, women become more tied to their children and become their primary carers. It is not uncommon to find a young *adivasi* mother carrying a head load of 35–40 kilograms on her head whilst a baby is tied to her back or lying under the shade of a ledge or, increasingly rarely, a tree. Children are also looked after by older family members but as soon as they are able to perform work, are employed in *khadans*. As workers in *khadans*, children often break stone into pieces with small hammers for their mothers who carry the load to the crushers. Dinu Soren of a village near Pakur wakes up before sunrise to clean her mud hut and to cook lunch for herself and her family of two children and an ageing father-in-law, tugs the children along to her work in the *khadan* some 10 km away from her village to work the entire day with only a short lunch break, and then goes back in the afternoon to complete her household chores such as feeding her goat, washing, and cleaning and preparing the family's food. Often, she has little or no food left at all after feeding her family, and she has to depend on *panta* or water-soaked leftover rice. Yet, Dinu's husband feels she is not working hard enough, and spending her time chatting to her manager. As this case shows, apart from family responsibilities, cultural barriers act as a major obstacle in pre-determining perceptions of women workers in quarries.

Kamins are generally considered by quarry supervisors to be more responsible than male workers. A quarry manager commented that women workers usually spend less time chatting to co-workers. They may take a break every now and then, to look after or to feed a baby, or to smoke a *bidi* (local tobacco leaf rolls), but usually one can get the most out of their labour. On the other hand, male workers may report to work drunk, and often drink in the absence of supervision. In addition to the family wage, women workers are also paid on a piece rate basis depending on the numbers of baskets they carry. However, it is interesting that women are the first to be sacked when it comes to the need for retrenchment; thus their jobs are more uncertain and insecure than their male counterparts. In *khadans*, women workers are treated as a dispensable group, although they are the ones who work the hardest and earn the least. This situation reflects an imitation of the prevailing attitudes towards women workers in formal mines, where their presence is virtually non-existent. Even when in the employ of a formal operator, they are the most invisible group to even the trade unions that are supposed to protect the interests of the "working class" as a whole.

The International Labour Organization, in its 1999 report, had noted (p. 34) the pitiful condition of women workers in small mines all over the world:

> Because mining is often seen as "men's work" it has influenced women's attitudes to it and the attitude towards women of other groups, including male miners, government agencies, banks and NGOs. Women's perceived inability to cope with much of the physical side of mining, superstition, implicitly extending bans on women working underground to other activities that are not prescribed, the fear that women's presence might lead to "indiscipline", and the general unsavoury atmosphere of much small-scale mining all affect their participation. Family responsibilities also restrict the time many women can devote to mining activity, reducing their productivity and earning capacity.

Unless greater gender sensitization occurs at all levels of the mining industry, there will be little change. Women workers will continue to be subjected to hard labour for minimal compensation and the chance of losing their occupations in the event of enhanced mechanization and formalization of *khadans*.

A recent result of large-scale expansion of quarrying in the Rajmahal region has been a rapid deterioration in the subsistence resources provided by the local ecosystem. This has had significant impact on the poor women of *adivasi* communities. Thus, abandoning the low-paying work in the *khadans* is no longer an option because other sources of livelihood are rapidly deteriorating or vanishing. The next section of the chapter will briefly discuss this problem, and show how it directly impacts the lives of poor women.

PHYSICAL AND SOCIAL CHANGES AND WOMEN IN *KHADANS*

By nature, mining and quarrying activities leave a conspicuous footprint on the land. Their effects are physically visible in a concentrated location, and occasionally, the resulting alterations are inevitable as well. Stone quarrying leads to a dereliction of land, transforming it into a wasteland devoid of valuable resources, and a dangerous and unhealthy tract, in turn, rendering it useless for future development. There is no doubt that because of a lack of government control, the stone quarries of the Rajmahal Traps have had devastating effects on the regional physical environment. However, the less visible impacts have been social in character, having a great bearing especially on the survival of poor indigenous women living in the region. The extensive quarrying industry of Pakur has not only destroyed the physical and biological ecosystems of the region but has also altered the region's entire social fabric. Derelict lands, huge quantities of airborne dust, abandoned pits, overburden dumps, waste dumps, devegetation, soil erosion and degradation, desiccation of springs, reductions in surface run-off, and encroachment upon agricultural lands are among the most significant the physical impacts of stone quarrying activity in Pakur (Das, 1993). A quarry owner may sometimes stop the work tentatively in one mine and move on to another that is determined to be more cost effective. Thus, within the *khadan* zone, there are both operational and abandoned *khadans*, resulting in the entire area appearing like a moonscape. As is the case with the physical environment, the culture and society of the region have also undergone fundamental alterations. According to the 1951 Census of India Report, which paints a picture of the country just after its independence, about 95% of the population of the region belonged to cultivating classes, whereas in 2001, only 13% were engaged in cultivation – identified as mainly a secondary source of incomes. The expansion of quarrying has largely affected the condition of agriculture by making it physically impossible. Apart from the *Raja* of Pakur, all entrepreneur-owners of stone quarries originate from outside of the region but have come to control nearly every aspect of the industry. They have immensely benefited from the industry and have created a new social fabric that is in tune with the mores of the *khadan* jobs. The work relationships of *khadans* have impacted upon the social and cultural relationships of indigenous communities. The new social relationships have followed those created at, and by, the *khadans*, have had the greatest impact on women, whose roles have been marginalized, and whose jobs have become dispensable.

The groups that have survived on the meager resources provided by the land, including water, fish and weeds from local ponds, and fruits and fuelwood from orchards have suffered the most (Minewatch, 1997; Howitt, 1996; Ghosh, 1996). The ecologies of the *adivasi* communities of the past have been the most upset by the expansion of quarrying in the last 30 years or so. The resource processes adopted by the indigenous communities were based on a judicious mixture of gathering, shifting cultivation, sedentary farming and artisan crafts. This concurrent use of several natural resources was in tune with the subsistence base provided by the region's environment. The indigenous communities of eastern India are well known for their egalitarian structure; however, the sudden and drastic changes occurring from the recent boom in the quarrying industry have led to a maldistribution of benefits, emergence of new interest groups, changes in material culture, imposition of social costs, and finally, alterations in the gender norms of *adivasi* cultures. The entrepreneurs have emerged as the new power groups in the regional economy, enjoying rather high profit levels as compared to their investments. Local communities have emerged as cheap labour providers in the new resource extraction economy but have failed to gain a substantial livelihood through these jobs because of the loss of their natural habitat that provided them with other sources of livelihood. The introduction of financial capital, organized sectoral allocation of responsibility, and labour and wages with associated strict formal differences have prevented local women from benefiting fully from the industry.

The traditional income-generating activities performed by the *adivasis* – the collection of forest products, silkworm rearing, lac growing and sabai grass cultivation – have all been affected. The pressures exerted by a declining resource base are visible when the local trains coming from Barharwa are inspected: bundles of fuel wood are not only packed in almost every compartment but the large loads are also found hanging on the outside of the train. This fuelwood, which originates from the nearby forests of Litipara, Hiranpur and Lakshmipahar, are carried by bullock carts to the domestic market of Pakur for sale.

Mainstream values have not only affected gender relations but have also altered *adivasi* perceptions of the environment. The forests that so far seem to have survived the expanding quarry industry are, in fact, now being destroyed in an effort to meet the basic needs of people. The habitat conditions required for silkworm rearing and lac growing industries have changed to such an extent that they, too, have declined; sabai grass cultivation is an indigenous source of sustenance that has entirely disappeared from the area. According to village *pradhans* (heads) of Pahariyas, sabai cultivation has ceased due to their lack of direct access to markets and the interference of *mahajans* or middlemen. However, the fact that the agricultural plots adjacent to the quarries have become strewn with dust and fragments of various sizes has deterred the expansion of agriculture. The powerful quarry owners are subsuming the interests of poor indigenous peasants. The social costs of the quarrying industry have fallen to the least benefited class in the *khadan* economy. The *khadans* have physically encroached upon agricultural land and invaded the psyche of the local cultivating classes. Though the *adivasis* are essentially marginal to the mainstream quarrying economy, their value systems have also changed. Thus, on the one hand, traditional alternative occupations have declined directly or indirectly due to the impacts of quarrying, whilst on the other, employment at quarries has become the only source of income.

In *adivasi* societies, women work alongside men instead of placing emphasis on their reproductive tasks, thus leading to a lower fertility level *vis-à-vis* the mainstream Indian population. As livelihood resources become scarcer, women find themselves disempowered and without many choices. Moreover, the new economy imposes new behaviour codes for women that emphasize their domesticity, and hence, those women who work in the *khadans* become objects of male lust. As *adivasi* communities cannot afford to keep their women within domestic spheres, they find it difficult to accept the fact that their women are equal in terms of labour value. Increased violence against women at home and at work has become common, and decreased security of livelihoods offers women no choice but to accept the very jobs that devalue their labour. As local perceptions of gender relations change and begin to reflect those of the mainstream society, women become obliged to surrender to the forms of exploitation offered to them by the quarrying economy.

Several authors have noted the separation of men and women's worlds in mining economies (see, for example, McDowell & Massey, 1984). Work at mines is commonly seen as a dangerous, hard and hazardous masculine job that contributes to a particular form of male solidarity and endows the manual labour with attributes of masculinity. The unequal economic and social relationships between women and men imposed by the social organization of the new mining economy makes the position of women subordinate both directly and indirectly. In the *pathar khadans* of Rajmahal, gender inequalities have also defined a place for women: at the very bottom of the mining economy.

IN SEARCH OF GENDER-INCLUSIVE POLICIES IN INFORMAL MINES

The formal mining sector has increasingly excluded women (Lahiri-Dutt, 2000; Metcalfe, 1987), whereas informal mining features a substantial female labour contingent. The case of women workers of the *pathar khadans* of Rajmahal encourages us to ponder about the position of women in mining as a whole, and to put forward a strong case for mainstreaming gender issues in mining. As we have seen, local environmental issues are inherently related with issues of women's work in informal mines. In developing countries with poor track records of policing and legal enforcement, informal mines present several challenges that are better seen holistically through a gender lens.

The case of women workers in informal mines points to the need for making the work of women more visible, especially in this sector, where they are still participants in the productive process. It helps us to develop a gender perspective in informal mining, enabling us to value the work of women workers of poor communities in the entire mining economy. Such a view ensures that women are not treated as an inherently "vulnerable group" needing protection and/or passive beneficiaries, and seen as important actors and stakeholders in mineral resource management. A gender approach also facilitates an understanding of the causes of the subordinate and vulnerable position of women, and the exploitations and difficulties they face in assuming their economic and social roles. Women's roles in informal mining and the realities of exploitation have to be placed in a broader context alongside those of men if they are to be identified, recognized and assigned value in planning and decision-making in the mining industry.

Involving women as well as men is not only desirable for solving practical problems and ensuring increased equity between women and men in mining, but it is also increasingly apparent that the full participation of women is essential for sustaining economic development. As noted earlier in this chapter, women's issues are better understood from a holistic perspective, which does not separate the physical and social changes in a mining region. Women of poor communities as resource managers at the household level can tell us a lot about sustaining local environmental quality.

Within the field of mineral resources management, although there is increased awareness that development cannot be sustained without the participation of local communities, gender has not yet been established as one of the most important socio-economic factors to be taken into account. There is still a significant lack of understanding of the roles and responsibilities of women in all areas of mining. In addition, most technical or engineering planners lack the necessary knowledge, skills and tools to incorporate gender analysis into their programmes. Concrete involvement with gender issues is still largely dependent upon individual awareness and commitment.

A major failing in all areas is the lack of attention given to gender in policies – at both the international and national levels – dealing with aspects of informal mines or formal mines. Laws and policies related to mining should therefore be revised wherever necessary to give women and men equal rights to land resources and representation to all mining activities at all levels. Documentation on mining projects often lack clear information on the involvement of women and the jobs they perform. The lack of attention given to women in documentation gives reason to believe that the actual importance placed on gender is, in reality, not significant. Gender strategies tend to be underdeveloped in mineral resource management, involving ad hoc activities or interventions that have no long-term benefit for women.

Women are usually seen as users of domestic fuel like coal, and as volunteers with unlimited resources of time and labour, and are not consulted at the time of preparing Environmental Impact Statements. They are seldom treated as managers of local natural resources, and as those who are most affected by resource deterioration. In mining, it is taken for granted that men have public roles at the community level, performing such tasks as consultation with management and public decision-making. Women are presumed to have domestic roles – namely, the collection and use of fuel and water, care and education of children. Technology, when present in mining, tends to displace women workers. This has led to a "welfare focus" in mining-related development initiatives, in which men obtain all the more "productive" and formal management roles – i.e. those involving skill development and monetary rewards.

In all forms of mining, communities and households are displaced physically and from traditional occupations, which change peoples' livelihoods completely or lead to a decreased access to subsistence resources. We have noted that these changes have gender implications since women and men make different contributions to household livelihoods. Failure to take gender roles and responsibilities into consideration can lead to displacement, particularly the women of proper groups and communities. The undervaluing of women's roles and needs can lead to development interventions such as formalization at the expense of women's interests. Women miners are not seen as economic actors in their own right but are a vulnerable and marginal group. The creation of a participatory environment for women in mining would provide a basis for their involvement in sustainable developmental planning and decision-making; a greater involvement of women in community consultations is one way of ensuring that their voices are heard. As mine workers, women require training and support to enable them to take part in consultation and negotiation processes. In mining, men inordinately dominate the organizational and institutional environments, whilst it is women's hard work that sustains the more informal forms of mining. It has now become essential to acknowledge women's work and plan more inclusively.

REFERENCES

Aich, P. (1996). Gender issues in mining: Effects of retrenchment. In A.K. Ghose (Ed.). *Small/medium scale mining: A global perspective*. New Delhi: Oxford and IBH Publishing Co. Pvt. Ltd.

Alfa, S. (1999). Child labour in small-scale mines in Niger. In N. Jennings (Ed.). *Child labour in small scale mining: Examples from Niger, Peru and Philippines*. Working Paper, Sectoral Activities Programme, International Labour Office, Geneva.

Burke, J. (1993). Asian women miners: Recovering some history and unpacking some myths. *Paper presented in the Women in Asia conference*, University of Melbourne, 1–3 October, 1993.

Chakravorty, S.L. (2001). Artisanal and small-scale mining in India, Report submitted to the MMSD, IIED and World Business Council for Sustainable Development, London.

Das, B. (1993). *Economic and ecological implications of Pakaur basalt quarries*. Unpublished Doctoral thesis submitted to The University of Burdwan.

Das, S.C. (1985). Common causes of accidents in opencast workings and its safety measures. *Mines Safety Week Celebration, Birbhum: West Bengal and Santhal Parganas: Bihar*. Directorate General of Mines Safety, Government of India, Dhanbad.

Dutta-Majumdar, N. (1955). The Santhal: A study in culture change. *Memoirs of the Anthropological Survey of India*, No. 2, pp. 21–40.

Fox, C.S. (1931). Contribution in *Memoirs of the Geological Survey of India*. Vol. L8, pp. 204–8.

Gausdal, J. (1960). *The Santhal khuts*. Oslo, pp. 9–12.

GDM (2000). International Stone Handbook. Giornale del Marmo, Faenza, Italy.
Ghosh, I. (1996). Changing land ownership situation and the displacement of peasantry in the Raniganj coalbelt, West Bengal. *Journal of Indian Geographical Foundation*, Calcutta.
Howitt, R., John, C. & Philip, H. (1996). *Resources, nations and indigenous peoples*. Melbourne: Oxford University Press, Australia.
Hunter, W.W. (1877). *A statistical account of Bengal*. Vol. 14, New Delhi: Reprint Legos Press.
Hunter, W.W. (1868). *Annals of rural Bengal*. Indian Studies Past and Present Reprint, pp. 84–125.
International Labour Organization (1999). *Report on social and labour issues in small-scale mines*. International Labour Office, Sectoral Activities Programme, TMSSM, Geneva.
Krishnan, M.S. (1961). (5th Edition) *Geology of India and Burma*. Madras: Higginbothams Pvt Ltd.
Kumar, A. (1986). *Tribal culture and economy: The Malpahariyas of Santhal Parganas*. New Delhi: Inter-India Publications.
Lahiri-Dutt, K. (2001). *Mining and urbanization in the Raniganj coalbelt*. The World Press, Calcutta.
Martinez-Castilla, Z. (1999). Child labour in traditional mining: Mollehuaca, Peru. In J. Norman (Ed.). *Child labour in small scale mining: Examples from Niger, Peru and Philippines*. ILO, Geneva.
McDowell, L. & Doreen, M. (1984). Coal mining and place of women: A case of nineteenth century Britain. In D. Massey & J. Allen (Eds.). *Geography Matters! A reader*. Cambridge: The Open University.
Metcalfe, A. (1987). Manning the mines: Organizing women out of class struggle. *Australian Feminist Studies*. Vol. 4, Autumn, pp. 73–96.
Minewatch (1997). *Women united and struggling for our land, our lives, our future*. Proceedings of the First International Conference on Women and Mining, Baguio, Bolivia.
Mukherjee, T. (1984). Satarkata. *Mines Safety Week Celebration, Birbhum, West Bengal and Santhal Parganas, Bihar*. Directorate General of Moines Safety, Government of India, Dhanbad, Bihar.
Nandy, A. (1996). Changing socio-economic profile of women working in small/medium mines in India. *Paper presented in Global conference on small/medium scale mining*, Calcutta, India, 2–4 December, 1996.
NISM (1993). News Bulletin No. 11, October, *The National Institute of Small Mines, India*, Calcutta.
NISM (1994). News Bulletin No. 12, January, *The National Institute of Small Mines, India*, Calcutta.
O'Malley, L.S.S. (1910). *Bengal District Gazetteers, Santhal Parganas*. Bengal Secretariat Press, Calcutta.
Pascoe, E.H. (1959). *A manual of the geology of India and Burma*. Vol. 2, pp. 975–79 and 1018–1019.
Participatory Research in Asia (PRIA) (2001). Bonded labour in stone quarries in central India. *Newsletter no. 1109*, New Delhi.

Raja, R. & Purushottam, A. (1962). Pitchstone flows in the Rajmahal hills, Santal Parganas, Bihar. *Records of the Geological Survey of India*, Vol. 91, Part II, pp. 341–7.

Rothermund, D. & Wadhwa, D.C. (Eds.) (1978 reprint). *Zamindars, mines and peasants*. New Delhi: Manohar Publications.

Roy Choudhuri, P.C. (1965). *Bihar District Gazetteers: Santhal Parganas*. Patna, Bihar: Secretariat Press.

Roy C., Rakhi (1996). *Gender and Labour in India: The Kamins of Eastern Coalmines, 1900–1940*. Calcutta: Minerva.

Sahu, N.K. (1992). Mine reclamation and afforestation: Different methods. *Lecture notes of Training Course on Management of Small Mines: Techno-Administrative Aspects,* Organized by TISCO, Noamundi, 1–10 February.

Somers, E.G. (1977). *The dynamics of Santhal traditions in a peasant society*. New Delhi: Abhinav Publications.

Wadia, D.N. (1976). *Geology of India* (4th edition). New Delhi: Tata McGraw-Hill Publishing Company.

Wolfe, E.R. (1966). *Peasants*. Englewood Cliffs, New Jersey: Prentice Hall.

26

A Perspective on Small-Scale Mining in India

MRINAL K. GHOSE

As can be discerned from the book thus far, the criteria used for categorizing mines as "small", "medium" and "large" varies from country to country; in short, there exists no universal yardstick. In India, the number of mining units considered small and medium-sized has increased substantially during the post-independence period (Rudra, 2002). Although there are no definitive categories of mine size in India, the mining and quarrying of industrial minerals and construction materials that occurs at a subsistence level throughout the country – *i.e.* operations producing relatively small quantities of minerals and employing few people – is generally considered to be "small-scale". These operations have an annual production capacity not exceeding 50,000 tonnes (Ghose, 2003a).

Worldwide recognition of the fact that small-scale mining can make a significant contribution to developmental objectives has been one of the principal motives behind recent industry formalization attempts worldwide (Noetstaller, 1994). When compared to large-scale mining, the industry's requirement in terms of minimum reserves, implementation time, and initial investments are small, skills and infrastructural requirements are moderate, and employment per unit output is high (Argall, 1978). However, as has been explained throughout this book, artisanal and small-scale mining makes important contributions to international mineral output, and provides employment to thousands of people residing in rural environments. Such has been the case of India, where artisanal and small-scale mining, as in most developing world settings, has become an important economic activity, largely because operations have moderate capital and infrastructure requirements, and short construction periods.

The purpose of this chapter is to briefly examine the impacts of small-scale mining in India, and to identify measures capable of improving conditions in the sector.

MINERAL POLICIES AND LEGISLATION IN INDIA

In Indian mineral legislation, there are no criteria that define "small-scale mining" *per se*. Consequently, no *precise* statistical data are collected, maintained and published for proper appreciation of the role of small-scale mining in the country's economy. However, the absence of a nationally-accepted definition has not prevented policy makers, researchers and Government officials from estimating the contribution small-scale mines – recognized locally as informal mining activity, and, as already explained, operations producing relatively small quantities of minerals and employing few people – make to the national economy. It has been reported (Chakraborty, 2002) that such mines constitute some 90% of India's operations, producing 42% of total non-fuel output and minor minerals taken together, and 6% of fuel-mineral output (coal). The Indian Government, which has devoted significant attention to small-scale industries (SSI) at the regulatory level, has paid little attention to its small-scale mining industry. In fact, not only has the Government largely ignored the industry from a policy-making standpoint but it has also failed to provide operators with the necessary guidance and support, in turn, causing the sector to develop in a haphazard manner. However, some progress has been made in the past 10 years, mainly among the more organized operations, which have begun to invest more funds in mine development and advanced technology; this has increased productivity and improved their position in what has emerged to become a competitive domestic market.

As there is no official definition of small-scale mining in India, there are no specific legal codes and regulations for its operations. However, it appears that indigenous policy-makers are using exemptions in mining policies as a basis for developing a working definition based on size and the nature of activities. For example, the *Mines Act*, 1952, does not apply to operations that use explosives, do not employ more than 50 persons, and are more than six metres in depth. Similarly, the *National Mineral Policy Statement* of 1993 does not apply to such operations. The likely reasoning behind using such "exemptions" as classificatory criteria for resident mines stems from the fact that there has not been any significant change made to the legal or policy framework during the past 10 years to indicate that there will be any effort to rationalize and regulate the existing/new activities within this sub-sector of industry.

Crude facts and figures indicate that informal, small-scale mining is practiced throughout India, and will continue for many decades to come (Ghose, 1990a,b). However, the techno-economic efficacy of all spheres of the industry's activities, from exploration, exploitation, management, and control, is in dire need of improvement (Chakraborty, 2002). The current contribution of Indian small-scale mining to global production is high in the case of certain minerals such as antimony (45%), calcium (50%), chromites (75%), clays (75%), feldspar (80%), fluorspar (90%), gypsum (70%), tungsten (80%) and vermiculite (90%).

Although mineral wealth rests with the State Government, the subject of "Mines and Mineral Development" is addressed within the national constitution of India. Parliament, therefore, has exclusive powers under the "Mines and

Mineral Development Rule", particularly with regard to assigning regulations – *as per* "Minor Minerals" – to the State Governments; it has been made law under Section 15F of the *Mines and Minerals (Regulation and Development) Act*, 1957. As a result, different states have framed rules under this provision, although country-wide, mineral rights are generally awarded in the following three ways: leases, quarry licenses and permits.

The *Industrial Policy Resolutions* of 1948 and 1956 recognized the important role of informal, small-scale mining to the national economy. A significant aspect of this resolution was that it did not place any hindrance on the operation of privately owned small mines, even in locations where large-scale operations were envisaged. Perhaps the most significant effort made to date to address the needs of small-scale mining from a regulatory standpoint was the inclusion of relevant passages in the recently declared *National Mineral Policy*. Specifically, Paragraph 7.12 of the Policy states:

> Small-scale mining with modest demand on capital expenditure and short lead time provides employment opportunities for the local population. Efforts will be made to promote small-scale mining of small deposits in a scientific and efficient manner while safeguarding vital environmental and ecological imperatives.

It further indicates that preference should be given to scheduled tribes for mineral concessions for small deposits in scheduled areas. With regard to the size of operation, Paragraph 7.2 emphasizes conservation and the development of scientific methods, and states that tenure, size, shape, disposition with reference to geological boundaries, and other conditions should be such as to favorably predispose leased areas to activities pursuing the systematic and complete extraction of minerals. The issue of environmental management in the industry is also adequately addressed (Ghose, 2001).

For the group of small mines involved in the extraction of minor minerals like building stones and sands, special provision has been made in the *Mineral Conservation and Development Rules* (MCDR) to relax the statutory qualification. The recently amended MCDR provides for the systematic planning of such mineral deposits through the use of a mine plan, and leaves implementation to those with lesser qualifications. The following operations are not considered under the purview of this Act:

- Mines that do not extend below the ground;
- Opencast workings where explosives are not used;
- Opencast workings that do not extend more than six metres below the ground; and
- Mines that do not employ more than 20 persons per day.

While the *Mineral Policy* decision clearly advocates leasing to local tribal groups, it is an accepted fact that statutory provisions are generally complicated in terms of language, and therefore, require parties to have a more thorough understanding of the implications of law. Most of the scheduled tribes for whose benefit the *Mineral Policy Resolution* has been implemented are not able to follow these provisions. More specifically, the provisions of all laws are not applicable in every case, and depend on the location and extent of deposit. Qualified experts can prepare a systematic mining plan, encompassing conservation, development, environment, and safety aspects, and the local people through persons experienced in mining can implement the annual action plan. Moreover, relaxation of legal provisions can be sought and granted by the statutory authority under the existing provisions of MCDR and MMR. However, what is truly needed in India is a government agency to assist individuals with obtaining a mining lease through a single-windowed system. The Government of Gujarat has taken steps in this direction through its Department of Industry, which has established the Industrial Extension Bureau.

A license to mine – in this case, on a small scale – must be obtained from the relevant state government. It is mandatory to first draft an environmental management plan (EMP) before commencing such projects (Ghose, 2001), as it helps to ensure that the potential environmental impacts of a project are assessed and incorporated into the early stages of development planning; the procedure for preparing an EMP has been accepted as a statutory requirement for granting a permit from an environmental angle (Ghose, 1997). For sanction for funding for all major projects, the Public Investment Board requires environmental clearance from, among others, the Department of Environment (DOE), and Ministry of Environment and Forests. All such mining projects must be cleared by the DOE to ensure that effective safeguards are in place to prevent environmental hazards. The DOE has issued guidelines for the preparation of an EMP report for mining projects. Finally, the Environmental Appraisal Committee for Mining (EAC-M), which was formed by the DOE, examines the EMP report before granting clearance for any project.

GEOLOGY

The Lesser Himalayan tracts of the lower reaches of the Himalayas, which have been fairly well explored, particularly during the last five decades, have proven to be a vast repository of several small, dispersed deposits of over 25 minerals and mineral commodities. These deposits are spread along a 3,000 km stretch, from Jammu and Kashmir in the northwest, to Nagaland and beyond in the east. They include limestone and dolomite, magnetite, phosphorite, talk-steatite, asbestos, gypsum, graphite, marble, rock salt, borax, barites, gemstones, fluorite, ochres, coal, oil and gas, various building materials, and metallic mineral resources of copper, lead, zinc, antimony, bismuth, cobalt, gold, chromium, tungsten tin, and iron ores. With the exception of oil and gas, the occurrences, prospects, and deposits of minerals and mineral commodities known to exist in the Himalayan region have proved to be invariably small by Indian standards, and therefore, amenable only to informal small-scale activity emphasizing mainly opencast methods (Rai, 1994).

Several small ancient pits, slag and mine dumps, and heaps, particularly in Uttar Pradesh, Sikkim, and the Nepalese Himalayas, bear testimony to the fact that mining, ore dressing, and smelting, for copper, and, to a smaller extent, lead, iron and gold, have long been carried out in region. The mining of gemstones, rock salt, borax and other minerals in the Western Himalayas appears to have been carried out for centuries. In recent years, quality resources of limestone, phosphorite, magnetite, talc, coal, oil and gas, and base metals have been identified in this region, and have attracted several private, public and joint sector organizations keen on working deposits on a small scale.

Most Himalayan mineral deposits are strata-bound, yet, small, tabular bodies, essentially controlled by a variety of stratigraphic, lithologic and structural factors. The majority of deposits are small and marked with small strike lengths and death extensions. The rocks hosting these deposits are generally highly stressed, as a result of complex geological folding, rejuvenated faulting, and thrusting. The prospecting and exploration of these deposits is circumscribed by the scarcity of observable outcrops, difficulties in distinguishing landslide or transported blocks of earth from *in situ* occurrences, and numerous problems associated with applying geochemical, biogeochemical and geophysical methods.

Drilling can also be challenging, as most deposits are covered with thick heterogeneous soils and talus covers, thereby requiring continuous casing of bore holes. Poor core recovery due to the highly disturbed nature of the strata, in turn, reduces the reliability of grade and reserve estimations. Numerous problems are encountered within the stereographic and structural correlations of mineralized zones, strata and veins – another factor reducing the reliability of grade and reserve estimations. Most ores, particularly nonmetallic deposits of phosphorite, magnesite, and coal, cause complex problems at the beneficiation stages, and are therefore associated with high processing costs.

When compared to underground mining, mechanization in resident opencast workings is less complex in design, configuration and operation. Some 70 of the 80 minerals (including minor ores) mined in India are extracted using small-scale means (Ghose, 1991). However, it has been a challenging task to categorize such mines because production is wide-ranging; certain operations have a daily output as low as tens of tonnes per day, whereas others have a daily production in the range of 150–200 tonnes. Moreover, operations have varying levels of capital injections, and some feature highly manual methods, while others are considerably mechanized. The variation in the nature of India's small-scale mining activities stems largely from the fact that each small-scale mining belt has unique techno-economic and socio-cultural characteristics.

The estimated number of small-scale mines operating in India during the 1985–1990 period is provided Table 26.1, and the locations of mineral deposits suitable for small-scale mining are identified in Figure 26.1.

SOCIO-ECONOMIC IMPACT OF SMALL-SCALE MINING IN INDIA

Small-scale mining is an important branch of India's mineral sector in terms of value, output, contribution to the economy, and employment. It has been

Table 26.1 Number of Small Scale Mines of some selected important minerals in India (1985–1990).

	Number of informal, small-scale mines (up to 50,000 t/annum)					Percentage of total number of mines				
Minerals	1986	1987	1988	1989	1990	1986	1987	1988	1989	1990
Asbestos	82	82	87	74	76	100	100	100	100	100
Bauxite	154	195	182	186	183	93	94	94	94	93
Baryte	52	46	51	51	45	100	100	100	100	100
Chromite	22	23	22	22	23	100	100	100	100	100
Coal	48	49	56	41	54	9	9	10	8	11
Dolomite	137	133	132	134	120	95	95	95	95	94
Feldspar	138	138	116	117	120	100	100	100	100	100
Fire clay	247	263	239	232	212	100	100	100	100	100
Graphite	31	40	51	51	50	100	100	100	100	100
Iron ore	243	259	237	238	206	71	72	73	73	68
Kaolin	180	198	182	183	170	100	100	100	100	100
Kyanite	13	15	15	10	9	100	100	100	100	100
Lime stone	486	563	569	525	546	77	79	79	76	77
Manganese ore	199	207	199	199	185	97	97	97	97	96
Mica	165	181	150	45	148	100	100	100	100	100
Ochre	93	113	91	87	91	100	100	100	100	100
Pyrophilite	40	45	43	44	44	100	100	100	100	100
Quartz	206	229	198	198	205	100	100	100	100	100
Silica sand	257	275	272	3012	274	100	100	100	100	100
Steatite	278	258	252	252	239	100	100	100	100	100

estimated that small-scale mines contribute about one sixth of the value of the world's non-fuel mineral output, although in many developing countries, output is significantly higher than this figure. As previously noted, in India, the output of some 3,000 small-scale mines accounts for 5% of national fuel mineral production (Ghose, 2003b). Moreover, it is commonly the only means of exploiting small deposits, and does not require substantial capital investment. In total, at least 500,000 people are employed at Indian small-scale mines (Carmen & Berger, 1990), which, collectively, represents some 50% of the country's mining labor force.

Other important benefits of small-scale mining in India include an ability to exist in remote areas with little infrastructure, to exploit otherwise uneconomic resources, and an ability to exist with economic overheads. Small-scale mining can usually be integrated readily into a wide range of social structures, which proves ideal in situations where the activity is seasonal. The ability of small-scale mines to generate employment and income in rural areas of the country has worked to stem urban migration. In addition, because they are generally locally owned, small-scale mines have provided more of a net gain to local communities and the national economy than large foreign-owned mines. At the same time, however, small-scale mining has been exorbitantly inefficient,

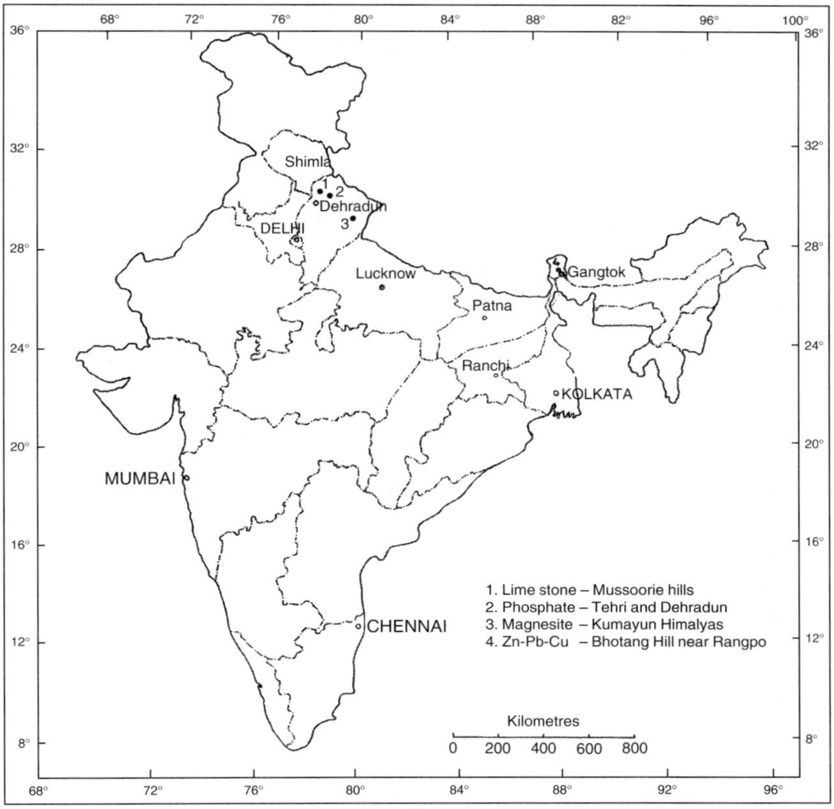

Figure 26.1 Himalayan mineral deposits suitable for small-scale mining.

featured poor working conditions, and has experienced its share of health, safety and environmental problems. Such problems have been compounded by the fact that much activity is carried out illegally, and has proven difficult to monitor and control. Moreover, because of widespread smuggling, there have been considerable losses to the miners themselves and the Government. It is still considered to be a disorganized industrial sector, often occurring in far-flung, isolated areas, disadvantaged by a lack of power and infrastructure.

To improve efficiency in the industry, the Government must establish mining centres that provide shared mining and processing facilities, as well as a practical mining advisory service that would provide information and training. Such a setup would enable small-scale miners to obtain free technical advice from mining engineers; gain access to drilling and blasting services; and utilize custom milling plants to process ore at competitive prices. One major factor that can inhibit the successful economic development of small-scale mines is the processing of ore at user-friendly prices. Artisanal miners must receive technical training and advice on the assessment of ore grades and mining practices so that processable mineral output can be maximized. Another important task of the Government is to create a geological base that provides small-scale miners with information regarding prospecting.

CONCLUSION: THE FUTURE OF SMALL-SCALE MINING IN INDIA

In the past 10 years, the status, role and importance of small-scale mining has improved in India. Although the value of all minerals extracted nationally has been ascertained statistically (Ghose, 2003b), no precise data have been compiled for small-scale mining. Similarly, the revenues derived from small-scale mine owners in the form of royalties, dead rents, and land rents, have not been properly maintained. Given the current level of disorganization in the industry, obtaining such data would be a time-consuming exercise.

In terms of legal reforms for small-scale mining, no significant progress has been made. The same applies for environment management; although the provision of Rule 31–41 of *Mineral Conservation and Development Rules*, 1988, is equally applicable to small-scale mining, its complete implementation is near-impossible. The Government, however, is in the process of developing a less rigorous EMP. Greater progress has been made in the area of technology, as many of the more organized operations have begun adopting mechanized equipment, which has undoubtedly improved production. Increases in production have stimulated increases in employment.

As this chapter has shown, small-scale mining provides a wealth of socio-economic benefits to the rural inhabitants of India, providing employment, income, adapting well within the confines of existing infrastructure, and helping to curb urban migration. However, the sector is still highly disorganized, and has failed to find a place in important national mineral policies. Given the obvious challenges associated with achieving heightened regularization, a practical next step is to educate mine owners and workers by establishing a mining centre with shared mining and processing facilities, and a suitable mining advisory service.

REFERENCES

Argall, G.O. Jr. (1978). Future of small-scale mining – important for the future, *Conference UNITAR*. New York, USA.

Carman, J.S. & Berger, A.R. (1990). Small-scale mining: a general review. In J.F. McDivitt (Ed.), *Small-scale Mining: A Guide to Appropriate Equipment* (pp. ix–xi). London: Intermediate Technology Publication.

Chakraborty, S.L. (2002). Issues and problems of Indian small-scale mines, *National Seminar on Policies, Statutes & Legislation*. January, CMRI, Dhanbad, India

Ghose, A.K. (1990a). Minerals, metals and developing countries – a forward look. *Proc. 14th Congress of CMMI*, Edinburgh, Institution of Mining & Metallurgy, London, pp. 249–255.

Ghose, A.K. (1990b). Technology trends and policies for solid energy materials in developing countries by 2001. Proc. 14th World Mining Congress, Beijing, Weitang, F. (Ed.) Pergamon Press, pp. 66–76.

Ghose, A.K. (1991). Towards an integrated plan for socio-economic development-the small-scale route. *Journal of Mines Metals and Fuels*, March, p. 7.

Ghose, M.K. (1997). Environmental management for disposal of spoils and tailings from the mines. *Environment and Ecology*, 15(1), 206–210.

Ghose, M.K. (2001). Management of topsoil for geo-environmental reclamation of coal mining areas. *Environmental Geology*, 40(11/12), 1405–1410.

Ghose, M.K. (2003a). Indian small-scale mining with special emphasis on environmental management. *Journal Cleaner Production*, 11, 159–165.

Ghose, M.K. (2003b). Promoting cleaner production in the Indian small-scale mining Industry. *Journal Cleaner Production*, 11, 167–174.

Noetstaller, R. (1994). Small-scale mining, practices, policies, perspectives. In A.K. Ghose, (Ed.), *Small-scale Mining – A Global Overview* (pp. 3–10). New Delhi: Oxford & IBH Publishing Co.

Rai, K.L. (1994). Geological and geo-environmental constraints in small-scale mining of Himalayan Mineral deposits – selected case studies. In A.K. Ghose (Ed.), *Small-scale Mining – A Global Overview* (pp. 51–64). New Delhi: Oxford & IBH Publishing Co.

Rudra, A.K. (2002). Policies, statues and legislation in small and medium mines. *National Seminar on Policies, Statutes & Legislation*, January, CMRI, Dhanbad, India.

27

Small-Scale Mining in India: Past, Present and the Future

AJOY K. GHOSE

Small-scale mining, which has thus far defied all attempts to be encapsulated into an apt generic definition, is a major segment of the global mining industry with extensive grassroots ramifications in national economies, especially of the developing world. Ever since the dawn of civilization, when the Paleolithic Stone Age man displayed his innate skills in shaping stone artifacts using the flint excavated by him, mining was essentially a small-scale entrepreneurial activity. Artisanal mining is a sub-set of small-scale mining, where, using either manual means or very simple and rudimentary tools, miners eke out a subsistence level of existence, mostly in the informal sector, outside the pale of legal or regulatory frameworks. Even if the definitions are fuzzy and overlapping, small-scale mining is important in its own rights as the only way to escape poverty in many parts of the globe. It has thus carved a distinct niche for itself for its economic value and also earned the dubious distinction of being "dirty, dangerous and disruptive".

Not much work has been undertaken in India into proto-history to trace the levels and origins of ancient mining activities; even the limited research efforts amply demonstrate that the level of mining as an industrial activity in the hoary past must have been way ahead of the rest of the world. Judged by today's perception, such mining must have been large in the-then contemporary scale of activities. In India, some 98% of minor minerals, some 90% of industrial minerals, and some 33% of coal are mined on a small scale, judged by the criterion of "tonnage produced".

In this chapter, we shall attempt a foray into the world of small-scale mining in India in the past, review the current status of developments, and seek to provide a configuration of the shape of things to come. Besides highlighting the major issues of small-scale mining, the focus will be on what needs to be done to optimize benefits flow from small-scale mining and mitigate its adverse impacts.

MINING IN ANCIENT INDIA

Even if it is difficult to establish clearly and conclusively through recorded evidence as to the origins of mining in ancient India, there are ample indirect evidences of a thriving mining sector right since pre-Harappan times (pre-4000 BC). The authors of *Rigveda* (circa 1200 BC) wrote about metals and gems. Discovery of artifacts from Mehrgarh in Baluchistan, and of pre-Harappan copper ore mining and smelting at Ganeshwar-Jodhpura, make evident the existence of such indigenous mining. The earliest C-14 date related to mining archeology in India is that of the Rajpura-Dariba lead-zinc mine near Udaipur, Rajasthan, which has been placed around 1260 BC; samples from gold mines of Hutti date to around 760 BC. Cementitious brass was first produced at the Harappan port of Lothal around 1500 BC. Kautilya's magnum opus, the *Arthashastra* (circa 300 BC), provides an insight into the state of the mining and minerals industry of that era. The text highlights the importance of the minerals industry, making reference to metals such as gold, silver, copper, lead, tin and iron, as well as alloys such as brass, bronze and bell metal. Mines for diamond and precious stones, quartz and mica are mentioned in the tome, which also provided a detailed hierarchy of management for the mining industry (Ghose, 2002). The *Arthashastra* described in detail the duties and responsibilities of principal mineral industry officials, such as the Chief Controller of Mining and Metallurgy, the Chief Superintendent of Mines, the Chief Superintendent of Metals, the Chief Master of the Mint, and the Chief Salt Commissioner, which make evident the existence of a thriving mining and minerals sector (Rangarajan, 1992).

The history of mining during medieval times is somewhat of a grey area, and one can reconstruct the mining industry scenario by gleaning from the writings of Tavernier, Alberuni and others, and from texts such as *Ain-I-Akbari* (Bagchi & Ghose, 1980). Diamond mining was pre-eminent in the period 1400–1700 AD. Tavernier, in his account of travels (1665–1669 AD), gave details of diamond mining on the banks of Krishna River in South India, which yielded the Hope Diamond and the Kohinoor. Lead and zinc mining had their heydays during the period 1400–1800 AD. Mochia Magra and Balaria deposits at Zawar, near Udaipur, were probably re-discovered during the reign of Rana Laksh Singh of Mewar (1382–1397) and were worked until the great famine of 1812–13. The mining of coal was undertaken on a local scale at many different places but the first organized efforts for coal mining were made around 1774, by Sumner and Heatly at Ethora, Damalia and Chinakuri in Ranigunj Coalfield. This marked a major milestone in the evolutionary development of India's coal industry.

SMALL-SCALE MINING – THE 20TH CENTURY OVERVIEW

Any discussion on small-scale mining in India, as of present, calls for a broad-brush review of the mineral industry as a whole to understand and appreciate the dimensions of the sector and its importance. It is equally necessary to define and delineate what constitutes small-scale mining *per se*. This poses an intractable definitional problem, as categorization could be based on employment, revenue, tonnage produced, and, in some cases, energy expenditure, in order to provide an envelope to differentiate between manual and mechanized mining. Some of these definitional issues have been examined elsewhere. However, in India, the attributes assigned to "small-scale" mines converge on such operations where exploitation is restricted to surface or near-surface deposits, or with output less than 100,000 tonnes per year (Ghose, 1994).

India's mineral industry ranks fifth among the countries in the world in terms of mineral production value, with an aggregate value of *Rs.* 614,210 million (some US$13.6 billion) during 2002–2003, excluding atomic minerals and minor minerals. The term "minor mineral" has been defined in Clause (E) of Section 3 of the *Mines and Minerals (Regulation & Development) Act*, 1957, and comprehends building stones, gravel, ordinary clay, ordinary sand, boulder, shingle, limeshell, kankar and limestone used in kilns for manufacture of lime, murrum, brick-earth, fuller's earth, bentonite, road metal, slate, marble, quartzite and sandstone for building material or for making road metals and saltpeter. Under the Act, the State Governments have been empowered to make rules for regulating the grant of quarry leases, mining leases and other mineral concessions in respect of minor minerals, whose production statistics are notoriously suspect besides widely varying legal regimes (Anon, 1999). While the exact tonnage of minor minerals produced in the country and their collective value are not known, the vast majority of such deposits are mined on a small scale, and, occasionally, informally. It is only the *Industrial Policy Resolutions* of 1948 and 1956 that provided the fillip to the emergence of giant mining monoliths in the form of SMEs (State Mining Enterprises), which, over the years, have attained "commanding heights" in the economy; the public sector contributes nearly 92% of the aggregate value of mineral production in the country, and, barring coal, through exclusively "large-scale" mining routes. The veering away of the Indian minerals sector from its "small-scale" face began following the launch of large projects in the State sector during the post-independence era, the aim of which was to achieve economies of scale.

Small-scale mining, by whatever yardstick that one could devise to categorize the activity, is pervasive in India's minerals sector; in sheer number, such mines account for nearly 95% of all mines in India, whether they are required to submit annual returns under the Mineral Conservation and Development Rules or are operating under the penumbra of the informal sector. Table 27.1 presents an analysis of the extent of small-scale mining amongst the mines reporting production volumes in the Annual Returns based on the yardstick of "average annual output/mine" as estimated by the author. The percentage of small-scale mines in the analysis for different minerals reflects the overwhelming dominance of small-scale mines in India's mining industry, and, perhaps, the low technological status of the industry as well. It is noteworthy that while India's coal industry ranks third in the global league table of coal-producing countries, with an aggregate production of 315 million tonnes, of the 574 coal mines in operation during 2000–2001, some 172 mines had an annual output less than 100,000 tonnes and

Table 27.1 Status of small-scale mining in India – 2000–2001.

Mineral	Production (t)	No. of Mines	% of small-scale mines
Apatite	11,117	2	100
Asbestos	14,516	28	100
Barytes	836,574	12	<5
Bauxite	7,893,000	174	93
Chromite	1,952,000	20	25
Coal	315,000,000	574	33
Dolomite	2,964,000	102	95
Felspar	157,897	54	100
Fireclay	348,000	84	100
Fluorite	43,802	6	100
Graphite	135,036	47	100
Gypsum	2,707,000	45	90
Iron Ore	79,210,000	202	70
Kaolin	791,000	126	100
Kyanite	4,385	4	100
Limestone	126,070,000	488	74
Manganese Ore	1,556,000	115	95
Silica sand	2,073,000	147	100

Source: Indian Bureau of Mines, 2001.

could therefore be rightfully classified as "small-scale" mines. The analysis as above necessarily excludes from its purview the innumerable artisanal mines operating in the country, some of which contribute significantly to regional, or even local, economies. Even today, juxtaposed with the coal mines of Eastern Coalfields Limited, a subsidiary company of the giant Coal India Limited, there are a plethora of artisanal mines in the informal sector reminiscent of mining in the early nineteenth century, at a time when the guiding tenet was "more hole, more coal". Besides rampant illegal coal mining from abandoned mines with significant safety risks, there are clusters of such small-scale operations in the informal sector comprising shallow pits/wells. We shall briefly examine the cases of coal mining in Meghalaya and marble mining in Rajasthan as exemplars of small-scale/artisanal mining.

Small-scale mining for coal in Meghalaya

Since 1975, there has been a spurt of small-scale/artisanal mining in the north-eastern state of Meghalaya, due largely to its proximity to the export market in Bangladesh. In the Garo Hills of Meghalaya, there are six coalfields with 117 million tonnes of proved reserves, some 40.8 million tonnes of indicated reserves, and inferred reserves of 300.7 million tonnes. The coals are of the low ash (3–30%), high moisture (4–25%), and high sulphur (1–4%) variety, with calorific values in the range of 6000–8500 kcal/kg. The west Darangiri Coalfield has witnessed a mushrooming of artisanal mines in the past decade; here, coal is extracted underground using a crude variant of bord and pillar mining with irregularly shaped and spaced pillars, using essentially manual methods. The coal seams outcrop on steep hill slopes and small patches are worked with entries extending to 200–250 m, provided that natural ventilation permits the operations to be carried out. Lighting is usually by candles. Although the mining operations are seasonal (as the operations are suspended during the rainy season), the aggregate annual output from Meghalaya is now estimated to be in the range of 3.5 million tonnes from 500 odd mines, with a significant impact on the local economy and employment. Understandably, the safety and environmental impacts are adverse, and surface areas of serious concern.

Marble mining in Rajasthan

The state of Rajasthan produces about 95% of the country's marble; it has carved a niche for itself in the global export trade with a share of about 5% of the world market. The total reserves of some 10 varieties of coloured marble in Rajasthan are estimated at 1100 million tonnes out of the country's total reserves of about 1200 million tonnes. The past decade has witnessed a major focus of interest on marble mining, and the production level has registered a significant leap from a level of one million tonnes to the current level of 3.5 million tonnes. During 1998–99, there were about 3475 quarry leases with a combined areal extent of 4245 ha; for a production level of 3.574 million tonnes, this works out to a statistical average of about 1000 tonnes per year per lease. However, there has been a modicum of effort made in mechanizing mining operations, which heretofore were largely manual, featuring jack hammers and jib cranes. By and large, the marble mining operations present a wide spectrum of sophistication in terms of the level of technology employed, ranging from purely artisanal operations at Makrana, to more planned large quarries around Udaipur. The proliferation in large, small and artisanal mining has inevitably given a spurt to the marble processing industry, which, today, can boast of some 1000 marble processing gang-saw units, and 350 automatic tiling plants in Rajasthan (Rathore, 2001). The medium and small-scale marble mining and processing industry provides employment for nearly 500,000 people. Waste rock from mining and slurry from processing plants, amounting to nearly 0.54 million tonnes per year, poses a significant problem as far as disposal is concerned.

SMALL-SCALE MINING – A CONSPECTUS OF THE CURRENT STATUS

As the discussions above have highlighted, small-scale mining is widespread in its ramifications in the Indian mineral industry scene, and, of the 84 minerals produced in India, some 36 are mined exclusively on a small scale. There are major, if not yawning, gaps in the database for minor minerals, such as building stones, aggregates and sand. There would be obviously no reported data on the extent of gold mining via the small-scale route or of high-valued gemstones. Even for some of the high-valued minerals, such as cassiterite, there is widespread artisanal mining that is not reflected in official statistics. According to official records, there is currently only one tin mine in India at Gonvindpal in Bastar district of the State of Chattisgarh, operated by the State Mining Corporation, the reported production of which was 12,979 kg during the period 2000–2001. In the adjoining mining belt in Malkangiri district of Orissa, there is a thriving business in tin production from eluvial placers and pegmatites. The author has seen underground operations some 10 m deep in pegmatite lodes, manned by intrepid artisanal miners working for cassiterite nuggets, some of which are as large as 8–10 kg in weight.

Despite the paradigm shift in the policy vis-à-vis small-scale mining over the past decade, not much concrete action has emerged. The *National Mineral Policy*, 1993 (for non fuel and non-atomic minerals), provided for a new emphasis on promoting the small-scale mining of small deposits in a scientific and efficient manner, at the same time, safeguarding vital environmental and ecological imperatives. The Policy recognized that small and isolated deposits "with modest demand on capital expenditure and short lead-time, also provide employment opportunities for the local population", and indicated that preference be given to Scheduled Tribes for mineral concessions for small deposits in Scheduled Areas. Regrettably, however, the policy environment as such has not been supportive of the development of small-scale mining, and, as a result, there have been serious delays and bottlenecks in the granting of concessions. A recent judgment of the Supreme Court has also dealt a blow to small-scale mining in Aravalli region. New winds of change appear, however, to be blowing to do away with some of the archaic rules that recognize the role and importance of artisanal and small-scale miners. At the Conference of State Mining Ministers in January, 2003, it was agreed that traditional gold and diamond mining be made legal with permits (of one year duration) by amending the *Mines and Minerals (Development and Regulation) Act*. The amendment would benefit artisanal gold and diamond mining communities in the states of Andhra Pradesh, Chattisgarh, Madhya Pradesh and Orissa. Likewise, recommendations of village level heads will need to be obtained before granting concessions for minor minerals in non-Scheduled areas (Business Standard, 2003). These augur well for a resurgence of small-scale and artisanal mining.

WHERE DO WE GO FROM HERE?

Any projections for the future of the small-scale mining sector must address the following key issues:

- Indian mineral endowment lacks large world-class "bonanza"-type deposits, and thus the focus should be on the economics of the wide spectrum of small deposits found all over the sub-continent. However, one must treat the preceding statement with some reservation, as the probability of discovery of "large" deposits cannot be ruled out.
- The winds of change in the economy, with a focus on market forces, will, in all likelihood, stimulate interest and activity in the sector, given supportive governmental actions and the removal of roadblocks.

- With increasing emphasis on the development of the rural sector, small-scale mining can play an expanding role, if not serve as an engine of growth.
- The configuration of the small-scale mining sector, which is exclusively dependent on brawn, is likely to be metamorphosed with significant inputs of appropriate high technology. Efforts in developing clean and smart modular mining systems, for each scale of operation, will lead to the emergence of modern and innovative digital mines, and the dissolving of the "technology curtains" dividing large and the small-scale mines. Such small-scale mining operations will be on the vanguard of developing the large mines of tomorrow.
- The share of small-scale mining will enlarge significantly, given its smaller ecological footprint and potential for employment in the rural sector. The burgeoning demand for minerals will also reinforce such a trend. One can envision significant growth of small-scale mining for gold, gemstones and diamonds, industrial minerals and even coal.
- Small-scale mining was, is, and shall continue to be, quite "big" in the Indian mineral industry context, offering, on a platter, multiple benefits to the economy.

REFERENCES

Anon (1999). *Digest of Indian Minor Mineral Laws.* Indian Bureau of Mines, Nagpur.

Bagchi, S. & Ghose, A.K. (1980). History of mining in India – circa 1400–1800 and technology status. *Indian Journal of History of Science*, 15(1), 25–29.

Business Standard, January 23, 2003, New Delhi.

Ghose, A.K. (1994). New configuration of small-scale mining for developing countries. In A.K. Ghose (Ed.), *Small-scale Mining – A Global Overview* (pp. 29–42). New Delhi: Oxford & IBH.

Ghose, A.K. (2002). *Kautilya, the art and science of mining, and the future of global mining industry.* First Kautilya Distinguished Lecture, The Institution of Engineers (India), *Journal of mines, metals & Fuels*, 50(12), 448–451.

Indian Bureau of Mines (2001). *Statistical profile of minerals 2001–2002*, Nagpur.

Rathore, S.S. (2001). Environmental Issues of Marble Industry – Overview of Rajasthan. In D.C. Panigrahi (Ed.), *Mine environment and ventilation* (pp. 539–546). New Delhi: Oxford & IBH.

Rangarajan, L.N. (1992). *Kautilya, The Arthashastra* (p. 868). New Delhi: Penguin Books India (P) Ltd.

28

Small-Scale Mining in China: Socio-Economic Impacts, Policy and Management

ZHONG ZIRAN

In China, the small-scale mining industry was born in the early-1980s, as a result of the implementation of economic reforms and the advent of an open door policy. It has experienced rapid growth over the last two decades, and has played, and will continue to play, a significant role in China's mineral industry.

Small-scale mining activities, which constitute one of the most important branches of the Chinese mineral industry, have helped the country meet its national demands for mineral commodities. The total output of solid mineral products from the small-scale mining industry accounts for about 72% of China's total mineral production. Moreover, the industry provides employment to well over six million people; effectively facilitates local economic development; and relieves poverty in the areas in which the mines are located. It has directly precipitated the development of transportation, building materials industries, metallurgical and processing facilities, chemical industries, electrical power industries, commercial services, and improvements to local infrastructure. China's small-scale mining industry, which now consists of few state-owned mining enterprises, first introduced market forces into the country's mineral sector; this, in turn, initiated a shift in operating mode in the country's mineral industry, from traditionally centrally-controlled planning to market-oriented operation.

The industry, however, has experienced, and continues to experience, a wealth of problems. The most significant of these include a shortage of identified mineral reserves, low recovery of mineral resources, obsolete mining technology and out-of-date equipment, poor environmental performance, a lack of sound or necessary mine safety facilities and measures, and a shortage of well-trained technicians and mine workers. Additional challenges are posed by an over-supply of mineral products such as coal and building materials, along with the restructuring of the mineral industry. In the wake of these challenges, the Chinese Government has, since 1998, worked to close illegal small-scale coal mines and restricted the yield of coal. Presently, the Government is reviewing existing policies in place for small-scale mining, and making greater efforts to incorporate small-scale mining activities under a legalized framework based on market mechanisms. This constitutes a part of a nationalized scheme aimed at promoting sustainable development in the country's entire mining industry.

A BRIEF OVERVIEW OF CHINA'S NATIONAL ECONOMY AND ITS MINERAL INDUSTRIES

Geography

China has a land area of over 9.6 million square kilometers, and a marine space of more than 3.0 million square kilometers. Topographically, it is situated high in the west and low in the east, having plateaus and high mountains in its western part, and mostly hills and plains in its eastern part. In southwestern China, there exists the Qinghai-Tibet Plateau, referred to as the "roof of the world", with Mount Qomolangma – the highest peak on the globe – at an elevation of 8,848 meters above sea level. In northwestern China, there lies the Taklimakan Desert, the second largest desert in the world, and the Turpan Basin with an altitude of 155 meters below sea level, which is the lowest depression in China and one of the well-known depressions in the world. In southern China, there are beautiful and peculiar karst landforms, and, in the north, there is a vast expanse of loess plateaus. The country has a complicated and varied geomorphology, with a crisscross pattern of magnificent mountains and rivers.

Economic background

As a result of economic reforms initiated in 1978, the Chinese economy became one of the most dynamic globally. During the past six years, GDP has increased annually at an average rate of 8.1%, amounting to over 9,593 billion RMB Yuan (over 1159 billion US dollars). In 1999, the country quadrupled its 1980 GDP.

In March 2001, the National People's Congress formulated the nation's tenth five-year plan (2001–2005) to help guide the national economy through the first five years of the new century. The plan set a target for doubling 2000 GDP by the year 2010, translating into an average annual economic growth of 7% or 12,500 billion RMB Yuan (1510 billion dollars), of which agriculture would account for 13%, industry 51%, and services 36%. Targets for the year 2005 include a GDP per capita of 9400 RMB Yuan (1135 US dollars), an unemployment rate of less than 5%, relevant stable pricing, and balanced international revenues and payment.

There is no doubt that, in order to achieve these goals, sufficient supplies of energy and raw materials from the petroleum, mining and mining-related industries must be extracted. This, however, is a challenging task, and would be quite unimaginable without active inputs from the small-scale mining industry.

China's mineral industries

China is one of the world's most important mining countries, both in terms of production and consumption. Its mineral sector has, and will continue to have, a positive impact on economic development and the welfare of its people. Moreover, domestic demand for such mineral commodities as copper, tungsten, tin, antimony, fluorite, and barite, has an important bearing on their international prices.

In 2001, 153,723 mining enterprises (including oil and natural gas development enterprises) were operating in China; of these, 500 mines were large-scale, 1254 were medium-scale, and 151,969 were small-scale. The number of operating mines peaked at more than 280,000 in the early-1990s, as the local government began supporting collectively-owned small-scale mines. Individual miners therefore flooded in, but their numbers have declined significantly in recent years, largely as a result of the implementation of restructuring policies and new mining legislation. Of the total number of mines operating in China, some five per cent are the state-owned, 175 have foreign investment, and others have investment interests from Hong Kong, Macao and Taiwan.

In 2001, China produced 4.35 billion tonnes of solid minerals, 165 million tonnes of crude oil, and over 27 billion cubic meters of natural gas, the total output value of which was 460.11 billion RMB Yuan (55.57 billion US dollars) and accounted for five per cent of GDP. Total earnings can be broken down as follows:

- The petroleum industry, 236 billion RMB Yuan (28.50 billion US dollars);
- The coal mining sector, 116.43 billion RMB Yuan (14.06 billion US dollars);
- The ferrous metal industry, 18.1 billion RMB Yuan (2.19 billion US dollars);
- The nonferrous metal industry 18.5 billion RMB Yuan (2.23 billion US dollars); and
- The precious metals industry 9.4 billion RMB Yuan (1.14 billion US dollars).

Figure 28.1 shows the output totals of China's mineral industry during the past thirteen years.

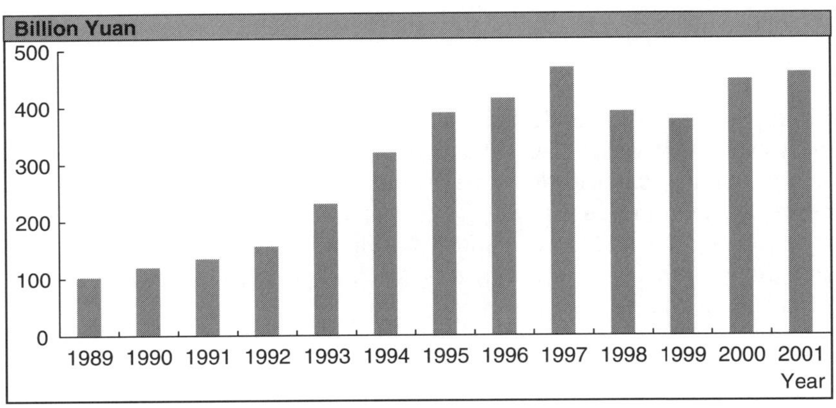

Figure 28.1 National output value of China's mineral industry, 1989 to 2001.

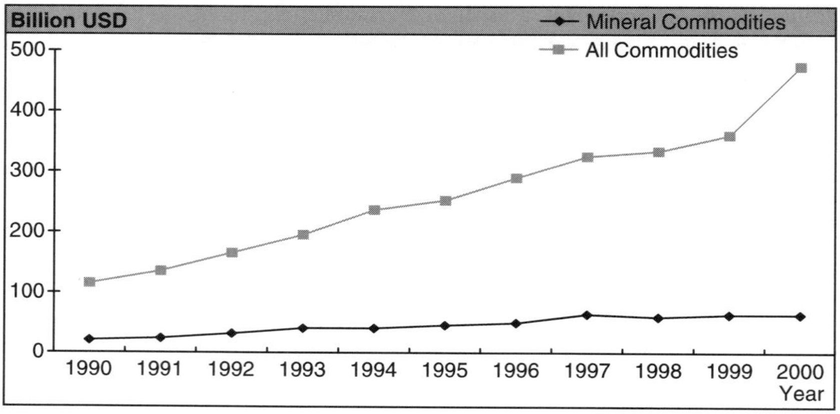

Figure 28.2 Historic trend of international trade of China's mineral commodities.

The import and export of China's mineral commodities have occupied a significant position in international trade in past two decades. In 2000, international trade of its mineral commodities amounted to over 100 billion US dollars; minerals also accounted for about 18% of the total number of commodities imported and exported. The trend of international trade of mineral commodities in the 1990s is shown in Figure 28.2.

DEFINITION OF SMALL-SCALE MINING

The definition of small-scale mines (SSMs) differs from country to country. In China, mining scales are divided into three categories: "large", "medium" and "small". These classifications are based on tonnage and/or ore volume.[1] For example, coal mines with a production of less than 300,000 tonnes per year are considered "small-scale"; those with production ranging between 300,000 and 900,000 tonnes per year, medium-scale; and those with production exceeding 900,000 tonnes per year, large-scale. Table 28.1 presents a more precise classification scheme of scales for selected segments of China's mineral sector.

Although these mining scales are not directly related to the ownership of mining operations, coincidentally, the majority of mining enterprises collectively owned by towns, villages and individual miners are "small-scale", as they conform to the classification criteria originally established (for small-scale mines) by the former Ministry of Geology and Mineral Resources (MGMR) in 1998. Some refer to these operations as "mini-scale mines", because several have mineral outputs significantly less than the upper production limits set for "small-scale" mines.

---

[1] For details, please refer to the Notification of the Ministry of Geology and Mineral Resources on the Classification of Scales of the Mines Construction issued on March 17, 1998.

Table 28.1 Classification of mining scales for selected minerals in China (unit: tonnes).

Mineral	Unit (ores)	Large-scale	Medium-scale	Small-scale
Coal	X100,000	>9	3–9	<3
Iron ores	X100,000	>20	6–20	<6
Copper	X100,000	>10	3–10	<3
Lead	X100,000	>10	3–10	<3
Zinc	X100,000	>10	3–10	<3
Bauxite	X100,000	>10	3–10	<3
Tungsten	X100,000	>10	3–10	<3
Tin	X100,000	>10	3–10	<3
Gold	X10,000	>15	6–15	<6
Phosphate	X100,000	>10	5–10	<5
Pyrite	X100,000	>5	2–5	<2

Source: The Circular of the Ministry of Geology and Mineral Resources on the classification of the mining scales of 1998.

Table 28.2 Number of different mining scales for selected minerals.

Mineral	Total	Large-scale	Medium-scale	Small-scale
Total	153,723	500	1,254	151,969
Coal	30,995	144	338	30,513
Iron ores	4,036	28	61	3,947
Copper	790	15	29	746
Lead	905	3	17	885
Zinc	1,191	3	22	1,166
Bauxite	274	3	4	267
Tungsten	190	1	21	168
Tin	269	2	12	255
Gold	1,806	40	82	1,659
Phosphate	512	12	23	477
Pyrite	805	1	6	798

Source: The Statistical Bulletin on the Development of Mineral Resources in 2001, issued by the Ministry of Land and Resources, 2002.

The structure of scales for selected minerals mined in China is shown in Table 28.2.

HISTORICAL DEVELOPMENT OF SMALL-SCALE MINING

Before 1978, China had a highly centralized planning system. Prior to 1982, all mineral exploration and development was organized, funded and highly controlled by the state. Corresponding mineral policies were integrated into the country's five-year plans. The mineral industry was one of the most complete sectors to adopt centrally-controlled planning as a basis for operation, which is why it is now difficult for its operations to both follow an open door policy and adapt to introduced market forces.

The small-scale coal mining industry was first given attention by Chinese leaders during the period of the Great Leap Forward (1958–1960). Coal production from small-scale mining had risen to over 34 million tonnes in 1958, but had sharply declined to eight million tonnes in 1962, and had further decreased to 5.68 million tonnes by 1966.

During the past two decades, China's small-scale mining industry has experienced the following three phases of development: (1) rapid and free expansion; (2) rectification; and (3) restructuring. Each is described more in detail in the discussion that follows.

Phase one: rapid and free expansion (1983–1986)

In China, small-scale mining began to expand rapidly in the early-1980s, at a time when the economy began growing as a result of the implementation of economic reforms and the advent of an open door policy (launched in 1978). An inadequate supply of raw materials and energy began restricting the development of the national economy. However, there existed a substantial amount of small-scale mineral deposits – which were widely scattered throughout the country – and thus, enormous potential for employment in rural areas.

In this regard, the Central Government decided to relax limitations toward mining activities conducted by (small-scale) mining enterprises collectively owned by towns, villages and individual miners. In March 1983, the State Council issued the *Circular on Eight Items of Measures for Speeding up the Development of the Small-Scale Coal Mining Industry.* In April 1984, the top leader of the Chinese Government, Mr. Hu Yaoban, requested to "make water to flow more quickly".[2] Thus, in January 1985, the Central Government made the declaration that economic entities with different ownership (state-owned as well as those owned by towns, villages and individuals), and enterprises operating on different scales (large, medium and small) could join the coal mining industry (Ye & Zhang, 1998).

Following the implementation of this new mineral policy, numerous small-scale miners commenced operation. In 1982, there were 16,000 small-scale coal mines, at a time when coal production was approximately 146 million tonnes. In 1983, however, the number of the small-scale coal mines had sharply increased by 30,000 to 46,000, resulting in an output increase to 170 million tonnes. By the end of 1985, there were 63,000 small-scale coal mines in operation, 51,000 of which were owned by towns and villages, and the remainder, by individuals. At the time, some 51% of all small-scale coal mines were operating without legal approval.

[2] Its intention was to accelerate the development of mineral resources located underground.

During this period, SSMs were expanding chaotically. Although many activities were based in areas amenable to small-scale mining, a considerable number of operators began considering the mineral deposits on which medium and large state-owned operations were based; this inevitably caused disputes between local communities and the management of state-owned mining enterprises (SOMEs), and caused additional tension between the local governments, which were in charge of SSMs and the former industrial departments that had taken charge of SOMEs. At the time, no national law regulated mineral exploration and mining operations. Since 1982, the former Ministry of Geology and Mineral Resources (MGMR) has been entitled to exercise duties to oversee the exploration and development of mineral resources, but at the time, it proved difficult to settle disputes without mining legislation.

Phase two: rectification of small-scale mining industry (1986–1998)

On March 19, 1986, China's first mining act – the *Mineral Resources Law of the People's Republic of China* – was passed at the 15th Meeting of the Standing Committee of the 6th National People's Congress, entering into force on October 1, 1986. Under Article 14 of the Law, the provincial standing committees of the People's Congress are entitled to promulgate provincial provisions on the management of mining enterprises collectively owned by towns, villages and individuals. Furthermore, the Law states that the former MGMR was responsible for issuing mining licenses to the mining enterprises approved by the State Council or the relevant industrial departments. Provincial bureaus of geology and mineral resources were responsible for issuing mining licenses to mining enterprises approved by the provincial governments. The bureaus of geology and mineral resources at prefecture and county levels were responsible for issuing licenses to the mining enterprises collectively owned by towns, villages and individual miners.

As the government agencies enforcing the mining laws and regulations, the former MGMR, its subordinate agencies at the provincial level, along with relevant government agencies, made great efforts to persuade the operators of the-then expanding SSM sector to adhere to mining legislation. For the same purpose, a series of mining rectification programs had been launched. In December 1995, the State Council issued the *Circular on Rectifying Mining Operation and Protecting the State's Ownership over Mineral Resources.* Vice Premier Zou Jiahua of the State Council chaired the rectification program, and personally directly the revamping of mining activities in Xiaoqingling gold mining district, which borders Lingbao County in Henan Province, and Tongguan County in Shaanxi Province. At the time, there was a chaotic gold mining rush in the region, resulting in the formation of hundreds of SSMs. MGMR, in collaboration with seven other ministries, oversaw the rectification of mining activities nationwide. The People's Governments at the provincial, prefecture and county levels were responsible for implementing the rectification program. As a result, 99.5% of mining enterprises collectively owned by towns, villages and individual miners, were subjected to a series of legal procedures for licensing purposes. Thousands of disputes between SOMEs and SSMs parties were settled under this framework of mining law and regulations.

During this period, in terms of number of operations and production, great growth occurred in the SSM sector. This was largely a result of high domestic demands for mineral commodities, and the favorable policy of the central government. In 1996, there were over 220,000 SSMs. At the time, coal production from SSMs amounted to 660 million tonnes, accounting for about 50% of the national total. Between 1995 and 1997, annual coal production lay in the range of 13.50–14.00 billion tonnes, and net exports amounted to about 30 million tonnes. Coal accounted for about 75% of primary energy production and consumption.

However, the new laws and regulations developed in the 1980s and 1990s to cover mineral rights (including exploration rights and mining rights), safety, and environmental protection, were not implemented effectively in the small-scale mining sector.

Phase three: restructuring of the mining industry (1998–present)

During the period 1995–1997, there was an oversupply of coal and a number of other minerals in China. From late-1997 to 1998, stockpiles grew to nearly 200 million tonnes, and prices in the domestic market plummeted. In early-1998, production from some SOMEs was suspended for up to two months to help alleviate the oversupply problem. Great efforts were made by government agencies at different levels to license and rectify an industry with numerous illegal and irrationally-location operations, low recovery rates, a poor safety record, and environmentally degrading practices.

The government reacted to both the short-term overproduction of coal and the industry's appalling long-term safety, environmental, and resource recovery record, by issuing the *Circular on Closing Illegal and Irrationally Located Coal Mines;* the Circular was implemented on December 25th 1998 by the State Council. The Chinese Government decided to reduce annual coal production by 250 million tonnes, and accomplished the task by banning 25,800 illegal small coal mines. The following operations were targeted for closure:

- All small coal mines that were operating without mining licenses and coal operation permits (banned before the end of 1999);
- All small coal mines that had commenced operation after 1 January 1997 and which were based within the mining areas of state-owned coal mines (banned before the end of 1999);
- All unlicensed small coal mines that had commenced operation before 1 January 1997 that were based within the mining areas of state-owned coal mines (banned before the end of 1999);
- All licensed and permitted small coal mines that had commenced operation before 1 January 1997 and which had negatively impacted the long-term development of state-owned mines as a result of working their land;

Table 28.3 Major economic indicators of the small-scale coal mining industry.

Year	Number of mines	Employment	Coal production	Output value	Profit
1996	over 80,000				
1997	over 70,000				
1998	51,100				
1999	43,659	2,588,526	430.64 Mt	32.43 BY	125.27 MY
2000	35,114	2,459,289	458.11 Mt	36.22 BY	857.78 MY
2001	30,513	2,108,095	385.06 Mt	34.58 BY	1781.03 MY

Note: Mt means million tonnes; BY means billion RMB Yuan; MY means million RMB Yuan (1 US$ = 8.28 RMB Yuan).
Source: The Statistical Bulletin on the Development of Mineral Resources in 2001, issued by the Ministry of Land and Resources, 2002.

- Small coal mines with neither mining licenses or coal operation permits that were operating beyond the mining regions of state-owned enterprises and which had failed to meet the legal requirements for licensing or permitting prior to February 1999 (banned before the end of 1999); and
- All coal mines exploiting high sulfur and high ash coal without effective measures to reduce sulfur and ash.

In addition, the Chinese Government banned all unlicensed mining enterprises exploiting tungsten, tin, antimony and rare earth metals.

As shown in Table 28.3, in the past six years, there has been a sharp reduction in the number of small coal mines in China. Between 1999 and 2001, annual outputs of coal from SSMs decreased by 45.58 million tonnes, while annual coal production from the SOME sector increased by 176.38 million tonnes; during this period, the number of SSMs operating in China had sharply declined by 71,723, and employment in the sector decreased by 709,422. However, in spite of these losses, the sector still managed to produce in excess of three billion tonnes of mineral product between 1999 and 2001. In fact, both production (in terms of value) and net profit in the sector have experienced significant growth during the past three years, as illustrated in Table 28.4.

It was estimated that annual domestic demand for coal has been 11–12 billion tonnes in recent years. Recent changes in supply and demand for coal has led to a sharp increase in coal prices. Between 1997 and 2001, the cost of a tonne of coal rose from about 110 RMB Yuan (13.29 US dollars) to nearly 200 RMB Yuan (24.15 US dollars).

CONTRIBUTIONS OF THE SMALL-SCALE MINING INDUSTRY TO CHINESE SOCIETY

In China, the small-scale mining industry has played an important role in supplying sources of energy and raw materials. As shown in Table 28.5, despite

Table 28.4 Major economic indicators of the small-scale mining industry.

Year	Number of mines	Employment	Ore production	Output value	Profit
1996	222,254				
1997	194,445				
1998	193,279				
1999	162,462	6,707,516	3065.21 Mt	92.32 BY	2.35 BY
2000	150,031	6,418,587	3313.44 Mt	96.78 BY	3.36 BY
2001	151,969	5,998,094	3140.75 Mt	98.47 BY	4.98 BY

Note: Mt means million tonnes; BY means billion RMB Yuan (1 US$ = 8.28 RMB Yuan).
Source: The Statistical Bulletin on the Development of Mineral Resources in 2001, issued by the Ministry of Land and Resources, 2002.

Table 28.5 Major indicators of the chinese mining industry, 2001 data.

Mining scale	Number of mines	Employment	Ore production	Output value	Profit
Total	152,301	8,749,705	4353.21 Mt	221.83 BY	9.19 BY
Large	475	1,687,908	827.85 Mt	88.35 BY	5.12 BY
Medium	1247	1,063,703	384.61 Mt	35.00 BY	-0.92 BY
Small	150,579	5,998,094	3140.75 Mt	98.47 BY	4.98 BY

Note: Mt means million tones; BY means billion RMB Yuan (1 US$ = 8.28 RMB Yuan).
Source: The Statistical Bulletin on the Development of Mineral Resources in 2001, issued by the Ministry of Land and Resources, 2002.

experiencing a period of prolonged rectification and mass closure, the small-scale mining industry still provides employment to some six million Chinese, and generates nearly five billion RMB Yuan (60.14 million US dollars) in profit for local governments and local communities. Output from small-scale mining accounts for about 72% of total mineral production in the country, and collectively, the industry has not only facilitated local economic development but has also helped to alleviate poverty (Table 28.6).

Small-scale mine production in China

The small-scale mining industry has produced over 50 per cent of China's mineral commodities in recent years. Table 28.6 compares the output from small-scale mining industry to that of the entire mineral sector.

During the period 1981 to 1998, the aggregate output from small-scale coal mines was over 7.5 billion tonnes, accounting for about 40 per cent of the country's total coal production, and 25% of national primary energy production[3] (Ye Qing, 1998). The small-scale coal mining industry greatly mitigated

Table 28.6 Contribution of small mining to China's mining sector, 2001 data (unit: %).

Mineral	Employment	Production	Sales revenues
Average	65	72	22
Coal	50	37	30
Iron ores	34	27	16
Copper	15	7	7
Lead	43	40	38
Zinc	42	35	29
Bauxite	52	52	32
Tungsten	20	37	17
Tin	39	25	28
Gold	29	32	14
Phosphate	35	43	36
Pyrite	36	46	34

Source: Based on the Statistical Bulletin on the Development of Mineral Resources in 2001, issued by the Ministry of Land and Resources, 2002.

shortages in energy supply, thus saving hundreds of billions of the State's RMB Yuan.

Introduction of new economic mechanisms

The occurrence and development of the small-scale mines owned by local communities precipitated the restructuring of state-owned mines, most of which were large and medium-scale. Since the early-1990s, SSMs have been in competition with SOMEs on the domestic market. Because of low production costs, flexible operation, and sales mechanisms, the small-scale mining industry has progressively taken up more shares on the mineral market. The competition between the parties has repressed the rise of mineral commodity prices, and has thus forced SOMEs to undertake reform measures to raise money, and to upgrade mine construction modes, operating systems, production management, and employment modes, for the purpose of improving their competitive capacity. Additionally, the rapid development of SSMs has prompted the government to introduce new mining legislation and new mineral resource management systems.

Full recovery of mineral resources in part of small-scale mines

A considerable number of SSMs exploit mineral resources from ore bodies abandoned by medium and large-scale SOMEs. Specifically, these resources

[3] The data are based on Ye Qing's publication (Ye & Zhang, 1998, p. 2–3) and the Statistical Bulletin on the Development of Mineral Resources 1996, 1997, 1998, issued by the former Ministry of Geology and Mineral Resources in 1997 and in 1998, and Ministry of Land and Resources in 1999, respectively.

are considered sub-marginal by the management of SOMEs, but are ideally suited for small-scale extraction. In 1995, the operations collectively owned by towns and villages reportedly mined some 156 million tonnes of coal from about 260,000 mining units abandoned by SOMEs. This group, which produced 27% of the sector's coal, owned 34 per cent of operating small-scale coal mines (Ye Qing, 1998, pp. 45–46).

Improvement of mining industrial layout

For historical and natural reasons, most of China's high energy-consuming enterprises are located in eastern coastal areas. Some 70% of China's coal consumption occurs in the eastern portions of the country, despite the fact that these areas provide only 17% of the country's supply. The majority of the coal mining operations located in the eastern part of the country are small, scattered, and complicated geologically, thus making them difficult to exploit using large-scale techniques. Most of the country's energy minerals, particularly coal deposits, are located in the western regions of the country. In fact, approximately 83 per cent of coal reserves occur in the middle and western parts of China; 75 per cent are found in Shangxi, Shaanxi, Inner Mongolia and Xinjiang Provinces alone. Local variations in supply and demand for coal commonly induce long distance transportation. Similar cases occur with other minerals.

Coal resources in eastern coastal areas, especially in nine provinces to the south of the Yantze River, are considered appropriate for small-scale extraction but unsuitable for mechanized mining. Coal production in eastern coastal areas was approximately 52 million tonnes in 1980, and amounted to over 240 million tonnes in 1998;[4] annual coal production in these areas had increased by 10 per cent over this time period. The production increase (108 million tonnes) achieved in the coal-rich Shangxi, Shaanxi and Inner Mongolia Provinces, underscored the need to construct two railways, a task that was absolutely impossible at the time because of financial and resource constraints. Thus, the development of the small-scale coal industry had helped to mitigate pressures from local consumers, and has improved coal distribution in the country overall.

The small-scale coal mines collectively owned by towns and villages are found in more than 1200 counties in 29 provinces, and have assumed the responsibility of supplying cheap coal to the less developed and remote rural areas of the country.

Promotion of local economic development

In China, small-scale mining has effectively facilitated local economic development and assisted in relieving poverty in many rural areas. In 1998, it generated

[4]The data are based on Ye Qing's publication (Ye & Zhang, 1998, p. 44) and the Statistical Bulletin on the Development of Mineral Resources 1996, 1997, 1998, issued by the former Ministry of Geology and Mineral Resources in 1997 and in 1998, and Ministry of Land and Resources in 1999, respectively.

4.36 billion RMB Yuan (526.57 million US dollars) in profits and contributed 44.07 billion RMB Yuan (5.32 billion US dollars) in tax revenues.[5] The industry has precipitated the development of transportation services, building materials operations, metallurgical processing facilities, chemical factories, electrical power plants and commercial services, and facilitates improvements to local infrastructure. It also accelerates economic development and improves the quality-of-life of many people in rural areas, not only through direct sales of minerals, but also through the provision of financial support to local agriculture, facilitated bridge and road construction, education, and the enactment of rural collective welfare projects.

COMMON PROBLEMS IN CHINA'S SMALL-SCALE MINING INDUSTRY

A number of problems have occurred in the industry. These have included, on different occasions, the proliferation of operations too small in scope, a shortage of proven mineral reserves, low recoveries of mineral resources, obsolete mining technology and out-of-date equipment, poor environmental performance, a lack of sound mine safety measures, and a shortage of well-trained technicians and mine workers.

The issue of scale

The average annual production of each small-scale mine was about 9,500 tonnes in 1998.[6] Many operations, however, are considered by the government to be too small in scale, as they fail to meet the cutoff requirements. Table 28.7 shows the mining scales of selected minerals. About 58% of small-scale coal mines have shafts producing less than 30,000 tonnes per year (Table 28.8).

Shortage of identified mineral reserves

Of China's 60 billion tonnes of coal reserves (not including those in Inner Mongolia and Xinjiang), approximately 65% are being mined. The reserves being tapped by small-scale mines account for only six per cent of identified reserves nationwide. Most operations are concentrated in the provinces of Shangxi, Shaanxi, Inner Mongolia and Xinjiang. However, coal production from the small-scale mining industry accounts for over 45% of national (coal) output. These operations, however, extract resources at rapid rates. The lives of many coal shafts are only five years, and almost all do not exceed 10 years.

[5] The figures are based on the Statistical Bulletin on the Development of Mineral Resources in 1998, issued by the Ministry of Land and Resources, 1999.

[6] The figure is based on the Statistical Bulletin on the Development of Mineral Resources in 1998, issued by the Ministry of Land and Resources, 1999.

Table 28.7 Output of selected minerals of each small-scale mine in 1998 (unit: tonnes, RMB Yuan).

Mineral	Ore production	Output in value
Average	9,500	325,398
Coal	6,793	414,286
Iron ores	13,430	409,836
Copper	2,973	293,916
Lead	2,839	673,968
Zinc	1,372	943,396
Bauxite	4,008	192,148
Tungsten	14,474	426,316
Tin	3,933	878,669
Gold	8,250	665,881
Phosphote	17,492	825,228
Pyrite	4,216	240,927

Source: The Statistical Bulletin on the Development of Mineral Resources in 1998, issued by the Ministry of Land and Resources, 1999.

Table 28.8 Production of small-scale coal mining shaft in 1995 (unit: tonnes, %).

	Number	Percentage in total	Production	Percentage in total
Total	72919	100	579 million	100
over 30,000	6646	9.1	245 million	42.4
10,000–30,000	15164	20.8	195 million	33.7
below10,000	51109	70.1	139 million	23.9

Source: The Report on National Small-Scale Coal Mines Survey by the State Planning Commission and the former Ministry of Coal Indsutry, 1997, p. 5.

Generally, most SSM operators lack formal geological information, which is why there is an element of high risk when initially investing in such operations.

Low recovery of mineral resources

A series of irregular mining methods, particularly manually techniques, are used at some 72% of China's small-scale coal mines. This, in turn, results in low recovery.

Illegal mining activities

About 10% of SSM have no legal status – *i.e.* they are operating without mining licenses. Mineral management authorities have made concerted efforts to ban all types of illegal mining activities, particularly those without licenses. However, in some areas, local governments and communities acquiesce in the fait accompli, directly and indirectly supporting illegal SSM activity. According to a survey undertaken in 1995, 11% of national coal output produced by the small-scale coal mining industry derives from mines operating without a license (Ye Qing, 1998). Illegal mining is one of the most serious problems in China's small-scale mining industry today, and a major reason why many people disregard the roles of SSMs altogether.

Negative environmental impacts

Small-scale mining activities have caused a wide range of environmental impacts. Solid wastes, including surrounding rock, overburden, tailings, and other potentially useful materials, are discharged uncontrollably from operations. Poisonous chemicals from these wastes often leak underground, thus leading to the contamination of underground water sources, and, eventually, surface rivers. Substantial quantities of water are consumed during extraction and ore processing activities at SSMs. Generally, this water is unsuitable for drinking, irrigation and recreation. The particular problematic group of operations are the irregular small-scale mining enterprises, which, as a result of utilizing obsolete and inefficient technology, cause a number of pollution problems and typically have low recovery levels.

Many open pit mining operations situated away from overburden have seriously degraded surrounding quality soils. Furthermore, extraction, conveyance of ores and mineral wastes, and ore processing activities generate substantial amounts of gaseous pollutants, which commonly contain metal-bearing dust and sulfur. Fugitive and point source pollutants migrate rapidly, eventually affecting areas beyond mining districts.

The environmental protection agencies have not been able to address effectively the environmental problems caused by SSMs. The general policy of environmental management is strongly bound to large-scale and medium-scale mining enterprises, but comparatively ineffective with respect to SSMs.

In the triangle region surrounding the borders of the Shangxi, Shaanxi and Inner Mongolia Provinces, which are endowed with coal and natural gas, the rapid development of small-scale coal mining activities has caused frequent flooding (during summer), severe water loss and soil erosion, and a deterioration of the ecological environment. During the late-1980s and early-1990s, these mines did not have effective environmental measures in place, and thus caused innumerable damages to river dams; this had impacted the exploitation of mineral resources and economic development, and had endangered areas downstream of the Yellow River. These circumstances were of major concern to the Chinese Government. The State Council decided to take urgent measures to rectify these operations and mitigate the environmental damages. It therefore instructed the departments concerned to develop and implement the *Program for Coordinated Development of Resources Exploitation and Environmental Protection* in the Region Bordering Shangxi, Shaanxi and Inner Mongolia Provinces.

Safety and health issues

According to the International Labour Organization (ILO), in China, over 6,000 fatalities occur at small-scale mines each year (ILO, 1999). This, in turn, translates into 4,500–5,000 fatalities at small-scale coal mines alone (according to 1997 statistics, this works out as 9.1 deaths per million tonnes of coal mined). According to China's statistic information, before 1991, the fatality rate exceeded 10 per million tonnes of coal mined; a peak fatality rate of 14.4 per million tonnes of coal mined occurred in 1987. Since 1995, however, the fatality rate has not exceeded 10 per million tonnes of coal mined.

Between January and July of 2002, 2,046 mine accidents occurred. There were 3,620 mine-related deaths recorded during this period, an increase of 4.8% from the corresponding 2001 period. An estimated 67% of deaths were a direct result of gas or coal dust explosions at SSMs. Six major accidents involving gas explosions took place at coal mines, of which, five were small-scale; one incident alone in Shangxi Province resulted in hundreds of deaths.

It is widely acknowledged that many accidents at SSMs go unreported, largely because the owners of illegal operations do not wish to draw attention to themselves. They frequently take no responsibility for paying compensation or making arrangements for social security for injury and death. Occasionally, owners of illegal small-scale mines sign contracts with employees that stipulate the provision of minimal compensation for death.

The causes of accidents at SSMs are varied, but can effectively be grouped under the following two broad categories: (1) "management/operation-related"; and (2) "equipment/work–environment-related" (Table 28.9).

The most frequently cited causes of accidents at Chinese SSMs are rock falls, subsidence, a lack of ventilation, misuse of explosives, lack of knowledge, lack of training, violation of regulations, and obsolete and poorly-maintained equipment.

Many small-scale coal mines were designed using outdated mining techniques, and lack safety equipment. Furthermore, a considerable number of SSMs do not comply with basic mining safety standards, and some illegal operations

Table 28.9 Major reasons for accidents at SSMs.

Management/operation	Equipment/work environment
Lack of awareness of safety	Rock falls; cave-ins; subsidence
Violation of regulations	Misuse of explosives
Negligence	Lack of ventilation
Lack of inspection	Unprotected equipment
Ignorance and lack of training	Poor access/exit
Poor management/supervision	Obsolete equipment
Anarchic exploitation of resources	Lack of maintenance
Over-exertion	Improper use of equipment
Substance abuse	Lack of or failure to use personal protective equipment
Lack of regulations	Poor working conditions and work practices

Source: Report of the International Labour Organization, 1999.

only have a single shaft. Gas explosions, coal outbursts and gas blow-outs caused about 70% of fatalities each year (ranged between 232 and 411) in Hunan Province between 1990 and 1997.

The overwhelming majority of mineworkers are peasants with basic education. They normally receive no formal training and lack the knowledge and skills to prevent accidents from occurring, and to deal with them when they do occur. In 1998, a lack of knowledge and insufficient gas detection equipment in Hunan led to the deaths of five people trying to rescue others following a fire.

The major health risks at Chinese SSMs have proven to be exposure to dust (silicosis); exposure to mercury and other chemicals; the effects of noise and vibration; the effects of poor ventilation (heat, humidity, lack of oxygen); and the effects of over-exertion, inadequate work space and inappropriate equipment.

LEGAL FRAMEWORK REGULATING SMALL-SCALE MINING

Development of laws and regulations

The following is a simple hierarchy of laws and law-making institutions that exist in China:

- The National People's Congress (the equivalent of a parliament or legislature), which holds the power to draft national laws;
- The State Council (the equivalent of a cabinet), which can draft implementing rules, regulations, decrees and orders;
- Subordinate Commissions and Ministries, which can issue orders, directives and regulations that are consistent with laws and State Council regulations; and
- The People's Congresses at the provincial and prefecture levels, which may adopt local regulations.

Generally, a law is first drafted by the relevant commission or ministry, under the guidance of the Law Office of the State Council, and the consultation of other relevant commissions and ministries. If agreeable, the law proposal will be submitted to the State Council for pre-review, and then to the National People's Congress or its Standing Committee for review and approval. Originally, a law proposal was reviewed twice, but today, it undergoes review three times. The State Council has a commanding influence over the development of laws and regulations, with substantial input from relevant Commissions and Ministries, particularly in the economic sector. Economic laws usually emerge from either the State Council or from a relevant commission or ministry. The State Council usually retains control over the draft and review process before the law is sent to the National People's Congress for approval. If the law is not approved, it is usually returned to the State Council for redrafting. Once the law is approved, either the State Council or the relevant department of government will be charged with drawing-up implementing regulations and measures. Local governments may then draft their own regulations.

During recent years, the National People's Congress has assumed a greater role in the drafting of legislation. The National People's Congress sometimes

Table 28.10 Selected laws and regulations on exploration and development of mineral resources.

Name of instrument	Year
Mineral Resources Law, amended	1986, 1996
Rules for the Implementation of the Mineral Resources Law	1994
Regulations on Registration for Exploration for Mineral Resources Using the Block System	1998
Regulations on Registration for Exploitation of Mineral Resources	1998
Regulations on Transfer of Exploration Rights and Mining Rights	1998

Source: Andrews-Speed et al. (2003).

empowers a special committee – for example, the committee on environment and natural resources, the committee on finance and economy, etc. – to directly draft law proposals without the input of the State Council and its member departments. Examples have included the *Proposal on the Law of Property Rights, and the Law on Contract of Collectively Owned Rural Land.* The Congress has also played a significant role in the development of laws via the State Council. For instance, the preliminary proposal on the *Amendments to the Mineral Resources Law* presented by the State Council initially involved only four articles. When entering into the process of legislation at the National People's Congress, a series of very important provisions, a topic of controversy among different commissions and ministries, had been included in the Amendments. Thus, the *Amendments to the Mineral Resources* Law passed by the Congress in August 1996 featured eighteen articles.

Laws and regulations governing small-scale mining activities

In China, there is now a substantial body of laws both directly and indirectly governing SSMs. These may be considered under the following six headings:[7]

- Laws and regulations on the exploration and development of mineral resources (Table 28.10);
- Laws and regulations on mine safety and labour (Table 28.11);
- Laws and regulations on environmental protection and land management (Table 28.12);
- Laws and regulations on township and village coal mines (Table 28.13);
- The Circulars of the State Council on Rectification of Mining Industries (Table 28.14);
- Laws and regulations governing mining taxation (Table 28.15).

[7] The tables of laws and regulations are intended to be illustrative, and cannot be considered to be complete or definitive.

Table 28.11 Selected laws and regulations on mine safety and labour.

Name of instrument	Year
Law on the Coal Industry	1996
Administrative Measures for Coal Production Licenses	1994
Rules for Implementation of the Regulations on the Management of Coal Production Licenses	1995
Regulations on Coal Businesses and Operations	1996
Law on Safety in Mines	1992
Regulations on Coal Safety Supervision	2000

Source: Andrews-Speed et al. (2003).

Table 28.12 Selected laws and regulations on environmental protection and land management.

Name of instrument	Year
Law on Water	1988
Law on Environmental Protection	1989
Law on Prevention and Control of Air Pollution	1987
Law on Water and Soil Conservation	1991
Law on Prevention and Control of Water Pollution, revised	1996
Law on Land Administration, amended	1986, 1998
Implementation Provisions for Law on Land Administration	1998
Regulations on Land Reclamation	1988
Implementation Measures for Regulations on Land Reclamation	1998

Source: Andrews-Speed et al. (2003).

Table 28.13 Selected laws and regulations on township and village coal mines.

Name of instrument	Year
Circular of the State Council on the Implementation of Industrial Management of Township and Village Mines	1986
Regulations on the Management of Township and Village Coal Mines	1994
Implementation Measures for the Regulations on the Management of Township and Village Coal Mines	1994
Regulations on Small Coal Mine Safety	1996
Law on Township Enterprises	1996

Source: Andrews-Speed et al. (2003).

The state's attitude toward small-scale mining

SSMs, including those collectively owned by towns, villages and individual miners, conduct their mining operations according to the *Mineral Resources Law* and relevant regulations. A specific chapter containing four articles under the *Mineral Resources Law* regulate small-scale mining activities.

Under the amended *Mineral Resources Law of 1996*, the same treatment is accorded to state-owned mines and non-state-owned small-scale mines with both foreign or domestic investment support.

Table 28.14 Circulars of the state council on rectification of mining industries.

Name of instrument	Year
Circular on Rectifying Individual Coal Mining	1991
Circular on Curbing Illegal Small-Scale Coal Mining and Maintaining Mine Safety	1993
Circular of Rectifying Mining Industries and Protecting the State's Ownership over Mineral Resources	1995
Circular on Rectifying Mining Activities in Xiaoqingling Gold Mining Area	1995
Circular on Closing Illegal and Irrationally Located Coal Mines	1998
Circular on Further Rectifying Mining Operation	2001

Table 28.15 Selected laws and regulations on mining taxation.

Name of instrument	Year
Regulations on Value Added Tax	1994
Regulations on Enterprise Income Tax	1994
Regulations on Resources Tax	1994
Regulations on Royalty (Mineral Resources Compensation)	1994

As stated in the *Mineral Resources Law of 1996*, the following policies apply to both collectively-owned mining enterprises and individual miners:

- The State having to adopt principles of vigorous support, rational planning and appropriate guidance and intensified supervision;
- The State assuming a role in directing and assisting collectively-owned mining enterprises and individual miners to raise unceasingly their technical levels, and to lift the rates of resources recovery and economic efficiency;
- That departments in charge of geology and mineral resources, geological exploration units and state-owned mining enterprises must, on the principles of vigorous support and mutual benefit, provide geological data and technical assistance with compensation to collectively-owned mining enterprises and individual miners; and
- That governments at and above the county level must direct and assist collectively-owned mining enterprises and individual miners to promote technological upgrading, improve the management of mining operations, and ensure safety in production.

Resources legally accessible to small-scale mines

As was prescribed in Article 34 of the *Mineral Resources Law of 1986*, according to Article 35 of the *Mineral Resources Law* of 1996, collectively-owned mining enterprises are encouraged to develop mineral resources within areas designated by the State. According to Article 38 of the Rules for *Implementation of the Mineral Resources Law of 1994*, these include:

- mineral deposits and occurrences unsuitable for large or medium-scale extraction;

- scattered small ore-bodies located close to the boundaries of mining areas licensed for state-owned mining enterprises, with approval by a competent department in charge of state-owned mining enterprises;
- remaining ore bodies within a closed state-owned mine, provided that the mining will be safe and not result in serious environmental consequences; and
- other mineral resources allowed to be exploited by collectively-owned mining enterprises pursuant to the state plan.

According to the updated mining law and regulations, collectively-owned mining enterprises and their restructured joint ventures may acquire mining rights for exploiting all mineral deposits, provided that staff have the necessary qualifications, and meet legal conditions and requirements.

Individual miners are permitted to exploit scattered and dispersed small mineral deposits and occurrences, as well as sand, stone and clay usable only as common building materials, and small amounts of minerals for their own personal use. Mineral deposits that are suitable for mining enterprises to exploit due to the relevant scale of reserves, the specific minerals for which protective mining policy is prescribed by the State, and other minerals prohibited by the State, cannot be mined by individuals. Individuals are prohibited from exploiting gold, tungsten, tin, antimony and rare earth minerals.

ACQUISITION OF MINING RIGHTS BY SMALL-SCALE MINERS

Acquisition of mining rights through the application/approval procedure

Pursuant to the *Regulations on Registration for the Exploitation of Mineral Resources of 1998*, a collectively-owned mining enterprise or individual miner, as a mining-right applicant, must check the availability of the mining areas being sought, and, if available, apply for demarcation of the mining area through the licensing authority based on an approved geological exploration report on mineral reserves in the area, prior to submitting an application (for a mining right) altogether.

An internet-based national mineral rights management information system, linking the end users at the Ministry of Land and Resources and its subordinate agencies at provincial, prefecture and county levels, has been developed to aid in the daily management of mineral rights, and for public acquisition of information on mineral rights. Similar management information systems on the programs for exploration and development of mineral resources being carried out at the national and provincial levels, have also been developed for guiding the granting of mineral rights, mine inspection, and for use by the public. As prescribed by the State Council, the granting of, and application for, mineral rights is subject to the programs for exploration and development of mineral resources.

In applying for a mining right, the applicant must present the following materials to the licensing authority:

1) an application for mining rights and a drawing or map showing the mining area;
2) documents of proof showing the qualifications of the mining-right applicant;

3) a plan for development and utilization of the mineral resources in the demarcated area;
4) approval documents on establishing the mining enterprise in accordance with the law;
5) an environmental impact statement on the mining of mineral resources; and
6) materials otherwise specified by the department incharge of Geology and Mineral Resources under the State Council.

Acquisition of mineral rights through public bidding

In addition to issuing mineral rights through the routine of application and approval, the licensing authorities may select bidding blocks, make public invitations for bids, establish and announce the bidding requirements, and set deadlines for bids, in accordance with the limits of authorities stipulated in the regulations on exploration rights, and the regulations on mining rights, respectively. Licensing authorities organize the evaluation of bids and select the best offer. Upon acceptance of a bid, the successful bidder must pay the requisite fees for exploration rights and reimbursement payments, or, as the case may be, fees for mining rights and reimbursement fees for mining rights. As prescribed, successful bidders must complete the registration procedures, after which they can obtain the exploration license or mining license, whereupon they can fulfill the obligations stipulated in the bid documents.

Special requirements for acquisition of mining rights by small-scale miners

Pursuant to Article 13 of the *Rules for Implementing the Mineral Resources Law*, an application for establishing a collectively-owned mining enterprise or privately-owned mining enterprise must meet the following requirements, in addition to the conditions as prescribed in relevant laws and regulations:
- having the information on mineral exploration required for mine construction;
- having a definite mining area approved without dispute;
- having the necessary funds, equipment and technical staff needed for the scale of the mine to be constructed;
- having a feasibility study report, and an appropriate mine design or mining plan in line with national industrial policies and other technical regulations; and
- the mine chief must have basic knowledge on mining operation, safety and environmental protection.

Pursuant to Article 14 of the *Rules for Implementing the Mineral Resources Law,* an application for individual mining must meet the following requirements:
- having a definite mining area approved and without dispute;
- having the necessary funds, equipment and technical personnel appropriate to the scale of mine being constructed;

- having corresponding information on mineral exploration and the approved mining proposal; and
- having necessary facilities and measures for safety and environmental protection.

Rights and obligations of legal small-scale miners

Pursuant to Article 30 of the *Rules for Implementing the Mineral Resources Law*, a mining licensee has the following rights:
- to conduct the mining operation within the valid term and the mining area prescribed by the mining license;
- to sell mineral products produced under the mining license;
- to construct production and living facilities within the mining area;
- to acquire the land use rights necessary for mine construction and mining operation; and
- other rights granted by laws and regulations.

Pursuant to Article 31 of the *Rules for Implementing the Mineral Resources Law*, a mining licensee must fulfill the following obligations:
- to conduct mine construction or operation within the valid term and mining area as designated by the mining license;
- to conduct efficient conservation and rational mining, and fully utilize mineral resources;
- to pay royalty and resources tax, and other taxes – *e.g.* corporate income tax, value added tax, etc. – pursuant to laws and regulations;
- to comply with state laws and regulations on labour, safety, water and soil conservation, land reclamation and environmental protection;
- to prepare and submit mineral reserve forms and progressive reports on mineral development according to relevant provisions; and
- to be subject to supervision and management from competent departments in charge of geology and mineral resources, as well as other relevant departments.

Security of mineral titles

In China, the right to mine is a critical precondition to investment in exploration. To guarantee the holder of a exploration right the power to apply for and obtain a mining right, the new *Mineral Resources Law* stipulates that, an exploration licensee has the privileged priority to obtain a mining right with respect to the mineral resources within the area under the exploration right. It means that, in the event of discovery, the holder of an exploration right seeking to mine a deposit in the exploration area, is guaranteed the right to mine, provided that he meets the conditions and requirements regarding, for example, funding and technical qualifications, mining plans, and environmental impact statements, prescribed by the *Mineral Resources Law* and the regulations on mineral rights. No person apart from the exploration licensee is permitted to apply to mine in an area of exploration. No government agency responsible for mineral rights management can issue a mining right to a person other than the exploration licensee.

New mining legislation considers the necessary time requirements for preparing feasibility studies, environmental impact statements, and financial arrangements. Most notably, a retention period between the exploration stage and the mining stage is prescribed in the *Regulations on Registration for Mineral Exploration.* After ascertaining an economic deposit within the term of an exploration license, the exploration licensee may apply for the retention of an exploration right with respect to the economic deposit within 30 days prior to the expiration of the exploration license. The maximum period of retention is two years. If there is a need to extend the term of retention, the exploration licensee may apply for an extension twice; each extension does not exceed two years. The area for retention of the exploration right should be that covering economic deposits.

Maximum period of mineral rights and renewal

In China, the maximum period of an exploration license for any mineral resource other than oil and gas is three years. The maximum period of an exploration license for oil and/or gas is seven years. If there is a need to extend the period of an exploration license, the exploration licensee can apply for the extension of the exploration license with the licensing authorities within 30 days prior to the expiration of the exploration license. Extension of an exploration license does not exceed two years each time.

The term of a mining license is decided in accordance with the magnitude of the mining project. If the scale of the project is large, the maximum term of the mining license is 30 years; for a project medium in scale, 20 years; and for a project small in scale, 10 years. If there is a need to extend the term of a mining license, the mining licensee can, within 30 days prior to its expiration, carry out the requisite procedures for extending the mining license with the licensing authority.

Payment for mineral rights

The payment for acquiring an exploration right or a mining right is equivalent to the rentals for mineral rights under the mining acts in the states of Australia, and for the dead-rents in Indonesia and India. The rates for the use of exploration rights are as follows: (1) 100 RMB Yuan per square kilometer per year for the first three years; and (2) 100 RMB Yuan per square kilometer added per year, commencing from the fourth year onward. However, the maximum rate cannot exceed 500 RMB Yuan per square kilometer per year. The rate of the fee for the use of mining rights is 1000 RMB Yuan per square kilometer per year.

Reimbursement fees for exploration rights and reimbursement fees for mining rights

If anyone applies for an exploration right or mining right to any blocks containing mineral deposits discovered at the State's expense, the applicant must pay reimbursement fees for an exploration right, in addition to the fee for using the exploration right as prescribed, or, as the case may be, pay the reimbursement fees for mining rights, in addition to the fee for using mining rights. Reimbursement fees must be paid in full or in installments, and must be evaluated by qualified organizations designated by the Ministry of Land and Resources. The results of the evaluation of the reimbursement fees must be confirmed by the Ministry of Land and Resources.

Transferability of mineral titles

A very important change in the new mining legislation is that exploration rights and mining rights are transferable. The holder of an exploration right may, after completing specified minimum exploration expenditures, transfer his exploration right. A mining enterprise holding a mining right may transfer its mining right because of merging or splitting ownership; it has entered an equity or cooperative joint venture; it has sold its assets, or when there has been some form of change with regard to ownership of assets.

Preconditions for transfer of mining right: Any mining licensee who wishes to transfer his mining right can only do so if: (a) one year or more has passed since the mining enterprise commenced operation; (b) the mining licensee has undisputed ownership of the mining right; (c) the fees for the use of the mining rights and the reimbursement fees for mining rights, royalty and the resources tax have been paid in accordance with relevant provisions of the State; and (d) other requirements otherwise specified by the Ministry of Land and Resources have been met.

Reporting and inspection

If the licensing authorities need to investigate the progress of the input of exploration expenditures and exploration work, the exploration licensee must report swiftly and accurately. Concealing relevant information or making false declarations constitutes a violation of the obligations of the exploration licensee. If suspected, an investigation cannot be refused.

At the request of the exploration licensees, the licensing authorities must keep confidential any materials or information deemed sensitive by the exploration licensee, including material submitted for the applications for the exploration licenses as well as materials concerning exploration results and financial reports. Licensing authorities are permitted to conduct and supervise matters regarding rational development and utilization of mineral resources, environmental protection and other obligations that must be fulfilled by mining licensees in their administrative areas. Mining licensees are required to report any relevant circumstances truthfully and submit an annual report to the licensing authorities.

Stricter enforcement of mining legislation

As already explained, mineral resources management authorities at different levels are expending great efforts to remove illegal small mining activities. During the past two years, ten of thousands of illegal small mines were banned. Most lacked the licenses to operate, had unacceptable environmental performance, and did not meet minimum safety requirements. It was realized that the enforcement of mining law and regulations needed to be improved to guarantee the interests of all legal holders of mineral rights. Thus, in accordance with new mining legislation, stricter legal liabilities, including administrative penalties and criminal responsibilities, have been imposed on the holders of mineral rights or any parties who have violated provisions under the amended *Mineral Resources Law* and the three items of Regulations. Legal liabilities have also been imposed on the staff of the governmental agencies in charge of supervising and administering mineral exploration and mining operations that illegally exercise their power. The department in charge of geology and mineral resources under the People's governments at higher levels are entitled to revoke exploration rights or mining rights granted illegally.

Stricter review and approval of mining plans

According to the *Circular of the Ministry of Land and Resources on Strengthening the Review of Mining Plans*, mineral rights management agencies can exercise a stricter assessment and approval of mining plans when reviewing applications for mining rights.[8] Efforts are now being made to determine whether or not applicants meet the requirements of rational mining scales and mining methods, environmental impact statements, and mine safety facilities and measures.

MINING TAXATION

Six tax laws and regulations were enacted in late-1993 and put in force from January 1, 1994, signifying a major reform in the country's tax regime. These include an enterprise income tax law, an individual income tax law (amended version of individual income tax law of 1980), regulations on value added tax (VAT), regulations on consumption tax, regulations on business tax, and regulations on resources tax. The most important taxes and charges are VAT, corporate income tax, resources tax and royalty. The regulations on VAT, consumption tax and business tax apply to all mining enterprises, including SSMs.

China's taxation restructuring in 1994

Restructuring of Turnover Taxes: The restructuring of turnover taxes was a key part of China's tax restructuring program in 1994. The new turnover

[8]It is derived from the Collection of Provisions on Mineral Rights Management, compiled by the Department of Mineral Development Management of the Ministry of Land and Resources, Geological Publisher, 1999, pp. 158–167.

taxes – including VAT, consumption tax and business tax – apply to both domestic enterprises and enterprises with foreign investment. The consolidated industrial and commercial tax that had been in place for enterprises with foreign investment up until the end of 1993, was removed. For domestic enterprises, including small-scale mines, *product tax* was replaced by *value added tax*. The rate of product tax was 5% for crude oil and natural gas, between five and eight per cent for phosphate rocks and sulfur, and three per cent for coal, ferrous metal ores and nonferrous metal ores.

The newly-introduced value added tax system imposes a heavier tax burden on the mineral industry, compared with the product tax previously applied. According to the *Provisional Regulations of the PRC on Value Added Tax*, which came into effect on January 1, 1994, there are three rates for VAT: (1) a 17% rate for sales and import of goods or labor services for processing and repairing; (2) 13% rate for sales and import of five kinds of goods specified in the Regulations on VAT; (3) and zero rate for the goods exported by taxpayers. The VAT payable represents the balance of output tax for a given period, after deducting the input tax for the same period. The Regulations on VAT disallows the offsetting of input tax on fixed assets from the output tax.

In the petroleum industry, a 17% VAT rate is applied to crude oil and natural gas. As for crude oil and natural gas produced by Chinese-foreign cooperative oil and gas fields, the VAT payable is 5% of sales revenues. The input tax cannot be deducted and is not be refundable. The value added tax rate is 13% for all solid mineral products, including coal, metals and non-metals.

Some 75% of VAT is distributed to the Central Government, and the remainder, to the local government.

Restructuring of corporate income taxes: Before the end of 1993, three different regulations were applied to enterprise income – namely, regulations for state-owned enterprises, collective-owned enterprises and private enterprises. A nominal rate of income tax (55%) was applied to state-owned enterprises (but was closer to 33%, as is explained below).

Following implementation of the *Provisional Regulations of the PRC on Enterprise Income Tax*, which came into effect on January 1 1994, a consolidated income tax of 33% was levied on all types of domestic enterprises, including state-owned, collective, private, joint venture, and joint-stock enterprises. Deductible items (for the purpose of calculating taxable income) include costs, expenses and losses as prescribed by the Regulations. The *adjustment tax* applied to large and medium-scale state-owned enterprises was removed as a result of the enterprise income tax restructuring program. In addition, the fiscal regime under which loans were to be repaid prior to the payment of enterprise income tax has been modified to parallel international practice.

Restructuring of the resources tax: The resources tax is levied on any operations producing mineral products – crude oil, natural gas, coal, metal and non-metal products – and salt in China. Any units/individuals, including domestic enterprises and enterprises with foreign investment that produce products on which resources tax is levied, pay resources tax. Revenues derived from resources tax revenues from offshore petroleum fields are allocated to the

Central Government; the balance of resources tax revenues are distributed to local governments.

Prior to January 1, 1994, resources tax was levied on a selected group of mineral products, including iron ores, crude oil and natural gas. It was based on excess profits, other than tonnage of crude ores or crude oil and gas; rates were generally very low. The newly-introduced resources tax system imposes heavy tax burdens on the mineral and petroleum industries, especially iron mining operations lacking follow-up processing operations.

Royalty (Mineral Resources Compensation): Apart from Chinese-foreign cooperative joint ventures involved in the exploitation of onshore and offshore petroleum, any party, including domestic mining enterprises and mining enterprises with foreign investment, involved in the exploitation of mineral resources, must pay royalties to the Chinese Government. The amount of compensation is determined from the revenues of mineral products; it ranges from 0.5 to 4% but on average, is 1.18%. Government mining agencies are responsible for overseeing the collection of such compensation. Revenues are then distributed equally among the Central Government and the relevant province or municipality. When autonomous regions are involved, the Central Government receives 40% of proceeds, and the autonomous region, 60%.

Integrated total tax burdens

The major taxes, charges and fees imposed upon China's mineral industries include value added tax, resources tax, royalties and mineral resources compensation, and corporate income tax. Other taxes and charges include city construction tax, land use tax, business tax, consumption tax, and an education surcharge. Based on a recent survey on mining taxation within state-owned mining enterprises, the following tax burdens were determined: 14% for petroleum; 12% for coal; 21% for iron ores; 6.5% for gold; 8.5% for nonferrous metals; 6.3% for fertilizer minerals; 11% for construction materials; and 1–6% for uranium. The rates of tax burdens are expressed as a percentage of sales income.

Contributions of selected taxes to total tax burden

In all categories of taxes and charges applicable to the mineral sectors, value added tax contributes substantially to total tax revenues. Resources tax constitutes a high percentage of total tax revenues derived from iron ore mines without follow-up processing operations. In a 1996 survey on mining taxation in China, the contributions of value-added tax, resources tax and royalty to total tax revenues was determined to be 66.52%, 11.44% and 5.80% respectively.

Mining taxation policy adjustments

Since 1994, a set of policies have been followed by the Chinese Government to mitigate tax burden in the mineral sector. For example, a reduced rate of 13%

of value added tax is applied to solid mineral products. Moreover, gold production has been exempted from value added tax since 1995, and, since January 1 1996, some 40% of the original rate of resources tax has been imposed on iron ore mines established before the end of 1993 that lack follow-up processing operations. In another example, 70% of the original rate of resources tax is levied on nonferrous metals. Efforts have also been made to reduce the tax burden of mining enterprises.

REGULATORY FRAMEWORK FOR MINING ENVIRONMENTAL MANAGEMENT

The basic laws governing environmental management in the mineral sector are the *Environment Protection Law* and the *Mineral Resources Law.* In addition, a law of prevention and control of water pollution, an air pollution control law, and more than 20 administrative regulations, have been enacted to regulate environmental protection.

Some general principles on mining environmental protection have been included in the *Mineral Resource Law.* In exploiting mineral resources, it is essential to observe the legal provisions on environmental protection to prevent the polluting of the environment. For mineral resources, special attention is paid to using land economically. In the case of cultivated land, grassland or forest land that has been damaged by mining, relevant mining enterprises are required to take measures to ensure that affected lands are made available for future use; common remediation measures include reclamation efforts, reforestation, and re-vegetation. Unless approved by a competent department authorized by the State Council, no one can exploit mineral deposits in nature reserves and important scenic spots designated by the State, major sites of immovable historical interest, and regions under state protection. If a mine is to be closed down, a report must be prepared with information about the mining operations, hidden dangers, land reclamation and utilization, and environmental protection, and an application for approval must be filed in accordance with the relevant state provisions.

According to the newly issued *Regulations on Registration for Exploitation of Mineral Resources*, when applying for a mining right, the mining-right applicant must present a plan for development and utilization of mineral resources in a demarcated area, as well as an environment impact statement. Licensing authorities are empowered to conduct, supervise and inspect matters pertaining to the rational development and utilization of mineral resources, environmental protection, and other obligations required to be fulfilled by the mining licensees.

INSTITUTIONAL FRAMEWORK FOR IMPLEMENTATION OF LAWS AND REGULATIONS GOVERNING SMALL-SCALE MINING

Institutional structure governing small-scale mines

There are five departments involved in the implementation of specific laws and regulations governing the small-scale mining industry. The first, the Ministry of Land and Resources and its subordinate agencies at provincial, prefecture and county levels, is responsible for overseeing the exploration and development of mineral resources. In collaboration with the government agencies in charge of environmental protection, they are also responsible for daily inspection and management of the mining environment, particularly the small-scale mining environment.

Secondly, the State Bureau of Environmental Protection and its subordinate agencies at provincial, prefecture and county levels, is responsible for overseeing the review and approval of EIS, and for monitoring and supervising the discharge of pollutants from mining operations.

Thirdly, the State Bureau of Safety Supervision and its subordinate agencies at provincial, prefecture and county levels, is responsible for overseeing mine safety and handling accidents at mines.

Fourthly, the Ministry of Labour and Social Security and its subordinate agencies at provincial, prefecture and county levels, is responsible for the management of labour and social security affairs. Nevertheless, they currently focus on resolving problems with unemployment and social security that occur at large and medium-scale SOMEs. However, less attention has been placed on small-scale mines. Prior to 1998, the former Ministry of Labour was in charge of mine safety supervision. Since 1998, the responsibilities related to mine safety have been reassigned to the State Bureau of Safety Supervision.

Finally, the State Tax Administration, and its branches at various levels, is responsible for collecting corporate income tax, value added tax, and resources tax.

Reorganization of agencies governing the mineral industry

During the past two decades, great changes have taken place in the Chinese Government as far as regulating resident mining activities is concerned. Under the traditional economic system, the former State Planning Commission (SPC) had an active role in setting priorities for the mining industry, approving minerals and energy exploration activities, and promoting operations in conjunction with various mineral-related ministries and provincial governments. Between 1950 and 1982, most mineral exploration and exploitation projects were organized and funded by the State. Certain mineral-related ministries were responsible for controlling the exploration and exploitation of specific minerals, energy development and production. These included (for ferrous metals) the Ministry of the Metallurgical Industry (MMI); (for nonferrous metals) the China National Nonferrous Corporation (CNNC); and (for petroleum on land territory) the China National Petroleum Corporation (CNPC). Finally, the Ministry of Geology, which, in 1982, was renamed the Ministry of Geology and Mineral Resources (MGMR), was responsible for organizing mineral exploration nationwide. More than ten ministries and national corporations were engaged in minerals and energy development activities.

Since early-1998, all functions in relation to mineral resources management have been concentrated within one ministry – *i.e.* the Ministry of Land and Resources (MOLAR). MOLAR was established in March, 1998 in accordance with the *Program on Reorganization of the State Council's Agencies* as approved by the 1st Session of the 9th National People's Congress and the Notification of the State Council on the Establishment of the Organizational Structure. MOLAR, as one of the member departments of the State Council, is in charge of the planning, management, protection and rational utilization of land resources, mineral resources, and marine resources. The Ministries of the Metallurgical Industry, Chemical Industry, and Coal Industry were abolished in 1998, and a limited range of functions on industrial management were reallocated to the newly-established State Bureau of Coal Industry. Further reorganization occurred in 2001, which effectively abolished the State Bureau of Coal Industry, replacing it with the State Bureau of Safety Supervision.

The functions that have been reallocated to MOLAR include: (a) the functions under the former State Bureau of Land Management; (b) the functions under the former Ministry of Geology and Mineral Resources; (c) the functions of the former State Oceanographic Bureau; (d) the functions under the former National Commission of Mineral Resources and its sub-agencies; (e) the functions of territory land planning, integral planning for land use assumed by the former State Planning Commission; (e) the functions in relation to mineral resources management assumed by the former Ministries of Metallurgical Industry, Coal Industry, Chemical Industry, the former China Nuclear Industry Corporation, and the former China National Nonferrous Corporation.

The restructuring of resident mineral industries occurred concurrently with the reorganization of government agencies. Almost all state-owned exploration units and mining enterprises, apart from the three national petroleum corporations and the national aluminum industrial corporation, have since been localized to provincial governments. The Central Government funds national geological surveys for which the China Geological Survey under MOLAR is responsible. It has also adopted a mineral policy to promote and enforce mining legislation in order to regulate the commercial exploration and development of mineral resources based on market mechanisms.

MINERAL POLICY AND THE RESTRUCTURING OF THE SMALL-SCALE MINING INDUSTRY

Restructuring policies and the mining industry

On August 9, 1999, the State Commission of Economy and Trade issued the *Catalog of Curbing the Unnecessary Repetition of Capital Construction Investment in Industrial and Commercial Circles.* According to this Catalog, the Chinese Government will attempt to stop, and prevent the development of, the following operations:

- projects prohibited by national laws and regulations;
- projects lacking technology and know-how;
- projects that will result in severe environmental pollution and unwise utilization of resources.

The Catalog stipulates that all listed projects cannot be approved by a competent government agencies, are not to be funded by banks and financial institutions, and cannot undergo procedures for review and approval by competent government agencies responsible for land management, urban planning, environmental management and customs.

The projects related to the mineral industry listed in the Catalog are as follows:

Steel and iron sector:

- coking process and equipment using indigenous methods.

Nonferrous metals smelting sector:

- new coarse copper smelter;
- new copper electrowinning;
- new copper processing;
- new aluminum electrowinning with a capacity of less than 100,000 tonnes per year;
- new aluminum processing;
- new lead smelter;
- new zinc smelter with a capacity of less than 50,000 tonnes per year; and
- all varieties of mining and separating projects at a mixed recovery of less than 60% of main product.

Coal mining sector:

- coal mining projects with shaft production not exceeding the following limits:
  - 150,000 tonnes per year in Shangxi, Shaanxi and Inner Mongolia;
  - 90,000 tonnes per year in Xinjiang, Gansu, Ninxia, Qinghai, Beijing, Tianjin, Hebei and in the northeast China and in the southeast China; and
  - 30,000 tonnes per year for extremely thin and unstable coal seams.
- projects whose desulfurization measures fail to meet the requirements of high-sulfur-containing coal mining projects, as prescribed by the State's environmental protection laws and regulations.
- new coal mining projects with recoveries less than 50% of the mine shaft.

Gold mining sector:

- independent cyanidation complexes with capacities of less than 10 tonnes per day;
- pyrometallurgical projects treating gold concentrates amounting to less than 50 tonnes per day;
- independent leaching projects with dumps treating less than 5000 tonnes per day;
- gold mining projects treating less than 25 tonnes per day of gold-containing hard rock;
- gold mining projects treating less than 200,000 cubic meters of placer gold.

Program for mineral exploration and development

The mineral resources programming system consists of a program for exploration and development of mineral resources at the national, provincial, prefecture and county levels. Special programs have also been implemented, including a national program for geological surveying and mineral exploration; a program for developing and protecting mineral resources; and a program for protecting the mining environment.

In April, 2001, the State Council approved the *National Program for Exploration and Development of Mineral Resources of 1999–2010.* Using the principles of sustainable development as guidance, the Program has the following four objectives:

– To raise domestic mineral availability by strengthening national geological survey funding (through the Central Government and local governments), and promoting commercial geological exploration for mineral resources through policy;
– To ensure adequate supplies of energy and minerals;
– To change the modes for developing and utilizing mineral resources; and
– To upgrade the mining environment.

The following contents have been included in the *National Program for Exploration and Development of Mineral Resources of 1999–2010*:

– National geological surveys and appraisals of mineral resources;
– Commercial exploration for mineral resources, both onshore and offshore;
– Mineral development;
– The introduction of foreign investment to the mineral industry of China;
– The import and export of mineral commodities;
– Exploration and development of mineral resources overseas;
– Conservation and rational development of mineral resources; and
– Protection of the mining environment.

As required by the State Council, the governments of 31 provinces, municipalities and autonomous regions worked out the provincial programs for exploration and development of mineral resources of 2001–2010, gaining approval from the Ministry of Land and Resources, the State Development Planning Commission, the State Economic and Trade Commission, and the State Bureau of Environmental Protection. All provincial programs for mineral exploration and development are working to control the total number of SSMs, particularly within the coal mining segment. These programs feature the following policy-related measures:

– delineation of the "three areas" – an area prohibited from being mined (*e.g.* nature reserves), a restricted area for operation (*i.e.* a region to be mined under specific conditions), and an area where mining operations will occur;
– a requirement that the scale of the mining operation matches the scale of the mineral deposit; and

- setting minimum requirements for the acquisition of mining rights, including (a) establishing minimum mining scales on a deposit-by-deposit basis; (b) setting minimum environmental requirements; and (c) establishing minimum requirements for mining equipment and techniques for safety, environmental and recovery purposes.

Program of rectification for mining operation

On November 3, 2001, the Administrative Office of the State Council issued the *Circular on Further Rectifying Mining Operation.* Its objectives are as follows:

- to sort out all exploration licenses and mining licenses issued, and correct the mistakes that have been made in the process of granting licenses;
- to put an end to illegal exploration and mining operations;
- to monitor mining enterprises in order to conduct mining activities pursuant to the approved mine design or mine plan, and to close the mines that do not meet the requirements for mine safety;
- to strictly enforce mineral rights management regimes, and to guarantee the grant of mineral rights subject to exploration and development programs for mineral resources, and legal requirements for resources recovery, mine safety and environmental protection;
- to reinforce mineral development management to monitor exploration licensees and mining licensees, and comply with legal obligations; and
- to reinforce mine inspection, and monitor any party in violation of laws and regulations (to be investigated for administrative penalties or, as the case may be, crime liabilities pursuant to laws and regulations).

Policies on closing mines and restricting yield

As a result of the closure policy implemented in 1998, over 50,000 illegal and irrationally-located small-scale coal mines were closed. However, some problems still exist because closure and restriction policies have not been effectively implemented in certain areas.

PROSPECTS FOR THE SMALL-SCALE MINING INDUSTRY

China's industrialization and modernization need small-scale mining

As was illustrated in Table 28.6, in the past two decades, the small-scale mining industry has played an important role in supplying the coal and raw materials necessary for economic and social development in China. The country is at an initial stage of industrialization and modernization, characterized by high consumption of primary energy and raw materials. High demands for coal and minerals are needed to support such development. Table 28.16 shows predicted demands for mineral products in China.

Table 28.16 Predicted demand for selected minerals (unit: tonnes).

Mineral	2005	2015
Coal	1.2–1.3 billion	1.6–1.8 billion
Iron ores	400 million	500–550 million
Copper	1.7 million	1.8–2.1 million
Lead	0.6 million	0.65–0.70 million
Zinc	1.26 million	1.3–1.4 million
Bauxite	3 million	3.5–4.0 million
Tungsten	10,000–12,000	15,000
Tin	30,000	68,000
Gold	300	320–350
Phosphate	24 million	38–40 million
Pyrite	20 million	30–32 million

Source: Based on the research report on the prediction of demand for major minerals in 2005 and 2015, Cao Xingyuan, 1999. Some are adjusted.

During the past five decades, coal has been a primary source of energy in China, accounting for over 70% of the country's energy production and consumption. This share has risen to 75% since the early-1990s (The Institute of Industrial Economics, 1997). It is predicted that the demand for coal will continue to be at a level of 1.2–1.3 billion tonnes of production per year. Predicted future high demands for mineral resources will provide additional opportunities for the small-scale mining industry.

Characteristics of resources suitable for small-scale mining industry

Small and medium-scale mineral deposits account for approximately 89% of all deposits identified nationwide, which provides a solid resources base for the development of the country's small-scale mining industry.

Regional economic development policies advantageous to the development of the small-scale mining industry

The Chinese Government has adopted a number of favorable policies to encourage and support the development of small-scale enterprises collectively owned by towns and villages. It has also sought to accelerate mineral resources development in western areas.

Improved infrastructure benefiting small-scale mining industry

Since the early-1990s, the constructing and renovating of railways and accompanying infrastructure has been given high priority in national economic development; more than ten main railway lines have been constructed. It is predicted that operating railways amount to 68,000 kilometers, carrying a rail freight of 1.8 billion tonnes annually (increasing by 210 million tonnes per year). It is projected that the rail freight of coal being shipped from Shangxi, Shaanxi, and Inner Mongolia to the outside, will increase by 140 million tonnes each year. A number of highways are also under construction. The improvement of transportation, in turn, will precipitate the development of both the state-owned mining industry and the small-scale mining industry.

CASE STUDY – LUQUAN'S APPROACH TO SUSTAINABLE PRACTICES IN THE SMALL-SCALE MINING INDUSTRY

Luquan (Deer Spring) is a city located in eastern Shijiazhuang, the capital of Hebei Province, some 300 kilometers from Beijing to the south. It is well endowed with limestone and other building materials, and high quality cement. As a result of a "mining rush" that occurred in the early-1980s, the number of small-scale operations grew at a prodigious rate; there were soon 290 limestone extractive sites in operation. Following a 10-year period of mining rectification, the number of small-scale limestone mining sites still exceeded 240, the output value from which amounted to 1.04 billion RMB Yuan (125.36 million US dollars) in 1995, some 24% of that of the county. However, legal, technical, environmental and safety requirements seem to have been ignored at most mining sites. Moreover, many of the county's operations were irrationally located, too small in scale, featured outdated techniques and equipment, and were environmentally degrading. Shijiazhuang City, which is located some 20 kilometers away, was suffering from severe air pollution produced by Luquan's small-scale limestone mining operations and cement factories.

The Luquan Bureau of Geology and Mineral Resources, under the guidance of the Hebei Provincial Bureau of Geology and Mineral Resources, and support from the city government, undertook several measures in an attempt to reverse the situation completely.

First, a program was designed and implemented for the limestone extraction industry (applicable from 1996 to 2010), which covered both mineral development and mining environmental protection. Following implementation of the program, six limestone extraction areas were demarcated; six different areas were deemed non-operating areas; five protective areas were created for limestone; four reserve areas were established for limestone mines; three bases were designated for cement burden; and two protective areas were designated for mineral water.

The program is based on the following principles:

- the control of limestone ore production;
- giving the "green light" to large-scale mining operations, and closing small-scale sites;
- rationalizing the location of mines;
- guaranteeing the exploitation of high quality ores and the production of high quality products;

- matching the mining scale of an operation to the scale of a reserve to be extracted; and
- guaranteeing that the requirements for environmental protection are met.

To this end, nine mining rights have been set up for large and medium-scale limestone mines; the six limestone extraction areas have been designated for nine mining blocks (one for each mine).

The second step was the exercising of the strict enforcement of mining laws and regulations. The Luquan Bureau of Geology and Mineral Resources succeeded in its attempts to get all mining rectification and mine closure specifications under a legal framework.

The third step was the provision of satisfactory service to mining enterprises, including small-scale operations. The Luquan Bureau of Geology and Mineral Resources raised money for researching and developing mining and processing techniques, in order to improve resource recovery, environmental performance and economic efficiency at mining operations and cement factories. The Bureau did its best to offer technical assistance and provide funding for selected activities. Its efforts were praised by the management of a number of mining operations and cement factories.

The fourth step was the implementation of a mining rectification program, and the restructuring of the mining industry. Small-scale mining operations and cement factories have since been encouraged to merge, and restructure by means of negotiation for stock sharing into joint ventures. The owners of small-scale mining operations and cement factories are now able to reap economic rewards through shareholding in joint ventures rather than strictly through direct ownership of mining operations or cement factories. Another important factor has been the introduction of public bidding and the auctioning of mining rights to mineral deposits discovered or identified using the State's fund. The public bidding and auctioning events have been consistent with the program on the exploration and development of mineral resources. Since May 2001, the Bureau has conducted public biding for mining rights on ten separate occasions, in the process acquiring 13.82 million RMB Yuan (1.67 million US dollars) in payments. As a result, 51 small-scale limestone mining operations have merged into several large-scale mining joint ventures.

The efforts of the Luquan Bureau of Geology and Mineral Resources in recent years have led to a more efficient and environmentally friendly mining industry. In 2001, the output of Luquan's mining industry, contributed by 154 mining enterprises, totaled 2.1 billion RMB Yuan (253.62 million US dollars) and accounted for 33% of the region's GDP.

CONCLUSIONS

In China, the small-scale mining industry has greatly contributed to socio-economic development both nationally and locally. It has constituted, and will continue to constitute, an important share of the country's mining industry. The Chinese Government continues to undertake efforts to regulate, guide and encourage the development of the small-scale mining industry. It is expected to continue to play a vital socio-economic role for many decades to come.

REFERENCES

Andrews-Speed, P., Yang, M., Shen, L. & Cao, S. (2003). The regulation of China's township and village coal mines: a study of complexity and ineffectiveness. *Journal of Cleaner Production*, 11(2), 185–196.

Cao, X. (1999). *The Research Report on the Prediction of demand for Major Minerals in 2005 and 2015.*

International Labour Organization (ILO) (1999). *Social and Labour Issues in Small-Scale Mines*. Report for discussion at the Tripartite Meeting on Social and Labour Issues in Small-Scale Mines, Geneva.

Song, R. (1997). *China Mineral Resources Report of 1996*. Geological Publisher.

The Department of Mineral Development Management of the Ministry of Land and Resources (1999). *The Collection of Provisions on Mineral Rights Management*. Geological Publisher.

The Institute of Industrial Economics of the Academy of Social Sciences (1997). *China's Industrial Development Report of 1997*. Economic Management Publisher.

The Ministry of Land and Resources (1999). *Mineral Resources Information of China*.

The Ministry of Land and Resources (1999) *China Mineral Resources Report of 1997/98*. Geological Publisher.

The State Planning Commission and the former Ministry of Coal Industry (1997). *The Report on National Small-Scale Coal Mines Survey*.

Ye, Q., Zhang, B. (1998). *China Collectively-owned-by-Town- and village Coal Mining Industry*. The Coal Industrial Publisher.

Ziran, Z. (1997). Overview of national mineral policy in China: Opportunities and challenges for the mineral industries. *Resources Policy*, 23(1/2). Elsevier Science Ltd., 1997.

Ziran, Z. (1999). Small-Scale Mining Activities in China – Contributions, Problems and Policy, the *Proceedings of the APEC/GEMEED Workshop on Small- and Medium-Scale Mining*. Cairns, Australia, 5th–8th October 1999.

Ziran, Z. (1999). Environmentally Sustainable Development of the Mineral Industries in China toward 21st Century, the *Proceedings of the APEC/GEMEED Workshop on Metals and Environment of the 3rd Environmental Cooperation Workshop on Sustainable Development of Mining Activities*, Cairns, Australia, 5th–8th October 1999.

Ziran, Z. (2002). New Century Opportunities for the New Mineral Industries in China – an Overview of the Mineral Industries and National Mineral Policy, prepared for the Asian Mining Yearbook 2001, to be published by the Australia based *Asian Journal of Mining*.

APPENDIX A: BACKGROUND INFORMATION ON ECONOMIC INDICATORS, MINING LEGISLATION AND FISCAL REGIME IN CHINA

1. Geography	
1) land area:	9,600,000 sq. km
2) population with growth rate:	1.276 billion (2001), 0.695% (2001)
2. Macro-economic Indicators	
1) GDP – purchasing power parity:	9,593.3 billion RMB Yuan (2001)
2) GDP – average annual growth rate:	8.1% (1996–2001)
3) GDP – per capita:	7,543 RMB Yuan (2001)
4) Inflation rate – consumer price index:	0.7% (2001)
5) Unemployment rate:	officially 3.6% in urban areas (2001)
6) Exchange rate – per US$1:	Yuan (¥) per US$1—8.2770 (2000); 8.2789 (1998); 8.2796 (1997), 8.2898 (middle rate in 1997), 8.3142 (1996), 8.3514 (1995), 8.6187 (1994), 5.7620 (1993) note: beginning 1 January 1994, the People's Bank of China quotes the midpoint rate against the US dollar based on the previous day's prevailing rate in the interbank foreign exchange market
7) Exports – total values:	$266.16 billion (f.o.b., 2001)
8) Imports – total values:	$243.61 billion (c.i.f., 2001)
9) Number of newly approved enterprise with foreign investment:	26,140 (2001)
10) Amount of actually introduced foreign capital investment:	$49.68 billion (2001)
11) Total number of enterprises with foreign investment	391,232 (1979–2001)
12) Total amount of actually introduced foreign direct investment:	$393.51 billion (1979–2001)
13) National reserve of foreign currency:	$212.17 billion (December, 2001)
3. Mining Legislation	
1) Mining Law:	Mineral Resources Law, No. 36 (March 19, 1986) and No. 74 (amended on Aug. 29, 1996)
2) Implementation Rule:	Regulations for Registering to Explore for Mineral Resources, No.240 (Feb. 12, 1998); Regulations for Registering to Mine Mineral Resources, No. 241 (Feb. 12, 1998); Regulations on Transfer of Exploration Rights and Mining Rights, No. 242 (Feb. 12, 1998); Implementation Rules for the Mineral Resources Law, No. 152 (March 26, 1994); Regulations on the Management of the Collection of the Mineral Resources Compensation (Royalty), No. 150 (Feb. 27, 1994)
3) Relevant Laws and Regulations:	Constitution of the People's Republic of China; Criminal Law; Law on Foreign-Capital Enterprises (April 12, 1986); Law on Chinese-Foreign

	Contractual Joint Ventures (April 13, 1988); Law of the People's Republic of China on Chinese-Foreign Equity Joint Ventures (Sept. 20, 1983); their implementation rules; Environmental Protection Law; Relevant Tax Regulations, etc.
4) Government Contract and Agreement:	None
5) Government Authority:	
Ownership of Minerals	State (MRL,[9] Art. 3)
Regulatory Government Authority	Ministry of Land and Resources
Granter of Titles	Ministry of Land and Resources and Department in charge of geology and mineral resources under the people's government of provinces, autonomous regions, municipalities [10] (MRL , Art. 12, 16; EMR,[11] Art. 4 ; MMR , Art. 3).
International Arbitration	Regulated by Art. 110 of the Regulations for the Implementation of the Law of the People's Republic of China on Chinese-Foreign Equity Joint Ventures (Sept. 20, 1983)
6) Restriction of Mineral Activity on:	
Foreign Direct Investment	Same treatment as domestic investment (EMR, Art. 37, 38; MMR, Art. 28, 29)[12]
Commodities	None
Application Area	State planned mining area, mining area of great value to the national economy, and the specified minerals for which protective mining policy is prescribed by the State (MRL, Art. 17); harbors, airports, national defense zone, important industrial district, large scale conservancy works, municipal engineering installations of cities and town, railways, highways, important rivers and embankments, natural reserves, important scenic spots, historical relics, or scenic beauty (MRL, Art. 20)

[9] Mineral Resources Law.

[10] The Ministry of Land and Resources shall examine, approve, register, and license to an exploration project which straddle two or more administrative areas , or is the sea area, or is from an enterprises with foreign investment applicant , or project for which those minerals listed in the Appendix attached to the regulations on exploration (EMR, Art. 4).

[11] Regulations for Registering to Explore for Mineral Resources Using the Block System.

[12] Regulations for Registering to Explore for Mineral Resources prescribes that "These Regulations shall be applied to any foreign investment in exploring for mineral resources" in Article 37. Regulations for Registering to Mine Mineral Resources prescribes that "These Regulations shall be applied to any foreign investment in mining mineral resources" in Article 28.

7) Coverage of the Mining Law;	Geological survey, exploration, mining, selling minerals[13]
8) Exploration Stage:	
Title Acquirement	Apply to the licensing authorities[14] (EMR, Art. 4;) or public bidding (EMR, Art. 16)[15]
Permitting Termination	40 days (EMR, Art. 8)
Necessary Conditions for Permitting	National plan for geological exploration or exploration contract, implementation proposal for exploration, certificate for the qualification of exploration unit, documents of proof showing the source of the funds (EMR, Art. 6)
Title(s) Exclusivity	Yes (EMR, Art. 9)
Initial Term for Exploration	EL:[16] 3 years (EMR, Art. 10) RER:[17] 2 years[18] (EMR, Art. 21)
Renewal(s)	2 years each time (EMR, Art. 10) RER: 2 years × 2 (EMR, Art. 21)
Extent of Concession	40 basic unit blocks[19] for metal, none-metal, and radioactive minerals[20] (EMR, Art. 3)
Shape of Concession	Polygon (EMR, Art. 3)
Reporting Obligation	Yes (EMR, Art. 18, 24, 25)
Relinquishment Obligation	None
Transfer of Concession	Yes, but need to be examined and approved within 40 days from application by the licensing authorities[21] (MRL, Art. 6.2; EMR, Art. 22.3; TR,[22] Art. 3, 4, 5, 7, 8, 9, 10)

[13] Minerals products to be purchased exclusively by designated units may not be purchased by any other units or individuals (MRL, Art. 34).

[14] Applicants shall pay Reimbursement Fee for Exploration Right or Mining Right of mineral deposits discovered by the State in any blocks, and the Reimbursement Fee will be based on costs incurred by the State (EMR, Art. 13; MMR, Art. 6, 9). The Exploration or Mining Fee and Reimbursement Fee may be reduced or exempt (EMR, Art. 15; MMR, Art. 12).

[15] The licensing authority shall give priority to any exploration project listed in the first category of the National Geological Exploration Plan (EMR, Art. 8).

[16] Exploration License

[17] Retention of Exploration Right, which can be applied by the exploration licensee within the term of a valid exploration license from the need to defer the present development of the deposit or from the present technical support (EMR, Art. 21).

[18] The licensee shall continue to pay the Exploration Fee (EMR, Art. 21).

[19] Basic Unit Block is longitude 1′ × latitude 1′ (EMR, Art. 3).

[20] The State shall adopt a Unified Block Registration System (MRL, Art. 12).

[21] Exploration license may be transferred to others after completion of the stipulated minimum exploration expenditure and subject to approval in accordance with law (MRL, Art. 6.1).

[22] Regulations for Transferring Exploration Rights and Mining Rights.

Concession Holding Fee	1–3 year-100 RMB yuan per sq. km annually; 4-year-100 RMB yuan per sq. km added per year; and the highest amount shall not exceed 500 RMB yuan per sq. km per year as annual Exploration Fee[23] (EMR, Art. 12).
Minimum Expenditure Obligation	1 year- 2,000 RMB yuan per sq. km; 2 year- 5,000 RMB yuan per sq. km; and 3 year- 10,00 0 RMB yuan per sq. km each year thereafter (EMR, Art. 17).
9) Security of Tenure:	
Exclusive Rights from Exploration to Mining	Yes[24] (MRL, Art. 6.1)
Accompanied Documents to the Application	Environmental Impact Assessment (EIA) report[25] (MMR,[26] Art. 5.4), exploration report[27] (MRL, Art. 13), mine design and development plan[28] (MRL, Art. 29, 30; MMR, Art. 5.3)
Approval termination	40 days (MMR, Art. 6)
10) Mining Stage:	
Maximum Extent of Concession	Not specified[29]
Initial Term for Mining	Large scale project- 30 years; medium scale project – 20 years; and small scale project- 10 years (MMR, Art. 7)
Renewal(s)	possible (MMR, Art. 7)
Concession Holding Fee	1,000 RMB yuan per sq. km per year as mining rights use fee (MMR. Art. 6, 9)
Minimum Production Requirement	Not specified
Mining Rights Transfer	Yes[30] (MRL, Art. 6.2; TR, Art. 3, 4, 6, 7, 8, 9, 10)
Mortgagability	It is permitted.

[23] The Exploration Fee may be reduced or exempt if the minerals or areas are encouraged for exploration by the State (EMR, Art. 15).

[24] Exploration licensees have the privileged priority to obtain mining right to the mineral resources in the exploration area (MRL, Art. 6.1). However, the authority shall examine its application of the mining area, mining design or plan, production technique, and safety and environmental protection measures before the approval (MRL, Art. 15).

[25] Environmental Impact Statement (EIS) report shall be presented to the licensing authorities prior to submitting the application for the mining rights (MMR, Art. 4).

[26] Regulations for Registering to Mine Mineral Resources.

[27] The mineral resources approving agency shall be responsible for examining and approving the prospecting reports to be used for the purpose of mine construction designing (MRL, Art. 13).

[28] The recovery rate and impoverishment rate in mining and recovery rate in ore dressing shall meet the design requirements (MRL, Art. 29), and minerals of commercial value shall be comprehensively mined and utilized in accordance with a unified plan to avoid waste (MRL, Art. 30).

[29] Three dimensional spaces for mining (MMR, Art. 32).

[30] Mining companies may transfer its mining right to others, subject to approval in accordance with law (MRL, Art. 6.2).

Compensation Requirement	Yes[31] (MRL, Art. 32)
EIA Requirement	Yes (MMR, Art. 5)
4. Fiscal Regimes	
1) Corporate income tax:	33%
2) Royalty:	Gold, 4%[32]; copper, 2%; and zinc, 2% on sales revenue
3) Import duty:	None for enterprises with foreign investment under specific conditions (Regulations for the Implementation of the Law of the People's Republic of China on Chinese-Foreign Equity Joint Ventures, Art. 71)
4) Dividend withholding tax:	None
5) Tax holiday:	5 years tax will apply if a mine will operate for more than 10 years (1, 2 year-100%; 3–5 year-50%), tax holidays does not apply to mines producing gold, other precious metals or rare earth metals
6) Foreign external account:	Allowed
7) Government equity requirement:	None
8) Export duties on minerals:	None for enterprises with foreign investment; gold can not be exported
9) Exchange control:	Free from major restrictions
10) Value added tax:	Actual tax rate is about 6–8% as a percentage in sales revenues for domestic mining enterprises, gold is exempt; zero rate is applied to the enterprises with foreign investment
11) Tax stabilization:	None
5. Environmental Regulations	
1) Environmental law:	Environmental Protection Law, Law on the Prevention and Mitigation of Water Pollution, Law on the Prevention and Mitigation of Air Pollution, Regulations on the Control of Noise Pollution, Regulations on Land Reclamation, Interim Measures on the Collection of Fee for Excessive Release of Pollutants, Measures on Environmental Protection Management on Capital Construction Projects
2) Adopted international law:	Antarctic-Environmental Protocol, Antarctic Treaty, Biodiversity, Climate Change, Desertification, Endangered Species, Hazardous Wastes, Law of the Sea, Marine Dumping, Nuclear Test Ban,

[31] Damages caused from mining mineral deposits to the production and livelihood of other persons shall be liable to making compensation and adopt necessary remedial measures (MRL, Art. 32).

[32] Gold must be sold to the Peoples Bank at a price which is equal to or higher than the international market rate, published by the central government.

	Ozone Layer Protection, Ship Pollution, Tropical Timber 83, Tropical Timber 94, Wetlands: signed, but not ratified: none of the selected agreements
3) Environmental administration for mining:	Ministry of Environment, Ministry of Land and Resources
4) Inspection and monitoring agency:	Ministry of Land and Resources (MMR, Art. 14)
5) Emission/effluent standard:	As specified by relevant provisions and rules
6) Community consultation obligation:	Not specified
7) Pollution taxes:	Levies for excessive release of pollutants
8) Mine close/reclamation bond:	None[33] (MRL, Art. 21)

[33] Need approval about mining operation, hidden dangers, land reclamation and utilization, and environmental protection (MRL, Art. 21).

29

The Impact of, and Responses to, the Closure of Small-Scale Coal Mines in China: A Preliminary Account

PHILIP ANDREWS-SPEED, GUO MA, XUNPENG SHI, AND BINGJIA SHAO

The last ten years or so have seen an increased concern in a number of features of the mining industry, two of which are of particular relevance to this chapter. The first is the management of the socio-economic impact of large-scale mines, and planning for their closure. The second is the effective regulation of small-scale mines. However, little attention has been devoted to the overlap of these two issues – that is, managing the *closure* of small-scale mines so as to minimise their negative socio-economic impact.

The aim of this chapter is to provide a preliminary analysis of the issue of managing the closure of small-scale mines,[1] using coal mines in China as an example. It provides a provisional account of the impact of, and some responses to, the enforced closure of large numbers of small-scale coal mines in China during the period 1998–2002. The results are provisional for two reasons: first, the closure campaign is very recent, and therefore, it is too early to judge fully the impact of closures, and the success or failure of responses to these closures; second, detailed field investigations by the authors have only just begun.

This chapter comprises four main sections. The first reviews current approaches to planning for the closure of large-scale mines, and identifies a number of features that distinguish small-scale mines in the context of mine closure. The second section provides background to the closure of small-scale coal mines in China, and describes the nature of the closure campaign. The third

[1] This chapter does not dwell on the definition of the term "small-scale mine", but rather follows the approach of Kumar and Amaratunga (1994), who include both artisanal mines and larger mines owned by small firms. The critical feature is that the output from such mines is small relative to the output of large-scale mines.

section outlines the overall impact of the closure campaign, and describes the general nature of responses to it. The final section presents one case study from Chongqing in south-west China.

APPROACHES TO MITIGATING THE SOCIO-ECONOMIC IMPACT OF MINE CLOSURE

Managing the environmental impact of mining, and the rehabilitation of mined land after the end of mining operations, has been a major concern for governments and mining companies for more than twenty years. In many countries, mining companies are required to draw up rehabilitation plans before operations can commence, and must put in place financial mechanisms to ensure that reclamation is actually implemented (*e.g.* Redgwell, 1992; Kuhne, 1992; Kahn et al., 2001). Special funding mechanisms have also been developed to clear up mine sites that have already been abandoned (*e.g.* Brook, 1994; Meyer et al., 1995).

Such formalised and tightly regulated approaches are generally absent with respect to addressing the local social and economic impact of mine closure. Only in the last ten years has the significance of this policy vacuum been identified. The first part of this section briefly reviews current approaches to this problem, which tend to be focused on large commercial mines. In contrast, little attention has been paid to the requirements of the small-scale mining sector. The second part of this section identifies some of the distinguishing features of small-scale mining that necessitate a rather different approach to planning for the social and economic consequences of mine closure.

Large-scale mines: current approaches

The closure of a mine will almost always have an immediate and significant negative economic impact on the local community, unless the mine has been operating in such a way as to minimise contact with the community. In simple terms, the negative economic impact takes the form of the removal of some or all of the positive contributions made by the mine during its life. These might include the following: employment in the mine, spin-off economic activity, local tax revenue, physical infrastructure and social services. The removal or deterioration of these economic components of life around the mine site will not only affect the economic strength of the community but substantial economic decline will also almost certainly result in a range of social problems (Rocha & Bristow, 1997; Mining Minerals and Sustainable Development, 2002a). Common symptoms include unemployment, crumbling infrastructure, failing social services, and rising crime.

It is widely agreed that mining operations should be designed in order to maximise their long-term benefit to the local community, and that the phrase "long-term" includes the period after mine closure (Veiga et al., 2001; Mining Minerals and Sustainable Development, 2002b). This parallels the realisation that care must be taken at a national scale to improve the chances of mining providing a basis for sustainable economic development rather than just a short-term flow of cash that is dissipated (Mining Minerals and Sustainable Development, 2002a).

For these very reasons, planning for mine closure needs to become an integral part of the planning for the whole life of the mine, and the socio-economic dimensions of mine closure must be addressed equally to its environmental components (Clark & Clark, 1999; Mining Minerals and Sustainable Development, 2002c; Jackson, 2002). Such a long-term approach has two advantages. First, by the time the mine is closed, the mining company and the higher levels of government will necessarily have lost interest in the region. Secondly, if their economic outlook is reasonable, workers and communities are less likely to object to the closure of the mine (Mining Minerals and Sustainable Development, 2002a).

The last few years have seen a spate of publications that broadly converge around a set of common themes (*e.g.* Rocha & Bristow, 1997; Roberts et al., 2000; Veiga et al., 2001; Hoskin, 2003; World Bank, 2002; Mining Minerals and Sustainable Development, 2002a, b; Chilean Copper Commission, 2002; Laurence, 2002). Two important dimensions of the proposed approaches for planning for closure include who should be involved, and what measures can be taken.

In simple terms, the central government and the mining company are identified as key players. The former must put in place policies and regulations and ensure implementation; the latter must provide some or all of the funding, and play an important role in implementation. Other parties that need to be involved from the earlier stages of planning onward include employees, local people, local government, local businesses, land owners, and, where relevant, NGOs. Most of these parties are local to the mine and can be regarded as expressing different concerns of the local community. So, aside from the NGOs, effective mine closure policy is seen to rely on a triangle of interests: central government, or, at least, a high level of government, the mining company, and the local community.

Concerning what *should* be done, there exists a wide range of ideas, different combinations of which may be used, depending on the circumstance. Measures include: developing long lasting infrastructure; stimulating alternative economic activity during the mine's operation in order to diversify the local economy; educating and training employees and locals to develop new skills; and encouraging the continuation of local subsistence activities.

These are relatively new concepts for the mining industry. Most countries have yet to put in place regulations, procedures and norms to address the socio-economic challenges posed by large-scale mines (Chilean Copper Commission, 2002).

Small-scale mines: distinguishing features

If the subject of managing the socio-economic impact of closure is new for large-scale mines, it is positively embryonic for small-scale mines. A review of the recent literature revealed only a small number of publications that address the problem of closure when applied to small-scale mines. These publications take two forms. The first addresses mine closure in general, and makes reference to one or more cases of small-scale mines, but without clearly identifying the specific problems of small-scale mines (*e.g.* Laurence, 2002). The second type focuses on the regulation and management of small-scale mines in general, and makes passing mention of the closure problem, but fails to address in any detailed way the management of mine closure (*e.g.* Hollaway, 2000; Mining Minerals and Sustainable Development, 2002d).

The special problems of formulating and implementing policies for the effective management of small-scale mines have been discussed at length elsewhere (*e.g.* Kumar & Amaratunga, 1994; Burke, 1997; Solomon, 1997; International Labour Organisation, 1999; Bugnosen et al., 1999). In our view, three factors distinguish small-scale from large-scale mines when it comes to managing mine closure:

- The legal and regulatory framework;
- The availability of funds; and
- The relative powers and interests of the parties.

In the case of large-scale mines, the legal and regulatory framework in most countries is at least moderately well established and functional. Regulatory systems and financial mechanisms have already been implemented to manage mine closure. To date, these procedures and funds have been directed at the physical reclamation of land and water, but such approaches could be adapted to ensure that adequate financing is available to address the socio-economic challenges described above. Key components for success include the financial strength of the mining company, and the transparency of the flow of economic rents. Finally, the central government, the mining company, and the local parties form a robust triangle of parties, within which interests can be distinguished, debate can proceed, and responsibilities can be allocated. An exception occurs where the mining company is state-owned, in which case, the interests of government and the mining company may be difficult to disentangle.

These favourable conditions do not exist for small-scale mines. The sector is permeated by weak legal and regulatory regimes, in which either the laws in place are inappropriate, or the regulatory institutions themselves are incapable of effective implementation. These weaknesses apply to the closure process as much as they do to the management of operating mines. The relatively impoverished nature of both mine owners and the local governments that tax mines, together with weak regulatory regimes, ensure that funds are rarely accumulated to address mine closure issues – environmental or socio-economic. Local governments are unlikely to have the vision to diversify the economy unless encouraged by higher levels of government.

Finally, the triangle of interested parties is quite different from that of the large-scale mines. The "mine owners", in many cases, are local people, local companies, or, indeed, the local government itself. The involvement of higher levels of government tends to be rather limited; the responsibility of regulation, therefore falls to local government. Thus, the triangle of interests is heavily distorted and focused on the "local" corner, which has three consequences. First, there are no "big" players with power and money involved who can provide guidance and funding. Second, the interests of higher and lower levels of government may be quite different. The higher levels may want to close the mines for environmental or other reasons, while the local government may want to keep the mines open for economic reasons. Third, conflicts of interest amongst local parties may prevent decisive action.

It might be argued that the closure of small-scale mines is a local problem that should be delegated to local government. However, without active support, advice and encouragement from above, few local governments will have the skills and the will to manage mine closure in a structured manner.

THE CLOSURE OF CHINA'S SMALL-SCALE COAL MINES

Background to the closure campaign

China's production and consumption of coal was the largest in the world during the 1990s. Between 1995 and 1997 annual production was in the range of 1,350 to 1,400 million tonnes, although net exports amounted to just 30 million tonnes. Coal accounted for nearly 75% of primary energy production and consumption (Thomson, 1996; Sinton & Fridley, 2000). Some 45% of this production – that is, more than 650 million tonnes – came from 75,000–80,000 so-called Township and Village Coal Mines ("TVCMs"), a large proportion of which were owned and controlled by local governments at the township and village level. A substantial minority of these mines – typically the smaller ones – were privately-owned. Others were owned by a variety of state companies and agencies, including the army and prison service.

Small-scale coal mines were found in almost every one of the country's 32 provinces, regions and municipalities; their collective output, individual capacity and workforce were highly variable. The percentage of provincial coal output derived from TVCMs within a single province varied from 20% to 80% (Andrews-Speed et al., 2002a). The average output from these coal mines ranged from as much as 25,000 tonnes per year in major coal mining provinces, to as little as 4,000 tonnes per year in provinces with fewer operations. The largest township mines had a capacity in excess of 100,000 tonnes per year. At the other end of the spectrum lay thousands of artisanal mines with outputs of a few hundred tonnes per year.

TVCMs began to be overtly encouraged by all levels of government at the beginning of the early-1980s, when China faced a crisis in energy supply (Wright, 2000). Laws and regulations, such as they were, were ignored and flouted in the interest of economic growth. Even new laws and regulations drafted in the 1980s and 1990s to cover safety, environmental protection, and licensing, were not implemented widely in the small-scale coal mining sector until the late-1990s (Andrews-Speed et al., 2002b).

In addition to providing much-needed energy, small-scale coal mines provided a basis for local employment and development in many poor and remote areas of China. Possibly as many as four million people were employed in the sector at the time of peak production. Other benefits from the growth and geographical spread of TVCMs included the following (Andrews-Speed et al., 2002b; Zhong, 2003):

- The exploitation of small, complex, remnant or otherwise economically-marginal, coal deposits that were of no interest to the state-owned mining companies.
- A diminished need for coal transportation, as coal was being exploited closer to the centres of demand.
- The substitution of coal for firewood, which reduced the destruction of forests in areas where coal had not been previously available.
- TVCMs providing economic competition for larger state-owned coal mines.

Although the size of some of China's TVCMs exceed most peoples' definition of a "small-scale mine" (Hollaway, 1986; Kumar & Amaratunga, 1994; Burke, 1997), these operations nevertheless suffer from many of the same problems: large numbers of illegal operations; irrational locations; low recovery rates; poor safety; and substantial environmental damage (Smil, 2000; Wright, 2000; Horii, 2001; Gunson & Jian, 2002). Of these negative attributes, it is arguably the environmental effects that are likely to have the most widespread, and long lasting, repercussions.

From the early-1980s to the mid-1990s, the central government's attitude towards TVCMs may be characterised as encouraging, yet, ineffective. Production grew at a prodigious rate but legal, technical, environmental, and safety requirements seemed to be ignored in most mining areas. Short-lived and half-hearted rectification campaigns failed to have any significant impact on the behaviour of miners or local governments.

From late-1997 to 1998, there was a sudden drop in demand for energy in China; the coal industry was affected more than other energy industries. This fall in demand for energy – especially for coal – is best explained by a general economic slowdown resulting from the Asian crisis; a decline in output from energy-intensive industries; closures of inefficient state factories; a general increase in end-use efficiency; and some substitution for coal by gas (Sinton & Fridley, 2000). Stockpiles of coal grew to hundreds of millions of tonnes, and prices in the domestic market plummeted (Sinton & Fridley, 2000; Wright, 2000). In early-1998, production from some state-owned mines was suspended for two months in order to ease the oversupply problem. By the middle of the year, it became clear that this was not a temporary phenomenon; drastic action was needed to protect the interests of state-owned mines into which large amounts of state investment had been poured.

The execution of the closure campaign

A plan was announced by the central government to close some 25,800 illegal and "irrational" mines (mainly TVCMs) by the middle of 2000 in order to reduce output by 250 million tonnes per year. It was reported that 33,000 mines had been closed by the end of 1999, which had effectively reduced annual output by 300 million tonnes.[2] By late 2001, the total number of TVCMs had reportedly been reduced to 23,000, resulting in a reduction in annual production from 620 million tonnes in 1997, to 200 million tonnes.[3] In addition to TVCMs, a number of larger mines (near exhaustion) have been closed and some enterprises made bankrupt.

Official documents clearly identified and categorised TVCMs targeted for closure. These included:

a. Mines with neither mining nor production licenses.
b. Mines opened within the areas of state-owned mines since 1st January 1997, which, by law, could not have been granted a mining license.
c. Mines opened within the areas of state-owned mines before 1st January 1997, which had mining licenses but lacked production licenses.
d. Mines opened within the areas of state-owned mines before 1st January 1997, which had mining and production licenses but had a negative impact on the state-owned mine.
e. Mines exploiting coal with high sulphur and ash content, and without appropriate countermeasures.
f. Mines operating outside the areas of state-owned mines, which had mining licenses but no production licenses.

Compensation would be paid only to those mines in category (d), as they represent the only full legitimate mines on the list (Zhang, 1999). Mines in category (f) were given the opportunity to raise their technical and safety standards, and apply for a production license before the end of February 1999. The smaller mines in this category were encouraged to merge with other mines to provide the financial and technical benefits of scale (Wang, 2000).

The level of local government responsible for the implementation of the closure campaign was the county. The county governments faced a conflict of interests. On the one hand, closure of some TVCMs provided protection for county-owned state mines; on the other hand, TVCMs provided an important source of employment and revenue from taxes and fees. As a result, it is likely that the actual extent of mine closure and production abatement is significantly less than that reported. Official figures show a sharp divergence from 1998 between rapidly declining coal production and only modestly declining coal consumption (Sinton & Fridley, 2000). The magnitude of this divergence suggests that coal supplies were drawn from unreported coal production – presumably from TVCMs – as well as from the large stockpiles (Sinton, 2001).

Investigations by the central government in 2001 revealed that the number of illegal coal mining operations had grown sharply, as had the number of related deaths and injuries.[4] As a result, the central government issued a circular on

[2] China Energy Efficiency Bulletin, June 2000.
[3] China Energy Efficiency Bulletin, October/November 2001.
[4] South China Morning Post, 5th July 2001.

June 13th 2001 ordering the suspension of small-scale coal mining operations, both those run by state-owned companies and those run by township and village enterprises. Only the licensed mines that had passed the inspections of the relevant authorities were permitted to recommence operation.[5]

In November 2001, five coal mine accidents occurred over a span of one week in Shanxi Province; the death toll was nearly one hundred.[6] All small-scale coal mines in the province were ordered to stop production and the Ministry of Land and Resources issued "Instructions for Further Strengthening the Order in Mineral Resources Operations".

Planning for the environmental consequences of TVCM closure

The closure of any coal mine requires a period of planning and significant effort in execution in order to minimise future environmental impacts. This is recognised in China, for large areas of coal mining provinces are already despoiled by the effects of mines closures carried out without due care and attention (Zhang & Peng, 2000). Despite recognition of this need, the TVCM closure campaign was carried out without any coherent environmental planning. In the specific cases studied in the field in 1999, those responsible for the closure of TVCMs referred to hasty implementation involving dynamiting shafts or adits, blockage with cement or bricks, removing machinery, cutting off utility supplies, destroying buildings, and erecting a sign declaring the mine closed. At that time, the authorities were not addressing the potential for future subsidence above underground mines, or disturbance or pollution of underground and surface waters (Chang, 1999). Further, no plans had been developed to deal with waste tips.

Planning for the socio-economic consequences of TVCM closure

At least three parties suffer economic loss when a small-scale coal mine is closed. The owner of the mine loses investment and future cash-flow. The local government loses tax revenue, in addition to any losses it sustains as partial or whole owner of the mine. Finally, the workers lose their livelihood, whether they are local people or migrants, as do those whose employment was indirectly related to the mining activity – in the fields of transportation, equipment manufacture and utility services.

The government policy mentioned compensation for the owners of only one category of legal mine which were closed (see above, category (d)).[7] Provincial governments were charged with formulating compensation policies. In Jiangsu, for example, the government decided that RMB 100,000 yuan[8]

[5] BBC Monitoring, Asia Pacific, 18th June 2001.

[6] South China Morning Post, 17th November and 26th November 2001.

[7] The term "legal mine" refers to a mine which possesses a valid mining licence and a valid production licence, and also applies the required environmental measures.

[8] The exchange rate generally lies in the range RMB 8.2–8.3 Yuan for US$ 1.

should be paid for each 10,000 tonnes of capacity closed. However, in 1999, few provinces had declared a compensation policy, let alone actually paid out money (Chang, 1999). No compensation appears to have been offered to owners of mines in other categories. Thus, private individuals and local governments, many of whom invested in good faith, believing that their enterprises were legitimate, even if not legal, will have lost substantial amounts of money.[9]

Of greater significance is the challenge facing those local governments that have lost revenue from the mines, and find themselves with an enhanced level of unemployment. Although the official line might be that the redundant workers should be redeployed to other township and village enterprises ("TVE"), in most areas, this is unrealistic, as the TVE sector as a whole is experiencing difficult times as a result of both market competition and the administrative closure of small, polluting enterprises in other sectors.

Migrant underemployed rural workers formed a major part of the TVCM labour force in some provinces, and hundreds of thousands will have lost their jobs if the number of TVCMs closed is indeed as high as has been reported. The general view of government officials in coal mining areas appears to have been that these people are not their responsibility and that they should return to their villages.

The available evidence suggests that the negative socio-economic impact of the TVCM closure campaign has not featured highly in the list of priorities for higher levels of government, and that little, or no, systematic support mechanisms have been put in place to deal with the inevitable impact on communities (Shi, 1999). Indeed, it is the lack of attention to these issues that has led to the formidable and sustained resistance to the closure policy by local governments and communities, as is discussed in the next section.

SOCIO-ECONOMIC IMPACT OF THE CLOSURE CAMPAIGN AND SOME RESPONSES

National impact

It is impossible to gauge the *actual* effectiveness of the mine closure campaign, let alone describe its social and economic impact at a national scale. That being said, assuming that the campaign has been as successful as officially declared, certain preliminary rough estimates can be made of the local impact on government revenue and employment.

A legal TVCM is required to pay to the local government as much as Yuan 60–70 RMB per tonne of coal produced.[10] If official figures are taken at face value, the output of TVCMs has fallen by 420 million tonnes. Assuming that local government only collected 50% of the revenue it was entitled to (RMB 30 Yuan per tonne), income would have declined by more than Yuan 12 billion

[9] The scope for legal action against the state is rather limited in China.

[10] Xiao Zhang and Ai Min (2001). "Bare confessions from the owner of an illegal mine," *Shanxi Daily* 18th September 2001 (in Chinese).

RMB. This is just a small percentage (2%) of the total local government revenue of Yuan 640 billion RMB in 2000 (National Bureau of Statistics of China, 2001a). However, given that TVCMs are necessarily geographically concentrated, it is almost certain that some areas had experienced a substantial reduction in government revenue.

Documenting the number of people employed by TVCMs during the industry's peak is just as difficult, given the illegal status of many of the mines, the preponderance of migrant labour in some areas, and the seasonal nature of mining in other areas. Unpublished sources suggest that as many as four million people worked at TVCMs in the mid-1990s. Specific estimates are provided for a few provinces by Ye and Zhang (1998). This information indicates a productivity varying between 100 and 200 tonnes per man-year in most of the documented provinces, with a high of 370 tonnes per man year in the largest mining province, Shanxi. Given that Shanxi Province at that time accounted for some 35% of TVCM output across the country, these figures suggest an average productivity of 200 tonnes per man year. Annual production at 650 million tonnes would imply a total work force of about three million.

Based on these figures, a reduction of TVCM output from 620 to 200 million tonnes would have caused some 2.0–2.7 million people to lose their jobs; this assumes that average productivity remained the same. Given that the least mechanised and least productive mines were probably closed in preference to the more efficient mines, this number is likely to have been larger. This estimate does not include those who did not work at the mines but whose employment depended on TVCMs.

Two numbers have been derived: a loss of local government revenue of Yuan 12 billion RMB per year, and a loss of employment for more than two million people. The derivation of these figures explicitly assumes that the closure programme has been a success. Even if the closure programme has been less successful than reported, these figures nevertheless indicate its nature and the scale of threat it posed to local interests.

Responses across the country

In simple terms, three end-member types of response are possible from local communities:

1) Comply with the mine closure campaign and successfully develop alternative economic activity;
2) Comply with the mine closure campaign and fail to develop alternative economic activity; or
3) Fail to comply with the mine closure campaign and allow mines to continue in operation.

At the national level, the press was the only source of information available to the public on this subject. Newspaper articles in China have tended to focus on either success stories of Type 1 or serious accidents that seem to have been concentrated at illegal mines (Type 3). Few stories have emerged relating to those communities that have failed to diversify their economies following mine closure. In short, on the one hand, news of success and law-breaking is published, but, on the other hand, news of poverty and failure (Type 2) is not.

Concerning success stories, three approaches may be identified from the press. Those areas that have a favourable geographical position, have an already-diversified economy, and are relatively rich, and can afford to invest in new economic activities. Two cases illustrate this, the first, Jiangsu Province, which lies on the east coast of China, just north of Shanghai, and is one of the country's richest provinces (National Bureau of Statistics, 2001a). The province closed nearly all of its 166 TVCMs, most of which lay in a suburb of Xuzhou. As a result, 30,000 workers lost their jobs, and local government revenue declined by Yuan 350 million RMB per year. However, the provincial government gave Yuan 25 million RMB toward the cost of mine closure and to help compensate owners. Further, the Xuzhou city government agreed to give Yuan 10 million RMB per year to the suburb, from 2001–2004, for the purpose of assisting with the diversification of the local economy, particularly in the field of agribusiness.[11]

The second, Zhejiang Province, lies just to the south of Shanghai and is the fourth richest province in China in terms of per capita GDP (National Bureau of Statistics of China, 2001a). Baixian town was an important coal production base in the mid-1990s with 36 mines employing 3,000 people. Twenty-four of these mines have been closed, resulting in a decrease in coal output of 60,000 tonnes per year. Some 2,000 workers have been transferred to other economic activities, and both their salaries and local government revenues are rising.[12]

A contrasting success story comes from Jiangxi Province in central China, one of the poorest in the country, where large numbers of TVCMs have been closed in Fenyi County. Here, the township and county government has been active in building upon existing economic activities, particularly in the agricultural sector. Jute production and processing was already a major business in the county and the local government invested a further Yuan 10 million RMB in two additional jute projects. Elsewhere in the county, local governments have given loans to peasants to diversify into sheep herding, forestry and fruit production.[13]

A third approach has been reported from Ningxia Province in north-west China, which has an equivalent GDP per head to Jiangxi, but has very poor quality agricultural land. Here, the closure of TVCMs has been offset by investment in larger mines owned by government at the county and higher levels. This

[11] Xiao Liming and Du Zhengjie, *Legal Daily* 21st June 2002, p. 2 (in Chinese).

[12] Jin Jiabao, "The days without coal mines," *China Coal Daily* 23rd February 2002, p. 3 (in Chinese). The apparently low productivity of thirty tonnes per worker per year suggests that many of the mine workers were part-timers.

[13] Xu Jinpeng, Zhan Guoqiang and Yu Bin, "New changes in Fenyi County, Jiangxi Province: new economic development whilst closing TVCMs," *China Coal Daily* 11th June 2002 (in Chinese).

approach has resulted in an overall increase in total coal output in the province, and in total revenue from the industry for government and workers, as well as an improvement in safety. [14]

Elsewhere, local government officials appear unable or unwilling to develop alternative economic activities (response Type 2). For example, 44 mines were closed in Jingxian city in Jiangxi Province, with a total loss of ten thousand jobs and of 80% of local government revenue.[15] A second example comes from Shanxi Province, the main coal production region of China. In Zuoyun county, which lies near the main coal centre of Datong, TVCMs produced 10 million tonnes of coal each year and contributed more than 80% of local government revenue. Local officials could see no way forward if the mines were closed.[16]

In such circumstances, it is hardly surprising that local government officials either turn a blind eye to illegal coal mining or provide active support. Recent years have seen an increase in the number of press reports in both China and in the west of accidents at TVCMs, a substantial number of which were found to be operating illegally.[17]

Few general conclusions can be drawn from such a limited and random selection of stories. Assuming they are sustainable, the successful projects illustrate the ingenuity and determination of local governments in some locations to react positively to the TVCM closure campaign. However, their ability and willingness to act in this way will, to a great extent, depend on local economic, social and political circumstances.[18] Also noteworthy is that such success has been achieved despite a lack of forward planning.

In more remote and less wealthy areas more dependent on TVCMs for economic development, rapid diversification of the local economy has been impeded by larger obstacles, and, in many places, little progress has been made. These communities are paying the price because of the sudden and unplanned implementation of the TVCM closure policy, and a lack of attention paid at any time by the higher levels of the Chinese government to planning for the closure of small-scale mines.

---

[14] Zhang Guanquan and Yang Zhen, "Information from Ningxia shows that closing small mines benefits development," *China Coal Daily* 9th March 2002, p.1 (in Chinese).

[15] Li Jingying, "A town sitting on an explosive cask," *China Youth Daily* 10th September 2001 (in Chinese).

[16] Chi Yuzhou, "Gas, small mines and the fate of Zuoyun county", *Economic Observer* 10th December 2001 (in Chinese).

[17] For example: "Gas explosion in coal mine,"*Mining Journal* 27th April 2001, p. 314; "92 feared dead in mining blast," *South China Morning Post* 24th July 2001; "Gas blasts kill 37 in two coal mines," *South China Morning Post* 17th November 2001; "Mine where 115 workers killed 'ordered to stop production 7 times'," *South China Morning Post* 8th July 2002.

[18] For a detailed discussion, see Andrews-Speed et al. (2002a).

A CASE STUDY OF CHONGQING

The evidence presented in the previous section is too sketchy to allow for a balanced appreciation of local government responses to the TVCM closure campaign. As a result, the authors have embarked on a field study, which will involve investigations at township and village levels. This section reports on one case study from Chongqing Province.

Background to Chongqing's TVCMs

Chongqing used to be the largest city in Sichuan Province in south-west China. It was elevated to Municipality status, equivalent to that of a Province, in 1996.[19] The population of Chongqing Province in the year 2000 was thirty million. At that time, the annual per capita GDP was Yuan 5160 RMB, compared to the national average of Yuan RMB 7080 (National Bureau of Statistics of China, 2001a). In the past, the economy of Chongqing relied on heavy industry, as this was one of Mao Zedong's strategic industrial areas located well away from the coast and from Russia to the north. Structural reforms to China's state industries have caused extreme hardship in the province, and consequently, a large number of factories lie idle.

Coal dominates the energy sector in Chongqing, accounting for some 80–85% of primary energy production and consumption (National Bureau of Statistics of China, 2001b; National Bureau of Statistics of China, 2001c). The other main indigenous sources of primary energy within Chongqing are natural gas and hydro-electricity. The province is a modest net exporter of coal and natural gas, and an importer of electricity (National Bureau of Statistics of China, 2001b; National Bureau of Statistics of China, 2001c).

TVCMs have traditionally produced about 50% of Chongqing's coal output. In 1997 and early-1998, some 4,500 TVCMs were active, producing approximately 15 million of the province's 28 million tonnes of coal (Administration of State Coal Industry, 1998; Chongqing Municipal Government, 1999). Thereafter, information for coal output diverges greatly. On the one hand, the Chongqing Statistical Yearbook (National Bureau of Statistics of China, 1999, 2000, 2001c) and the National Energy Statistics Yearbook (National Bureau of Statistics of China, 2001b) report a total output steady at 28–30 million tonnes. On the other hand, the Coal Industry Yearbook (Administration of State Coal Industry 1999, 2000; Administration of State Work Safety, 2001) records a reduction to about 20 million tonnes in 1998 and 1999, and provides no data for 2000. The Chongqing Yearbook (Chongqing Municipal Government, 2000, 2001) documents a reduction of 11 million tonnes over two years, to an annual output of 16 million tonnes in 2000. Unpublished data from the Chongqing Coal Bureau show a total of 1,233 TVCMS active in 2001 with an annual output of

[19] Officially, China is a unitary state, but substantial political and economic powers are delegated to different levels of government. In simplified form, these are (in descending order): Province, Prefecture or City, County, Township.

9.7 million tonnes, and a total annual capacity of 22 million tonnes. However, output from TVCMs may have been as low as 6.5 million tonnes in 2000 (Chongqing Municipal Government, 2001).

These figures reflect a vigorous and successful implementation of the TVCM closure campaign. If they are to be believed, a total of some 3,300 TVCMs have been closed since 1998. The initial closure resulted in a reduction in annual output from 15 million tonnes to possibly fewer than seven million tonnes, with a subsequent rise to nearly 10 million tonnes in 2001. It is understood that this subsequent increase in TVCM production has been the result of officially-sanctioned upgrading of some of the remaining TVCMs.

The case of Tianfu township: description

Tianfu township, which lies some 50 km north of the city of Chongqing in a mountainous area covering some 54 square kilometres, has a population of 47,000. Here, coal mining has provided an important source of wealth since the late Ming Dynasty (Seventeenth Century). The quarrying of construction materials is the township's other main industry; its agricultural output is low because the land is of poor quality. Coal output reached a peak of two million tonnes per year during the Second World War, when it was mainly transported down the nearby Jialin River for industrial use. From 1950 to 1980, only a few mines remained in operation, and production stayed at a low level, mainly for local use. Since the early-1980s, the national drive to raise coal production from TVCMs resulted in local production increases; output in the township soon reached its previous peak, achieved when 32 mines were active. These mines, all of which were underground, were collective enterprises owned at the township or village levels. Some 95% of mine workers were local farmers, some of whom had worked at the mines for more than 10 years.

The closure campaign in the late-1990s resulted in the closure of 14 mines and the conversion of the remaining 18 mines into joint-stock companies. All mines with an annual capacity of less than 10,000 tonnes have been closed, regardless of their legal status. This limit will rise to 30,000 tonnes by 2003. Total annual coal output for the township fell to 100,000 tonnes in 2001 and is expected to rise to 200,000 tonnes in 2002, though the maximum allowed by the county government is only 90,000 tonnes.

A number of parties have suffered adversely from the closure campaign, including:

- The local government and local population;
- The mine owners;
- The mine workers;
- The local farmers; and
- Local energy users.

The mine closures have significantly changed the structure and scale of the local economy. During the 1980s and 1990s, coal mining accounted for some 70–80% of the township government's revenue. This has fallen to 20–25%, and the revenue to the township government has declined by three million Yuan RMB per year. This will have reduced the capacity of the local government to invest in new infrastructure and facilities in the township.

No compensation has been paid to the mine owners – previously, the collective enterprises – despite the fact that the local government claims that they were all operating legally. The owners have thus seen their investments in the mines, and related infrastructure, go to waste.

The total number of workers employed at the mines fell from 1,200 during peak periods to 700–800 in 2002, with the average wage being Yuan 500–1,000 RMB per month. Not only have some 400–500 workers lost their jobs, but they are also owed unpaid wages by the mine owners. As previously noted, the land is too poor for inhabitants to return to agriculture and so they pester the township government for financial assistance or alternative employment. The only apparent help available to them is a subsidy lasting a maximum of eight years for the planting of trees. However, the farmers are not attracted by the short-term nature of the subsidy and the consequent uncertainty of subsequent income.

The local farmers were accustomed to receiving compensation from the coal mines for the use of their land. The closure of the mines had hit them in two ways. First, they no longer receive this compensation payment; second, the lack of any reclamation work means that the land still cannot be used for agriculture.

Finally, the reduction in coal output has decreased coal supplies within the township, and some people have resorted to felling trees for their household energy uses, thus further damaging the environment.

It is against these negative impacts of the closure campaign that two positive outcomes are being sought. First, the higher levels of government are making substantial efforts to raise the standards of the remaining small-scale coal mines. Regulations relating to coal washing and the filtering of water after washing have been enforced with greater vigour, and there has been a greater take-up of training opportunities for workers and technical staff in the mines. Second, most of the remaining mines have been permitted or encouraged to invest in improving their facilities and increasing their capacity, though it seems that financing for these investments has had to come from friends and relatives of the owners, rather than from banks.

The case of Tianfu township: analysis

Tianfu Township provides an example of a community that has long been heavily dependent on small-scale coal mining for its economic livelihood. The sudden and radical closure campaign has had a substantial negative economic impact. It is evident that no planning to mitigate economic or environmental impacts had been carried out prior to the mine closures. Almost the entire population of the township will have been affected in some way, either directly or indirectly. The local government appears to lack the resources and the encouragement to diversify the local economy. Indeed, its history, its location, and the poor quality of the land are working against it. The only good news is that the remaining mines will soon have the capacity to raise their output once again, though whether they are permitted to by the county government is another question. The outlook for local employment at these mines is unclear. If their upgrading includes substantial mechanisation, this may, indeed, reduce the need for workers.

Despite the fact that the environmental standards of operating mines should improve in the near future, no resources have been identified to reclaim the land destroyed by mining over the previous three hundred years. The community will continue to suffer the negative effects of this damage, presumably in the form of polluted water and land.

Overall, Tianfu provides an example of a sudden mine closure campaign implemented with little regard to local social, economic or environmental consequences. It is evident that the township government cannot address these consequences without assistance from a higher level of government; however, this assistance does not appear to be forthcoming.

CONCLUSIONS

The socio-economic challenges posed by the closure of large numbers of small-scale mines have yet to be addressed in a systematic manner by academics, advisors, governments and international agencies. The necessarily close interdependence between small-scale mines and their local communities means that a wide-ranging closure campaign may have an even greater local socio-economic impact than the closure of a large mine. Just as regulating small-scale mines is more difficult and costly than regulating large-scale mines, as is managing the socio-economic impact of the closure of small-scale mines. The absence of direct input from major mining companies and the central government in most small-scale mining provinces often leaves a vacuum in terms of authority, experience and finance for effective planning.

The case of small-scale coal mines in China shows a mine closure campaign which, officially, has resulted in the closure of 60–70% of operational mines and an equivalent reduction in output. The campaign was enacted with no planning for the mitigation of environmental or socio-economic impacts. If the closure campaign was indeed as successful as officially reported, then crude estimates suggest that more than two million people may have lost their jobs, and that local government revenue from the sector may have declined by more than ten billion Yuan.

A provisional evaluation of press reports and one case study at the township level has revealed three types of response by local government in China. In those areas with natural economic or political advantages, the government has been able to redeploy mine workers in other economic activities, or into larger mines that have been upgraded. In other areas, which lack such opportunities or where local government has failed to develop them, two types of responses have been identified: the first involves a defiance of government policy and the continued operation of mines that should have been closed; the second is characterised by inaction and increasing poverty. It is impossible to determine the scale and distribution of such responses across the whole of China, but press reports indicate that a significant proportion of responses lie in the latter two categories.

The failure to plan systematically for the closure of the country's small-scale coal mines and the apparent lack of any concerns at higher levels of government for the resulting adverse socio-economic impacts, has undoubtedly resulted in a legacy of local economic deprivation, which could take years to address. Conversely, in those locations where local government continues to defy the official campaign, illegal mining will continue with the consequent environmental damage and appalling death rate.

REFERENCES

Administration of State Coal Industry (1998). *China Coal Industry Yearbook 1998* (Beijing: China Coal Industry Publishing House).

Administration of State Coal Industry (1999). *China Coal Industry Yearbook 1999* (Beijing: China Coal Industry Publishing House).

Administration of State Coal Industry (2000). *China Coal Industry Yearbook 2000* (Beijing: China Coal Industry Publishing House).

Administration of State Work Safety (2001). *China Coal Industry Yearbook 2001* (Beijing: China Coal Industry Publishing House).

Andrews-Speed, P., Zamora, A., Rogers, C.D., Shen, L., Cao, S. & Yang. M. (2002a). A framework for policy formulation for small-scale mines: the case of coal in China. *Natural Resources Forum,* 26, 43–52.

Andrews-Speed, P., Yang, M., Shen, L. & Cao S. (2002b). The regulation of China's township and village coal mines: a study of complexity and ineffectiveness. *Journal of Cleaner Production*, 11, 185–196.

Brook, D. (1994). Reclamation of abandoned underground mines in the United States. *Mineral Planning*, 61, 21–26.

Bugnosen, E., Twigg, J. & Scott, A. (1999). Small-scale mining legislation and regulatory frameworks. *Minerals and Energy*, 14(2), 35–38.

Burke, G. (1997). Policies for small-scale mining: the need for integration. *Journal of Mineral Policy, Business and Environment*, 12(3), 11–14.

Chang Yuyao (1999). Problems and countermeasures in the process of closing up the mines and restricting the yield. *Energy Bases Construction*, 3, 15–16 (in Chinese).

Chilean Copper Commission (2002). Research on mine closure policy. In *Mining for the Future*, Paper No. 44 (London: International Institute for Environment and Development).

Chongqing Municipal Government (2000). *Chongqing Yearbook 2000* (Chongqing: Chongqing Yearbook Publishing House, in Chinese).

Chongqing Municipal Government (2001). *Chongqing Yearbook 2001* (Chongqing: Chongqing Yearbook Publishing House, in Chinese).

Clark, A.L. & Clark, J.C. (1999). The new reality of mineral development: social and cultural issues in Asia and Pacific nations. *Resources Policy*, 25, 189–196.

Gunson, A.J. & Jian, Y. (2001). Artisanal mining in the People's Republic of China. In *Breaking New Ground: Mining, Minerals and Sustainable Development*, Paper no. 74 (London: International Institute for Environment and Development).

Hollaway, J. (1986). The small-scale mining sector in Africa; restructuring for profitability. *Natural Resources Forum*, 10(3), 293–297.

Hollaway, J. (2000). Lessons from Zimbabwe for best practice for small- and medium-scale mines. *Minerals and Energy*, 15(1), 16–22.

Horii N. (2001). Development of small-scale coal mines in market transition and its externality, In N. Horii & S. Gu, (Eds.). *Transformation of China's energy industries in market transition and its prospects* (Chiba, Japan: Institute of Developing Economies) pp. 23–62.

Hoskin, W.A. (2003). Mine closure – the 21st Century approach. Avoiding future abandoned mines. In E.A., Bastida, T.W. Waelde, & J. Warden-Fernandezel (Eds.), *International and Comparative Mineral Law and Policy.* London: Kluwer.

International Labour Organisation (1999). *Social and Labour Issues in Small-Scale Mining* (Geneva: International Labour Office).

Jackson, R.T. (2002). Capacity building in Papua New Guinea for community maintenance during and after mine closure, in *Breaking New Ground: Mining, Minerals and Sustainable Development*, Paper No. 181 (London: International Institute for Environment and Development).

Kahn, J.R., Franceschi, D., Curi, A. & Vale, E. (2001). Economic and financial aspects of mine closure. *Natural Resources Forum*, 25, 265–274.

Kuhne, G. (1992). Abandonment and reclamation of energy sites and facilities: Germany. *Journal of Energy and Natural Resources Law*, 10(1), 4–20.

Kumar, R. & Amaratunga, D. (1994). Government policies towards small-scale mining. *Resources Policy*, 20(1), 15–22.

Laurence, D.C. (2002). Optimising mine closure outcomes for the community – lessons learnt. *Minerals and Energy*, 17(1), 27–34.

Meyer, P.B., Williams, R.H. & Yount, K.R. (1995). *Contaminated Land – Reclamation, Redevelopment and Reuse in the United States and the European Union.* Cheltenham: Edward Elgar.

Mining Minerals and Sustainable Development (2002a). Minerals and Economic Development, in *Breaking New Ground: Mining, Minerals and Sustainable Development*, Chapter 8. London: International Institute for Environment and Development.

Mining Minerals and Sustainable Development (2002b). Mine Closure Working Paper in *Breaking New Ground: Mining, Minerals and Sustainable Development*, Paper No. 34. London: International Institute for Environment and Development.

Mining Minerals and Sustainable Development (2002c). Local Communities and Mines, in *Breaking New Ground: Mining, Minerals and Sustainable Development*, Chapter 9. London: International Institute for Environment and Development.

Mining Minerals and Sustainable Development (2002d). Artisanal and small-scale mining, in *Breaking New Ground: Mining, Minerals and Sustainable Development*, Chapter 13. London: International Institute for Environment and Development.

National Bureau of Statistics of China (1999). *Chongqing Statistical Yearbook 1999* (Beijing: China Statistics Press).

National Bureau of Statistics of China (2000). *Chongqing Statistical Yearbook 2000* (Beijing: China Statistics Press).

National Bureau of Statistics of China (2001a). *China Statistical Yearbook 2001* (Beijing: China Statistics Press).

National Bureau of Statistics of China (2001b). *China Energy Statistical Yearbook 1997–1999* (Beijing: China Statistics Press).

National Bureau of Statistics of China (2001c). *Chongqing Statistical Yearbook 2001* (Beijing: China Statistics Press).

Redgwell, C. (1992). Abandonment and reclamation obligations in the United Kingdom. *Journal of Energy and Natural Resources Law*, 10(1), 59–86.

Roberts, S., Veiga, M. & Peiter, C. (2000). *Overview of Mine-Closure and Reclamation in the Americas* (Vancouver: International Development Research Centre).

Rocha, J. & Bristow, J. (1997). Mine downscaling and closure: an integral part of sustainable development. *Minerals and Energy*, 12(4), 15–20.

Shi Xunpeng (1999). Analysis, consideration and proposals concerning the programme to close mines and restrict the output. *Coal Economic Research*, 6, 18–21 (in Chinese).

Sinton, J.E. (2001). Accuracy and reliability of China's energy statistics. *China Economic Review*, 12, 373–354.

Sinton, J. & Fridley, D. (2000). What goes up: recent trends in China's energy consumption. *Energy Policy*, 28, 671–687.

Smil, V. (2000). China's energy and resources uses: continuity and change. In Edmonds, R.L. (Ed.). *Managing the Chinese Environment* (pp. 211–227). Oxford: Oxford University Press.

Solomon, M.H. (1997). Small and mid-scale mining in South Africa: beyond the rhetoric. *Journal of Mineral Policy, Business and Environment*, 12(3), 23–30.

Thomson, E. (1996). Reforming China's coal industry. *The China Quarterly*, 147, 726–750.

Veiga, M.M., Scoble, M. & McAllister, M.L. (2001). Mining with communities. *Natural Resources Forum*, 25, 191–202.

Wang, X. (2000). Always emphasise the importance of the task in mine closure and production reduction. *Coal Enterprise Management*, 2, 9–10 (in Chinese).

World Bank (2002). *Its Not Over When Its Over. Mine Closure Around the World* (Washington DC: World Bank).

Wright, T. (2000). Competition and complementarity: township and village mines and the state sector in China's coal industry. *China Information*, 14(1), 13–129.

Ye, Q. & Zhang, B. (1998). *Township and Village Coal Mines in China.* Beijing: Coal Industry Publishing House (in Chinese).

Zhang, B. (1999). Text of speech at a meeting on the closure of illegal and irrationally located coal mines, 11th November 1998. In *Selected Documents on Mine Closure and Output Reduction in China's Coal Sector* (Beijing: Coal Industry Publishing House, in Chinese), 38–49.

Zhang, X. & Peng, D. (2000). Investigations on ecological restoration and mine-site land rehabilitation in Shanxi Province of China. In Lu Xinshe (Ed.). *Mine Reclamation and Ecological Restoration for the 21 Century* (pp. 108–111). Beijing: China Coal Publishing House.

Ziran, Z. (2003). Small-scale mining in China: socio-economic impacts, policy and management. In this volume, Chapter 25.

30

Informal Gold Mining in Mongolia: Invaluable Social Safety Net and Environmental Hazard

William Murray

Gold mining has a long history in Mongolian territory, dating back to at least the eleventh century – longer than the Mongols have been identified as a distinct people. However, informal gold mining is a recent activity in modern Mongolia, barely acknowledged by Government policy or the donor community. Although this chapter is based on extensive fieldwork and discussions with many miners and concerned Government officials, the topic is not documented and the numerical estimates are subject to wide margins of error.

The chapter discusses the history and prospects of informal mining in relation to some of the major forces that have impacted Mongolian society since a transition to a market economy commenced in 1990. These forces are chiefly the privatization of pastoral and arable agriculture; privatization of industries and services; ever-widening and deepening poverty; emergence of the informal sector; and Government attempts to develop a modern mining industry.

COUNTRY BACKGROUND

Mongolia is larger than the combined areas of France, Germany, the United Kingdom and Italy but its 2.5 million people occupy the territory at an average density of only 1.6 persons per km^2 nationally, and 0.7 per km^2 in rural areas. It has a dry continental climate with great diurnal and seasonal temperature variations, typically $\pm 40\,°C$ over the year, and annual precipitation ranging from over 450 mm in the north to less than 200 mm in the Gobi deserts in the south.

It was a communist state from 1924 until 1990 but has since been a democracy and in transition towards a market economy. The communist administration carved the territory into 18 provinces, each with a capital town, and some 20 rural districts, and cooperatized the traditional nomadic pastoral economy.

During the last decades of communism, there was a strong trend towards urbanization and industrialization, accompanied by a provision of excellent social and education services, with Ulaanbaatar developed as the primate city; however, a third of the population remains nomadic.

Following the end of communism and withdrawal of COMECON subsidies, which were equivalent to a third of GDP, much of the economy collapsed. The livestock, arable, and urban industrial and service industries were privatized. In the latter cases, this led to significant shedding of labour, and to the emergence of an informal economy, now accounting for approximately one third of GDP (Bikales et al., 2000). In the former case, however, it led to increases in labour, as herds were privatized. There was a significant urban-to-rural migration of people opting for a sustainable, if rudimentary, rural livelihood in preference to abject urban poverty. The herding population more than doubled, and, in 1999, the national herd increased by more than a third to almost 34 million head. Since then, however, a series of summer droughts and appalling winter conditions have combined with the collapse of support for the industry to completely reverse the increase in the number of animals, leaving many families destitute; increasing rural poverty has added to the severe urban poverty. In 1998, more than a third of the population were below the national poverty line of about £17 per person per month (NSO, 1999), itself about half the commonly used international yardstick of one (US) dollar per day. The indications are that the situation has deteriorated further since 1998.

GOLD MINING

Mongolia is rich in mineral reserves. Gold mineralization occurs widely in lode deposits, bulk-tonnage disseminated mineralised zones, and placers. Geological research has identified ten main gold metallogenic provinces, although exploration is incomplete, and exploitable deposits may occur in all provinces. Total reserves are probably of the order of 3,000 tons. Fields of gold pits from the Manchurian colonial period – and possibly earlier – are widespread but panning for gold has never been a major activity of nomads.

Mongolia's largest mine is the copper (Cu) and molybdenum (Mo) concentrate state-owned Mongolian-Russian joint venture plant at Erdenet, commissioned in 1978; it has been the main pillar of the export economy during transition, with exports of £148 million of copper concentrate in 2001, double that of gold. A number of gold mining joint ventures with countries in the Soviet Block were established in the 1970s and 1980s, and have continued to operate during transition. In 1992, eight of these produced only 775 kg of gold (Sermier, 2002, p. 113). However, one of the post-communist Government's most successful policies has been the promotion of mining, and gold exports, in particular, have increased since 1993 to rival those of Erdenet; in 2002, gold exports were £118 million compared with £138 million for copper concentrate.

Recognizing the pivotal importance of the mining sector, the Government launched its "Gold" Programme in 1992, which led to a regeneration of placer mining. It was strengthened by the *1997 Minerals Law*, regarded as one of the most favourable in the world for exploration, production and foreign investment, and by the "Gold 2000" campaign. Exploration licences are valid for an initial three years, extendable up to seven years, and are granted, subject to annual fees, for sixty years, extendable by forty years.

The principal placer areas mined to date are Zaamar, Bayanhongor, Tolgoit, and elsewhere in the North-Central one-fifth of Mongolia's territory. About 130 licensed enterprises are currently operating; exploration has also greatly intensified. Mongol Gazar, Altandornod, and the Shijir Alt Mongolian-Russian joint-venture companies dominate placer gold mining and together account for more than 45% of production. A main feature of both the long-established and more recently-established mines is that, with very few exceptions, they use inefficient technology that recovers only 30–70% of gold, leaving the balance in waste, which is very extensive, and predominantly un-rehabilitated. Contrary to regulations and agreements, environmental restoration undertakings are commonly flaunted. The mines are capital-intensive and employ foreign and city labour, rather than recruiting locally. For example, the current development of the Boroo Gold mine is expected to cost £45 million and employ 200 people. Thus, formal gold mining provides minimal local economic benefits but potentially has massive negative environmental impacts.

The 12.1 tons of production in 2002, valued at US$145 million, brings total production during transition to the equivalent of £800 million in 2003 prices.[1] These are very large numbers in the context of Mongolia's small economy, and have been vital in terms of preserving the macro economic stability of the national currency, and, by providing a sound borrowing customer base of gold mining companies, in critically assisting the development of commercial banking. Gold production is not fully accounted for in GDP estimates; moreover, gold exports are not included in the published trade data, much to the detriment of the trade balance, which has been increasingly negative since 1996 (see Fig. 30.3).

Frequent bullish announcements about combined world class gold and copper prospects made by Ivanhoe Mines Ltd, Rio Tinto, Bell Coast Capital Corp, Maximum Ventures Inc. and other exploration companies continue to inflate international interest in exploiting Mongolia's mineral wealth.

INFORMAL GOLD MINING

Inspired by the recent successes of formal gold mining, and attracted by the results of its inefficiency, the newly-developed informal economy has seized on illegal gold mining as a salve for the poverty caused by widespread rural and urban unemployment (see Murray, 2003 for more detail on these socio-economic forces).

[1] Throughout the chapter a world price of 12 euro per gram, prevailing in January 2003, is used.

Figure 30.1 On-site services at Uyanga District, Overhangai Province.

The most rudimentary of technology prevails, undeserving of the adjective "artisanal", although with very distinct regional variations. Figure 30.1 illustrates a typical scene following the shutdown of an Erel Company mine for the winter; Figure 30.2 illustrates relatively sophisticated panning techniques by Mongolian standards. Group and family mining predominates, with frequent occurrences of patterned behaviour on a larger scale, although this does not normally amount to cooperative effort but rather parallel replication of small group activities. Informal gold miners have acquired their own collective name, as is the case in several other countries, such as the *galamsey* in Ghana (Hilson, 2002); in Mongolia, they are colloquially referred to as *ninjas* because the green plastic pans carried on their backs are reminiscent of some popular turtle cartoon characters. Although gold is the principal product extracted, informal miners also work coal and fluorspar.

Informal gold mining began in the mid-1990s in the north and central provinces of Selenge, Darkhan-Uul and Tov, where formal gold mining is most extensive. Inefficient licensed mining has left some 150 tons of "waste" gold to date, excluding pre-transition activity; this has been the foci for informal miners, who know exactly where they can find gold and are unrestrained by the costs of exploration and primary mining works. Whilst the majority engages in placer mining, increasing numbers are now developing small adits and shafts to explore small, high grade lode-gold deposits.

After a relatively slow start, the activity is now expanding very rapidly and, in addition to the symbiotic relationship with licensed mining, miners are

Figure 30.2 Fine separation panning, Galuut District, Bayanhongor Province.

increasingly working in other areas. In Bayanhogor Province, for example, informal mining began as recently as 2000 in association with the Bulgarian-Mongolian Duvunt mine in Galuut District, with only a few dozen miners. Two summers later, there were some 4,000 miners in this District, and mining had spread more than 300 km to the extreme south of the Province, where several hundred households who had lost their livestock due to harsh weather conditions had migrated to commence mining in areas where there was no licensed mining. Similarly, in Overhangai Province, there is intensive informal mining

Table 30.1 Estimated composition of informal miners.

Proportion of total	Types of miners
55%	• Destitute or near destitute families having lost their livestock due to adverse weather or arable employment by privatization • "Redundant" urban residents unable to make a living in urban areas They rarely manage to make more than a poor living from mining.
30%	• Occasional miners supplementing incomes to pay for education, purchasing a vehicle, etc.
7%	• Full time "professional" miners
8%	• Mining only to satisfy requirements for alcohol

Note: Based on field observations and discussions with concerned Government officials.

being carried out by several thousand miners in two Districts where there is no formal mining. This new mining is guided by "clues", such as old place names incorporating the word "gold", of which there are many, and by the expanses of ancient gold mining pits.

Although it is possible for the five storey high licensed Russian dredges to operate throughout the year, they cannot be started up during winter. The foreign managers and technicians need a winter break, and consequently, many months of production are sacrificed, and most mining companies operate only in the milder weather from April to October. However, thousands of informal miners have no reluctance about working throughout the very cold winter, burning wood, tyres, dung, and benzene to warm themselves, and to thaw the permafrost and soil to release gold-bearing deposits and melt the ice of ponds and watercourses to facilitate panning. The closure of a mine is, unless security is unusually good, the signal for the immediate appearance of swarms of informal miners, eager to exploit any exposed mining opportunity.

Table 30.1 presents estimates of the proportions of miner types in Mongolia. They come from all strata of society. Most have lost the basis of their family's livelihood because of urban or arable privatization, or misfortunes to their livestock. Mongolians are generally well educated, and many miners are professional people. There are mature people of both sexes, and women and children are well represented (approximately 40% of the workforce is female). Residents in close proximity to mining opportunities, including both urban residents and herdsmen, are very active.

Gold buyers pay about 80% of the world price for gold, although it may be less if there is little competition, and as much as 20% more if there are convenient border crossings to China. The usual 25% margin on the field price is reasonable, and the street price in Ulaanbaatar – *i.e.* with the addition of transport costs and with maximum competition – is only 2 or 3% higher. Individual incomes vary greatly, depending on skill, energy and luck but many make barely enough to live on. Estimates based on field discussions are that about

25% earn less than MNT 6,000[2] per person per day, 60% between MNT 6,000 and 10,000, and 15% MNT 10,000–15,000; this translates into a mean of some MNT 7,400 (£7.50) per day.

Gold buyers are present on site, and in district centres, provincial centres and licensed mining camps but higher links in the dealing chain are uncertain. The Bank of Mongolia (BOM) has been almost exclusively the official purchaser and exporter to date, and purchases approximate licensed production, with little room for the addition of illegal production. BOM pays the world market rate, in accordance with Article 37 of the *1997 Law on Minerals*, subject to a 7.5% tax, which serves as a deterrent to some sellers. Demand from the Mongolian jewellery industry is modest, and it appears that the vast majority of informal gold is aggregated in Ulaanbaatar, and smuggled principally to Korea and China.

SOCIAL SAFETY NET

The scale of informal gold mining is generally under-estimated. The Mineral Resources Administration estimate that in July 2002, there were at least 10,000 miners[3] but field surveys suggest that the true figure is very much higher than this, and probably ranges between 50,000 and 100,000 at the August peak; thousands work throughout the winter. The IFC[4] estimate collective illegal mining production between 0.5 and 1.0 ton of gold per year but this also appears to be an underestimate. There are at least 50,000 miners active for six months of the year and 12,500 at other times; a mean income of £7.5 per day generates 7.5 tons of gold per year with a value of £107 million. This is approaching the volume of formal production of 12 tons, and actual figures may be significantly higher than these assumptions and could quite conceivably exceed formal production in the next few years. Any inaccuracy in estimation of the current scale of activity is likely to be only temporary, given the rapid rate of growth resulting from widespread poverty and unemployment, and the almost inexhaustible mining opportunities in association with licensed mines and in other areas.

As a novel source of household income, informal gold mining is performing an invaluable role as a social safety net for tens of thousands of people. The poverty of recent years has been deep and widespread but in the few areas where informal gold mining has been active for a number of years, it has clearly had beneficial impacts on livelihoods. It is providing relatively lucrative employment for tens of thousands of people in rural areas. As such, it may be worth some £85 million each year, an amount unmatchable by donors and Government through poverty alleviation programmes, not least of all because it has no associated heavy administrative costs, and indeed has made no call at

[2] MNT indicates the national Mongolian currency, the togrog.

[3] Cited in IFC, 2002.

[4] IFC, 2002.

all on scarce Government expenditure. With its low weight, high value, and non-perishability, informal gold mining is immune to the tyranny of distance that thwarts many of Mongolia's development ambitions. The low technology is cheap to purchase and operate, and there are big added value margins – much greater than for livestock herding, which it is beginning to rival, being already worth about one third of that industry.

ENVIRONMENTAL IMPACTS

Informal gold mining does, however, have strong negative impacts on the environment. The most obvious of these is land degradation, and although reworking and tunnelling of the waste of licensed mines has only limited additional impact, and is, in fact, used as an excuse by companies for inaction on mandatory rehabilitation, the consequences associated with the development of new areas for informal mining are very serious. Although Mongolians have a deep respect for the environment, with even the turned-up toes of their boots reputedly designed to minimise disturbance of the ground, much customary behaviour has been put under great pressure by the depth of the new poverty. Newly mined areas are currently without any control whatsoever. For example, one interviewee complained that his regular nomadic wintering place had been completely destroyed by miners.

Impacts on watercourses are also exceedingly important. Water use is profligate where it is plentiful, although careful use for panning in arid areas demonstrates that, in fact, large quantities are not required; mining has therefore been blamed for the recent drying up of many important watercourses. The impact of thousands of miners panning in close proximity has a severe effect on the hydro ecology. The intensity of mining settlement threatens all combustible material in the vicinity used for heating and cooking requirements; widespread de-vegetation is caused by these direct uses, as well as by accidental forest fires. Thus, the rapid and uncontrolled spread of the activity has major implications for Mongolia's fragile environments. There are numerous secondary impacts, including damages to livestock and wildlife watering, and, more generally, habitat depletion, loss of biodiversity and the spreading of desertification.

Environmental conditions, in the form of droughts and harsher winters, have accelerated the demand for mining by killing off millions of head of livestock; this more intense activity is further accelerating environmental deterioration.

SOCIAL IMPACTS

Mercury contamination from informal mining is another environmental problem experienced in the hard rock gold areas of north-central Mongolia but its human health consequences are more serious, with hundreds of mainly untrained miners, including many children, using it to separate gold from uneconomic particles. This has resulted in serious impairments, including birth defects preventing speech and even walking (Saijaa, 2002). More widespread health risks arise from the collapsing of unsupported tunnels in placer mining areas, with several dozen deaths each year.

The mining is illegal and miners are marginalized, with the exception of a few cases, where local governors, because of the extent of involvement of the local community and/or the depth of local poverty have recognised them in attempts to introduce some regulation. Miners operating outside their home areas are not registered locally and cannot avail themselves of school or health facilities. There is no effective public administration, including law and order, in the mining areas, and criminal activity, as well as antisocial behaviour – particularly, drunkenness – are common. Arrangements for potable water and sanitation in overcrowded, makeshift conditions are often inadequate, and are likely to result in public health crises, with cholera outbreaks understood to have already occurred.

The relationships with local resident communities and licensed miners vary. Some companies manage to maintain effective barriers against informal operators; some harass informal miners, with varying degrees of success in mitigating their activity, by beating, fining, locking up, and confiscating tools, whilst others try to reach an agreement with them by tolerating activities in parts of the licensed area or charging nominal mining fees. When illegal miners are dispersed by guards, they sometimes resort to working at night, which is even more dangerous than by day; in some cases, children spend several days underground, sustained only by tea and curds (PTRC, forthcoming). Miners do provide new markets for the local herders' dairy products and meat, but they often occupy, and sometimes ruin, the best pasturelands that adjoin the rivers; there are direct conflicts where demand for mining opportunities outstrips supply.

Although the activity is rural, many miners are urban residents, and it is helping to slow the rate of migration to the capital city. However, Ulaanbaatar is the pre-eminent focus for investment and much of the profits from mining are invested there in property and taxi vehicles.

GOVERNMENT REACTION TO INFORMAL GOLD MINING

In 2001, the Ministers of Infrastructure and Trade and Industry issued a joint Interim Regulation on *Gold Extraction By Non-Industrial Means*, which permitted informal mining on already worked and unworked, licensed land, subject to licensee's agreement, during the following 12 months. This has not been renewed, and responsibility is now left entirely to the local administrations.

Despite the obvious relevance of informal mining to much needed poverty relief, the Government reaction to date has been decidedly muted, although it is now creeping into the most recent policy documents (*e.g.* the *Interim Poverty Reduction Strategy Paper* of October 2002). The reluctance to deal firmly with the issues may be related to a desire to preserve the good international image of the industry; however, it leaves local Government with extremely serious problems, and no regulatory framework or guidance on how to manage them. It is more likely that the Government has not fully appreciated either the scale of the problems or the massive benefits that might easily be captured: (i) its direct local social safety net impacts and multiplier effects fit well with the development policies for rural areas, where conditions continue to deteriorate; (ii) if properly channelled, it would add at least £8 million to the Treasury each year through the gold tax alone; and (iii) the mining is practiced entirely by Mongolian nationals, and its rural nature helps the nation to overcome concerns about properly occupying its thinly populated territory in the light of its massive Chinese and Russian neighbours. Slight reservations about increasing dependence on it are that it makes scant use of the high quality human resource character of Mongolia's under-used labour, and, like all the export commodities on which Mongolia depends, is subject to the vagaries of world markets.

If annual production is in the range of 5–15 tons per year then the world price value minus 7.5% gold tax and 80% for miners leaves 12.5% or some £8–23 million which, if the activities can be formalised, could be applied to a variety of possible uses. Community-based involvement is essential to formulating satisfactory solutions, and finance could, in part, perhaps support a mechanism for accelerating much needed fiscal decentralization and community Government (*i.e.* by creating a financial basis for providing improved field services). Other essential ingredients include wealthy and well-equipped companies keen to compromise with informal operators, and increasing Government recognition of the need to accept and formalise activities.

Table 30.2 identifies some of the problems and possible approaches to solutions. In the case of Mongolia, it appears that there ought to be two basic sets of solutions, dealing respectively with and without the willing involvement of licensed mining companies. It is critical that informal miners are organized at least sufficiently to perform as stakeholders in forming, and, more importantly, implementing, agreements. Although gold buyers pay high prices to miners, there is enough of a margin to develop mutually beneficial policies when miner purchases, Government tax, and handling charges are deducted from world prices. Formulae could be developed to allocate this appropriately to mining-related services, environmental improvements, other local area improvements and contributions to a centralized fund for distribution to provinces in which gold is not present.

Inclusion of formal and informal gold exports in the published trade figures and national accounts could well, for example, eliminate the adverse balances of trade that have been increasing in recent years, and strengthen the Government's arguments with the IMF over monetary policy. Figure 30.3 indicates that Mongolia has had negative trade balances since 1996, during which time the excluded gold exports have ranged between £52 million and £118 million annually. The graph indicates the situation adjusted for the inclusion of these formal exports. The situation is complicated by the fact that imported services, having a negative impact, are also excluded from the trade figures, so these too are included in the adjusted figures. Although individually large, they do not have a dramatic net impact. The Figure also indicates the effects of

Table 30.2 Towards possible solutions to informal mining problems.

Problem situation	Basis of solution	Identifying/providing access to mining locations	Provision of field services	Infrastructure improvements	Environment		Initial gold buying	BOM Intermediary and mark-up allocation payment
					Good operating practice	Rehabilitation		
1. Associated with a co-operating licensed company	Symbiotic relationship should be developed between company and community	Community/ Mining Co. For a company fee? Appropriate panning and "residential", as well as mining, locations must be identified	Companies (or other open competitors) provide services for a fee. (Services = security, health and safety, possibly education etc.)	Company/ community/ district cooperate/ finance in accordance with a plan. (e.g. improved roads, bridges, water supply, public facilities)	Special non-watercourse panning areas; rehabilitate as part of process; arrange fuel sources etc.	By company in accordance with agreed restoration plan. District/ Province approve rehabilitation plan and contribute finance as agreed	Company, with rotating Government on-site represen-tation, buys	Company pays directly to district amounts owed, after short delay for getting cash from BOM. Balance to BOM/Ministry of Finance.
2. Independent of licensed company (because company does not want to cooperate or is not present)	District/ mining community are principal parties, but need funds to be effective	Community/ District. For a fee? Appropriate panning and "residential", as well as mining, locations must be identified	District or community organizes/ purchases for a fee	Community and District	More respect for environment "Miners' Code of Conduct"?	District/ community prepare plan. Province approves. District/ community finances	Existing buyer system.	Buyers sell to District Government appointed agent e.g. local Bank office, the new Mongolian Wholesale network, at higher price than currently. Agent pays locally and balance to BOM/Ministry of Finance

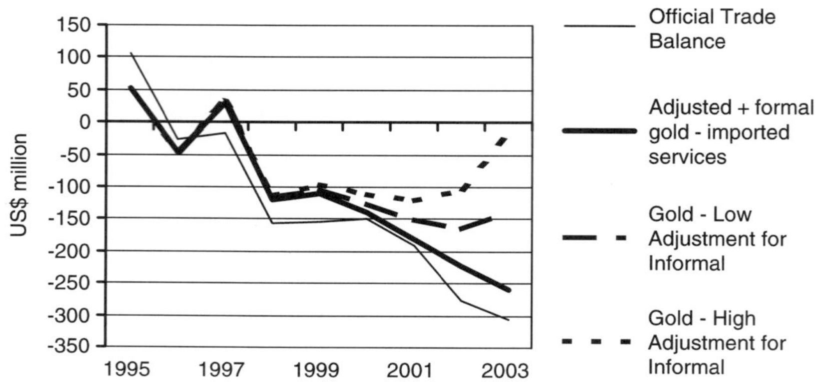

Figure 30.3 Potential impact of gold exports on Mongolia's trade balance.

including higher and lower estimates of informal exports. These are crude estimates only but are the best available, and they indicate that it is quite conceivable that the trade balance could be defined as being positive in the near future.

Formalizing the illegal exports would also be sufficient catalyst for the establishment of a Mongolian gold refinery, for which the Government has been trying to attract private investment for some years. At present, legal gold is sent by BOM to Russia, Japan or England, with transport costs significantly eroding its value. The break-even threshold for a refinery is about 10 million tons per year, so that the addition of illegal to legal processing would make the investment very attractive.

CONCLUSIONS

Informal gold mining in Mongolia is already large, is increasing rapidly, perhaps now justifying the word "rush", and will continue for several decades. It is functioning as a social safety net on a much grander scale than anything the Government and donor community can implement, and at much lower cost, but is creating problems of health and safety, civil order, child labour and adverse environmental impacts. Different levels of Government, particularly the lower ones, are, with some degree of desperation, trying to meet the demands of the situation, but a coordinated response has not yet been formulated; nor has Government sought the assistance of the donor community, which is almost completely oblivious to the phenomenon. Whilst it is necessary to recognize and deal with the immediate and direct problems as a matter of urgency, it would clearly be advantageous for GOM to seek donor assistance in dealing properly with the issues because potential benefits are massive in terms of benefits to accounted GDP and the treasury, as well as to regional development policy and possible effective decentralization of Government. Aided by a transitional atmosphere still expecting change and prepared to experiment, this may yet provide the incentive for energetic Government attention, perhaps with the opportunity to demonstrate innovative solutions from which other countries can learn.

REFERENCES

Bikales, B., Khurelbaatar, C. & Schelzig, K. (2000). *The Mongolian Informal Sector*. Manuscript, Development Alternatives Inc. for the Economic Policy Support Project, United States Agency for International Development, Ulaanbaatar, 2000.

Hilson, G. (2002). The environmental impact of small-scale gold mining in Ghana; Identifying problems and possible solutions. *The Geographical Journal*, 168, (1).

International Finance Corporation (IFC). *Mining and Petroleum Sector Profile*, Ulaanbaatar, September 2002.

Murray, W. Informal Gold Mining and National Development – the Case of Mongolia. *International Development Planning Review* forthcoming May 2003.

National Statistical Office (NSO) and UNDP. *Living standards measurement survey 1998*. Ulaanbaatar, August 1999.

Population Teaching and Research Centre (PTRC), National University of Mongolia. *Assessment of the child labour situation in gold mining*, commissioned by the International Labour Office, forthcoming.

Saijaa, N. (2002). *Mercury poisoning*. State University Environment and Health Resources Centre, manuscript in Mongolian, Ulaanbaatar, 2002.

Sermier, C. (2002). *Mongolia, empire of the steppes*. Airphoto International Ltd, Hong Kong.

31

Managing Mercury Pollution Emanating From Philippine Small-Scale Gold mining Activities: An Economic Analysis

DANILO C. ISRAEL AND JASMINDA P. ASIROT

As has been mentioned throughout this book, small-scale gold mining is an activity that relies heavily on manual labor and features simple implements and methods. Although it is a humble form of livelihood, it contributes significantly to gold production and rural employment in the Philippines. However, small-scale gold mining has been the target of strong opposition in the country in recent years, mainly because of its environmental and social side-effects, the most significant of which are induced by mercury pollution.

Numerous studies (*e.g.* Veiga 1997a, b; Akagi et al., 1995; Aula et al., 1995; Malm et al., 1995; Porvari, 1995; Barbosa et al., 1995; Guimaraes et al., 1995; Veiga et al., 1995; Veiga & Meech, 1995) have investigated the environmental impacts associated with mercury use at small-scale gold mines. High levels of mercury concentrations have been found in the hair and blood samples of miners and people in many communities in close proximity to activities, as well as within the fish, soil sediments, and forest and river ecosystems in mining regions.

In the Philippines, several studies (*e.g.* Mahinay et al., 1998; Bacani et al., 1996; Breward, 1996; Balce & Cabalda, 1992; Williams et al., 1995) have looked into the environmental impacts of operations in Diwalwal, the largest small-scale mining region in the country. High concentrations of mercury pollution have been found on-site as well as downstream. For example, mercury loads in some sections of the Agusan River, into which Diwalwal drains, are already considerable. Moreover, the water within the mining site itself has higher concentrations of mercury than the water in many other gold rush areas around the world (Williams et al., 1995).

Most of the studies that examine the environmental impacts of mercury pollution in the small-scale gold mining industry are technical in nature. Few touch on the economic aspects of the problem, and, of those that have, only do so in a superficial manner. Of what has been done, none have targeted the Philippine small-scale gold mining industry.

The main objective of this chapter, therefore, is to examine the state of small-scale gold mining in the Philippines, and to quantitatively assess the economic cost of the mercury pollution emanating from its activities. As a corollary goal, the study discusses related environmental and developmental problems associated with the industry. The chapter concludes by prescribing measures to both address these problems and promote better management in small-scale mining.

The study reported in this chapter used both secondary and primary data. Surveyed were key informants, representatives from Local Government Units (LGUs), and various small-scale gold miners and processors at two study sites. The secondary sources of data used to complement these findings included data extracted from the Mines and Geosciences Bureau (MGB), the World Bureau of Metal Statistics, National Statistical Coordination Board (NSCB), and the existing body of literature on small-scale mining.

REVIEW OF SMALL-SCALE GOLD MINING IN THE PHILIPPINES

Philippine gold production vis-à-vis global gold output

Global gold production has increased in recent years, rising from 1.2 thousand metric tons in 1980 to 2.3 thousand metric tons in 1997. During the 1980–1997 period, the world's leading producers of gold were South Africa, which contributed 34.54% of total production; the former USSR, which added 14.39%, and the USA, which contributed 11.07%.

Gold in the Philippines also increased during this time period (Table 31.1). Output rose from 20 metric tons in 1980, to 33 metric tons in 1997, although national output accounted for only a small percentage of world production. On average, annual gold output in the country is but 1.5% of world production. Nevertheless, gold is the most important mineral produced in the Philippines in terms of value (Table 31.2), accounting for 32.54% of the country's mineral production between 1980 and 1997.

Philippine exports of gold have increased over the years, both in terms of volume and value (Table 31.3). Between 1980 and 1997, volumetric output increased from 12.7 to 15.7 metric tons, and FOB value rose from about P1.78 billion to P5.0 billion. The top importing countries of Philippine gold during this period were Japan, followed by the United Kingdom, United States, Taiwan, South Korea, China and North Korea. Exports to Japan have decreased over the years, while exports to the United Kingdom have risen. More recently, exports to the United States and Taiwan have ceased, while exports to China and North Korea have been intermittent.

During the 1980–1997 period, the average annual gross value added (GVA) in mining and quarrying was only 1.56% of national Gross National Product (GNP). Thus, the GVA in gold mining was 0.6% of GNP.

Table 31.1 World annual gold production, by country, 1980–1997 (in metric tons).

Year	South Africa	USSR	United States	Philippines	Others	Total
1997	493	237	338	33	1,166	2,266
1996	495	214	312	30	1,080	2,130
1995	522	213	313	27	1,021	2,097
1994	584	229	326	27	974	2,140
1993	620	236	331	25	914	2,125
1992	609	226	330	26	935	2,125
1991	601	220	294	26	905	2,046
1990	601	302	295	25	864	2,086
1989	606	304	201	30	718	1,858
1988	619	278	201	30	664	1,792
1987	602	260	155	33	552	1,602
1986	638	273	116	35	477	1,540
1985	671	272	76	33	430	1,481
1984	680	269	63	26	382	1,420
1983	680	268	61	25	359	1,392
1982	664	266	46	26	330	1,331
1981	656	262	43	24	272	1,256
1980	672	258	30	20	207	1,188
Average	612	255	196	28	680	1,771
Percent	34.54	14.39	11.07	1.57	38.43	100.00

Note: Data for USSR for the years 1992–1997 were only for Russia, Armenia, Georgia, Kazakhstan, Kyrgyzstan, Tajikstan, Ukraine and Uzbekistan.
Sources: World Bureau of Metal Statistics (January 1987–December 1998) and Mines and Geo-Sciences Bureau (1980–1997)

Table 31.2 Mineral production of the Philippines, 1980–1997 (in million pesos).

Year	Gold	Copper Concentrate	Sand and Gravel	Salt	Coal	Others	Total
1997	9,909	2,792	10,062	7,250	1,187	1,880	33,080
1996	9,855	3,364	8,733	6,078	956	2,093	31,079
1995	8,484	5,786	6,601	3,851	1,266	2,013	28,001
1994	8,966	5,521	4,050	3,542	1,298	1,209	24,586
1993	7,926	6,262	3,389	3,128	1,582	1,152	23,439
1992	7,189	6,641	2,400	2,194	1,828	4,411	24,664
1991	8,177	7,677	2,257	2,178	1,421	2,349	24,059
1990	7,116	7,891	2,194	1,803	1,060	2,305	22,369
1989	7,925	7,948	2,072	1,801	1,042	2,379	23,168
1988	8,844	7,953	1,207	443	1,028	2,285	21,760
1987	9,352	6,141	1,392	413	931	1,277	19,506
1986	8,395	5,461	1,221	391	1,149	3,421	20,037
1985	6,088	5,630	1,071	365	1,509	6,323	20,987
1984	4,773	4,970	1,142	341	1,134	5,163	17,522
1983	3,822	4,047	1,022	318	393	4,039	13,641
1982	2,651	3,446	858	219	189	3,937	11,301
1981	2,642	3,782	691	213	64	4,470	11,862
1980	2,785	4,409	613	204	59	4,750	12,821
Average	6,939	5,540	2,832	1,930	1,005	3,081	21,327

Source: Mines and Geosciences Bureau (1980–1997).

Table 31.3 Philippine gold exports, 1980–1997.

Year	Quantity (in metric tons)	FOB value
1997	15,747	5,013,627,916
1996	10,888	3,533,324,525
1995	12,328	3,619,530,939
1994	13,864	4,407,254,262
1993	13,419	3,978,926,107
1992	13,564	3,739,674,617
1991	4,747	1,922,197,914
1990	6,149	1,597,459,480
1989	6,900	1,903,645,183
1988	6,807	2,075,991,452
1987	6,513	1,902,579,202
1986	6,775	1,579,657,385
1985	7,394	1,365,966,994
1984	9,138	1,790,102,037
1983	11,417	1,367,483,371
1982	15,535	1,594,430,803
1981	14,713	1,693,904,187
1980	12,747	1,784,746,890
Average	2,999.8	2,492,805,737

Source: Mines and Geosciences Bureau (1980–1997).

The Philippine small-scale gold mining sector

There are two types of ore reserves relevant to small-scale gold mining: indicated and inferred (Javelosa, 1997). Indicated reserves are those for which tonnage and grade are computed using specific measurements, samples or production data. Inferred reserves are those for which quantitative estimates are based; in the majority of cases, quantitative estimates are based on broad knowledge of the geologic character of a given deposit.

In 1992, the Philippines had an estimated 60.8 million metric tons of indicated reserves, and 180.4 million metric tons of inferred reserves (Table 31.4). The country's indicated reserves had a mean metal content of 87.59 metric tons and appeared to be ideal for small-scale extraction. In contrast, inferred reserves had a mean metal content of 0.58 metric tons, which suggested that they were unsuitable for small-scale mining. The largest indicated and inferred reserves were found in the Cordillera Autonomous Region (CAR).

Small-scale mining covered a total claim area of about 4,939 hectares in 1992 (Table 31.5). The largest claim areas were in CAR, Region XI and Region XII. Region VI had the smallest claim area, followed by Region X and Region IV; in total, there were 135 mine sites. The largest sites in terms of total claim area were found in the provinces of Sultan Kudarat, Davao del Norte and Nueva Ecija, while the smallest were located in Agusan del Norte, Ilocos Sur, Bohol, Surigao del Norte, Agusan del Norte and Cebu (see Fig. 31.1).

Because a substantial portion of the small-scale mining activities that occur in the Philippines is illegal, accurate production data are hard to find, and of the

Table 31.4 Estimated geologic mineral reserves of the small-scale mines in the Philippines, by region, 1992.

Region	Claim area (Ha)	Indicated reserves: Volume (000 m³)	Reserve (000 MT)	Metal content (Mean MT)	Inferred reserves: Volume (000 m³)	Reserve (000 MT)	Metal content (mean MT)
CAR	1,266.89	79,214	21,050	9.55	237,534	62,939	0.05
I	137.50	8,593	2,276	11.64	25,780	6,830	0.05
II	320.00	19,999	5,297	8.10	59,999	15,895	0.04
III	306.28	19,142	5,072	0.07	49,927	13,229	0.12
IV	56.00	3,498	924	0.16	10,498	2,780	0.06
V	333.00	20,811	5,511	17.46	62,436	16,539	0.05
VI	5.00	312	82	2.10	937	248	0.01
VII	332.50	20,781	5,505	3.95	62,343	16,515	0.05
VIII	199.62	12,475	3,304	3.56	37,427	9,913	0.05
IX	343.00	21,439	5,678	3.82	64,313	17,040	0.03
X	39.50	2,466	651	0.52	7,404	1,955	0.05
XI	1,080.00	67,500	17,884	11.17	202,500	53,657	0.04
XII	520.00	32,500	8,611	25.04	97,500	25,833	0.03
Total	3,672.40	229,516	60,795	87.59	681,064	180,434	0.58

Source: Javelosa (1997).

Table 31.5 Small-scale gold mining claim areas and sites in the Philippines, by region, 1992.

Region	Claim areas (Ha)	Percent	Sites Number	Percent
CAR	1,266.89	25.65	39	28.89
I	137.50	2.78	5	3.70
II	320.00	6.48	9	6.67
III	306.28	6.20	3	2.22
IV	56.00	1.13	8	5.93
V	333.00	6.74	16	11.85
VI	5.00	0.10	2	1.48
VII	332.50	6.73	6	4.44
VIII	199.62	4.04	10	7.41
IX	343.00	6.94	9	6.67
X	39.50	0.80	10	7.41
XI	1,080.00	21.87	11	8.15
XII	520.00	10.53	7	5.19
Total	4,939.29	100.00	135	100.00

Source: Javelosa (1997).

estimates that have been published in secondary sources, most are wide-ranging. The production estimate made for 1992 (including both seasonal and continuous operations but excluding alluvial gold panning) was 6,826 kilograms (NSCB, 1999; Santelices, 1997), with underground small-scale gold mining accounting for approximately 25% of this total.

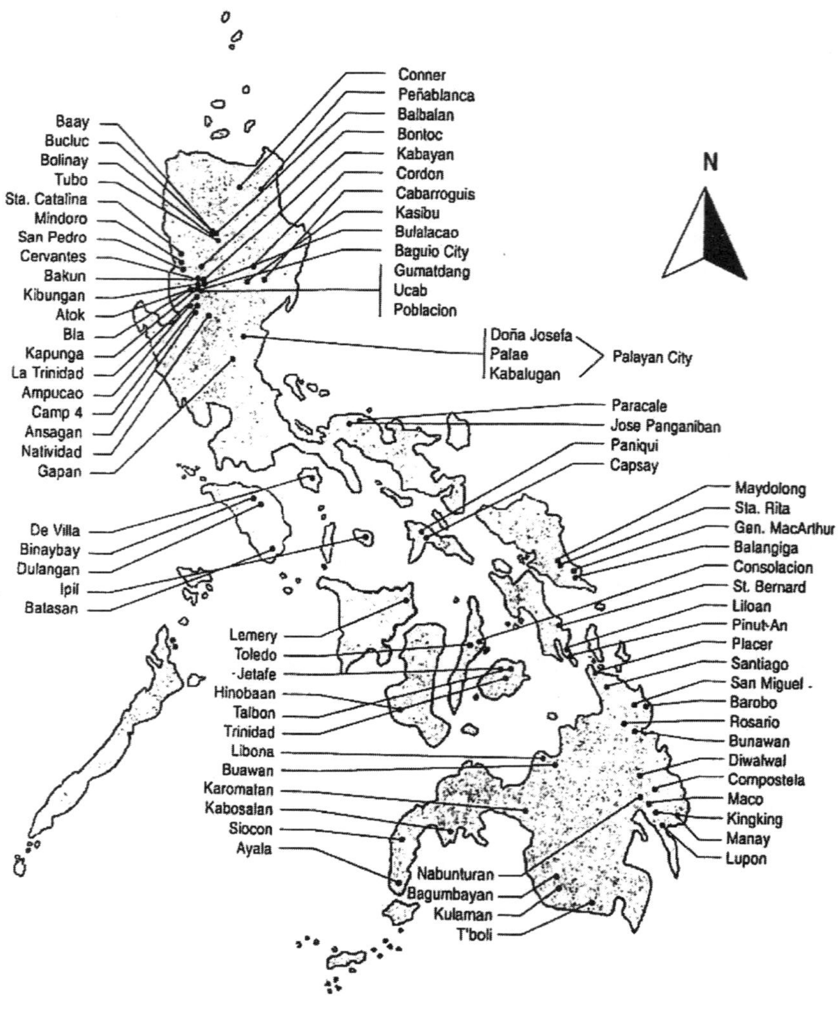

Figure 31.1 Locations of important small-scale gold mining regions in the Philippines.

Studies have cited much higher production levels, however. Stewart (1994), for example, estimated that between 1985 and 1986, approximately 10,000 tons (of gold) were produced in the industry. Quoting local sources, Dhar (1994) asserted that approximately 150 tons are produced by local small-scale miners each year, of which only 25–26 tons are sold to the Central Bank because of pilferage into the black market. However, these figures should be treated with caution as they imply that overall gold mining in the Philippines is dominated by small-scale activities.

The exact number of small-scale miners in the country is also not known. Bayle (1995) estimated 250,000 to be directly involved in small-scale gold mining activities, while Dhar (1994) figured that between 400,000 and 500,000 were involved, including those employed in the backward and forward linkages of the industry. While these figures are wide-ranging, they nevertheless manifest the importance of small-scale mining as a source of employment in the rural upland areas of the country.

Small-scale mining is not an important public revenue-generating sector for the government at present because of its largely illegal nature. Viewed in a more positive light, the activity should become a solid tax base when fully licensed, given the large number of people and economic activities dependent upon it.

Relevant laws and institutions

The earliest mining law in the Philippines was the *Commonwealth Act 137*, which was promulgated in 1936. This legislation had no separate provision for small-scale mining, since the activity was not practiced extensively at the time. It had taken effect for many years until the martial law era, during which time it was amended through Presidential Decree (PD) 463, otherwise known as the *Mineral Resources Decree* of 1974. Like its predecessor, this Law did not have separate provisions for small-scale mining.

In 1984, PD 899 established small-scale mining as a new dimension in mineral development, and defined it as a specific activity. Succeeding orders based on this law stipulated, *inter alia*, the rules and regulations governing the granting of small-scale mining permits, and the mandatory sale of recovered gold to the Central Bank and its authorized representatives.

During the term of President Corazon C. Aquino, the Congress of the Philippines passed Republic Act (RA) 7076 or the *People's Small-Scale Mining Act* of 1991. This Law established the People's Small-Scale Mining Program, and identified the (small-scale mining) areas that could be opened under it.

During the administration of President Fidel V. Ramos, RA 7942 – otherwise known as the *Philippine Mining Act of 1995* – was passed. The Law stipulated that small-scale mining would continue to be governed by the provisions of RA 7076 and PD 1899, and their implementing rules and regulations.

Institutions

Prior to the implementation of PD 899, no single government agency was responsible for managing small-scale mining, since it had not been recognized as a formal economic sector. Those engaged in the activity at the time did so without any government interference. With the passing of the Law, the Mines and Geosciences Bureau (MGB) was given the authority to administer small-scale mining concessions.

In the early-1990s, the administrative supervision of small-scale mining remained with the MGB under RA 7076. Shortly afterward, through a series of administrative orders, the authority to grant small-scale mining permits was devolved to Governors or Mayors, upon clearance from the regional offices of the Department of Environment and Natural Resources (DENR), under which the MGB administratively falls.

As part of its current reorganization, the MGB created the Small-Scale Mining Section under its Mining Environment and Safety Division. The functions of this new unit include the environmental assessment of small-scale mining areas, and policy formulation related to the environment and small-scale mining activities.

MERCURY POLLUTION EMANATING FROM RESIDENT SMALL-SCALE MINING ACTIVITIES

Gold amalgamation

Mercury amalgamation is a popular technique in small-scale mining regions because it is simple and requires relatively low investment. After ore is crushed manually with sledgehammers, the resulting particles are fed into a grinding facility, such as a rod mill or ball mill. Lime and water are then added to the ore, after which grinding commences. After several hours, the mill is turned off and mercury is mixed with the fine ore. Eventually, the milled ore is placed in a large basin, and the heavy metal alloy is allowed to settle to the bottom. Water is again added to the milled ore to remove the slurry, leaving behind the amalgam.

The amalgam is then collected and placed in a fine cloth, which is squeezed to free excess mercury. Borax is often added to the amalgam as a cleaning material to remove impurities, and then blowtorched in a circular clay pot to separate the gold from the remaining mercury. The final product is 14–16 carat sponge gold.

Amalgamation, up until the point where the impure amalgam is produced, is performed by the miners themselves, or by workers employed at rod mills or ball mills.

The health impacts of mercury

Mercury is often unintentionally spilled as a result of careless handling. Moreover, it is discharged, together with other wastes, into inadequate tailings ponds, or, in the worst of cases, discarded *directly* into rivers and waterways. Vaporized mercury can also be released into the atmosphere when amalgam is blowtorched and refined.

Once in the environment, mercury is dangerous because of its potential adverse impact on human health. Mercury discharged into rivers and waterways is often transformed into methylmercury, and then ingested by aquatic species that are later consumed by people. Once inside the human body, it can trigger neurological disturbances, as well as problems within the reproductive organs (Veiga, 1997a). Typical symptoms of mercury poisoning include visual constriction, numbness of the extremities, and hearing and speech impairments. The long-term effects of excessive exposure include metabolic impairments, and damages to the nervous system. The exhibition of exaggerated emotional responses, muscular tremors and gingivitis, are typical symptoms of such deterioration.

The release of mercury into the atmosphere during blowtorching also poses a risk to human health. The activity is usually performed in open containers and enclosed houses. Thus, the inhalation of vaporized mercury is likely to be high among people refining gold.

CASE STUDIES

Diwalwal

Figure 31.1 shows the newly-established province of Compostela Valley, the municipality of Monkayo, and the barangay of Diwalwal (the official name of the barangay is actually Mt. Diwata but was renamed to "Diwalwal" by folklore). In 1997, Diwalwal had a population of 9,490 people and a land area of 729 hectares. Although it is not known how many of its inhabitants are, in fact, small-scale miners, most are nonetheless involved in mining-related activities; many have also organized themselves into associations.

Mining activities in Diwalwal began in September 1983, when near-surface gold deposits were first discovered by a group of miners. Production reached a peak in 1985, and although it subsequently declined, output has remained significant. During its peak, the regional population had reportedly reached 50,000 (ECS, MTD 1986).

Initially, mining in Diwalwal was carried out on a small scale, emphasizing the exploitation of narrow and shallow tunnels using manual and artisanal methods. However, local mining techniques have since become more sophisticated. Several of today's operations feature bigger, deeper and well-developed tunnels; use heavy machinery and explosives for ore extraction; and deploy mine cars and related equipment for ore transport. They also employ engineers and other technical personnel. Apart from amalgamation, some processing operations in Diwalwal use the carbon-in-pulp (CIP) methods.[1] In 1988, there were 57 CIP plants, and approximately 220 rod mill and ball mill operators.

Diwalwal is highly environmentally-sensitive because of its location. The barangay is located at the upper part of the Mamunga River watershed, and surrounding creeks connect to the Mamunga and Navoc Rivers, both of which drain into the Agusan River some 24 kilometers away from the mine site. Thus, the potential for mercury pollution extends far beyond the site, a case in point being reported contamination in the provinces of Agusan del Sur and Agusan del Norte, which are crossed by the Agusan River and Butuan City. Related incidents of contamination have been widely reported in the local media (Table 31.6).

---

[1] The CIP method of gold processing involves adding lime and water to crushed ore, and then transferring the resulting slurry into a container called a *repulper*. Cyanide is added to the "pulp", which is further agitated in a series of tanks. Using an air compressor, activated carbon is added to the last tank, which, through a series of chemical processes, separates gold from the pregnant cyanide solution.

Table 31.6 Selected newspaper reports on mercury pollution in Diwalwal, Monkayo, Compostela province, 1997–2000.

Newspaper	Title/Content	Date
Manila Bulletin	Solon wants Diwalwal mining probed – Environmental problems arising from illegal mining in Diwalwal had reached alarming proportions. Cyanide, mercury and other mine tailings had turned Naboc River in Monkayo into a milky dead body of water.	January 30, 2000
Manila Standard	Kids exposed to mercury poisoning – Children and farmers living near Diwalwal are excessively exposed to mercury poisoning in the illegal gold mines of the mountain. So are lowlanders and their vegetable farms downstream of Monkayo, Diwalwal in Davao del Norte.	February 23, 1998
Manila Bulletin	House committee rejects move to lift suspension of mine permits – About 20 million tons of tailings heavily contaminated by mercury around Diwalwal are now polluting the streams, rivers and bays of East Mindanao.	October 3, 1997
The Philippine Star	Diwalwal kids, workers have high mercury, cyanide levels in blood, urine – DOH study – The Department of Health has reported that schoolchildren and workers exposed to mining operations in Diwalwal were detected to have high levels of mercury and cyanide in their blood and urine.	July 9, 1997
The Philippine Journal	DENR monitors four Southern provinces for mercury pollution – DENR is closely monitoring the provinces of Zamboanga del Norte, Cagayan de Oro, Davao del Norte, and Negros regarding the extent of mercury pollution in their particular waterways. The massive use of mercury by small-miners in the Diwalwal gold rush area resulted in mercury contamination in rivers, streams and bays, and the poisoning of residents.	June 28, 1997
Manila Times	Rivers near mining sites have high mercury levels – A number of rivers in Mindanao near gold mine areas are contaminated with mercury due to uncontrolled and improper use of the chemical in mining activities. The British Geological Survey warned that the river below the Diwalwal gold rush area has become a potential health risk due to high levels of mercury caused by the dumping of mine tailings or wastes.	April 21, 1997

Panique and Tugos

Panique has a land area of 10 hectares, and, in 1998, had a population of 5,775 people. For residents, the major sources of livelihood include small-scale gold mining, farming, fishing, and vending; mining, however, is Panique's chief economic activity. There are presently 825 households in the barangay, and key informants estimate that each household has an average of two members who are miners. There are approximately 160 ball mills and four CIP operations in Panique. Many small-scale miners have long been operating in areas within the mining claim of the Atlas Mining Company – a large-scale mining firm – that has recently been taken over by Base Metal Mineral Resources Corporation.

Tugos has a land area of nine square kilometers, and, in 1997, had a population of 3,625 people. As is the case in Panique, mining is a major source of livelihood for the inhabitants of Tugos. There are presently some 100 small-scale gold mining tunnels in the area, although only 35 are in operation. Each site is estimated to employ between 11 and 20 miners. There are approximately 35 ball mills and 10 CIP operations in Tugos, only half of which were reported to be functioning. A segment of the small-scale mining operations based in Tugos falls within the mining claim of the United Paragon Mining Corporation.

Case study methods

Until only recently, there had been no mercury studies conducted at any of the three stipulated locations. A brief survey was therefore conducted between April and June of 1999, for the purpose of determining the general occurrence of mercury pollution within the small-scale gold mining regions identified. Because of time and resource constraints, short interviews were conducted. The data gathered were then relayed back to local government officers and informants.

The respondents included both small-scale gold miners engaged in basic amalgamation activities, and operators of ball mills employing amalgamation practices; they were selected randomly. The survey gathered key demographical, environmental, institutional and financial data related to mercury pollution. It particularly sought to elicit local peoples' perceptions of mercury.

Demographic characteristics of small-scale miners

A total of 95 small-scale miners (45 in Panique and 50 in Tugos) were surveyed, a substantial number of whom were married and residing in close proximity to mining activities with their families (Table 31.9). At the time of the survey, most were involved in processing activities, had family-members who were also involved in mining-related works, and had attained at least an elementary education.

The majority of miners noted having heard of people contracting illnesses as a result of mercury exposure; some had also mentioned having been repeatedly exposed to mercury during the course of mining (Table 31.7). A substantial number of miners at one site in particular mentioned that their operations were in close proximity to a water body, where significant siltation, sedimentation, and a loss of fishery resources have occurred since mining activities

Table 31.7 Environmental information from small-scale miners in Panique, Aroroy, Masbate and Tugos,

Survey area	Have you heard of people getting sick due to mercury exposure in your area?				Have you been exposed to mercury during the course of your mining activity?				Is your mining a water body?	
	Yes	No	No response	Total respondents	Yes	No	No response	Total respondents	Yes	No
Panique										
Frequency	27	16	2	45	7	30	8	45	4	41
Percentage	60.00	35.56	4.44	100.00	15.56	66.67	17.78	100.00	8.89	91.11
Tugos										
Frequency	22	28	0	50	12	38	0	50	30	20
Percentage	44.00	56.00	0.00	100.00	24.00	76.00	0.00	100.00	60.00	40.00
Total										
Frequency	49	44	2	95	19	68	8	95	34	61
Percentage	51.58	46.32	2.11	100.00	20.00	71.58	8.42	100.00	35.79	64.21

Table 31.8 Institutional information from small-scale miners in Panique, Aroroy, Masbate and Tugos,

Survey area	Are Local Government Units monitoring small-scale mining activities in your area?				Is the National small-scale in your area?	
	Yes	No	No response	Total respondents	Yes	No
Panique						
Frequency	34	8	3	45	1	43
Percentage	75.56	17.78	6.67	100.00	2.22	95.56
Tugos						
Frequency	36	13	1	50	9	40
Percentage	72.00	26.00	2.00	100.00	18.00	80.00
Total						
Frequency	70	21	4	95	10	83
Percentage	73.68	22.11	4.21	100.00	10.53	87.37

commenced. They also suggested that increased concentrations of mercury have resulted in health-related problems among miners and other people.

Institutional information

Some miners noted that LGUs were monitoring activities within their respective areas, while others revealed the exact opposite (Table 31.8). However, key informants reported that LGUs do not rigorously monitor mercury pollution overall. Moreover, most miners mentioned that national government agencies have not been actively involved in monitoring, educating and regulating local small-scale gold mining activities. Specifically, a number of interviewees noted that local and national governments have paid minimal attention to the need for monitoring mercury use in small-scale gold mining areas. This neglect has undoubtedly exacerbated the mercury pollution problem throughout the country.

Paracale, Camarines Norte, 1999 data.

Area close to		Have you noticed significant siltation and sedimentation in the water body since mining started?				Have you noticed significant fishery loss in the water body since mining started?			
No response	Total respondents	Yes	No	No response	Total respondents	Yes	No	No response	Total respondents
0 0.00	45 100.00	3 75.00	1 25.00	0 0.00	4 100.00	3 75.00	1 25.00	0 0.00	4 100.00
0 0.00	50 100.00	23 76.67	7 23.33	0 0.00	30 100.00	18 60.00	12 40.00	0 0.00	30 100.00
0 0.00	95 100.00	26 76.47	8 23.53	0 0.00	34 100.00	21 61.76	13 38.24	0 0.00	34 100.00

Paracale, Camarines Norte, 1999 data.

Government involved in mining related activities		Are there Non-Government Organizations in your area?			
No response	Total respondents	Yes	No	No response	Total respondents
1 2.22	45 100.00	1 2.22	43 95.56	1 2.22	45 100.00
1 2.00	50 100.00	33 66.00	16 32.00	1 2.00	50 100.00
2 2.11	95 100.00	34 35.79	59 62.11	2 2.11	95 100.00

Financial information for small-scale miners

Some 74 of the small-scale miners surveyed – 34 in Panique and 40 in Tugos – were involved in amalgamation activities. However, the average volume of ore produced by individual miners varied significantly between the two case study sites (Table 31.9). The majority reported that they processed ore at ball mills, although a notable number reported processing ore at mills owned by those who possess the rights to the very tunnels they had gathered the ore from. Detailed production cost and returns data were difficult to generate from the miners but their estimated average annual gross and net incomes were positive, yet, low. Key informants also indicated that the average incomes of small-scale miners were low but sufficient to sustain their rural standards of living.

Because of subsistence earnings, it is highly likely that miners will be unwilling to pay for the environmental damages caused by the mercury discharges

Table 31.9 Demographic information from small-scale miners in Panique, Aroroy,

	Civil status				Are you living alone	
Survey area	Single	Married	No response	Total respondents	Living alone	Living with family
Panique						
Frequency	11	34	0	45	20	21
Percentage	24.44	75.56	0.00	100.00	44.44	46.67
Tugos						
Frequency	10	40	0	50	17	19
Percentage	20.00	80.00	0.00	100.00	34.00	38.00
Total						
Frequency	21	74	0	95	37	40
Percentage	22.11	77.89	0.00	100.00	38.95	42.11

	Are your family members also involved in mining-related activities?				Highest educational	
	Yes	No	No response	Total respondents	Elementary	High School
Panique						
Frequency	13	32	0	45	28	14
Percentage	28.89	71.11	0.00	100.00	62.22	31.11
Tugos						
Frequency	38	12	0	50	31	17
Percentage	76.00	24.00	0.00	100.00	62.00	34.00
Total						
Frequency	51	44	0	95	59	31
Percentage	53.68	46.32	0.00	100.00	62.11	32.63

emanating from their activities. A more realistic approach would be to convince them to pay for equipment that helps to ensure their personal protection from mercury pollution, provided that it involves only small deductions from personal earnings.

Environmental information for ball mill operators

In total, 45 ball mill operators were surveyed – 25 from Panique and 20 from Tugos. The small-scale miners involved in amalgamation at ball mills were generally not required by the (ball mill) operators to use gloves when handling mercury and other chemicals (Table 31.10). Moreover, all operators mentioned having tailings ponds that contain wastes but key informants noted that facilities were typically inadequate to handle discharges. This was evidenced by the degree of tailings-related contamination. Specifically, of the operators whose

Masbate and Tugos, Paracale, Camarines Norte, 1999 data.

or with your family in mining area?			Are you also involved in processing activities?			
Not living in site	No response	Total respondents	Yes	No	No response	Total respondents
4	0	45	32	13	0	45
8.89	0.00	100.00	71.11	28.89	0.00	100.00
14	0	50	50	0	0	50
28.00	0.00	100.00	100.00	0.00	0.00	100.00
18	0	95	82	13	0	95
18.95	0.00	100.00	86.32	13.68	0.00	100.00

attainment			Are you a member of a cooperative?			
College	No response	Total respondents	Yes	No	No response	Total respondents
3	0	45	0	45	0	45
6.67	0.00	100.00	0.00	100.00	0.00	100.00
1	1	50	0	50	0	50
2.00	2.00	100.00	0.00	100.00	0.00	100.00
4	1	95	0	95	0	95
4.21	1.05	100.00	0.00	100.00	0.00	100.00

sites were situated close to waterbodies, the majority mentioned that significant siltation, sedimentation and a loss of fishery resources have occurred.

Although most ball mill operators reported that the blowtorching of amalgam occurs outdoors, all noted that mercury retorts were not used. Many recalled incidents of people becoming ill as a result of mercury contamination. It is also important to note that some of the ball mill operators, who, at the time of the survey, were not as directly involved in gold processing as small-scale miners and workers, admitted that they themselves were also exposed to mercury pollution.

Financial information for ball mill operators

Detailed data on costs and returns proved difficult to obtain from the ball mill operators themselves. On average, operators at one site were found to have

Table 31.10 Financial information from small-scale miners also doing amalgamation processing in Panique,

Survey area	Average ore production per month				Where is the ore processed?			
	In kgs.	Number responded	No response	Total respondents	Tunnel owners' ball mill	Other ball mill	No resonse	Total respondents
Panique	742.2							
Frequency		34	0	34	16	16	2	34
Percentage		100.00	0.00	100.00	47.06	47.06	5.88	100.00
Tugos	349.3							
Frequency		40	0	40	14	26	0	40
Percentage		100.00	0.00	100.00	35.00	65.00	0.00	100.00
Total								
Frequency		74	0	74	30	42	2	74
Percentage		100.00	0.00	100.00	40.54	56.76	2.70	100.00
Average	545.8							

higher annual gross and net incomes than those operating at the other surveyed site. Since ball mill operators are generally low income earners, they are also unlikely to assume responsibility and thus pay for the environmental damages their activities have caused. As is the case with small-scale miners, however, they may be more willing to provide payment toward relatively minor expenditures related to the prevention of mercury pollution.

ECONOMIC VALUATION OF MERCURY POLLUTION IN SMALL-SCALE GOLD MINING REGIONS

Valuation methods

There are different methods available for estimating the economic costs of mercury pollution resulting from small-scale mining activities. One popular technique is the *loss of income approach*, which, if applied to this case study, would consider the incomes forgone by the people affected by mercury pollution as a measure of the economic cost of the problem. This method, however, is demanding since it requires a thorough physical accounting of the dose of mercury produced by the sampled small-scale mines over time, as well as the *exact* responses – in terms of morbidity or mortality – taken by the affected population. It also requires accurate income data from the individuals affected over time, in order to convert mortality and morbidity cases into economic values.

Another potential valuation approach, which can be used together with the *loss of income approach*, is the *productivity change method*, which, in this case, would estimate the economic value of the aquatic resources lost as a result of mercury pollution. This method is likewise tedious to apply because data on the physical dosage of mercury into the environment, and the exact volume of fish killed and rendered inedible as a result of its dispersion over time, are needed. In addition, historical price data on fish species in the locality are required to convert lost fishery output into monetary values.

Aroroy, Masbate and Tugos, Paracale, Camarines Norte, 1999 data.

Average annual gross income				Average annual net income			
In pesos	Number responded	No response	Total respondents	In pesos	Number responded	No response	Total respondents
29,568				22,648			
	34	0	34		34	0	34
	100.00	0.00	100.00		100.00	0.00	100.00
30,664				24,929			
	40	0	40		40	0	40
	100.00	0.00	100.00		100.00	0.00	100.00
	74	0	74		74	0	74
	100.00	0.00	100.00		100.00	0.00	100.00
30,116				23,789			

Because of an acute shortage of data, both methods were not used in this study. More specifically, there are no available epidemiological and fishery productivity studies for small-scale mining that can provide, on the one hand, accurate dose-response estimates between mercury pollution, and, on the other hand, data on human health and fish productivity.

Another possible method, which could be used to measure the economic costs of mercury pollution caused by small-scale mining activities, is *contingent valuation*. Briefly, this approach requires asking the affected population their willingness to pay to mitigate mercury pollution. The contingent valuation technique also has its share of limitations, however. In particular, it is inaccurate to apply the technique to individuals who are poor and whose willingness to pay for mercury pollution mitigation is low.

Therefore, for practical purposes, this study uses the *defensive expenditures approach* to measure the economic costs of mercury pollution caused by resident small-scale miners. This involves measuring the cost of implementing the necessary facilities and equipment to prevent mercury pollution from occurring, and then using these results to estimate the economic costs of the problem. It is important to note, however, that the *defensive expenditures approach* is limited in that it can only measure the costs associated with future mercury pollution, and not the impacts of past and existing contamination.

Determining the economic costs of preventing mercury contamination

The cost of hand gloves

Survey results reveal that 86% of small-scale miners are involved in gold processing (Table 31.9), which – provided this ratio is correct – means that there are 215,000 pairs countrywide. This total, however, does not include the undetermined number of workers employed at ball mills and rod mills.

The current price of good quality industrial rubber hand gloves ideally suited for gold processing, capable of lasting at least one year, is, at most, P500.00 per

Table 31.11 Environmental information from small-scale ball mill operators in Panique, Aroroy, Masbate and Tugos, Paracale, Camarines Norte, 1999 data.

Survey area	Do you require gloves as protective equipment in the handling of mercury and other chemicals?				Is there a tailings pond in the processing area?				Is your processing area close to a water body?			
	Yes	No	No response	Total respondents	Yes	No	No response	Total respondents	Yes	No	No response	Total Respondents
Panique												
Frequency	0	25	0	25	25	0	0	25	18	7	0	25
Percentage	0.00	100.00	0.00	100.00	100.00	0.00	0.00	100.00	72.00	28.00	0.00	100.00
Tugos												
Frequency	1	19	0	20	20	0	0	20	14	6	0	20
Percentage	5.00	95.00	0.00	100.00	100.00	0.00	0.00	100.00	70.00	30.00	0.00	100.00
Total												
Frequency	1	44	0	45	45	0	0	45	32	13	0	45
Percentage	2.22	97.78	0.00	100.00	100.00	0.00	0.00	100.00	71.11	28.89	0.00	100.00

Table 31.12 Marginal costs and returns of the use of a mercury retort in gold processing, 1999.

Item	Quantity	Value/piece	Total value
Marginal costs			
Purchase price of retort[1]	1 piece	P40,000/piece	P40,000.00
Repair and maintenance[2]			8,000.00
Total costs			P48,000.00
Marginal returns			
Retrieved mercury[3]	262,500 grams	P0.371 per gram	97,387.50
Retrieved gold[4]	1,750 grams	P260 per gram	455,000.00
Total returns			P552,387.50
Net income			P504,387.50

Notes
1. The retort is depreciated fully at the end of its economic life.
2. The total repair and maintenance cost during the entire life of the retort is 20% of its purchase price. No additional labor cost in the operation of the retort.
3. The economic life of a retort is 500 hours. Per batch of use takes 30 minutes. Therefore, 1,000 batches can be accommodated during its whole economy.
4. If the retort is not used, blowtorching results to the loss of about 2% of the processed gold. The use of the retort will result to 100% gold recovery. It is assumed that the recovered gold per batch is about 25% of the amalgam, or 87.5 grams of 350 grams load per batch. The 2% of this or 1.75 grams, are saved due to the retort. Given 1,000 batches, total gold saved is 1,750 grams.

pair; to satisfy the entire mining populace would require an expenditure of P107.5 million. However, individually, gloves are affordable, as their cost would constitute but a small percentage of net income (Table 31.12). Furthermore, the actual price of gloves should be significantly lower if they are used carefully, which would help to ensure that they last longer than the assumed economic life of one year.

Cost of mercury retorts

There are mercury retorts sold in the market not exceeding more than US$1,000 per unit or P40,000. The marginal cost and returns analysis for retort use during its entire economic life is shown in Table 31.12.

The marginal costs of using a retort in gold processing include purchase price, repair costs and maintenance. There is no marginal labor cost associated with the operation of a retort as it can be incorporated in blowtorching activity. In a case where 100% of mercury and gold is recovered, a retort actually saves its user approximately half a million pesos (in terms of recovered gold and mercury) during its entire economic life. It is only in situations where the efficiency rate of the retort is less than 8.6% that a retort is economically inefficient.

If retorts are acquired and shared among groups of miners, expenses will undoubtedly be less. According to key informants, the mercury-to-gold ratio in amalgam is approximately three to one (see Table 31.12). If the 1992 level of gold production (6286 kg) is, in fact, an accurate representation of present-day production, the total amalgam produced by the surveyed miners would be in the range of 27,304 kilograms. As each retort can process 350 kg of amalgam in its economic life, the total number of pieces required would be 78. At a price of P48,000, including maintenance costs, the total projected cost would be P3.7 million.

As retorts may be less than 100% efficient, and may be unsuitable to be shared because of the distance between mine sites and individual processing operations, more would be required. However, even if the cost is magnified ten times, the total cost at P37 million is still small in relation to the cleanup costs posed by excessive mercury pollution. Key informants in the two case study areas also explained that mercury retorts could be locally produced at a much reduced price.

Cost of tailings ponds

Based on Santelices (1994), the NSCB (1998) estimated the volume of small-scale mine tailings produced at Philippine small-scale mines, as well as the impoundment cost per unit of tailings for the period 1988–1994. Using simple average annual growth rates, these data were expanded to cover the period through to 1999. An additional variable, "total cost of impoundment", was calculated by multiplying the total volume of tailings generated by the impoundment cost per unit. In 1999, the total amount of tailings generated by small-scale mining was 9.2 million metric tons, while the cost of impoundment per metric ton was P107. Using these data, the total cost of impoundment was P986.5 million.

The cost of impounding the tailings from small-scale mining would be shared by mill operators, CIP operators and miners all over the country. Although the total quantity of tailings produced at amalgamation operations is not known, they should nevertheless share a substantial part of the cost, as they are the country's main gold processors.

The average volume of wastes produced per month by the mills examined in the two case study sites is not known because the miner responses were inconsistent, and, therefore, unreliable. However, the average cost of impoundment of P107 is certainly an accurate assessment, and costs could be significantly less if improvements were made to existing tailings ponds.

Final assessment of total costs of preventive equipment and facilities

Table 31.13 summarizes the total (annual) costs of the equipment and facilities needed to prevent future mercury pollution contamination from small-scale gold mining countrywide. The total estimated costs are as follows: hand gloves, P125 million; mercury retorts, P37 million; and tailings ponds, P789 million. The total cost, therefore, would be in the range of P933.5 million.

Table 31.13 Estimated costs for mercury pollution prevention in the Philippines.

Item	Cost (million pesos)
Hand gloves	107.5
Mercury Retorts	37
Tailings Ponds	789
Total	P933.5

RECOMMENDATIONS

Although the laws in place for regulating mercury pollution at small-scale gold mines are adequate, accompanying monitoring and enforcement is weak. The following recommendations are therefore made:

License all small-scale gold mining and processing operations, and impose, as a licensing requirement, membership in a cooperative. Licensing provides legal status to miners and processors, helps to organize them into cooperatives, and facilitates common efforts for improved environmental management. An added advantage of cooperatives is that they promote better marketing of the gold produced by the miners.

Earmark licensing proceeds for the establishment and operation of a small-scale mining monitoring and enforcement unit within the management framework of LGUs. Together with other relevant national and local enforcement units, this office would function to apprehend violators and impose appropriate penalties upon them.

Develop an effective system to enforce the use of gloves, mercury retorts and tailings ponds in the industry. Miner cooperatives should impose a fee upon their members for the purchase and distribution of hand gloves and other protective equipment. Processor cooperatives could impose additional – albeit, affordable – fees for the use of mercury retorts and appropriate tailings ponds.

Active participation of Non-government Organizations (NGOs) and other responsible members of the local population in monitoring and enforcement activities. Having these watchdogs on board would lower the cost of environmental management for LGUs, and would improve the monitoring environment overall.

Strengthen the Small-Scale Mining Section of the Environment and Safety Division of the MGB. This section has staff at both its national and regional offices. Its reinforcement is therefore essential because it is the national government office that oversees small-scale mining, and which liaises with LGUs.

Conduct educational and awareness campaigns on mercury pollution (a responsibility that should be assumed by the national government, LGUs and NGOs). Generally, miners are not fully knowledgeable about the health risks associated with mercury use. Heightened education and awareness is therefore a key to making them more voluntarily compliant.

Involve international organizations in the fight against mercury pollution, particularly the promotion of preventive technologies. A case in point is the

United Nations Industrial Development Organization (UNIDO) project that promotes the use of mercury retorts among miners in Diwalwal.

These actions, however, are geared specifically toward preventing *future* mercury pollution in the Philippines. To ameliorate existing mercury pollution problems, which would be an extremely costly task to undertake, and a subject beyond the scope of this chapter, the government must emphasize the dredging of affected rivers, soils and associated waterways.

CONCLUSIONS

To recapitulate, the study reported in this chapter examined selected financial aspects of mercury management in the Philippine small-scale gold mining industry. To date, a number of landscapes and waterways have been contaminated by mercury originating from small-scale gold mining activities. The key to preventing mercury pollution is twofold. First, the government must improve mercury monitoring practices and begin enforcing applicable environmental legislation. Secondly, a series of "hardware" changes are needed, including the increased use of protective gloves and mercury retorts, and the construction of robust tailings ponds.

REFERENCES

Akagi, H., Malm, O., Kinjo, Y., Harada, M., Branches, F., Pfeiffer W. & Kato, H. (1995). Methylmercury pollution in the Amazon, Brazil. *The Science of the Total Environment*, 175(2).

Aula, I., Braunschweiler, H. & Malin, I. (1995). The watershed flux of mercury examined with indicators in the Tucurui Reservoir in Para, Brazil. *The Science of the Total Environment*, 175(2).

Bacani, M.M., Jr., Zabala, R.R. & Aler, J.M. (1996). *Small-scale gold mining operations in Mt. Diwata, Monkayo, Davao del Norte.*

Balce, G.R. & Cabalda, M.V. (1992). Impact of mineral resources development on the environment – Philippine setting. *Prospects for Mining to the Year 2000.* Proceedings of the Seminar on Prospects for the Mining Industry to the year 2000, New York, 1989.

Barbosa, A., Boischio, A., East, G., Ferrari, I., Goncalves, A., Silva P. & Da Cruz, T. (1995). Mercury Contamination in the Brazilian Amazon. Environmental and Occupational Aspects. *Water, Air, and Soil Pollution*, 80, Nos. 1–4.

Bayle, R.L. (1995). *Small-scale mining in the Philippines: Situation and potentials for community-based upland resource management.* Paper Presented at the Third National Conference on Research in the Uplands Held on September 4–9, 1999 at SEARSOLIN, Cagayan de Oro, Philippines.

Breward, N. (1996). *Mercury and other heavy-metal contamination associated with gold mining in the Agusan River Catchment, Mindanao, the Philippines.* British Geological Survey, Overseas Geology Series Technical Report WC/96/61/R, Keyworth, Nottingham, United Kingdom.

Dhar, B.B. (1994). Small-scale mining in the Philippines with special reference to gold rush: Problems and issues. In K. Ghose, (Ed.), *Small-Scale Mining A Global Overview*. Proceedings of International Conference on Small-Scale Mining Organized by Mining, Geological and Metallurgical Institute of India at Calcutta from 3 to 5 October, 1991.

Environmental Control Section, Mining Technology Division (1986). *An Environmental Study of the Small Scale Gold Mining Activities at Diwalwal, Monkayo, Davao del Norte*. Mines and Geosciences Bureau, Department of Environment and Natural Resources.

Guimaraes, J., Malm, O. & Pfeiffer, W. (1995). A simplified radiochemical technique for measurement of net mercury methylation rates in aquatic systems near goldmining areas, Amazon, Brazil. *The Science of the Total Environment*, 175(2).

Javelosa, R.S. (1997). *Estimation of the geologic mineral reserve of the small-scale gold mines in the Philippines*. National Statistical Coordination Board. Makati City.

Malm, O., Castro, M., Bastos, W., Branches, F., Guimaraes, J., Zuffo C., & Pfeiffer, W. (1995). An assessment of mercury pollution in different gold-mining areas, Amazon, Brazil. *The Science of the Total Environment*, 175(2).

Mahinay, M.S., Manguray, A.C., Pico, V.T. & Ruiz, R.J. (1998). Assessment of mercury pollution in gold processing plants and riverine environment in Mt. Diwalwal. *KIMIKA* 14(2).

Mines and Geosciences Bureau (1980–1997). Mineral News Service.

National Statistical Coordination Board. (1999). Special Studies on Small-Scale Gold Mining Activity. ENRA Report 3c.

——(1998). Report on Environmental Degradation Due to Economic Activities I: Agriculture, Fishery and Forestry, Manufacturing, Mining.

——(1993) and (1999). *Philippine Statistical Yearbook*.

Porvari, P. (1995). Mercury levels of fish in Tucurui Hydroelectric reservoir and in River Moju in Amazonia, in the State of Para, Brazil. *The Science of the Total Environment*, 175(2).

Santelices, E.B. (1994). *Estimation of production, tons mined and tailings generated by the small-scale gold mining activity*. Study Conducted as Part of the Overall Effort of the National Statistical Coordination Board (NSCB) through the Environmental and Natural Resources Accounting (ENRA) Programme of the Integrated Environmental Management for Sustainable Development (IEMSD).

Stewart, D.F. (1994). Prospects and problems: Small-scale gold mining in Papua New Guinea and the Philippines. In K. Ghose, (Ed.), *Small-Scale Mining A Global Overview*. Proceedings of International Conference on Small-Scale Mining Organized by Mining, Geological and Metallurgical Institute of India at Calcutta from 3 to 5 October, 1991.

Veiga, M. (1997a). Mercury in Artisanal Gold Mining in Latin America: Facts, Fantasies and Solutions. Paper Presented at the UNIDO – Expert Group Meeting – Introducing New Technologies for Abatement of Global Mercury

Pollution Deriving from Artisanal Gold Mining, Vienna, Austria, July 1–3, 1997.

——(1997b). Introducing new technologies for abatement of global mercury pollution Phase II: Latin America. In United Nations Industrial Development Program Document. Final Version.

Veiga, M., Meech, J. & Hypolito, R. (1995). Educational measures to address mercury pollution from gold-mining activities in the Amazon. *Ambio,* 24(4).

Veiga, M. & Meech, J. (1995). Gold mining activities in the Amazon: Clean-up techniques and remedial procedures for mercury pollution. *Ambio,* 24(6).

Williams, M., Apostol, A. & Miguel, J. (1995). *Mercury Contamination in Artisanal Gold Mining Areas of Eastern Mindanao, Philippines: A Preliminary Assessment*. British Geological Survey Technical Report WC/95/72/R, Overseas Geology Series.

World Bureau of Metal Statistics. January 1987–December 1998. World Metal Statistics. Ware, Hertz, England.

32

The Evolution of Mining Policy in Developing Countries: Seven Generations in Indonesia's Contract of Work System

HAMID ETEMAD AND KAMALEDDIN S. SALMASI

In developing countries, mineral policies have experienced several cycles of major change over the past three to four decades that have affected small-scale mining (SSM). Each cycle has been a reaction to socio-economic change, and has resulted in a different perception (on the part of the main players in the global mining industry) of the prevailing environment. Under the pressure of global competition, large mining enterprises – mainly, multinational enterprises (MNEs) – have lobbied for progressively larger concessions and greater privileges from local governments. They have pointed to more competitive offerings from other countries, developing and developed alike, and have often threatened to search for greener pastures elsewhere. To the host government, on the other hand, mineral resources represent a national heritage, successfully preserved for many generations. In the past, this concern, combined with the ownership legacy of subsoil mining resources, often gave governments the impression that they possessed greater bargaining power than they actually had, leading many to peruse aggressive policies of taxing away large portions of the proceeds from mining activities, particularly those operated by large foreign companies. The balance between these perceptions and reality has fluctuated many times, each wave of changes impacting the direction of mining policy.

Consider, for example, the 1950s and 1960s, during which periods government regulations restricted foreign investment, required higher levels of national ownership and control, and (in the late-1960s and early-1970s), placed limits on the repatriation of profits (Walde, 1983; Daniel, 1993). By the mid-1980s, however, several factors, such as the worldwide economic recession of 1981–82, the worsening of real terms of trade, a dramatic reduction in the intensity of metal use (due to search for, and actual substitution of, composites for metals in response to the volatility of mining policies and metal prices) had precipitated a slow paradigm shift; the emerging paradigm would focus more on private capital initiatives and the privatization of public mineral enterprises (Morgan, 1994). This trend led to a gradual change of 1960s and 1970s policies, with policy emphasis shifting from increased state participation and control of mining industry,[1] toward privatization and denationalization.[2]

Similarly, the impact of policy change on the SSM sector has not been unidirectional (*i.e.*, it has followed similar direction to that of large enterprises but with some notable exceptions). In certain countries, SSM enterprises are a part of the informal sector, and therefore, remain relatively immune to formal policy change and challenge (ILO, 1999). In other countries, however, governments view the sector's operations as smaller-scale enterprises that need to be protected. In such cases, the laws and regulations in place for mining are not equally applied or as tightly enforced, for the purpose of providing the sector with necessary flexibility. The discussion presented in this chapter, while recognizing that it has taken on a multitude of definitions, treats small-scale mining as the highly labour-intensive, low-tech, rudimentary branch of the industry.

The initial working hypothesis of this chapter is that the historical cycles of conflictive relations between international mining companies and host governments, as briefly highlighted above, have gradually come to a peaceful rest. More specifically, instead of playing brinkmanship to gain a few percentages from the other party, the large mining enterprises and host governments have come to manage their differing orientations with much less conflict and more cooperation than ever before.

Following this introduction, the chapter will briefly review the relevant theoretical underpinnings of mining investments in general, as well as SSM in terms of its operational risks and rational returns. This review, which is anchored heavily on mining policy in developing countries, seeks to show that the emerging cooperative policies are superior to the conflictive practices of the past, regardless of the size of mining operation. The chapter uses Indonesia's cooperative and flexible system of "Contract of Work" to illustrate the case. It proceeds to suggest that a shift in paradigm for policy formulation in mining operations, from a "zero-sum" to "constant and positive sum type of games", and their corresponding operating policies, are long over-due. The implications of this shift are then discussed and concluding remarks are offered.

LITERATURE REVIEW

The history of mineral policies in developing countries

Mines are highly capital-intensive, of high risk, and have long lead and exposure times (Bilodeau & Davidson, 1992). These conditions take on special characters

[1] There is an increasing de-mystification in the ownership providing for control paradigm. In fact, there is no strong evidence to suggest that ownership is necessary for control. Unfortunately, developing country policy makers are still viewing ownership as the necessary condition for control.

[2] De-nationalization is a term used to signal the reversal of previous processes of nationalization or take-over of otherwise privately held mining enterprises.

of their own in different developing countries, where the mining sector has been viewed historically as an engine of growth, contributing substantially to society, government revenues and foreign exchange receipts.

During the 1950s and 1960s, rapid growth was experienced in the mining sector in both developed and developing world economies, largely because of the increased worldwide demand for ores and metals fuelled largely by the general development of heavy industries and elevated production of capital goods (Bomsel, 1990). The increased economic benefits of, and returns to, the mineral industry attracted attention, and led to increased government involvement in the control and operations of mining projects. By 1980s, State-Owned Mining Enterprises (SMEs), which had collectively produced negligible quantities of mineral in the early-1950s, were responsible for the production of almost one third of the world's minerals (Radetzki, 1985a). Mineral output from the SME sector was even higher in developing countries. For example, in 1981, government equity holdings in Lesser Developed Countries (LDCs) reached 41% in bauxite, 58% in copper, and 62% in iron ore (Radetzki, 1985).

In the late-1960s and early-1970s, many national governments began envisioning, and thus, promoted, increased state control of mining: government regulations were introduced to restrict foreign investment; there existed greater national ownership and control; and limits were placed on the repatriation of profits (Walde, 1983; Daniel, 1993). These changes exposed foreign investors operating in the mining sector to even higher real and perceived risks; higher commensurate rewards were therefore required but did not materialize (mainly due to the oppressive and unstable regulatory environment). Nationalization of mining operations in, for example, Chile, Ghana, Guyana, Bolivia, Zambia and Zaire were direct manifestations of higher desires for the local control of mining operations on the part of host countries. Thus, in the 1970s, Foreign Direct Investment (FDI) began decreasing in the mining sector because of higher perceived risks.

However, several other factors, such as the world economic recession of the early-1980s, the worsening of real terms of trade, and a dramatic reduction in the intensity of metal use (due to the substitution of composites for metals) contributed to an eventual paradigm shift. The paradigm that emerged in the 1980s was that of the promoting private investment in mining. More specifically, there was a promotion of capital investment to public funds, and the privatization of publicly-held mineral enterprises (Walde, 1988; Morgan, 1994). The combination of softening demand (due to both recession and substitution) and a shift in the philosophical stance worldwide in favour of private sector and market regulation (Etemad, 1999), reversed the previous trend of high state participation and control of mining. The higher efficiencies associated with private sector operations in a free market regime facilitated privatization and denationalization, including a reliance on others (*i.e.*, inter-dependent mining operations by national and international enterprises) to slowly constitute a different *modus-operandi*.

The manifestation of the policy reversal, necessitated by the shift, is clearly detectable in the ensuing revisions of mineral industry investment laws and regulations enacted in different countries, beginning in the late-1970s. Primary examples include: Columbia, Chile, Egypt, Argentina, and Indonesia (1977); Cameroon (1978); Nigeria, Burundi, China, the Philippines and Mauritania (1979); Sudan and Malaysia (1980); Djibouti (1981); Liberia (1982); Algeria, Burkina Faso and Ivory Coast (1985); Ghana and Zaire (1986); and Nepal and Guinea in 1987 (Gabre-Maryam, 1989; Morgan, 1994). Some ninety countries revised their mining codes during the 1980s and 1990s, and many others are still in the process of introducing major amendments (Eggert, 1997).

While reasserting permanent sovereignty over national resources, the evolving legislations were invariably designed to ease, if not reverse, the policies of the past. They certainly served as mitigation against the higher perceived risks, uncertainties and complexities inherent in mining. Typical mining legislation of the present provided, for example: (a) a set of competitive financial incentives, in contrast to past punitive provisions (e.g. including, but not limited to, longer tax holidays, lower tax rates, accelerated depreciation, lower royalties in the initial years of operation, exemptions from withholding taxes on dividends and interests, tax credits for domestic purchases, and guarantees for foreign exchange remittance of profits); (b) fair and equitable treatment encouraging, as opposed to hampering, FDI , codified as "national" or "non-discriminatory" treatments (*e.g.* tax treatment for FDI, equitable compensation in the event of expropriation, settlement of disputes through international arbitration bodies instead of domestic courts); and (c) streamlining of the bureaucratic procedures specific to mining operations (*e.g.* codification and simplification of regulations pertaining to licenses and permits, the relationship between the license-holder/operator and the state on one side, to the relationships between the land owners and the mineral rights holders on the other). All of these revisions were, and still are, aimed at reducing perceived risk and re-attracting foreign investment in order to achieve real increases in mining activity. Although there is no substantial information on the impact of such stiffening and softening cycles on mining regulations relevant to SSM, it is safe to assume that operations were not entirely immune.

Particular characteristics of mining operations

An extensive literature on the state of the mining industry world-wide (*e.g.* Kumar & Amaratunga, 1994; World Bank, 1996; Eggert, 1997) points to several inescapable facts: (a) that mining, which was formerly an isolated industry with its own laws, regulations and anomalies, has become an integrated part of global trade and investment systems; (b) that exploration investment, which, at one time, was concentrated in a handful of mainly developed countries, has slowly shifted to the developing world; (c) that the industry is still controlled by a small group of multinational mining enterprises based in mainly industrialized countries; (d) that mining projects in developing countries are competing for the same resources responsive to, and governed by, global market forces; (e) that SSM operations, which used to be shielded and hence, completely protected, from most global market influences, have involuntarily become an integrated part of the worldwide supply chain of mineral commodities, and are now forced to play by global market rules; and (f) that the host governments, which used to protect most of their national industries, are also compelled to play by rapidly-emerging global investment provisions, such as those constituted by the World Trade Organization or advocated by now defunct Multilateral Investment Systems (MIS) supported by most highly-industrialized countries in the OECD. As a direct result, host countries and mining projects are compelled to reform their operating conditions to globally-acceptable operating standards, including competitive risks and returns, in order to operate normally (*e.g.* raise capital in global capital markets, respond to global opportunities, and deal with potential threats just like other enterprises in the global system, irrespective of location and size).

"Level playing field," "national treatment" and "reciprocal treatment" are therefore no longer abstract concepts, and have, in fact, become a part of the normal operating vocabulary of investment, helping to guide worldwide operations regardless of industry-type and location. Small-scale enterprises in general and small-scale mines in particular were favoured by the sins of omission on commission (*i.e.* not included in mining legislation, and favoured, if not protected, by their national governments in the past for well-justified reasons at the time). However, sustaining such a discriminatory system proved exceedingly difficult. Not only did foreign operators demand similar treatment under, for example, a "level playing field" or "national treatment," but their large national counterparts also expected nothing less. Consequently, and, in spite of their will, small-scale mines are trusted into, and are affected by, market forces extending far beyond their control. Despite the superimpositions of these difficult operating conditions, from which they were traditionally protected, small-scale mines generally thrive in developing countries (Etemad & Salmasi, 2001).

Characteristics of small-scale mining in perspective

Small-scale mining operations have become extremely important, particularly where capital and expertise is constrained – the case in most developing countries. Their collective benefits generally materialize much faster and easier than those of large-scale projects, largely because of the ease of a nation supporting greater numbers of small to medium-sized mining operations; larger operations require heavy capital investment as well as skilled manpower, infrastructure, and associated services. Small-scale operations tend to engage a large number of mineral services suppliers that would otherwise not exist in the presence of large-scale operations. They also have smaller social costs and impose much smaller environmental disturbances than their larger counterparts. They are able to begin operation more quickly and through a variety of business organizations, from individual proprietors and partnerships to medium-sized companies. Small-scale operations are known to be flexible in their hiring practices, and even offer seasonal employment opportunities to local farmers and others, whereas larger mining operations cannot cope with varying and transient labour forces. When properly developed and controlled, SSM can provide many of the benefits brought about by large-scale mining but at lower costs. Primary examples include contributions to foreign earnings through exports, and the promotion of local entrepreneurship as well as regional development. Although they may not achieve the economies of scale and utilize the more costly sophisticated technology associated with large-scale mining, small-scale mines can easily attain other national objectives at lower financial, environmental and temporal costs; such has often been sufficient justification for their protection in the past.

Entrepreneurship both in the mining sector and indigenous regional developmental plans is among the main national developmental priorities of most mineral-endowed countries. In contrast, most large operators, especially foreign investors, frown upon the limits placed on local participation rates or regional development requirements. Conversely, small-scale mines, generally because of their smaller size, are much more sensitive to their operating environment than their large-scale counterparts. Among the most prominent factors impacting operations directly include policies pertaining to mining law, royalty and taxes, export regulations, and foreign investment laws (Kumar & Amaratunga, 1994; World Bank, 1996). These factors are capable of tilting the balance in favour, or against, small-scale mining operations.

THE CASE STUDY OF THE CONTRACTS OF WORK

According to the Constitution (as well as the Mining Law) of Indonesia, all minerals in the form of natural deposits in the country are part of the national wealth of the people and controlled by the state; therefore, only the state can hold exclusive mining rights. However, Indonesian mining companies and citizens may conduct mining based on mining authorizations known as "Kuasa Pertambangan" (KP). Foreigners are not allowed to hold KPs and any holder does not own the land but is authorized to conduct mining on the land (Directorate General of Mines, Ministry of Mines and Energy of Indonesia, 1998).

The Contract of Work (CoW) was developed as a solution to permit foreign investors to develop mineral properties and share in the benefits of such activities, while at the same time, adhering to the words and the substance of the Constitution of Indonesia. The nation maintains ownership of the mineral rights while contracting out the work to the local or foreign investor. What is particularly noteworthy about the CoW system is that the contract is highly regarded, and valued, despite the fact that mineral rights are not granted to the investor. In international mining circles, it is rare for an investor to work on property for which it does not hold a firm title. Under the CoW system, the company is simply appointed by the government as the sole contractor to explore, evaluate and develop mineral resources according to provisions (of the CoW) and within a designated area; the project is to be financed wholly by the company at its own risk.

A typical CoW represents an agreement forged between the government or a SME, and the contractor (an independent foreign firm or a joint venture with Indonesians) to carry out some, if not all, stages of mining, from geological surveys to extraction and marketing. The Mining Law does not specify the content of the CoW. Responsibility lies with the Department of Mines for non-fuel minerals, and with the state-owned company, PT Tambang Batubara, for coal. To execute the CoW within the mining sector, government have established some SMEs. Private enterprises may also act as mine operators or contractors for a state enterprise. However, contractors are required to bear all risks associated with the mining venture (EIU, 1998; Price Waterhouse, 1997).

The CoW is a remarkable document in that it covers just about every legal aspect of exploration and mining that must be covered in an agreement between the state and the private contractor. It has two critical provisions that make it particularly attractive to the investor: (i) the "conjunctive title" which gives security of tenure to the investor; and (ii) the *lex specialis*, which specifies the treatment of a given contract by the Government. The latter states that the contract terms pronounce the expected Modus-Operandi and override the general law (Sigit & Yudonarpodo, 1994). For example, the income tax provisions that are set out in the CoW take precedence over the law on income tax. Generally speaking, the tax rules included in the CoW reflect the general tax rules in effect at the time the contract is signed, although there are some exceptions. One noteworthy point is that a typical CoW limits future volatilities by, for example, freezing the taxation rules for the duration of the contract, again, with some exceptions. The downside risk to this is that the company may not always be permitted to take advantage of the more favourable changes in the general law as it evolves.

A critical examination of conditions and terms of agreements in various CoWs reveals systematic changes made to Indonesia's policy on foreign investment in the mineral sector over the past three and-a-half decades. It also portrays the actual process of planning, change, and implementation in the policy environment for foreign investment. A comparison of the provisions of the various generations of CoWs is provided in Figure 32.1, which points clearly to the evolutionary path of a progressively more cooperative policy environment, from one generation of CoW to the next, in the past three to four decades.

To date, seven generations of CoWs have been implemented in the mineral sector of Indonesia (January 2003). The first generation of CoWs was introduced in April 1967, and led to the development of the first copper-gold mine in Irian Jaya. In the second generation, 1968–1974, 16 CoWs were signed, of which six culminated in operating mines. The third and fourth generations of CoWs (covering the periods of 1976–1984 and 1985–1990, respectively) required the contractor to offer shares in the company to local enterprises, commencing with 5% in the third year of production, and increasing up to 15% by the fifteenth anniversary of the contract.

Retrospectively, the CoW program recognized the role of Indonesian SSMs, mandating them to participate in major mining operations. Furthermore, the Indonesian policy formulation provided for an equal footing for SSMs in terms of partnership, specialization and a division of labour, as opposed to the traditional practice of total separation between SSMs and large mining companies. The protection of environment was also regulated for the first time in this generation. The debt-to-equity ratio requirement for issuing common shares was also

Description		Generations 1-2	Generations 3-4-5-6-7	The Current Generation
Physical Operating Conditions	Exploration Period	18 months (G.1) 3 years (G. 2)	1 year + 2 years extension	The Same
	Extraction Period	30 years		The Same
	Construction Time Limit	6-8 years		The Same
	Protection of Environment	Not regulated	Regulated based on the rules and regulations (During the G. 4-7, law was not changed)	New restricted regulation
Financial Conditions	Financing	In G. 1, Not regulated. In G. 2, company under the Cow can publish shares when its Debt to Equity (D/E) ratio is 70 /30	Company under the CoW can publish shares with different D/E ratios: 60/40 (G. 3), 75/25 (G. 4), 5/1 for investment equal or less than $US 200 million, and 8/1 for investment more than $US 200 million (G. 5-7). All rates are fixed at the outset	Basically the same
	Depreciation Rate	12.5% Straight Line (SL)	10-50% (SL) For different assets	Basically the same
	Tax Holiday	3 years	Not regulated	
	Import Duty	Free	Free for the first 10 years	
	Royalty	Royalties are fixed by the Minister of Mines for each CoW Generation, and are the same for all CoWs. (Rates have not changed for G. 4 to G. 7)		Increase in royalties
	Regional Tax	Free	Regulated according to the stages of mining, type of the assets, and area of the operation	Unclear, will be based on the region's decision
	Export Tax	Free	Applies only for copper	
	Company Income Tax	Different rates for different years of operation (G.1), or different levels of income (G. 2 to 7). Tax incentives for investing in remote areas from the fifth generation. Tax cuts in the sixth and the seventh generations. All rates are fixed at the outset.		Not clear yet
	Foreigner Individual Tax	Similar to the tariff stage in origin country	For G. 3, based on tax regulation in Indonesia, and For G. 4 to 7, free if the foreigner works more than 183 days a year in Indonesia	
Rights and Requirements	Security of Tenure	Not regulated in generation 1	The company has sole right over the CoW area (G. 2 to 7)	Probably will remain the same
	Transferring the Right	Transferable with Government's approval (G.1, 3 to 7), Transferable with partner's approval (G. 2)		

Figure 32.1 Summary comparison of seven generations of CoWs, and current situation.
Source: Compiled from (Aspinall, 2000).

changed. Other changes included modified depreciation rates, and regional and export taxes (see Fig. 32.1); nearly all of these changes were incorporated into the fourth generation of CoWs.

The fifth generation of CoWs (1991–1994), also known as the *Frontier Contracts,* aimed at attracting foreign investment to the remote areas of the eastern regions of Indonesia. More tax incentives were offered to foreign companies during this period. The sixth generation (signed en-masse in 1997) was primarily intended to accommodate the provisions of the 1994 *Tax Law*, which came into effect in January 1995. The main feature of the law was the reduction of corporate tax to a maximum of 30% from the previous 35%. Other incentives included the acceleration of depreciation allowance, and a lower withholding of taxes for dividends.

The seventh generation of CoWs (signed en-masse at a ceremony in February 1998) is very similar to the sixth. However, it provides for greater government control of the contractor's performance throughout each stage of mining activity.

NEW REGULATIONS AND POTENTIAL IMPACT ON FUTURE VARIANTS OF THE CoW

Mining investment in Indonesia experienced stability from 1967 to 1997 – uncharacteristic behaviour of its mining sector – because investors were convinced that the CoW framework provided a conducive environment for investment, and that its terms would be respected by the government. Several Ministerial Decrees declared that a CoW was *lex specialis* and that the bureaucracy administered it according to its terms. However, at the beginning of the sixth generation of CoWs, the rules had changed.

Mining investors are currently watching carefully the ongoing review of several laws and regulations in Indonesia as they unfold. If they are not amended to encourage mining investment, the growth of the Indonesian mining industry may begin to experience a slow down or decline from its present position. The following laws and regulations are currently under revision:

1) The Law on Regional Autonomy (1999)
2) The Law on Forestry (Law # 41, 1999)
3) The Regulation on Mineral Royalties (1999)
4) The Toxic Waste Regulation (PP # 18 and 85) issued in 1999
5) The PP (regulation) # 75, 2001

The issuance of Laws # 22 and 25 on January 1st, 2001, which provide autonomy to the regions to negotiate and administer a CoW, appear to be a departure from some 35 years of the exemplary evolution of the CoW governance structure. The autonomous regions do not have the manpower, expertise, or finances in place to deal with the complexity of the mining industry's administrative demands. In some cases, there have been calls for provinces and regencies to enact additional taxes on mineral production at the regional level (Wahju, 2002), which would undoubtedly be viewed as a reversal of the past trends.

Although the mining industry supports the principle concept of regional autonomy in Indonesia, companies are still in need of more information and additional clarification on future action in the event of regional autonomy.

IMPLICATIONS AND CONCLUDING REMARKS

The case study of Indonesia clearly indicates that attracting investing companies to the mineral sector is dependent both on the quality of available resources and, perhaps even more importantly, on the general socio-political and legal environment of the country as well as the proper enforcement of policies by local authorities, mostly outside of the investor's control. Once investment is in place, the investor, regardless of size and nationality, is at the mercy of both the essence of policies and enforcement. While large enterprises can tolerate deviation for a longer period of time, SSM enterprises find themselves almost helpless. As a result, SSM enterprises tend to cooperate more readily with the local authorities and remain more flexible to avoid losses potentially catastrophic to their existence.

As for the bargaining power side of the equation, while the relationship of the investing MNE, with its home government as well as its international standing (in terms of the true strength of OSA and the size, structure and control of internal market and its global network), play significant roles, SSM enterprises find themselves much less powerful to engage in conflictive situations. They generally adopt a cooperative and much less confrontational stance than their larger counterparts. Furthermore, their flexibility, expertise, and specialization – attributes often associated with industries operating on smaller scales – enable SSMs to master work in their favour. In fact, local SSMs are capable of providing smaller, yet, highly effective, business propositions to local authorities that their larger and especially international counterparts cannot. For example, SSMs can establish a web of *strategic alliances* with local buyers and suppliers, including mining services, labour and even value-added processing services, and thereby deliver much to the local economy. SSMs can also be more socially- and locally-responsive, which enables the delivery of higher benefits than those formally required by a typical contractor. These are some of the areas towards which both the governments and SSMs can readily aspire to.

As evidenced by the large number of foreign companies attracted to Indonesia prior to the move towards regional autonomy, not only has the CoW system served the country well in terms of developing world-class mining operations but it has also proven that indigenous SSM enterprises can operate symbiotically and synergistically with larger MNEs. Most CoW companies have Indonesian partners and a substantial portion of the profits remains in Indonesia. Moreover, the expenditures of CoW companies have created valuable opportunities for small local suppliers; in addition to providing income taxes and royalties, these smaller companies provide a stable employment base and the solid beginnings of a local mining and mineral services industry in Indonesia.

The positive response of foreign mineral investors illustrates the efficacy of the CoW system in Indonesia. The stability and predictability of contract, the credibility and good will that the Indonesian government has accumulated over the past 35 years, combined with the quality of the country's natural resources, have made mining successful in the country. These must be preserved if the country is to continue to build upon its now-marketable track record of economic successes.

This study confirms that it is not the body of a very rich mineral deposit alone that entices a firm into making a long-term investment but rather the general investment climate and the cooperative nature of the local or national authorities, which magnifies (or diminishes) the value of the ore body and attracts (or discourages) the investment. In fact, the impact of local authority can enhance a weak case or weaken a very strong case. Indonesia serves as a vivid example of such regulatory impact. In spite of some thirty-five years of stability in the policy environment, the political turmoil of the past three years, along with fundamental changes in government policies and regulations, mineral investment in Indonesia has declined dramatically. While even richer deposits in unexplored regions are becoming available, the shift of authority from the central government to provincial governments and regencies, which have pronounced their short-term schemes for rapid development, has added an element of risk previously absent. In direct response, the investing community has adopted a "wait-and-see" attitude until political stability replaces the turmoil of the past three years.

The inevitable requirement of successful investment appears to be cooperative and trustworthy relations between the investing enterprise and the government. Such relations can only occur if both acknowledge each another as interdependent entities, are open and candid in discussing problems, and are committed to finding joint solutions for joint action, as opposed to relying on unilateral action based on their own bargaining (in the case of MNEs) or regulatory powers (in the case of host governments). The end result for both parties should be the attainment of mutual goals, at the same time, accommodating the needs of the other party. However, trust takes time to build; activities associated with early exploration and development can play an important role in the building of trust.

In terms of mining, the ultimate goal of a government is optimal exploitation of a mineral resource, which provides maximum benefit to a country both economically and socially. The ultimate goal of the company, on the other hand, is to obtain an acceptable rate of return from its operation comparable to that of an investment elsewhere, at the same time, being recognized as a respectable corporate citizen of the country in which the operation is located. Although conflictive on the surface, especially when viewed within the zero-sum frameworks that prevailed in the past, mining enterprises and governments have many common interests. They must, however, search for, and build upon, mutually interdependent, if not symbiotic, synergistic relations (Etemad, 2003) to achieve them. Changing policies, the introduction of an element of volatility and insecurity, the lack of direction and ignorance of the other side's aspirations and problems are all potentially damaging to trust on both sides. Credibility, fairness, trust and a host of other concepts relating to the general phenomenon of confidence in the other party are the catalytic ingredients of symbiotic and synergistic relations in this context.

Stability is another important factor. The governments lacking basic policies, clear direction, and which continue to deal with different companies inconsistently, are not attractive options to foreign investors, as they contribute to the heightening of perceived risks. This is why it is important that decisions are made fairly and transparently, especially when it comes to enforcing one party's goals and objectives through the development of mining policies. These are among the fundamental prerequisites of attracting the investor and preserving trust. On the other hand, companies must also ensure that their positions are realistic, and take the host nation's fundamental objectives, along with cultural and social morals, into account, and respond to the prevalent values and aspirations during the course of their tenure in the country.

Finally, the concept of rule of law – a universally accepted principle upon which fair transactions are based – appears indispensable. The rule of law is achieved through constitutionalism. All citizens and governments should stand equal and be subjected to such rule of law, which basically means that actions of government will have to conform to the internationally-acceptable body of legislation, and that individual authorities will not be permitted to prejudge a person or property. Such rule of law helps to ensure fairness, equity, transparency, and security of property and individual rights, irrespective of size, social position or international bargaining power. Any change in contractual provisions raises the perceived risk of mining investment to higher levels. This becomes even more important to SSMs where their powers are limited, resources are constrained, and sustainability is at the mercy of policy fluctuations and associated losses.

The creation and maintenance of a predictable policy and regulatory environment is therefore conducive to building mutual trust and joint action, particularly within the countries in which the stability and rule of law as defined above – perceived as lacking or difficult to maintain – becomes indispensable. In the final analysis, the principles that allow for building trust and facilitate increased cooperative and joint policy actions, as opposed to conflictive and unilateral actions and reactions, may help parties achieve mutually beneficial goals and objectives in the longer term, even in situations where the law proves unaccommodating in the short term.

REFERENCES

Aspinall, C. (2000). *Summary comparison of contracts of work from first generation to seventh generation*. Clive Aspinall's Report.

Bilodeau, M.L. & Davidson, J.I. (1992). *Special characteristics of mining activity: Implications for taxation international seminar on mining taxation, CIM and UNDTCD.* Montreal, Canada, September.

Bomsel, O. (1990). *Mining and metallurgy investment in the Third World: The end of large projects.* Organization of Economic Co-operation and Development (OECD), p. 13.

Daniel, P. (1993). *Evaluating state participation in mineral projects: Equity, infrastructure and taxation*. In Foreign Investment and Joint Venturing in the Mining Sector, Selected Papers, United Nations, p. 23.

Directorate General of Mines (1998). *The progress of mining contractual agreement in Indonesia.* Ministry of Mines and Energy, Republic of Indonesia, pp. 1–11.

Economist Intelligence Unit (EIU) (1998–99). *Country Profile – Indonesia.* London.

Eggert, R.G. (1997). National Mineral Policies and The Location of Exploration. *Proceedings of the Sixth Annual Professional Meeting of the Mineral Economics and Management Society (MEMS),* February 27–March 1, pp. 1–8.

Etemad, H. (1999). Globalization and small and medium-sized enterprises: Search for potent strategies. *Global Focus* (formerly *Business and Contemporary World*), 11(3), 85–105.

Etemad, H. &. Salmasi, K.S. (2001). The rugged entrepreneurs of Iran's small-scale mining. Forthcoming *Journal of Small Business Economics,* Issue 1, Spring.

Etemad, H. (2003). Managing relations: The essence of international entrepreneurship. In H. Etemad & R. Wright (Eds.), *Globalization and entrepreneurship: Policy and strategy perspectives.* Northampton, MA: Edward Elgar Publishing (Forthcoming).

Gabre-Maryam, T.B. (1989). Mineral legislation in developing countries. In *Mining Policies and Planning in Developing Countries*, United Nations, pp. 25–37.

International Labour Organization (ILO) (1999). Social and labour issues in small-scale mines. *Report for discussion at the Tripartite Meeting on Social and Labour Issues in Small-Scale Mines*, Geneva, 17–21 May.

Kumar, R. & Amaratunga, D. (1994). Government policies towards small-scale mining. *Resources Policy*, 20(1), 15–22.

Morgan, C.A. (1994). Privatization of state mining companies. *Minerals Industry International*, July pp. 13–17.

Price Waterhouse (1997). *Doing business in Indonesia.* pp. 1, 5, 6.

Radetzki, M. (1985). *The state copper industry in Zambia.* State Mineral Enterprises, Resources for the Future, pp. 129–131.

Radetzki, M. (1985a). *State Mineral Enterprises: An Investigation into their Impact on International Mineral Markets.* Resources for the Future, p. 28.

Sigit, S. & Yudonarpodo S. (1994). *Legal aspects of the mineral industry in Indonesia.* In Mineral Development in Asia Pacific Into the Year 2000, *Proceedings of the 4th Asia Pacific Mining Conference*, Jakarta, pp. 76–82.

Wahju, B.N. (2002). *Indonesian mining industry in the period of transition, between 1997–2001.* Presented in International Convention, Trade Show Investors Exchange, Prospectors & Developers Association of Canada (PDAC), Toronto, Canada, March 10–13.

Walde, T. (1983). *Permanent sovereignty over natural resources: Recent developments in the mineral sector.* Natural Resources Forum, United Nations, p. 240.

Walde, T. (1988). Third world mineral investment polices in the late-1980s: From restriction back to business. In *Mineral Processing and Extractive Metallurgy Review*, 3, 123–126, 174–177.

World Bank, Occasional Paper No. 6 (1995). *A Summary of the Proceedings of the International Roundtable on Artisanal Mining,* Organized by the World Bank, Washington, D.C. May 17–19, Industry and Energy Department, Barry, M. (Ed.), April 1996.

33

A Review of Small-Scale Mining Activity in Papua New Guinea (PNG)

GEOFF CRISPIN

Papua New Guinea (PNG) occupies a unique position in the developing world. Here, gold extraction and processing is almost the exclusive subject of small-scale mining (SSM) activities; miners are widespread throughout the country. Every single one of the country's 19 provinces has identified gold deposits or prospects for village level miners. Such widespread activity has had an impact both locally and nationally on economic and social systems, as well as the development path of the country.

Since the early part of the last century, the importance of SSM to rural communities and rural economic activity has increased. In the long term, the sustainability of these activities and on-going positive contributions to development are subject to a number of management parameters. These include limits to resources, environmental impacts, occupational health and safety issues (especially the use of mercury and future possibility of cyanide usage), legal and legislative frameworks, relationships with existing or proposed large mines, the role of women and children (See Fig. 33.1), and the level of mechanization in mining processes.

The sustainability of large mining projects has been defined by a number of parameters, some of which are also relevant to the SSM community. These are significant if SSM is to make a positive contribution to the sustainable development of PNG. The most relevant parameters include (Hancock, 2001, p. 121):

1) Finding, extracting, producing, adding value to, using reusing recycling, and, where necessary, disposing of mineral and metal products in the most efficient, competitive and environmentally responsible manner, and, where possible, using best practices.
2) Respecting the needs and values of all resource users, and considering those needs and values in government decision-making processes.
3) Maintaining or enhancing the quality-of-life and the environment for present and future generations.

Figure 33.1 Family using a basic sluice box for gold in Korunga Creek, Morobe Province.

4. Securing the involvement and participation of stakeholders, individuals, and communities in decision-making processes.

Creating a management climate for SSM in a sustainable fashion – based on these parameters – is an ongoing challenge for regulators and other important people involved in SSM.

PNG has a diverse and complex social structure, with over 700 language groups contained within a population of approximately five million. This cultural diversity is further complicated by geographical parameters; it ranges from the rugged Highlands areas, to the riverine-based Sepik and Fly River basins, to the island groups, including New Ireland, New Britain, Bougainville, various coastal groups, and the numerous communities in Milne Bay Province.

These national parameters almost certainly mean that any attempt to quantify and define the extent and impact of SSM will be a complex and difficult task. The SSM industry has been accepted into the mainstream by the Department of Mining (DoM) as a legitimate means of generating income, especially in rural and remote areas, provided that it is practiced with concern for the environment and safety.

This chapter provides a thorough documentation of small-scale gold mining activity in PNG, describing its main characteristics, its function in the rural economy, and the main assistance-related projects commissioned on its behalf.

SMALL-SCALE MINING IN PNG

SSM has been carried out in PNG since the nineteenth century. Since country independence in the mid-1970s, an increasing number of people have engaged in this occupation throughout its rural areas. It provides either a regular income or seasonal income to rural people who have been progressively marginalized. Specifically, government administration has become more difficult due to reductions in financial returns, which has led to a degradation of infrastructure, such as roads, transport, medical facilities, and educational services.

Using the Human Poverty Index, the Asian Development Bank (ADB) rated PNG at the same level as Burundi and Mali. Some 31% of the population is living below the poverty line of US$1 per capita per day. The richest 10% of the country accounts for 36% of consumption, and the poorest 50% account for just 20% of consumption. The report further suggests that there is a dualistic pattern of growth, with part of State revenues being used to support the growth of a large public service. The subsistence sector has shown little growth (ADB Report, 2000). The challenge at hand is to ensure that all sectors of the population participate in promoting sustainable economic growth.

The SSM sector had long been ignored by many rural development programs but since the Government of PNG began recognizing its importance as part of the country's small business sector, and made its development and management part of national policy, efforts have been made to identify and quantify the sector. Initially, a number of donor agencies and the private sector were supporting the innovative outreach program commenced by the DoM. The Australian Agency for International Development (AusAID) was the first to provide technical assistance and personnel training. The European Union (EU), through the SYSMIN vehicle, the Japanese Social Development Fund (JSDF), and Asian Development Bank (ADB) has also begun to administer assistance. All are aiming to provide industry assistance in a variety of ways, including the promotion of increased environmental awareness, and, in some cases, attempting to measure the effects of both current and past mining activities on various SSM communities.

There are large several large gold mines in PNG, such as Ok Tedi, Porgera, and Lihir, and smaller projects such as the Tolukuma and the soon-to-be-closed Misima. The Kainantu Gold Mine may become operational in the next two years. There are no medium-sized mines at present. The focus has always been on large mines and their economic contribution.

New polices were introduced by DoM to recognize the role of SSM, and its contribution to the rural economy. These policies attempt to manage both the industry and its impact on rural communities. The PNG DoM Draft 5 Year Plan includes, and emphasizes, the following:

> Broad policy statement by the government covering the Small Scale Mining (SSM) sector is the original PNG Government's Eight Point Plan which emphasizes developing the rural based industries with participation by PNG citizens. SSM is a legally recognised economic activity. Promotion of SSM should be done in an environmentally friendly manner beneficial to both present and future generations.

Specifically, it seeks:

a) To promote mining as a means of stimulating national development.
b) To promote and facilitate development of SSM in order to stimulate rural economics.
c) To facilitate and ensure mining operations are carried out in a manner that is technically, socially and environmentally responsible.

Some of the Provincial Governments have been agitating for control of SSM to be divested to the various regional administrations. This would mean up to 19 different mining regulatory frameworks for SSM, which could generate very complex management regimes for the entire country.

LEGAL FRAMEWORK[1]

There are three main categories of operators in SSM in PNG. They are as follows:

1) Mechanized miners – less than 1% of the miners
2) Semi-mechanized miners – less than 10% of the miners
3) Grassroots or village based miners – more than 90%

The miners within these three categories are expected to have registered formal mining leases – preferably Mining Leases (ML) and Alluvial Mining Leases (AML) – granted under the *Mining Act 1992*. Most of village level miners operate on unregistered, customary-owned land, and are recognized as formal miners according to provisions under the *Mining Act 1992*. In some areas where traditional land is being mined by outsiders, they are expected to come to an agreement with traditional landowners and pay a tribute. In such cases, a lease is expected to be registered by the landowners.

The simple, mechanized miners are not permitted to mine without proper registration because of the nature of their operations. These provisions may be changed in the future to permit traditional landowners to use small static machinery without formal registration processes.

As of 31st December 2000, the tenement registry had records for 113 AMLs and 367 MLs. For the convenience of administration, each lease is

[1] The discussion that follows is summarized from Susapu and Crispin (2001, pp. 6–11).

represented by only one miner. In particular situations, especially under earlier arrangements for business or family connections, a lease can be registered under two or three names.

Another group of miners operate on traditional or customary land; these individuals mainly use unsophisticated tools, and had been issued identification cards by the SSM Branch to enable them to mine, possess and sell gold. The system was discontinued when gold dealing was deregulated in 1987. Prior to the enactment of the *Mining Act 1992*, the number of registered miners in this group were as follows:

Mining Tenement Holders	615
Miners with ID Holders	1700
RGC Tributes	167

There is no complete record to indicate that these were the only people mining gold at the time. It may have well been that many more people were mining, and that any gold recovered was sold by the person holding the appropriate registration on their behalf; alternatively, miners could have been paid salaries (or given a share of the gold recovered), with the arrangement that all recovered gold was passed to the landowner/leaseholder.

In 1987, operations producing (alluvial) gold were deregulated to allow any person to buy and market gold. Prior to deregulation, only people possessing either a gold dealer's permit or a lease were authorized to deal in gold. Small-scale miners are no longer subjected to any sectional tax, apart from VAT. Few, if any, pay income tax on what they earn, and are therefore free to enjoy benefits from their gold proceeds.

However, a licence is required to export gold out of the country. The export licence is issued on annual basis by the bank of PNG through its Exchange Control Department. As of December 2002, only four such licences had been granted (Bank of PNG, 2002)

In accordance with the provisions of mining legislation, operators are required to submit monthly production returns to DoM. However, as most miners do not submit returns regularly, figures do not accurately reflect overall production in the SSM sector. The gold production statistics, therefore, more accurately reflect the volume of gold being exported by authorized dealers. Generally, miners utilize existing marketing facilities – namely direct sales to commercial banks, individual gold dealers or Metals Refining Operations (MRO) Ltd. (Metals Refining Operations, 2002).

ECONOMIC PARAMETERS

The proliferation of small-scale gold mining has changed the economic structure in the areas in which it is practiced. The once-predominant barter/subsistence agricultural framework has since been replaced by a cash-based or money-based system. In the Wau area, for example, it is estimated by DoM officers that 75% of the total population is involved in mining. Here, agricultural pursuits have diminished to the extent that coffee, which formerly had economic importance in the area, has all but ceased, although recently (2002), a large company did revive some of the industry. Part of the diminution of agricultural production is due to declining infrastructure maintenance, particularly roads. Diminishing prices have also played a part in the decline of the coffee industry in the area.

People still maintain gardens for domestic use but the cycle of life in many communities now revolves around mining activity, whether it be seven days a week or on a seasonal basis. There is less time available for agriculture, as the returns from gold mining are more favourable and instantaneous.

The reliance on mining has changed the micro economies where gold buyers have come to play an important part. They are authorized to buy gold on a cash basis on any day of the week. The availability of cash has changed local economic activity, as people now depend more on store-bought goods that can be readily obtained with the cash accrued from gold mining. In Wau, for example, until only recently, there were seven large trade stores supplying goods to the local community. This is many more than would normally be expected in similar populations in rural PNG.

The gold mined in PNG is rarely 100% pure. In the Wau area, for example, gold purity can be as low as 45–50%, contaminated with silver. The highest purity of alluvial gold reported to the DoM during the survey period was 97.1% (gold recovered at Ioma in Oro Province).

In the Wau/Bulolo area, gold purity varies so much that 10–15 different (varying) set prices are used. Gold recovered from Sandy Creek, for example, can be 85–90% pure, whilst Eddie Creek gold can be as low as 48% pure. Each of the buyers knows exactly where the gold comes from, and is capable of making estimations close to the value of the gold submitted. Prices vary for nuggets, flake gold, and amalgam cakes, the latter tending to have higher losses through heating because of mercury or other contaminants, which results in lower paid prices.

The major gold buyers in the Wau/Bulolo area receive information each day on the gold price and the Kina[2] value against the US dollar, so that prices accurately parallel the fluctuations in the world market. In more remote areas, however, this is not the case. There have been many examples of people taking advantage of remote miners' lack of knowledge about price movements.

In early December 2002, the price of refined gold was approximately K42 per gram, and the price for Sandy Creek gold (85% purity) in the Wau area was approximately K26 per gram. Miners, therefore, received approximately 75% of the refined gold value – a good return compared to miners in more remote areas, who receive only 20–25% of the market value of refined gold.

[2] The PNG currency is the Kina, which, in December 2001, was equal to approximately AUD $0.50. The ADB Report on Micro-finance in PNG (2001) indicated that in more than two thirds of the country's provinces, the average monthly income is clustered around K461 per year, and the monetary income is decreasing compared to non monetary income.

Based on official sales, the cash flow generated from the SSM of gold is estimated to be around K16 million/year from the Wau/Bulolo area. This volume of cash generates the need for the availability of cash for transactions both with the gold buyers and the retail outlets. It has also changed the way society operates, and because people now use cash, they need somewhere to store surplus money other than "under the bed" or "buried in a tin outside". Financial services are part of the project commenced by the ADB in the Wau area, as there are no banks in Wau, and a 60 km round trip to Bulolo to access the only bank branch in the area opens the way for criminal activity.

The estimated mining population of 60,000, despite being acknowledged as a conservative estimate, nonetheless represents about 1.25% of the population of PNG. The SSM sector has considerable economic impacts in PNG, with average earnings estimated to be in the range of K250 per month. Income estimates from gold production – based on informal observations – have been made by DoM officers. During 1999, they ranged upward to about K550/month (at gold prices, \$260-70USD/oz, and Kina parity of K1 =US\$0.35–40). These estimations have completely changed in 2002–3, and are discussed more in detail later in this chapter.

It has been estimated that including support systems and related activities, another 420,000 people depend indirectly on SSM in PNG. Thus, including both direct and indirect participation, some 9–10% of PNG's population is supported by this activity (European Union Interim Report of SYSMIN eligibility of PNG, 2001).

ILLEGAL EXPORTS

The DoM informal survey carried out over the period 1998–2001 indicates that the amount of gold recovered by small-scale miners exceeds that passing through official channels. In many cases, gold is being transported to major centres without the issuance of export permits. In the more remote areas of the country, it is easier to move gold without official records. In West Sepik, during the recent political unrest, gold was sold at a premium price in neighbouring Indonesia – exceeding that provided by commercial channels.

Based on these observations, and officially recorded production, it has been estimated by a number of sources that the country's actual gold production figure is at least double, if not, triple, what official figures indicate. In 2001, production was estimated to be between K150 and K180. These numbers are an educated guess, as it is very difficult to gather accurate data in the remote, rural areas of PNG, because of the scattered population and the rugged terrain.

ADDED VALUE

Refining a great majority of the gold recovered by SSM would help to retain maximum value in the country. Further economic gains could be made if downstream processing occurred in larger quantities. For example, jewelry is currently being manufactured on a small scale in PNG. If this could be expanded, then a greater national return could be generated from PNG's resources.

DISTRIBUTION OF MINING POPULATIONS IN PNG

In 2001, the estimated number of people involved in small-scale mining in PNG was 50,000–60,000, with a possible upper limit of between 80,000 and 100,000 people (Table 33.1). This estimate was derived from a basic head count conducted by DoM officers, who traveled around the country over a three year period (1998–2001). This head count revealed that approximately 30% were school-aged children, and 20% were women.

When the Program began in 1998, it was estimated that the mining population was between 15,000 and 20,000.

These figures, however, are not complete, as not all of the mining areas were visited, due to remoteness and rugged terrain. Even so, it is not always possible to accurately estimate the numbers of people working at any one location, as they are usually spread out and can be hidden away in small creeks and valleys beyond the major centres. It has been suggested that up to 80,000 might be a more accurate census figure (Tongo & Crispin, 2001).

Table 33.1 Estimated mining population in PNG.

Province	Estimated mining populations
Bougainville	900–1,000
Central	2–300
East New Britain	50–100
East Sepik	10–12,000
Eastern Highlands	1000
Enga	4–5,000
Gulf	3–4,000
Madang	2–3,000
Manus	20–50
Mine Bay	1,000
Morobe	15–20,000
National Capital District	Uncertain
New Ireland	4–500
Oro	4–500
Southern Highlands	2–300
Simbu	Uncertain
West New Britain	A few only
West Sepik	3–5,000
Western	50–100
Western Highlands	50–100
	4,200–53,950

Source: Susapu and Crispin 2001.

There are three main clusters of SSM activity in PNG. They are based in Wau/Bulolo in Morobe Province; Enga Province in the Highlands area; and both the East and West Sepik Provinces (Fig. 33.2). The Wau/Bulolo area has been mined since the early part of the last century, and increasingly by small-scale means since the mid-1970s.

Much of the activity in Enga Province centres around the huge Porgera Gold Mine, where miners utilize the waste from the mine to exploit unrecovered gold reserves as well as recovering both alluvial ore and hard rock mined ore.

In both Sepik Provinces, mining communities are widespread; there are a number of major centres, as well as smaller communities throughout the region, through to the Indonesian border to the west, the border with Madang Province to the east, and Western and Enga Provinces to the south.

MINING TECHNIQUES

Observations carried out by DoM over the period 1998–2001 indicate that more than 90% of miners use gold panning techniques or sluice boxes without any mechanical aids. Moreover, less than 10% use semi-mechanized equipment, and less than 1% use fully mechanized equipment such as washing plants and bulldozers.

Non-mechanised mining

The non-mechanized miners start at a very simple level, using their bare hands to pick up pieces that are easily visible. Banana leaves have been used to capture gold; they are either placed in streams or are used as very basic panning devices. Gold pans can be carved out of wood or plastic, and metal pans can be purchased from suppliers. One of the main restrictions on more widespread use of both simple and semi-mechanized equipment in the PNG SSM sector is a lack of appropriate supplies, as well as deficient gold mining knowledge and skills.

The use of sluice boxes varies from area-to-area. In the Wau-Bulolo region, about 60–70% use sluicing equipment. In other areas, the percentage varies enormously from very few (because of lack of knowledge or use of floating dredges) to just about everyone. Sluice boxes range from simple wooden construction with no riffles or mats, to manufactured galvanised iron and lightweight aluminium featuring a lift out riffle ladder and purpose-supplied matting (Fig. 33.3).

The simpler sluice boxes are made out of all woods, and can be a straight through rectangular model, or can feature a large area designed to contain gravel and a narrow channel through which the gravel can be washed. The shape of the box is almost totally individualized, and often relates to the size of timber available, rather than to suggested effective gold recovery proportions.

The simplest sluicing technique involves the laying of steel-reinforcing mesh on the ground, and later washing the apparatus until only the coarsest gold is retained. This can be combined with cut Kuni grass, which acts as a mat. The grass is simply burned to recover the gold trapped within at times

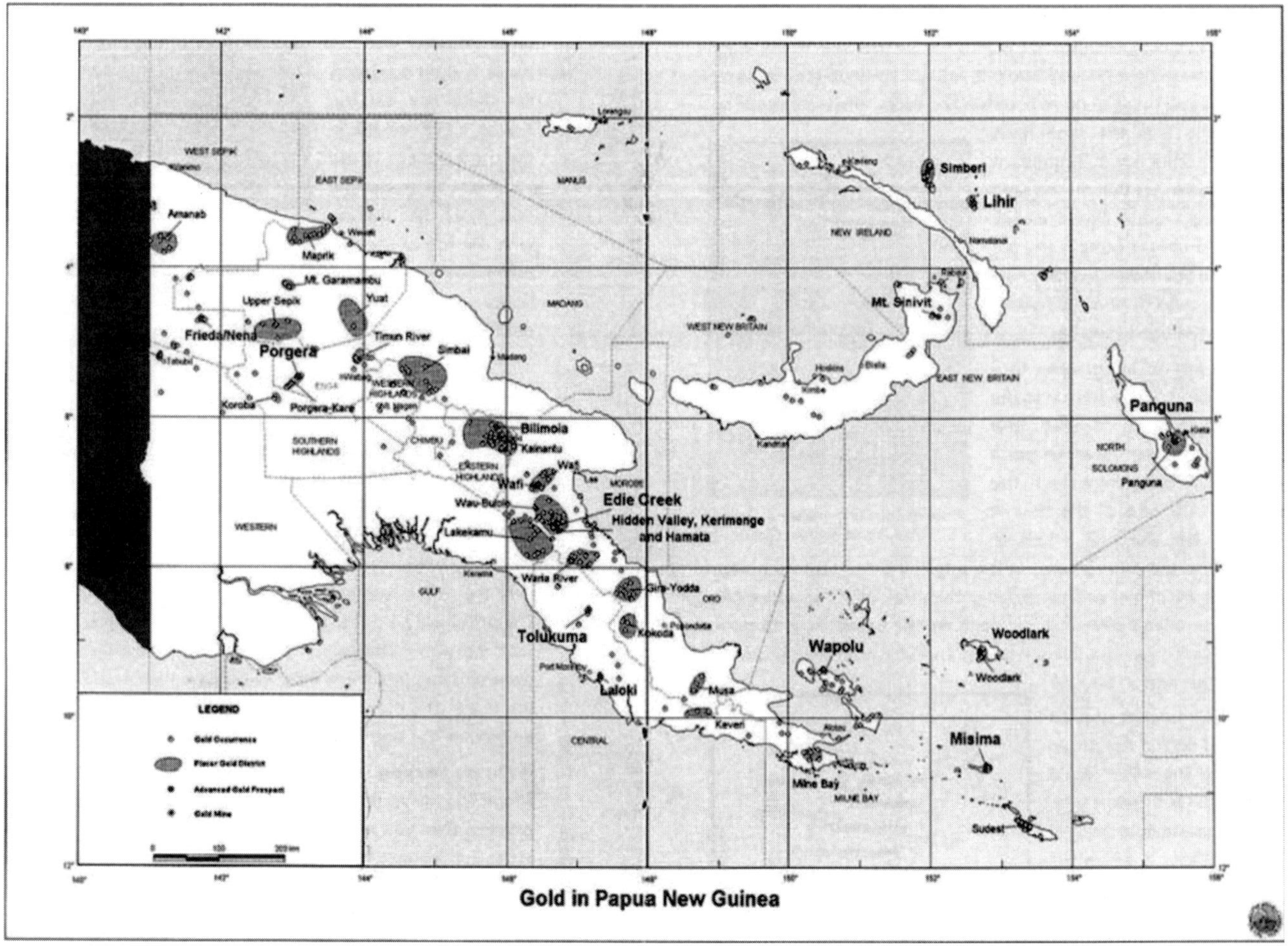

Figure 33.2 Major alluvial goldfields in PNG. *Source*: Gavu et al., 1999a–c; Gavu et al., 2001a–d.

Figure 33.3 Multiple sluice boxes, Watut River, Morobe Province.

when the miner decides to check his gold. The resultant fine back sand and ash can then be panned to separate the gold with or without mercury.

Many have no riffles, or have one at the top and/or the bottom of sluice. These boxes are fed by water that has been diverted from a flowing stream, the velocity of the water, in many cases, being insufficient to affect quality separation of the gold passing through. Many people do use such devices, however. The inefficiency of many sluicing techniques results in significant gold passing through to rivers, providing a resource for those working downstream. These people are usually mining the flood gold deposited after every rise in the river following heavy rains. This can occur a number of times throughout the year, which means "your patch" in the river bed can be mined over and over again.

Sophisticated gravity-fed water systems are used by a significant proportion of the miners in the Wau/Bulolo area because of the steep topography, which allows for stream diversion through water races down the mountainsides to the mining areas. Polypipe is also used by those who can afford it at the local cost of about K30/metre. Water is also collected in small dams, and then transferred to the work area. Sometimes, the water is used directly from the water race with sufficient pressure to run a jet set/monitor/musket, and to wash the gravel into a sluice box. In times of low rainfall, when water volume is reduced, all of the people sharing the water race must ration their work schedules to share available water between various working groups.

In areas such as the Waria Valley in Morobe Province, large boulders must be removed to access underlying gold. When it is hot, boulders are winched away or, fires lit and when it is hot the stone is broken down using water. They can also be moved by simple mechanical means.

Semi-mechanised mining

Semi-mechanised mining occurs where a number of types of machinery have been introduced. For example, water pumps are used to help with gold recovery. In particular, they are used for removing water from holes in riverbeds to allow the mining of the deeper river gravels where the gold has settled. In most cases, the use of a pump results in 200–400% more gold yield than that recovered using pans and simple sluice boxes.

Pumps can also be used for transferring water to dams, or to directly run a musket to wash gravel into a sluice box. The pumps that operate a musket are usually diesel, whereas models used for water transfer are mainly petrol. Petrol pumps are much cheaper and lighter, although the fuel is more expensive, and the models require continuos attention to best maintain spark plugs and fuel feed.

In the Sepik Provinces, Madang Province, and a few other isolated areas around the country, floating dredges are used. They provide the best option in these areas because of the large readily available water supply, and the fact that the gold is still found in reasonable quantities in rivers and stream beds. These dredges range from about a 2″ diameter hose to a 6–8″ diameter hose. These are supplied by retailers and come mostly from the United States, although one local manufacturer was making them in Wewak for a period of time.

These dredges can be combined with simple snorkelling gear, such as face masks, gloves and wetsuits to direct the suction inlet along the river bed, which best ensures the collection of as much free gold from the gravels as possible. In some areas, individuals also use goggles to free dive to collect individual flakes of gold under the water.

Mechanised mining

This type of mining, whilst still occurring on a small scale, can involve larger machinery such as bulldozers, trummels, jigs, excavators, wash plants, front end loaders, and large trucks (Fig. 33.4).This type of mining is only practiced by a small number of people throughout PNG, mainly by ex-patriots, in conjunction with local partners. Several of the larger operations are mining hard

Figure 33.4 Trummel and jigs at Eddie Creek, Morobe Province.

rock deposits, processing the original source of the gold in the host rock, as well as larger alluvial deposits.

An example of a PNG national with a mechanised operation is one man who supports his extended family of about 30 people. In addition to the family, he employs about 10 groups of tributers who work the tailings on his leased plot, paying him a percentage of the gold recovered as rent. The estimated number of people dependent upon this one operation, including women, children, and elderly people, is 200.

Hard rock mining

Unlike alluvial mining, which occurs throughout PNG, hard rock mining is carried out in a limited number of areas. Primary examples include: Mt Kaindi in the Wau/Bulolo area, pockets in and around the local mountains of Morobe Province, Bilamoia in the Kainantu area (Eastern Highlands Province), the Mt Kare and Porgera regions in Enga Province, and in and around the Tolokuma Gold Mine in Central Province. There are other clusters of hard rock activity in the Sepik provinces and, in all likelihood, other places around the country that the DoM is not yet aware of.

Hard rock mining techniques in PNG are variable. In some areas, the gold veins are exposed by machinery, and then worked by hand to recover the gold-bearing portion (Fig. 33.5). The host rock is altered so that it can be reasonably

Figure 33.5 Gold vein exposed by a bulldozer and then worked by hand at Mt Kaindi.

soft to allow for this technique to be used. Machinery is also used to mine the ore where a processing plant is available to match the throughput.

At Bilamoia in Eastern Highlands Province, very sophisticated tunnels have been constructed by hand, featuring well-engineered wooden supporting frames for both the sinks and the drives. Whilst hard rock operators are one of the more safety-conscious groups of miners, many are actually operating illegally on plots leased to a mining company (an issue discussed more in detail later in this chapter). Some miners have also been tunnelling illegally into the surrounding areas that major mining companies are exploiting, which has led to a number of deaths.

Where hard rock mining is carried out by hand, gold is recovered by pounding mineral in a dolly pot, after which the crushed sediment is panned. Rarely is mechanical crushing carried out, the exception being larger operations.

In some areas, unsupported tunnels are driven following any gold vein that is uncovered or suspected to be present. Although not strictly hard rock mining, this tunnelling can be quite common within alluvial deposits where the drive is following the bottom of the wash; here, most of the gold accumulates. In some areas of PNG, such tunnelling has caused the destruction of public roads and deaths among miners.

ENVIRONMENTAL AND HEALTH ISSUES

Environmental degradation

Mass small-scale alluvial mining activities along rivers, creeks and banks have caused damage in many of PNG's riverine environments. In Wau/Bulolo, for

Figure 33.6 Unconsolidated sediments being reworked in Koranga Creek, Morobe Province.

example, many operations are working tailings from old mines in already-degraded environments. In other areas, unstable sediments are being worked, which collapse during periods of high rainfall, causing flash flooding (Fig. 33.6).

In areas containing placer gold, such as Mt Kare, miners dig pits of various depths to rework sediments. Most of these workings are relatively small and shallow, and can have a detrimental effect on the environment.

In places of intensive mining activity, such as the Eddie Creek Merri Creek areas around the old Namie Mine in Wau, the landscape has been completely transformed into a degraded area of pits and waste. The acid soils that have been exposed are difficult to revegetate. Constant re-working of soils slows and prevents the regrowth of vegetation. The lack of vegetative cover means that when heavy rains do occur, there is greater chance of erosion and landslips.

In some areas where restoration has been carried out, such as in the old Namie Mine pit in Wau, the combined pressure of small-scale mining following vein gold along the old pit walls and from the people utilizing planted trees for firewood, has made some of the restoration unsuccessful over the long-term.

Digging around the roots of large trees along riverbanks eventually causes them to fall into the watercourses. This has caused problems, especially during periods of regular flooding, when tree trunks are swept downstream. These floods cause damages to bridges, and induce changes to water flow when they become entangled and cause blockages. The loss of tree cover also increases the chance of soil erosion in the high rainfall and steep-sloped environments.

The outreach program of DoM has also raised the issue of environmental degradation from mining activities, and the need to prevent damages to both private property and public lands. It is continually emphasizing that damages to surrounding environments can adversely impact future generations, their ability to garden and feed themselves, and peoples' health. This has been demonstrated through to use of pictures and VCDs showing areas that have been affected by past mining activity and not rehabilitated. This includes areas where current mining practices do not follow regulations.

In PNG, the majority of gold deposits worked by small-scale miners are alluvial, in which particles are separated from waste using simple mining equipment and techniques. In most cases, miners apply mercury to riffles in sluice boxes, boil box compartments, or in gold pans to separate gold from the final black sand. Mercury poses a hazard to the environment as well as to human health. In the Wau/Bulolo area, where dredge mining commenced in the late-1920s and continued to the 1960s, it is still common for bulldozers to uncover large puddles of mercury. Recently, a field trip to Bouganville (October 2001) by MRO revealed that the same situation was occurring around the alluvial centre on the island as a result of the mining carried out earlier this century.

The hydraulic sluice operators occasionally dislodge naturally-formed amalgamations of mercury and alluvial gold. Moreover, mercury is insoluble and often remains in river sediments. It is not unusual for river water to flow clear of mercury whilst high levels of mercury lie on the riverbed (Tiller, 1990).

Recent studies have shown that mercury may be emitted to the environment as a result of forest burning. In the Amazon Basin, it has been argued that up to 50% of the mercury present in waterways and subsequent mercury poisoning in people originates from this source (Veiga et al., 1998; Pearce, 1999). This factor could also be of significance in PNG because of the drought during 1987–88. Specifically, large areas of forest were burnt in and around the Wau/Bulolo mining area, which may have led to an increase in the levels of mercury and/or methyl mercury circulation in the environment.

Small-scale mining activity has intensified in the rural communities of PNG but an ongoing problem is that the Government's medical staff working at the community level has not been properly trained to recognize the symptoms of mercury poisoning, and have no drugs to treat it. The relevant authorities and many local communities have been made aware of this, and are being pressured to address this shortfall. Some communities have made an effort to try and pursue this lack of knowledge and to ensure that their community health workers are aware of the potentiality of this problem. This enthusiasm is tempered by the lack of resources in rural areas.

Mercury

The main hazard to small-scale miners is posed by mercury usage in the mining process, as discussed above. At present, some four tonnes of mercury are sold annually to small miners. There may also be illegal imports but there are no figures available to determine possible levels. Figures regarding the quantity of mercury imported cannot be used as reliable estimators of gold production country-wide. Based on these figures, it would mean that only two grams of mercury arc used to recover one gram of gold, which is untenable for several reasons. First, evidence from other countries indicates that at least four grams of mercury arc used to recover one gram of gold, and, in certain places, it is even higher. Secondly, there are a number of people who do not use mercury. In PNG, people are still recovering nugget gold or flakes (40% according to DoM informal reports). Finally, because an accurate small-scale mining census is not yet available, the quantity of recovered gold has not been accurately recorded. If the assumed total gold production (both official and unofficial) was taken into account, then the amount of mercury used would amount to less than the 1:1 theoretical formula, meaning that PNG would have the lowest usage of mercury/gram of gold in the world. Whilst there are some specific parameters in PNG (e.g. low levels of technology), it would appear that on the surface, there should be a higher level of mercury consumption than the import figures quoted.

The use of mercury and creation of awareness of its possible dangers is a major concern. Various authors (Pearce, 1999; Krabbenhoft et al., 1999; Veiga et al., 1998; Shanley, 1999; Hruschka, 2001) have shown that mercury poisoning has become a major problem in gold mining communities in a number of developing countries. In PNG, mercury is used in panning to extract gold from black sand. More specifically, it is mainly used to capture the very fine gold remaining in the black sand after panning. Typically, the mercury is rubbed into the black sand using bare hands as a means of ensuring that no gold is missed (Fig. 33.7). Since the miners handle mercury without gloves, they absorb it through their skin, especially if they have cuts, abrasions or open sores on their hands. The amalgam is then heated to evaporate the mercury to leave a purer gold residue. Most people heat the amalgam to drive off the mercury in their huts or houses. This is done to prevent others, including criminals, from seeing what is being done and how much gold is being recovered.

Occasionally, the amalgam is placed on a knife or other household utensils, such as plates or cups, and then held in the fire until "cooked". These same utensils are often used for normal household activities. Widespread mercury poisoning cannot be proven at this stage but it would appear from hearsay evidence that headaches and illnesses result from this activity. The current AusAID project surveying mercury levels in the mining community of Wau/Bulolo is aiming to explore this possibility. The results should be available sometime in 2003.

Figure 33.7 Rubbing the mercury into the black sand to capture fine gold.

When the amalgam is burned outside the living quarters, it is quite often carried out over an open fire surrounded by a number of people trying to keep warm; these people likely inhale the fumes.

Mercury stories

The following stories concern mercury usage in PNG, effectively illustrating the variety of situations in which it is used and misused. These cases also show the importance of ongoing education concerning the use of mercury, and the dangers associated with its use. These examples are summarized from Susapu and Crispin (2001).

Case #1

In the first visit to the Middle Sepik area in Nov 1999, the author visited Bisario, near the headwaters of the Crossamarie River (Fig. 33.8). Here, gold buyers had given the local people mercury to capture fine gold but had not informed them of the dangers. The intention was for the gold to be returned to the gold buyers because the mercury was being supplied free of charge. The outreach program, through its workshops, helped the people to understand the dangers of mercury, and how to reduce its exposure to both children and pregnant women. Many people soon expressed their concern and anger at the lack of information and the deception or lack of full information being supplied by the people who had given them the mercury in the first place.

Figure 33.8 Outreach somewhere along the Sepik River in the Sepik Provinces.

Case #2

This foregoing story has been repeated in a number of areas in PNG. Many times, however, the dangers of mercury have been explained (*e.g.*, in Waria River Valley, Porgera, Kainantu, Milne Bay, and many communities in and around the Wau/Bulolo), and, as a result, practices have changed and people are now more aware. The main catalyst has been a video that shows the effects of mercury on the people in Minimato Bay in Japan and in small-scale mining communities in Brazil. The video has been put together by DoM, with the assistance of AusAID.

Case #3

Another practice that occurs in some areas is the burning of amalgam cake in huts where people live. The main reason given is to ensure that criminals cannot see what is being done, and how much gold is being processed. The dangers of this practice have since been explained (accumulation of mercury and mercury vapours in the huts), and some people have changed their practice. Emphasis has always been placed on its possible effects on children and pregnant women.

Case #4

The various methods of burning amalgam have been explained in all of the areas visited by representatives of the DoM/AusAID education outreach program. The various options that can be used, from full retorting to use of the "tin fish tin" method of recycling mercury (Described in detail in *Mercury-Safe*

Uses in Small Scale Alluvial Mining published by the Department of Mining 1999), have been discussed. Briefly, the "tin fish tin" method involves use of two different sized cans of a popular brand of tinned fish, which are used to contain mercury vapour (Figs 33.9–33.13). The apparatus is therefore widely available, and costs nothing, apart from the initial cost of buying the tinned food. This is an effective means of recycling and an appropriate technology.

Case #5

It has also been emphasized that people can save money by recycling mercury. Traditionally, people have burned the amalgam cake wrapped in leaves, or on a tin exposed to the atmosphere. People often burn the amalgam cake while other people are sitting downwind from the smoke. It has been pointed out on many occasions that this is dangerous, and many people have since taken this on board and now try to keep upwind from the fire. Allied with this is the habit of smoking cigarettes whilst burning the amalgam cake, something that can be further complicated by sitting in smoke to keep warm. Again, many times this has been pointed out, with people appearing to accept the advice once the dangers are explained.

Case #6

In one village called Sambio (Morobe Province), several people were burning their amalgam cakes on the blade of a knife that was subsequently used for preparing food. After a visit from the outreach team, who explained the dangers

Figure 33.9 Wrapping amalgam cake in leaves.

Figure 33.10 Placing wrapped amalgam cake in the smaller tin.

Figure 33.11 Placing the larger tin upside down on top of the smaller tin.

Figure 33.12 After the "cooking" the larger tin is removed.

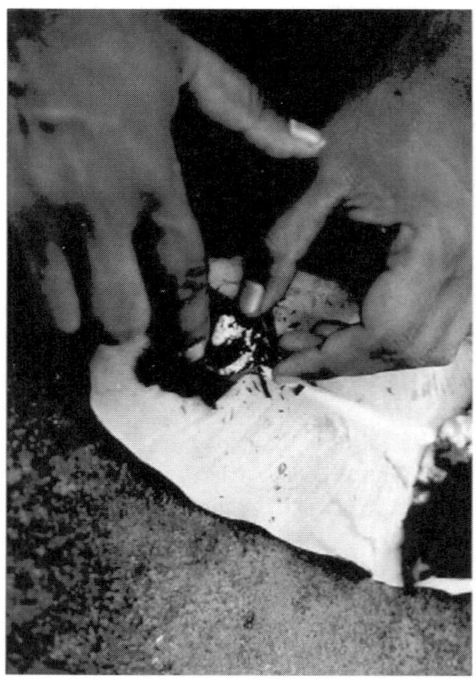

Figure 33.13 Removing the remains of the wrapping from the "cooked amalgam cake".

of this, the people indicated that they would stop such practice and begin using a different method to burn the amalgam.

Case #7
In the Simbai Valley (Madang Province), no mercury was being used to recover gold because they only try to recover nugget gold. This is changing, however, as information on mercury is filtering through to this isolated valley. The outreach team carried out mercury awareness programs so that miners would be aware of the dangers before it is widely used in the valley. It may prevent some of the health risks associated with using mercury.

Case #8
In the river below the giant Porgera Gold Mine, mining practices have changed, according to the local mining community. Previously, the mine people were able to capture gold of larger size. Now that the tailings from the mine cover the original gravel beds, the gold is much finer, and requires mercury to capture it.

Case #9
AusAid has funded a baseline mercury level study in the Wau/Bulolo area. Up until the beginning of the study, there had been no large-scale investigations of mercury levels in people. Results should be available later in 2003. If a problem is identified, further investigations will be carried out to determine if contamination is being caused by naturally-occurring mercury, mercury from previous mining, mercury from current practices, or mercury originating from forest burning during the major drought in 1997.

Studies carried out in PNG on mercury contamination of miners
A number of studies have been carried out in Wau/Bulolo investigating possible mercury contamination. The most detailed is a study by Mallard and Hugman (1996), who collected water samples, freshwater fish from the remains of old dredging ponds, and urine samples from the population. They carried out a survey of people for mercury poisoning symptoms, and used a questionnaire to gather information. The main conclusions of the study were that a more extensive survey is needed to establish the degree of exposure of miners to mercury poisoning, and to determine the risks to their health.

Several other studies have been carried out (*e.g.* Blowers, 1998; Tetsuji, 1997; Subasinghe & Okada, 1998) but with inconclusive results. These studies were intended as pilot work but the methodologies could be questioned in each, especially with regard to the number of samples (which are small), and the way the sampling was carried out. Hearsay evidence of local villagers indicated that samples in one study were taken from any person in the vicinity of the study area at the time, which meant that the sample could have contained individuals not involved in mining. A population of some 15,000–20,000 miners needs far more than 40–50 people to provide an accurate representation of the population.

The population, whilst concentrated in the valley and surrounding hills, is still well scattered. Thus, a representative sample of *all* of the communities is needed for reliable conclusions to be drawn.

The studies recommend that further work be carried out that especially focuses on the mining population. The largest study to date is currently being carried out (funded by AusAID) in the Wau/Bulolo area, largely because of the recommendations put forward and questions raised by the mining community. The full results should be available in early-2003.

TECHNICAL ASSISTANCE TO PNG FOR SSM

One of the most important outcomes of the aforementioned groundbreaking project in PNG was the tripartite nature of both the funding and the inputs into the program. All three sectors have provided inputs in the forms of professional-level advice and support, financial contributions, as well as co-ordination and support with the planning of logistics. This unique combination provided a strong support base for what was a very successful project, which has motivated other agencies continuing to build upon the positive outcomes.

AusAID

AusAID was the first agency to get involved with small-scale mining in PNG, and did so in 1998–2001 with the implementation of a technical assistance program. The program aimed to address the following:

- Assisting economic development in rural areas and helping to alleviate poverty;
- Addressing environmental and health issues in rural areas;
- Developing the role of women both socially and economically; and
- Investigating the involvement of children in economic activities.

The Program was developed through the SSM Branch of DoM for the purpose of reviving and expanding its role in mining in rural areas *as per* Government policy.

A technical adviser was provided to assist with changes made to organizational infrastructure and personnel development. Initially, the main aims were to carry out a survey to determine the extent of SSM across PNG, define the working population, and to investigate the extent of the industry's economic contribution to the overall economy of the country.

Part of the initial program was to develop a comprehensive educational program for the SSM community. To this end, a series of videos (later changed to VCDs) and booklets written in *Tok Pisin* and English were developed, and a travelling educational road show was inaugurated. The topics of the educational material included:

1) Introduction to Small-Scale Mining techniques;
2) Mercury – Safe use in Small-Scale Mining;

3) Introduction to Mechanized Small-Scale Mining;
4) Occupational Health and Safety in Small-Scale Alluvial Gold Mining;
5) Restoration of the Landscape and the Environment;
6) Economics of a Small-Scale Mining Business; and
7) Gold – From Gold Pans and Sluice Boxes to Bars and Jewellery.

An eighth section was planned on the legal structures and responsibilities but was not completed, as changes are being planned for the near future. These educational materials are available from the DoM and the MRO in Port Moresby.

The initial aim was to visit all SSM sites throughout the country over a period of two years (later extended to three). However, this proved to be almost impossible, given funding limitations and the ever-increasing number of mining communities documented in some very remote and relatively inaccessible areas.

The initial SSM population figure in 1998 was 15,000–20,000 people but by the end of 2001, it had increased to 50,000–60,000, based on substantiated results. It has been suggested that if the population is completely documented, that this figure could rise to between 80,000 and 100,000 (Crispin, 2002).

The on-site staff of the SSM Branch in Wau was increased from three to five. This included the first woman to be employed at a professional level in SSM. As part of this process, various people have been seconded to DoM in Wau to gain experience in SSM techniques and regulation, and to help build a number of professional level experienced people as quickly as possible. As a result, both the private sector and the government department were able to benefit from pooled resources – again, essential for a rapid building of the system.

Survey of mercury awareness and techniques

An evaluation of the effectiveness of the mercury awareness program was undertaken as part of the AusAID project in 2001 (see Crispin, 2003). This was undertaken to provide systematic feedback about the outreach program, and to decrease reliance upon informal discussions with people. A survey was developed, which covered only the knowledge and mercury usage techniques. In view of limited time and funds, it was decided that the mercury awareness among village-level miners would be the first priority. A pilot survey was carried out to test the validity of the survey, after which certain adjustments were made.

Initially, the evaluation was carried out in the Wau/Bulolo area, as it contains the most concentrated population of miners. People were interviewed at three villages visited by representatives of the outreach program over the previous two years. The findings were compared to those obtained from three villages that had not been visited by representatives of the outreach program. A control group was used from outside the mining province but which was still within the general area to enable effective comparisons of basic mercury knowledge between the communities. Questions were asked in *Tok Pisin* by the staff of DoM and seconded people from private industry. However, some people preferred to be asked questions in English. Some of the survey results that relate specifically as to whether or not the message about mercury was getting through are discussed below.

The results of the survey showed that more than 70% of the miners had been mining for over five years. Some of the miners had been working for as many as 25 years. Moreover, all interviewees in both the visited and unvisited villages used mercury. In the Wau/Bulolo area, people are well aware of mercury and its uses. For the most part, the gold recovered is fine, requiring mercury for its recovery from panned black sands.

Question 17: "Do you cook mercury in your hut?"
In visited villages, some 80% (20/25) reported not cooking mercury in their hut, compared to 35% (7/20) in the unvisited villages. It appears that the awareness program may have made a difference in persuading people not to cook amalgam in their living areas. The change from cooking in the house to outside was one of the major thrusts of the educational program, once it was established that the practice was widespread.

Question 19: "Do you recycle mercury?"
The results of this question were the same –25% (6/24) in the visited villages, and 25% (5/20) in the unvisited villages. It is interesting that no change occurred, in spite of the efforts of the educational program. Making people aware of the potential for savings and reduction in environmental and health problems through mercury recycling would be the next logical step. Those who did recycle mercury used the "tin fish tin" method. Retorts are not used (Fig. 33.14), despite an extensive educational push for their use. This suggests that more educational work is required in order to get people thinking about other ways to recycle mercury.

Retorts cost about K100, which may be an obstacle to their use, as most would have to think carefully about spending such an amount of money from a limited income on one item without a perceived instantaneous benefit. Those miners who have higher cash flows, or who can "see" the business sense in reusing mercury may need to be targeted and shown the monetary gains that can take place.

Informal discussions with the miners revealed two further reasons why retorts are not widespread. The first relates to the gold sponge being burned within a closed atmosphere in the retort, in which case it can change colour and must be cleaned afterwards. This is significant because gold is bought on the basis of colour and density: the more silver contamination, the less golden the colour, and, therefore, the less money received. Near Wau, the price can be as little as K8/gm for amalgam containing 50% silver and up to K18/gm for gold of 90% purity (according to 1999 prices). The education team may need to spend more time both explaining and demonstrating how this occurs, and what must be done about it.

Secondly, some people believe that mercury can only be used once. It has a bright silver colour, and, when used, becomes less shiny. People may therefore discard it, believing that it is less effective. An interesting result from questions regarding the toxicity of mercury both in the home and the workplace is that people view mercury as suitable for use in the workplace but dangerous at home.

Figure 33.14 Miner using a small amalgam retort.

Some 64% (16/25) of interviewees from the visited villages indicated that the information they have been given has changed the ways in which they use mercury, compared to only 20% (4/20) of those interviewed from unvisited villages. The pilot study "yes" response was 50% (4/8), and 25% (1/4) of the people from another group of miners from Milne Bay Province said they would change the way they use mercury. The knowledge about mercury present in the unvisited villages comes from both general community awareness and information from the education program that has filtered through to the community. Whilst not recorded in the survey response sheets, discussions afterward with residents from visited villages indicated that miners had made some decisions regarding their mercury practices. They indicated that they had decided to change mercury practices that affected both children and pregnant women. An example given in several cases was that they now ensured that people kept away from smoke during the burning of amalgam, and they checked the wind direction before starting the fire.

This was reinforced during a visit to Kobiac village (unvisited village), where the headman spoke to the villagers and the outreach team; he indicated having spoken to the people from Mrs Booth (a visited village). As a result of the information received from Mrs Booth, Kobiac village miners have changed their mercury practices, although there has been no attempt made to use retorts or to recycle mercury.

Another factor regarding mercury use and recycling is that only a low percentage in both groups of villages indicted that "mercury smoke" can be captured and reused. This indicates that the educational program will need to re-examine the ways this message is being communicated to the villagers, and make some adjustments to its presentation.

Another major effort aims at making people aware of how to safely store mercury. It was recommended that people always store mercury under water. The survey results showed that in the visited villages, some 60% of miners stored mercury in the recommended way, compared to only 15% surveyed in the unvisited villages. Both groups of villagers did, however, mention storing the mercury away from children.

The lack of awareness of some of the problems with mercury may be a result of the limited communication skills of the team, or a function of the educational levels of the miners themselves, the latter possibly inhibiting an understanding of the issues (especially among the older generation). There may also be a financial constraint regarding the purchase of a retort, which could be beyond the priorities of the limited budget of the majority of miners.

The survey instrument appears to have succeeded in identifying some of the differences between the visited and unvisited villages but further refining of the questionnaire may allow more of the details to be drawn out as the outreach program continues. Nearly all people (one exception) wanted the program to come back to their village. There is probably an entertainment motive behind this view but it does indicate that education is being taken seriously in a number of villages, and that the benefits of using mercury carefully are beginning to be understood. The outcomes generated by the initial survey have pointed some areas of the educational program that will need to be adjusted. If these shortcomings can be addressed initially and updated as the program develops, then there is a strong chance that an ongoing environmentally-aware SSM sector in PNG can be managed.

AusAID has also jointly funded a micro finance project with the Asian Development Bank (ADB) that commenced in late 2002, the details of which are examined later in the discussion.

SYSMIN

European Union (EU), through its SYSMIN program, has been developing a project to assist SSM, and building on the work of AusAID. The *Draft Proposal for Institutional Strengthening*, produced by the SSM Branch of the DoM in 2001, reads:

> To increase the insertion of Papua New Guinea into the global economy whilst reinforcing the sustainability of its development and securing/developing the sources of budgetary income required to provide essential services to the population and to fight against poverty. To enable the SSM sector to increase its inputs into the PNG economy whilst enhancing social and cultural life.

The Project will continue the work of the outreach program, and the training of support staff. It will also promote the development of infrastructure support, as professional staffing is expanded. Another major aim is to develop vocational training colleges in the major SSM centres around the country. The priority is to refurbish the Wau centre, and to construct new centres at Porgera in the Highlands, and Wewak in the Sepik area. It may also be necessary to expand into the islands, possibly constructing a centre in Milne Bay or New Ireland. The Porgera centre will also include a women's development centre.

This five year program should commence during 2002–3.

The Japanese Social Development Fund (JSDF)

The JSDF has also prosed assistance for SSM in PNG, which, according to a draft proposal, would aim to:

> Improve rural incomes and reduce rural poverty by supporting Intergrated Sustainable Development of Communities affected by or involved in SSM.

This project may commence during 2003.

The Asian Development Bank (ADB)

The ADB has initiated a project supporting the development of microfinance in PNG. A small part of this project involves the conducting of a pilot project to test micro finance products in Wau, the location of one of the country's SSM centres. This project is extremely important for the SSM community, as there is no Bank in Wau, and, therefore, no facilities providing financial services in the area. The income from SSM in this area is in the order of K16–18 million. This income is cash based and occurs all year-round, unlike other rural pursuits such as cropping, which is very seasonal.

It has been documented by the ADB that the major banks do not return savings to small rural communities. A community bank, however, may be able to provide savings services, as well as loans, to ensure that profits are returned to the community to assist in its development, rather than that of the larger mining companies in Port Moresby and Lae.

World Bank

The World Bank has been running a project on institutional strengthening in DoE. Small-Scale Mining constitutes one part of the general effort, which aims at developing the Mining Department.

Metal Refining Operations (MRO)

Metal Refining Operations (MRO), the privately owned gold refinery in Port Moresby, has provided support for the development of SSM in the country. There is a profit motive to this support, in that the more gold that goes through the refinery, the more money the company makes. There is, however, a congruence regarding some of the roles of the donors and the private sector.

The aim of the donor agencies is to help build economic infrastructure, especially in rural areas, with an added concern over the environment and occupational health and safety issues, which helps to better attack the issue of poverty more comprehensively. In the case of MRO, aims coincide with those of donor agencies. A well-educated, and organized, SSM sector will provide income to many the rural villagers, and by building organized infrastructure, the outcome will help to provide a basis for the established refinery or a similar endeavor. The suppliers of equipment will also add to the distribution of wealth generated by the SSM sector, and skilled jobs created within manufacturing channels will add to the contribution of SSM to the economy of the country. Another important challenge is reducing the transit of unofficial gold mined through the system, which would improve PNG national accounts and official export earnings.

Personnel from the SSM section of the refinery business have been seconded to the SSM Branch of DoM for training purposes. This has resulted in an increase in the numbers of experienced people in SSM over a short period of time. Assistance was also given with sharing the costs of the outreach program, with books and videos being used as support material. Logistical support has been provided to DoM to assist with the implementation of the project.

Mining companies

Various mining companies have assisted with the development of the SSM program in PNG. They have provided transportation, accommodation, and access to mines or prospects to carry out the outreach program. Financial assistance has also been provided to the program. Some personnel have also been seconded to DoM program for training in SSM techniques and management practices.

Provincial governments

Several provincial governments assisted with the aforementioned program by providing transport, manpower and finances. These administrations were those who had a SSM policy drawn up within their overall provincial plans.

SSM AND LARGE-SCALE MINING RELATIONS IN PNG

In PNG, there are large-scale mines, as well as washing plants and small-scale crushing plants that directly employ between 10 and 15 people. At the beginning of 2001, there were five large-scale mines operating in PNG. Exploration has virtually come to a standstill, with little activity apart from work aimed at expanding the reserves of known deposits. The industry's relationship with SSM is becoming increasingly important, as more people are being made aware of the value of minerals and the community-level desire to share more of this perceived value. During negotiations, traditional ownership of land is often perceived to be more important than national Government regulation. In the end, both sides sill have to be accommodated to ensure maximum benefit from these resources.

Relationships with exploration companies

There is a lack of awareness amongst village people that the State owns the minerals in the ground. Many still believe *mi papa bilong graun – i.e.* that "the minerals belong to me". Compensation is demanded for all activities but because there is little differentiation made between exploration and mining activities at the villager level, the spreading of awareness of this distinction should become part of any ongoing education program.

One exploration company tried to steer artisanal miners away from hard-rock deposits, assisting them with work on alluvial gravels. The company provided expert assistance with operations design, the locating of appropriate equipment, and production. This initially enabled local people to gain income from alluvial gold, and, at the same time, allowed the company to work the hard rock deposits. However, over time, maintaining this division proved to be very difficult, as miners began entering the leased area illegally, following the gold veins through the rock. They began using excavators to remove large amounts of soil, which is a clear breach of the mining regulations on mining and exploration leases.

The company attempted to ban the use of mercury on its lease because of fears of problems developing with its use, and wanted to stop information about mercury being disseminated because it was felt that the information may cause greater problems. Downstream from the camp, the alluvial miners were processing the gravels using mercury but were unaware of the dangers.

Representatives from the outreach program of DoM visited the site and explained the dangers associated with mercury; awareness was soon aroused in the community. The company decided that it was best to allow the information to be passed on but still banned the use of the material on their lease.

SSM and large-scale miners

In PNG, mining company interactions with the SSM communities have varied. In one case, small-scale miners were excavating the back of a mountain containing a gold deposit leased to a company. Several miners were killed when blasting commenced (after ignoring repeated warnings of the danger). The danger extends to the waste dump, with people getting in the way of trucks and the dumping of waste rock. There is a cut off grade for the large mines; thus, there will always be some discarded material containing gold worth recovering by small-scale means. In addition to working the primary deposit, artisanal miners are working within the stream below the tailings outlet, and are recovering gold.

The intrusion of miners into areas controlled by large-scale mining companies will continue to be a problem. It will be impossible to stop people from mining gold for themselves, without resorting to vigorous interventions. The ultimate solution may be for the people to be assisted with working alluvial gold deposits, leaving the major hardrock deposits for the major companies. As indicated above, some companies are already attempting to address this issue, searching for a way to peacefully resolve the issue.

Misima mine closure

On Misima Island in Mine Bay Province, a gold mine is now entering the closure phase. The mine closure plan emphasizes the provision of assistance to the community to enable it to better cope with the changed economic conditions induced by the mine's closure. The mine operators have mainly pursued agricultural options as a means of income generation but the financial returns are restricted because of the isolated location, distance from markets, and limited arable agricultural land.

However, small-scale mining has the potential to replace a major portion of the royalties paid out each year (which, on average, total 1.7 million Kina/year). This would spread across the community of Milne Bay. Gold is found on many of the islands in Milne Bay, as well as areas on the mainland.

Talks were held with the business development arm of Misima Mines regarding the issue of working with the small-scale mining community. Many of the mine's employees will be returning to their traditional lands following closure, and the skills they have already gained from working in the mine will probably be utilized to generate an income to replace lost wages.

Two of the gold miners who brought gold to be sold during the last trip by DoM and AusAID to Milne Bay (in October 2001) were employees of the mine. They indicated that they realized that their employment was coming to an end, and that they needed to find alternative sources of income. However, they have become part of the cash-based economic system, which means reverting to the village life, working as a subsistence food producer, no longer holds a strong appeal.

A wider view of the region must also be taken into account, as such a reduction in monetary input will lead to diminishing levels of secondary economic activity. Because of the widespread gold occurrence across the area, SSM could replace this financial input to some degree, and continue to support economic activity across the Province.

It should be pointed out that environmental issues are an important part of the scenario for island-based SSM. Pollution of fishing grounds from high sediment loads in streams added to by mining may affect reef systems, and, therefore, have an impact on food sources. These environmental problems could be managed by the local communities through the use of low key appropriate technologies such as settling ponds. This is extremely important, especially where soils are clay-based, which may lead to sediments being held in suspension for longer periods of time, or carried greater distances in streams and the sea.

The sustainability of SSM in Milne Bay can be reinforced with the full cooperation of all stakeholders. The acceptance that the promotion of SSM as part of the mine closure plan would be an important breakthrough for the future. It may also convince other mining companies operating in the country that SSM has to be an integral part of any mine closure plan (Susuapu & Crispin, 2001).

UPDATE: DECEMBER 2002

A recent trip to PNG has enabled officials to update some of the key information about SSM. The most significant economic factor has been an increase in the price of gold, from about US$270/oz to about US$320–350/oz, combined with a drop in the relative value of the Kina, from about US$0.30–0.35 to US$0.23–0.25.

This has effectively precipitated a 40% increase in the returns from gold from the SSM community, which is significant because the mining community is probably the only group keeping up with inflation in PNG. Those who have fixed incomes or salary will have found it much more difficult to cope with rising prices.

It was suggested by a number of local sources that the number of people mining gold in the Wau/Bulolo area is increasing, reflected chiefly in the increasing numbers of "settlers" (those from other areas) moving into the area. Some gold buyers suggested that the average amount of gold/miner was decreasing, whilst others admitted having not yet experienced such a change.

The amount of alluvial gold that is now passing through the official system has increased. The combination of MRO-refined gold production and exported, unrefined gold from other export licence holders could have an annual value of K150,000,000 (based on export records). This price estimate is based on the price as of December 2002, taking into account increases in the price of gold and the decline in the value of the Kina. When illegal sales of gold are added, the official figure could exceed K300 million. This is well in excess of previous estimates of approximately K150–180 million.

The ADB Micro Finance Project is now being implemented, as well as a pilot scheme in Wau, to service the local community.

WHERE TO NOW?

In PNG, SSM activity is now a recognized economic activity that generates income for rural populations. The government policy on SSM has been difficult to support, having been implemented with limited funds and manpower resources.

The AusAID-funded technical and education assistance program was the first attempt made to provide comprehensive assistance to the SSM sector. As a result of this initial program, it has been shown that, despite the financial constraints, SSM functions can be developed, promoted and managed. This has been accomplished by soliciting resources from within PNG, through external donor sources, and with the resources of DoM.

DoM's vision and long-term goals have been documented in a Draft Five-Year Development Plan (D5YDP) covering the 2002–2006 Period. The expansion of proposed funded programs under ADB, JSDF and the EU's SYSIM Fund will be guided by the priorities contained within the 5YDP document. DoM has an overall plan to become a corporate body within the next five years. If this corporatization becomes a reality, and a levy is charged on all gold exports, the long term sustainable development of SSM can quite conceivably be internally financed.

It is recognized that mining activity, whether large- or small-scale, has the ability to substantially change the physical landscape, and, therefore, has the potential to cause long-term environmental impacts. It also has the potential to adversely impact the health and safety of a mining community. If SSM is to continue to make an important and sustainable contribution to rural economic development in PNG, improved management of the activity is needed.

The assistance of the donors has been incorporated into long-term planning, to ensure that the assistance is targeted and has a direct effect on capacity building within DoM. The outcomes are targeted towards achieving a sustainable SSM Branch, with professional staff working towards improving the working conditions, economic outcomes and infrastructure support of rural-based communities. These aims can be achieved by emphasizing mining skills training, business skills, environmental awareness, occupational health and safety issues (including the safe use of mercury), financial services support that assists in achieving environmentally sustainable lifestyles, and poverty alleviation initiatives in rural areas.

REFERENCES

Bank of PNG (2002). Private communication.

Blowers, M. (1998). *Mercury survey*. Private communication.

Crispin, G. (2003). Environmental management in SSM in PNG. *Journal of Cleaner Production*, 11, 175–183.

Gavu, V., Tongo, G. & Lole, H. (1999a). *Introduction to SSM Techniques*. Department of Mineral resources, Port Moresby.

Gavu,V., Tongo, G. & Lole, H. (1999b). *Mercury-safe uses in small-scale alluvial gold mining*. Department of Mineral Resources, Port Moresby.

Gavu, V., Tongo, G. & Lole, H. (1999c). *Introduction to mechanised small-scale gold mining*. Department of Mineral Resources, Port Moresby.

Gavu, V., Tongo, G. & Lole, H. (2001a). *Environment and rehabilitation*. Department of Mining, Port Moresby.

Gavu, V., Tongo, G. & Lole, H. (2001b). *Gold: From sluice boxes and pans to bars and jewelery*. Department of Mining, Port Moresby.

Gavu, V., Tongo, G. & Lole, H. (2001c). *Economics of SSM*. Department of Mining, Port Moresby.

Gavu, V., Tongo, G. & Lole, H. (2001d). *Occupational health and Safety in SSM*. Department of Mining, Port Moresby.

Hancock, G.E. (2001). Mining and Sustainable Development: The Status of Mineral Policy in PNG. *Proceedings of the GEM 2001 Conference*, pp. 117–133 AUSIMM (port Moresby Branch).

Hruschka, F. (2001). *Mercury uses by small scale miners in Peru*. Private communication.

Krabbenhoft, D.P. & Rickert, D.A. (Online). *Mercury containination of Aquatic Systems* (Acessed February, 1999). US Department of Interior, US Geological Survey, Fact Sheet FS-216-95.

Mallard, A.E. & Hugman, S.J. *The impact of mercury on the Bulolo River valley*. ReportCt/13 Department of Chemical Technology, University of Technology. Lae PNG.

Metals Refining Operations (2002). *Production reports*. The staff of the Dar.

Mercury (Online), *Material Data Safety Sheet*. Http://www.atstdr.cdc.gov/tfacts46.html (Acessed January 2001), University of Utah.

Pearce, F. (1999). A nightmare revisited. *New Scientist*, February 6th.

PNG Department of Mining (Mineral Resources), Various informal reports, unpublished during the period 1998–2000. Authors usually officers of the Department.

Shanley, A. (1999). Mercury effluent threatens Bolivian waterways. *Chemical Engineering*, 106(11), 64.

Subasinghe, G.H.N. S. & Okada, S. (1998). *The Role of Mercury in alluvial Gold Mining: Handleing, Toxicity, Pollution and Possible Alternatives*. Report, Department of Mining Engineering, University of Technology, Lae, PNG.

Susapu, B. & Crispin, G. (2001). *Country Study Report on Small Scale Mining In Papua New Guinea*. Mining Minerals and Sustainable Development Reports London.

Tetsuji, K. (1997). Private communication, 8/11/97.

Tiller, D. (1990). *Mercury in the freshwater environment. The contamination of waterbodies in Victoria as a result of past mining activities*. Department of Natural Resources.

Tongo, G. & Crispin, G. (2001). *SSM Update, Proceedings of the GEM 2001 Conference*, pp. AusIIM (Port Moresby Branch).

Veiga, M.M. Meech, J.A. & Onate, N. (1995) (Online). *Mercury pollution from deforestation*. http//water.ugs.gov/public/wid/FS_216-95.html. Accessed February 1998.

Part V

Latin American Case Studies of Artisanal and Small-Scale Mining

34

Introduction Part V

GAVIN M. HILSON

The artisanal and small-scale mining activities of Latin America began receiving considerable attention during the heyday of the Brazilian *garimpeiro* gold rushes some 15–20 years ago. At the time, issues such as land use competition in the Amazon Basin, mercury contamination, and associated crime, made headlines worldwide. The environmental problems of resident activities were frequently weighed heavily against the economic importance of the *garimpeiro* activities, which, up until only a few years ago, produced some 90% of the country's gold.

The history of Latin America is intimately associated with the prospecting and extraction of gold, silver and gemstones on a small scale; mining has been practised in the region for centuries. The Andean zone of South America, for example, has a 400 year-old mining tradition. Similarly, Bolivia and Mexico have both supported silver mining since the mid-16th century, when the Spanish first exploited trace quantities of minerals using rudimentary techniques. Brazil, which, again, has received a considerable amount of attention in the literature, was the world's leading producer of gold some 250 years ago as a result of intensive artisanal activity.

Today, artisanal and small-scale mining activities are widespread throughout Latin America (Fig. 34.1). To provide an indication of the size of the industry today, a now-outdated United Nations' report entitled *Developments in Small-Scale Mining* (UN, 1996) estimated there to be over one million small-scale and artisanal miners in Latin America, most of whom are operating in Brazil, Bolivia, Venezuela and Chile (Table 34.1). However, recent research has revealed that there are active artisanal and small-scale mining contingents in many of the lesser-developed Latin American countries. For example, in Suriname, which has a population of only 400,000, an estimated 10,000–15,000 people are involved in small-scale gold mining (de Kom et al., 1998). In neighbouring Guyana, at least 14,500 small-scale mining permits and 1800 licenses for dredging have been issued by the Geology and Mines Commission (Colchester et al., 2002).

In this, the final section of the book, a series of Latin American case studies of artisanal and small-scale mining are presented. The first case, by Andrea

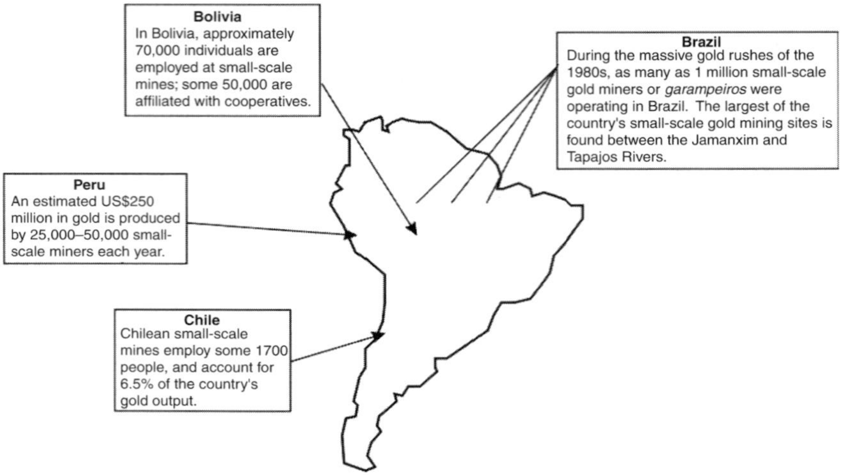

Figure 34.1 Important small-scale mining regions in Latin America.

Table 34.1 Small-scale mining employment estimates in Latin America.

Country	Estimated employment
Argentina	5,800
Bolivia	70,000
Brazil	1,000,000
Columbia	100,000–200,000
Ecuador	92,000
Guyana	10,000–20,000
Haiti	4,500
Mexico	20,000–40,000
Peru	20,000
Suriname	10,000–15,000

Sources: UN, 1996; ILO, 1999; Ávila, 2000.

Seeling, reports on fieldwork conducted in Peru. The author aims to generate a series of livelihood profiles for a sample of artisanal mining households in the community of La Eugenia in the Sur Medio of Peru. The research, which is structured around the UK Department for International Development's Sustainable Livelihoods Framework, addresses the many technical, socioeconomic, legislative, organizational and environmental aspects of the industry.

Marieke Heemskerk and Rachael van der Kooye then describe the dynamics of artisanal gold mining in Suriname. The authors show how resident operators, despite constituting a small percentage of the total number of miners working in the Trans-Atlantic region, have had profound economic, political, land-use and environmental impacts in Suriname; the industry is a source of livelihood for many of the country's impoverished people.

In Chapter 37, Luiz Drude de Lacerda, in a case study of Brazil, provides an in-depth analysis of the impacts mercury emanating from small-scale gold mining activity has had on the environment. The author makes a link between atmospheric emissions of mercury from gold mining with fish contamination and human exposure, using the large amount of data now available for the Brazilian Amazon.

In Chapter 38, Håkan Tarras-Wahlberg, B. Lundberg, A. Bermeo, and R. Hoffner examine the socioeconomic impacts of artisanal and small-scale mining in Ecuador. In their case study analysis, the authors use data from the recently-completed Mining Sector Development Project, *Proyecto Desarollo Minero y su Control Ambiental* (Prodeminca), executed by the Ecuadorian Government, with the support of the World Bank, and Swedish and UK Governments. The authors complement these data with findings from extensive socio-economic and environmental investigations recently undertaken in the country's major mining centres.

In the final chapter, Sergio Castro provides a comprehensive overview of artisanal and small-scale mining in Chile. The author examines production and employment in the sector, and examines both its socioeconomic and environmental impacts. The industry's institutional support framework is also described.

REFERENCES

Ávila, E.C. (2000). *La llamada pequeña minería: un renovado enfoque empresarial*. División Recursos Naturales e Infraestructura, Santiago, Chile.

Colchester, M., La Rose, J. & James, K. (2002). *Mining and Amerindians in Guyana: Exploring Indigenous Perspective on Consultation and Engagement within the Mining Sector in Latin America and the Caribbean*. Executive Summary, http://www.nsiins.ca/download/Guyana/Guyana_Exec_Summary_eng.pdf

de Kom, J.F.M., van der Voet, G.B. & de Wolff, F.A. (1998). Mercury exposure of maroon workers in the small scale gold mining in Suriname. *Environmental Research*, 77, 91–97.

ILO (1999). *Social and labour issues in small-scale mines*. Report for discussion at the Tripartite Meeting on Social and Labour Issues in Small-Scale Mines, International Labour Organization, Sectoral Activities Programme, International Labour Office, Geneva.

United Nations (UN) (1996). *Recent Developments in Small-Scale Mining*. Economic and Social Council, United Nations, New York.

35

Report on Field Research into the Socio-Economic and Social Impact of Artisanal and Small-Scale Mining in Peru

ANDREA SEELING

The main purpose of the research reported in this chapter was to generate a set of detailed livelihood profiles of a sample of artisanal mining households in Peru, in order to increase general understanding of their work and lives. In particular, the work encompassed an analysis of key documents on socio-economic and social impacts of small-scale mining on affected communities, especially in the working region of the GAMA project (managed by Projekt-Consult) in Southern Peru. Based on this analysis, suggestions were made as to how micro-level small-scale mining interventions – in general as well as project-specific – might be improved in order to optimise socio-economic and social benefits. On a broader level, the investigation sought to provide lessons about the usefulness of the Sustainable Livelihoods (SL) approach to livelihood research in the artisanal and small-scale mining (ASM) sector.

Artisanal and small-scale mining is still perceived by many as an activity steeped in illegality and delinquency. It continues to have a critical social welfare function, with its low barriers to entry and ability to absorb large amounts of labour. It serves as an important cash-generating complement to agricultural activity in some countries, and in others, such as Peru, as the ultimate safety net in times of economic stress. These considerations have raised the question as to whether there is intrinsic value in preserving artisanal and small-scale mining as a "way of life", or whether to begin to see small-scale mining more as a means rather than an end in and of itself – *i.e.* in terms of its capacity to contribute to the development of more sustainable livelihoods, economic growth and development, and to the improvement of quality-of-life in rural areas. The main challenge was ascertaining what impact artisanal mining has on the livelihoods of the people involved in the sector, and how it contributes to poverty reduction.

The research was structured around the UK Government's Department for International Development's Sustainable Livelihoods Framework in order to capture the complexity of the issues surrounding those engaged in artisanal mining: environmental, technical, social, cultural, organisational, legal and economic. The main outcome of the research was a set of detailed livelihood profiles of a sample of artisanal miners in the mining community of La Eugenia in the Sur Medio of Peru, and an increased general understanding of their work and lives.

The research has confirmed some important facts and assumptions about the sector. First, there is no generic profile of an artisanal small-scale miner in Peru; motivations, backgrounds, skill levels and education vary greatly between and within communities. Secondly, living conditions for women and children in artisanal mining communities are particularly harsh. The research revealed an alarming level of alcohol abuse and domestic violence, and no support structures to deal with these problems. Thirdly, working conditions in La Eugenia are expectedly unsafe: most people wear absolutely no protective clothing whatsoever, even when handling highly toxic materials such as mercury or cyanide. Fourthly, and typically, infrastructure services are as good as non-existent; there is an acute lack of water and sanitation services, waste management, roads, communications and affordable electricity. Finally, the level of trust among community members, and with it, effective community organisation, is extremely low.

The research also confirmed that any approach to improve the situation of those involved in artisanal mining must be cross-sectoral. Some of the core strategies include the generation of alternative livelihood opportunities, bearing in mind that any programme methodology has to reflect the specific needs of women and must allow for diversity in their circumstances, and the formalisation of artisanal mining. The government of Peru has already taken an important step in formalising the sector by passing Law No. 27651, *Formalisation and Promotion of Small-scale and Artisanal Mining*, in January 2002.

SUMMARY OF LIVELIHOOD ANALYSIS IN LA EUGENIA

Research methodology

"Artisanal (small-scale) mining is the most primitive type of mining, characterised by individuals or groups exploiting deposits – usually illegally – with the simplest equipment."[1] As a livelihood strategy, small-scale and artisanal mining comprises economic, social, cultural, political and environmental aspects. In order to take account of this diversity and the highly context-specific nature of people's realities, the Sustainable Livelihoods (SL) framework, devised by the UK's Department for International Development (DFID), was selected as the main

[1] World Bank IEN Occasional Paper No. 6 – Regularising Informal Mining, editor B. Mamadou.

analytical tool. SL approaches first became prominent in DFID in 1998 and were originally developed as a means by which to study rural people's livelihoods. As described in DFID's SL Guidance Sheets (1999),

> a livelihood comprises the capabilities, assets (including both material and social resources) and activities required for a means of living. A livelihood is sustainable when it can cope with and recover from stresses and shocks and maintain or enhance its capabilities and assets both now and into the future, while not undermining its resource base.[2]

Sustainable Livelihoods Analysis (SLA) is, above all, people-centred and takes a holistic view of the factors that cause poverty, including lack of assets, lack of access to services, institutional inadequacies and vulnerabilities. SLA, therefore, offered a promising approach to structuring research on Peru's artisanal and small-scale mining sector's current contribution to poverty reduction, starting with the premise of looking at the full context of the capabilities, assets, and sources of vulnerability of the community in question. SLA also offered a way of elucidating the links between the micro and the macro-levels, looking at the market and policy environments.

The SL approach is not described in detail in this chapter, as it is assumed that the reader will have a level of knowledge of SL concepts, although this is not totally necessary in order to understand the main findings of the research. However, Table 35.1 provides a brief summary of the core concepts of the sustainable livelihoods approach and should be read with an appreciation of the issues surrounding artisanal mining communities (*e.g.* social aspects, regulation, technology, investment and the environment).

The *SL Framework* (see Fig. 35.1) provides a structure and focus for thinking about systemic change and a mechanism to ensure that the voices and concerns of the poor – men, women and children of all ages and abilities – are heard. It is a practical analytical tool for understanding livelihoods systems and strategies in all their complexity, and, as such, needs to be highly context-specific. The framework makes explicit the relationships between poverty and vulnerability by analysing people's livelihood assets: social capital, human capital, natural capital, financial capital and physical capital.

Data collection

The field research used for the study described in this chapter was carried out in the artisanal mining community of La Eugenia, located in the district of Arequipa in the Sur Medio of Peru (see Fig. 35.2).

The site is located at an altitude of approximately 1,800 metres, with an annual average day-time temperature of 22/23 degrees C. At night, temperatures drop by about 10–12 degrees C. Rainfall, and hence, vegetation, are virtually non-existent

[2] DFID, 1999, Sustainable Livelihoods Guidance Sheets, DFID: London, 1.1.

Table 35.1 SLA core concepts.

Concept	Outline
Vulnerability context	This "frames the external environment in which people exist. People's livelihoods and the wider availability of assets are fundamentally affected by critical **trends** as well as by **shocks** and **seasonality** – over which they have limited or no control."
Livelihood assets	The study of livelihood assets "seeks to gain an accurate and realistic understanding of people's strengths (assets or capital endowments) and how they endeavour to convert these into positive livelihood outcomes. The approach is founded on a belief that people require a range of assets to achieve positive livelihood outcomes; no single category of assets on its own is sufficient to yield all the many and varied livelihood outcomes that people seek." The study of assets is broken down into categories of human, social, natural, physical and financial capitals.
Policies, Institutions and Processes (formerly "Transforming structures and processes")	"Transforming structures and processes within the livelihoods framework are the institutions, organisations, policies and legislation that shape livelihoods." The SLA examines them in terms of the effect they have on livelihood assets and the degree to which they include or exclude a group of people and provide them with a sense of well-being.
Livelihood strategies and outcomes	*Strategies* is "the overarching term used to denote the range and combination of activities and choices that people make/undertake in order to achieve their livelihood goals." The SLA seeks to promote choice, opportunity and diversity. *Outcomes* are "the achievements or outputs of *Livelihood Strategies*." It must be recognised that the outcomes that people pursue may be very different from those of the researcher.

Source: Rouse, J. & Ali, M., 2001, Waste Pickers in Dhaka, WEDC, Loughborough University, p. 7.

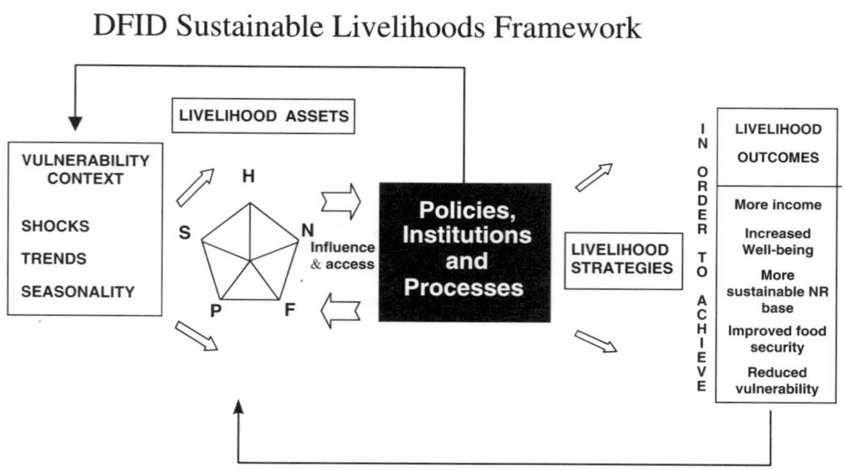

Figure 35.1 DFID SL framework.

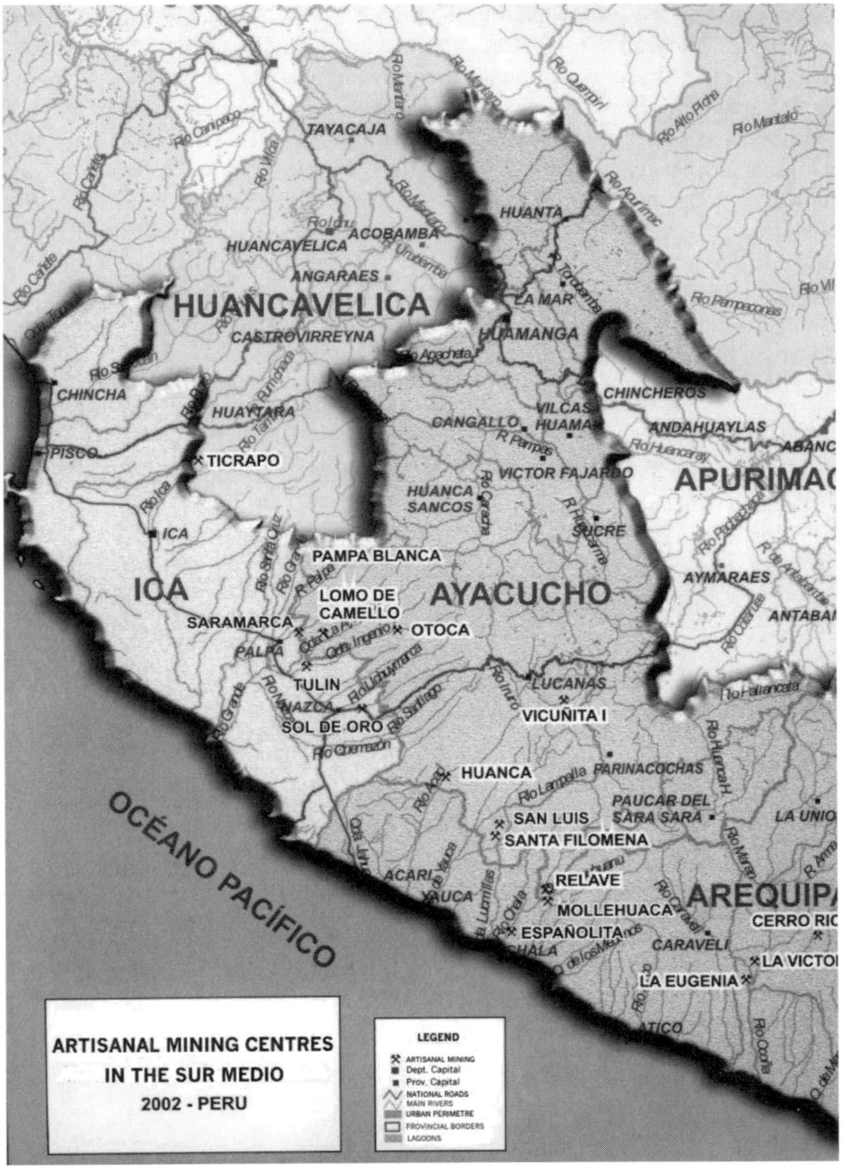

Figure 35.2 ASM communities in the Sur Medio of Peru.

in the region, which means that mining activity is not subject to seasonal variability. The area is not connected to the national power and/or telecommunications network, and there are no water and sanitation services in La Eugenia. The German development agency, GTZ, tried to install pipelines in the community approximately five years ago, but the project was later abandoned for reasons unknown. The nearest town, Caravelí, can be reached by dirt road in approximately three hours; apart from that, there are no other major roads or settlements in the area. The community has been in existence for decades, and population numbers have fluctuated greatly over time. Currently, La Eugenia's total population is estimated at just under 1,000 (Table 35.2 below gives a detailed breakdown). Strictly speaking, La Eugenia comprises five individual "barrios" – San José, Santa Rita, Santa Rosa, Bocamina, and Eugenia/Polvorín, all of which are quite different in size and character. The main services (medical post, nursery, school, telephone/radio post, satellite dish and TV, electricity generator, churches, shops, restaurants, and bars) are located in San José, and people move freely between barrios in order to secure work or to socialise.

Table 35.2 Eugenia population estimate.

Population estimation by age groups La Eugenia

Age	Number
<1	17
1	17
2	17
3	17
4	17
5	11
6	11
7	11
8	11
9	11
10–14	43
15–19	79
20–24	167
25–29	142
30–34	118
35–39	84
40–44	57
45–49	37
50–54	32
55–59	14
60–64	13
65 y+	18
Total	944

Women of child-bearing age	
Not pregnant	165
Births	19
Pregnant	15

Source: RED CAMANA – CARAVELÍ: 2002.

Whilst La Eugenia is located in one of the priority intervention regions of Proyecto GAMA, the project has not actually been active in this community. The reason for choosing to conduct field research in La Eugenia was to obtain data that was as "unbiased" and crude as possible in order to capture the livelihood conditions that form the reality for the majority of Peru's artisanal miners. Before commencing the research, the community was made well aware that there would be no immediate technical or other assistance as a result of the study. Ideally, the data collected in La Eugenia would have been compared with data from one of GAMA's project communities; however, time and resource limitations did not allow for this interesting exercise. The data was collected during a period of four weeks (16 days desk research and literature review in Lima, and nine days field research in La Eugenia). The data collection, therefore, focused on depth rather than on breadth (a total of 13 in-depth individual household surveys were conducted). The following methods were employed:

- household selection based on *level of poverty* (determined through wealth ranking exercise);
- stakeholder analysis to disaggregate by salient variables such as status within the community, ethnic group and gender, and to elicit the varying impacts of mining activity on the livelihoods of different groups; the stakeholder analysis also helped to identify key informants;
- semi-structured interviews and focus group discussions to build up an understanding of:
  - main economic and livelihood activities, and coping strategies
  - key constraints for these livelihood activities
  - intra-household allocation of income, tasks and responsibilities
  - endogenous and exogenous risks and shocks
  - trends
  - prevailing structures and processes;
- the findings were cross-checked with findings from participatory, community level, exercises and survey samples for representativeness.

RESULTS

The results of the research will be presented following the SL framework in order to make them more accessible to the reader. The framework does not attempt to provide an exact representation of reality. Further, it is a way of thinking about the objectives, scope and priorities for development, and to help understand and analyse the livelihoods of artisanal mining communities. Since people, rather than the resources they use or the governments that serve them, are the priority concern of the approach, we start by analysing the assets and strengths, rather than the needs, of the members of the community. These were established using tools such as focus group discussions, semi-structured interviews, and general observation and conversation with community members.

Figure 35.3 Pallaqueo.

Livelihood assets

Human capital

Human assets include the ability to labour, skills and knowledge/information, health, creativity, and control over decisions.

The level of skills in the community varied greatly – from the professional miner with over 30 years' experience to the absolute lay person with only minimal previous exposure to the job. Over half of male respondents expressed a desire to improve their skills and knowledge, particularly with regard to mining techniques, and health and safety at mines. Women, on the other hand, considered that little or no skill was required for carrying out their jobs: "pallaqueo" (sorting ore for its gold content in preparation for the pulverisation process – see Fig. 35.3) and "quimbaletar" (amalgamation process in huge man-powered grindstones where the pulverised mineral is treated with water and mercury – see Fig. 35.4), which can both be learned within the space of a few weeks or hours respectively. This does not necessarily indicate that women have a low opinion of themselves, especially as they all identified the hard physical labour involved in these activities. All female respondents did, however, wish to develop additional skills, ranging from confectionery to computing, in order to escape the drudgery of their daily existence and to increase their income.

The level of education amongst respondents was equally varied, ranging from university graduates to illiterate. In general, the literacy rate in the community seems to be quite high (probably 90%). Education, either for themselves or for their children, was high on the respondents' list of priorities, as it is still perceived

Figure 35.4 Quimbaletar.

as one of the most promising ways out of poverty. Nevertheless, schooling was something of an issue in the community. There is only one primary school in La Eugenia, which basically comprises three classes:

a) first and second grade, 33 children between 6 and 7½ years;
b) third and fourth grade, 22 children between the ages of 8 and 12; and
c) fifth and sixth grade, 22 children between 10 and 12 years.

If parents want their children to progress to secondary education, they have to send them to live with relatives, friends, or in "pensiones" in bigger cities. Only a few households can afford to do this. Most parents considered their children's education substandard, due not least to the frequent absences of the teachers (who explained that during their absences they are "either visiting their families or attending training courses themselves") and the lack of adequate teaching materials and books. When asked about the level of school attendance in the community, the teachers affirmed that all children in the community attend class at all times. This, however, could not be seen and contradicted the views of some parents, who admitted to keeping their older children at home to look after younger siblings, or to help out with the "pallaqueo".

Whilst education as a whole was perceived as important, even those with high levels of education lamented the lack of employment opportunities in the cities, one of the most cited reasons for entering the ASM sector in the first place. It seems clear that many artisanal miners would take up training opportunities if they were on offer, and most, even the least well-off, indicated that they would be willing to pay for these opportunities, provided they were affordable and offered real employment prospects for the future.

Table 35.3 Eugenia general morbidity, 2001.

Causes of morbidity	Cases
Common cold	62.5
Parasitis	65.7
Light concussion	21.5
Food poisoning	26.9
Other acute respiratory illnesses	50.6
Measles	60.5
Amebiosis	1.07
Gastroenteritis	1.07
Pneumonia	1.07
Causes of maternal morbidity	
Urinary tract infections	4.32
Inflammation of the pelvic	2.16
Causes of infant morbidity	
Acute respiratory infections	59.5
Common cold	8.65
Pneumonia	4.32

Rate = Nr × 1000/PT (total population).

Table 35.4 Eugenia: general mortality.

Causes of mortality	1997	1998	1999	2000	2001
Severe concussion		1	1		
Cholera		1			
Burns			1		
Suicide			1		
Politraumatism					2.15
Asphyxia		1			
Heart attack		1			

Using rates as solicited for each year (×1000).

Whilst there were surprisingly few accidents in the community that can be directly related to mining activity (given the low level of safety in the mines, it was expected that many more serious injuries would occur as a result of falls in mine shafts, collapse of underground workings, falling stones, and misuse of explosives), the level of respiratory problems was relatively high, especially among children (see Tables 35.3–35.5). These are, to a large extent, likely to have been caused by contaminated dust. All members of the community expressed a great concern over the level of contamination caused by mercury and other chemicals used in the amalgamation process. They are, however, lacking information on the *exact* effects on their health and protective measures available to them. "Sabemos que estamos contaminandonos. Pero que podemos hacer? Tenemos que trabajar y no hay otro trabajo." ("*We know that we are contaminating ourselves. But what can we do? We have to work and there is no*

Table 35.5 Eugenia: infant mortality (>1 year).

Causes of mortality	1997	1998	1999	2000	2001
Pneumonia		1			
Respiratory insufficiency			1		
Asphyxia				1	

Using rates as solicited for each year (×1000).

Figure 35.5 Handling mercury without mask and protective clothing.

other employment." Sra Dora de la Cruz, cited during semi-structured interview held on 11 July 2002). The usage of retorts in La Eugenia is virtually unknown or considered too expensive so that mercury is handled without any protective clothing whatsoever (see Fig. 35.5), and children were observed to be playing with the pretty, silvery liquid – unreprimanded by their nearby parents.

Another major health problem is contamination from waste – both household and human. Sanitary facilities are as good as non-existent (there is one disused latrine for the children in the school), and most respondents acknowledged to using the "open hill" system, even though women complained about harassment (both verbal and occasionally physical) by men. Household waste is disposed of at random, often only a few metres from dwellings, with pigs roaming freely in the village (see Fig. 35.6). Due to these conditions, diarrhoea, especially among infants, is common throughout the community, a fact confirmed by health officials in the area. Villagers also reported outbreaks of dysentery and cholera in the past. Other injuries related to household waste are cuts to hands and feet from broken glass and sharp tin can edges.

Figure 35.6 Dumped rubbish in La Eugenia.

Health care in the community is very basic. The medical staff (one trainee doctor and two nurses) are not authorised to perform surgery, nor does the doctor have access to sterile surgical instruments, and the nearest fully equipped hospital is in Camaná, an eight hour journey away. In cases of transferral to the hospital, patients have to find their own transport (often by foot or hitching a lift on the back of a truck). The cost of a consultation for adults is three soles (US$0.90), and children up to the age of 17 are free-of-charge, as they fall under the national *Seguro Medico Escolar*. However, if a person cannot afford the price of a consultation, the doctor will not charge them. In general, the medication available in the post is generic; it is not branded, and has, in some cases, been donated by the hospital in Camaná; patients have to buy their medication from the medical post at a nominal cost, but many can still only afford to buy drugs for one or two days rather than the whole course of a treatment.

Means of contraception (the pill, injections, coils, condoms) are available free-of-charge; however, relatively few women solicit the family planning services provided by the medical post. It seems that the majority of women do not know enough about family planning or do not see the advantage to it. There are also a number of cases were women are prohibited from using birth control by their husbands or partners. Cases of HIV/AIDS were not reported, although prostitution is not unknown in the community (these tend to be women from outside La Eugenia, who come into the village at the weekends, when the bars are full of men with disposable cash to spend on alcohol and women). Excessive alcohol consumption was considered to be one of the major problems in the community; some respondents even estimated that about half of all men spend all their hard-earned cash on drink. For comparison, the price of one 0.75 l bottle of beer is five soles (US$1.49) as opposed to a 220 l canister of fresh water, which is

13–14 soles (approx. US$4.18). Alcohol abuse was also seen to be the main cause of domestic violence (both against women and children), a phenomenon that is alarmingly widespread in the community. (During the time of the research, one woman was beaten so severely by her partner that a neighbour took it upon himself to alert the doctor, which, according to information gathered during women focus groups and individual interviews, is not a common reaction. Usually, domestic violence is considered "family business" and is draped in silence, but once women felt comfortable with the – all-female – research team, they were eager to share their experiences and air their concerns openly.)

There is no trained midwife in the community, and most women give birth in the medical centre. Some women are, however, afraid to use the centre and prefer to give birth in the privacy of their homes. There was one case involving a woman who had given birth to all three of her children on the floor of her hut, by candlelight, and assisted only by her husband – the most recent delivery only a few weeks before the survey was carried out.

The extremely hard living conditions and inability to improve one's situation affect people's, and especially women's, mental health. Many respondents admitted to being depressed a lot of the time, which does have negative effects on their ability to labour and the well-being of their dependants.

Physical/productive capital

This category comprises, for example, shelter, transport, energy, land, cattle, and consumer durables such as household appliances, tools, and clothing.

As already explained above, access to infrastructure such as roads, transport, electricity, communication, waste management and sanitation is extremely limited. Electricity is generated and distributed in San José only, but even there, many households cannot afford the one sole (US$0.30) per day needed to operate an electric light bulb. The main means of lighting are therefore candles, battery powered torches and lanterns, and carbide lamps, which are also used in the mines. Cooking is performed with gas or kerosene, and, in some cases, wood, but the latter is not easily available in the community.

There are four two-way radios in Eugenia, by means of which it is possible to establish telephone connections. These radios transmit signals that are picked up by a radio in Arequipa. The owner of this radio makes the phone call and via the two-way-radio connection the person in Eugenia can communicate with the person they wanted to call. Another service that has been in existence in the community for a relatively short time is Rural Telephone. There is only one phone of this type, which is located in one of the shops in Eugenia. The person who wants to make a phone call needs to buy a pre-paid telephone card from the shop owner, who charges one sol (US$0.30) above the usual selling price per card in order to cover for his expenses. The shop owner is also the one who receives the calls, a service for which he also charges one sol (US$0.30), which usually the people do not pay. The telephone uses a satellite signal and is powered by a solar panel, which is installed outside the shop. On a cloudy day, therefore, it is impossible to make or receive calls as the panel does not generate enough energy. The rural telephone company has offered to improve this service in two ways: by installing two-way communication lines (*i.e.* that one person does not have to wait until the other person has finished speaking); and by making the service available 24 hours a day. This could be an excellent entry point for a government or NGO scheme to improve people's access to information (*e.g.* the price of gold) and knowledge on a variety of things. Some NGOs are already installing rural telephone systems in rural communities in Peru (ITDG, for example), and the MEM may be well advised to explore ways of linking in with one of their programmes.

Dwellings are almost exclusively made of straw matting (see Fig. 35.7), which needs to be ordered and transported by truck from Caravelí, and plastic sheeting for the roof. With temperatures dropping quite sharply at night, the buildings do not offer much protection against the cold. Most of the huts are divided into two parts: kitchen and living-cum-sleeping space. Hygiene standards inside the huts are poor, as dust and dirt enter through every crevice, and water is too scarce and expensive to allow for regular cleaning. Only a few buildings are made of stone (school, medical post, nursery) or have concrete floors. The number of residents per dwelling varies (there is no typical household size, as many people live with relatives, extended family or friends), but in general, living conditions are very cramped.

There was widespread consensus amongst community members that access to land in La Eugenia was relatively easy, except in the centre of San José, where housing prices were deemed prohibitive, especially for commercial property such as a shop or a bar (apparently, prices can rise up to US$1,000). La Eugenia, which is in the district of Urazqui, is recognised as a "centro poblado menor", where everybody has the right to construct a dwelling. Finding a mine to work in, however, is a more difficult enterprise. One either needs to be introduced by a friend or relative to be allowed to enter a working mine (the fear of "informants" or "spies" is widespread), or one has to reinstate an abandoned mine.

Levels of wealth within the community vary greatly, with the most prosperous group being merchants (*e.g.* shopkeepers, bar or restaurant owners,

Figure 35.7 Straw huts in La Eugenia.

water vendors, and traders in explosives and/or mercury). Absolutely everything has to be imported into the community; the richest shopkeepers tend to be those who own their own truck. Villagers complained fiercely about the high cost of living in La Eugenia and the fact that shopkeepers can demand almost any price they want for their goods. (For example, according to one focus group, S/.1 will buy 5–6 kg of onions in Arequipa, whereas the same amount will only buy 1 kg of onions in Eugenia.)

The level of wealth amongst *miners* seems to differ in line with the kinds of tools they have access to: the more sophisticated the tool, the greater the production. Those miners who have nothing but their hands to work with, will usually sell their labour on a day-to-day basis to various mine owners or pulverize ore for others.[3] With one remarkable and extremely successful exception of a female mine owner, all other mines in this study were owned and run by men. Amongst the poorest in the community are the elderly and members of female-headed households. This is not surprising given the low-income activities they are restricted to (women NEVER enter into the mines – partly for the extreme physical strength that is required to extract ore, and partly for cultural beliefs: women in mines are deemed "bad luck").

Because of the lack of vegetation, pigs are the only livestock that is kept in the community, except for the occasional chicken. The people who keep pigs raise them, usually on household refuse, in order to sell them in the big cities rather than for personal or local consumption. Due to the high level of soil contamination in La Eugenia, there have been outbreaks of various diseases among pigs in the past, such that large numbers have had to be culled. The community is currently trying to take concerted action against keepers who let their pigs roam freely in the village, as the fear of illness is widespread.

Natural capital

Includes assets, or access to assets, such as water, land, wildlife, biodiversity, and the environment in general.

For obvious reasons, access to land and water are of key importance to artisanal miners. The mineral deposits in the area form the basis of livelihood strategies, either directly or indirectly, of approximately 1,000 people in La Eugenia. Some members of the community, who have lived at this site for many years, have noticed a decline in the quality and quantity of the deposits over recent years. Whether or not, or to what extent, this is true is beyond the scope of this research; however, on the whole, miners did not seem to think that the deposits were anywhere near exhaustion.

[3] There are a number of ball mills in the community that can be hired for S/.2,5 per bucket if the material is processed in a "quimbalete" that does not belong to the mill owner, or for S/.1,5 per bucket if the quimbalete of the mill owner is also used for amalgamation. Some miners prefer to process their own ore, others prefer to hire a day-labourer who gets paid S/.20 per day plus one refreshment (usually a bottle of fizzy pop). The maximum number of buckets of material that one person can process in a day is eight.

As mentioned previously, the region is extremely dry and vegetation virtually non-existent. Water, both for production and household purposes, is brought in to the community by lorry twice a week. Water comes in two types: *agua salada* (salt water) and *agua dulce* (fresh water). A 220l canister of salt water costs between S/.7–8 and a canister of the same size of fresh water between S/.13–14. In general, salt water is used for production, cleaning and washing, but poorer households also use it for cooking and drinking. As an example, a family of four may use a canister of salt water per week, and a canister of fresh water every two weeks.

In terms of nutrition, people are completely dependent on what is provided in the shops; because of the infertility of the soil and the lack of water, wildlife and livestock, they cannot supplement their diet with fresh meat or home-grown vegetables or crops. Whilst the researcher did not come across any acute cases of malnutrition, the quality and variety of the diet obviously leaves a lot to be desired. Children in particular suffer from vitamin deficiency due to a lack of fresh produce (fruit, in particular) and fresh milk.

Given the fact that there is no waste management system to speak of, the atmosphere in the village is filled with unpleasant smells, and in summer, mosquitoes become a real plague. Furthermore, the prevalence of household waste has attracted rats in great numbers, which can transmit diseases. But more importantly, the contamination caused by waste and toxic dust (mercury, cyanide) which is carried everywhere, is a constant health hazard, especially for children.

All women that participated in the survey who spend their days "pallaqueando", worried about the intensity of the sun, to which they, and often their children, are exposed without any real protection. Combined with the ever-present dust, this makes for very oppressive work conditions indeed. Respondents complained about frequent eye infections, headaches and skin irritations. It seems

Figure 35.8 Women *pallaqueando* in blazing sunlight.

that when talking about a "mining environment", the realities of those who don't spend their days underground, but are nonetheless heavily involved in the mining activity, are often not sufficiently taken into account when designing project interventions.

Social capital

Social assets comprise household relations, membership of groups and communities, relationships of trust, social networks, and access to institutions.

> "La unica solución es la unidad del pueblo." *(Dr. Edgar Zela, interviewed 5 July 2002)*

The diversity of the population of La Eugenia is representative of the country as a whole. People come from all-over Peru: the coast and the highlands, the north and the south, rural and urban centres. The length of time that people stay in the village also varies greatly: some have been living in the village for over 20 years, whilst others, who may originally have intended to spend only a few weeks or months, have ended up staying several years, and some stay only a few months and then leave again. Reasons why people are motivated to join the ASM sector in the first place, and choose La Eugenia as their base, differ from person-to-person, which is also reflected in the kinds of livelihood activities they carry out.

With such diversity and fluctuation, it is difficult to create a sense of unity and common purpose in the community. It seems that the lack of active support networks within the community, however, is one of the main problems in La Eugenia. All 13 interviewees (five women, eight men) complained about the degree of disorganisation and mistrust among community members. This was echoed in focus group discussions, where many deplored the absence of women's associations – *e.g.* for the provision of childcare facilities – and voiced their frustration over the community's inability to deal effectively with the problem of dumped rubbish. People do not tend to forge close relationships with their neighbours and work colleagues, and many complained about the amount of gossip and badmouthing in the community. When asked to whom they would turn to in an emergency or if they simply needed help, over two thirds of the interviewees (in both individual surveys and focus groups) responded that they would turn to their immediate family – either within or outside Eugenia – and no-one else. This atmosphere of distrust and isolation fosters socials ills such as domestic violence, alcoholism, and depression. The people are well aware of the benefits that organising themselves might have, and various attempts at forming associations have been made. As discussed in 29.2.2 below, the "Frente de Defensa", is one such attempt where people have recognised that only as a group will they stand a chance to defend themselves against foreign invasions. Other examples of successful associations are limited to the following:

- Asociación Minera Sta Rita,
- Asociación Minera San José (both associations of miners who intend to form a mining enterprise in the future in order to gain legal status),

- Empresa Minera Victoria (the only mining enterprise in La Eugenia so far that has obtained legal status), and
- the Comite del Vaso de Leche.[4]

In addition, there are a number of religious groupings in the community who have their own "churches" and can and do act as a support network in case of shock. These include:

Israelitas	30–50 adherents
Evangelistas	10–12 adherents
Testigos de Gehova	4 adherents

There is no church for the large majority of Catholic community members in the village, a fact worth mentioning as priests can often act as initiators of concerted community action among their congregation.

Some 80% of households were described as "family units" – *i.e.* a couple, married or cohabiting, with a number of children, including from previous relationships. A large proportion of women and men also has dependent children outside Eugenia and regularly send remittances to support them (*e.g.* for rent, food, clothing and education). Of the remaining 20%, many are single or female-headed households. There is more than likely a strong correlation between the high number of long-term couple relationships and the low incidence of HIV/AIDS in the community, a disease rampant in many African mining communities where men tend to live separated from their wives and family for months or even years on end.

Most women did not think that they have the same status as men within the community, a viewpoint not necessarily shared by their male compatriots, although there are some who would like to see women become more integrated in the community. On the whole, community members are not used to trusting each other or working together – this became particularly apparent during some of the participative exercises conducted within the framework of the research. On one occasion, when the villagers were asked to draw a map of their community, the failure to communicate amongst each other nearly turned this into a futile exercise. In addition, female participation during this exercise was almost zero.

Financial capital

Credit, savings, access to micro-credit, and banking facilities

"There are good months and there are bad months" – this is a reference to the fact that incomes among artisanal miners in La Eugenia vary greatly not only

[4] Government scheme under which each child in Peru is entitled to one glass of milk per day; the committee was founded in 1997 and is comprised of two women. On an irregular basis, the government supplies a quantity of a dried milk powder, supplemented with cereals, which the committee distributed among the parents of school-aged children. The product is actually harmful to some children as it can cause diarrhoea, and many dislike the extremely sweet taste. The product often ends up being fed to pigs.

from household to household, but also from month to month. Money earned depends on the activity carried out (extraction, quimbaletar, pallaquear, etc.), size of the mine and quality of the mineral, access to tools (hammer, pneumatic drill, pulleys, etc.) and explosives. Incomes from "pallaqueo" are usually in the region of 200 to 600 soles (US$60 – 180) per month (typically, a person – usually women – earning less than 300 soles (US$90) "pallaqueando" would have to supplement their income with other activities such as quimbaletar, tool/machine hire, cutting hair, giving lessons, or shop keeping, bars and restaurant assistance). Incomes from mining extraction can range from 500 to 6,000 soles (US$150 – 1.790) or more per month. Some 6,000 soles per month is, of course, an exceptional income, which is achieved only by people who own a particularly lucrative mine, can afford a generator and electrical equipment, form part of a mining association, and/or have a number of day labourers working for them. It is impossible to generalise how income is distributed within households (due to their diverse composition); however, within the sampled family units, income is combined and spent jointly on household expenses such as food, water, energy, personal hygiene, medicine, transport and education (where applicable). Very little is spent on clothing, and, in most cases, people are unable to save money from one month to the next. Frequently, monthly expenditure exceeds monthly income, in which case, people are either able to compensate their shortcomings with earnings from the previous month(s) or they buy basic necessities (both for working and living) on credit. Borrowing money from friends or relatives for larger investments also seems to be common practice.

Whilst most of the respondents expressed a wish for access to credit, there is also a certain degree of scepticism, and even apprehension, associated with taking out a loan from "unknown" sources. Two thirds of respondents would want to use a credit for buying mining equipment or consumables (pneumatic drill, compressor, explosives) or gold. The rest would like to take out a loan in order to leave the sector, for example to go back into agriculture, open a shop or to simply buy a house or a piece of land in the city. Banks and micro-credit institutions are, however, extremely reluctant to lend to artisanal miners because of the volatility and semi-legality of their activity, and possibly because of the sector's bad reputation in terms of social conflict and environmental damage. In any case, given the nature of people's requirements in La Eugenia, it may be more useful to instigate a tool-hire or -leasing programme rather than a credit scheme.

Banking facilities in La Eugenia are non-existent. Most miners sell their final product (gold) locally, either by using the same buyer all the time or shopping around for the best price among several buyers.[5] The buyers then sell the gold on to intermediaries in Arequipa (a 10-hour bus or truck journey away) or other big cities. Occasionally, the miners prefer to sell their product in the

[5] In general, gold buyers are shopkeepers at the same time, also selling explosives and/or mercury. Buyers obtain the global price of gold from the radio or sometimes the press (often days old).

bigger cities themselves, which allows them to obtain a better price and to deposit their earnings in a bank straightaway. However, the journey is expensive, long, strenuous and potentially dangerous, not only because road conditions are precarious, but also because buses and/or trucks from the region frequently fall victim to armed robberies. All respondents expressed the wish to receive up-to-date gold prices on a daily basis from an independent source rather than from the local buyers.

Political capital
People's ability to influence policy decision-making and the political process.

Whilst the SL framework does not include this type of asset, it was judged suitable to incorporate this category because, if harnessed wisely, artisanal and small-scale miners possess a potentially very influential capital – their sheer number and the contribution their production makes to the national economy. It would, for example, be very difficult for the government to ignore the demands of a united ASM labour force, given the many thousands of people who are involved in the sector either directly or indirectly. Not only does the ASM sector contribute to the regional and national economy, but it can also provide a motor for enhancing a country's poverty reduction strategy. After all, the sector does represent an important source of employment and income for the miners, their families and the community as a whole, even if this may – as is the case in La Eugenia – to a large extent be due to the lack of employment opportunities elsewhere. Having recognised the sector's growing importance as a "safety net" for the poor (not to mention the fact that it contributes over 7% of the country's total gold production), artisanal miners all over Peru have formed associations in order to improve their situation and to force the government to give the artisanal mining sector its rightful legal status. In September 2001, the First Convention of the Artisanal Miners of the Sur Medio was held, with over 30 nominated delegates from associations in Ica, Arequipa and Ayacucho and another 30 plus participants from Puno, Moquegua and other mining areas. The aim of this gathering, which was held under the auspices of Proyecto GAMA, the World Bank, the ILO, and the Dirección Regional de Energia y Minas Ayacucho, was to put forward proposals for recognizing artisanal mining as a legal activity. The outcome of this concerted action was the promulgation of Law No. 27651, *Formalisation and Promotion of Small-scale and Artisanal Mining*, on 24 January 2002, which created the basic conditions that will allow the formalisation of artisanal miners, becoming "artisanal mining producers", either as regular concession holders in free areas or as contractors in the new modality of artisanal exploitation agreements in consensus with existing concession holders. This law is the first step in the official recognition of the sector, which was made possible only because artisanal miners recognised and made use of their political capital. Since then, a second Convention was held between 30 June and 1 July 2002 in Ayacucho, the aim of which was to consolidate the organisation and to drive forward the development of artisanal mining enterprises.

Vulnerability context

Critical trends, shocks and seasonality over which people have limited or no control

Whilst it would seem understandable if artisanal miners were mostly concerned about shocks related to their health and that of their families, one of the most cited threats amongst respondents was the ever-present risk of becoming the victim of so-called "invasiones", where gangs of bandits external to the community (usually from the nearby Nazca coastal region) attack the village and occupy particularly prosperous mines by force. So great is the fear of this danger that the community has recently organised a night watch rota, in which all male community members are obliged to participate (Anybody who misses their turn three times is expelled from the community). At the moment, women are not allowed to take part in this scheme, although many have expressed an interest to do so, not only within the framework of this research but also publicly during previous community reunions.

The risk of ill health resulting from accidents and contamination at work is due significantly to the lack of safety measures applied. The use of mercury is an integral part of the gold recovery process: once ore has been excavated (using explosives, drills, hammers and sheer manpower) and sorted (pallaqueo), it undergoes a series of poundings (hammer, ball mill) into fine powder. Subsequently, the mineral is mixed with water and mercury in a quimbalete (usually 15–20 minutes for every 33 kg of mineral, depending on the quality and quantity of gold contained therein). The resulting amalgam is then separated by heating. Health risks occur at all stages of this process. Illness or injury resulting in an individual being unable to work, for however long or short a period, would obviously result in a loss of earnings that may have serious implications not only for the individual him- or herself but also for his or her dependants. Most participants in the survey stated that they had no savings or other capital assets (e.g. property, land, machinery) that could be used/sold in emergencies, e.g. for medical expenses or to substitute income deficits. Most of the respondents involved in mining activities considered their work potentially dangerous and are well aware of the risks, particularly with regard to the long-term effects of mercury contamination. However, living and working in Eugenia for most people is a conscious decision that has been taken under full acceptance of the risks. That said, all interviewees would like to see their living and working conditions improved.

Given the region's geomorphology and agro-ecological conditions (as described in earlier sections), mining activity in La Eugenia is not subject to seasonal variability; it is carried out full-time and all year round. The region is, however, prone to earthquakes, with the last major tremor, the Atico Earthquake, hitting the region in June 2001. Although tragic, the death toll was fortunately fairly low considering the magnitude of this earthquake, at the time, the largest one to have occurred in the world in the past 25 years. The damage was localised in quite a small area in and around Atico (see Fig. 35.2), and no structural damage to either buildings or mine shafts was recorded in La Eugenia. However,

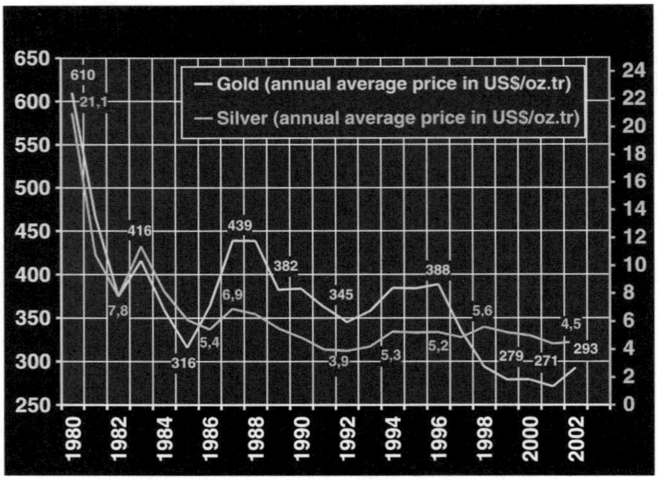

Figure 35.9 Gold and silver prices, 1980–2002.
Source: APEMIN Project Presentation, 2002.

many people originating from the Atico area, and who are now living in Eugenia, have been affected by the disaster in that they have lost most or all of their possessions (including homes). Miners who were working underground at the time of the earthquake described their fear at the time when the earth suddenly started shaking and rocks started falling. Miraculously, none of the pits collapsed and there were no casualties. The school teaches children what to do in the case of an earthquake, and the medical post also gives out advice. However, these measures are only negligible and could not mitigate the effects of a more localised tremor. Medical supplies and facilities are insufficient and would take a long time to reach the community in an emergency because of the area's inaccessibility.

La Eugenia produces an estimated 600 kg of gold per annum. Without exception, this is being sold on the national market, which is subject to global gold price fluctuations. Whilst the price of gold is generally considered more stable and profitable than, say, the price of copper, there have been serious price variations in recent years (see Figs. 35.9 and 35.10). Financial instability related to global trends is thus a serious risk to which artisanal miners are exposed at all times. In addition, the uncertainty over the quality and quantity of their production does not foster a climate for long-term investments or even mere planning ahead in the medium-term.

Due to their often illegal and clandestine activity, artisanal miners in general have little or no recourse to the law. In La Eugenia, there is no permanent judicial or police representation, so that villagers are often forced to take the law into their own hands. As described above, women and children are particularly vulnerable to domestic violence, and there is a constant risk of being verbally or even physically abused by men. Rape is not unknown either, and

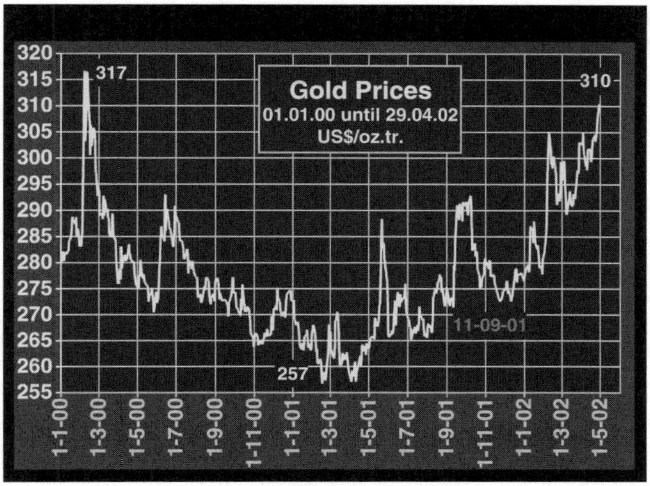

Figure 35.10 Gold prices 1/1/00–29/4/02.
Source: APEMIN Project Presentation, 2002.

villagers told of a particularly tragic and brutal case involving a 14–year old girl in the relatively recent past. It is very likely that because of the stigma attached to rape victims and the lack of appropriate authorities in the village, many cases of sexual assault go unreported.

Policies, institutions and processes

These include institutions, organisations, policies and legislation that shape livelihoods

The influence of *Policies, Institutions and Processes*, both positive and negative, and direct and indirect, are related to all elements of the SL framework. They are, for example, linked invariably to the *Vulnerability Context* as the existence or lack of processes (policies) can affect trends or cushion the impact of external shocks in a number of ways (fiscal policy impacts on economic trends, health policy impacts on population trends, etc.).

Peru is one of the few countries where the government has taken action to recognize and promote artisanal mining activities within a legal framework, thus not only increasing people's sense of self-control, but also laying the groundwork for sustainable development of the sector. It is a well-known fact that many environmental and social problems only exist because of the sector's illegality and the authorities' unwillingness to "get involved": lack of mining and processing know-how, lack of financial and material resources to install mitigative measures, no clear system of land titles, and the lack of access to acceptable tribunals all mean that miners are constantly at risk of losing their possessions and future gains. The absence of a legal framework also makes them vulnerable to all types of corruption and criminal behaviour, thus perpetuating the illegality-poverty continuum.

Policies to extend social services into the areas where artisanal mining is carried out are currently not in place in Peru. However, these could improve people's well-being enormously, whilst the provision of social safety nets could significantly reduce people's vulnerability. The community's social structures and informal institutions are, to a great extent, described under *Social Capital* and will therefore not be reiterated here.

There is only one form of local authority to speak of in La Eugenia. The community has an elected *Teniente Gobernador* (Governor), whose functions include maintaining order, liaising with the district authorities in Camaná, dealing with indictments and other honorary tasks. The current incumbent, Roman Pequeña Aguilar, a 29-year old with no previous governance experience, was elected to post by popular vote only a few months ago and for an indefinite period. Should the community grow dissatisfied with his performance, they will remove him from office and elect a new Teniente Gobernador. However, when talking to community members from barrios other than San José, it transpired that most people did not actually know Sr Aguilar or were not aware of (or interested in) his functions as Governor. A great deal of awareness-raising is necessary in order to ensure better community relations and broader community involvement in planning and decision-making. But the community cannot solve these problems alone. As discussed below, building capacity may be part of the solution.

It is probably fair to say that the government does not really endeavour to play a key role in the social and economic life of artisanal mining communities, and it is therefore seen in a rather negative light. On more than one occasion, it was explained that "we [the miners] work for our fatherland but not for the government". People do not expect anything from the government, and it is only now, with the creation of various mining associations and the Empresa Victoria, whose president, Manuel Reinoso, was heavily involved in the two Conventions of the Artisanal Miners of the Sur Medio, that they have become aware of their rights, and understand the potential benefits of closer co-operation with the government. This is an important step to reducing levels of mistrust on both sides. At the moment, however, the necessary structures of governance for the industry and administrative norms do not exist, both at the local and global levels.

Livelihood strategies

Range and combination of activities and choices that people make/ undertake in order to achieve their livelihood goals.

In view of the adversities described in earlier sections, why do people still choose to become involved in the ASM sector? This was one question central to this research, as the reasoning behind present livelihood activities might reveal a lack of options and alternatives elsewhere, and possibly suggest future intervention strategies.

The lack of employment opportunities, both in the cities and the rural sector, was cited most frequently as the prime motivation behind entering the ASM sector and for *remaining* involved in it. Besides, entry into the sector is relatively easy as it requires few skills, little previous knowledge and hardly any initial capital. Many of the respondents had already spent several years in La Eugenia and/or other mining sites and were not envisaging moving in the foreseeable future. More women than men were eager to leave La Eugenia and to carry out alternative, less strenuous, livelihood activities, whereas men, in general, wanted to become more proficient in their work in order to increase their output.

This division in preference is not surprising, as living conditions for women (and children) tend to be much more difficult in the very male dominated and male oriented ASM communities. At present, diversification opportunities for both men and women in La Eugenia are extremely limited; out of the thirteen respondents, six men out of eight and one woman out of five replied that they were not involved in any other income-generating activities besides mining. Four women admitted to having to supplement their income from mining with other menial tasks such as cutting hair or helping out in shops and restaurants. Only one of the interviewed women had had enough start-up capital to invest in a small business (a billiard table). Table 35.6 shows the typical daily activities of one female and one male respondent – within families, looking after children and carrying out the housework (cooking, washing, cleaning, etc.) are almost exclusively the tasks of women.

The factors that determine the wealth of a household are not necessarily determined by access to natural resources (*i.e.* land, minerals, water), which is basically the same for all community members, but rather differences in financial and human capital. Access to financial capital makes it possible to purchase better mining equipment and consumables or to become a member of a mining association or mining enterprise. Improved human capital would put people in a better position to recognise and defend their rights, organise themselves, and to better manage existing ventures or to set up new ones. In general, the different livelihood strategies that people pursue are dependent on their financial status and their ability to take risks.

The dry climate in the region and the lack of water does not allow for any agricultural activities from which people could either supplement their income or enhance their food security. Furthermore, the access to markets (other than within the community) is extremely limited. At the moment, the only products that are sold outside the community are gold and a few pigs.

The poorest people in the community – *i.e.* those who do not possess their own home and have no regular employment – are entirely dependent on the goodwill of others in the community, who will put them up for a few days or weeks. However, this kind of assistance is not limitless, and those who cannot support themselves over a long period of time will find that they exhaust the generosity and obligations of their friends, neighbours or relatives. Without outside aid, it is very unlikely that these people will ever be able to enhance their asset base (financial, human, social, physical), and to improve their livelihoods.

Many families compromise the education of their children by sending them to help with the "pallaqueo", whereas others make huge sacrifices in order to send their children to secondary or higher education outside the community, even though career prospects through education in Peru are generally poor.

Table 35.6 Typical daily activity charts for women and men.

Time	Activity
Dionisia Muñoz	
6:00	Get up
6:00–7:00	Have a wash, cook breakfast and lunch, wash clothes
7:00–7:20	Breakfast
7:20–7.40	Continue washing clothes, clean the house
7:40–8:20	Look after the children
8:40	One child goes to kindergarden
9:00–13:00	Pallaqueo
13:00–13:30	Lunch at home with the children
13:30–14:30	Continue washing clothes
14:30–17:00	Pallaqueo (with children)
17:00	Go home, carrying mineral
17:15–17:30	Have a wash
17:30–18:00	Prepare dinner
18:00–18:30	Have dinner with children and husband (if he's already home)
18:30–21:30	Have a rest, go to watch TV, chat with neighbours/friends
21:30 ó 22:00	Go to bed
Raul Martinez	
5:30–6:00	Wake up and listen to radio
6:00	Get up
6:10–6:15	Go to work
6:15–8:00	Work (transport rocks in wheelbarrow)
8:00–8:30/9:00	Return home to have breakfast
9:00–13:00	Work
13:00–14:00	Return home to have lunch
14:00–17:00	Work
17:00–17:30	Descansar y luego regresar a casa
17:30–17:45	Regresar a casa
17:45–18:30	Have a break before going home
18:30–18:40/19:00	Dinner
19:00–20:30	Read
20:30–21:00	Listen to radio
21:00–22:00	Get ready for bed

The richest households in the community rely mainly on commerce for their livelihoods. These are often far-sighted people who recognised a business opportunity, and are willing to put up with the hardships and discomforts of life in a remote mining site such as La Eugenia. Here, people tend to obtain much better prices for their products than they would in the bigger cities, where competition obviously is much fiercer. Their access to markets is also favourably affected by their wealth as they can afford their own means of transport.

Livelihood outcomes

Achievements or outputs of livelihood strategies

Understanding the priorities, objectives and ambitions of artisanal miners is vital when trying to design more effective interventions that address the specific needs of all involved, as well as broader development objectives. People's reasons for entering the ASM sector have been discussed in previous sections of this chapter. In summary, ASM is considered a real, and sometimes the only, alternative to unemployment or labour market instabilities. For many, it is thus merely a means of survival or an opportunity to leave the vicious circle of poverty through their own work and in the form of self-employment, whereas for others, mining constitutes a "vocation", a distinct form of life they choose for a number of reasons not necessarily related solely to subsistence.

The lack of access to information and training opportunities, for both men and women of all ages, was an issue of great concern to all participants in the survey. Even a very simple and inexpensive initiative, such as establishing a library in the school, could otherwise make a difference to people's quality of life. Another problem is the inability to save money that could be invested in alternative income-generating ventures; a basic finance or credit scheme could be an effective means to help people improve their circumstances.

DISCUSSION

The findings of this survey have not necessarily revealed any new trends or revolutionary surprises. They have, however, confirmed some important facts and assumptions about the sector and, as such, serve as a means to improve the understanding of a relatively little-known form of life and its important contribution to poverty reduction. In doing so, it is hoped that the research will also help to disperse a number of prejudices that stand in the way of meaningful assistance to the sector.

The first point that has been confirmed is that there is no generic profile of an artisanal small-scale miner in Peru: variations exist from region to region, and even among and within communities, with regard to the size of mines and type of products extracted, techniques and tools employed, activities carried out by individuals, as well as their level of involvement (*e.g.* full-time, seasonal, migrant, etc). Individuals' motivation for becoming involved in the ASM sector also varies greatly; some do it to supplement income from other sources, whilst for others, mining is a "vocation" or a family tradition. Yet, others see it as an "easy" option, but for most, it represents the only alternative to being unemployed. In brief, artisanal miners cannot be treated as a single consistent group; their livelihoods are highly context-specific, which is one of the reasons why it is so difficult to plan effective broad-scale interventions in the sector.

Secondly, evidence to confirm that living conditions for women and children are particularly harsh in ASM communities was expected in its abundance. However, the seeming prevalence of domestic violence and alcohol abuse in La Eugenia was striking and should be the subject of further research.

Thirdly, another disturbing, albeit expected finding, was the dangerous working environment in La Eugenia. In the majority of instances, miners work with no protective clothing whatsoever, not even hard hats or boots. Tunnels and pits are often poorly ventilated and lit with carbide lamps subjecting miners to respiratory problems. Miners are not equipped with first aid skills to help potential casualties and the nearest fully equipped hospital is an eight hour journey away, giving seriously injured casualties slim chances of survival. Despite these adverse conditions, the researcher was surprised at the low level of accidents reported.

There is widespread use of mercury in La Eugenia – however, the dangers of contamination are not sufficiently known; in fact, some miners falsely believe that handling mercury with bare hands is harmless as it does not enter through the skin's pores. Problems in La Eugenia arise with raised mercury levels in soils and air pollution from fumes released during the heating of the amalgam. These are well-known dangers, and countless aid projects and government efforts have been launched to deal with these problems. The question to be addressed, therefore, is why the problems persist in La Eugenia. Clearly, it is not about a lack of standards and regulations, but rather weaknesses in implementation and propagation of these standards.

Typically, there are no public or private structures to provide adequate infrastructure services such as water and sanitation, transport, energy, health care, or basic education to artisanal and small-scale mining communities, despite the fact that these services may be relatively well established in other regions of the country. In this respect, La Eugenia is no exception. The regional authorities are unable to intervene, or may have the impression that they are not obliged to intervene, in such a vast and uncontrollable development process.

On a community level, the lack of trust among community members and the absence of organisational skills was striking. This is an area where outside agencies could achieve great impact with relatively little, albeit long-term, input. A capacity building project to improve people's ability to formulate their demands, and to mobilise their resources, may be an effective channel for action on a number of issues. In La Eugenia, it was obvious that people need support to help them relate to each other in ways that enable them to address problems in the community and to bring about positive change in a sustained and sustainable manner. In particular, effective community organisation could be a means to:

- promote discussion of, and finding solutions to, common concerns (such as waste disposal and sanitation);
- offer mutual support and solidarity (*e.g.* on the subjects of alcoholism and domestic violence);
- enhance self-esteem and collective confidence (necessary for successful communication with outsiders);
- increase community members' participation in the political process (*e.g.* in the fight for formalisation of the sector);
- gain access to and negotiate with government and other officials through collective action;
- be a forum for learning.

The latter in particular can be achieved through improving people's access to information, whether through formal education, training (*e.g.* in management skills), public meetings or the media, and by establishing intra- and inter-community networks to foster horizontal lesson-learning and knowledge-sharing. In any case, capacity-building and social organisation need a long-term strategy and not a "quick-fix" project approach. An intervening agency must aim to invest in new and middle-level leaders, recognise and build on people's existing practical and organisation skills as well as their diverse backgrounds, and allocate prime responsibility for assessment, decision-making, planning, implementation and evaluation activities, to community members.

A capacity-building approach could also enable the community to overcome discriminatory practices, which are currently limiting women's quality-of-life. It appears that many people within the community felt "threatened" by the subject of gender equity whenever it was openly addressed, some choosing to ignore inequalities altogether or putting them down to "culture". It is hence important that any effort to help women better organise themselves – *e.g.* around economic activities, or in order to campaign for better health and community services – be accompanied by efforts to raise men's gender awareness to avoid conflict in the future. Community radio stations and audio-visual materials (videos, theatre plays, etc) could help bring the interests of women to the fore. It must also not be forgotten that for women to participate effectively and to lighten the burden of their daily existence, support must be provided for their domestic work such as caring for children and household chores. Furthermore, intra-household resource distribution is determined by individual household members' bargaining power. This, in turn, is affected by (a) control over resources, such as assets (physical or human capital), (b) influences that can be used to control the bargaining process, (c) access to interpersonal networks (social capital), and (d) basic attitudinal attributes, such as self-esteem and self-confidence. Finding the appropriate indicator of bargaining power is therefore an effective entry point for the design of interventions, bearing in mind that bargaining varies both within and between communities.

Finally, how suitable and useful was the SL Framework as a research tool for analysing the livelihoods of artisanal miners and the contribution of their activity to poverty reduction? In general, it can be said that the framework was constructive in that it helped to organise thoughts and work out the research plan in advance of the field work, *e.g.* in preparing the semi-structured interview and focus group questions and in selecting relevant participative exercises. However, this is also where the framework can be restrictive. A great deal of flexibility is required on the part of the researcher in order to capture the diversity and sometimes covert dynamics of people's livelihoods. This means keeping an open mind at all times, listening to what people are saying, and being willing to rework the framework whenever necessary, whether by adding certain categories (as in this case "political capital"), or by disregarding or paying less attention to others (seasonality, for example). Inevitably, the researcher will also bring his or her own gender bias to the investigation. Women respondents, for example, have most certainly discussed the issues of domestic violence and gender inequalities so openly because the researcher was female herself and showed a clear interest in those topics. Whilst showing an interest in a subject matter is not necessarily a bad thing, the researcher has to be aware of this and avoid asking "leading" questions. As the SL framework is not meant to be an *exact* reflection of reality, results are not always quantifiable and can be prone to the researcher's over-interpretation and personal assumptions.

However, the methodology applied was well-suited in that focus group discussions and other participatory exercises were helpful in establishing the vulnerability context of artisanal miners' livelihoods quite precisely within the very short time available. The individual semi-structured interviews gave a clear picture of the micro-context: how assets and activities differ within the community, not least along gender lines, and what social relations exist (intra-household and extra-household) in La Eugenia. The macro-context, however – i.e. the structural, historical and institutional elements – was more difficult to discern. Here, the framework did not provide any clear guidance on how to structure the analysis, and the methodology was unable to identify or fully appreciate the inequalities of power, which surely contribute to the persistence of poverty in ASM communities in Peru. Much more research, probably using a different methodology, would also be necessary in order to obtain satisfactory information regarding artisanal miners' operating environment, namely, the government's commitment to poverty eradication or its ability to think cross-sectorally.

In summary, the SL approach provides a solid starting point for involving as many stakeholders as possible, for building up an understanding of the sector's main problems, for developing a vision, and for thinking about ways to find the best ways to deliver development outcomes.

CONCLUSION

What can be concluded from this particular study? First, to reiterate: this research has not revealed any revolutionary new findings about the artisanal mining sector as a whole, nor are the subsequent recommendations for an improvement of the livelihoods of those engaged in artisanal mining meant to be a blueprint for success. The research has, however, confirmed that artisanal mining makes a considerable contribution to poverty reduction, and, as such, is a form of life worth preserving and promoting. It is high time that long-standing misconceptions and prejudices surrounding the sector (environmental degradation, delinquency, etc.) are put right, and that artisanal miners are given the respect and dignified living conditions they deserve.

In summary, artisanal mining in Peru is characterised by the following:

- geographical dispersion and atomisation of operations;
- spontaneous and temporary occupation of areas of exploitation; this is reflected also in the type of productive activities and in the population of the centres;
- women and children are involved in mining operations;
- artisanal mining communities have hardly any environmental training or education; however, they can better associate with the problems of environmental pollution via the effects it has on people's health;

- represents activity of economic subsistence and is often the only alternative to unemployment;
- productive activities that are generally informal;
- productive activities that are carried out by individuals or in very small groups, which has led to the disintegration of the processing industry and, in some cases, triggered dependency chains of third parties;
- organisations that deal with production and basic social services that are still in their infancy;
- people's various origins and ethno-cultural roots influence the social and productive fragmentation; and
- miners often have a conservative and distrustful attitude towards unfamiliar technologies, which is often reinforced by their lack of knowledge of other alternatives.

These characteristics show that interventions in the small-scale mining sector have to encompass technical, social, organisational, economic, legal and cultural aspects, whilst taking account of the differences that exist between, and even within, artisanal mining communities.

Moreover, interventions must happen not only at the micro level (*i.e.* within artisanal mining communities), but must also include action at the macro level (regulative and administrative policy). Changes at the macro level are indispensable if efforts at the micro level are to be successful. A priority for future work will therefore be to develop:

- a better understanding of overall governance structures and their effect on the livelihoods of artisanal miners; and
- improved ways to understand the relationships between the micro and the macro, in order to enable us to pinpoint where constraints to the development of more sustainable artisanal mining livelihoods lie.

RECOMMENDATIONS

1. No definitive statistics exist for the ASM sector: figures for Peru range between 25,000 small-scale miners (official government estimate), 40,000 mining households (compromise reached by Proyecto GAMA), and unofficial estimates of approx. 50,000 miners in the Puno region *alone*. National surveys of the demographics of small-scale artisanal mining are required urgently. Without exact figures, the importance of the sector and its contribution to poverty reduction both on a national and global scale remains guesswork, and the effectiveness of interventions will be limited. This is especially true as the contribution of the ASM sector to a country's entire economy is deemed big enough to warrant special attention only if:
   - it can guarantee the government budgetary resources for poverty reduction programmes and it has the potential to act as a powerful catalyst in the development of the private sector in the region; and
   - a country has ASM sites whose population is essentially migrant, lives in vast territories and produces revenue at a subsistence level, and where

there is no direction and no social or environmental protection. If these communities constitute a total population of 50,000 individuals or more, a government might envisage investing in a poverty reduction strategy because the social and environmental consequences, as well as ensuing cultural and political conflicts, could be explosive.

2. Traditionally, the one group that has benefited the least from interventions in the ASM sector is women, in particular, ultra-poor women, despite the pivotal role they play in supporting (or, in many cases, heading) the household. Women's contribution to household income generation and food security is, at best, overlooked – and often dismissed – by outsiders who generally base their assumptions on the notion that there is only one (male) decision-maker and income-earner in the ASM household. Research on many counts, however, has shown that the most effective way to improve a household's well-being is to concentrate development efforts on women. Given the extent and importance of women's contribution to artisanal and small-scale mining, efforts to improve productivity can only be effective if they are based on women's perceptions, needs and participation. One way of ensuring that women's interests are represented in policy-making and implementation is to enhance existing, or to establish new, women's organisations or co-operatives so that these become mechanisms for true empowerment. It is important to remember that the most effective groups are built on trust and may take quite a long time to form – imposing deadlines (*e.g.* born by funding or reporting pressures) must therefore be avoided at all cost. In a similar vain, it is essential that even the most disadvantaged are given an opportunity to be involved in decision-making processes. This may be quite a culture shock in the still very male-dominated, technically focused domain of artisanal and small-scale mining, especially in a Latin American context, not only within the communities themselves, but also on the side of implementation bodies.

   Programme methodology and services have to reflect the specific needs of the target group, and flexibility must be provided to allow for diversity in women's circumstances (single, co-habitant, married, number and age of children, level of education, etc.). Experience has shown that focusing on income-generating activities alone is not enough to provide women with economic empowerment in the true sense. They need to be accompanied by appropriate awareness and confidence raising efforts, education – for example, literacy or vocational training – and social sector assistance. Furthermore, specific safety net linkages need to be explored, as well as ways in which these linkages can contribute to the sustainable livelihoods of the poorest of the poor.

3. Interventions need to be highly context specific due to the high level of variation between and within mining communities.

Micro-level intervention

Poverty is often caused by a lack of, or closed access to, choices and options, and is therefore reduced in line with expansion of those choices and options. The creation of alternative income-generating opportunities in ASM communities is therefore one of the areas in which outside agencies can provide assistance at the micro level. This, of course, does not mean that diversification possibilities are endless; labour market and environmental conditions, as well as entry constraints (such as lack of start-up or investment capital) might place severe restrictions on the range of opportunities available. Furthermore, outside agencies can assist in the diversification process "by taking action to improve information, encourage mobility and reduce regulatory restrictions" (Ellis, 1999).

Meso and macro level intervention

A necessary need in the future is job creation, in order to prevent sharp increases in unemployment that is anticipated certainly in relation to the expected numerical growth of the labour force over the next 20 years. Artisanal and small-scale mining already provides a viable alternative to unemployment for a large number of people in Peru, and with careful and efficient management of both the natural resource base, and improved living conditions, the sector is capable of absorbing an even larger work force.

Policy-makers also need to take into account urban-rural linkages, as these facilitate financial and skill transfers, and provide access to markets for the outputs of artisanal and small-scale mining enterprises. A *local economy perspective*, or *meso* economy perspective, with the aim of regenerating the local economy, could be a useful reference frame in which to forge livelihood strategies for a scaled-up attack on poverty. The *local* is now defined by rural-urban linkages, and affected in places by a process of "rural-urbanisation". In Latin America, for example, rural spaces are characterised "by the growth of towns and medium-sized cities and by stronger ties between them and their rural hinterland, through non-agricultural trade, transportation systems, and a wide range of services oriented toward production, consumption and recreational needs" (Berdegué et al., 2000, p. 2). Exogenous activities such as investment in manufacturing and light industry can act as an engine in bridging traditional sectoral or rural-urban divides, and in creating a dynamic local and regional economy. A useful approach in identifying policy changes that could have a very wide positive impact is sub-sector analysis, which studies the vertical structure of a given economic activity (in this case, artisanal and small-scale mining) from input supply through the production process and selling of finished products. As argued by Dawson (1997), "[t]his permits the identification of (1) bottlenecks constraining the economic activity; (2) niches in which small producers can have a comparative advantage; and (3) specific interventions which can ease the bottlenecks and facilitate the exploitation of niche opportunities".

Macro-level intervention

Whilst micro-level interventions should focus on providing support (for example, in the form of business development services) and training for both mining and non-mining enterprises, and meso-level interventions on local economy regeneration, the focus at the macro level should be on policy and institutional reforms and removal of obstacles to ASM production. Some specific policy suggestions for reform include:

- Real employment creation and small enterprise development on a national scale.
- Promoting diversification and value-added mining production as well as increasing processing capacity.
- Recognising the need for government to play a greater role in determining and promoting investment opportunities for private business.
- Explore linkages between the development of the mining sector and regional development strategies.
- Development of infrastructure projects (roads, water and sanitation, electrification, waste management, telecommunications, etc.) that serve ASM communities.
- Social organisation and social services provision (health and education).
- Policies that take into account environmental impacts and security issues.
- An emphasis on land reform and institutional reforms to increase democratic practice and accountability.
- Generation and communication of information about the industry to a number of stakeholders (aid agencies, NGOs, government officials) in order to reduce levels of mistrust.
- Development of norms and structures of governance for the industry at the global level, with a clear set of incentives to encourage compliance.

It seems that the latter is perhaps the greatest challenge to facilitating a transition toward a sustainable ASM sector, as well as ensuring that institutional arrangements are acceptable to all of the actors involved.

REFERENCES

APEMIN (Apoyo a la Pequeña Explotación Minera) (2002). Project Presentation, Oruro, Bolivia, http://www.apemin.com.bo

Berdegué, J.A., et al. (2002). Policies to Promote Non-Farm Rural Employment in Latin America. *Natural Resource Perspectives*, No. 55, London: Overseas Development Institute.

RED CAMANA – CARAVELÍ. (2002). Data provided by Dr Edgar Zelada, Eugenia.

Dawson, J. (1997). Beyond credit – the emergence of high-impact, cost-effective business development services. *Small Enterprise Development*, London: Intermediate Technology Publications, Vol. 8, No. 3, pp. 15–25.

DFID (1999). *Sustainable Livelihoods Guidance Sheets*. London: DFID.

Ellis, F. (1999). Rural Livelihood Diversity in Developing Countries: Evidence and Policy Implications. *Natural Resource Perspectives*, No. 40, London: Overseas Development Institute.

Rouse, J. & Ali, M. (2001). *Waste Pickers in Dhaka*. WEDC, Loughborough University.

Weber-Fahr, M., Strongman, J., Kunanayagam, R., McMohan, G., & Sheldon, C. (2001). *Mining for Poverty*. Washington: World Bank, http://www.worldbank.org/poverty/strategies/chapters/mining/min0409.pdf

World Bank. (1996). World Bank IEN Occasional Paper No. 6, *Regularising Informal Mining*. B. Mamadou (Ed.), Washington.

36

Challenges to Sustainable Small-Scale Mine Development in Suriname

MARIEKE HEEMSKERK AND RACHAEL VAN DER KOOYE

Over the past two decades, mining for gold has been a major catalyst for ecological, political, demographic and socio-economic change in the Amazon rainforest (Cleary, 1990; MacMillan, 1995; Schmink & Wood, 1992). This chapter describes these dynamics for the small Amazon country of Suriname (Fig. 36.1). No-one knows *exactly* how many people are mining for gold in Suriname. Our estimates suggest that there are between 20,000 and 25,000 small-gold miners active in at any given time.[1] However, some believe that there may be twice as many (Quick et al., 2001).

Even though Suriname miners constitute a small proportion of the total number of miners in the Trans-Amazon region, their activities are reshaping economics, politics, land-use, the natural environment, and social relations in Suriname. Small-scale gold mining provides foreign exchange to the economy of Suriname and a livelihood for many of the rural poor. These economic benefits are offset by an increase in violent crime; a malaria epidemic; the spread of sexually transmitted diseases; mercury pollution of the aquatic ecosystem and forest peoples; and a degraded forest ecosystem (De Kom, Van Der Voet, & DeWolff, 1998; Healy, 1996; Heemskerk, 2001; Peterson & Heemskerk, 2001; Pollack et al., 1998; Van der Kooye, 1997).

[1] The Central Bank of Suriname bought, in the past five years (1997–2001), an average of 5,391 kg gold/yr, and has estimated that this accounts for 25% of the national gold production (Central Bank, pers. com., 1999 and 2002). These figures suggest that about 21,500 ± 2,000 kg gold are produced in Suriname annually. Observations from Suriname suggest that a small-scale miner who works year-round produces, on average, 1 kg/gold per year (Veiga, 1997). Hence, it is likely that there are roughly between 20,000 and 25,000 miners actively mining for gold in Suriname at any given time.

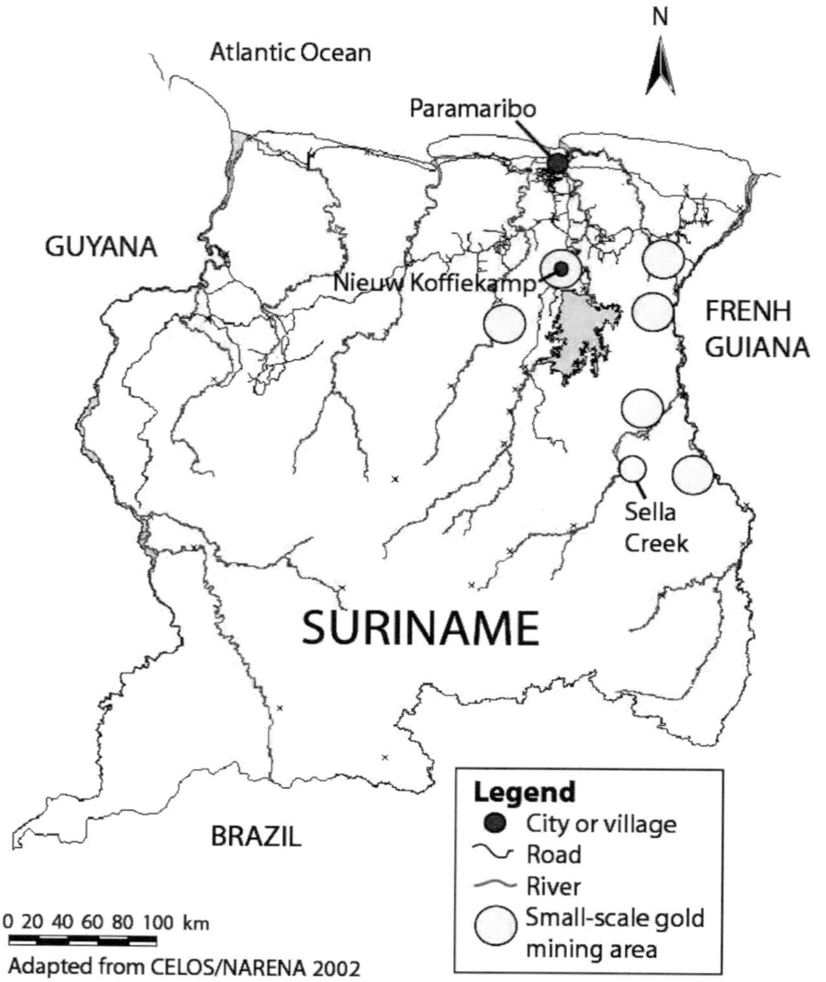

Figure 36.1 Map of Suriname with the main small-scale gold mining areas and sites mentioned in this chapter.

For the purposes of this discussion, the term "small-scale gold mining" is used to describe both manual and mechanized mining characterized by:

- Informality – that is, a large degree of autonomy of national social, legal, and economic regulations; and
- A labor force that is not formally trained in mining.

Small-scale gold miners in Suriname work on dredges in rivers, pan river-sediments and mine tailings, and use hydraulic machines of different degrees of power. This chapter is primarily concerned with hydraulic, land-based gold mining, which is by far the most common extraction process in Suriname. This method is explained in more detail in Box 1. Because miners follow trails of

Box 1. Small-scale mining methods.

Hydraulic gold miners work through several stages. Miners start by exploring possible sites, often near the site of another miner who is known or believed to have hit a good location, following the projected direction of the ancient streambed. Rudimentary prospecting methods consist of digging one or several holes of about two metres deep. The contents are washed with a *batea*, a circular metal pan with a pointed bottom. If gold is encountered, a site is deemed suitable.

After site selection, a forest area of about one ha is cleared from trees and under-story. A generator-powered mining machine is placed at the prepare site, and empowers two types of hoses. One or two power hoses divert high pressure water to remove first the top layer of sand and clay, and later the gold bearing layer of soil (see illustration). The soil-water mixture is pumped through the suction hose into a sluice-box. This piece of mining equipment consists of a series of tilted wooden boxes. Gold particles and other heavy minerals are trapped behind riffles and/or a metal screen, and in the fine mat that covers the bottom of the sluice box. The mine tailings – gravel, sand, and clay from which gold has been (partly) removed – flow into either an abandoned mining pit or adjacent forest.

After two or three weeks of work, the sluice-box is “washed”. Gold is recovered by washing the screen and the mat with water, meanwhile applying mercury that chemically binds with gold but not with the other heavy minerals that have been retained. Gold and mercury combine in a ratio of 1:1 to form an amalgam. At last, mercury is separated from the gold by evaporation. The most cost-effective and healthiest way to burn off mercury would be to use a closed system, such as a retort, which recovers mercury for re-use. In our experience, however, most miners simply heat the gold mercury amalgam in a batea, either in the open air or in a closed space.

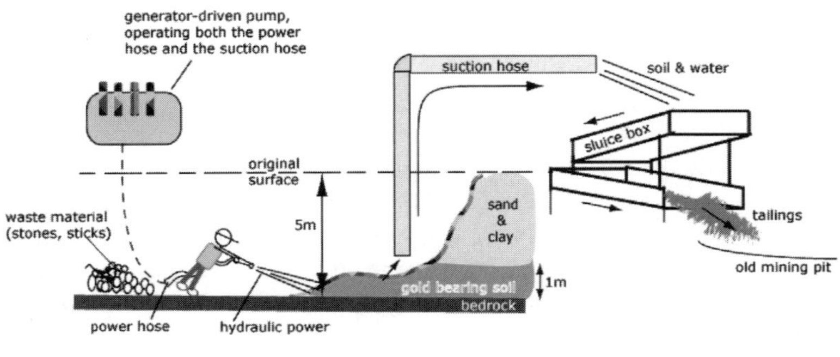

Mining engineer Dehlberg (1984) provides detailed descriptions of various scale-scale gold mining methods, based on experience in Suriname.

Figure 36.2 Hydraulic mining in the Suriname forest, seen from the air.

secondary alluvial deposits through the forest, this type of mining is also called strip mining (Fig. 36.2).

Our analysis builds on primary data from anthropological fieldwork and investigative journalism, supplemented with information from secondary sources (Heemskerk, 2000; Van Der Kooye, 1997). In addition to gold miners and their families, interviews were conducted with officials from the Suriname government, and institutions in Suriname, including the Central Bank, the Organization of American States, the *Cooperativo de Garimpeiros*, local Non-Governmental Organizations, and the University of Suriname in Paramaribo. The following sections introduce the study country, Suriname, and its mining population. Next, the chapter describes labor relations and organization, jobs and earnings in mining areas, and health and environmental impacts. The chapter concludes by describing the challenges to sustainable gold mining development in Suriname, reflecting upon policy interventions that might steer this development in a positive direction.

SURINAME

Suriname, formerly Dutch Guyana, is situated north of Brazil between French Guiana and Guyana. Its 425,000 people live almost entirely along the coast (ABS, 2000). Minimally impacted tropical rain forest covers the remaining 80% of the country. This forest houses and provides sustenance to forest peoples: Amerindians (est. 10,000 to 22,000 people) and Maroons (est. 45,000 people). (Kambel & MacKay, 1999). Maroons are the descendants of runaway African slaves, who established independent communities in the rainforest in the sixteenth and seventeenth centuries.

Suriname's forest peoples have maintained relative political, social and economical sovereignty within the nation state. However, their access to land, natural resources and self- determination is vulnerable because the government has refused to recognize their individual or group property rights (Kambel & MacKay, 1999). According to Suriname law, all land and subsoil resources within the territory of Suriname belong to the state. Human rights lawyers have contested this law. Referring to international conventions ratified by the Suriname state, including the United Nations and Organization of American States declarations on the rights of indigenous peoples, they argue that:

> Suriname has definite and substantial obligations under international human rights law to recognize and respect indigenous and Maroon rights ... to be free from discrimination; to own and use their lands and resources; to participate in decision-making; and to practice their cultures. ... (Kambel & MacKay, 1999: 173).

Today, Maroon and indigenous rights are threatened by logging and mining concessions that overlap with their territories and villages (Healy, 1996; Heemskerk, in press).

Suriname's gold deposits are part of the Guiana Shield, a geological greenstone formation that covers 415,000 km^2 of Venezuela, the Guyanas, and Brazil (Veiga, 1997). These deposits have been mined informally and formally throughout Suriname's history (Bubberman, 1977). However, the current number of people involved, the amount of gold extracted, and the social and ecological impacts of mining are greater than during earlier mining activities (Heemskerk, 2001). Because most small-scale gold mining in the country occurs informally and quasi-legally, estimates of its production are speculative. These estimates suggest that annual gold production increased from about 30 $kg\,yr^{-1}$ in 1985 (Gemerts, 1986) to between 20 and 25 tonnes in the late-1990s.[1]

Virtually all gold is extracted by *small-scale* gold miners. Suriname Maroons make up approximately one quarter of the mining population, and the remaining three-quarters are Brazilian gold miners called *garimpeiros* (Veiga, 1997). They usually work on small-scale (<200 ha) concession areas that belong to individuals who obtained a government small-scale mining permit. Miners officially have to pay a fee for working on someone else's concession, but poor infrastructure and control mechanisms prevent this rule from being enforced in the more isolated parts of the forest.

Depressed gold prices discouraged large-scale mine development throughout the 1990s, but price recovery motivated multinational companies to restart their activities in 2002. The Canadian gold exploitation company, Cambior Inc., established the operational corporation Rosebel Goldmines to construct a gold mine in the area of Gross Rosebel (Fig. 36.1). The mine is anticipated to be operational by the beginning of 2004.[2] Given prior hostilities between GSR, Maroon communities in the concession area, and local gold miners, conflict can be expected as the mining company develops the site (Box 2).

WHO IS INVOLVED AND WHY?

Maroons

Maroons have historically mined for gold and provided services in the gold mining industry, but the role of mining in the household economy has changed. In the early days, small-scale gold mining was conducted infrequently and temporarily when cash was needed. Today, by contrast, it has become the primary source of subsistence for many Maroon households in eastern Suriname.

Maroons are working in gold mining more than any other ethnic group in Suriname in part because they have been disproportionally impacted by the rise in rural poverty, increased economic instability, and the 1986–1992 civil war (Heemskerk, 2001). Maroon miners themselves explained their occupational choice by the lack of other jobs that earned sufficient income to sustain a family, and their limited formal education (Heemskerk, 2002). Historic tension between Maroons and urban people, who hold the keys to alternative jobs, poses an additional barrier to entering the national labor market (Price, 1995). Moreover, several Maroons said they preferred informal work and life in the forest rather than the city.

Garimpeiros

The arrival of *garimpeiros* in the 1980s was crucial in modernizing small-scale mining methods and management in Suriname. Most *garimpeiros* had many years of mining experience before coming to Suriname. Their exodus out of Brazil into the larger Amazon area was a response to efforts by the Brazilian government to regulate, limit, and control small-scale mining (MacMillan, 1995; Schmink & Wood, 1992). *Garimpeiros* said that they had become gold miners for largely the same reasons as Maroons: increasing costs of living,

[2] The investments will be US$95 million, while an annual production of 6700 kg gold with a value of US$66 million/yr is expected. The company anticipates that it will employ 600 people for construction and operation of the mine.

Box 2. The Koffiekamp conflict.

In 1994, the Canadian company Golden Star Resources Ltd. (GSR) concluded a Mineral Agreement with the government, granting it exclusive rights to explore the Gross Rosebel concession. The Maroon village of Nieuw Koffiekamp, which has a population of 500–800 people, was – and still is – located centrally in the company's 17,000 hectare concession. In accordance with the mining agreement, GSR wanted to expel local small-scale miners because they "hindered the exploration activities of the company". At first, GSR wanted the entire village to be removed because it was situated too close to the planned mine location. Officials said they were concerned that exploitation activities would threaten the safety of the villagers.

Villagers of Nieuw Koffiekamp feared and refused relocation, in particular because they still experienced trauma from a previous relocation in 1965. In that year, their historic villages were flooded to make place for a dam to generate electricity. Ironically, the villagers were relocated on a large heap of gold, which became the Gross Rosebel Concession.

More informed than in the past and backed by human rights lawyers, Maroons voiced the opinion that they had a right to determine how the land and resources around the villages should be divided, managed, and used. In early 1995, Nieuw Koffiekamp residents complained that they were intimidated by armed guards and that their subsistence activities, including small-scale gold mining, were being restricted by GSR security personnel and armed police units, including the paramilitary Special Police Support Group. Villagers also said that these parties were firing ammunition at them to keep them away. These allegations have been substantiated by Suriname's main human rights organization.

The conflict was never resolved but withered away in the late-1990s, when gold prices reached such a low point that it became unprofitable for GSR to further invest in mine development. Recovery of gold prices in 2001 has motivated GRS to restart its activities, to great distress of Nieuw Koffiekamp residents. GSR has stated that it no longer foresees removal of the entire village in the process of mine construction. Instead, company officials have declared that community development is an integral part of their mining project. As of today, it is unclear what this means, and whether the project plan is acceptable to local Maroons and gold miners.

Sources: Healy 1996; Kambel and MacKay 1999; Van der Kooye 1997.

coupled with unemployment and minimal formal education. They had come to Suriname because, as they lamented, protected areas in Brazil had been closed off to *garimpeiros*. The few leftover mining places were exhausted or overpopulated. Suriname had seemed attractive for its relative lack of government control, and exaggerated rumors about its richness.

Better organized than local miners, many *garimpeiros* are part of a cooperative where they exchange information, sell gold, and find labor and legal assistance. Because foreigners cannot obtain a small-scale mining concession, legal *garimpeiros* need to work with a Suriname associate or pay a Suriname concessionary about 5–10% of their earnings for the privilege of working on his or her concession. Some *garimpeiros* felt deceived by promises of rich gold fields. They complained about being exploited by Maroons and targeted by criminals. Others were pleased by the relative safety in Suriname mining areas and tranquillity of its capital city. A significant number of Brazilians in this latter group have settled in Suriname, learned some of the local Creole dialect, and sent for their families or begun relationships with Suriname women. Today, entire neighborhoods in Paramaribo are ethnically Brazilian, featuring Brazilian restaurants, music, stores, and services. It is unclear how much of *garimpeiros'* earnings flow back to Brazil. While Brazilians said that few *garimpeiros* sent money to their families at home, others asserted that they did so whenever possible.

Local and foreign miners

Brazilians and Maroons work together in a brittle symbiosis. The lack of a common language, history, and culture creates distrust, tension, and occasional conflicts. Notwithstanding, each group usually respects the others' skills and knowledge, and realizes their mutual dependency. Brazilians tend to admire the Maroons' extensive knowledge of the river system and natural resources, and depend on them for access to transport, food, lodging, and mine sites in the isolated and inhospitable Suriname interior. Nevertheless, *garimpeiros* prefer to work with and for other Brazilians who, in their eyes, have more advanced mining skills, experience and knowledge. Moreover, *garimpeiros* find Suriname mine operators cheap. The food in Suriname mining camps, in particular, the lack of vegetables and fresh meat, is a source of continuous complaints. "I only worked for three days there", recounted one *garimpeiro*: "when you work with the blacks [Maroons] you are always hungry".

Maroons acknowledge the mining expertise and work ethic of Brazilian miners. One Maroon mining boss said that he preferred working with Brazilians because he had found Maroons less reliable. His (Maroon) relatives would unexpectedly travel to their home villages for festivities, rituals, and funerals, or to visit their wives and children. Moreover, Maroon laborers objected to rigid work schedules and inflexible orders from their kinsmen. Brazilians, who do not have family or home communities nearby, would labor 12-hour days for months in a row. Yet, *garimpeiros* are not trusted like family. Maroon miners are convinced that, if given the chance, Brazilian team members would hold back a couple of grams of gold here and there. For this very reason, mine operators always have at least one Maroon miner on the team to supervise the foreigners.

The mining economy

Monetary transactions in Suriname mining areas typically occur in gold, but there are large differences in payment systems and earnings between, and within, professional groups. Miners who work with hydraulic equipment typically form teams of six laborers, a boss, an overseer, a cook, and occasional temporary laborers.

The machine-owner typically supplies food, shelter, and equipment in exchange for 70% of gold recovered. Surveyed mine operators in Eastern Suriname had invested approximately US$20,000 to start up an operation (6-inch hydraulic unit), and spent an additional US$5,000 per month to keep their operations going (N = 21, SD = 421). Despite these expenses, mine operators reported higher average net earnings than people in other professions (N = 21, Mean = 344 g gold/month, SD = 364). These earnings ranged from a low of 14 g to over 1 kg/month – reported by the most successful operators (1 g gold ~9 US$, 1999). These well-off entrepreneurs augmented their income by running stores and bars with satellite TVs and videos to entertain their laborers and those of surrounding camps.

Pit workers divide the remaining 30%, which translates to 5% per laborer if there are six people working in the pit. Among surveyed pit workers, 39% had earned between 20–40 g/month in the month prior to the interview, which amounts to US$180–360 at an average of US$9 per gram of gold (in 1999). Over a quarter (27%) had earned less than that. Pit workers earned higher wages in more advanced operations that featured bulldozers, back-hoe excavators, and heavier pumps. Several pit workers in such operations reported earnings of more than 150 g (US$1350)/month. In 2002, the Cooperativo de Garimpeiros in Suriname estimated that the average *garimpeiro* earned between US$500 and US$1,500 a month (personal communication, October 2002).

Cooks typically receive fixed wages that they augment by performing services. In the late-1990s, in the Sella Creek mining area, cooks usually earned 60 gram of gold per month – the equivalent of US$540. The earnings of shop owners and commercial sex workers were more variable and because they are often paid in credit, they risk receiving only about half of what they are owed. Nevertheless, sex workers earned some of the highest wages, and often lent money to gold miners.

Regardless of one's profession, mining wages compared favorably to wages in formal jobs. In 2002, low and unskilled workers in the city made approximately US$100 per month, and college-educated professionals, such as nurses and administrators, made about US$200 per month. An additional benefit of working in the mining area is that the mine operator will buy food and provide housing. Hence, one can save all the money one earns. With today's inflated prices for food and housing, an uncertain, yet, on average, better mining income, may provide more security than a salary paid in devaluated Suriname guilders.

HAZARDS AND HEALTH RISKS AT SMALL-SCALE GOLD MINES

Life in the mining area is emotionally and physically demanding. Miners live for months in a row in mining camps that are based a long way from their home. Their working methods and environment expose them to chemical contaminants, heat stress, ergonomic problems, unsafe equipment and mine structures, unsanitary conditions, malaria, and unsafe sex. Unbalanced diets, long work hours, and alcohol consumption further decrease the body's natural resistance mechanisms to disease (Walle & Jennings, 2001).

Significant mining-related environmental damage was observed in six investigated gold mining areas. Due to mining waste disposal, most creeks and rivers near important Suriname gold mining areas were extremely turbid. A recent study found that water transparencies varied from 0 to 50 cm in some creeks, to 50 to 70 cm in larger rivers. Only two of Suriname's larger rivers showed transparencies higher than 100 cm (Quick et al., 2001). As a result, villagers situated near gold fields lack safe drinking water, and in some places, women walk for hours to find clean water. Furthermore, hydraulic mining has shifted creek banks, and the residues have killed trees by flooding the forest. Some abandoned mining pits have evolved into quick sand areas that trap wildlife. Other pits that have filled with water have become habitat for the malaria-infected mosquito. Its epidemic form and increasing drug-resistance have made malaria a leading cause of illness and death in forest communities, and the most severe mining-related public health hazard.

Infection with sexually transmitted diseases, including HIV/AIDS, threatens the health of commercial sex workers and their clients, as well as that of their families. A recent report on HIV/AIDS in Suriname by a consortium of international organizations was foremost characterized by a lack of information (UNAIDS et al., 2002). In particular, there are no HIV/AIDS data on Maroons and small-scale miners, who are the groups most at risk.

Media and scientific attention toward the Amazon gold rush has foremost focused on miners' use of mercury and resulting pollution of aquatic resources, forest peoples, and gold miners (MacMillan, 1995). Estimating how much mercury miners release into the environment is difficult, as is estimating its net effect on human and ecosystem health. Miners reported using about one kg of mercury per kg of gold extracted. Given an annual gold production of 20,000 to 25,000 kg and low mercury recycling rates, over 20,000 kg of mercury may be released into Suriname's air and waterways per year. This figure corresponds with estimates that 20,000 kg of mercury is imported annually (Pollack et al., 1998). None of the Brazilians or Maroons interviewed and witnessed used retorts (closed systems to recycle mercury). The only precaution they took was to stay upwind when burning the mercury-gold amalgam.

Studies comparing mercury-exposed Maroon miners with non-exposed Maroons report some differences in mercury levels in blood, hair, and urine samples (Cordier et al., 1998; De Kom et al., 1997; Mol et al., 2001; Pollack et al., 1998). Quick et al. (2001) found that one third of surveyed carnivorous fishes in Suriname rivers had mercury levels above the safety cut-off standard of 0.5 mg per kg. Due to the relatively recent onset of the Suriname mining boom, chronic mercury contamination in people is not yet apparent in most places. Of the 28 miners surveyed by De Kom et al. (1997), only one individual had a urine mercury level exceeding World Health Organization (WHO) standards. Pollack et al. (1998) found that even though mercury levels in exposed women and children near mining areas were higher than those in the control group, the average Hg levels were comparable with mean total mercury reference values in industrialized countries. Notwithstanding, classic symptoms of mercury poisoning, such as trembling and nausea, were observed in mining areas.

Because unhealthy conditions in the mining area reinforce one another, it is difficult to single out the effects of *any* one disease. Headaches, nausea, dizziness, and faintness can indicate malaria but also mercury contamination, alcoholism, and overexposure to sun. People whose liver has been affected by excessive alcohol consumption have less resistance to malaria, and people who frequently contract malaria are more susceptible to the effects of mercury contamination.

SMALL-SCALE GOLD MINERS, LARGE-SCALE MINERS, AND THE GOVERNMENT

Suriname's Ministry of Natural Resources has little grip on small-scale gold miners, few of whom bother to seek legal endorsement of their activities. Many Maroon miners feel it is unnecessary to apply for a permit to work on lands that they traditionally view as theirs', even though they do not have legal titles. Moreover, the bureaucratic application procedure is complicated, slow, and believed to favor government allies. Maroons also have little means to meet the legal requirements to have an office in Paramaribo, and to supply the Geological Mining Service with a written report of exploitation and exploration findings every three months (Government of Suriname, 1986).

Despite non-conformance to national regulations, most miners will adhere to local rules about stakes in the mining area. Tight kin-relations and strong social control allow many conflicts among Maroons to be solved informally. When conflicts do get out of hand, they are brought before a council of Maroon elders, which allocates responsibility and punishment. While the tribal legal system successfully mitigates local disagreements, it has little power to settle conflicts between gold miners, and the government and large scale companies.

In 1995, conflict escalated between the village of Nieuw Koffiekamp and the Canadian gold exploration company Golden Star Resources, which was backed by the Suriname government (Van der Kooye, 1997). This conflict illustrates the issues at stake in regulating the Suriname gold mining sector; its development is described in more detail in Box 2. Today, the Nieuw Koffiekamp conflict is being replicated in the Nassau Mountains (Fig. 36.1). In 2002, the Suriname Aluminum Company (Suralco), a joint-venture between the state of Suriname and the US-based bauxite company ALCOA, acquired a gold concession in the Naussau mountains. The concession area overlaps with the traditional homelands and economic use zone of the Paramaka Maroons, who were not informed or consulted. In the summer of 2002, the government forcefully removed local gold miners from the concession to keep them from interfering with the company's exploration activities. Like the miners of Nieuw Koffiekamp, Paramaka miners feel that Suralco's mining activities violate their traditional rights.

LESSONS FROM ABROAD

Examples from other countries suggest that it is possible to develop the small-scale mining industry in a way that minimizes social conflict, environmental degradation, the spread of disease, and illegality. In some Amazon countries, large-scale mining corporations have developed partnerships with small-scale gold miners (Barry, 1996; ILO, 1999). One such arrangement in Venezuela allows small-scale miners to mine alluvial gold on the concession of Las Cristinas (a joint venture between the Corporación Venezolana de Guyana and Placer Dome Venezuela) with the technical advice and assistance of company engineers (Jeffrey Davidson, pers. com. September 2002). Mine managers from Las Cristinas worked with community leaders to select a working area that was both acceptable to small-scale miners and of little economic interest to the company. Las Cristinas further agreed to tolerate gold panners on its main property provided they would not interfere with exploration activity. By working with, rather than against, local miners, the company is hoping to reduce the possibility of invasion by miners who previously worked on its concession.

Another promising direction is the certification of eco-gold or Green Gold, that is, gold that is produced in an ecologically responsible manner. An international group of ecologists, miners, activists and small traders is currently exploring viable mining methods and market potentials (Coalition for Green Gold, 2002). Its aim is to certify gold that is extracted without cyanide or mercury, with methods that minimally impact forest cover and replace the organic topsoil after mining. One Green Gold pilot project is currently being developed in the Choco rainforest in Colombia. Its aim is to develop a system of certification through an alternate marketing system that empowers artisanal miners to conserve and protect their lands.

POLICY RECOMMENDATIONS AND CONCLUSIONS

Transformation of Suriname's current small-scale mining sector into a more sustainable enterprise will require small-scale miners, concessionaries, the Suriname government, transnational mining companies, and international agencies to enter dialogue and collaborations. Our policy recommendations for each of these stakeholder groups, and relations between them, are summarized in Table 36.1 and outlined below:

Maroon small-scale gold miners

Maroon gold miners would stand stronger if they were to organize themselves into associations or cooperatives, like many of their colleagues elsewhere. In the recent past, Suriname Maroon miners have organized when confronted with direct threats to their livelihoods. In 1994, when Golden Star Resources (GSR) wanted to remove approximately 2000 small-scale gold miners from its exploration concession, Maroon miners created an association to negotiate

Table 36.1 Policy recommendations for improving the state of the industry.

- Organization of Maroon small-scale gold miners into Miners' Associations.
- Holding concessionaries liable for social and environmental mining damage on their concession areas.
- Establishment of partnerships between large-scale mining companies and small-scale miners that include training in more sustainable small-scale mining methods by mine engineers.
- Implementation of stricter mining and environmental laws that suit the Suriname context; allocation of human and financial resources to enforce these laws.
- Fostering collaboration and dialogue between all stakeholders; recognition of indigenous and Maroon land rights.
- Commitment of donor countries and institutions to provide financial and professional support for legal, institutional, and technological transformation of the small-scale mining industry.
- Continued research on the driving forces and consequences of small-scale gold mining in Suriname; Evaluation of the perspectives, motivations, and future visions of local gold miners and their families.

with GSR and the Suriname government. Their negotiations ended with the assignment of another, acceptable, gold mining area to local miners. In 1995, a similar conflict occurred at the Lawa River between local Maroon gold miners and a consortium of Grassalco (Suriname) and the Canadian mining company Canarc. Organized in an association, Maroon miners won the right to continue mining on part of the concession.

Examples from Brazil, Venezuela, Peru, and Bolivia show that by providing basic health care and education services and regulating alcohol consumption and prostitution in mining sites, Miners' Cooperatives can improve the well-being of miners and their families. Cooperatives in these countries have helped small-scale gold miners negotiate agreements with the government and large-scale operations; acquire concession rights and information; and invest in newer, more efficient, and cleaner technology. Unlike these examples, Maroon miners' associations usually have fallen apart when threats have vanished. At this moment, only one group of gold miners from the village of Nieuw Koffiekamp has organized in the Koffiekamp Collective. The Collective has obtained permission to mine on the concession of GSR, as long as its members are not hindering the company's exploration activities. Small-scale miners elsewhere in Suriname could gain from following this example.

Environmental and health effects associated with current small-scale mining methods jeopardize the future of Maroon communities near mining areas. Because miners' families that live in these communities are most at risk, Maroon miners may have a personal stake in mining with cleaner methods. Indeed, some local communities are actively seeking out ways to reduce environmental damage. For example, one Maroon village received a grant to develop a demonstration site where gold miners could be trained in environmentally sound mining methods. It is difficult to motivate small-scale miners to change their practices, as long as environmental impacts do not visibly affect them, and alternative income-generating options are rare.

Suriname concessionaries

At present, concessionaries are not held liable for adverse social and environmental impacts resulting from mining in the areas which they have acquired legal mining titles. As the main parties responsible for, and the beneficiaries of, mining on their concessions, it should become possible to oblige concessionaries to clean up abandoned mine sites. However, such a regulation only makes sense if law-enforcement authorities have the capacity to control and enforce mining legislation. Specifically, laws are more likely to be respected if violators can be fined or otherwise punished (for example, by confiscating their concessions).

Stricter regulations and more serious consequences for not following them may motivate concessionaries to encourage small-scale miners to use environmentally sound mining techniques, as it will ultimately be cheaper to reduce environmental impacts than to clean up damages. In more positive terms, if green gold certificates can guarantee higher market prices for gold, certification could make the promotion of green mining on one's concession ultimately more profitable.

Large-scale mining companies

Large-scale gold mining companies in Suriname can learn from the Venezuelan example in their attitude towards small-scale gold miners. The story of Las Cristinas teaches that working with local miners decreases sabotage by local people, lost work days due to conflict, the need to hire armed guards, negative public opinion, and international court cases. In addition, partnerships can benefit local miners by supporting technical and financial inputs that transform small-scale mining operations into more efficient and stable enterprises.

Violent removal of local miners and villagers fixes the trespassing problem in the short term, but offers no *real* solution. In order to develop more beneficial and durable company-community relations, large-scale mining companies must recognize community rights to ancestral lands and resources. In addition, mining multinationals have an opportunity to contribute to community development – and thus appease stock holders – by helping small-scale miners use more productive, safe, and environmentally responsible gold mining techniques. By facilitating small-scale miners in these ways, large-scale companies will also contribute to environmental protection in and around their concession areas.

The Suriname government

The Suriname government is responsible for steering collaboration between the above-mentioned parties. Recognition of traditional land rights would be a first step towards achieving this goal, because social disruption is likely to follow multinational activities on Indigenous and Maroon lands, such as in the area around Koffiekamp and the Naussau Mountains. Moreover, a better relation with its interior populations may improve the image of Suriname among donor countries, on which it depends to settle its national accounts.

The Suriname government has announced a multidisciplinary approach to accomplish this goal, incorporating the Ministries of Justice and Police and

Defense (to fight crime), the Ministry of Public Health (to reduce health hazards), the Ministry of Regional Development (to incorporate local communities in policy-making), the Ministry of Foreign Affairs (to regulate *garimpeiros*), and the Ministry of Labor, Technological Development, and Environment (to improve labor conditions and reduce environmental pollution). To date, the human and financial resources needed to effectively implement this policy have been lacking. Hence, the government is turning a blind eye to small-scale gold mining practices.

International agencies

More sustainable development of Suriname's small-scale mining sector requires implementation and active enforcement of laws that discourage the use of mercury, that punish deforestation and award reforestation, and that encourage development and transfer of cleaner mining techniques. Donor countries and Non-Governmental Organizations (NGOs) can assist in this process by offering financial and professional assistance to ground personal in the forest (e.g. rangers, police), vehicles and other equipment, and bureaucratic infrastructure. Training of concessionaries, small-scale gold miners, and government officials could help stakeholders work more efficiently.

Secondly, outside funding and knowledge can help design and implement infrastructural changes. The Suriname mining sector could benefit from the assistance of mining engineers who have tested and implemented green mining techniques elsewhere. Building water purification installations and locating groundwater sources can improve access to clean water. Moreover, the health and well-being of forest peoples is likely to benefit from improved access to communication (e.g. radio, phone), quality education, and a strong health care system. International assistance is important because it is unlikely that the Suriname government will be able to finance the above mentioned developments.

At present, several NGOs (PAHO, World Wildlife Fund) are proposing small-scale gold mining related projects. These seek to: increase technological efficiency; improve the working and living conditions of gold miners; improve the health of miners and communities nearby mining areas; and reduce mercury contamination and river turbidity.

Policies aimed at greening the mining sector will be more likely to succeed if they incorporate the social and economic concerns of small-scale gold miners and their families. Reliable data about the forest and its inhabitants is currently sparse. It is not known how many locals and foreigners are mining; how many families rely on mining income; where miners work; how much they find; what their money is spent on; and how these issues are changing over time. The development of a database with demographic and socioeconomic data on gold miners and their families is a first step towards the design of new mining policy.

In addition to a base-line survey, research is needed to answer pressing questions about the causes and consequences of the Suriname mining boom. Suriname and foreign studies continue to document the technical, environmental, and health aspects of the small-scale mining industry. For example, Suriname researchers are studying the impacts of mercury and sedimentation of creeks and rivers. The Suriname Medical Mission (Medische Zending) and PAHO are planning a survey on HIV/AIDS infection and transmission, and scientists from the University of Suriname are designing more benign mining techniques. International funding and governmental logistical support will help bring these projects to completion.

In addition to the mentioned fields of interest, the socio-cultural and economic lives of small-scale gold miners merit more attention. Without incorporating the opinions, beliefs, and life realities of local miners, policy interventions aimed at changing the behavior of these miners are doomed to fail.

REFERENCES

Barry, M. (1996). *Regularizing Informal Mining. A Summary of the Proceedings of the International Roundtable on Artisanal Mining*. May 17–19, 1995. Occasional Paper #6, World Bank, Industry and Energy Department, Washington, DC. URL: http://www.worldbank.org/html/fpd/mining/m3_files/art/arthome.htm (Access date: November 10, 2002).

Bubberman, F.C. (1977). De gouden draad. *Suralco Magazine*, 9(3), 14–19.

Cleary, D. (1990). *Anatomy of the Amazon Gold Rush*. University of Iowa Press, Iowa City, 245 pp.

Cordier, S. et al. (1998). Mercury exposure in French Guyana: Levels and determinants. *Arch. Environ. Health*, 53, 299–303.

De Kom, J.F.M., Van Der Voet, G.B. & De Wolff, F.A. (1998). Mercury exposure of maroon workers in small-scale gold mining in Suriname. *Environmental Research*, 77(Section A), 91–97.

Gemerts, G. (1986). *Goudmijnbouw in Suriname, Seminar*. University of Paramaribo, Paramaribo, Suriname.

Coalition for Green Gold (2002). *If it ain't green, it ain't gold*. Rainforest Information Centre (RIC). URL: http://www.rainforestinfo.org.au/gold/ecomine.htm (Access date: November 10, 2002).

Government of Suriname (1986). *Mining Decree E-58 of May 8, 1986*, Containing General Rules for Exploration and Exploitation of Minerals. E-58, Republic of Suriname, Paramaribo.

Healy, C. (1996). *Natural resources, foreign concessions and land rights: A report on the village of Nieuw Koffiekamp, Organization of American States (OAS), special mission to Suriname*. Unit for the Promotion of Democracy, Paramaribo, Suriname.

Heemskerk, M. (2000). *Driving forces of small-scale gold mining among the Ndjuka Maroons: A cross-scale socioeconomic analysis of participation in gold mining in Suriname*. Ph.D. Thesis, University of Florida, Gainesville, FL, 194 pp.

Heemskerk, M. (2001). Do international commodity prices drive natural resource booms? An empirical analysis of small-scale gold mining in Suriname. *Ecological Economics*, 39, 295–308.

Heemskerk, M. (2002). Livelihood decision-making and environmental degradation: Small-scale gold mining in the Suriname Amazon. *Society and Natural Resources*, 15(4), 327–344.

Heemskerk, M. (In press). *Scenarios in Anthropology*. Reflections on Possible Futures of the Suriname Maroons. Futures. Special issue: Futures of Indigenous Peoples., Scheduled to appear Fall 2002.

International Labour Organization (ILO) (1999). *Social and Labour Issues in Small-Scale Mines*. Report for discussion at the Tripartite Meeting on Social and Labour Issues in Small-scale Mines., International Labour Organization, Sectoral Activities Programme, Geneva. Chapter 5: Legislation: A path to sustainable small-scale mining? & Chapter 6: Large-scale and small-scale mining: Cooperation or confrontation? URL: http://www.ilo.org/public/english/dialogue/sector/techmeet/tmssm99/tmssmr.htm#5. (Access date: November 10, 2002)

Kambel, E.R. & MacKay, F. (1999). *The rights of indigenous peoples and Maroons in Suriname*. International Work Group for Indigenous Affairs, Copenhagen, Denmark.

MacMillan, G. (1995). *At the end of the rainbow? Gold, land, and people in the Brazilian Amazon*. Methods and Cases in Conservation Science Series. New York, NY: Columbia University Press.

Mol J.H., R.J., Lietar, C. & Verloo, M. (2001). Mercury contamination in freshwater, estuarine, and marine fishes in relation to small-scale gold mining in Suriname, South America. *Environmental Research*, 86(2), 183–197.

Peterson, G. & Heemskerk, M. (2001). Deforestation and forest regeneration following small-scale gold mining in the Amazon: The case of Suriname. *Environmental Conservation*, 28(2), 117–126.

Pollack, H., Kom, J.D., Quik, J. & Zuilen, L. (1998) (n.d.). *Introducing retorts for abatement of mercury pollution in Suriname*. HWO Consultants NW, Paramaribo, Suriname.

Price, R. (1995). Executing ethnicity: The killings in Suriname. *Cultural Anthropology*, 10(4), 437–471.

Quick, J.A.A. (2001). Mercury in the Surinamese environment. *Interactie*, 5, 29–37.

Schmink, M. & Wood, C.H. (1992). *Contested frontiers in Amazonia*. New York: Columbia Press.

UNAIDS, UNICEF, PAHO & WHO (2002). *Epidemiological fact sheet on HIV/AIDS and sexually transmitted infections*. Suriname.

Veiga, M.M. (1997). *Artisanal gold mining activities in Suriname,* UNIDO (United Nations Industrial Development Organization), Vancouver, Canada.

Walle, M. & Jennings, N. (2001). *Safety and health in small-scale surface mines: A handbook*. International Labour Organization, Geneva, Switzerland.

Van der Kooye, R. (1997). *Porknokkerij in de Media*. Berichtgeving en Effecten van Dagbladberichtgeving over kleinschalige goudwinning in Suriname, 1994–1996. Bachelors Thesis, Akademie voor Hoger Kunst- en Cultuuronderwijs, Paramaribo.

37

Human Health Implications of Mercury Usage in Small-Scale Gold Mining Activities in the Brazilian Amazon

LUIZ DRUDE DE LACERDA

In many tropical countries where gold is found in significant amounts in near-surface mineralized deposits, small-scale mining is seen by thousands of people as a fast route to prosperity, due to growing poverty, a lack of alternative employment, climate change and its impacts on crop production, and a "*get-rich-quick*" mentality, a common feature of a "globalized" world (Veiga, 1997; ILO, 1999; Bonzongo et al., 2002). Small-scale gold mining involves individuals or organized small groups of prospectors exploiting deposits, sometimes illegally, with the simplest equipment, and is a significant source of income, particularly in areas lacking alternative employment such as in most of the developing tropical regions. For example, in the Brazilian Amazon during the peak of the activity in the 1980s and early-1990s, over 50% of the offered job positions were in small-scale gold mining. This activity, however, causes a host of environmental and social problems such as pollution, diseases, child labor, and regulatory/ legal issues, which calls for an interdisciplinary evaluation of its costs and benefits. In light of recent knowledge, however, the most serious impacts of such activities is its propensity to affect people and ecosystems not directly involved with gold mining and mostly displaced in time and space from affected areas (Lacerda & Salomons, 1998).

Small-scale gold mining uses mercury (Hg) amalgamation as the main technique of gold concentration. This technology is at least 3,000 year old, and has escaped the waves of innovation that swept comparably ancient practices such as those used in historic agriculture or medicine (Bonzongo et al., 2002). Due to its characteristics of being cheap, portable and reliable and able to be operated by a single individual prospector, the amalgamation technique spreads very fast wherever available gold resources abound (Lacerda & Salomons, 1998). In amalgamation, gold in concentrated ore sludge is mixed with Hg^0 to form amalgam, which is then heated to release Hg^0, leaving the gold behind.

The process is nearly entirely performed outdoors, resulting in direct emissions to the environment. In addition to the direct introduction of metallic elemental Hg^0 into aquatic systems and tailings, this mining technology releases significant amounts of Hg^0 vapor into the atmosphere, contributing to regional, and possibly, global pollution. Estimates from many parts of the world agree that between 50 and 65% of the Hg used for amalgamation is lost to the atmosphere during roasting of the amalgam, frequently performed in the mining site proper, and during smelting of the gold in small towns where it is commercialized (Lacerda, 2002).

The environmental fate and impact of Hg related to both historic and current small-scale gold mining activities has been extensively investigated in many areas, particularly in the South American Amazon region (*e.g.* Pfeiffer & Lacerda, 1988; Malm et al., 1995; Nriagu et al., 1992; Akagi et al., 1995; Aks et al., 1995; Palheta & Taylor 1995; Eve et al., 1996; Lacerda & Marins, 1997; Kehrig et al., 1998; Lodenius & Malm, 1998; Malm, 1998; Boischio & Cernichiari, 1998; Lacerda & Salomons, 1998; Castilhos et al., 1998; Barbosa et al., 1995; Ware, 1998; Grandjean et al., 1999; Santos et al., 2000; Lechler et al., 2000; Boischio & Henshel, 2000; Harada et al., 2001; Maurice-Bourgoin et al., 2000; Amouroux et al., 1999; Kom et al., 1998; Nico & Taphorn, 1994; Veiga, 1997; Gulleb et al., 1997; Velásquez et al., 1997); in North America (*e.g.* Gill & Bruland, 1990; Nriagu, 1994; Bonzongo et al., 1996a,b; Wayne et al., 1997; Mastrine et al., 1999; Lyons et al., 1999;) and Africa (*e.g.* Ikingura & Akagi, 1996; Ikingura et al., 1997; Bonzongo et al., 1996c, 2002; Harada et al., 1999; van Straaten, 2000a,b).

Notwithstanding the now enormous amount of data, most of the studies were devoted to local, present effects, and few relate to the important biogeochemical processes linking the atmospheric emissions and its consequential impact on remote areas, including remote human groups potentially exposed through environmental, rather than occupational, routes (Rofhus & Fitzgerald, 1995; Lacerda 1997c, 2002).

Unfortunately, the reality is that, artisanal gold mining is on the increase in gold-rich-developing countries, and the lack of safe and environmentally sound production methods are causing adverse effects to human health and severe contamination of the environment. One could then conclude that catastrophes are in the making in these gold-rich-developing countries. In this chapter, the link between atmospheric emissions of Hg from gold mining with fish contamination and human exposure is discussed, using the large amount of results available for the Brazilian Amazon.

THE IMPORTANCE OF SMALL-SCALE GOLD MINING AS A SOURCE OF MERCURY IN BRAZIL

In Brazil, due to the enforcement of strong regulations for industrial point sources of Hg during 1970s and 1980s, and the consequent decrease of their Hg emissions to the environment (Lacerda, 1997b), small-scale gold mining became the major environmental source of Hg, in particular to the atmosphere (Moreira & Pivetta, 1997, Lacerda, 1997b). During the late-1980s, gold mining was responsible for over 80% of the total Hg input to the environment in the country, with emissions varying from 100 to 180 $t\,yr^{-1}$ (Ferreira & Appel, 1991). Even with the present day decrease in mining operations, this source still contributes to nearly 30% (about 11 to 30 $t\,yr^{-1}$) of the total Hg inputs to the atmosphere in Brazil (Lacerda, 2002). Table 37.1 resumes the estimated contributions from different anthropogenic sources to the total Hg load to the environment in Brazil during three distinct periods in the past 30 years. Among the most important sources are energy generation, with emissions mostly to the atmosphere; the chlor-alkali industry; the electro-electronics and chemical industry, with major emissions to soils and waters; and small-scale gold mining, with significant emissions to both the atmosphere, and soils and waters.

Fossil fuel combustion is presently the principal source of Hg to the atmosphere on a global scale. In North America, for example, this source contributes about 100 tons per year, some eight times higher than the 2002 estimates for Brazil (Pirrone et al., 1998). Notwithstanding the relatively small emissions when compared to more industrialized countries, an ongoing energy matrix change in Brazil will significantly influence Hg emissions at the country level (Lacerda & Vaisman, 2002). One major reason is the continuous reduction of Hg emissions from other industrial sources, which have been more significant for Hg than for other metals. The relative atmospheric Hg contribution from the power generation industry will significantly rise in the coming years, and, by 2005, may reach about 15% of total Hg emissions to the atmosphere (2 to 12 $t\,yr^{-1}$). However, higher proportions can be attained if a faster decrease in

Table 37.1 Estimated Hg emissions from major sources in Brazil in the late 1970s, during the peak of gold mining between 1986 and 1989 and today ($t\,yr^{-1}$). Emissions were estimated based on emission factors and consumption parameters for major sources; for details of estimates one is referred to (Pfeiffer & Lacerda, 1988; Ferreira & Appel, 1991; Lacerda & Salomons, 1998; Lacerda & Marins, 1997; Lacerda, 1997a,b; Veiga, 1997).

Source	1970's	1986–1990	2002
Energy generation	<1	<1	2–12[1]
Chlor-alkali industry	90–110	8–20	14–20
Electro-electronics	1–3	5–14	<5[2]
Dental & Pharmacy	25–31	8–16	<8[2]
Paints & Chemistry	30–44	8–32	<8[2]
Gold mining	22–33[3]	100–170	11–30
Total	168–221	129–252	48–83

[1] Includes new coal and oil fired energy generation plants to be in operation by 2005 (Lacerda & Vaisman, 2002).

[2] Recycling of batteries and lamps are expected to reduce emissions to less than the minimum estimates of the 1980s.

[3] Very rough estimate based on re-sale of mercury assumed to gold mining, few if any, data available on consumption parameters.

small-scale gold mining occurs, a possible scenario resulting from improved socio-economic conditions in the country and near-exhaustion of easily mined deposits (Maron, 1999). Furthermore, most emission control measurements will certainly be more effective when applied to other industrial sources, where there is an enormous difficulty in further reducing Hg emissions from power generating plants (Pires et al., 1997).

The chlor-alkali industry (mercury electrolytic cell users) was a significant source of Hg to the environment in Brazil during the 1970s, consuming 60% of the country's total mercury demand in 1979. Ten years later, it accounted for only 7.4% of the total atmospheric Hg emissions (Lacerda, 1997b). Today, it is still one of the major sources of mercury emissions in Brazil, although total banning of this technology is expected to occur within the next decade. Using the parameters given by Moreira and Pivetta (1997) and ABICLOR (2001), emissions from this source are estimated to be in the range of 14–20 $t\,y^{-1}$.

Small-scale gold mining was responsible for about 80% of the total Hg emissions to the environment in Brazil in 1989, an average of about 136 $t\,y^{-1}$ (Lacerda & Marins, 1997), though it has shown a marked decrease in the past few years. Presently, small-scale gold mining is responsible for the release of 11–30 metric tons (Lacerda, 2002) or about 12 tons per year if official gold production numbers are used for estimating Hg emissions (Maron, 1999). This decrease is due to a drastic reduction in the number of gold mining operations in the Amazon following the exhaustion of most of its easily mined deposits. Notwithstanding, small-scale gold mining remains the most significant source of Hg to the atmosphere in Brazil.

Other Hg sources to the environment in Brazil, such as the chemical and electro-electronics industries have remained relatively constant throughout the past 10 years, and no significant decrease or increase is expected to occur in the near future. More recently, the recycling of Hg in fluorescent bulbs, and special disposal of batteries, have become common practices in the country, helping to reduce Hg emissions from this group source.

If only atmospheric emissions are considered, small-scale gold mining is the most significant polluter of Hg, followed by chlor-alkal: production and energy generation.

ATMOSPHERIC DEPOSITION OF Hg IN THE AMAZON

Atmospheric deposition of Hg has demonstrated its potential for assessing the impacts of natural and anthropogenic activities on the global Hg cycle (Vandal et al., 1993; Nriagu, 1996; Martínez-Cortizas et al., 1999). Atmospheric deposition of Hg is a key step in this elemental biogeochemical cycle, due to the dominant role of Hg gaseous species. Atmospheric deposition rates are linked to soil Hg concentrations (Grigal et al., 1994); concentrations in remote lakes, waters and sediments (Engstrom & Swain, 1997); and the Hg content of freshwater (Lacerda, 1997c) and marine fish (Rolfhus & Fitzgerald, 1995). Thus, estimating Hg atmospheric deposition rates and their variability is a fundamental step toward modeling Hg transport and accumulation in the environment. Further, the atmospheric pathway may cause human exposure to Hg in populations far from original sources.

Surveys on Hg concentrations in lake and peat bog sediment, and ice cores have shown significant changes in pre-historic deposition rates during glacial periods, due to changes in aridity and temperatures, and consequent changes in soil cover (Vandal et al., 1993; Martínez-Cortizas et al., 1999; Rasmussen, 1998). During historic times, deposition rates increased in concentrations relative to background values in many parts of the world, in particular, because of gold and silver mining (Nriagu, 1994). The most significant changes, however, have occurred during the past 60 to 80 years, and have been a result of industrialization (Nriagu, 1996; Pirrone et al., 1996, 1998).

Estimates of the anthropogenic deposition of Hg in North America and Europe during the present century show increasing deposition rates, with a peak of maximum deposition occurring in the 1960s and 1970s. More recently, a relative decrease in deposition rates has been observed due to emission control policies implemented in the industrialized nations in the last two decades (Pirrone et al., 1996, 1998, 2001; Iverfeldt et al., 1995; Swain et al., 1992). In general, pre-industrial Hg atmospheric deposition in remote areas of the northern hemisphere ranges from five to 10 $\mu g\,m^{-2}yr^{-1}$, whereas present day rates range from 10 to 25 $\mu g\,m^{-2}yr^{-1}$, although during peak deposition periods in the 1960s and 1970s, it had reached over 100 $\mu g\,m^{-2}yr^{-1}$ (Iverfeldt et al., 1995; Engstrom & Swain, 1997; Mason et al., 1994). These high atmospheric deposition rates have been linked to Hg pollution in remote areas of the northern hemisphere (Fitzgerald et al., 1998). Consistent data for South America have only recently been generated, notwithstanding its significant anthropogenic contribution of Hg to the atmosphere, in particular, from small-scale gold mining.

Natural emissions of Hg in Brazil are most probably very small, since no volcanoes or significant Hg-ore deposits exist in the country. Therefore, natural emissions are solely from re-volatilization of deposited Hg from soils and waters, from long-range atmospheric transport. This reemission of deposited Hg is highly influenced by climatic changes and changes in land use (Vandal et al., 1993; Martínez-Cortizas et al., 1999; Lacerda et al., 1997c,d). Changes in Hg atmospheric deposition rates are due to natural and anthropogenic causes. Among the natural causes, climatic changes, volcanic emissions, and changing vegetation cover are the major factors affecting deposition rates (Martínez-Cortizas et al., 1999; Shotyk et al., 1998). Among the anthropogenic causes, mining and industrialization, emission control policies, changes in soil uses and technological changes, are among the most important controllers of deposition rates. In the Amazon region, however, atmospheric emissions are due to direct inputs from mining activities and from land use change (Lacerda, 1995, 1997c,d; Nriagu, 1994, 1996).

Generally, examination of lake sediment cores are the easiest means of recording past Hg atmospheric deposition. However, certain characteristics must exist in order to fully interpret results from this type of sample (Porcella,

1996). A suitable environment for recording past atmospheric deposition rates are those, such as headwater lakes, where the basin area to lake surface area ratio is very small, thus decreasing the importance of input from basins, which can be of only local scale. Such lakes must be remotely located, in order to best avoid direct inputs of trace metals from point sources, but such remoteness can create additional logistical problems as well (Blais & Kalf, 1993; Meili, 1995). Furthermore, this group of lakes typically has low sedimentation rates – *i.e.* in the order of 0.1 to 0.01 $cm\,yr^{-1}$ – which, in turn, decreases the temporal resolution obtained from cores. Notwithstanding this, remote headwater lake sediment cores are still regarded as the best material to analyze past deposition rates of Hg. Taking the above into consideration, some studies (Lacerda et al., 1999, Santos et al., 2001; Cordeiro et al., 2002) have drawn a consistent picture of past atmospheric deposition in the Amazon region that can be used as background values for detecting change in present-day deposition rates resulting from small-scale gold mining activity. Figure 37.1 shows the location of the studies reviewed here, while Figure 37.2 presents a consistent set of data from a lake located at São Gabriel da Cachoeira, effectively illustrating how Hg deposition rates have varied in the region over time (Santos et al., 2001). The lake is located at the Hill of Six Lakes at latitude 0°16′N and longitude 66°41′W, in the Amazon lowland of northwestern Brazil. It is surrounded by dense tropical rain forest, in a hot humid climate. All lakes from the Hill occupy small, closed, steep-sided basins with flat or shelving bottoms under 7 to 15 m of water.

Holocene deposition rates show two periods clearly marked, notwithstanding the within-period variability, as seen from the data of São Gabriel da Cachoeira lake cores (Fig. 37.2). The two periods of different Hg deposition are separated by a hiatus of sedimentation rate during the last glacial maximum (LGM). Prior to the LGM at about 18,000 BP, Hg deposition rates averaged $2.8 \pm 1.5\,\mu g\,m^{-2}yr^{-1}$. After the LGM until about 2,000 BP, Hg deposition rates increased by a factor of two to three, averaging $8.6 \pm 2.6\,\mu g\,m^{-2}yr^{-1}$ (Santos et al., 2001). Results from the other sites shown in Figure 37.1 are rather similar, with pre-LGM Hg deposition rates between 1.7 and $4.2\,\mu g\,m^{-2}yr^{-1}$, and post-LGM between 3.8 and 11.2 (Lacerda et al., 1999), and these are, most probably, the background deposition for the entire Amazon region. What is interesting to note is that these values are between two and four times lower than the reported pre-LGM Hg deposition in the northern hemisphere during the same period (Martínez-Cortizas et al., 1999). Large landmasses relative to sea and larger Hg deposits in the Northern hemisphere relative to the southern hemisphere probably explain these results. Post-LGM deposition rates, however, are only slightly smaller than those reported for the northern hemisphere.

The periods of highest Hg deposition during the Holocene were characterized by drier, colder climates with higher frequency of forest fires, as shown by coal and pollen distribution data (Colinvaux et al., 1996; Turcq et al., 1998; Absy et al., 1991). Forest fires are an important source of Hg to the atmosphere in the Amazon, due to enhanced re-emission of deposited Hg from increasing albedo and microclimatology at the soil level, as well as from volatilizing Hg present in the biomass itself (Haygrarth & Jones, 1992; Lacerda, 1995; 1997c,d).

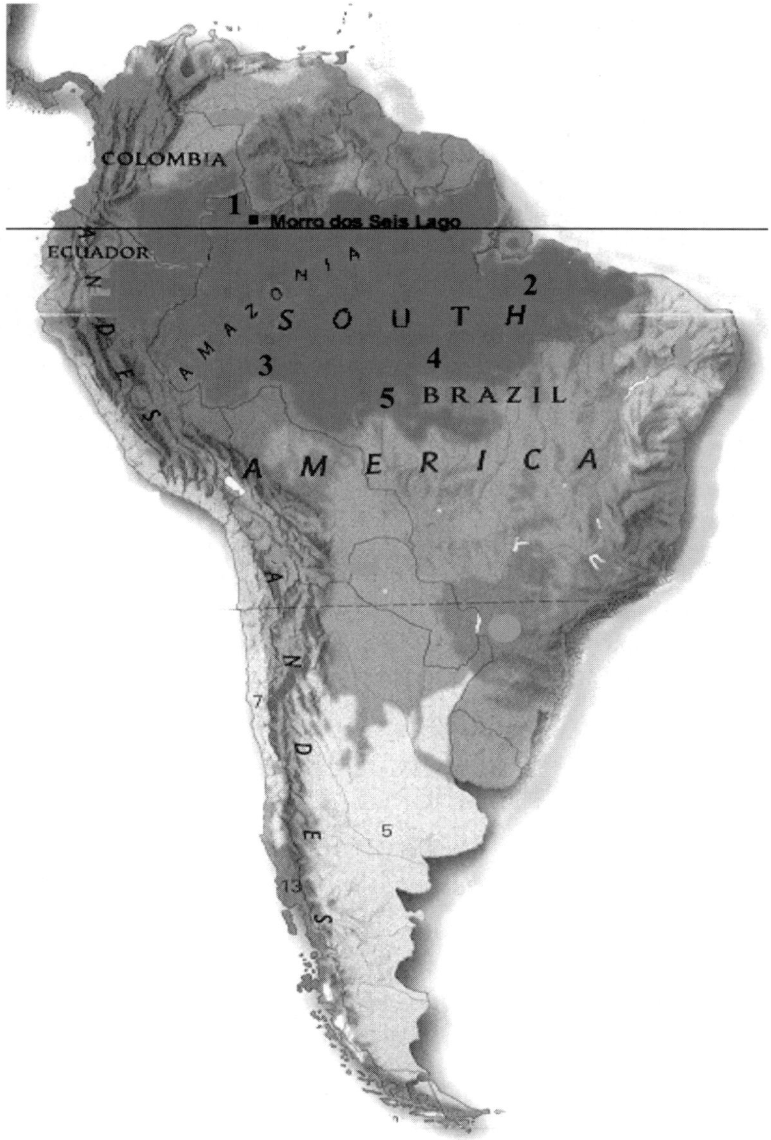

Figure 37.1 Location of major studies on historical recording of atmospheric mercury deposition in the Amazon Region, Brazil. 1. São Gabriel da Cachoeira, 2. Carajás Mountains, 3. Madeira River Prospecting area, Rondônia, 4. Alta Floresta, 5. Poconé, Northern Pantanal.

Furthermore, black carbon ashes may nucleate and reduce Hg residence time in the atmosphere, increasing its deposition (Rasmussen, 1998; Artaxo et al., 1993). A good correlation between black carbon distribution and Hg deposition rates were obtained, at least for one site at Maranhão State, eastern Amazon

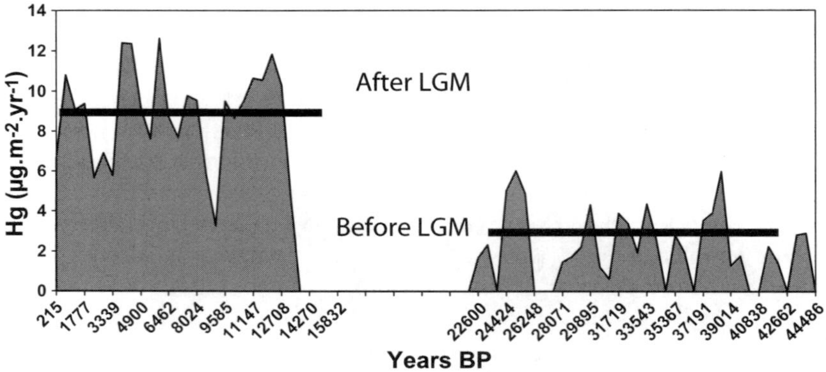

Figure 37.2 Mercury accumulation rates recorded in lake sediment cores from São Gabriel da Cachoeira, Central Amazon, adapted from Santos et al. (2001). See Figure 1 for location of the site.

(Lacerda et al., 1999) as well as for recently-deposited lake sediments in Alta Floresta, southern Amazon (Cordeiro et al., 2002) (see Fig. 37.1 for location of the sites). Although still based on very few studies, increasing Hg deposition during dry periods in the Holocene have been reported, at least for Europe (Martínez-Cortizas et al., 1999) and Antarctica (Vandal et al., 1993).

ATMOSPHERIC Hg DEPOSITION OVER THE AMAZON REGION DURING THE PAST 1000 YEARS

Figure 37.3 shows Hg deposition rates and other paleo-environmental parameters in a sediment core from Lake CSN93.3, in the Carajás Mountains (see Turcq et al. (1998) and Lacerda et al. (1999) for details of lake characteristics and Fig. 37.1 for location). Sediment is mostly composed of organic matter (Org-C = 46%) deposited under a constant and low regime at a rate of 0.03 $cm\,yr^{-1}$. Since the lake has a very small basin and no fluvial inputs, Hg deposition is mostly from the atmosphere. The distribution of the fluxes of organic matter, chlorophyll derivatives and charcoal particles in the core, show no relationship with Hg deposition rates, strongly suggesting that variations in Hg deposition is due to variation in Hg inputs to the lake, rather than to recycling. Also, natural events, such as periods of more intense biomass burning, represented by the flux of charcoal particles, which seemed to have influenced Hg deposition during the Holocene, or change in lake productivity, as indicated by organic matter and chlorophyll derivatives fluxes, have no effect on Hg distribution. Therefore, we may assume that any Hg variability in the core is due to anthropogenic inputs.

Up to year 1600AD, Hg deposition rates are similar to those recorded for the pre-Columbian times (up to 3 $\mu g\,m^{-2}yr^{-1}$). However, from 1600AD until the present, Hg deposition showed a marked increase to present day values ranging from nine to 12 $\mu g\,m^{-2}yr^{-1}$. Natural and man-made emissions to the

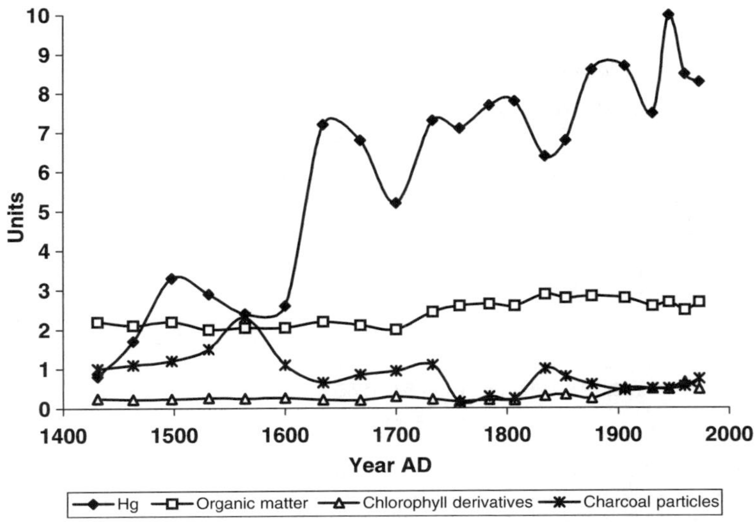

Figure 37.3 Atmospheric mercury deposition ($\mu g\,m^{-2}yr^{-1}$) over the Amazon Region and distribution of major sediment components: Chlorophyll derivatives (SPDU $dm^{-2}yr^{-1}$); organic matter (kg $M^{-2}yr^{-1}$) and charcoal particles ($10^3\,cm^{-2}yr^{-1}$); obtained from the analysis of remote lake sediment cores at the Carajás Mountains, modified from Lacerda et al. (1999).

atmosphere are likely to enter the global cycle if they are in the form of Hg (0), but deposit locally or regionally if they are oxidized (Hg (II) (Porcella 1994), since between 65% and 85% of Hg emissions from gold and silver mining enters the atmosphere as Hg (0) – vapor (Pfeiffer & Lacerda, 1988), and emissions from the Spanish silver mines and, more recently, the present Amazon gold rush, are likely to reach at least a regional scale and easily affect the entire Amazon basin (Lacerda & Salomons, 1998).

A comparison of Hg distribution in Lake CSN93.3 with the estimated Hg emissions from the Spanish silver mines, the major Hg source to South America prior to 1900AD (Nriagu, 1994; Lacerda et al., 1999), shows that Colonial Spanish mining was the major source of Hg to these Amazonian lakes. Mercury deposition in the profile increases from assumed background values to about 7.0 to 9.0 $\mu g\,m^{-2}yr^{-1}$ in the period 1700–1840, when the largest annual emissions of Hg from Spanish colonial silver mines occurred. Silver mining in South and Central America between AD 1580 to 1900, resulted in an average annual loss of Hg of 612 tons (in a range of 292 to 1,085 $t\,yr^{-1}$) and totaled about 196,000 tons, whereas in North America, during the gold rush of last century, the total amount of Hg emitted into the environment is reportedly 61,380 tons (Nriagu, 1994). This result strongly suggests that remote areas of the Amazon region have been subjected to anomalous Hg deposition for the past 500 years.

The importance of gold mining as the major source of Hg to the atmosphere in the Amazon, and its relationship with Hg accumulation in surface

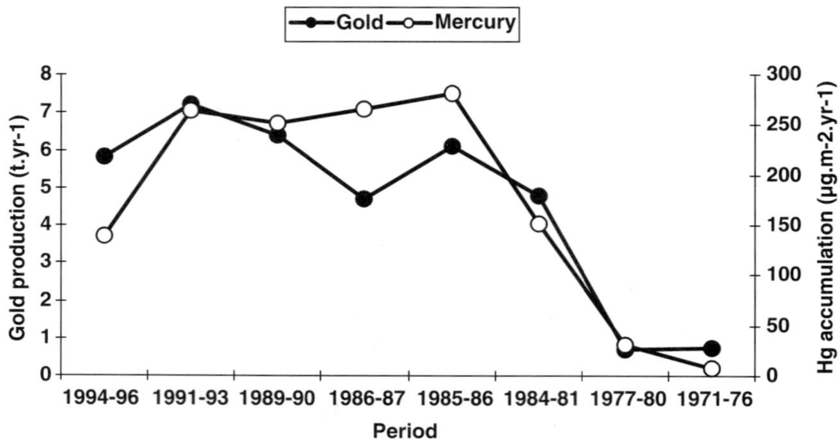

Figure 37.4 Estimated gold production in Alta Floresta mining district, Southern Amazon and mercury deposition rates measured in dated sediment cores from a nearby lake, adapted from Cordeiro et al. (2002).

environments, is demonstrated in Figure 37.4, which shows a covariation between Hg accumulation rates and gold production in an important gold mining region in Alta Floresta, Southern Amazon (based on original data from Lacerda et al. (2001) and Cordeiro et al. (2002)).

COMPARING ATMOSPHERIC Hg DEPOSITION WITH BRAZIL WITH OTHER WORLD REGIONS

Pre-historic deposition of Hg in the Amazon, is somewhat lower than the values reported for the northern hemisphere (Table 37.2), and also appears to have been influenced by regional paleoclimatic changes. The mechanism involved, however, may include changes in vegetation cover and forest fires, rather than simple changes in overall temperature. Mercury deposition during the past millennium, notwithstanding the few results evaluated, seems to have been strongly influenced by colonial mining activities. Presently, although with deposition rates similar to those reported for other remote areas in the northern hemisphere, the estimated Hg deposition rates are higher than the region's pre-historic background, suggesting the importance of anthropogenic sources even in such remote environments.

Whereas a constant increase in Hg deposition has been verified for the Amazon region in recent years, as shown, for example, in the cores of Figure 37.3 and from diverse published studies (see Lacerda et al., 1999 for a review), Hg atmospheric deposition measured in industrialized Brazil, such as along the industrialized Rio de Janeiro-São Paulo axis (data from the Itatiaia Mountains in Table 37.2), provide a similar figure to that reported for North America and Europe, with values showing a significant decrease, after a peak in the 1970s, in response to specific emission control measurements implemented in the past 15

Table 37.2 Reported present (1990's) Hg deposition rates from diverse remote industrial/urban affected areas of the world, compared to Brazilian Lakes ($\mu g\,m^{-2} yr^{-1}$).

Site	Region	Hg deposition	Author
North Sea	Remote	10.8	Mason et al. (1994)
Scandinavia	Remote	10.0	Iverfeldt et al. (1995)
Northern Minnesota, USA	Remote	10.4–15.4	Steines and Anderson (1991)
Lake Michigan, USA	Remote	12.0	Mason and Sullivan (1997)
Rural Mid-west, USA	Remote	12.5	Swain et al. (1992)
Upper Mid-west, USA	Remote	1.8–8.8	Engstrom and Swain (1997)
Mid-continental, USA	Remote	24.5 ± 7.2	Benoit et al. (1994)
Norway	Remote	9–30	Rekolainen et al. (1986)
Finland	Remote	25–50	Mason and Morel (1993)
Carajás Mountains	Remote	8–12	Lacerda et al. (1999)
Itatiaia Mountains	Remote	15–20	Lacerda et al. (1999)
Sepetiba Bay, Brazil	Industrialized	76	Marins et al. (1996)
Finland	Industrialized	370	Mason and Morel (1993)
Denmark	Industrialized	30–200	Lacerda et al. (1991)
Poconé, Central Brazil	Au mining operation	90–120	von Tumpling et al. (1995)
Poconé, Central Brazil	Au mining operation	67–151	Lacerda et al. (1991)
Penido Vello, Spain	Hg mining operation	56 ± 25.7	Martínez-Cortizas et al. (1999)
Jabone, SE Brazil	Metropolitan	23–37	Lacerda et al. (1999)
Penido Vello, Spain	Holocene	1.6–8.5	Martínez-Cortizas et al. (1999)
Amazon	Holocene	1.7–5.6	Lacerda et al. (1999)

years. The different Hg deposition patterns, particularly during the last two decades, are due to different source categories of Hg for the two areas (point industrial sources in the Southeast vs. non-point gold mining sources in the Amazon).

An interesting analogy can be traced with the economic development of Brazil. Economists of the 1980s used the expression "*Bel-India*", to characterize the unequal economic growth distribution between the industrialized south and southeast and the underdeveloped north and northeast regions of the country. The results presented on Hg deposition rates clearly divided the country into two different regions based on pollution sources and characteristics. In the industrialized southeast, Hg emissions are mostly from industrial origins, and following the same trend observed in industrialized nations, these sources have begun to be controlled effectively, thus resulting in a strong reduction in emissions, as verified in the Itatiaia Mountain (Lacerda et al., 2001). Conversely, in the northern region, Hg deposition has occurred as a result of increased emissions from artisanal gold mining.

THE LINK BETWEEN Hg ATMOSPHERIC DEPOSITION, FISH CONTAMINATION AND HUMAN EXPOSURE

Although humans may be subjected to Hg via a myriad of sources and routes, the principal pathway of exposure to this toxic metal is through consumption of Hg-contaminated fish (Porcella, 1994). Numerous studies have shown that, the majority, if not all, of Hg that is bio-accumulated in food chains, takes the form of methyl-Hg (Grieb et al., 1990; Bloom, 1992; Watras & Bloom, 1992; Kim, 1995). Thus, people who consume fish from Hg-impacted aquatic systems – namely, fish eating populations – are at serious risk.

Mercury concentrations in Amazonian fish are frequently found to be independent of the local Hg emissions from gold mining. To understand this apparent paradox, one must trace the major pathways of Hg from gold mining to the fish populations themselves.

Assessing environmental impacts from small-scale gold mining

A summary of the major pathways and behaviour of Hg emissions released into the environment from small-scale gold mining is presented in Table 37.3. Due to the specifics of the process, some 30% of the Hg emitted by the local industry occurs in mine tailings as low-reactive elemental metallic Hg, while about 70% is released into the atmosphere as Hg^0 (vapor), which forms as a result of amalgam burning and gold purification processes. Details on the Hg mass balance of a typical operation can be found in Pfeiffer and Lacerda (1988).

Table 37.3 Major pathways and behavior of the different mercury emissions from small-scale gold mining in the Amazon.

Property	Mining operation	Amalgam burning and gold smelting
Fate	Tailings, from soil mining, and river sediments, from dredges.	Rural, mostly from amalgam burning, and urban atmosphere, from gold smelting.
Chemical species	Hg^0 metallic.	Hg^0 vapor, from relatively low temperature amalgam burning. Hg^{2+} and Hg-particle from gold smelting.
Mobility	Low, dependent on transport of Hg-containing materials.	High, dependent on chemical reactions and atmospheric precipitation.
Residence time	Long, decades to centuries.	Short, days to months.
Reactivity	Low in oxic surface environments, small fraction may be volatilized to the atmosphere.	High under the conditions present in tropical atmosphere, where acidity and high ozone during the dry season accelerates oxidation.
Methylation capacity	Very low.	High, deposition is of readily methylated Hg^{2+}.
Bioaccumulation and toxicity	Very low.	High as Hg^{2+} to extremely high as methyl-Hg.

Mercury vapor oxidizes very quickly in the tropical atmosphere, and oxidation occurs even quicker in the presence of ozone and soot, which persist during, and following, forest fires (Hacon et al., 1995; Lacerda, 1995). This results in a relatively short residence time in the atmosphere and high atmospheric deposition rates between two and five times higher than background rates recorded for the Amazon; deposition rates are over 100 times higher in areas located close to mining sites or tailing deposits (Andrade et al., 1988; Tumpling et al., 1995, Lacerda & Salomons, 1998; Marins et al., 2000).

Upon reaching surface ecosystems as highly reactive Hg^{2+}, biological entities within tropical soils and aquatic ecosystems begin cycling the metal. It is rapidly mobilized, particularly in the extensive floodplains of large tropical rivers and lakes, where high organic matter content, bacterial activity and optimum hydrochemistry favour high rates of Hg methylation (Guimarães et al., 1998). Highly toxic methyl-Hg formed under such conditions is carried to rivers and accumulates in various biota. Recent studies of Hg distribution in fish obtained in the Amazon region, show that, notwithstanding the general decrease in gold mining activity, carnivorous species still contain high concentrations of Hg (70–98% as methyl-Hg) similar to those measured during the early 1990s, when gold mining had reached its peak (Lechler et al., 2000). Also, higher Hg concentrations are found in such environments as floodplains, floodplain lakes, and artificial reservoirs, where methylation capacity is at its maximum (Aula et al., 1994; Lacerda et al., 1994).

The ecology of most tropical fish species also influences Hg bioaccumulation. Many Amazon fish species are characterized by a migratory behavior to flood plains during the flood season (six to eight months annually), when fish are probably most exposed to high methyl-Hg concentrations because of optimal methylation conditions in those flooded areas compared to the main channel of rivers (Lacerda, 1997c).

The ecological cycle of Hg in Amazonian aquatic systems also affect humans. Prospectors are seldom contaminated with Hg; the highest Hg content in these individuals is generally found in blood and urine, is associated with occupational exposure, and is mostly in an inorganic form (Cleary et al., 1994; Lacerda & Salomons, 1998; Grandjean et al., 1999; Harada et al., 2001). Prospectors are mostly immigrants from other Brazilian regions with diets based on meat, rather than fish. Riverine populations, including Indians, and particularly those residing along black water rivers and flood plains, have the highest Hg content among human groups in the Amazon. Very high Hg concentrations have been measured even in areas where gold mining is small or absent (Silva-Forsberg et al., 1999). The highest Hg concentrations found in these human groups are present in hair as methyl-Hg (Malm et al., 1995; Akagi et al., 1995; Barbosa et al., 1995; Boischio & Cermnichiari, 1998); this exposure occurs through the food chain within a population that depends on fish as its only source of protein (Clarkson, 1997).

This pattern of Hg distribution among human groups in the Amazon is quite perverse, since the most exposed populations are those not directly linked to the gold mining activity itself, and which will never receive any profit from it.

Apart from being a significant source of methyl-Hg, tropical aquatic ecosystems may also enhance the cycling of Hg through reduction and re-emission into the atmosphere. Hg volatilization fluxes from surface waters in the French Guiana Amazon forest were reportedly much higher than those measured in temperate aquatic environments, suggesting that tropical aquatic systems are important sources of Hg^0 to the atmosphere (Amouroux et al., 1999). The potential Hg^0 reduction to Hg^{2+} capacity of these environments seems very high, representing a major pathway for the dynamic cycling of Hg in tropical environments (Amouroux et al., 1999). Similar results have been recently reported for Hg re-emission from soils, in particular when the original forest cover is substituted by agriculture or pasture (Lacerda et al., 2001).

Land use changes and mercury remobilization in the Amazon

A second important component of the tropical rain forest Hg cycle includes the Hg deposited on soils. In principle, the residence time of Hg in tropical soils is in the order of hundreds to thousands of years, thus allowing the build up of relatively high concentrations even when natural atmospheric deposition is the major Hg source (Silva-Forsberg et al., 1999). Leaching of Hg accumulated in Amazon soils is accelerated by the conversion of tropical forests into pasture and/or agriculture (Roulet et al., 1996, 2000, 2001; Lacerda et al., 2001). Also, Hg degassing, although still poorly studied in tropical regions, is probably accelerated by forest soil conversion, as suggested by preliminary results (Table 37.4) from comparing Hg concentrations and distribution in forest and pasture soils submitted to similar Hg input in southern Amazon (Lacerda et al., 2001).

Figure 37.5 proposes theoretical behavior of Hg in tropical rain forest soils, based on a model proposed by Hesterberg et al. (1992) and modified by Stigliani (1995). The graph shows the response of tropical rain forest soils under different uses to increasing Hg input. The slope of the line at a given point $\Delta Q/\Delta C$ defines the binding capacity of soil conditions for Hg. Thus, in natural tropical rain forest soils at the beginning of the contamination event, the binding capacity is quite high ($\Delta Q_1/\Delta C_1$). As the contamination progresses, this capacity declines ($\Delta Q_2/\Delta C_2$), until a maximum sorption capacity is reached, where the binding capacity is zero – i.e. any small additional increase in Hg input (ΔQ_2) results in large leaching from soils (ΔC_2) and probably enhances Hg degassing fluxes.

Under different soil uses, Hg leaching and degassing from soils increases. Converting tropical rain forest soils into pasture results in large remobilization of

Table 37.4 Mercury concentrations (ng g^{-1}) in soils under tropical forests and pasture developed in the same soils, from southern Amazon Brazil, after Lacerda et al. (2001).

Land use	Far from mining sites	Close to mining sites
Forest	25–140	150–210
Pasture	9–25	30–42

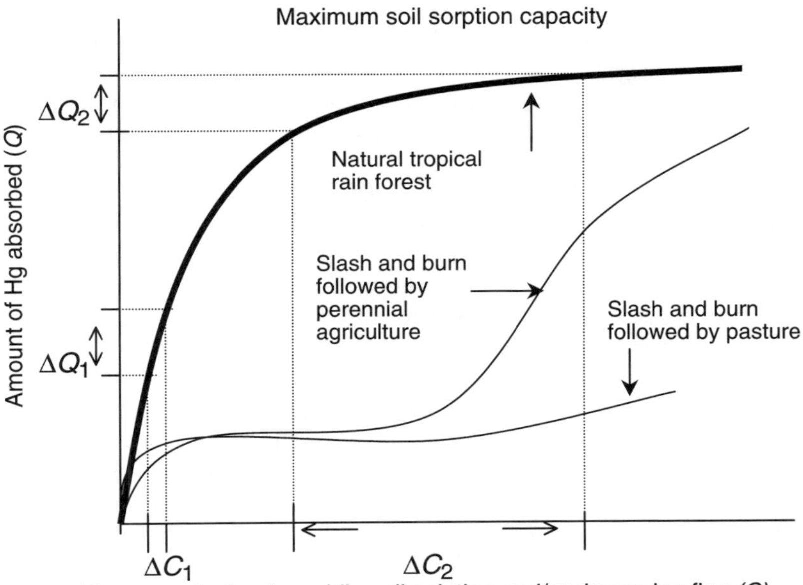

Figure 37.5 Theoretical behavior of Hg in tropical rain forest soils under different processes of land use changes.

Hg, either by increasing degassing fluxes as suggested by results obtained in Alta Floresta in Southern Amazon (Lacerda et al., 2001), or by increasing erosion suggested by mass balance and geochemical tracer studies performed along the Tapajós River, a tributary of the Amazon River in Central Amazon (Roulet et al., 1996, 2000). Deforestation, followed by agriculture, may eventually recover some Hg binding capacity after a period of low sorption capacity, due to the recovering of soil properties. This eventual increase in sorption capacity may be enhanced under perennial crops such as cocoa, but may be kept very low if annual crops are preferred.

This theoretical framework, however, needs testing, but preliminary results, although still very site-specific, highlight the importance of forest conservation practices to minimize Hg remobilization in tropical rain forest ecosystems.

REFERENCES

ABICLOR (2001). *Anuário estatístico 2000*. Associação Brasileira da Indústria de Álcalis e Cloro-derivados, São Paulo, p. 24.

Absy, M.L., Cleff, A., Fournier, M., Martin, L., Servant, M., Sefeddine, A., Silva, M.F., Soubies, F., Sugio, K., Turcq, B. & Van der Hammen, T. (1991). Mise en évidence de quatre phases d'overture de la forêt dense dans le sud-est de l'Amazonie au cours des 60.000 derniéres annés. Premiére comparaison avec d'autres régions tropicales. *CR Acad Sci Paris Ser II*, 318, 1645–1652.

Adimado, A.A. & Baah, D.A (2002). Mercury in human blood, urine, hair, nail, and fish from the Ankobra and Tano river basins in Southwestern Ghana. *Bull Environ Contam Toxicol*, 68, 339–346.

Akagi, H., Malm, O., Kinjo, Y., Harada, M., Branches, F.J.P., Pfeiffer, W.C., & Kato, H. (1995). Methylmercury pollution in the Amazon, Brazil. *Sci Tot Environ*, 178, 85–95.

Aks, S.E., Erickson, T.B., Branches, F.J.O., & Hryhorczuk. D.O. (1995). Blood-mercury concentrations and renal biomarkers in Amazonian villagers. *Ambio*, 24, 103–105.

Amouroux, D., Wasserman, J.C., Tessier, E., & Donard, O.E.X. (1999). Elemental mercury in the atmosphere of a tropical Amazonian forest (French Guiana). *Environ Sci Technol*, 33, 3044–3048.

Andrade, J.C., Bueno, M.I.M.S., Soares, P.V., & Choudhuri, A. (1988). The fate of mercury released from prospecting (Garimpos) near Guarinus and Pilar, Goiás (Brazil). *An Acad Bras Ciênc*, 60, 293–303.

Artaxo, P.F., Gerab, M.L., & Rabello, C. (1993). Elemental composition of aerosol particles from two background monitoring stations in the Amazon basin. *Nuclear Instru Methods Phys Res*, B75, 277–279.

Aula, I., Braunsweiller, H., Leino, T., Malin, I., Porvari, P., Hatanaka, T., Lodenius, M., & Juras, A. (1994). Levels of mercury in the Tucurui reservoir and its surrounding are in Pará, Brazil. In C.J. Watras & J.W. Huckabee (Eds.), *Mercury pollution: Integration and synthesis* (pp. 21–40). Boca Raton: Lewis Publ.

Barbosa, A.C., Boischio, A.A.P., East, G.A., Ferrari, I., Gonçalves, A., Silva, P.R.M. & Cruz, T.M.E. (1995). Mercury contamination in the Brazilian Amazon, environmental and ocuppational aspects. *Water Air Soil Pollut*, 80, 109–121.

Benoit, J.M., Fitzgerald, W.F. & Damman, A.W.H. (1994). Historical atmospheric mercury deposition in mid-continental U.S., as recorded in an ombrotrophic peatbog. In C. Watras & J.B. Huckabee (Eds.), *Mercury pollution: Integration and synthesis* (pp. 234–239). Boca Raton: Lewis Publ.

Blais, J.M. & Kalf, J. (1993). Atmospheric loadings of Zn, Cu, Ni, Cr, and Pb to lake sediments: The role of catchment, lake morphometry, and physico-chemical properties of the elements. *Biogeochemistry*, 23, 1–22.

Bloom, N.S. (1992). On the chemical form of mercury in edible fish and marine invertebrate tissues. *Can J Fish Aquat Sci*, 49, 1010–1017.

Boischio, A.A.P. & Cernichiari, P. (1998). Longitudinal hair mercury concentration in riverside mothers along the Upper Madeira River (Brazil). *Environm Res Sect A*, 77, 79–83.

Boischio, A.A.P. & Henshel, D. (2000). Fish consumption, fish lore, and mercury pollution-risk communication for the Madeira river people. *Environm Res Sect A*, 84, 108–126.

Bonzongo, J.C., Heim, K.J., Chen, C., Lyons, W.B., Warwick, J.J., Miller, G.C., & Lechler, P.J. (1996a). Mercury pathways in the Carson River-Lahontan Reservoir System, Nevada, USA. *Environ Toxicol Chem*, 15, 677–683.

Bonzongo, J.C., Heim, K.J., Warwick, J.J. & Lyons, W.B. (1996b). Mercury levels in surface waters of the Carson River-Lahontan Reservoir System,

Nevada: Influence of historic mining activities. *Environ Pollut*, 92, 193–201.

Bonzongo, J.C., Donkor, A.K., Nartey, V.K. & Lacerda, L.D. (2002). Mercury pollution in Ghana: a case study of environmental impacts of artisanal gold mining in sub-Saharan Africa. In L.D. Lacerda & R.E. Santelli (Eds.), *Facets of Environmental Geochemistry in Tropical and Sub-Tropical Environments*. Berlin: Springer Verlag (in press).

Castilhos, Z.C., Bidone, E.D. & Lacerda, L.D. (1998). Increase of the background human exposure to mercury through fish consumption due to gold mining at the Tapajos river region, Para State, Amazon. *Bull Environ Contam Toxicol*, 61, 202–209.

Clarkson, T.W. (1997). The toxicology of mercury. *Critical Rev Clin Lab Sci* 34, 369–403.

Cleary, D., Thornton, I., Brown, N., Kazantzis, G., Delves, T. & Worthington, S. (1994). Mercury in Brazil. *Nature*, 369, 613–614.

Colinvaux, P.A., de Oliveira, P.E., Moreno, J.E., Miller, M.C. & Bush, M.B. (1996). A long pollen record from lowland Amzonia: Forest and cooling in glacial times. *Science*, 274, 85–88.

Cordeiro, R.C., Turcq, B., Ribeiro, M.G., Lacerda, L.D., Capitâneo, J., Oliveira da Silva, A., Sifeddine, A. & Turq, P.M. (2002). Forest fire indicators and mercury deposition in an intense land use change region in the Brazilian amazon (Alta Floresta, MT). *Sci Tot Environ*, 293, 247–256.

Engstrom, D.R. & Swain, E.B. (1997). Recent declines in atmospheric mercury deposition in the Upper Midwest. *Environm Sci Technol*, 31, 960–967.

Eve, E., Oliveira, E.F. & Eve, C. (1996). The mercury problem and diets in the Brazilian Amazon: planning a solution. *Environm Conser*, 23, 133–139.

Ferreira, R.H. & Appel, L.E. (1991). Mercury: sources and uses in Brazil. In *Inter Symp on Environmental Studies on Tropical Rain Forests*, pp. 207–216. Manaus, 8–11 November, 1991.

Fitzgerald, W.F., Engstrom, D.R., Mason, R.T. & Nater, E.A. (1998). The case for atmospheric mercury contamination in remote areas. *Environ Sci Technol*, 32, 1–7.

Gill, G.A. & Bruland, K.W. (1990). Mercury speciation in freshwater systems in California and other areas. *Environm Sci Technol*, 24, 1392–1400.

Grandjean, P., White, R.F., Nielsen, A., Cleary, D. & Santos, E.C. (1999). Methylmercury neurotoxicity in Amazonian children downstream from gold mining. *Environm Health Perspect*, 107, 587–591.

Grieb, T.M., Driscoll, C.T., Gloss, S.P., Schofield, C.L., Bowie, G.L. & Porcella, D.B. (1990). Factors affecting mercury accumulation in fish in the upper Michigan peninsula. *Environ Toxicol Chem*, 9, 919–930.

Grigal, D.F., Nater, E.A. & Homann, P.S. (1994). Spatial distribution of mercury in an east-central Minnesota landscape. In C. Watras & J.B. Huckabee, (Eds). *Mercury pollution: Integration and synthesis.* Boca Raton: Lewis Publ.

Guimarães, J.R.D., Meili, M., Malm, O. & Brito, E.M.S. (1998). Hg methylation in sediments and floating meadows of a tropical lake of the Pantanal wetland, Brazil. *Sci Tot Environ*, 213, 165–175.

Gulleb, A.C., Schenck, C. & Staib, E. (1997). Giant otter (*Pteronura brasiliensis*) at risk? Total mercury and methylmercury levels in fish and otter scats, Peru. *Ambio*, 26, 511–514.

Hacon, S., Artaxo, P., Gerab, F., Yamasoe, M.A., Calixto, R.C., Lacerda, L.D. & Conti, L.F. (1995). Atmospheric mercury and trace elements in the region of Alta Floresta in the Amazon Basin. *Water Air Soil Pollut*, 80, 273–283.

Harada, M., Nakachi, S., Cheu, T., Hamada, H., Ono, Y., Tsuda, T., Yanagida, K., Kizaki, T. & Ohno, H. (1999). Monitoring mercury pollution in Tanzania: relation between head hair mercury and health. *Sci Tot Environ*, 227, 249–256.

Harada, M., Nakanichi, J., Yasoda, E., Pinheiro, M.C.N., Oikawa, T., Guimarães, G.A., Cardoso, B.S., Kizaki, T. & Ohno, H. (2001). Mercury pollution in the Tapajós River basin, Amazon, mercury level of head hair and health effects. *Environm Intern*, 27, 285–290.

Haygrarth, P.M. & Jones, K.C. (1992). Atmospheric deposition of metals to agricultural surfaces. In D.C. Adriano (Ed.), *Biogeochemistry of Trace Metals* (pp. 175–244). Boca Raton: Lewis Publ.

Hesterberg, D., Stigliani, W.M. & Imeson, A.C. (1992). *Chemical time bombs: linkages to scenarios of socioeconomic development.* Executive Rep 20 (CTB Basic Deoc 2), IIASA, Laxenburg, p. 321.

Ikingura, J.R. & Akagi, H. (1996). Monitoring of fish and human exposure to mercury due to gold mining in the lake Victoria goldfields, Tanzania. *Sci Tot Environ*, 191, 59–68.

Ikingura, J.R., Mutakyahwa, M.K.D. & Kahatano, J.M.J. (1997). Mercury and mining in Africa with special reference to Tanzania. *Water Air Soil Pollut*, 97, 223–232.

ILO (1999). *Social and labour issues in small-scale mines.* Report for discussion at the Tripartite Meeting Social and Labour Issues in Small-scale Mines, International Labour Organization, Geneva, 17–21 May, pp. 1–107.

Iverfeldt, A., Munthe, J., Brosset, C. & Pacyna, J.M. (1995). Long-term changes in concentration and deposition of atmospheric mercury over Scandinavia. *Water Air Soil Pollut*, 80, 227–233.

Kehrig, H.A., Malm, O., Akagi, H., Guimaraes, J.R.D. & Torres, J.P.M. (1998). Methylmercury in fish and hair samples from the Balbina reservoir, Brazilian Amazon. *Environ Res Sect A*, 77, 84–90.

Kim, J.P. (1995). Methylmercury in rainbow trout (*Onchorhynchus mykiss*) from lakes Okareka, Okaro, Rotomahana, Rotura and Tarawera, Nort Island, New Zealand. *Sci Tot Environ*, 164, 209–219.

Kom, J.F.M., Voet, G.B. & Wolf, F.A. (1998). Mercury exposure of Maroon workers in small-scale gold mining in Suriname. *Environ Res*, 77, 91–97.

Lacerda, L.D. (1995). Amazon mercury emissions. *Nature*, 374, 21–22.

Lacerda, L.D. (1997a). Global mercury emissions from gold and silver mining. *Water Air Soil Pollut*, 97, 209–221.

Lacerda, L.D. (1997b). Evolution of mercury contamination in Brazil. *Water Air Soil Pollut*, 97, 247–255.

Lacerda, L.D. (1997c). Atmospheric mercury and fish contamination in the Amazon. *J Braz Assoc Adv Sci*, 49, 54–57.

Lacerda, L.D. (2002). Updating global Hg emission from small-scale gold mining and assessing its environmental impacts. *Environm Geol* (in press).

Lacerda, L.D. & Salomons, W. (1998). *Mercury from gold and silver mining: a chemical time bomb?* Berlin: Springer-Verlag, p. 146.

Lacerda, L.D. & Vaisman, A.G. (2002). Estimates of mercury emissions to the atmosphere due to prospected changes in the Brazilian energy generation matrix. *Regional Environ Changes*, in press.

Lacerda, L.D. & Marins, R.V. (1997). Anthropogenic mercury emissions to the atmosphere in Brazil: The impact of gold mining. *J Geochem Explor*, 58, 223–229.

Lacerda, L.D., Salomons, W., Pfeiffer, W.C. & Bastos, W.R. (1991). Mercury distribution in sediment profiles of remote high Pantanal lakes, Central Brazil. *Biogeochemistry*, 14, 71–77.

Lacerda, LD., Bidone, E.D., Guimarães, A.F. & Pfeiffer, W.C. (1994). Mercury distribution in fish from the Itacaiúnas-Parauapebas River system, Carajás regions, Amazon. *An Acad Brasil Ciên*, 66, 373–379.

Lacerda, L.D., Ribeiro, M.G., Cordeiro, R.C., Sifeddini, A. & Turcq, B. (1999). Atmospheric mercury deposition over Brazil during the past 30,000 years. *J Braz Assoc Adv Sci*, 51, 363–371.

Lacerda, L.D., Ribeiro, M.G., Souza, M. & Ayres, G.A. (2001). *Distribuição de mercúrio em solos e sedimentos lacustres na região de Alta Floresta, Mato Grosso, Brasil.* Seria Tecnologia Ambiental CETEM/CNPq, Rio de Janeiro, p. 65.

Lechler, P.J., Miller, J.R., Lacerda, L.D., Vinson, D., Bonzongo, J-C., Lyons, W.B. & Warwick, J.J. (2000). Elevated mercury concentrations in soils, sediments, water, and fish of the Madeira River basin, Brazilian Amazon: a function of natural enrichments? *Sci Tot Environ*, 260, 87–96.

Lodenius, M. & Malm, O. (1998). Mercury in the Amazon. *Rev Environ Contam Toxicol*, 157, 25–52.

Lyons, W.B., Welch, K.A. & Bonzongo, J.C. (1999). Mercury in aquatic systems in Antarctica, *Geophys Res Lett*, 26, 2235–2238.

Madsen, P.P. (1981). Peat bog records of atmospheric mercury deposition. *Nature* 293, 127–130.

Malm, O. (1998). Gold Mining as a source of mercury exposure in the Brazilian Amazon *Environ Res Sect A*, 77, 73–78.

Malm, O., Castro, M.B., Bastos, W.R., Branches, F.J.P., Guimaraes, J.R.D., Zuffo, C.E. & Pfeiffer, W.C. (1995). An assessment of Hg pollution in different gold mining areas, Amazon Brazil. *Sci Tot Environ*, 175, 127–140.

Marins, R.V., Andrade, J.B., Pereira, P.A.P., Paiva, E.C. & Paraquetti, H.H.M. (2000). Sampling techniques for the assessment of anthropogenic vapour and particulate mercury in the Brazilian Amazon atmosphere. *J Environ Monit*, 2, 235–238.

Marins, R.V., Silva Filho, E.V. & Lacerda, L.D. (1996). Atmospheric mercury deposition over Sepetiba Bay, SE Brazil. *J Braz Chem Soc*, 9, 77–81.

Maron. M.A.C. (1999). Ouro. *Sumário Mineral*, 19, 78–79.

Martínez-Cortizas, A., Pontevedra-Pombal, X., García-Rodeja, E., Nóvoa-Muñoz, J,C, & Shotyk, W. (1999). Mercury in Spanish peat bog: Archive of climate Change and atmospheric metal deposition. *Science*, 284, 939–942.

Mason, R.P. & Morel, F.M.M. (1993). An assessment of principal pathways for oxidation of elemental Hg and the production of methyl-mercury in Brazilian waters affected by gold mining. pp. 413–416. In J.J. Abrão, J.C. Wasserman & E.V. Silva Filho, (Eds.), *Proc Intern Symp Environmental Geochemistry in Tropical Countries,* Universidade Federal Fluminense, Niteroi.

Mason, R.P. & Sullivan, K.A. (1997). Mercury in Lake Michigan. *Environ Sci Technol*, 31, 942–947.

Mason, R.P., Fitzgerald, W.F. & Morel, M.M. (1994). The biogeochemical cycling of elemental mercury. *Geochim Cosmochim Acta*, 58, 3191–3198.

Mastrine, J.A., Bonzongo, J.C. & Lyons, W.B. (1999). Mercury concentrations in surface waters from fluvial systems draining historical precious metals mining areas in southeastern U.S.A. *Appl Geochem*, 14, 147–158.

Maurice-Bourgoin, L., Quiroga, I., Chincheros, J. & Coureau, P. (2000). Mercury distribution in waters and fishes of the upper Madeira rivers and mercury exposure in riparian Amazonian populations. *Sci Tot Environ*, 260, 73–86.

Meili, M. (1995). Pre-industrial atmospheric deposition of mercury: uncertain rates from lake sediment and peat cores. *Water Air Soil Pollut*, 80, 637–640.

Moreira, J.C. & Pivetta, F. (1997). Human and environmental contamination by mercury from industrial uses in Brazil. *Water Air Soil Pollut*, 97, 241–246.

Nico, L.G. & Taphorn, D.C. (1994). Mercury in fish from gold-mining regions in upper Cuyuni River system, Venezuela. *Fresenius Environ Bull*, 3, 287–292.

Nriagu, J.O. (1994). Mercury pollution from the past mining of gold and silver in the Americas. *Sci Tot Environ*, 149, 167–181.

Nriagu, J.O. (1996). A history of global metal pollution. *Science* 272, 223–224.

Nriagu, J.O., Pfeiffer, W.C., de Souza, C.M.M. & Mierle, G. (1992). Mercury pollution in Brazil. *Nature*, 356, 389.

Palheta, D. & Taylor, A. (1995). Mercury in environmental and biological samples from a gold mining area in the Amazon region of Brazil. *Sci Tot Environ*, 168, 63–69.

Pfeiffer, W.C. & Lacerda, L.D. (1988). Mercury inputs into the Amazon Region, Brazil. *Environ Technol Lett*, 9, 325–330.

Pires, M., Fiedler, H. & Teixeira, E.C. (1997). Geochemical distribution of trace elements in coal: modeling and environmental aspects. *Fuel*, 76, 1425–1437.

Pirrone, N., Keeler, G.J. & Nriagu, J.O. (1996). Regional differences in worldwide emissions of mercury to the atmosphere. *Atmospheric Environ*, 30, 2981–2987.

Pirrone, N., Alegrini, I., Keeler, G.J., Nriagu, J.O., Rossmann, R. & Robins, J.A. (1998). Historic atmospheric mercury emissions in North America compared to mercury accumulations in sedimentary records. *Atmospheric Environ*, 5, 929–940.

Pirrone, N., Costa, P., Pacyna, J.M. & Ferrara, R. (2001). Mercury emissions to the atmosphere from natural and anthropogenic sources in the Mediterranean region. *Atmospheric Environ*, 35, 2997–3006.

Porcella, D.B. (1994). Mercury in the environment: Biogeochemistry. In C. Watras & J.W. Huckabee (Eds.), *Mercury pollution: Integration and synthesis* (pp. 3–19). Boca Raton, Florida: Lewis Publishers.

Porcella, D.B. (1996). *Protocol for estimating historic atmospheric mercury deposition. Technical Report* – TR – 106768, Electric Power Research Institute, Palo Alto, California.

Rasmussem, P.E. (1998). Long-range transport of trace metals: the need for geoscience perspectives. *Environm Geol*, 33, 96–108.

Rekolainen, S., Verta, M. & Liehu, A. (1986). The effect of airborne mercury contents in some Finish forest lakes. *Publ Water Res Int Finland*, 65, 11–20.

Rolfhus, K.R. & Fitzgerald, W.F. (1995). Linkages between atmospheric mercury deposition and the methylmercury content of marine fish. *Water Air Soil Pollut*, 80, 291–297.

Roulet, M., Lucotte, M., Rheault, I., Tran, S., Farella, N., Canuel, R., Mergler, D. & Amorim, M. (1996). Mercury in Amazon soils: accumulation and release. In S.H. Bottrell (Ed.) *4th Inter. Symp. Geochem. Earth's Surface*, Ilkely, pp. 453–457.

Roulet, M., Lucotte, M., Canuel, R., Farella, N., Courcelles, M., Guimarãe, J.R.D., Mergler, D. & Amorim, M. (2000). Increase in mercury contamination recorded in lacustrine sediments following deforestation in central Amazon. *Chem Geol*, 165, 243–266.

Roulet, M., Luycotte, M., Canuel, R., Farella, N., De Freitos Goch, Y.G., PachecoPeleja, J.R., Guimarães, J.-R.D., Mergler, D. & Amorin, M. (2001). Spatio-temporal geochemistry of mercury in waters of the Tapajós and Amazon rivers, Brazil. *Limnol Oceanogr*, 46, 1141–1157.

Santos, E.C., de Jesus, I.M., da Silva Brabo, E., Loureiro, E.C.B., da Silva Mascarenhas, A.F., Weirich, J., Camara, V.D.M. & Cleary, D. (2000). Mercury exposures in riverside Amazon communities in Para, Brazil. *Environ Res Sec, A* 84, 100–107.

Santos, G.M., Cordeiro, R.C., Silva filho, E.V., Turcq, B., Lacerda, L.D., Fifield, L.K., Gomes, P.R.S., Hausladen, P.A., sifeddine, A. & Albuquerque, A.L.S. (2001). Chronology of the atmospheric mercury in lagoa da Pata basin upper Rio Negro of Brazilian Amazon. *Radiocarbon*, 43, 801–808.

Shotyk, W., Weiss, D., Appleby, P.G., Cheburkin, A.K., Frei, R., Gloor, M., Kramers, J.D., Reese, S. & Van Der Knaap, W.O. (1998). History of atmospheric lead deposition since 12,370 ^{14}C yr BP from a peat bog, Jura Mountains, Switzerland. *Science*, 281, 1635–1640.

Silva-Forsberg, M.C., Forsberg, B.R. & Zeidmann, V.K. (1999). Mercury contamination in humans liked to river chemistry in the Amazon Basin. *Ambio*, 28, 519–521.

Steines, E. & Anderson, E.M. (1991). Atmospheric mercury deposition of mercury in Norway: temporal and spatial trends. *Water Air Soil Pollut*, 56, 391–404.

Stigliani, W.M. (1995). Global perspective and risk assessment. In W. Salomons & W.M. Stigliani (Eds.), *Biogeodynamics of pollutants in soils and sediments* (pp. 331–343). Berlin: Springer Verlag.

Swain, E.B., Engstrom, D.R., Brigham, M.E., Henning, T.A. & Brezonik, P.L. (1992). Increasing rates of atmospheric mercury deposition in midcontinental North America. *Science*, 257, 784–787.

Tumpling, W.V., Wilken, R.D. & Einax, J. (1995). Mercury contamination in the northern Pantanal region, Mato Grosso, Brazil. *J Geochem Explor*, 52, 127–134.

von Tumpling, W., Wilken, R.D., Einax, J. (1996). Mercury deposition resulting from setting on fire of grasslands – estimation of the annual mercury deposition of the tropical northern pantanel region, central Brazil. In *Proceedings of the International Conference on Mercury as a Global Pollutant*, Hamburg.

Turcq, B., Sefeddine, A., Martin, L., Absy, M.L., Soubies, F., Suguio, K. & Volkmer-Ribeiro, C. (1998). Amazonia rainforest fires: A lacustrine record of 7,000 years. *Ambio*, 27, 139–142.

Vandal, G.M., Fitzgerald, W.F., Boutron, C.F. & Candelone, J.-P. (1993). Variations of mercury deposition to Antarctica over the past 34,000 years. *Nature*, 362, 621–623.

van Straaten, P. (2000a). Human exposure to mercury due to small-scale gold mining in northern Tanzania. *Sci Tot Environ*, 259, 45–53.

van Straaten, P. (2000b). Mercury contamination associated with small-scale gold mining in Tanzania and Zimbabwe. *Sci Tot Environ*, 259, 105–113.

Veiga, M.M. (1997). *Introducing new technologies for abatement of global mercury pollution in Latin America*. UNIDO/UBC/CETEM-CNPq, Rio de Janeiro, p. 94.

Velásquez, G.Á.S., Fenzi, N., Guillen, S.M. & Nascimento, F. (1997). Contaminação por Hg, Pb e CN^- gerada pela garimpagem dde ouro sobre os Rios Sucio, Mico e Sinecapa, Nicarágua, América Central. *Geochim Brasil*, 11, 121–131.

Ware, G.W. (1998). Mercury in the Amazon. *Rev Environ Contam Toxicol*, 157, 25–52.

Watras, C.J. & Bloom, N.S. (1992). Mercury and methylmercury in individual zooplankton: Implications for bioaccumulation. *Limnol Oceanogr*, 37, 1313–1318.

Wayne, M.D., Lechler, P.J., Gill, G.A., Warwick, J.J. & Lyons, W.B. (1997). Mercury concentrations in the Carson River, Nevada: modern influence of historic mine wastes. *Water Air Soil Pollut*, 92, 391–399.

38

How Beautiful is Small-Scale Mining? Evidence from Small-Scale and Artisanal Gold Mining in Ecuador

N.H. Tarras-Wahlberg, B. Lundberg, A. Bermeo, and R. Hoffner

This chapter presents a case study of the Ecuadorian small-scale and artisanal mining industry, with the aim of evaluating its socio-economic impacts. Mining in Ecuador has long been near-synonymous with gold mining, which has a history stretching back to pre-colonial days. Some very substantial mines existed in the late-19th and early-20th century. At present, however, Ecuador does not have any large mines, and comparisons with existing definitions (*e.g.* Spiropolous, 1989; Noetstaller, 1987; Davidson, 1989; DSE and UN/DTCD, 1991) suggest that all current mining activities are either "small-scale" or "artisanal". Although operations have diminished in recent years,[1] an estimated 20,000 people are still directly involved in the industry.

The data presented in this case study were collected during the recently-completed Mining Sector Development Project, *Proyecto Desarollo Minero y su Control Ambiental* (Prodeminca), which was executed by the Ecuadorian Government, with the support of the World Bank, and Swedish and UK Governments. The project was initiated in 1994 and completed in 2000. Its main aim was to stimulate the development of a modern and environmentally-sustainable mining sector. As part of this work, extensive socio-economic and environmental investigations were undertaken in Ecuador's major mining centres.

---

[1] The decrease in the number of miners is a result of recent economic problems. It is especially believed that the rise in costs associated with the recent adoption of the US dollar as the official Ecuadorian currency may have adversely affected the financial viability of many small mining ventures.

AN OVERVIEW OF ECUADOR'S MAIN GOLD MINING DISTRICTS

Although there are numerous pockets of small and/or semi-permanent mining activities scattered throughout Ecuador, the country has four major (gold) mining regions. They are as follows: (i) Portovelo-Zaruma; (ii) Ponce Enríquez; (iii) Nambija; and (iv) Santa Rosa (Fig. 38.1). Each exhibits similar characteristics in terms of organization and mining techniques, largely because the Ponce Enríquez, Nambija and Santa Rosa gold fields were all discovered and first exploited by miners originating from the Portovelo-Zaruma district, the former centre of industrial mining.

Portovelo-Zaruma

The mining district surrounding the twin cities of Portovelo and Zaruma is Ecuador's oldest and most important gold mining centre. The two cities are situated on the western side of the Eastern Cordillera, at altitudes of 800–1,500 masl. In addition to mining, the areas surrounding Portovelo-Zaruma support low intensity agriculture and cattle grazing; the region is sparsely populated overall. The mining district itself is situated at the headwaters of the Puyango River, which flows in a south-westerly direction, and empties into the sea in Peruvian territory.

Spanish colonialists began mining the area in the mid-1500s; in 1549, the city of Zaruma was established (Zhapan, 2001). Small-scale mining continued

Figure 38.1 Map of southwestern Ecuador, showing the location of its four main mining areas.

in Zaruma until the late-1800s, when international mining companies first established operations in Portovelo. The most important of these was the South American Development Company (Sadco), an American firm. Up until the mid-1900s, Sadco operated a large-scale mine in Portovelo. The mine was then taken over by the Ecuadorian Government, which controlled its operation for the next three decades. In 1984, the mine was invaded by local miners, many of whom had lost their jobs as the fortune of the original mine had diminished. Thereafter, mining and processing assumed its present artisanal and small-scale character. Today, there are close to 400 small mines in the area, more than 80% of which are rudimentary and collect fill material from the old Sadco mine (SGAB-Prodeminca, 2000). Typically, following crushing and grinding, ore is gravity concentrated; the gold is subsequently recovered from heavy mineral concentrates using mercury. Cyanidation plants of various types and sizes treat the gravity and amalgamation tailings.

In the mid-1990s, the Bira Mining and Processing Operation was established in central Zaruma; the gold bearing veins exploited were first identified by Sadco. Bira's mines, which are considered to be the most modern in Ecuador, produce about 150 tonnes of ore daily, and have an annual gold output in the range of 28,000 oz (0.9 tonnes). The Bira Processing Plant is a modern, state-of-the-art, cyanidation plant.

By 1999, there were 65 plants for crushing, grinding and amalgamation, and about 80 cyanidation plants in Portovelo-Zaruma. Data from 1994 indicate that, nearly 11,000 people were directly involved in mining activities in the district (Predesur, 1994). Gold production in the region was estimated to be 78,000 ounces in 1999, of which 50,000 was produced by small-scale and artisanal operators. Bira accounted for approximately one third (28,000 oz) of this total.

The gold deposits in Portovelo-Zaruma have generated international interest in recent years. A Chinese venture started a comparatively large and modern mining operation in the mid-1990s but the endeavour faltered due to insufficient ore. More recently, the Canadian junior mining company, IAM Gold, conducted exploration work in the area and acquired a number of concessions, the likely aim of which is to attract investment interests from foreign large-scale mining companies.

Nambija

The high gold price that prevailed in the early-1980s, combined with the problems experienced at the Portovelo Mine, prompted many miners to begin prospecting for gold in other parts of the country. One area that drew considerable attention was Nambija, which is situated on the eastern side of the Andean Cordillera in what is still a relatively remote part of the rainforest. The Nambija deposits were known in the pre-colonial period, and their re-discovery in 1981 initiated an intense gold rush. The deposit was considered world class, and, as a result, a number of international mining companies attempted to secure it. However, the overall frenzy of the local gold rush prevented its acquisition. The area was instead invaded by tens of thousands of illegal miners, and the deposit was intensively worked by a number of informal companies. The mining district remained completely lawless for a number of years, until the Government intervened and brought in the military to control activities. At about the same time, roads were constructed through the jungle to make Nambija more easily accessible to the authorities, the miners themselves and visitors.

At the time of the mine's re-discovery, the Nambija village had only 250 inhabitants. By 1985, however, its population had grown to approximately 20,000; the Nambija area soon became Ecuador's chief gold-producing region. Ore extraction was accomplished using both manual and semi-mechanized methods. Gravity concentration and mercury amalgamation methods were used to process gold particles. In contrast to the other Ecuadorian gold mining areas, cyanidation techniques were never widely utilized in the Nambija area. Intense mining, in combination with poor mine planning, resulted in the exploitation and rapid exhaustion of the region's most accessible ore deposits, after which, miners began experiencing technical difficulties with extracting ore buried at greater depths. Large underground chambers were opened up but soon became unsafe, forcing miners to abandon their work. There were several accidents caused by collapsing adits, and, by the mid-1980s, activities were on the decline. In 1989, Nambija's population was about 8,000 but by 1991, it had declined to 4,000; in 1996, only 1,600 people remained in the region. At present, the gold mining activities in the area are modest and still on the decrease.

Major mining companies, which could not enter the area during the time of the early-1980s gold rush, have since attempted to orchestrate large-scale exploitation of remaining gold resources. However, the instability of the ground, as well as a combination of technical and social problems, have hampered such initiatives. Thus, Nambija serves as an example of small-scale mining, which, due to the number of groups and people engaged, became a large-scale activity but one which lacked the necessary planning and control. As a result of uncontrolled activity, this world class deposit was exploited in such a haphazard manner that a considerable amount of resources are now impossible to recover. The area is likely to remain a small-scale/artisanal district until accessible ore has been completely mined out.

Ponce Enríquez

The gold deposits of the Ponce Enríquez Mining District, located some 40 kilometres northeast of the town of Machala, were discovered in 1983. They occur where the Western Andean Cordillera rises above the coastal plain; here, small-scale exploitation is carried out along a number of narrow veins. A series of small rivers and streams drain the area. These flow through downstream banana plantations and mangrove swamps, the latter of which have been largely cut down and converted into ponds for shrimp aquaculture. Both the banana and shrimp industries are of great importance to the Ecuadorian economy but potentially conflict with the mining activities that occur upstream.

In 2001, the Ponce Enríquez area had a total population of approximately 8,700 people, the majority of whom resided in its two main mining villages: (1) Bella Rica, which is situated at 700–1,100 masl height, and five kilometres west of Ponce Enríquez; and (2) San Gerardo, a smaller village situated 15 km northwest of Bella Rica at an altitude of 800–1,600 masl. Gold was first discovered in Bella Rica, and, by 1997, about 1,500 individuals were engaged in mining and mineral processing in the area, and 50 processing plants (milling, amalgamation and cyanidation) were in operation. In the same year, 750 people were reportedly working the gold deposits in San Gerardo, where some 20 processing plants were also in operation.

Major prospecting companies have evaluated selected portions of these areas but have not been able to prove that reserves are economically viable for large-scale exploitation. At present, it is only economically viable to extract gold using simple, semi-mechanized methods similar to those deployed in the Portovelo/Zaruma region.

Santa Rosa

The gold fields of Santa Rosa were discovered in the early-1980s. It is the smallest of Ecuador's four major mining districts. The region's gold is exploited mainly in narrow veins. However, up until 1996, Odin Mining, a joint Ecuadorian-Australian Venture, was exploiting alluvial sediments. Presently, the dominant operator in the area is Eminsa, its production ranging between 500 and 1000 oz/gold per annum. In total, the mining area has some 200 active miners. All mining activities are carried out upstream of the town of Santa Rosa, which holds a sizeable population of some 40,000 people, 10% of whom reside in an area directly downstream of the mining district.

MINING AND PROCESSING TECHNIQUES

Mining activities in the Portovelo-Zaruma, Ponce Enríquez and Nambija districts are strikingly similar, featuring the following five main processes: (1) mining proper; (2) gravity processing; (3) pan washing; (4) amalgamation; and (5) cyanidation (see Table 38.1).

Table 38.1 Summary of the five major activities carried out in the gold mining districts of southern Ecuador.

Process	Description
Mining proper	The extraction of gold bearing ore from rock.
Gravity processing	The ore is first crushed and ground in mills, and the ground ore is then passed through jigs or chutes from which a heavy mineral concentrate is obtained.
Pan washing	The heavy mineral concentrate is further concentrated by washing in a pan.
Amalgamation	The heavy mineral is mixed with mercury, which draws the gold out of the concentrate to form a gold-mercury amalgam.
Cyanidation	The waste products from the above processes are re-treated in a solution of sodium cyanide, which dissolves most of the remaining gold. The gold is subsequently precipitated out of solution onto either zinc dust or organic carbon.

Most of the mined ore is hard rock, and is extracted underground and brought to the surface via shafts and adits. Some alluvial gold mining has occurred in rivers, although it is of comparatively less importance. Most operations produce between 0.5 and two tonnes of ore daily; only a handful process more than five tonnes of ore per day. Individuals, small groups and co-operatives operate the mines, as there are few formal companies involved in ore exploitation. However, Bira's mines, which are mechanized, produce about 200 tonnes of ore daily.

In contrast to the mainly primitive mining activities, most processing operations have reached at least a modest level of mechanization; some are quite sophisticated. Ore is typically transported from the mines via trucks or mules to a large number of mills where, following comminution, it is gravity concentrated. Most operators use filt-clad chutes over which the ground ore is passed as a dilute slurry. Often, mercury is added to the crushed ore before it is milled, and then passed over the chutes, as this is believed to increase retrieval efficiencies. The chutes have riffles, where gold and other heavy minerals are deposited. This heavy mineral concentrate is further concentrated in pans, and eventually recovered and mixed with mercury. The latter is accomplished manually by rubbing the mercury into the concentrate, using round stones in a pan. The amalgam is thereafter retrieved, and burned several times to vaporise the mercury, thus producing the final gold product.

The heavy mineral concentrate obtained is treated further via cyanidation. The following two methods are used: (i) percolation; and (ii) agitation. Percolation plants are the most rudimentary and most common. In percolation plants, cyanide solution is added to a pool filled with ore, and the cyanide-ore mix is left for a few days whilst the gold dissolves. Thereafter, the pregnant cyanide solution is drained from the pool, and the gold is retrieved by precipitating it onto zinc dust. Agitation is performed in tanks that are filled with ore and cyanide solution, and the resultant slurry is continuously mixed using rotor blades. This method is faster than percolation and also allows for the optimization of the slurry's cyanide concentration, which allows better control of the process, and, ultimately, better efficiency. The pregnant cyanide solution is then withdrawn, and the gold is precipitated out on either zinc dust or activated carbon.

LEGAL AND ORGANIZATIONAL ISSUES

Small-scale miners and the law

The *Ecuadorian Mining Law*, 1991 (and its subsequent reform in 2000), governs the legal context for mining activities. The Law aims to attract foreign investment to the country. Some of its salient features include:

- The State awarding mining titles and providing administrative protection admissions, dispossessions, invasion or any other form of disruption;
- The commercialisation of minerals is free;
- All natural or juridical persons may freely carry out prospecting;

- Mining exploitation and exploration rights are exclusive rights, distinct from property titles of surface land;
- Exploration title holders may freely convert their areas into exploitation titles; and
- Mining activities cannot be levied with additional taxes or contributions that are not specifically established in the Mining Law.

In terms of environmental protection, mining ventures are required to prepare an Environmental Impact Assessment (EIA) that details how environmental impacts will be mitigated, and which includes plans for mine closure. Further, miners must submit annually an environmental audit that shows the degree to which the stipulations of the EIA are being followed. Miners are held responsible for both direct contamination and environmental damages that occur after mine closure, although bonds to ensure reclamation are not required by law. Moreover, the contamination or degradation of natural resources, be it deliberate or by negligence, is deemed a criminal offence.

Although the legislative framework is probably sufficient for controlling resident mining activities, they are inadequately enforced. In fact, small-scale/artisanal mining in southern Ecuador was, for a long time, carried out completely illegally. Countless invasions of concessions have occurred, particularly in the Nambija area, where the situation became completely lawless, thus prompting military intervention. The situation has since improved in Nambija and other areas, although illicit undertakings still occasionally occur. A major change was achieved through a strong effort by the national Government to organize and legalize operations. This was done by accepting the invaders as *de facto* owners but forcing the many individual companies to organize themselves into co-operatives – associations of small companies formed with the purpose of serving as the legal representative and responsible body in relation to legislation and the authorities. This has normalized the situation radically, despite the fact that many of the cooperatives are still too small to cope efficiently with the obligations laid down in laws and regulations.

This weakness, in combination with the large number of organizations that must be controlled, requires wide-ranging supervision from state authorities. This was confirmed by Carvajal et al. (1997), who noted that the public institutions controlling the mining activities are, in fact, often too weak to do so. The authors explain that, mining authorities are caught in a vicious cycle of cause and effect whereby they are unable to enforce regulations because of a lack of operational resources. This, in turn, leads to illegal operation, poor environmental and health and safety standards, and a loss of fiscal revenues. A lack of revenue limits the ability of authorities to perform their regulatory function, and therefore, perpetuates uncontrolled mining (Noetstaller, 1987).

Organisation

Although all mining activities in Ecuador are small in scale, the mining community is itself heterogeneous with regard to activities, level of mechanization,

Table 38.2 Characteristics of the three main groups of miners operating in Ecuador (adapted from SGAB-Prodeminca, 2000).

	Activities		
	Artisanal	Small-scale/semi-mechanised	Small-scale/industrial
Legal status	Informal	1/3 legal; 2/3 informal	Legal
Activities	Mining and mercury amalgamation	Processing, and only rarely mining	Mining and processing
Work conditions	• Artisanal • Minimal safety measures • Informal groups of 1–2, and rarely more than 10 people	• Semi–mechanised • Few safety measures • Work performed in small entities with about 10 people	• Industrial processes • Acceptable working conditions
Economic situation	• Miners own few tools • Only obtain about 40% of available gold extracted • No capital • Subsistence mining • Small or no investments needed	• Firms may own a few mills/pools for cyanidation and 1–2 trucks • Some capital and limited investments made	• Full ownership of industrial equipment • Substantial capital and investments

and capital investments (Table 38.2). SGAB-Prodeminca (2000) identified three main groups within the Portovelo-Zaruma area: (i) artisanal miners; (ii) small-scale/semi-mechanised miners; and (iii) small-scale/industrial operations. The *artisanal miners* are almost entirely engaged in the extraction of ore, although some also perform simple amalgamation activities. Most of the *small-scale/semi-mechanised* group is concerned with processing ore purchased from artisanal miners; only a select few have their own mines. The *small-scale industrial operations* group, on the other hand, is engaged in all activities, from mining to the processing and smelting of gold.

The individuals comprising these three groups can be found in each of the main mining districts but in varying abundances. In Portovelo-Zaruma, for example, approximately 70% of the people involved in local mining activities are considered to be informal small-scale miners who mine using rudimentary tools and techniques, whereas only 5% of miners are recognized as “industrial”. In Nambija, there is a larger proportion of informal miners, whereas in the Ponce Enríquez area, there is proportionally more semi-skilled and skilled miners.

Mine workers in groups (i) and (ii) usually belong to small companies. Many of these companies form co-operatives or mining associations, which, as explained earlier, have been established in response to pressure brought on by the authorities who require miners to associate in larger entities. These entities/co-operatives are usually quite fluid, allowing new members to join freely and existing members to exit with ease.

Members of the *small-scale/semi-mechanised group* (*i.e.* group ii) have become increasingly well organized. There now exists a small number of tightly-knit groupings or associations of such small-scale/semi-mechanized companies and cooperatives. These associations have also made some effort to agree on joint measures of environmental protection (Cenda, 1996).

In addition to the three main groups, there are also gold buyers, as well as individuals involved in mining during intermittent periods. The gold buyers purchase gold extracted by the artisanal and small-scale/semi-mechanized miners, and usually live within the districts themselves. There also exists a group of people, referred to locally as the *jancheros*, who make an income out of "picking" among waste rock and other discarded materials left behind by the "real" miners. Groups of *jancheros* are usually comprised of groups of women and/or school children who work during school holidays or breaks from household duties. Similarly, there are individuals who derive incomes from the simple panning of gold in rivers. They, along with the *jancheros*, often travel vast distances to the mining districts for the sole purpose of supplementing their incomes during a few months of the year.

ENVIRONMENTAL AND HUMAN HEALTH RELATED IMPACTS

The environmental impacts associated with gold mining in southern Ecuador have been well documented (SES, 1999a; SGAB-Prodeminca, 2000; Tarras-Wahlberg et al., 2000; Tarras-Wahlberg et al., 2001). Resident mining and processing activities are causing a variety of environmental problems, including deforestation and accelerated erosion, noise and dust pollution (from heavy transport), and land destabilization (from the myriad of poorly planned and maintained mine adits). The most serious environmental problems, however, result from discharges – cyanide, metals, and large amounts of silt – from tailings into river systems. According to SGAB-Prodeminca (2000), the small-scale semi-mechanised group of miners are responsible for most of this pollution.

The Puyango River System has been severely affected by the mining activities in Portovelo-Zaruma. There is essentially no aquatic life present up to 20 km downstream of the mining district. Further downstream, some life reappears, although evidence suggests that it has been severely reduced up to 160 km downstream of the mining area, well into Peruvian territory. The situation in the Siete River, downstream of the Bella Rica Mining District near Ponce Enríquez, is similar: aquatic life in the downstream river system has virtually disappeared all the way through to the river delta. Fortunately, there is no evidence of any significant adverse effect on the banana and shrimp farming activities that occur downstream, the primary reasons being that shrimp farmers use seawater in their dams and banana estates utilize water from unpolluted rivers. Tarras-Wahlberg et al. (2001) argue that the management of the waste products originating from mining and processing activities must be radically improved if the ecological health of the river system is to be restored. It is suggested that a near-complete mitigation of environmental problems can be achieved by constructing adequate impoundments for process tailings and associated mine wastes.

Tarras-Wahlberg (2002) considers the environmental problems that are caused by mining activities in the Portovelo-Zaruma district. In this paper, it is shown how the Government, aided by international development organizations, has made a concerted effort to both define local environmental problems and identify solutions. It is also proven that, mine waste and tailings can be confined at a reasonable cost. However, the complex topography of the mining district forces tailings management to be communal, whereby all operators are connected to one central tailings impoundment. This, in turn, implies two things: (i) that a large number of operators must agree to pool resources to bring such a facility into reality; and (ii) that in order to be able to make such a long term investment, the miners will need to begin moving towards bigger, mechanized and longer-term sustainable operations that are based on proven ore reserves. Sadly, experiences in Prodeminca indicate that the idiosyncrasies that exist within the Ecuadorian mining community make it exceedingly difficult to convince miners to participate in communal tailings management, even in cases where measures are subsidized. To date, there exists only one example of miners pooling resources with the Government to construct a tailings impoundment. This is a small and temporary pilot-scale facility that has been constructed in the Ponce Enríquez area. Negotiations between the central Government, as represented by Prodeminca, and local miners concerning the construction of a larger and more sustainable facility, have been ongoing. However, doubts exist as to whether or not the miners could realistically advance such a project, either with or without technical and financial assistance. The problems relate only in part to a perceived lack of financial resources to construct tailings dams. More important has been the failure to obtain a commitment from Ponce Enríques miners to invest jointly in environmental protection. Furthermore, a lack of pressure from Government and local communities has encouraged miners to believe that, in effect, they can continue with current practices.

At present, there are only two examples of small-scale gold mines that have achieved acceptable environmental performance. The first is Bira in Portovelo-Zaruma, and the other, Eminsa in Santa Rosa. Although both Bira and Eminsa are considered small-scale operations, in the Ecuadorian context, they feature more modern technology and are much larger than most of the country's mining operations. This implies that poor environmental performance of the smaller operations may be partly related to the fact that they are small in scale, and, as a result, must use technologies more susceptible to causing environmental degradation. Hence, a key component in achieving an environmentally acceptable and more sustainable form of mining in the district appears contingent upon promoting large-scale and more technologically-advanced forms of mining and mineral processing.

Adverse health impacts resulting from mining and processing activities are well known. The most common of these relate to miners' working environment, although there may also be health problems resulting from contamination from mining and mineral processing. To gain a better understanding of these issues in an Ecuadorian context, extensive health related investigations were undertaken in Prodeminca.

In general, questions of health and safety are almost totally ignored within the Ecuadorian small-scale mining sector, in spite of the substantial efforts made by the Government and within international development projects to raise the level of consciousness among miners and mine managers. It is true that certain improvements have been achieved by supporting work within mining cooperatives and women's groups but the overall gains have been quite modest. The main reasons for this are primitive and perilous working conditions, especially within underground mines; a shortage of financial resources; as well as the prevailing culture of "manliness", particularly among young men, which gives such efforts low priority.

The most serious problems are: (1) accidents caused by rock falls and mud slides in the poorly planned mines; and (2) silicosis. In 1993, one large mudslide/rock fall in the Nambija district buried and killed as many as 300 people. Smaller rock falls have claimed, and continue to claim, lives in various Ecuadorian mining regions. Further, medical investigations have shown that between 10 and 15% of Ecuadorian miners suffer from silicosis, which is caused by the continuous inhalation of the dust-filled air inside and around a mine.

SES (1999b) investigated the possible effect of mercury and other metal exposure by sampling blood and urine from a large number of miners. Mercury exposure was found to be significant among the gold buyers but not within the wider mining community. Some of the gold buyers were found to display symptoms of mercury poisoning, including finger tremor, co-ordination problems, and prolonged reaction time. The content of other metals in blood and urine was generally low, with the exception of lead, which was found in elevated concentration in the blood of miners operating in Portovelo-Zaruma. The cause of this is unknown, although it is speculated that, the lead-enriched dust in air, and the provision of, up until only a few years ago, potable water obtained from the local contaminated river, maybe the cause of lead exposure.

SOCIO-ECONOMIC EVALUATION

The present small-scale/artisanal gold mining activities in southern Ecuador emanate from the large-scale Sadco Mine in Portovelo-Zaruma. Triggered by lay-offs at the mine, and stimulated by rumours of "great gold finds" in Brazil and other countries, miners began making trials in areas allegedly containing gold. The result was a gold rush in Nambija, and, a decade later, less dramatic efforts in Ponce Enríquez and adjoining areas. The original driving force was a shortage of employment. Today, the majority of people engaged in this occupation are males aged between 15 and 30. Nearly all of these individuals work as day labourers earning subsistence wages; few operators have managed to enjoy lavish lifestyles.

The investments made by the three different groups of miners identified earlier (i.e. the artisanal, small-scale/semi-mechanised, and small-scale/industrial) vary considerably. Carvajal et al. (1997) provide estimates of such investments. Artisanal miners are, in most cases, entirely dependent upon manual labour, and therefore, require only minimal investments. A small group of artisanal miners require US$500–1,000 to establish a base camp, purchase simple tools, construct sluice boxes, and invest in an electric generator. Operations comprising the *small-scale/semi-mechanised* and *small-scale/industrial* groups have been established as a result of investments made by outsiders – either local entrepreneurs, politicians or military people. Small-scale/semi-mechanized miners are typically organized into small companies and/or co-operatives, which contain between 10 and 80 people. The equipment needed may include explosives, air compressors, a truck, a mill and/or a simple processing plant. Investments needed range between US$10,000 and US$30,000.

Small-scale industrial operations are legal entities that are associated with the Ecuadorian Chamber of Mines, and which include a variety of technically-skilled personnel such as geologists, engineers and administrators. The Bira Mine has several hundred employees, and the handful of other industrial operators in the Ponce Enríquez and Portovelo-Zaruma areas have close to one hundred staff. Such operations rely on initial investments in the range of US$100,000.

The incomes generated by gold mining activities are substantial. To reiterate, in 1999, it is estimated that, gold production in Portovelo-Zaruma alone was 78,000 ounces, of which 50,000 (1.6 tonnes) were produced by small-scale and artisanal operators. Assuming that each ounce fetched a gold price of US$275, the total value of the region's gold was US$21.5 million (SGAB-Prodeminca, 2000), a substantial figure considering that, the district's total GDP was estimated at US$59 million during the same year (Tarras-Wahlberg, 2002). The situation in the Ponce Enríquez and Nambija districts is similar, although the absolute incomes generated are smaller. Hence, it is clear that, the mining activities generate incomes that are, at least at the regional scale, very important indeed.

However, substantial portions of the incomes generated by gold mining are not reinvested in the gold mining districts themselves (SGAB-Prodeminca, 2000). Rather, the money earned is commonly spent elsewhere, as many successful mine owners have invested their money in the country's big cities (especially Guayaquil) or abroad (in the US).

Nevertheless, in this respect, each of three mining districts differs. Zaruma, for example, is a mature city with a true mining culture and a relatively stable population; many of its families have lived in the area for several generations. Thus, numerous Zaruma miners have invested locally, which can be seen in the form of newly constructed private houses. Furthermore, the local Bira Company employs permanent miners and provides certain social benefits and healthcare. Conversely, Bella Rica, San Gerardo and most of Nambija, are transient settlements, where virtually all investments are made only to sustain the mining activities themselves. In general, small-scale mining activity does not provide payments that enable workers to make appreciable investments, and most of the economic gains of the company owners and investors are invested outside of the areas.

The mining communities also support a range of supporting businesses and other activities. Numerous merchants make a living from selling equipment and consumable goods to miners. Additional clandestine activities have also agglomerated in these areas, and crime and prostitution have, at times, been rampant; crime and prostitution is, however, noticeably less pronounced at present than during "gold rush" periods. Again, the city of Zaruma, with its more stable social fabric and long mining history, has experienced the fewest socio-economic problems.

CONCLUSIONS

What, then, are the merits and drawbacks of artisanal and small-scale mining? How beautiful, or sustainable, is this way of living (c.f. Schumacher, 1973; Beckerman, 1995), which certain people, placed at an agreeable distance from the activities, tend to regard as an almost romantic occupation? In reality, as described above, the problems are manifold and grave. To start with, small-scale/artisanal mining is decidedly different from industrial mining in several fundamental respects. This is, at least, if we compare the well-regulated and managed, international industrial mining of today, with the likes of small-scale mining in southern Ecuador. Actually, most of the differences, as well as many of the problems, stem from precisely the fact that the individual small-scale operations are small. *Small* does not only mean a reduced level of production and technology but also implies small resources to cope with all of the different obligations related to mining, not the least of which is the protection of the environment. A summary of the main technical, economic and social factors involved is presented in Table 38.3.

The overview illustrates that there are three main benefits associated with small-scale mining: (i) the possibilities it offers to exploit mineral resources that are otherwise uneconomical to recover; (ii) creation of job opportunities; and (iii) revenues for those involved. These are certainly weighty factors in a poor economy, even if these activities may occasionally hinder the rise of industrial ventures and provide very little tax income to the community.

However, more serious drawbacks include such negative factors as the extensive pollution of nature, the very poor health and safety conditions under which the work is carried out, and various social problems such as criminality and prostitution. It is certain that many of its negative effects can, to certain degrees, be mitigated, but always at a considerable cost and rarely entirely. This leads to the conclusion that any proposals for promoting or subsidising small-scale and artisanal mining ventures must be analysed carefully and critically. Small-scale and artisanal mining may provide a source of livelihood but there is usually little beauty associated with it. Therefore, it should not be promoted as a business under normal circumstances but only in situations of socio-economic alert, when no other reasonable alternatives are to be found.

Table 38.3 Schematic comparison between the conditions of small-scale/artisanal mining and typical large-scale, industrial mining.

	Small-scale mining	Large-scale, industrial mining
Effective utilization of natural mineral resources:		
Small deposits	Good	Poor
Big deposits	Poor	Good
Approach to exploitation	Chance and opportunity	Systematic
Prospecting/exploration	Little to none	Extensive
Commitment	Short-term	Long-term
Requirement for public supervision	High	Modest – high
Payment of taxes	Minimal	Yes
Compliance with legislation	Unsatisfactory	Satisfactory
Rate of employment	High	Relatively low
Social development and professional improvements	Poor	Good
Concern for health and safety	Low – very low	High

What, then, are the future prospects of the small-scale and artisanal miners in Ecuador? The problems associated with these activities are now well known to the Ecuadorian authorities, and initiatives have therefore been ongoing to promote the formation of larger mining entities and firms. Miners' environmental obligations are now defined in legislation, and a public authority has been formed, which is charged with environmental supervision and control. Considering that the small-scale and artisanal miners will, in general, have difficulties in complying to such environmental legislation, it seems inevitable that the small-scale miners of Ecuador will find it more and more difficult to operate. At the same time, there is a clear Government policy to promote foreign investment in the mining sector, in order to enable the establishment of a larger-scale, and more technologically advanced form of mining in the country. Considering that Ecuador now has a relatively attractive mining policy, it is not unlikely that international mining companies may again establish operations in the country. This would probably lead to improved overall earnings, although the potential for employment within the sector will most likely decrease. As always, all such developments will be closely governed by commodity price fluctuations, especially with regard to gold.

REFERENCES

Beckerman, W. (1995). *Small is stupid: blowing the whistle on the Greens.* London: Duckworth.

Carvajal, M., Espín, J., Pereira, J. & Rivadeneira, J. (1993). Perspectiva socioeconómica de la pequeña mineria y la mineria artesanal. Quito: Prodeminca, Ministerio de Energía y Minas (in Spanish).

Cenda (1996). Estudio colectivo de impacto ambiental y plan de manejo ambiental para las plantas de beneficio de mineral aurífero ubicadas en la Vega del río Calera/Salado. Loja: Fundación Cenda (in Spanish).

Davidson, J. (1989). Small scale gold mining for developing countries – an international professional development seminar. *SMI Bulletin*, 1, 8–10.

DSE and UN/DTCD (1991). *Mining and the environment – The Berlin guidelines*. United Nations' Department of Technical Co-operation for Development, and the Development Policy Forum of the German Foundation for International Development. London: Mining Journal Books.

Noetstaller, R. (1987). Small-scale mining: a review of the issues. Technical paper no. 75, *Industry and Finance Series 23*, World Bank, Washington.

Predesur (1994). *Estadísticas socioeconómicas de la region sur*. Loja: Predesur (in Spanish).

Schumacher, E.F. (1973). *Small is beautiful: a study of economics as if people mattered*. London: Blond & Briggs.

SGAB-Prodeminca (2000). Plan maestro ambiental para el distrito minero Portovelo – Zaruma y la cuenca del río Puyango. Quito: Prodeminca, Ministerio de Energía y Minas (in Spanish).

Spiropolous, J. (1989). Small-scale miners association. *SMI Bulletin*, 1, 6–7.

Swedish Environmental Systems (SES) (1999a). *Monitoreo ambiental de las áreas mineras en el sur del Ecuador entre 1996 y 1998*. Quito: Prodeminca, Ministerio de Energía y Minas (in Spanish).

Swedish Environmental Systems (SES) (1999b). Exposure to and toxic effects of mercury in the Ecuadorian gold mining industry. Quito: Prodeminca, Ministerio de Energía y Minas (in Spanish).

Tarras-Wahlberg, N.H., Flachier, A., Fredrikson, G., Lane, S.N., Lundberg, B. & Sangfors, O. (2000). Environmental impact of small-scale and artisanal gold mining in southern Ecuador: implications for the setting of environmental standards and the management of small-scale mining. *Ambio*, 29, 484–491.

Tarras-Wahlberg, N.H., Flachier, A., Lane, S.N. & Sangfors, O. (2001). Environmental impacts and metal exposure of aquatic ecosystems in rivers contaminated by small-scale gold mining: the Puyango river basin, southern Ecuador. *The Science of the Total Environment*, 278, 239–261.

Tarras-Wahlberg, N.H. (2002). Environmental management of small-scale and artisanal mining: the Portovelo-Zaruma goldmining area, southern Ecuador. *Journal of Environmental Management*, 65/2, 165–279.

Zhapan, J.C. (2001). Historia de la Mineria en el Austro del Ecuador. Quito: Ministerio de Energía y Minas (in Spanish).

39

The Socio-economic Impacts of Artisanal and Small-Scale Mining in Chile

S.H. CASTRO

From a socioeconomic point of view, small-scale mining (SSM) is a very important activity in Chile. The sector includes around 10,000–15,000 direct and indirect workers, taking into account the artisanal miners and work sites dedicated to the extraction and/or processing of gold, copper, silver and coal. In 2001, SSM was responsible for 6.04% of national gold production, 0.85% of copper production, and some 25% of coal production. Metallic SSM in the North receives State support through ENAMI (National Mining Company), which maintains a promotional policy of purchasing power sustained through a system of tariffs and subsidies. Small-scale coal mining in the South receives State assistance through CORFO (Production Promotional Corporation), whose programs are mainly directed at social aspects and labor reinsertion. The present work reviews the general characteristics of SSM in Chile, highlighting production levels through 2001, the institutional support framework, production technology, environmental impacts, and its socioeconomic characteristics.

SCOPE

In Chile, small-scale and artisanal mining has historically been linked to the North (Regions II–VI), where it has developed around deposits of gold, silver and copper, and the South (Regions VIII and X), where it has developed around coal deposits (Castro & Sánchez, 2003). Figure 39.1 shows Chile's chief mining regions.

SSM constitutes an important activity from a social point of view, as it provides employment to a vast segment of the population in zones where alternative work availability is very low. However, deficient company conditions, the impact on the economic and social reality of the workers (Chaparro, 2000), and related environmental problems (Sánchez & Enríquez, 1996) are sources of concern for Chilean and international agencies (ECLAC, World Bank, etc.).

Figure 39.1 Locations of important small-scale mining regions in Chile.

For decades, the State has maintained permanent technical and financial support policies for small-scale and artisanal miners that have enabled them to subsist even during times of crisis in mineral product prices. However, these efforts have been extremely costly for the State, and have not fully resolved the central socio-cultural problems of the local community. More recently, training and technological transfer activities have been undertaken (Lagos et al., 2001; Borregard et al., 2001) but have failed to address issues of deficient safety, hygiene, health, education, and training – key aspects for socioeconomic improvement in SSM.

In the case of coal, after 150 years of operation, the Lota Mine, owned by the large State company, ENACAR, was closed on 15 April 1997. Some 1,100 workers were left unemployed, and the associated economic and social impacts in the zone were grave. This was an example of the closing of a work site by the State for social reasons; the mine's subsidy had ended, and the government elected to invest resources in social protection and labor reinsertion plans rather than continuing an unprofitable mining activity. Small-scale and artisanal coal mining has continued, nevertheless, due to the lack of alternative employment.

Work in the sector continues, playing a dynamic role in many local Chilean economies. SSM has played a historic function in Chile for more than two centuries. It should be noted that small-scale and artisanal mining pioneered mineral exploration; many deposits currently being exploited on a large scale were first discovered by small-scale miners and *pirquineros* (artisanal miners).

CHILEAN MINING CLASSIFICATION

Although Chilean mining is highly diversified, the following three distinct size divisions are commonly identified within the sector:

- Large-scale mining (production >10,000 tons Cu content/year);
- Medium-scale mining (processing capacity >200 tons of ore/day); and
- Small- and artisanal-scale mining (processing capacity <200 tons of ore/day).

According to SERNAGEOMIN's Resolution No. 0408 (15 January 1998), small-scale mining is considered to be activity that has less than 200,000 man hours worked during the course of one year (with an average of 80 workers during one year).

The mining sector is also divided into State-owned (CODELCO and ENAMI) and private (*Escondida, Collahuasi, Los Pelambres, Candelaria, Disputada, Cerro Colorado, Quebrada Blanca, El Tesoro, Mantos Blancos, Michilla*, etc.) companies. Both private and State-owned mining activities are important employment generators; estimates indicate that over one million people in Chile are economically tied to mining.

Chile is divided into thirteen geopolitical regions (numbered from north to south), but from strictly a geographic perspective, can be divided into the following three large zones: the northern zone (Regions I–IV), home to the Atacama desert; the central zone (Regions V–VIII); and the southern zone (Regions IX–XII). Mining activities are distributed throughout the entire country. The largest copper and gold deposits are found in the North, some copper and small gold deposits can be found in the central zone, whereas coal and gold deposits are the most important mineral resources in the South (Regions VIII, X, and XII).

SSM PRODUCTION IN CHILE

Metallic production statistics (Table 39.1) indicate that Chilean SSM in 2001 produced 40,675 metric tons of Cu content (0.85% of total production); 2,578.1 Kg of Au (6.04%); and 11,694.2 Kg of Ag (0.87%).

Table 39.1 Chilean metallic production in 2001 by mining sectors.

Mining-scale	Cu content (tons)	Mo content (tons)	Au (Kg)	Ag (Kg)	Pb content (tons)	Zn content (tons)
Large-scale	4,432,463	33,492	20,040.9	1,052,404.9		
Medium-scale	292,924		20,053.6	284,567.7	1,193	32,762
Small-scale	40,675		2,578.1	11,694.2		
Total	4,766,062	33,492	42,672.6	1,348,666.8	1,193	32,762

Source: Anuario de la Minería de Chile, 2001.

Table 39.2 SSM gold production in 2001 by regions and product types (kg).

Region	Metal *doré*	Concentrates	Gold bars	Concentration ores	Direct smelting ores	Total
II	203.7			1.8		205.5
III	1,103.9	92.3	0.5	120.6	17.8	1,335.1
IV		150.7	31.3	45.0	5.0	232.0
V		218.6	5.9	159.4	11.3	395.2
Total	1,307.6	461.6	37.7	326.8	34.1	2,167.8

Source: Anuario de la Minería de Chile, 2001.

Small-scale gold production made up 2,167.8 Kg, or 84.1%, of total SSM production (2,578.1 Kg), which includes the gold produced by small-scale copper and silver mining operations. The number of people employed directly in metallic SSM was an estimated 3,605, including 2,120 in copper mining and 1,485 in gold mining (Anuario de la Minería de Chile, 2001).

Because of the percentage of its contribution to national metallic production, gold mining is the most important. Found principally within Regions III and IV, gold mining frequently produces silver and copper as by-products. Table 39.2 shows gold production in SSM by region and product type. Five product types are identified: metal *doré*, concentrates, gold bars, concentration ores (low grade ores of Cu, Au, Ag, Pb, Zn, etc., which are destined for processing into concentrates at concentration plants), and direct smelting minerals (high grade Cu and/or Au, that may also contain Ag, which are directly destined for national smelters). It can be seen that metal *doré* was the main product (60.32%), followed by Au in concentrates (21.29%).

Tables 39.3 and 39.4 show SSM copper production and associated gold and silver by-products. Production in 2001 reached 39,974 tons of Cu content. There were six principle product types: electrowinning cathodes, precipitates, concentrates, direct smelting ores, concentration ores, and leaching ores.[1]

[1] It should be clarified that, of the total copper produced by SSM (40,675 tons), 701 tons (1.72%) originates from small-scale gold mining.

Table 39.3 SSM copper production in 2001 and noble metals by-products.

	Concentrates				Direct smelting ores			
Region	(tons)	Cu content (tons)	Au (Kg)	Ag (Kg)	(tons)	Cu content (tons)	Au (Kg)	Ag (Kg)
II					648	177	1.1	29.3
III	13,278	2,876	97.1	1,631.9	1,285	238	4.5	18.5
IV	3,977	1,094	21.5	277.0	230	56	2.3	56.9
V	44,614	11,792	100.2	4,967.7	124	22	3.8	10.2
Total	61,869	15,762	218.8	6,876.6	2,287	493	11.7	114.9

Source: Anuario de la Minería de Chile, 2001.

Table 39.4 SSM copper production in 2001 and noble metals by-products.

		Concentration ores				Leaching ores		Precipitates	
Region	EW Cathodes	(tons)	Cu content (tons)	Au (Kg)	Ag (Kg)	(tons)	Cu content (tons)	(tons)	Cu content (tons)
II		99,511	2,805	41.6	1,147.9	142,284	4,831	7	3
III		114,130	1,659	64.5	103.8	208,161	5,350	1,000	619
IV		78,539	1,621	61.8	135.0	95,884	2,370	60	39
V		9,091	254	5.1	156.0	650	18	3,208	2,431
Total	1,719	301,271	6,339	173	1,542.7	446,979	12,569	4,275	3,092

Source: Anuario de la Minería de Chile, 2001.

Table 39.5 Coal production in the Region VIII from 1991–2001.

Year	Coal production (thousand tons)
1991	1,341.53
1992	997.78
1993	792.68
1994	480.82
1995	361.55
1996	233.60
1997	134.94
1998	116.61
1999	82.24
2000	110.04
2001	123.32

Source: Anuario de la Minería de Chile, 2001.

The highest level of Cu production was in concentrates from flotation plants (39.43%), followed by leaching minerals (31.44%). Cathode production in small-scale copper mining was extremely low, constituting only 4.3% of production.

Chilean coal is produced in Regions VIII, X, and XII. Small-scale and artisanal mining is concentrated in Region VIII, and, to a lesser degree, in Region X. In 2001, net production in Region VIII had reached 123,320 thousand tons, which constituted 21.7% of national production. On the other hand, production in Region X had only reached 21,943 tons, barely 3.86% of the national total.

Coal production in Region VIII decreased throughout the 1990s, mostly as a result of the closing of the ENACAR-Lota work sites. Currently, only residual artisanal activity remains in the zones of Arauco, Lebu, Curanilahue, and others. The closing of the mines in Lota had a tremendous social impact in the Lota and Coronel areas; the government was forced to implement plans that moved the miners into other productive work sectors.

CHILE'S SSM SECTOR

SSM is a traditional activity that generates employment for the low-income population. It is concentrated in the North (Au-Cu) and South (coal) of Chile (Table 39.6). Small-scale and artisanal miners may work in their own installations or those of someone else. Although SSM mineral processing capacities can reach 200 metric tons of ore per day, 98% of the sector has an ore extraction capacity below 25 metric tons/day. Small-scale miners may produce ores for direct commercialization or process them themselves in small concentration plants.

The following two subdivisions coexist within the Chilean SSM sector: (1) formal small-scale mining, in which mines or plants operate with some degree of mechanization; and (2) artisanal-scale mining, in which ore is extracted from small mines, and, in some cases, processed by small artisanal *trapiches* for gold amalgamation. These artisan miners are called *pirquineros*, and are a particularly poor group of the Chilean population. There is no formal organization of these miners; in other words, they do not form a legally constituted company. One the main characteristics of the artisanal sector is a lack of operational continuity – *i.e.*, the miners only work when metal prices and/or the local weather make the activity feasible.

Small-scale miners and *pirquineros* sell their product to ENAMI agencies, which act as promotional and regulatory bodies, processors, and exporters. ENAMI has a legal obligation to process the products extracted by small producers. ENAMI often receives mineral product from artisanal miners who sell to intermediaries such as formal SSM plants or mines. The SSM sector abides by the State's promotional policies, selling the majority of its product to ENAMI on a tariff system. There is a high level of interdependence between small-scale miners and *pirquineros*, with the former hiring artisanal miners for ore supply. Both small-scale miners and *pirquineros* can sell their mineral products directly to ENAMI's agencies.

In order to characterize SSM, it is necessary to distinguish between mines and plants. Statistics from 1994 identified a total of 1,626 mines and 281 plants in the SSM sector (Table 39.7).

Table 39.6 Main provinces and municipal areas where SSM and artisanal-scale mining are located.

Region	Mining activity	Provinces	Municipal areas
II: Antofagasta	Cu-Au	Tocopilla, Antofagasta, and El Loa	Tocopilla, Antofagasta, and Taltal
III: Atacama	Cu-Au	Chañaral, Copiapó, and Huasco	Chañaral, Diego de Almagro, Copiapó, Tierra Amarilla, Huasco, Freirina, and Vallenar
IV: Coquimbo	Cu-Au	Elqui, Limari, and Choapa.	Andacollo, Ovalle, Combarbalá, and Illapel
V: Valparaiso	Cu-Au	Valparaiso	Quilpué and Casablanca
VIII: Bio-Bio	Coal	Concepción and Arauco	Coronel, Lota, Arauco, Curanilahue, Lebú, and Los Alamos

Table 39.7 Number of small-scale mines and processing plants in Chile.

Region	Locations	Plants	Mining operations	Total
I		3	27	30
II	Taltal	11	191	202
III	Vallenar, Diego de Almagro, Copiapó, Chañaral	99	641	740
IV	Ovalle, Andacollo	139	588	727
V	Quilpué	20	95	115
VI	Rancagua	1	55	56
RM		8	29	37
Total		281	1,626	1,907

Source: Subterra Ingenieros Limitada, 1994.
Note: RM denotes "metropolitan region".

INSTITUTIONAL FRAMEWORK

The different organisms and institutions that constitute the institutional support framework for Chilean mining sector are identified in Figure 39.2. These are described in the discussion that follows.

Mining ministry

As the representative of the Executive in the mining sector, the Ministry is in charge of designing and carrying out relevant governmental policies. Like all other Chilean Ministries, a regional structure functions via a Regional Ministerial Secretary (SEREMI), who is in charge of coordinating tasks with the main offices operating out of Santiago. This Ministry is responsible for the National Geology and Mining Service, SERNAGEOMIN, whose functions are described later in this chapter.

In 1991, the Environmental Unit of the Mining Ministry was created to foster environmentally sustainable mining activities and promote their introduction

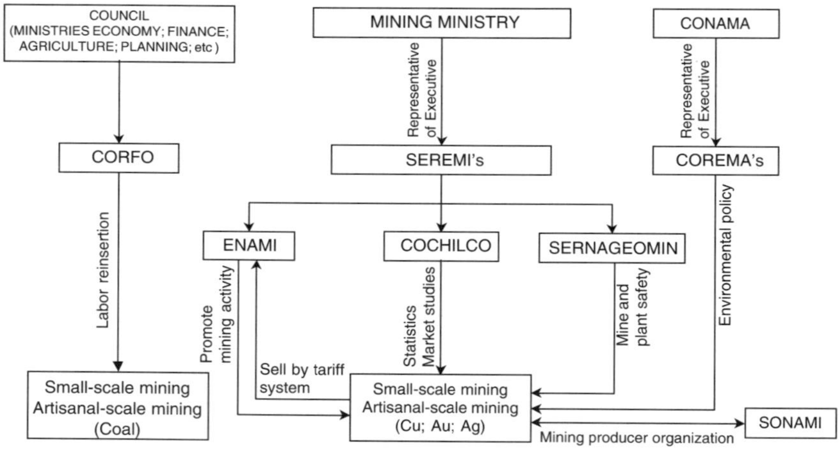

Figure 39.2 Institutional support framework for SSM in Chile.

into the regulatory framework that is currently being established in Chile. This unit works in conjunction with CONAMA.

ENAMI (National mining company)

Several decades ago, the State created the Mining Credit Union, which later became ENAMI. Its purpose is to stimulate mining activity and contribute to the sustainability of work sites, and small and medium mining companies. ENAMI has two main areas of activity: production and promotion. Its current goal is the improvement of mining production and productivity, allowing small- and medium-scale miners access to international markets. ENAMI's role in the SSM sector is established by law; it is required to buy all SSM production on a tariff system and offer direct subsidies to alleviate fluctuations in the price of copper.

The promotional work of ENAMI includes technical assistance and credit programs that are carried out through two instruments:

1) Mining promotion funds are destined for the discovery of reserves, the financing of mining operations, and the design of mechanisms that will increase production levels and lower plant operation costs.
2) Commercial promotion funds have a two-fold purpose: (a) the provision of access to treatment and metal sales markets for small and medium miners, and (b) price-stabilizing services in these sectors.

Hence, ENAMI does not *have* mines, but rather buys raw materials from the medium- and small-scale sectors for processing in its own plants or those of others. It produces concentrates and precipitates that, together with direct smelting minerals, are fed into its smelters.

ENAMI has several mineral flotation plants (*Planta Matta, Taltal, Salado, Vallenar*, and *Panucillo*) and two smelters (*Paipote* and *Ventanas*, which includes an electrorefining plant).

SERNAGEOMIN (National geology and mining service)

A public entity dependent on the Mining Ministry, SERNAGEOMIN supervises mining property, exploration, and geology services. According to Supreme Decree 72 of the Mining Ministry and Article 71 of the *Sanitary Code*, SERNAGEOMIN is also responsible for regulating the presentation of projects for sterile dumps and the disposal of industrial and leaching residues. In 1992, its Department of Environmental Management was activated. Its work is fundamentally aimed at work site safety control, and it is in charge of authorizing the operational conditions of mines, plants, and tailing deposits, as well as the storage of waste mining ores and leaching residues.

SERNAGEOMIN has six regional offices (Regions I to V and Region VIII) through which it channels its supervision. These offices also consult with CONAMA as to mining-related Environmental Impact Evaluation material.

SERNAGEOMIN keeps a cadastre of tailing deposits on which it bases its supervision. Its work in recent years has been geared toward new projects in plant safety, dams construction, and the supervision of operating installations.

SONAMI (National mining society)

One of the two oldest guilds in Chile, SONAMI represents the interests of its over five thousand members, including large, medium, and small mining producers. Its Environmental Commission works in three main areas: training, regulation, and international tendencies. The training directed mainly at the small- and medium-scale mining sectors focuses on environmental aspects rather than general aspects. The first phase seminars and workshops concentrate on the diffusion of relevant information.

COCHILCO (Chilean copper commission)

COCHILCO is an independent, fundamentally technical, organism, specialized in copper mining. Its functions include the elaboration of, and assistance with, implementing policies; the supervision of compliance with specific legislation in the private and public sectors; and the use of strategic evaluations in safeguarding the interests of the State in its mining ventures. COCHILCO's scope includes sectoral studies, legal consulting, and environmental consulting. In recent years, its work has focused on market studies for different mining products and aspects related to large-scale State and private mining.

CONAMA (National environmental commission)

This entity defines reference terms for environmental impact studies and analyzes impact declarations according to the Basic Environmental Law.

CONAMA is the foremost authority on environmental policy in Chile, and operates in all thirteen regions of the country through Regional Environmental Commissions, or COREMAs. The COREMAs must assure compliance with the *Basic Environmental Law* and establish the lines of environmental policy at the national level. Their fundamental role is to coordinate the work of the Ministries and Public Services (all of which operate with their own technical structures), grant permissions, and supervise compliance with the corresponding norms.

CONAMA is in charge of the Environmental Impact Evaluation System established by the *Basic Environmental Law*. It manages the mechanism that generates environmental emission standards and regulates the management of natural resources.

SMALL-SCALE MINES

Underground mines, quarries, open mines, panning sites, and waste mining ores are exploited in Chilean mining. However, underground mines are the main form of exploitation, constituting 92% of all work sites and 95% of all production.

SSM reaches from Region I to Region VI, with the greatest number of mines being located in Regions III (39.4%) and IV (36.2%). Region II follows in importance, but is a distant third. The large-scale mines are found in Region II, whereas the smallest are in Region IV. Five municipal areas host nearly 40% of all known mines: in Region III, Vallenar, Diego de Almagro and Copiapó; and in Region IV, Ovalle and Andacollo (see Table 39.6).

There are few gold panning sites in the north due to the scarcity of water. Those that do exist are distributed from Copiapó to the southern limit of the northern zone. Most are found in the Coastal Mountain Range, specifically, in the Valle de Copiapó, Andacollo, Casablanca and Marga-Marga. Only 26 such sites exist in Regions I to VI, and, generally, their production levels are low. The municipal areas having the greatest presence of panning sites are Andacollo (Region IV), Quilpué (Region V), and Vallenar (Region III) (Sánchez & Enríquez, 1996).

Seventy-six percent of all SSM work sites are dedicated to copper extraction, and they produce a similar percentage of the total SSM copper production. The gold mines are generally smaller and supply only 11% of the total extracted mineral. Very few poly-metallic mines exist. No mines registered in the SSM sector directly exploit silver deposits. Molybdenum is only produced as a by-product of large-scale copper mining.

THE SMALL-SCALE PLANTS

A large number of small ore processing plants, usually with production capacities of below 50 metric tons/day, process gold (42%) and copper (37.5%) ores. Very few plants are devoted to poly-metallic, silver, and non-metallic ores.

Table 39.8 Concentration and leaching plants in the Atacama Region (Region III).

Type of product	No. of plants
Cu concentrates	32
Au concentrates	33
Ag concentrates	1
Cu precipitates	11
Total	77

Source: Navea, 2002.

The majority of plants are located in Regions IV (50%) and III (35%). Plants in Region II barely make up 6% of the total. Region IV's plants have the greatest production capacity (50.5%), followed by those in Region III (25%).

Andacollo is an important auriferous mining district, and therefore, has the greatest number of concentration plants (21.2%), which are mostly dedicated to processes of amalgamation with mercury. The municipal area of Copiapó, a traditionally important mining area in the North and home to ENAMI's Fundición Paipote, follows in importance. Although Taltal and Chañaral are the municipal areas with the greatest mining extraction capacity, they register very low participation in the total number of plants.

Andacollo and Copiapó remain the municipal areas with the greatest plant processing capacity. Many small plants in these municipal areas are dedicated to obtaining metallic gold by amalgamation or flotation.

Most artisanal plants are dedicated to the production of amalgamated gold, although they also benefit some from copper. Copiapó and Andacollo are the municipal areas with the greatest quantity of plants.

Table 39.8 provides a tabulation of concentration and leaching plants in the Atacama Region.

CHARACTERISTICS OF SMALL-SCALE AND ARTISANAL-SCALE MINING IN CHILE

The characteristics of Chile's SSM sector are practically universal. There is a low level of mechanization, and, consequently, a large percentage of SSM work is manual and burdensome. Knowledge of deposit reserves is minimal or non-existent. SSM operators have a low technical level and professionals are absent. Safety and hygienic conditions are precarious, and there is a deficient use of the mineral resources due to selective exploitation of high grade ores.

Found mostly in zones offering few non-mining alternatives, SSM is characterized by a direct relationship between the degree of poverty and the precarious nature of the work. Small-scale miners are highly marginal and cannot permanently incorporate themselves into the market; they generally do not own mining property. Most work is manual although, in some cases, there is a degree of mechanization.

The small artisanal miner is defined as an informal worker having little or no patrimony and operating using rudimentary techniques. They work in groups of up to eight people, and their average production per person does not surpass one ton per day. Artisanal miners are a middle-aged group (average age = 40 years), with low education levels and early exclusion from the national educational system.

An estimated 2,800 artisans make up this segment of the mining industry. It should also be noted that an important number of informal work sites are found in Regions III and IV, although Region V also registers a relatively important activity level. The municipal areas having the highest number of productive work sites are Diego de Almagro, Andacollo, Vallenar, Copiapó, Tocopilla, Chañaral, and Taltal. All of these, with the exception of Taltal, are work sites that have very small production capacities.

In this segment, nearly everyone works in an installation that is not their own and with a certain degree of individualism. The work methods used are unsafe, manual, and result in low productivity. The miners are associated with informal organizations. Incomes are low and unstable, and the work is highly seasonal and related to the employment alternatives offered within the zone. Given the characteristics of the deposits, the type of exploitation selected is governed by situational metal prices. Miners of the SSM sector must commercialize their product through third parties. Since many do not have direct access to ENAMI, they sell their products to medium-scale miners or other small miners larger than themselves.

Extraction occurs primarily through mines. Workers are rarely inclined to participate in waste mining ores or work in small processing plants; quarries are even more unusual. Most of the few identified panning sites are located in the municipal areas of Andacollo, Illapel, and Vallenar (Regions IV and V).

SSM and artisanal miners are highly exposed to work accidents, hygiene problems, and professional illnesses. In general, there is little access to the health system, and, according to the responses of the miners, a decline in medical coverage has occurred over time. In the case of Region I, this decline is further complicated by very distant medical attention centers.

PRODUCTIVE PROCESSES

In Chilean SSM plants, five main processes are employed for the treatment of copper, gold, and, to a lesser extent, silver. These are:

- Flotation
- Flotation–amalgamation
- Amalgamation
- Leaching
- Simple ore extraction

In general, these processes are carried out using old technology, have inefficient yields, and cause environmental degradation.

Gold production from SSM comes from underground mines, waste mining ores, and panning sites. The process of gold amalgamation with mercury is very widespread. It is found at 26% of all work sites, and is largely concentrated within the artisanal segment. Artisanal miners sometimes rent *trapiches* for grinding and amalgamating the mineral from which they obtain metallic gold for rapid commercialization.

Some plants use a combined flotation–amalgamation process. Artisan *trapiches* are used for grinding high grade ores and coarse gold amalgamation. The ores are combined with mercury in the launder of the *trapiche*, and, after the set running time, the Au-Hg amalgam is recovered and filtered to produce a paste with around 60% Au. This amalgam is distilled at a high temperature and metallic gold is recovered. During distillation, emissions of Hg vapors are

Figure 39.3 Typical *trapiche* or Chilean mill.

Figure 39.4 Chilean *pirquineros* of the Atacama desert.

Figure 39.5 *Pirquinero* of artisanal coal mining in the Lebú-Arauco Province (Region VIII).

produced. If they are not condensed safely, these vapors pose a serious health risk for the operators. Artisanal miners usually handle mercury carelessly, with direct skin contact and exposure (through inhalation) to toxic vapor.

Tailings disposal is usually conducted in dams that do not comply with basic construction norms. The contamination of underground water and safety issues in winter are some problems resulting from the use of this technology.

Figures 39.3 through 39.5 illustrate certain mine production activities in Chile.

PRINCIPAL PROBLEMS IN THE SSM SECTOR

The principal problems associated with SSM Chile are as follows:

a) Low technological level applied to the exploitation of deposits in SSM operations;
b) High exploitation costs;
c) Low productivity level;
d) Unknown availability of mineral reserves;
e) Low training level; and
f) Traditional work organization.

These are examined in this section of the chapter.

Environmental impacts

The environmental impact in both small-scale mines and plants affects water, soil, air, the natural environment, and human life. Water contamination from effluent liquid emissions containing dissolved heavy metals, acid, and processing reagents, is a chief environmental problem. Another environmental risk is posed by active and abandoned tailing deposits.

The most severe environmental impact is wastewater discharged from processing plants. Consider the following (Castro et al., 1998):

a) At flotation and amalgamation plants, tailing ponds usually contain water with Hg and Cu concentrations exceeding the legal norm.
b) At leaching plants, wastewater discharged exhibits an acid pH with high concentrations of Fe, Cu, and As.
c) Around half of all tailing dams present wall instability or leakage, or were abandoned without planning. This is especially true of dams close to urban zones.
d) There is a high risk of inhaling toxic mercury vapors while distilling gold amalgam.

The SSM sector has no environmental management whatsoever and a very limited financial capacity for investing in improved practices. In this segment, and especially at the most artisanal of work sites, the environmental risk linked to the manipulation of mercury in the gold amalgamation process is of particular importance. Vapors of Hg and mercury-rich amalgamation tailings constitute a serious environmental risk.

Gold extraction using mercury involves an Au-Hg amalgam decomposition step, which is usually carried out by two methods: (a) direct open-air burning in a pan using a blowtorch; and (b) placing the amalgam in a retort, then heating it until the volatile mercury is condensed in a cooler tube. A sponge-like gold *doré* containing about 20 g of mercury per kg of gold is produced and can be sold in the local market or melted to produce metallic gold. Significant air pollution by volatile Hg is produced by the first method, which is employed preferentially by artisanal miners. This has a major ecological and human health impact. Inhaled Hg is oxidized in the lungs, forming blood-soluble Hg(II) complexes, which can readily penetrate the brain producing erethism (exaggerated emotional response), and can cause gingivitis, and muscular tremors (Veiga et al., 2002).

It is presumed that many small – and even medium – plants in Chile dump their effluents into nearby rivers or directly into the ocean.

An estimated 658 tailing dams exist between Chile's Regions II and VII, along with other mineral deposits such as: leaching residues deposits, low grade ore deposits, and filtered tailing deposits. Region V hosts 57% of all the country's registered dams (Borregard et al., 2001).

In general, the ground conditions, water, scarce rain level, and the typical desert landscape of the zone add to the characteristics of the concentration processes to determine a general perception of a low level of environmental impact derived from SSM activities. However, there is a lack of quantification and administration of the environmental damage caused by SSM.

SOCIOECONOMIC OVERVIEW

In spite of the problems caused by SSM, it fulfills an important social role in Chile. It offers employment in remote rural regions where there is frequently little productive activity, and constitutes an important labour alternative in sectors affected by poverty or scarce work options. In this sense, it has a dynamic effect on local economies. Nevertheless, SSM offers low income levels, is unstable (dependent on climatic factors and metal prices), lacks capital, and is socially and legally conflictive (compliance of wage norms, benefits, work site installation permissions, mining concessions, etc.). From a socioeconomic point of view, SSM presents high indexes of poverty; few possibilities in the areas of education, health, and labor alternatives; and a cultural mining tradition in which movement to other activities is not auspicious in the short term.

A variety of information regarding socioeconomic variables tells of a poor sector in which reinsertion in other work activities is difficult. It should be noted that the group dedicated to gold amalgamation is the hardest to identify since many of these miners commercialize their product through intermediaries. This type of mining has unique characteristics, including legislation allowing the free commerce of gold and no control of any kind on the commercialization of mercury.

A first look reveals two types of small miners: (1) those whose presence in the sector is intimately linked to fluctuations in the price of copper, and, therefore, is itinerant; and (2) those whose main activity is mining and who remain in this area despite difficult price conditions. The first case consists of a generally young group working in other activities and the unemployed who live in mining zones or migrate to them, taking advantage of periods of high prices by linking themselves with the sector through very short-term mine rental contracts.

The second case consists of older miners, whose persistent presence is explained by a series of cultural aspects that form the mining tradition, including a low aversion to the risky nature of the work, and, in some cases, the lack of alternatives for subsistence. In general, this type of mining is characterized by the following, with limitations being more acute at the more artisanal work sites:

- A property situation that does not permit development;
- A high percentage of renters operating with very short-term contracts;
- Little knowledge of exploitable reserves; and
- Highly limited or no access to financing.

This type of miner works with high grade ores. The elevated costs result from the use of antiquated and rudimentary technology, royalty payments to property owners, and the large distances traveled to the concentration plants, among others.

In Chile, there are no existing indigenous groups associated with mining activity. Nevertheless, the *Atacameño* and *Quechua* people are demanding that the Chilean State mining companies turn over property and water rights in the area based on ancestral use. They also demand compensation in training, employment, and environmental protection plans for future generations.

Given that workers in this sector move periodically to other temporary activities, it is difficult to quantify the number of people that work in artisanal mining in the North of Chile. In 1993, the amount was estimated to be between 7,500 and 9,100 people, whereas in 2000, it was estimated to be around 6,000.

A considerable reduction in the jobs of several thousand workers took place during the 1990s.

There are two statistical sources regarding mining work: (1) those of SERNAGEOMIN and COCHILCO, which quantify the direct number of workers in mining companies; and (2) those of INE (National Statistics Institute), which measure the total number of direct and indirect workers linked to mining activity. The values found in INE's statistics are therefore higher than those of SERNAGEOMIN. Thus, considering both direct and indirect work, labor estimates in 1999 reported around 15,000 people (Table 39.9) in small-scale and artisanal mining (Lagos et al., 2001).

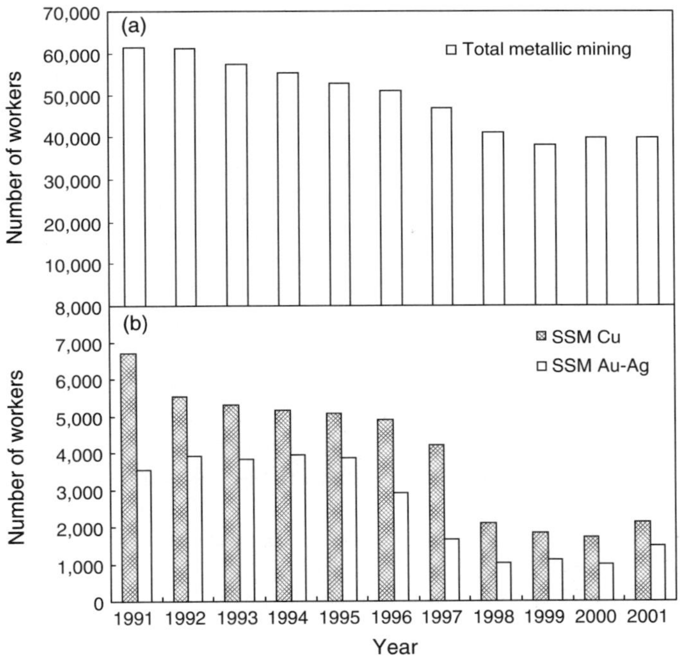

Figure 39.6 Number of workers by year in the mining sector: (a) Total metallic mining; and (b) SSM sector classified by Cu, and Au-Ag activities.
Source: Estadísticas del Cobre y otros minerales, 2001.

Table 39.9 Total mining employment (thousand people) by Region (direct and indirect).

Year	Region III	Region IV	Region V	Region VI	Region VIII	Region X	Total
1992	14.30	11.60	6.80	8.50	12.10	1.40	87.30
1995	16.00	10.10	7.40	8.70	7.70	0.60	89.50
1997	15.50	11.10	7.90	7.70	4.10	1.40	92.60
1998	11.08	7.46		9.58			75.59
2000	10.47	7.52	8.64	7.90	2.72	0.66	70.31

According to statistics from 2001 (SERNAGEOMIN), around 2,120 of Chile's small-scale and artisanal mining workers were employed in copper mining activities. Around 1,485 worked in gold mining. This is of a total of 3,605 workers in the sector, or 9.5% of all copper and gold mining workers. At the same time, these figures represent around 7.6% of the total workers in the Chilean mining and combustible sector, including copper, gold, iron, lead-zinc, non-metallic mining, and combustibles.

Given their capacities, the majority of the SSM plants have between one and 10 workers. Large plants can have 30 to 40 people per work site. Given the mobility existing in the sector, however, the information regarding the number of workers that operate at mining work sites is imprecise.

Region III, or the Region of the Atacama, has one of the greatest concentrations of small-scale and artisanal mining activities. The number of workers that can be identified in this area is around 2,010. Table 39.10 shows the number of workers and mines of the artisanal and SSM sector distributed by municipal areas (<3,000 tons/month of production) at Chañaral, Copiapó and Huasco, in Region III (Atacama) (Navea, 2002).

The educational level of small and artisanal miners is low and age-dependent. Average schooling for miners in the 14–29 year age range is eight years; for those in the 30–45 year range, it is six years; and for those in the 45–65 year range, it is only five years. A total absence of education can be found in 3.9% of the miners (Muñoz, 1999). The majority of the miners are in the 26–40 year age range and have a family group of between four and seven members.

The State has created training programs for this sector. Through ENAMI, a program called PAMMA (Artisanal Mining Assistance and Modernization Program) was instituted from the early-1990s until 2001. The estimated investment of the program was US$10 million. A new program, called Training and Technological Transfer to Small Artisanal Mining, was launched in 2002.

Table 39.10 Number of small-scale and artisanal workers in the Atacama Region (Region III).

Municipal area	Number of workers
Chañaral	410
Diego de Almagro	250
Total Chañaral Province	660
Caldera	90
Copiapó	625
Tierra Amarilla	302
Total Copiapó Province	1017
Huasco	42
Freirían	66
Vallenar	205
Alto del Carmen	20
Total Huasco Province	333

Total Atacama Region (Region III): 2,010 workers.

Another noteworthy initiative is the program for Environmental Management in the Small and Medium Scale Mining Industry (ECOMIN), whose objective is to increase positive effects of SSM in the Chilean economy, reduce environmental problems, introduce environmental management tools, and develop and disseminate clean and appropriate technology. Union organizations exist in formal small- and medium-scale mining, such as the National Society of Mining (SONAMI), and the Confederation of Working Miners.

FINAL REMARKS

SSM is a traditional activity that generates employment for the low-income population of Chile. The direct employment formally registered in the metallic SSM in 2001 was around 3,600 workers, 9% of the total workers contracted in the country's metallic mining industry. However, including artisanal miners and indirect employment, between 10,000 and 15,000 people are tied to this sector.

SSM and artisanal-scale miners are mainly distributed in Regions III and IV, and VIII of the country, and, in 2001 were responsible for 6.0% of national gold production, 0.85% of copper production, and around 25% of coal production.

In the Northern Zone, SSM exhibits a high interdependence with artisanal-scale mining, and both sell the majority of their mineral products to ENAMI (State-owned company) on a tariff system. SSM is characterized by low income levels, instability, and very limited financial capacity. However, in spite of the problems presented, SSM plays an important societal function. To improve sustainability of this segment, technical, legal and financial assistance is necessary.

REFERENCES

Anuario de la Minería Chilena (2001). Gobierno de Chile. Ministerio de Minería-INE-SERNAGEOMIN.

Borregard, N., Dufey, A., González, P., Kewin, P. & Newbold, J. (2001). *Sustentabilidad del sector minero:¿qué rol puede jugar un fondo de sustentabilidad?* CIPMA report.

Castro, S.H. & Sánchez, M. Environmental viewpoint on small-scale copper, gold and silver mining in Chile. *Journal of Cleaner Production*, 11(2), 207–213.

Castro, S.H., Sánchez, M.A., Vergara, F. & Oyaneder, E. (1998). Water and slurry wastes assessment in the Chilean copper industry. In M.A. Sánchez, F. Vergara & S.H. Castro (Eds.), *Environment & innovation in the mining and mineral technology* (pp. 1005–1016). University of Concepción-Chile.

Chaparro, E. (2000). *La Llamada Pequeña Minería: Un Renovado Enfoque Empresarial*. CEPAL/ECLAC, Serie: Recursos Naturales e Infraestructura, Santiago de Chile, Julio.

Lagos, G.E., Blanco, H., Torres, V. & Bustos, B. (2001). *Minería y minerales de Chile en la Transición hacia el desarrollo sustentable* MMSD-Chile. CIPMA.

Muñoz, H. (1999). *Diseño del programa de capacitación y transferencia tecnológica a la pequeña minería artesanal.* Ministerio de Minería. Santiago-Chile.

Navea, J. (2002). *Proyecto de Transferencia Tecnológica para la Pequeña Minería de la Tercera Región* – SEREMI de minería III Región, Ministerio de Minería.

Sánchez, J.M. & Enríquez, S.M. (1996). Impacto ambiental de la pequeña y mediana minería en Chile. In *Environmental Study of Artisanal, Small and Medium Mining in Bolivia*, Chile and Perú, Dec.

Subterra Ingenieros Limitada. *Desarrollo de un patrón de análisis ambiental de la pequeña minería.* Ministerio de Minería-Unidad Ambiental. Comisión Nacional del Medio Ambiente CONAMA.

Veiga, M.M. & Hinton, J.J. (2002). Abandoned artisanal gold mines in the Brazilian Amazon: A legacy of mercury pollution. *Natural Resources Forum* 26, 15–26.

Index